ICE Handbook of Concrete Durability

emerald PUBLISHING
ice Publishing

ICE Handbook of Concrete Durability

A practical guide to the design of resilient concrete structures

Second edition

Edited by
Marios Soutsos
Queen's University Belfast

Published by Emerald Publishing Limited, Floor 5, Northspring, 21-23 Wellington Street, Leeds LS1 4DL.

ICE Publishing is an imprint of Emerald Publishing Limited

Other ICE Publishing titles:
Concise Guide to Reinforced Concrete Design to Eurocode 2
Patrick Purcell. ISBN 978-0-7277-6572-7
Designing and Constructing Prestressed Bridges
Jiří Stráský and Radim Nečas. ISBN 978-0-7277-6385-3
Design of Fire-resistant Concrete Structures
Kristian Dahl Hertz. ISBN 978-0-7277-6444-7

A catalogue record for this book is available from the British Library

ISBN 978-0-7277-6375-4

Cover photo: Welbeck Street car park, London. Magnus Bjermo/Shutterstock

Commissioning Editor: Michael Fenton
Development Editor: Cathy Sellars
Production Editor: Sirli Manitski

Typeset by Nova Techset
Index created by Sunrise Setting
Printed and bound by CPI Group (UK) Ltd, Croydon, CR0 4YY

Contents

About the editor

Marios Soutsos has been Professor of Structures/Materials in the School of Natural and Built Environment at Queen's University Belfast (QUB) since 2012. Prior to this, he was a lecturer/reader in Structures and Concrete Technology at the Centre for Engineering Sustainability, School of Engineering, University of Liverpool. He has had industrial experience in Cyprus, Libya, Saudi Arabia and Bahrain prior to returning to academia. His principal research experience is in construction materials, and his current research interests include high-strength concrete, cement replacement materials, chemical admixtures, concrete rheology, the use of recycled demolition aggregate in concrete products, repair materials, heat of hydration effects in concrete structures, as well as alkali-activated binders. Research funding sources include the Engineering and Physical Sciences Research Council (principal investigator for two EPSRC research grants and co-investigator for two EPSRC/EU research grants), the Department of the Environment (co-investigator for the DoE funded Cardington concrete frame project), The Onyx (Veolia) Environmental Trust, the Northwest European Regional Development Fund Programme, Carbon Trust Applied Research Grants, FP 7 Collaborative Research and Invest Northern Ireland.

He is an author or co-author of more than 120 technical publications (including more than 40 refereed journal papers and 80 refereed conference publications). He is a senior editor for *Construction and Building Materials* (Elsevier) and member of the editorial advisory board for *Magazine of Concrete Research* (ICE). He is a co-editor for two other books: *Properties of Fresh and Hardened Concrete Containing Supplementary Cementitious Materials* (Springer) and *Construction Materials: Their Nature and Behaviour, Fifth Edition* (CRC Press). He has contributed chapters to three other books: 'Chapter 18: Special concretes' in *ICE Manual of Construction Materials* (Thomas Telford), 'Chapter 1: Introduction: Key issues in the non-destructive testing of concrete structures' in *Non-destructive Evaluation of Reinforced Concrete Structures, Volume 1* (Woodhead Publishing) and 'Chapter 3: Estimation of on-site compressive strength of concrete' in *Non-Destructive Assessment of Concrete Structures: Reliability and Limits of Single and Combined Techniques* (Springer).

List of contributors

Chris Atkins, Mott MacDonald, UK
Salim Barbhuiya, University of East London, UK
PA Muhammed Basheer, University of Leeds, UK
John Bickley, JA Bickley Associates, Canada (†)
Klaas van Breugel, TU Delft, The Netherlands
Jorge de Brito, Universidade de Lisboa, Portugal
John Broomfield, Broomfield Consultants, UK
Peter A Claisse, Coventry University, UK (†)
Oguzhan Copuroglu, TU Delft, The Netherlands
Dave Cullen, Jacobs, UK (†)
Philip Ebbatson, HKA, London, UK
R Doug Hooton, University of Toronto, Canada
Kenneth C Hover, Cornell University, Ithaca, USA
Jeremy Ingham, Consultant, London, UK
Iqbal Johal, Galvanizers Association, UK
Fragkoulis Kanavaris, Arup, UK
Jon Knights, Jon Knights Materials Consulting, UK
Konstantin Kovler, Technion – Israel Institute of Technology, Israel
Robert C Lewis, Ferroglobe Plc, UK
Desmond Makepeace, Galvanizers Association, UK
Bryan Marsh, Arup, UK
Alexander Mezhov, Technion – Israel Institute of Technology, Israel
John Myers, Missouri S&T, USA
Antonio Nanni, University of Miami, USA
Sreejith Nanukuttan, Queen's University Belfast, Northern Ireland
Alan Poole, AB Poole & Associates, UK
John L Provis, University of Sheffield, UK
John Reddy, Ecocem Materials Ltd, Republic of Ireland
Alan Richardson, Northumbria University, UK
Marco Rossini, University of Miami, USA
Eric Schlangen, TU Delft, The Netherlands
Asia Shvarzman, Antex Western, Winnipeg, Canada
Kosmas Sideris, Democritus University of Thrace, Greece
Rui V Silva, Universidade de Lisboa, Portugal
Robert Viles, Fosroc International Limited, UK (†)
Don Wimpenny, Jacobs, UK
Kai Yang, University of Leeds, UK
Guang Ye, TU Delft, The Netherlands

Acknowledgements

The editor would like to thank the following for their contributions to the first edition of this book:

Mustafa Sahmaran
David Cochrane
Douglas Gremel
Lindon Sear
Michael D Connell
Scott Humphreys

The following contributors have sadly passed away since the publication of the first edition or during the preparation of the second edition:

John A Bickley
Dave Cullen
Peter A Claisse
Robert Viles
Odd E Gjorv
Ken Day

Their contribution of knowledge, experience and guidance extended well beyond the confines of this book and their passing is an irreplaceable loss, keenly felt across the concrete community.

About this book

Significance of durability for reinforced concrete construction

Far more concrete is produced than any other manufactured material. It is the basic material for the construction industry, which employs 7% of the workforce worldwide and over half in some countries. One tonne of concrete is produced every year per head of population on the planet.

Over 100 years ago in 1907, Knudson showed in his paper entitled 'Electrolytic Corrosion of Iron and Steel in Concrete' that the passage of a small current through the reinforcement in concrete would cause corrosion. His opinion that stray currents were the cause of the problem was shared by most researchers at the time, but within a few years, Rosa, McCollum and Peters had concluded that 'the presence of chlorides always facilitated trouble' ('Electrolysis of Concrete', 1912). At the time this was written, the chlorides that did the damage were almost always from seawater. It was not until half a century later that winter salting of roads became widespread and soon after that, the extensive deterioration of highway structures was observed. This corrosion is the most significant of a wide range of different problems that affect modern structures, and these problems are discussed in this book.

Clients now expect their structures to last longer, with service lives of over 100 years being demanded in some cases. To avoid the liability for long-term repairs, they are looking for a 'warranty' from construction companies and, in many cases, give them contractual responsibilities for long-term maintenance. Contractors can, therefore, no longer bid based on the cheapest solution. They must consider the long-term durability of the structure to avoid spending their profits on repairs.

Durability is a very difficult area for design. A structural engineer can decide on the size and shape of components and carry out the necessary calculations (or get a computer to do them) to determine their load-bearing capabilities with great accuracy. No such process is possible for durability. The mix design of concrete can be specified and the exposure conditions determined (often not very accurately), but the only things that these affect directly are the properties of the matrix, such as the hydrate structure, the pore size and the fluid chemistry. These affect only the transport properties, and it is these that finally determine the durability of the structure. The link between what we can control and the result we get is, therefore, extremely complex and can be determined only by experimental or probabilistic methods.

To add further complications to a complex problem, how we make and use concrete is currently going through a series of fundamental changes. These have been driven partly by costs but primarily by environmental concerns. Cement production is responsible for approximately 5% of the world's carbon dioxide emissions. To put this into context, note that carbon dioxide emissions for concrete are 20 times less per tonne than for steel, but the quantities of concrete used are vastly greater. Thus, there is enormous pressure to replace as much of the cement as possible with other materials. The use of such materials and their effects on durability are discussed in detail in this book.

The final issue is that of workmanship. Despite many advances in quality control, the plans and specifications produced by the designer are often not accurately followed. The depth of cover concrete, which protects the steel from the exterior environment, often ends up less than that specified. Also, the curing, which should prevent the concrete from drying out after it is poured, is often not effective. Drainage systems for road bridges are notorious for failing to keep the damaging salt water away from vulnerable concrete surfaces. These factors combine to cause even the best designs to fail if inadequate provision is made to prevent, or allow for, poor construction.

In the past, there was a clear division between the structural engineers, who prepared the designs, and the concrete technologists, who often became involved only when durability problems arose. This came about due to conflicting pressures on the syllabus in civil engineering courses. However, designers are now finding that durability considerations have become a central part of the process and must be considered at every stage from concept to completion. Those wishing to study the topic have found the relevant information to be fragmented, partly due to the huge volume of research that has been published in the numerous journals that cover the area. Much of the information is also presented in terms of complex physical and chemical theories. This book is a compilation of essays by a selection of leading practitioners covering the full range of durability topics at a level that can be understood by engineers without the need to study the science in great detail.

Structure and contents

This book has 16 chapters with contributions from 39 practitioners and academics. The contents of the chapters are as follows.

1. Introduction

This chapter introduces the reader to concrete durability and aims to convey the need for all those involved in the design or construction of reinforced concrete structures to have a basic understanding of all the degradation processes.

2. Pore structure and transport processes

This chapter describes how the characteristics of the pore structure of cement-based materials play a fundamental role in governing the transport processes in concrete, and hence, its durability.

3. Physical deterioration mechanisms

This chapter reviews the range of mechanisms that result in the physical degradation of concrete. Specific deterioration types and mechanisms described include abrasion, erosion, cavitation, frost, exfoliation and fire.

4. Chemical deterioration mechanisms

This chapter reviews the range of mechanisms that result in the chemical degradation of concrete. Specific deterioration types and mechanisms described include weathering and leaching, acid and alkali attack, sulfate attack, delayed ettringite formation, salt attack, alkali–aggregate reaction and corrosion of the steel reinforcement.

5. Material selection to improve durability

This chapter describes construction materials that can be used to improve the durability of reinforced concrete structures. These include chemical admixtures, mineral admixtures (pulverised fly ash, ground granulated blastfurnace slag, condensed silica fume (microsilica) and metakaolin) and alternative reinforcing materials to conventional steel reinforcement (black bars) (epoxy-coated steel reinforcement, galvanised steel reinforcement, stainless steel reinforcement and glass-fibre-reinforced polymers).

6. Construction processes to improve durability
This chapter describes how, for concrete to be durable, the designer's intentions must be translated into the actual structure. The concrete must be produced to give a consistent final in situ product, allowing for any variations in the constituent materials, production conditions and site practices.

7. Design aspects that can reduce risks from deterioration mechanisms
This chapter describes the considerations needed at the design stage to maximise the protection that can be achieved both using suitable design aspects and through appropriate materials measures.

8. Durability performance tests
This chapter describes tests that cover two main areas, namely physical and chemical material tests. It will assist the designer and specifier in choosing the correct test to determine durability for a given service life, which is related to the expected overall durability of the structure.

9. Performance-based specifications for concrete
This chapter mainly deals with performance specifications from the North American perspective, although the international literature has also been reviewed. The chapter concludes that a multi-step plan is needed, with possibly nine steps, to realise the transition from prescription to performance specifications.

10. Modelling and predicting the effects of deterioration mechanisms
This chapter describes and discusses the potential of some numerical models in tackling specific threats to durability. The deterioration of concrete is often a very slow process in which chemical, physical and mechanical mechanisms proceed simultaneously and may interact as well. Thus, modelling is very complex.

11. Case studies: Durability problems, repair strategies or proper consideration in durability design?
This chapter presents case studies of durability problems that have arisen in major structures to raise awareness of the extent and cost of repairs that may be needed for the structures to remain operational. One hundred years or more of service life is achievable with careful selection of concrete mix proportions and constituent materials as demonstrated by the East Sea Bridge in Shanghai.

12. Repair methods
This chapter describes a wide range of options for repairing and rehabilitating damaged reinforced concrete structures. Selecting the optimum combination of repair techniques requires a well-designed and executed condition survey.

13. Non-destructive testing and structural health monitoring
Several in situ tests for assessing the mechanical, permeation, corrosion and integrity characteristics of concrete in structures are reviewed in the first part of this chapter. The second part focuses on sensing techniques, which are classified as electrical or optical.

14. Durability of concretes based on non-Portland cements
This chapter addresses the durability of concretes based on non-Portland cements, with a particular focus on four key binder types: calcium aluminate, calcium sulfoaluminate, alkali-activated and magnesia-based cements.

15. Durability of concrete made with aggregates recycled from construction and demolition waste
This chapter reviews the durability of concrete made with aggregates recycled from construction and demolition waste. It also presents the design of slabs and beams made from reinforced recycled aggregate concrete in conformity with Eurocode 2.

16. Lessons learnt by expert witnesses involved in construction disputes
This chapter discusses, through examples, the recurrent themes that have been identified as contributing to the causes of disputes involving a failure of design or workmanship for construction projects involving concrete structures.

Soutsos M
ISBN 978-0-7277-6375-4
https://doi.org/10.1680/icehcd.63754.001

Chapter 1
Introduction

Marios Soutsos
Queen's University Belfast, UK

1.1. Potential durability problems

Concrete is, due to its versatility, comparative cheapness and energy efficiency, of great and increasing importance for all types of construction throughout the world. Concrete structures can be durable and long lasting but to be so, due consideration needs to be given at the design stage of the effect on the concrete of the environment that the structure will be exposed to. Degradation can result either from exposure to the environment exposed, for example, frost damage, or from internal causes within the concrete, such as the alkali–aggregate reaction. It is also necessary to distinguish between degradation of the concrete itself and loss of protection and subsequent corrosion of the steel reinforcement. ACI Committee 201 [1] defines concrete durability as: 'Its resistance to deteriorating influences which may through inadvertence or ignorance reside in the concrete itself, or which are inherent in the environment to which it is exposed.'

Initially, concrete was regarded as having an inherently high durability, but more recent experiences have shown that this is not necessarily the case, unless durability design forms an integral part of the design and construction process. There is a need to consider all potential deterioration mechanisms at the design stage before selecting and specifying an appropriate concrete mixture from a durability perspective. Prescriptive specifications for concrete based on a permissible maximum water-to-cement ratio and minimum cement content have received much criticism in recent years. They may even have inadvertently allowed designers and contractors to avoid considering or implementing all the available information required for a sound design for durable construction.

Unexpected maintenance and repairs arising very early in the specified service life of structures have led to enormous financial burdens to clients. The US Department of Defense funded a major research project entitled 'Concrete Durability – a Multibillion-Dollar Opportunity' to determine whether the problem with concrete durability was a technical or an institutional problem [2]. It concluded that

> Most of the knowledge exists which, if properly applied, would produce durable concrete. The lack of proper application may be attributed to a lack of knowledge by practitioners and the system's failure to make durability the responsibility of the organization which can most directly provide it – the contractor.

It is not so much the lack of knowledge within the design team but the failure to recognise the magnitude of potential durability problems, which may incur high maintenance and repair costs. If structures are not properly maintained, subsequent strengthening is not always an option, so that they may have to be demolished. This was the case with the Churchill Way flyovers in Liverpool, which were demolished between September and December 2019. The flyovers were closed in September 2018 pending a structural safety review, which found that they were 'no longer adequate to carry vehicles or pedestrians'.

Figure 1.1 Multistorey car park in Colchester that was pulled down in 1995 because concrete corrosion problems were beyond economic repair

Hopefully, such structures showing signs of structural distress will be identified in good time and demolished safely rather than collapsing suddenly with loss of life. The latter was unfortunately the case with the Morandi Bridge in Genoa, a section of which collapsed in August 2018 killing 43 people and leaving 600 people homeless. The collapse also exposed the state of disrepair of Italian motorways in general, but especially those in Liguria, a region of north-western Italy, whose capital is Genoa.

Some examples of durability problems that have appeared in the *New Civil Engineer*, the monthly journal of the Institution of Civil Engineers, in the past decades and durability problems reported in newspapers and websites are briefly described to convince the reader that durability needs to be considered seriously and at the design stage of the construction project.

1.2. Concrete condemned: The Queen Street car park in Colchester

A multistorey car park in Colchester was pulled down in 1995 because concrete corrosion problems were beyond economic repair (Figure 1.1) [3].

The Queen Street car park was put up only in 1971 under a design and build contract by the now defunct contractor Shears-Neal. By 1985, evidence of corrosion in precast concrete units on the in situ concrete frame prompted Colchester Borough Council to commission consultant Eastwood & Partners to report on the structure. Eastwood found that 3% of units were affected and made repair recommendations to prolong its life by 5 years. In 1992, the car park, which sits over a live bus station, was closed for safety reasons. Eastwood reported that 40% of the units were then affected. Refurbishment costs were put at £1.5 million, rebuild at £3 million and demolition at £350,000. The council opted for demolition.

1.3. Repairs to a heavily used bridge in Runcorn carried out without disrupting traffic

Problems on the 34-year-old 27-span crossing with its striking 330-m central lattice steel arch were first identified in 1989 [4]. The three-span bridge itself, with its concrete deck, was in reasonably good condition, thanks to owner Cheshire County Council's £1 million a year maintenance programme. However, all the concrete approaches were in a less favourable state (Figure 1.2). Each is made up of four in situ longitudinal deck beams carried generally on a single central pier with an integral cross head. The ever familiar story of road salt seeping down through leaking deck expansion joints to attack beams, cross heads and piers was all too evident on both sides of the Mersey.

Figure 1.2 Concrete approaches to the Runcorn Bridge

The worst damage was beneath the Widnes approaches, which were widened in 1977 by adding a separately supported 5-m-wide strip of deck. Here the main culprit was a 500-mm-wide longitudinal infill slab that connected the original and extended sections of the deck.

Flexible joints supporting both sides of the infill were leaking, allowing chloride-rich surface water to run down. Chloride levels in the rectangular beam, which is up to 2 m deep, approached 2% by weight of cement. The concrete had spalled and delaminated and the link steel had been attacked. However, the four layers of the main densely packed 50-mm rebar were relatively unscathed.

Cathodic protection was installed in the chloride-riddled approach spans, which avoided the widespread demolition and reconstruction of the damaged spans for one of the River Mersey's most strategic crossings. Importantly, this was the first full-scale application of a micro-concrete gunite overlay, which claimed to solve the historic package of problems surrounding such specialised sprayed-concrete applications.

1.4. Terminal operation: Miracle cure for the Marsh Mills Viaducts

What to do about the Marsh Mills Viaducts became a major preoccupation of engineers in the West Country for more than 15 years once it was realised that the elegant concrete structures were condemned to a lingering but terminal decline [5]. The revelation that the Marsh Mills Viaducts were afflicted by the alkali–silica reaction (ASR), which would inexorably burst it apart, had come as a shock. The industry had assumed that ASR was a technically interesting cause of deterioration to concrete overseas but generally of only academic interest in Britain. The ASR deterioration of structures, such as Charles Cross car park in Plymouth and the foundations of electricity substations in the South West, had previously been dismissed as freak incidents. Cracking on the viaduct, which had already been monitored for 18 months when the situation was headlined by *New Civil Engineer*, was dismissed at the time by engineers at Devon County Council as something 'rather superficial'.

The trouble was caused by alkali-rich cement from the nearby Plymstock Works, which was used in combination with sea-dredged aggregates, and aggravated by road de-icing salt. Moisture is required for ASR, which produces an expansive gel that can burst concrete structures apart. The internal expansion causes a characteristic map cracking effect on the surface of a structure.

Discovery of the full extent of the problems at Marsh Mills prompted a nationwide examination of other highway structures. Many were found to be in trouble to a greater or lesser degree and several were replaced. As well as Marsh Mills, there were several other reinforced-concrete bridges on the 1969/70 vintage, grade-separated A38 highway between Exeter and Plymouth. Measures adopted were to observe and contain the problem with remedial works, such as weather shields, to extend the working life of structures until such time as replacements could be built.

Miracle cure? Bold innovation won Hochtief the contract to replace Plymouth's concrete cancer-crippled Marsh Mills Viaducts at a design and build cost of £12.25 million. The idea was simple. Its execution, however, was nerve-wracking. Traffic diversions could be avoided almost entirely, figured Hochtief and its designer Tony Gee & Partners; just assemble the new viaduct decks on temporary supports beside the old structures while building the permanent foundations and piers beneath them. Then, traffic could be transferred away for a few hours while each new viaduct was slewed sideways onto the permanent supports (Figure 1.3). Slewing in the viaducts involved the biggest such bridge jacking operation ever attempted. Each slip road deck was some 400 m long, weighed about 5250 tonnes and was supported on bearings sliding on tracks set on seven or eight intermediate piers. Just for good measure, the viaducts were each set out on a curve with a severe gradient and a crossfall.

Motivation for this extreme solution came from the lane licence charges imposed by the Highways Agency. Overnight closure of any two lanes of the A38 would have cost the contractor £5000, rising to £18 000 a day at a weekend and a thumping £25 000 a day during the week. In effect, Hochtief saved these charges and spent the money instead on extensive temporary works. The crippled viaducts carried slip roads for the A38 up to 12 m above the Plym Valley. They were heavily trafficked and tightly constrained by obstacles, including the main railway line to Penzance, the River Plym and buried services, including most of the trunk gas, water and electricity supplies to Plymouth.

Figure 1.3 Replacing the Marsh Mills Viaducts [5]

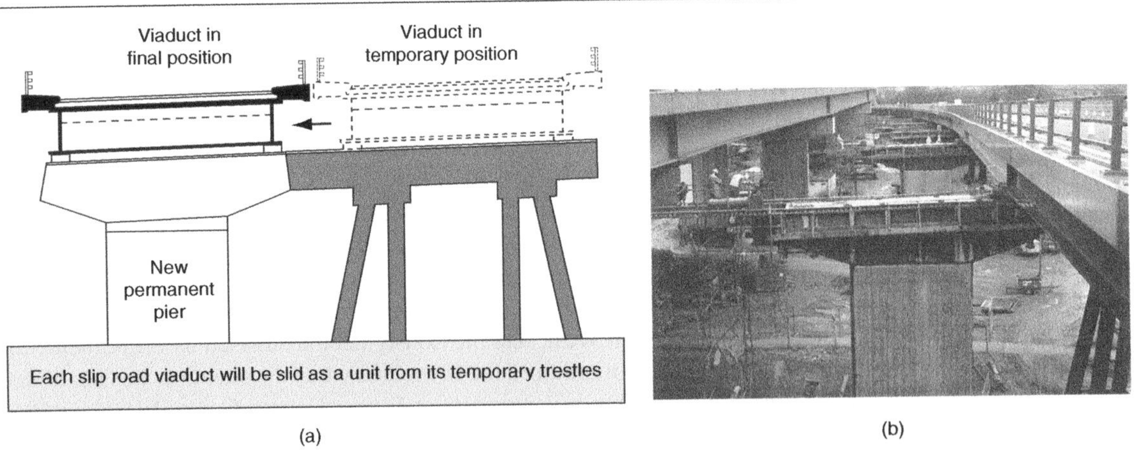

Figure 1.4 The collapse of the Ynys-y-Gwas Bridge led to a ban on grouted tendons [6]

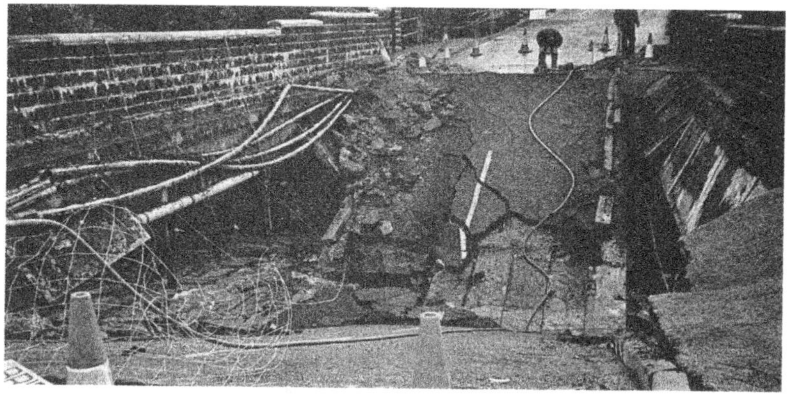

1.5. Lasting effect: The collapse of the Ynys-y-Gwas Bridge

Bridge owners have for many years been concerned about the corrosion of prestressing cables and the difficulty of inspection. These concerns were highlighted in December 1985 with the sudden collapse of a 32-year-old 18.3-m-span post-tensioned segmental road bridge in South Wales (Figure 1.4) [6]. The failure of the Ynys-y-Gwas Bridge was directly caused by tendons corroded by chlorides from de-icing salts. The salt penetration was eventually attributed to a combination of inadequate tendon protection, poor workmanship and ineffective deck waterproofing. Other key factors identified included the lack of an in situ top slab and joints opening under load.

Although possibly the most newsworthy, this is by no means the only bridge to have had problems. In September 1992, the Department of Transport's concern as an owner and client led to the announcement of a temporary ban on the commissioning of any new bridges of the grout duct post-tensioned type until specifications had been reviewed. Construction of some bridges already designed using bonded internal prestress was allowed to continue. The Department of Transport's decision, in effect, laid down a challenge to the UK concrete bridge industry to put its house in order and to demonstrate that it had done so. The response by the Concrete Society, supported by the Concrete Bridge Development Group, was to set up a working party in June 1992 to study the problem and prepare recommendations. In May 1994, the working party held a seminar that summarised the position at that time. Detailed discussions started with the Highways Agency in April 1995 with a view to making use of the revised design and construction procedures to allow a phased reintroduction of bonded post-tensioned bridges. Concrete Society Technical Report 47: *Durable Bonded Post-tensioned Concrete Bridges* was published in 1996 [7].

1.6. Thaumasite test for the Cotswolds

Fears that thaumasite sulphate attacks against concrete could hit thousands of house foundations prompted the government to commission a major in situ research project in 1998 [8]. The research took place at a site in the Cotswold village of Shipston-on-Stour, where cases of thaumasite in house foundations were discovered as long ago as 1990. The tests monitored 176 concrete specimens over periods of 3 and 10 years. Some 43 mix types were used with four different aggregates and ten different binders. A combination of precast and cast-in-place specimens were placed in two 11-m-long by 2.5-m-deep trenches.

The research was a clear indication of the government's increasing fears over the possible scale of the problem. The Highways Agency revealed ten cases of thaumasite on bridges on the M5 in Gloucestershire (Figure 1.5). A further 27 structures were understood to have been identified as vulnerable to thaumasite attack on country roads in Gloucestershire.

Figure 1.5 Thaumasite attack against a bridge in Gloucestershire

The subject of thaumasite attack was in the spotlight for almost a year [9–12]. The government's expert group produced, after an 8-month investigation, a guidance document on thaumasite attacks [13]. Articles on thaumasite, however, continued to appear months after its publication [14–17].

1.7. Joint detailing key to flat-slab collapse of Pipers Row car park

The Health & Safety Executive published its report into the 1997 collapse of the Pipers Row multistorey car park in Wolverhampton in 2004 (Figure 1.6) [18,19]. The research identified that lift slabs were susceptible to punching shear if the concrete had deteriorated in the slab/column zone.

Pipers Row was a punching shear failure and occurred when there was only a dead load, implying that there had been an apparent major reduction in strength. The failure was initiated by local deterioration of the concrete, confirmed both by examination of the debris and by subsequent rigorous structural analysis. The concrete in the collapsed top slab was of lower quality than elsewhere in the structure. It had been exposed to the elements for over 30 years and undoubtedly went through many freeze–thaw cycles. Frost action

Figure 1.6 Collapse of Pipers Row multistorey car park in Wolverhampton [18]

Figure 1.7 Deterioration of the concrete on (a) the parapet walls and (b, c) concrete deck of Fornello Viaduct

(a) (b) (c)

led to the concrete becoming friable all the way into the slab near an internal column and as far as the top steel, eventually leading to a complete loss of bond and anchorage. Rigorous analysis of Pipers Row clearly showed that the undeteriorated as-built strength had a safety margin of at least 1.5 against the worst possible in-service load.

1.8. Fornello Viaduct of the Orte–Ravenna (E45) highway

The European route E45 goes from Norway to Italy through Finland, Sweden, Denmark, Germany and Austria. It has a length of about 5190 km, and it is, thus, the longest north–south European route. Fornello Viaduct is part of the mountain highway connecting Orte to Ravenna in Italy. It was built in the early 1970s and has four 32-m spans of width 9.5 m in each direction. The deck consisted of four prestressed concrete girders and a reinforced-concrete slab of thickness 200 mm. Figure 1.7 shows the extent of the concrete deterioration in the parapets and reinforced-concrete slab. Studies of the deck indicated that the primary causes of deterioration were not only corrosion of the steel reinforcement but also freeze–thaw attack [20]. The studies were carried out in 2008 after it was decided that the reinforced-concrete deck should be demolished and replaced.

1.9. The Churchill Way flyovers in Liverpool were demolished in 2019

The Churchill Way flyovers were built in the 1960s and opened in the 1970s as part of Liverpool's abandoned inner ring road scheme. The flyovers linked Lime Street to both Dale Street and Tithebarn Street and ran directly behind the city's museums on William Brown Street. Remedial works were previously carried out in the 1980s, in 2005 and in 2013. They were closed in 2018 and a review was carried out which showed that they were 'no longer adequate to carry vehicles or pedestrians'.

The flyovers had been poorly built. Wooden moulds had been used to build the hollow-core post-tensioned concrete girders. In some parts, the wooden moulds were overloaded from the weight of the fresh concrete and broke open. Therefore, there was more concrete than was required, which increased the self-weight of the structures. The lack of access holes that allowed entry into the hollow cores of the girders meant that this went unnoticed. Moreover, the wooden formwork inside the flyover decks had rotted, and there were serious problems with drainage. The concrete cover was not adequate. For example, some cables and steel reinforcement were too near the surface. Corrosion of these led to concrete spalling. The structure was, as a result, 'overstressed' and cracks had appeared over some of the supports [21–23].

Liverpool Council decided that it would have been too costly and difficult to repair or replace the flyovers. They estimated that it would have cost £60 million to rebuild whereas demolition would have cost only

Figure 1.8 Removal of Liverpool's condemned Churchill Way flyovers [22]

£6.75 million. The latter was considered to be the safest and most economical option, even if the road network in the area had to be improved to help the traffic flow. New bridges were required to replace the pedestrian footbridges that crossed the busy main roads.

Demolition was a complex job since the flyovers were multi-span post-tensioned concrete girders (Figure 1.8). Taking down one part of any structure could have caused damage elsewhere. The concrete girders were cut in situ, lowered to the ground using heavy-lifting equipment and then cut into smaller sections and transported off site. This was to minimise the dust, noise, vibration and environmental impact to the surrounding area.

1.10. Collapse of the Morandi Bridge in Genoa in 2018

A 250-m-long deck section of the Polcevera Viaduct, more familiarly known as Genoa's Morandi Bridge, collapsed on 14 August 2018 killing 43 people (Figure 1.9) [24]. The viaduct was 1.1 km long and had three 90-m-tall towers. One of the three towers collapsed. The foundations of the tower were being strengthened

Figure 1.9 Collapse of the Morandi Bridge in Genoa in 2018 [25]

(a) (b)

at the time of the collapse. The report into the collapse of the bridge has not yet been released, despite that several years have since passed. However, it does appear that the foundations of the collapsed tower were not the cause, as they appeared to be stable. The bridge was 50 years old and significant repairs to the tower and stay cables had been identified as being needed. The repairs and strengthening were expected to be like those carried out in the 1990s for one of the other two towers. This included strengthening of the cables encased in concrete with new steel cables strapped to the surface of the concrete and replacement of the steel anchor blocks. This would have required 55 tonnes of steel. The reinforced concrete of the deck and the two towers was to be repaired, and this was estimated to require 1200 m^3 of new concrete. The contract notice was issued on 28 April and the deadline for the tenders was 11 June 2018. There is no evidence that the contract had been awarded.

Traffic volumes were very different in the 1960s, and concepts such as the durability or lifespan of a building were not commonly discussed. Therefore, there are thousands of 1960s and 1970s bridges that are in the same state, as no bridge designer foresaw the long-term durability issues back then. The collapse of the Morandi Bridge also brought into the spotlight other smaller incidents, which nonetheless indicated that the transport infrastructure in the area needed urgent maintenance. These problems were as follows [26, 27].

- In 2016, a flyover close to Milan collapsed under the weight of a truck. One person died.
- In 2017, a bridge collapsed near Ancona killing two people.
- In 2019, a motorway bridge fell onto the A6 following a landslide.
- In December 2019, the ceiling of a tunnel collapsed on the A26 not far from Genoa. Luckily, there were no victims.
- In April 2020, a 260-m bridge near the town of Aulla, which is between Genoa and Florence, collapsed and two people had to be taken to hospital.
- Six 40-m spans of the Albiano Bridge collapsed, highlighting maintenance failings [27].

Autrostrade per l'Italia, the operator of the motorways in the region, acted quickly to repair both its roads and its image. The new Genoa bridge opened only 2 years after the tragic collapse. The operator announced a 4-year investment budget that was three times the previous budget. It was expected that this would reach €7.5 billion spread over investment and maintenance. Unlike the 1960s, today we focus much more on how to extend a building's life and we put in place a maintenance guide to keep structures in good condition.

1.11. Design-build-operate-maintain

The expectation of the owner of a structure is that it will require only very little or no maintenance during its design life. Owners have realised that the cheapest option for constructing a structure may work out to be an expensive option in the long run, so they have sought ways of minimising their project risks. Formerly, the design-bid-build delivery system was the norm, where the owner entered into separate contracts for the design and construction of a project. However, they then adopted design-build delivery systems, where from inception to completion, only one organisation was liable to the owner for defects, delays and losses. Streamlining the delivery system also reduced the delivery time for completion by forcing consultancy/design teams and contractors/construction companies to form collaborations and complete the separate tasks at the same time, thus working in parallel. This system is used to minimise the project risk for an owner and to reduce the delivery schedule by overlapping the design phase and construction phase of a project. However, this approach does not incorporate 'life-cycle costing'. The benefits of life-cycle costing are particularly important, as most infrastructure owners spend more money maintaining their systems than on expansion. In addition, the life-cycle approach removes important maintenance issues from the political vagaries affecting many maintenance budgets, with owners often not knowing how much funding will be available to them from year to year. In such cases, they are often forced to spend what money they do have on the most pressing maintenance needs rather than on a more rational and cost-effective, preventive approach. Major infrastructure projects have now moved to design-build-operate(-maintain) or 'turnkey' procurement, which for example, the US Department of Transportation – Federal Highway Administration defined as an integrated partnership that combines the design and construction responsibilities of design-build procurements with operations and maintenance.

The advantage of the design-build-operate(-maintain) approach is that it gives responsibility for usually disparate functions – design, construction and maintenance – to a single entity. This allows the private partners to take advantage of several efficiencies. The project design can be tailored to the construction equipment and materials that will be used. In addition, the design-build-operate(-maintain) team is also required to establish a long-term maintenance programme up front, together with estimates of the associated costs. The team's detailed knowledge of the project design and the materials utilised allows them to develop a tailored maintenance plan that anticipates and addresses needs as they occur, thereby reducing the risk that issues will go unnoticed or unattended and then deteriorate into much more costly problems.

Few structures fall down in the UK, but when they do, the consequences and ramifications are huge. Avoid the complacency, as it leads to tragedy.

1.12. Conclusions

A basic understanding of all the degradation processes is needed by all involved in the design or construction of reinforced-concrete structures. Potential problems can be eliminated, or at least minimised, by due consideration to durability criteria in the design and specification of new structures. Ignorance of, or lack of attention to, such criteria in the past has led to high maintenance costs for owners of infrastructure. It is to be hoped that today's practitioners will learn from these lessons and reduce the need for such remediation in the future.

REFERENCES

1. ACI Committee 201 (2001) Guide to durable concrete. American Concrete Institute, Farmington Hills, MI, USA, Report ACI 201.2R-01, p. 41.
2. National Research Council Committee (1987) *Concrete Durability – A Multibillion-Dollar Opportunity*. National Academy Press, Washington, DC, USA.
3. Anon. (1995) Concrete condemned. *New Civil Engineer*, 30 Mar.: p. 6.
4. Coull ZL, Atkin CP, Lambert P and Chrimes J (2003) *The Evolution of Reinforced Concrete Repair Techniques at the Silver Jubilee Bridge*. Latincorr, Universidad de Santiago de Chile, Chile.
5. Anon. (1995) Terminal operation – miracle cure. *New Civil Engineer*, 23 Mar.: pp. 18–21.
6. Woodward RJ and Williams FW (1988) Collapse of Ynys-y-Gwas Bridge, West Glamorgan. Part 1. *Proceedings of the Institution of Civil Engineers* **84A**: 635–669.
7. Concrete Society Working Party in collaboration with the Concrete Bridge Development Group (1996) *Durable Bonded Post-tensioned Concrete Bridges*. Concrete Society, Crowthorne, UK, Technical report 47, p 64.
8. Anon. (1998) Thaumasite test for Cotswolds. *New Civil Engineer*, 21 May: p. 3.
9. Anon. (1998) Sulphate attack hits M5 bridges. *New Civil Engineer*, 2 Apr.: p. 3.
10. Anon. (1998) Industry ignored warnings on thaumasite threat. *New Civil Engineer*, 9/16 Apr.: p. 3.
11. Anon. (1998) More sulphate casualties come to light. *New Civil Engineer*, 23 Apr.: p. 8.
12. Anon. (1998) Crash damage clouds M5 thaumasite case. *New Civil Engineer*, 11 June: p. 8.

13. Thaumasite Expert Group (1998) *The Thaumasite Form of Sulphate Attack: Risks, Diagnosis, Remedial Works and Guidance on New Construction*. Department of the Environment, Transport and the Regions, London, UK, p. 180.
14. Anon. (1999) Experts cast doubt on Thaumasite report. *New Civil Engineer*, 28 Jan.: pp. 3–4.
15. Anon. (1999) No need to panic. *New Civil Engineer*, 11 Feb.: pp. 12–13.
16. Anon. (1999) New thaumasite evidence. *New Civil Engineer*, 4 Mar.: p. 6.
17. Anon. (1999) Contractors bear thaumasite responsibility. *New Civil Engineer*, 15 Apr.: p. 6.
18. Somerville GS (2004) Joint detailing key to flat slab. *New Civil Engineer*, 8 Jan.: p 12.
19. Wood JGM (1997) *Pipers Row Car Park, Wolverhampton – Quantitative Study of the Causes of the Partial Collapse on 20 March 1997*. Health and Safety Executive, London, UK, http://www.hse.gov.uk/research/misc/index.htm.
20. Tittarelli F, Ruello ML, Capuani D and Aprile A (2017) Damaged concrete viaduct in an Italian highway: Concrete characterization and possible strengthening techniques by FRP applications in comparison. *IOP Conference Series – Materials Science and Engineering* **245**(**3**): 32–69.
21. Milne O (2017) Flyover won't be given special protection from demolition. *Liverpool Echo*, 9 Sep., https://www.liverpoolecho.co.uk/news/liverpool-news/flyover-wont-given-special-protection-13595960 (accessed March 2021).
22. Houghton A (2019) Churchill Way flyovers WILL be demolished – everything you need to know. *Liverpool Echo*, 22 Feb., https://www.liverpoolecho.co.uk/news/business/churchill-way-flyovers-demolished-everything-15869473 (accessed March 2021).
23. Sholli S (2019) Liverpool flyovers' catalogue of faults. *New Civil Engineer*, 5 Mar.
24. Mattioli G (2019) What caused the Genoa bridge collapse – and the end of an Italian national myth? *The Guardian*, 27 Feb., https://www.theguardian.com/cities/2019/feb/26/what-caused-the-genoa-morandi-bridge-collapse-and-the-end-of-an-italian-national-myth (accessed March 2021).
25. Botti D, Kerr S and Marcolini B (2018) Genoa bridge collapse: Video shows moment road fell. *The New York Times*, 14 Aug., https://www.nytimes.com/video/world/europe/100000006054214/genoa-bridge-collapse-video-shows-moment-road-fell.html (accessed March 2021).
26. Pinna M (2020) Italy's crumbling motorways: How the Genoa bridge collapse exposed a national scandal. *Euronews*, 23 Mar., https://www.euronews.com/2020/02/28/italian-motorways-a-national-scandal (accessed March 2021).
27. Horgan R (2020) Latest Italy bridge collapse highlights maintenance failings. *New Civil Engineer*, 9 Apr., https://www.newcivilengineer.com/latest/latest-italy-bridge-collapse-highlights-maintenance-failings-09-04-2020/ (accessed March 2021).

Soutsos M
ISBN 978-0-7277-6375-4
https://doi.org/10.1680/icehcd.63754.013

Chapter 2
Pore structure and transport processes

PA Muhammed Basheer
University of Leeds, UK

Salim Barbhuiya
University of East London, UK

2.1. Introduction

Concrete durability depends largely on the ease (or difficulty) with which fluids in the form of liquid (water), gas (carbon dioxide and oxygen) or ions (chlorides and sulfates) can migrate through the hardened concrete mass. Concrete is a porous material, and hence, moisture movement can occur by flow, diffusion or sorption. Generally, the overall potential for moisture, gas and ion ingress in concrete by these three modes is referred to as its permeability, but permeability does not strictly represent all the three mechanisms of transport. The durability of concrete is also influenced or controlled by the number, type and size of pores present. It is believed that capillary voids of 50 mm in diameter, referred to as macropores, are detrimental to strength and impermeability, whereas voids smaller than 50 mm, referred to as micropores, affect drying shrinkage and creep [1]. Any water in pores bigger than 50 nm behaves as free water and plays an important role in the durability of concrete. Therefore, to improve durability, a clear understanding of the characteristics of the pore structure and permeation properties of concrete is very much essential.

2.2. Pore structure of concrete

The properties of concrete are strongly affected by how the pores of various sizes are distributed within it. Concrete with the same total pore volume can exhibit entirely different properties, depending on whether it contains a small number of large pores or a large number of small pores. Concrete is a porous material, and its pore structure is extremely tortuous, complex and spatially inhomogeneous. For example, the cement paste close to aggregates, that is, in the interfacial transition zone, has a different pore structure than the bulk cement paste. In addition, fine and coarse aggregate particles in concrete may have their own pore systems, which might be completely different from the pore system of the cement paste. The pores present in concrete may also vary in size, shape and origin. There are various types of pores in concrete, such as

- pores in the hydrated cement paste (HCP)
- pores in the aggregates
- pores associated with the interfacial transition zone
- water-filled capillary voids
- voids due to construction deficiencies, for example, honeycombing due to poor compaction
- internal discontinuities in the HCP.

Figure 2.1 shows schematically the sequence in the formation of pore structure in HCP as the hydration proceeds. This involves the replacement of water that separates individual cement grains in the fluid paste

Figure 2.1 Schematic diagram of the structural development of pores in HCP: (a) initial mix, (b) after 7 days, (c) after 28 days and (d) after 90 days. (Reproduced from [2], © 2002 Taylor & Francis)

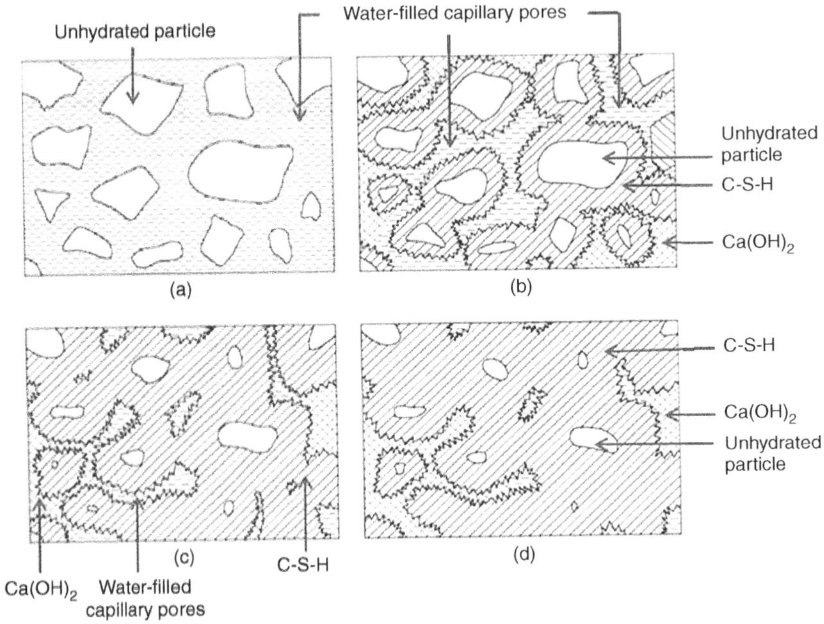

(Figure 2.1(a)) with solid hydration products. Owing to their lower specific gravity (~2.0 compared with 3.2 of cement), the hydration products occupy a greater volume than the original cement compounds and, thus, form a dense matrix. The hydration products also bind the residual cement grains together over time, as illustrated in Figures 2.1(b)–2.1(d). Although calcium sulfoaluminates crystallise as separate phases, for simplification these are included as part of calcium silicate hydrate (C-S-H) in Figure 2.1.

HCP is produced when cement chemically reacts with water. The HCP contains C-S-H gel, crystals of calcium hydroxide ($Ca(OH)_2$), calcium sulfoaluminate (ettringite), minor residues of the original unhydrated cement and residues of the original water-filled spaces in the fresh paste. There are four main types of pores in HCP, namely

- gel pores
- capillary pores
- hollow-shell pores
- air voids.

The features and significance of each type of pore in the transport processes in concrete are given in Table 2.1, which gives three methods for classifying pores by size.

2.2.1 Gel pores

Gel pores are inherent pores in C-S-H gel. Although Powers assumed that the gel pores have a width of 18 Å (1 Å = 10^{-10} m) [3], Feldman and Sereda [4] suggested that these spaces may vary from 5 to 25 Å. These pores do not play any significant role in the flow of water through concrete, because pores with a radius of less than 500 Å do not contribute to the water permeability [5–9]. However, they may contribute to drying shrinkage and creep [4].

Table 2.1 Classification of pore sizes according to the general classification by IUPAC and concrete science terminology

According to IUPAC [12]		According to Mehta [1]		According to Mindess *et al.* [13]		
Name	Diameter	Pore type	Size range	Name		Diameter
Micropores	<2 nm	Inter-particle space between C-S-H sheets	1–3 nm	Inter-layer between micropores	Gel pores	<0.5 nm
				Micropores Small capillaries		0.5–2.5 nm
Mesopores	2–50 nm	Capillary pores (low water-to-cement ratio)	10–50 nm	Medium capillaries	Hollow-shell pores Capillary pores	2.5–10 nm 10–50 nm
Macropores	>50 nm	Capillary pores (high water-to-cement ratio)	3–5 μm	Large capillaries		50 nm–10 μm
		Entrained air voids	50 μm–1 mm	Entrained air		0.1–1 mm

2.2.2 Capillary pores

Capillary pores are the spaces not filled by the solid components of HCP. Therefore, the volume and size of the capillary voids depend on the distance between unhydrated cement particles in the freshly mixed cement paste and the degree of hydration [10,11]. Unlike gel pores, these pores are assumed to have a major effect on transport processes but only a minor effect on hydration rates. In well-hydrated and low water-to-cement ratio pastes, the capillary voids may range from 100 to 500 Å, whereas in high water-to-cement ratio pastes, at early ages of hydration, the capillary voids may be as large as 3000 to 5000 Å.

2.2.3 Hollow-shell pores

Hollow-shell pores are closed, distinct pores, which formed from the original cement grain boundaries. Their size ranges from 1 to 15 μm. At low water-to-cement ratios, they can be larger than the capillary pores by more than two orders of magnitude [12].

2.2.4 Air voids

Air voids in concrete are due either to air entrapped during casting or to intentional entraining when using an air-entraining admixture. Entrapped air cavities may be as large as 3 mm, and entrained air voids may range from 50 μm (5×10^5 Å) to 200 mm (2×10^9 Å). They are much bigger than the capillary voids and have a significant role in the permeability of concrete.

As discussed, the range of pore sizes in HCP is very wide, from 1 nm to 1 cm. A comparison of the different pore size ranges according to the International Union of Pure and Applied Chemistry (IUPAC) and using concrete scientists' terminology is presented in Table 2.1.

2.3. Pore structure of the transition zone

A preferential orientation of large crystals of calcium hydroxide and calcium sulfoaluminate occurs around aggregates due to a local increase in the water-to-cement ratio close to them. A water film forms around the aggregates in fresh compacted concrete and bleed water collects underneath flaky coarse aggregates, which account for this variation in the water-to-cement ratio. As a result, the concrete is more porous close to the aggregates, with larger capillary voids closer to the aggregates. Therefore, the properties of the transition zone significantly affect the permeability of concrete.

2.4. Techniques used to study the pore structure

A wide variety of techniques have been used to characterise the pore structure of concrete. Some of the methods have access only to the open pores, whereas others have access to both open and closed pores. Each of these techniques is most suitable for a specific size range. The various techniques used in characterising the pore structure of concrete are

- mercury intrusion porosimetry
- gas adsorption
- displacement methods
- thermoporometry
- nuclear magnetic resonance
- small-angle scattering
- optical microscopy
- scanning electron microscopy.

Details on the suitability of these test techniques along with the range of pore sizes over which each technique is applicable can be found in [12].

2.5. Factors influencing the pore structure

The main factors that influence the pore structure of concrete include the water-to-cementitious materials (w/cm) ratio, degree of hydration, supplementary cementitious materials, chemical admixtures and curing conditions. In addition, the type of cement and its age can also influence the pore structure of concrete. The pore structure of HCP and, hence, the concrete largely depend on the w/cm ratio and the degree of hydration. In general, the higher the w/cm ratio for a given degree of hydration, the higher will be the volume of larger pores in HCP [14].

The addition of various supplementary cementitious materials, such as pulverised fuel ash (also known just as fly ash), ground granulated blastfurnace slag, microsilica (also known as silica fume) and metakaolin, refine the pore structure of concrete mainly by producing secondary C-S-H. For microsilica and metakaolin, the fine nature of the particles also contributes to pore refinement in concrete. The pore structure of HCP also changes greatly when different chemical admixtures are used. This is mainly due to the change in the degree of hydration of the cement. At a constant w/cm ratio, the addition of a superplasticiser has variously been reported to refine the pore structure [15–17]. In particular, Khatib and Mangat [17] demonstrated that the inclusion of a superplasticiser decreased the total intruded pore volume of HCP.

Curing concrete properly is widely recognised as a necessity for ensuring the adequate field performance of concrete structures. Not only is it important to minimise evaporation of the concrete mix water but also to provide a source of external or internal curing water to replace that consumed by chemical shrinkage during the hydration of the cement. A cement paste, mortar or concrete cured under sealed conditions will self-desiccate, resulting in the creation of coarse capillary pores within the microstructure. For a w/cm ratio greater than approximately 0.42, there is sufficient water in the mix such that complete hydration of the cement can be achieved theoretically without supplying additional water to the cement paste. Even if complete hydration is achievable, the lack of additional curing water can still result in the creation of relatively large pores within the final microstructure. In this case, the addition of curing water ensures that all pores remain filled with water and are eligible as locations for the precipitation and growth of hydration products during curing.

2.6. Transport processes in concrete

The durability of concrete depends largely on the resistance it offers to the penetration of fluids (such as water, carbon dioxide, oxygen and salts) from its service environment into and through its matrix. As concrete is a porous material, the transport of these substances can occur by capillary-suction-driven flow (absorption), concentration-driven flow (diffusion) or pressure-driven flow (permeation). Generally, the overall potential for the ingress of substances into concrete by these four modes is referred to as its permeability.

- *Adsorption* is the process by which ions or molecules from a gas, liquid or dissolved solid adhere to the surface of concrete pores [18].
- *Absorption* is the process by which concrete takes in a fluid due to capillary suction in pores. This fluid fills the spaces within the material [18,19].
- *Diffusion* is the random motion of free molecules or ions in a saturated pore system under the influence of a concentration gradient [18,20]. Diffusion can be in either gaseous or ionic form. A typical example of gaseous diffusion is the diffusion of carbon dioxide into concrete, which can lead to the carbonation of the concrete. In contrast, a typical example of ionic diffusion is chloride diffusion, which can cause corrosion of the reinforcing bars embedded in the concrete. Diffusion can be either steady or non-steady. More details on this can be found elsewhere [21].
- *Permeation* refers to a fluid passing into and through the body of the concrete under a pressure difference [18,20,22]. A typical example is that occurring in deepwater marine structures or the bases of retaining structures.

2.6.1 Fundamentals of transport processes

Rose [19] distinguished six stages in the transfer of a fluid flux through a porous medium. These are presented in Figure 2.2 from (a) to (f). In addition to these stages, Report 31 of the Concrete Society [18] outlined two more stages, which are the ionic diffusion through (g) a partially saturated or (h) a fully saturated porous system. For simplicity a single pore with a neck at each end only is presented in Figure 2.2. The vapour flux is incident from the left in (a) to (f), and an ionic flux is incident from the left in (g) and (h). The various stages presented in Figure 2.2 are explained below.

The first stage (a) is adsorption. Until this is complete, a vapour flux cannot be transmitted, and 'conductivity' has no meaning, even though the flux will move to adsorption sites as vapour. This does not prevent surface diffusion in the adsorbed phase. After the initial adsorption, the second stage (b) is one of unimpeded vapour movement, where the vapour behaves like an ideal gas. This is expressed by Fick's first law for diffusion:

$$J_x = -D_x \frac{\delta c}{\delta x} \tag{2.1}$$

where J_x is the flux of fluid at position x in the x direction from the origin, D_x is the diffusion coefficient at position x and $\delta c/\delta x$ is the concentration gradient at position x. (Fick's second law of diffusion relates the rate of change of concentration gradients by assuming that the diffusion coefficient is independent of position.)

The third and fourth stages occur when the necks contain liquid, either (c) without or (d) with a thin film of significant thickness on the walls of the cavity. In stage (c), the system is impervious to an inert gas and pervious to liquid only by a process of distillation, in which the necks act as short circuits for vapour movement. The process is described as liquid-assisted vapour transfer, the assistance arising because liquids shorten the effective path length for vapour diffusion.

Figure 2.2 Idealised model of the transfer of liquids and ions within concrete [18,19]

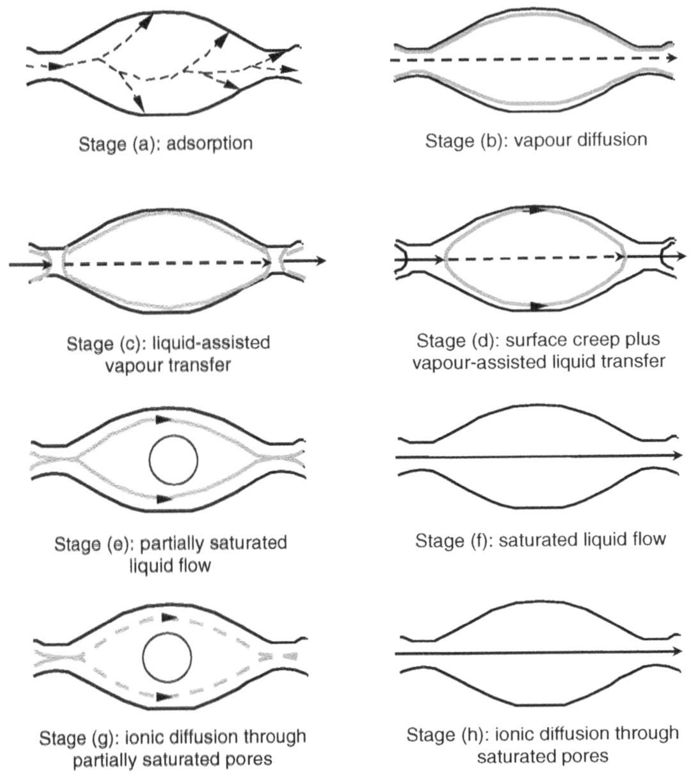

In stage (d), surface creep occurs. That is, the flow is in thin liquid films, in which there is vapour-assisted liquid transfer. Eventually, there is a transition to the fifth and sixth stages, where there is liquid flow, with stage (e) representing the unsaturated condition and stage (f) representing the saturated condition. There is an important difference between (d) and (e), best understood by ignoring the arrows and treating the diagram as depicting a dynamic equilibrium. In (e), the air–water boundary has the same curvature everywhere, but in (d), the curvature away from the necks is largely determined by the shape of the solid surface.

As flow occurs in a pore due to the pressure differential across the boundary (meniscus) in stage (e), the rate of flow is given by Washburn's equation [23]:

$$v = \frac{r\gamma}{4d\mu}\cos\theta \tag{2.2}$$

where v is the rate of flow, r is the capillary radius, γ is the surface tension, d is the depth of penetration of the fluid, μ is the fluid viscosity and θ is the contact angle.

The flow in stage (f) – the saturated condition – is due to a high-pressure head across the pore. Therefore, the rate of flow is governed by Darcy's law. For a non-compressible fluid and a saturated porous medium, Darcy's law states that the steady-state rate of flow is directly proportional to the hydraulic gradient:

$$v = \frac{Q}{A} = -Ki = -K\frac{\delta h}{\delta l} \tag{2.3}$$

where v is the apparent velocity of the flow, Q is the flow rate, A is the cross-sectional area, i is the hydraulic gradient, δh is the head loss over a flow path of length δl and K is the coefficient of permeability or hydraulic conductivity.

Note that the application of Darcy's equation to the flow through porous media is based on the following assumptions.

- Complete saturation has been achieved.
- The flow is laminar and viscous.
- Equilibrium flow conditions have been established.

A more rational concept of permeability, which is independent of the fluid properties and dependent purely on the characteristics of the porous body, is intrinsic permeability. This is expressed as:

$$v = \frac{Q}{A} = -\frac{k}{\mu}\frac{\delta p}{\delta l} \tag{2.4}$$

where μ is the viscosity of the fluid, k is the intrinsic permeability of the porous medium and δp is the pressure loss over the flow path of length δl.

In addition to the stages from (a) to (f), ionic diffusion can occur in stages (e) and (f), as shown in stages (g) and (h). This is superimposed over the other transport processes already explained with reference to these figures. Ionic diffusion is also governed by Fick's first law. It can be seen in Figure 2.2 that the moisture condition of the concrete influences various stages of the transfer of fluids through it.

2.6.2 Methods used to study transport processes

The primary test methods used to study the transport processes in concrete can be broadly classified into three categories: absorption tests, diffusion tests and permeation tests. The various methods under these categories are listed in Figure 2.3. Detailed descriptions of these test methods can be found in the literature [23–25].

2.6.3 Factors influencing transport processes

The factors influencing various transport processes in concrete are presented in Table 2.2. It is worthwhile to mention here that the interaction among the factors identified in Table 2.2 is complex, as some of the variables are interdependent. For instance, an increase in the cement content normally means a reduction in aggregate content. A detailed discussion on how all these factors influence the transport processes in concrete is beyond the scope of this chapter. More information on these can be found elsewhere [24–26].

It can be seen in Table 2.2 that both the constituent materials and treatment given subsequent to concreting accounted for majority of the variations in transport properties reported in publications. Both the ambient temperature and moisture condition were found to be factors influencing transport processes. Therefore, tests for measuring the transport properties of concrete should take account of these environmental parameters for all three transport processes.

Figure 2.3 Classification of test methods used to study the transport processes in concrete

Absorption tests	Tests for water absorption capacity			
	Sorptivity tests	Sorptivity from water absorbed		
		Sorptivity from water penetration depth		
	Absorptivity tests	Surface absorptivity tests	Initial surface absorption test	
			Autoclam sorptivity test	
			Stand pipe sorptivity test	
		Drill-hole absorptivity tests	Figg water absorption test	
			Covercrete absorption test	
Diffusion tests	Gas diffusion tests	Water vapour transmission test		
		Water vapour transpiration test		
		Oxygen diffusion test		
	Ionic diffusion tests	Steady-state diffusion test		
		Non-steady-state diffusion tests	Immersion test	
			Ponding test	
		Electric field migration tests	Steady-state test	
			Non-steady-state tests	Rapid chloride permeability test
				NT Build 492 [25]
Permeation tests	Liquid permeability tests	Steady-state water flow test		
		Non-steady-state water flow test	Steinart guard ring test	
			Autoclam water permeability test	
		Water penetration test		
	Gas permeability tests	Constant head gas permeability test		
		Falling head gas permeability tests	Drill hole suction test	
			Drill hole overpressure test	
			Surface suction test (Torrent test)	
			Surface overpressure test (Autoclam air permeability test)	

2.7. Overview of mechanisms of concrete deterioration

The technical term for the damage done to concrete is deterioration, in contrast to corrosion, which is the damage done to metals. Based on the durability problems reported in various technical works in the literature [27,28], the main causes of deterioration of concrete are as follows

- abrasion, erosion and cavitation
- alkali–aggregate reaction, alkali–silica reaction and alkali–carbonate reaction
- chemical attack
  - acid attack
  - alkali attack

Table 2.2 Factors influencing transport processes in concrete

Factors	Subfactor	Percentage of references in nearly 150 publications
Constituent materials	Water content	16.55
	Cementitious materials content	7.20
	Aggregate size, type and content	5.76
	Admixtures and cement replacements	7.20
Method of preparation	Mixing	0.72
	Method of casting	0.72
	Compaction	0.72
	Trowelling	0.72
Subsequent treatment	Curing	16.55
	Protective treatments	0.72
	Age	9.32
Interdependence	Porosity	7.19
	Pore size distribution	4.32
	Strength	4.32
Test parameters	Test pressure and duration of test	4.32
	Ambient environmental conditions	3.60
	Moisture content	5.04

  - carbonation
  - chloride attack
  - leaching
  - salt attack
  - sulfate attack
  - delayed ettringite formation
  - thaumasite attack
- corrosion of reinforcement
- cracking
- freeze–thaw and frost damage.

Among the various causes listed above, carbonation, chloride ingress and leaching can all lead to the corrosion of reinforcement, whereas their chemical effect on the integrity of the concrete is less important. Similarly, the action of salt is not solely due to the chemical reaction with the concrete but also due to the crystallisation pressure in the capillary pores in the concrete [29]. Further, though corrosion is a chemical process, the deterioration of reinforced concrete is due to the reduction in the cross-section of the reinforcement as well as the physical effects of the corrosion products, which can crack the concrete. Various forms of non-structural cracks in concrete are also due to a combination of physical and chemical effects [30,31]. That is, it is difficult to classify the causes of concrete deterioration due to the interactions among various factors, but efforts have been made to group the various types as physical or chemical [27,28].

2.7.1 Physical deterioration mechanisms

The physical deterioration mechanisms of concrete include cyclic freezing and thawing, abrasion, erosion and cavitation [28,29]. Concrete deteriorates due to cyclic freezing and thawing because of the generation of hydraulic pressure in the capillaries, diffusion of gel water in these capillaries by freezing and osmotic pressure. The rate of deterioration due to freezing and thawing depends not only on the number of cycles of freezing and thawing and the exposure conditions but also heavily on the content and spatial distribution of the water in the pore structure of the concrete. For concrete to deteriorate due to cyclic freezing and thawing, a critical saturation must be reached and the exposure temperature must fall below what is required for ice to form. The result of cyclic freezing and thawing is extensive cracking of concrete, loss of mechanical properties and eventual spalling of the concrete. Freeze–thaw damage needs to be differentiated from the frost damage of newly placed concrete; the latter occurs when newly placed concrete is exposed to a temperature below 0°C and the water in it freezes, resulting in cracking and spalling of the concrete.

Both abrasion and erosion of concrete occur progressively [28]. Abrasion occurs due to the physical attrition on the surface of the concrete. In contrast, erosion occurs when concrete is in contact with continually moving water for a long period of time, resulting in the progressive leaching of calcium ions from it. This may induce dissolution of cement hydrates, so that the microstructure of the concrete becomes porous, which eventually increases its permeability. Cavitation occurs on a concrete surface when there are discontinuities or irregularities in the path of a high-speed water flow. If the bubbles collapse against a concrete surface, there is a very high-pressure impact over a tiny area of the surface, resulting in the removal of substantial volumes of concrete.

2.7.2 Chemical deterioration mechanisms

The effect of some of the most common chemicals on the deterioration of concrete, as summarised by ACI Committee 201 [27], was presented in the introduction to this section. A detailed description of these mechanisms is available in Chapter 16 of the *ICE Manual of Construction Materials*, 'Durability of Concrete' [28]. In most cases, the chemical interactions between the aggressive agents present in the external environment and the constituents of the cement paste determine the extent of the deterioration of the concrete [28,32]. An exception is the effect of salt used as a de-icing agent, in which case the effect is both physical and chemical. That is, the main cause of almost all chemical deterioration in concrete is the ingress of aggressive agents, except for the alkali–aggregate reaction, which is an internal process. Even in that case, the deterioration depends on the availability of moisture for the reaction products to expand [28,29]. These aggressive agents include water, chloride and sulfate ions, carbon dioxide, alkalis and acids.

2.7.3 Corrosion of reinforcement

The major form of deterioration of many reinforced-concrete structures is corrosion of the steel reinforcement, which is initiated when a change in the nature of the pore solution surrounding the steel occurs due to carbonation, chloride ingress or leaching of the pore fluid [33]. Chlorides can penetrate concrete due to absorption or diffusion. Carbonation occurs when carbon dioxide from the atmosphere penetrates the concrete and reacts with some of the cement hydration products, resulting in a reduction in alkalinity. Leaching of alkaline material from concrete occurs when the concrete is exposed to continually flowing water. Note that all these mechanisms are related to the ease with which a fluid or ions can move through the concrete pores. As stated previously, the effect of corrosion is both chemical (reduction in steel cross-section) and physical (cracking of concrete).

2.7.4 Cracking of reinforced concrete

Cracking indicates the presence of disruptive forces within a mass of material that exceed its tensile strength. These forces can be caused by the application of an external load or by internal changes in the material. For concrete, they can occur due to either of these causes or by a combination of the two while the concrete is in either a pre-hardened or hardened state [31]. To identify the causes of cracking, cracks formed in the hardened state are generally separated from those formed in the pre-hardened state. The various types in

Table 2.3 Cracks in reinforced-concrete structures

Crack propagation state	Type of cracks (cause of crack formation)
Pre-hardened state	Early frost damage
	Plastic settlement
	Plastic shrinkage
	Formwork movement
	Subgrade movement
	Constructional movement
Hardened state	Physical
	Drying shrinkage
	Shrinkable aggregate
	Crazing
	Chemical
	Alkali–aggregate reaction
	Corrosion of reinforcement
	Cement carbonation
	Thermal
	External seasonal temperature variations
	Freeze–thaw
	Early thermal contraction
	Structural
	Accidental overload
	Creep
	Design loads
	Foundation movements

each category are represented by a comprehensive family tree of cracks in Table 2.3. Cracks accelerate the penetration of aggressive substances into the concrete, which in turn aggravates any one or a number of other mechanisms of deterioration.

One of the common characteristics in almost all the processes is that they require the transport of aggressive substances from the exposure environment into the concrete [32]. In other words, concrete deterioration processes are highly dependent on the penetration of water or aggressive agents, such as chlorides, sulfates, alkalis, etc., via pores and microcracks inherent in its microstructure.

2.8. Influence of pore structure and transport processes on the durability of concrete

The overall effect of pore structure and permeability on the penetration of various aggressive substances and thereby their effect on the durability of concrete is visualised in Figure 2.4 [32]. The parameter at the tail of an arrow influences that at its head. The constituent materials of concrete, the manufacturing method and the subsequent treatment given to the hardened concrete all have a direct effect on the properties of the concrete in a structure. These factors also govern the pore structure and the formation of various microcracks in the concrete. The characteristics of the pore structure and the presence of various microcracks influence the transport processes. Therefore, these are represented in the top three blocks in the figure.

Figure 2.4 Interdependence of transport processes and the durability of concrete [32]

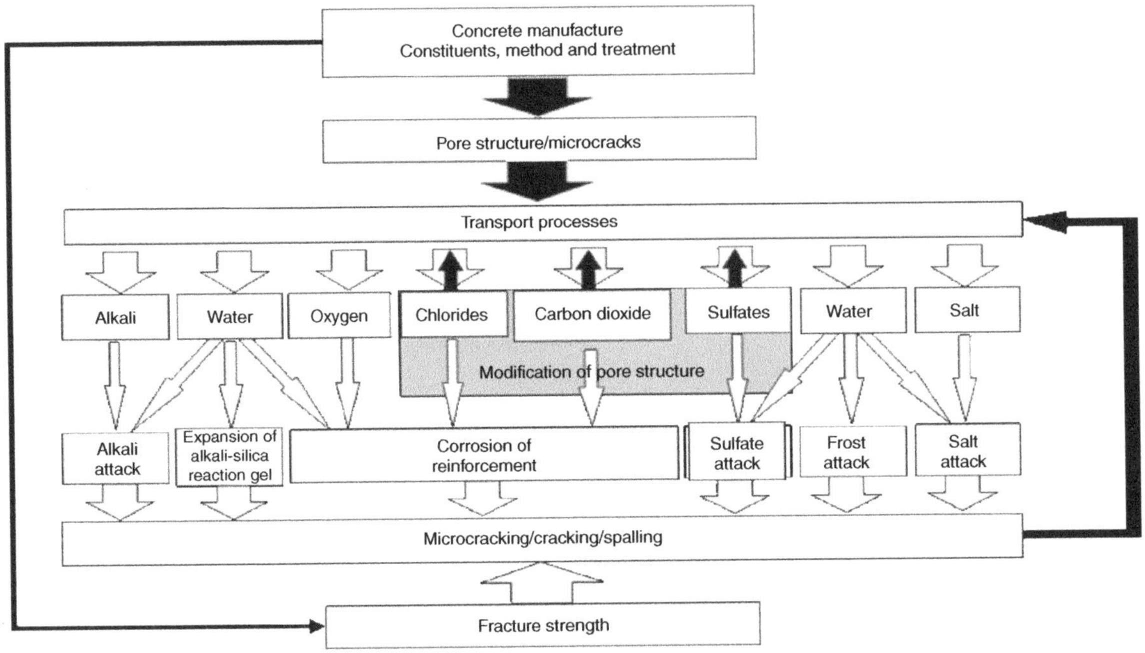

The following line of blocks in Figure 2.4 represents the penetration of various aggressive substances, which govern the different deterioration mechanisms. These mechanisms are shown in the next line, below the aggressive substances. The arrows between these two lines indicate the dependence between the penetration of an aggressive substance and the deterioration mechanisms. As a result of all the mechanisms, concrete cracks, which accelerates the transport processes. In a sulfate attack, a chloride attack or carbonation, a slight modification of the pore structure can occur, which can diminish the transport processes. However, under the prolonged action of the mechanisms, concrete will eventually crack.

Experimental evidence suggests that there is a correlation between the relevant transport properties and either the penetration of different aggressive substances or the mechanisms of deterioration. Typical correlations are presented in Figures 2.5–2.8. The dependence of carbonation on the gas permeability of concrete is illustrated in Figures 2.5 [34] and 2.6 [32]. In Figure 2.5, the depth of carbonation after 20 weeks in an accelerated carbonation test is presented against the intrinsic permeability in a logarithmic scale. Figure 2.6 illustrates the depth of carbonation for 27 different mixes after 14 days in an accelerated test plotted against the Autoclam air permeability. In both sets of data, there is a good correlation between the depth of carbonation and the gas permeability of concrete. The differences between them is considered to be due to the differences in sample preparation, testing conditions and the types of cementitious materials added to the concrete.

Long *et al.* [35] reported that there was a good correlation between Autoclam water permeability and the mass loss (expressed as a percentage of the initial saturated mass) after 300 cycles of the ASTM C666 Procedure A freezing and thawing test. The data presented in Figure 2.7 are based on tests carried out on 27 different concrete mixes.

The dependence of chloride penetration on the sorptivity of concrete can be seen in Figure 2.8 [36]. The data in this figure are based on ponding tests carried out on both surface-treated and untreated concrete

Figure 2.5 Relation between depth of carbonation and permeability [34]

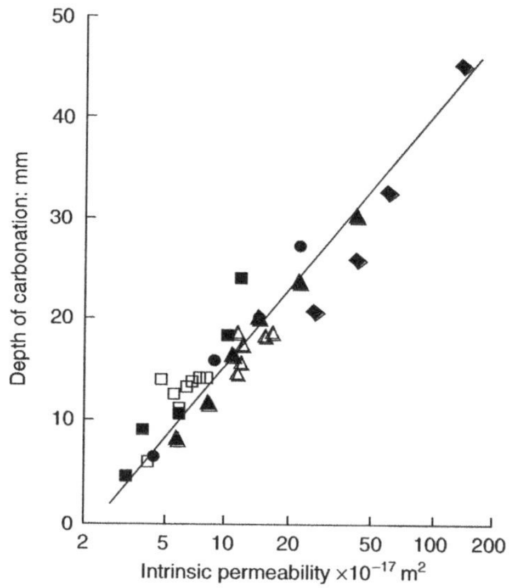

Figure 2.6 Relation between the depth of carbonation and the Autoclam air permeability index [24]

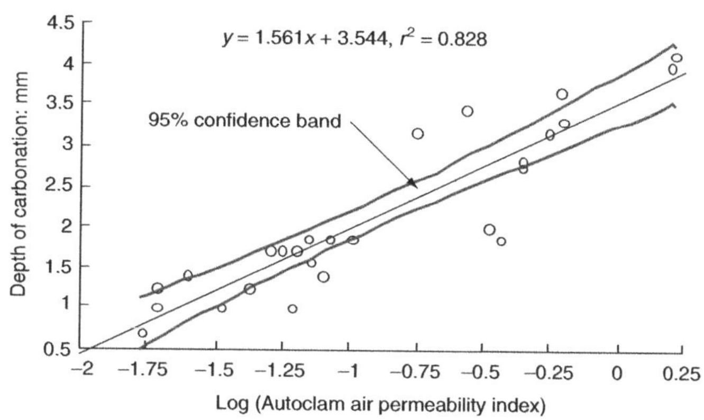

specimens. For the three untreated concrete specimens, there is a linear correlation between the chloride ingress and the sorptivity, so that a reduction in sorptivity due to surface treatment correspondingly resulted in a decrease in the chloride content. Basheer and Cleland [37] reported that the reduction in both the sorptivity and the chloride ingress was accompanied by a corresponding reduction in the rate of corrosion of steel bars embedded in the test specimens.

Whereas the dependence of various mechanisms of deterioration on permeation properties has been well documented by laboratory investigations, there are not much data from in situ investigations. Two graphs are presented in Figures 2.9 and 2.10. These report the dependence of both carbonation and chloride ingress on the air permeability of concrete, based on tests carried out with the Autoclam Permeability System on 11 concrete motorway bridges in Northern Ireland [24,38]. Although the effect of moisture was not completely accounted for in the air permeability results, large variations in both the depth of carbonation and chloride ingress corresponded to a similarly large variation in the air permeability of concrete.

Figure 2.7 Relation between mass loss due to freeze–thaw deterioration and the Autoclam water permeability index [24]

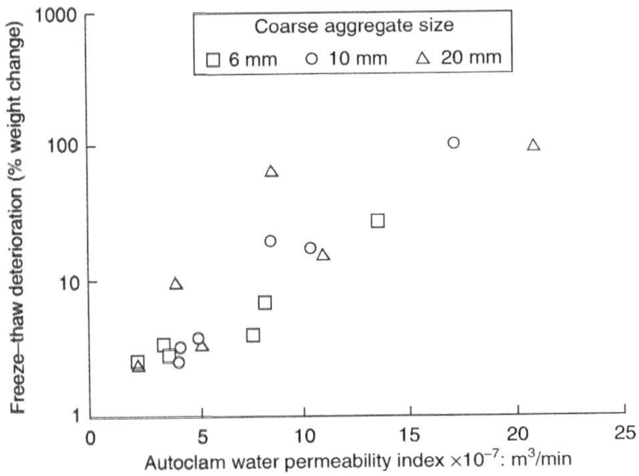

Figure 2.8 Relation between chloride ingress and the Autoclam sorptivity index [36]

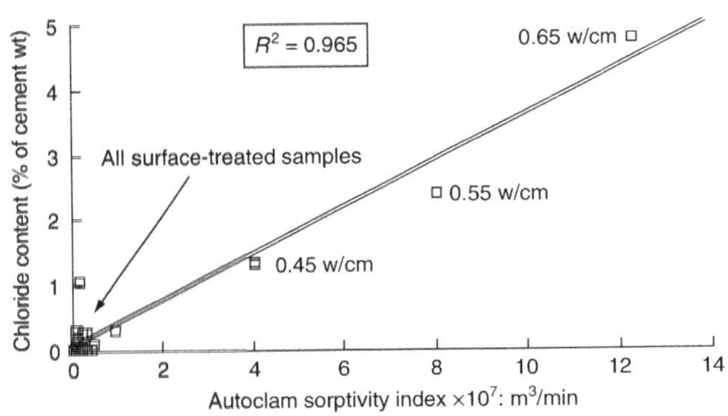

Figure 2.9 Dependence of carbonation on the air permeability of concrete bridges [32]

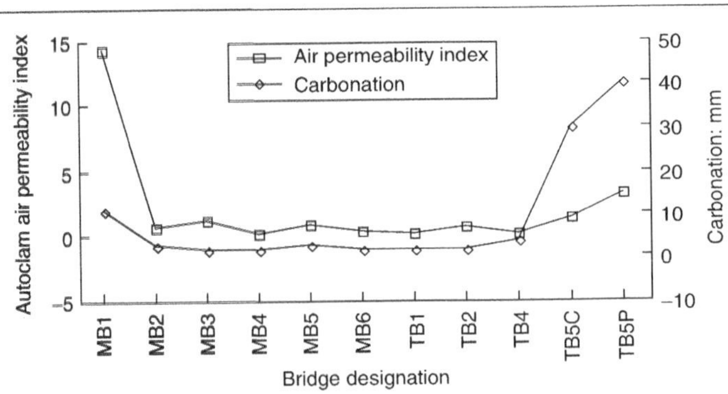

Figure 2.10 Dependence of chloride ingress on the permeability of concrete bridges [32]

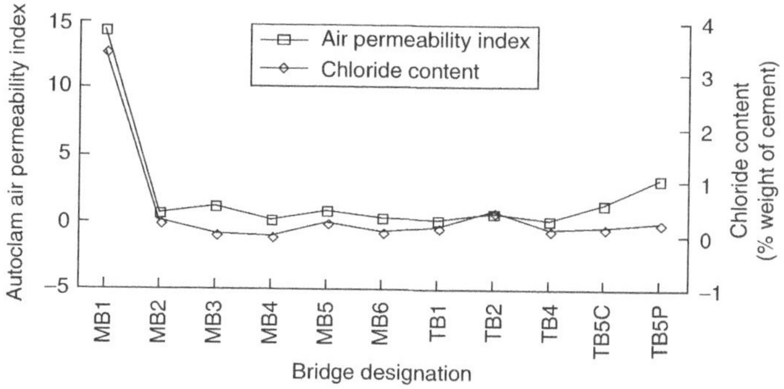

Recall that the conceptual model relating permeation properties to both the penetration of aggressive substances and the deterioration mechanisms (Figure 2.4) can be validated by data, in so far as the mechanisms of deterioration involve the transport of aggressive substances and that these initiate the deterioration. Also, it is relatively easy to show that there is a good correlation between a specific mechanism of deterioration and a related transport property. However, in reality, one mechanism may act as the catalyst for a number of other mechanisms, and the permeation properties alone may not be sufficient to estimate both the rate and the extent of deterioration. Therefore, assessing the durability of concrete requires that the permeation properties are used along with a knowledge of both the physical and chemical characteristics of the concrete.

2.9. Summary

The characteristics of the pore structure of cement-based materials play a fundamental role in governing the transport processes in concrete, and hence, its durability. Variations in the size of the pores in concrete and their influence on both transport properties and durability are highlighted in this chapter. Numerous test methods can be used to characterise the pore structure and permeability of concrete. Environmental factors, such as temperature and moisture variations, affect all transport processes, and hence, there is a need to consider their effects on the measurement of transport properties. Both laboratory and field data highlight the dependence of durability on transport properties. The properties of the constituent materials, the methods of manufacturing concrete and the subsequent treatment of concrete affect both the pore structure and the transport processes, and hence, to design durable structures, attention should be given to these factors.

FURTHER READING

Aligizaki KK (2006) *Pore Structure of Cement-based Materials: Testing, Interpretation and Requirements. Modern Concrete Technology No. 12.* Taylor & Francis, London, UK.

Kropp J and Hilsdorf HK (1995) *Performance Criteria for Concrete Durability.* RILEM Technical Committee 116-PCD, Bagneux, France. RILEM Report 12.

Mehta PK and Monteiro PJ (2005) *Concrete: Structures, Properties and Materials*, 3rd edn. McGraw-Hill, New York, NY, USA.

Neville AM (2011) *Properties of Concrete*, 5th edn. Wiley, London, UK.

Ramachandran VS and Beaudoin JJ (eds) (2000) *Handbook of Analytical Techniques in Concrete Science and Technology: Principles, Techniques and Applications.* Barnes & Noble, Lyndhurst, NJ, USA.

REFERENCES

1. Mehta PK and Monteiro PJ (2005) *Concrete: Structures, Properties and Materials*, 3rd edn. McGraw-Hill, New York, NY, USA.

2. Bensted J and Barnes P (2002) *Structure and Performance of Cements*, 2nd edn. E&FN Spon, London, UK.
3. Powers TC (1958) Structure and physical properties of hydrated Portland cement paste. *Journal of the American Ceramic Society* **41(1)**: 1–6.
4. Feldman RF and Sereda PJ (1968) A model for hydrated Portland cement paste as deduced from sorption-length change and mechanical properties. *RILEM Bulletin* **1(6)**: 509–520.
5. Diem P (1986) Determination of capillary absorptiveness of very dense concrete. *Betonwerk und Fertigteil-technik* **52(11)**: 719–724.
6. Mehta PK and Manmohan D (1980) Pore size distribution and permeability of hardened cement pastes. *7th International Symposium of the Chemistry of Cement, Paris, France*, vol. 3, pp. 181–185.
7. Midgley HG and Illuston JM (1983) Some comments on the micro-structure of hardened cement pastes. *Cement and Concrete Research* **13**: 197–206.
8. Nieminen P and Romu MI (1988) Porosity and frost resistance of clay bricks. *Proceedings of the 8th International Brick/Block Masonry Conference, Dublin, Ireland*, pp. 103–109.
9. Nyame BK (1985) Permeability of normal and lightweight mortars. *Magazine of Concrete Research* **37(130)**: 44–48.
10. Parrott LJ and Killoh DC (1984) Prediction of cement hydration. The chemistry and chemically related properties of cement. *British Ceramic Proceedings* **13(2)**: 41–53.
11. Parrott LJ (1985) Effect of changes in U.K. cements upon strength and recommended curing times. *Concrete: Journal of the Concrete Society* **19(9)**: 22–24.
12. Aligizaki KK (2006) *Pore Structure of Cement-based Materials: Testing, Interpretation and Requirements. Modern Concrete Technology No. 12*. Taylor & Francis, London, UK.
13. Mindess S, Young JF and Darwin D (2002) *Concrete*, 2nd edn. Prentice Hall, Englewood Cliffs, NJ, USA.
14. Klieger P and Lamond JF (1994) *Significance of Tests and Properties of Concrete and Concrete-making Materials*. ASTM, Philadelphia, PA, USA.
15. Mor A and Mehta PK (1984) Effect of superplasticizing admixtures on cement hydration. *Cement and Concrete Research* **14(5)**: 754–756.
16. Syal SK and Kataria SS (1982) Development and interaction of a concrete additive for improved performance and durability. *Cement Concrete and Aggregates* **4(2)**: 110–114.
17. Khatib JM and Mangat PS (2002) Influence of high-temperature and low-humidity curing on chloride penetration in blended cement concrete. *Cement and Concrete Research* **32(11)**: 1743–1753.
18. Report of a Concrete Society Working Party (1988) *Permeability Testing of Site Concrete – A Review of Methods and Experience*. Concrete Society, London, UK, Technical report 31.
19. Rose DA (1965) Water movement in unsaturated porous materials. *RILEM Bulletin* **29**: 119–124.
20. Feldman RF and Sereda PJ (1968) A model for hydrated Portland cement paste as deduced from sorption-length change and mechanical properties. *RILEM Bulletin* **1(6)**: 509–520.
21. Nanukuttan SV (2007) *Development of a New Test Protocol for the PERMIT Ion Migration Test*. PhD thesis, Queen's University Belfast, Belfast, Northern Ireland.
22. Mercer LB (1945) Permeability of concrete – 1. Theoretical considerations, laboratory test methods, details of experimental work with new apparatus. *Commonwealth Engineer* 349–357.
23. Ramachandran VS and Beaudoin JJ (eds) (2000) *Handbook of Analytical Techniques in Concrete Science and Technology: Principles, Techniques and Applications*. Barnes & Noble, Lyndhurst, NJ, USA.
24. Basheer PAM (1991) *Clam Permeability Tests for Assessing the Durability of Concrete*. PhD thesis, Queen's University Belfast, Belfast, Northern Ireland.
25. NordTest (1999) NT Build 492. Concrete, mortar and cement-based repair materials. Chloride migration coefficient from non-steady state migration experiments. NordTest, Espoo, Finland.

26. Kropp J and Hilsdorf HK (1995) *Performance Criteria for Concrete Durability.* RILEM Technical Committee 116-PCD, Bagneux, France. RILEM Report 12.
27. ACI Committee 201 (1982) *Guide to Durable Concrete.* American Concrete Institute, Detroit, MI, Report ACI 201.2R-77.
28. Basheer PAM and Barbhuiya S (2009) Durability of concrete. In *ICE Manual of Construction Materials, Section 3: Concrete* (Forde MC (ed.)). Thomas Telford, London, UK, vol. 1, pp. 169–183.
29. Neville AM (1996) *Properties of Concrete*, 4th edn. Wiley, London, UK.
30. Richardson MG (1987) Cracking in reinforced concrete buildings. *Concrete International* **9(1)**: 21–23.
31. ACI Committee 224 (1985) Cracking in concrete structures: Why concrete structures crack and what to do about it. *Concrete International* 26–30.
32. Basheer PAM, Long AE and Montgomery FR (1994) An interaction model for causes of deterioration of concrete. *Malhotra Symposium on Concrete Technology, SP 144.* American Concrete Institute, Farmington Hills, MI, USA, pp. 213–233.
33. Building Research Establishment (1982) *The Durability of Steel in Concrete. Part 1 – Mechanism of Protection and Corrosion.* Building Research Establishment, Garston, UK, BRE Digest 263.
34. Dhir RK, Hewlett PC and Chan YN (1989) Near-surface characteristics of concrete: Prediction of carbonation resistance. *Magazine of Concrete Research* **41(148)**: 137–143.
35. Long AE, Basheer PAM and Montgomery FR (1997) In-situ permeability testing – a basis for service life prediction. *Proceedings of the CANMET/ACI International Conference, SP 171, New Zealand*, pp. 651–670.
36. Basheer L (1994) *Assessment of the Durability Characteristics of Surface Treated Concrete.* PhD thesis, Queen's University Belfast, Belfast, Northern Ireland.
37. Basheer L and Cleland DJ (1995) Surface treatments for concrete quantifying the improvement. *Proceedings of the IABSE Symposium, San Francisco, CA, USA*, pp. 275–280.
38. Basheer PAM, Montgomery FR and Long AE (1990) In-situ assessment of durability of concrete motorway bridges. *International Symposium on the Durability of Concrete, Toronto, SP 131.* American Concrete Institute, Farmington Hills, MI, USA, pp. 305–311.

Soutsos M
ISBN 978-0-7277-6375-4
https://doi.org/10.1680/icehcd.63754.031

Chapter 3
Physical deterioration mechanisms

Chris Atkins
Mott MacDonald, UK

3.1. Introduction

Concrete is a porous material. Generally, the pore structure is produced as a result of excess water added when it was first placed. For a typical structure, the amount of water required to make the cement fully hydrated is approximately 30% of the total mass of cement. If a cement paste is mixed with this amount of water, it is quite stiff. If aggregate is added to make concrete, it would not be possible to adequately mix or place the concrete. The easiest solution is to add more water. This excess water is not chemically combined into the hydrate structure and causes voids to form. More water results in more voids, which in turn results in lower-strength concrete.

There are a range of mechanisms that result in the physical degradation of concrete.

- Conventional abrasion and cavitation can cause damage.
- Where concrete surfaces are subject to traffic, large hydrostatic pressures can arise when moisture is compressed in defects, and this can result in accelerated abrasion.
- Material within the pores nearest the surface can expand, which imparts an indirect tensile stress on the concrete and results in the surface scaling off.
- Finally, fire, and subsequent firefighting, can damage concrete.

This chapter reviews these mechanisms.

Note that a concrete structure often contains joints. These are typically made from a softer, less durable material, and so can be subject to more damage than the surrounding concrete. However, once these have degraded, the result is an exposed edge that is always more vulnerable to damage. If the concrete is founded on a granular sub-base, the passage of water through failed joints can lead to erosion of the material under the concrete with little or no indication that this is occurring until the concrete itself is sufficiently undermined that failure occurs. Any exposed joints and edges should always be given special consideration in areas prone to physical damage.

3.2. Abrasion and erosion

Abrasion or erosion can be defined as the wearing away of surfaces. It is commonly considered that it requires the action of a hard material on a softer one, but this is not the case. Given enough time, a soft material can slowly wear away a harder one. For example, at canal locks, cast iron posts are often provided to hold a barge in place while a lock is emptied or filled: the barge is held close to the wall of the lock by looping a rope around the post. Over time, these posts can become abraded by the rope. An example is shown in Figure 3.1.

Figure 3.1 Rope abrasion of cast iron

Similarly, stone steps in very old structures are often worn into an inverted bell curve by the passage of hundreds of thousands of feet. Concrete is no exception to this wear and abrades over time in any environment where other materials pass over it. Where it is subjected to a flowing environment, which may contain suspended solids, the resulting damage can be extensive, as shown in Figure 3.2.

Typically, abrasion is a very slow process, unless the concrete is subject to the other mechanisms discussed later. For conventional reinforced concrete, the abrasion is likely to be of most concern if it results in an uneven surface for traffic or if it significantly reduces cover to the reinforcement. Uneven surfaces often allow puddles of water to form, which can dramatically accelerate the abrasion process. When water is trapped while being subject to wheel loading, significant pressure is exerted in the horizontal plane. This can rapidly lead to the undermining of the surrounding surfaces or the opening up of existing defects,

Figure 3.2 Abrasion of concrete in a stilling basin

which, in turn, causes increasingly widespread surface degradation. The ability of water to penetrate cracks and disrupt concrete has led to the use of ultra-high water pressure as an effective means of removing concrete in a technique commonly referred to as hydrodemolition.

Repairing concrete can itself lead to physical degradation of the surrounding parent concrete. For areas subject to traffic loading or that require a rapid return to service due to operational constraints, it is tempting to use a high-strength material, on the basis that this will achieve the minimum strength required for a rapid return to service. Using material with a Young's modulus higher than that of the original parent concrete can result in abrasion of the concrete surrounding the repair. The repair material will act as a hard spot, and so the surrounding concrete may be subject to abrasion.

3.2.1 Avoiding abrasion

As with most concrete problems, the risk of abrasion can be minimised by properly placing and curing a high-quality concrete. Higher strengths will typically provide higher abrasion resistance, but the curing and placement of concrete is of equal importance. Anything that results in a soft or porous surface containing cracks, such as plastic settlement or shrinkage, will result in a concrete more susceptible to abrasion. Typically, industrial floors require a high-quality smooth finish. To achieve this, the surface is often heavily trowelled or power-floated. This draws water to the surface and can result in a smooth but weak surface. Thus, by attempting to achieve a smooth surface, the abrasion resistance may, in fact, have been compromised. This will become apparent shortly after the floor is subject to wheel loading, when it will begin to degrade and the surface will break up. Increasing requirements for greater joint spacing have improved the technology available for casting floors, but concrete will still shrink and be subject to thermal movements. In addition, clients are more stringent regarding acceptable tolerances for flooring, resulting in more queries at handover as to what is and is not an acceptable floor.

Even high-performance concretes can be subject to abrasion if the environment is severe enough. In this case, there are several commercial products known as 'dry shake' materials that can provide a significant enhancement of the surface properties. As the name suggests, these are dry products shaken onto the surface of wet concrete prior to finishing. They typically comprise a cement-based product and carefully selected aggregate material. The most hard-wearing materials are typically metallic and can provide resistance to wear by tracked military vehicles.

Fibre reinforcement can be used to improve the integrity of exposed concrete surfaces. The fibres are typically either polypropylene or steel and are included in the original mix. After placement, the resultant concrete contains randomly oriented fibres that enhance its tensile strength. The presence of fibres in a mix often means that the concrete must be placed with greater care, which in turn can lead to an improvement in the abrasion resistance of the concrete. Additionally, the fibres provide enhanced tensile properties in the cover, which can also improve the abrasion resistance of the concrete. The fibres also hold damaged concrete in place, to some extent.

3.2.2 Repairing abrasion damage

Once the surface begins to break down, achieving an acceptable repair is usually complicated because the areas subject to heavy abrasion are usually in constant use and there is no alternative other than to continue using the structure. As a result, several rapid-set materials have been developed that can be used to reinstate lost material. Due to the rapidity of the set, such materials should be placed by a contractor experienced with the specific product. Trials should be performed before commencement of the actual repair to confirm that the contractor is familiar with the material. However, as the main purpose of a repair is to fill a hole, corrosion protection requirements can be secondary. Consequently, several rapid-set materials do not have the level of alkalinity required to prevent the reinforcement from corroding.

The nature of abrasion often leaves a surface with a good mechanical key due to exposed aggregate. However, shallow repairs may need to be feather edged. Ideally, to achieve a successful repair, all edges should be saw cut to a depth of 10 mm, which provides a sharp edge to form the repair against.

Also note that to provide a repair that is structurally compatible with the existing structure, the material used should have a similar Young's modulus and thermal properties as the concrete being repaired and should exhibit little or no shrinkage. In areas where access times are not limited, a range of repair products can provide an effective repair.

3.3. Cavitation

Cavitation occurs when liquids undergo a reduction in pressure until the liquid is below its vapour pressure. This produces bubbles in the liquid, which subsequently collapse and produce shock waves. This can result in a very aggressive environment and extremely high rates of abrasion. This process is accepted as the mechanism for coastal erosion. Air pockets within the water are forced into fissures, and the continuing inflow of water results in the pockets collapsing and a range of shock waves being produced that result in degradation.

3.4. Frost

Frost damage is caused by a simple mechanism. Moisture within the concrete expands as it freezes, and as a result, tensile stresses form near the surface. Typically, water expands by approximately 9% when it begins to freeze. If this expansion exceeds the volume available in the pores, then the concrete will be damaged. This can result in the surface crazing and then scaling off. Repeated cycles result in further damage, as the surface opens more and more, allowing a greater volume of water to penetrate the concrete.

There are two critical factors. First, the water in large capillary pores is free to freeze, whereas the water in gel pores is adsorbed on the cement hydrates and so has a limited effect, although it will migrate into the capillary pores and contribute to the damage. Second, this is balanced against the air entrapped in voids, which is compressible and allows the expansion of the frozen water. Controlling frost attack is, therefore, normally achieved by using a low-permeability concrete to keep the water out and by using air entrainment to increase the compressible volume of the voids.

A range of products can achieve air entrainment. Typically, they are organic materials that produce a series of evenly dispersed discontinuous pores or bubbles within the concrete. These are relatively uniform in diameter. Generally, they are less than 0.1 mm in diameter, and the volume that is typically required for protection is of the order of 4–7% by volume of concrete. The bubbles are different from the air normally entrapped in concrete due to their size and because they are discontinuous. This provides two benefits. The first is that because the pores are discontinuous, the surface absorbency is reduced. The second is that because the pores remain filled with air and, hence, unsaturated, they can absorb the stresses induced by frost. However, as the concrete can absorb these stresses, placing air-entrained concrete by pumping is generally not normally feasible.

The amount of air entrained is influenced by several factors. An increase in mixing time or concrete temperature reduces the amount of air entrained. The type of cement and the cement content both influence the volume of air entrained, with high cement contents or cements with high specific surfaces resulting in problems with air entrainment. The presence of organic matter can also affect the performance of the air entrainment.

Standardised methods of determining air content are available, for example, BS EN 12350-7 and ASTM C231. Similarly, there are two standard tests for determining the freeze–thaw resistance of a concrete.

ASTM C666 describes two methods that both involve the application of rapid freezing. Currently, there is a European standard being drafted (PD CEN/TS 12390-9:2016). Basically, the standard methods subject a sample to a number of freeze–thaw cycles and compare tested and untested samples.

3.5. Exfoliation

Exfoliation is caused by a similar mechanism to frost damage, in that the pore structure becomes filled with a material that subsequently expands. In this case, it is a solution containing soluble salts. As the moisture content decreases, the solution becomes more and more concentrated until, ultimately, the soluble salts recrystallise. The recrystallised salts may be greater in volume than the original solution, and so tensile stresses are imparted into the concrete. Figure 3.3 shows an example of exfoliation.

Note that the surface of concrete that is exposed to the solution may not be the surface that suffers from exfoliation. If one face of a concrete surface is adjacent to a highly concentrated solution and the other is exposed to a drying environment, the salt solution can be drawn a significant distance to the drying face, where the recrystallisation takes place. This can be a problem in immersed tunnels. Typically, the solution on the exterior of the tunnel is contaminated with chloride, so that capillary absorption can cause corrosion as well as exfoliation.

First, note that concrete already contains a supersaturated solution of calcium hydroxide. However, calcium salts typically have a relatively low solubility, and so, as the moisture dries out, the salts that are precipitated do not expand by a large volume. For exfoliation to occur, a more soluble salt must usually be present. In many cases in natural environments, this is based on sodium chloride, but any highly soluble salt can produce a similar effect. Table 3.1 provides an illustration of the range of solubilities and densities of soluble salts. It is clear that when a saturated solution of ammonium nitrate dries out, there will be an increase in volume. For example, 1900 g will dissolve in 1 litre of water, but the volume of 1900 g of the solid is greater than a litre.

Figure 3.3 Exfoliation of a sea wall

Table 3.1 Solubilities and densities of various salt solutions

Chemical	Solubility: g/l	Density: g/l
Sodium chloride	359	2160
Calcium chloride (anhydrous)	745	2150
Calcium chloride (dehydrate)	745	835
Ammonium nitrate	1900	1700

This effect has been reported for concrete contaminated with sodium chloride and subsequently over-coated. Due to the nature of the pore structure, the cement contains other soluble products that will also precipitate.

Coatings or waterproofing for concrete are commonly vapour permeable to prevent osmotic blistering. As a result, this can cause the coated concrete to dry out. If the concrete is contaminated with a soluble salt, the concentration in the pores under the coating will increase until the salts precipitate. This can cause the surface to exfoliate, which results in coating failure.

As a general rule of thumb, all nitrates as well as sodium, potassium and ammonium salts are highly soluble. Due to its high solubility, significant problems can be encountered when trying to contain an ammonium nitrate solution in a concrete tank.

Also note that solubility typically increases with temperature. As a result, exfoliation may also occur due to a temperature change. Further, the density and, consequently, the volumetric change that occurs when compounds such as calcium chloride dry out depend on the amount of chemically combined water. This can also affect the solubility of a salt and can further complicate the temperature dependence of the solubility. For example, sodium sulfate decahydrate loses its chemically combined water at 32°C and forms a more soluble anhydrous salt.

To control the risk of exfoliation or frost attack, the basic approach is either to prevent significant salt ingress or to prevent the surface from drying out. Typically, a coating system protects against salt ingress but note that a vapour-permeable coating can increase the degree of drying out of the concrete and may, therefore, cause exfoliation if the substrate is heavily contaminated. As with any attempt to repair concrete, success requires a full understanding of the problem.

3.6. Fire

A coating of concrete is a basic way to protect steel structures from fire. After the great fire there, all steel structures in Chicago were coated with a basic form of concrete to provide fire protection. When cement hydrates, it chemically combines with a significant amount of water. In a fire, as the temperature increases, the hydrated compounds break down and release water, thus dissipating heat and energy. This means that although the surface of the concrete may have been subject to high temperatures, the depth of damage may be limited.

After a fire, the appearance of any structure can give the impression of total destruction. Due to the smoke, everything will be blackened. There will be areas where the concrete has cracked and may have spalled. The level of damage may appear significant, but due to the ability of concrete to dissipate heat, the actual damage may well be recoverable.

If the temperature of the concrete has not exceeded 300°C, the residual strength, on cooling, is commonly considered not to have been significantly affected [1]. The reduction in strength is often offset by factors that resulted in the concrete having a higher strength than required for the original structural capacity. However, caution is advisable since it is often assumed that the strength of concrete increases with time beyond that required by the design. For older types of cement, this was the case, as to achieve the required strength at 28 days, the ultimate strength needed to be higher. Today, cements are more finely ground and cement-producing companies have improved their quality control systems, which means that the factors of safety included to ensure that concrete meets the required 28-day strength are now lower. Thus, the ultimate strength achieved may not be significantly greater than that obtained at 28 days.

The temperature at which significant structural deterioration occurs is often considered to coincide with a colour change in many concretes, as the material develops a pink hue (Figure 3.4). This discolouration is due to the presence of ferrous salts in the aggregates and may not occur if these are not present. Importantly, since this colour change does not always occur, the level of heat damage and degradation of the cement should be confirmed using a petrographic analysis of samples. The strength is usually severely compromised in places where temperatures exceed 500°C.

The actual degradation curve depends on a range of factors, including the aggregate type. Limestone aggregate starts to decompose at a relatively low temperature compared with a siliceous aggregate but decomposes more gradually and over a greater temperature range. This obviously affects the performance of

Figure 3.4 Typical effect of heat on the compressive strength of dense aggregate concrete after cooling. Colour changes shown for concrete containing most types of siliceous aggregate may not apply to concrete with calcareous or crushed flint aggregates. (Reproduced from [1] with permission from the Concrete Society)

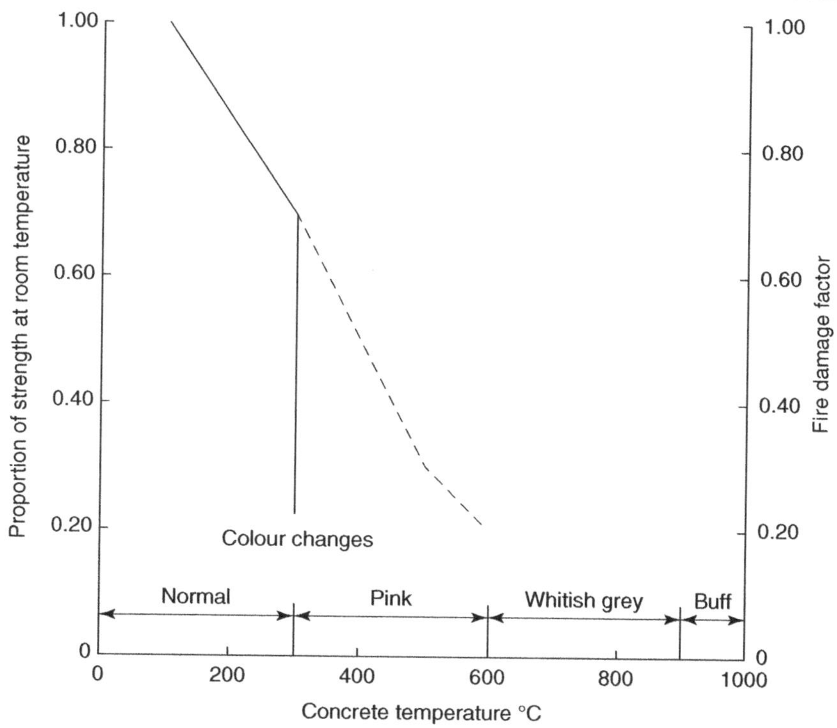

concrete in a fire. CS TR 68 provides a range of examples produced using a variety of software packages that illustrate the thermal properties of concrete [1]. In one example, a 380-mm-square column was subject to a standard furnace test for 1 hour. At a depth of 25 mm, the temperature reached 300°C. At 50 mm deep, the temperature was 150°C, and at 75mm deep, the temperature was only 100°C. In this case, only 25 mm of the concrete was likely to have suffered significant degradation and would require replacement. BS EN 1992-1-2 has tables for a range of member types of minimum dimensions that satisfy specified periods of fire resistance.

When concrete is subject to fire, the free water is heated and will expand. If the pore structure is relatively open, the pressure may be able to dissipate. In a dense impermeable concrete, however, the pressure cannot dissipate, so that the concrete may rapidly spall due to a phenomenon often termed explosive spalling. This degradation results in a more rapid loss of concrete and, therefore, an increased level of damage.

This mechanism has been postulated to have occurred in recent tunnel fires, such as that in the Channel Tunnel in October 2008, where it was suggested that the high-strength concrete underwent explosive spalling. Note that it was reported that this particular fire reached temperatures in excess of 1000°C and burned for 16 hours. It is unlikely that any concrete exposed to this temperature for this duration would survive. Explosive spalling is generally considered to occur early in the life of a fire, as the free water in the near-surface concrete expands rapidly. If the concrete cannot dissipate the pressure, it will spall. There is much debate about the use of direct fire protection, such as sprinkler systems, in tunnels. Statistically, a fire is a rare event, and so it can be difficult to justify the cost of installation and ongoing maintenance requirements for active fire control measures. The major cost associated with dealing with the aftermath of a fire is not that of repair. For any infrastructure, downtime is difficult to cost. For example, for a structure such as the Channel Tunnel, the reduction in service can be directly measured and costed. However, there will be additional lost revenue due to the reduction in public confidence that inevitably follows a major fire.

It is also important to consider the effects of heat on the steel reinforcement. Typically, heating steel up to 450°C and allowing it to return to room temperature naturally is unlikely to affect most common structural steelwork. Above 550°C, steel typically begins to deform plastically, and if heated still further, internal phase changes may occur, which may get locked in if the steel is subsequently quenched during firefighting. However, prestressed and post-tensioned steels will typically have undergone specific treatment to achieve a higher strength, which can be lost due to a fire. The increased temperatures will result in accelerated creep and stress relaxation in the tendons, which, ultimately, may result in their capacity reducing to that of conventional steel. This is important to understand when assessing the level of fire damage in a structure.

In addition to this, recent research [2] suggests that post-tensioned slabs designed with a 2-hour fire resistance may collapse in 66 minutes.

If it is considered that the structure has suffered to the point where its structural integrity is brought into question, it is necessary to carry out coring to obtain the strength of the concrete. Although surface techniques, such as a sclerometer or internal fracture test, can provide useful information about the strength of a concrete structure, a fire will obviously modify the surface concrete significantly, and so coring is necessary. For steelwork or embedded reinforcement, surface hardness tests provide information on the surface strength of the steel. However, this may not be representative of the entire section due to the fire and subsequent firefighting.

In general, repairing fire-damaged structures simply involves the removal and replacement of the deteriorated material, typically using cementitious repair products. Prior to undertaking any repair, it is necessary to confirm the integrity of the structure and the extent of the concrete requiring replacement. Concrete must be removed beyond the damaged area, which may affect the stability of the structure. It may, therefore, be appropriate to repair the concrete in sections. The repair material should, obviously, be of a similar structural capacity to that required and can be applied using a range of techniques (e.g. using sprayed concrete, flowable or self-compacting concrete or hand-applied repair products), depending on the extent and position of the repair. Note that some high-build repair mortars have a lower strength than similarly named products. In December 2008, European standard EN 1504 was introduced, which covers the products, processes and approaches to be used when carrying out a concrete repair.

3.7. Summary

Concrete degrades physically due to a limited number of causes. Typically, the simplest way of avoiding degradation is to gain a full understanding of the working performance required of the concrete and then selecting the right material. Choosing the right mix is only the start of the process, as installation and curing are of equal if not greater importance, since all the physical degradation mechanisms start at the surface of the concrete, and it is the surface that is most vulnerable to placing and curing effects.

As always, to carry out an effective repair, it is necessary to understand the processes involved. Standards such as EN 1504 provide a route to achieving a durable repair. When a coating is applied to address chloride contamination of concrete, the repair itself can sometimes result in further physical degradation of the concrete. Using inappropriate repair materials can also result in the repair causing physical damage to the original concrete.

REFERENCES

1. Concrete Society (2008) *Assessment, Design and Repair of Fire-damaged Concrete Structures.* Concrete Society, Camberley, UK, Technical report 68.
2. Kelly F and Purkiss J (2008) Concrete slab fire test results questioned. *New Civil Engineer*, 23 Oct.

Soutsos M
ISBN 978-0-7277-6375-4
https://doi.org/10.1680/icehcd.63754.041

Chapter 4
Chemical deterioration mechanisms

4.1. Introduction

Alan Poole, AB Poole & Associates, UK

Concrete, if correctly formulated, cast and cured, can endure for millennia in one environment but may completely degrade in a few years or even months in another. Thus, a 'durable concrete' is one that can resist the forces that would cause premature deterioration in the particular environment that it is exposed to. The assumption that using the right strength, cement content and water-to-cement (w/c) ratio will ensure that the concrete is durable can be misleading, as it must also possess properties appropriate to its local environment and to the external and internal factors that can influence its ability to resist deterioration.

With real-world concrete structures, investigations typically show that several mechanisms act together to produce the observed deterioration, although usually only one is the initiator or the predominant decay mechanism.

Although, as is described later, there are numerous deterioration mechanisms, they can be categorised according to two distinct causes. First, there are extrinsic deterioration mechanisms, in which external agencies, such as local environmental factors, are the causative factor. Second, intrinsic mechanisms arise from the use of inappropriate materials, mix designs or specifications, which allow deterioration to initiate and develop within the concrete.

Except for mechanical damage, all deterioration mechanisms are initiated by changes to the chemical composition or the mineralogical components in the concrete fabric. Thus, an investigation should first focus on a detailed petrographic and chemical study of possible changes to the chemistry and mineralogy of a concrete exhibiting signs of deterioration. As the deterioration progresses, physical damage, such as staining, cracking, spalling or misalignment of adjacent concrete elements or the breakdown and crumbling of the concrete fabric, provides the macroscopic evidence of deterioration and is often the first evidence of incipient decay and eventual failure.

4.2. Concrete as a material

Alan Poole, AB Poole & Associates, UK

Concrete is formed from a mix of stone or artificial aggregates and a cementitious material, usually Portland cement. Modern concretes often contain additions of mineral admixtures, most usually a ground granulated blastfurnace slag (GGBS) or pulverised fuel ash, and chemical additives, which modify the rheological and setting properties of the mix. The cement is hydrated by the addition of water. Hydration is a chemical reaction that causes the concrete to set into a durable stone-like material. The strength and durability of the resulting concrete is dependent on the mix design, the materials used and the amount of water added, usually measured in terms of the w/c ratio.

4.2.1 Aggregates

Some 70% to 80% of a typical concrete is composed of a carefully graded mix of coarse and fine aggregate (>4 mm), which is a natural gravel, crushed quarried rock, or natural or crushed rock sand. In recent times, increasing use is being made of artificial manufactured aggregate materials and crushed recycled concrete. Foamed natural materials, such as pumice, or artificially foamed or expanded aggregate materials are used to produce 'lightweight concrete'. Other specialised concretes may make use of aggregates selected for particular properties, for example, stability at high temperatures, or mix designs specifically formulated for particular applications, for example, to enhance thermal insulation.

Most natural aggregates are essentially inert within the set concrete, although the nature of their interfaces with the hydrated cement paste and their water absorption properties are significant factors influencing the strength and durability of the concrete. However, certain mineral constituents in some aggregate materials can lead to chemical reactions with the hydrated cement, external solutions or the pore fluids that are always present in a concrete. Numerous national and international specifications have been developed to avoid or restrict the use of such aggregates. Table 4.1 is a list of some of these mineral constituents that have the potential to cause intrinsic reactions and degradation of a concrete.

Table 4.1 Mineral constituents in aggregates with the potential for initiating concrete deterioration

Mineral	Typical occurrence	Comments
1. Mineral components that may be potential initiators of an internal deleterious sulfate attack		
Pyrites FeS_2	Low percentages present as small crystals in many aggregates.	May degrade to sulfuric or sulfurous acid and attack the cement matrix.
Marcasite FeS_2 Pyrrhotite $Fe_{1-x}S$	Low percentages present as small crystals in many aggregates.	May degrade to sulfuric or sulfurous acid and attack the cement matrix.
Chalcopyrite $CuFeS_2$	Rare component of mine waste and rock aggregates.	May degrade to sulfuric or sulfurous acid and attack the cement matrix.
Gypsum $CaSO_4 \cdot 2H_2O$	A rare contaminant of aggregates, more common in desert regions.	May cause significant expansion and cause concrete degradation.
2. Mineral components that may induce deterioration due to alkali–aggregate reactivity		
Opaline silica	A component of some cherts or a secondary recrystallisation of some siliceous microfossils. Used as cement for some sandstones and other siliceous rocks, e.g. greywacke.	Because of its amorphous and porous nature, usually highly reactive and need only be present in a small amount. May devitrify to cristobalite or tridymite.
Cristobalite SiO_2 Tridymite SiO_2	Rare components in siliceous, igneous metamorphic or sedimentary rock.	Highly reactive in alkali pore solutions.
3. Other aggregate components or contaminants that may result in weak or non-durable concrete		
Wood, coal, clay, rubber, glass, bone and organic materials	All aggregate contaminants.	Reduces the strength and durability of the concrete and produces areas of weakness in the concrete.

4.2.2 The cementitious component

The cement most commonly used for concrete is Portland cement, CEM I. For some applications, Portland cement with mineral admixtures is used. Five principal types are recognised in standard specifications, as listed in Table 4.2. These are further subdivided into 27 combinations based on the percentage and type of mineral addition. The range of chemical and mineralogical percentage compositions of Portland cement prior to hydration is shown in Table 4.3. Typically a few per cent of gypsum ($CaSO_4 \cdot 2H_2O$) is interground with the cement clinker as a set regulator. The gypsum provides the sulfate for the initial formation of ettringite from tricalcium aluminate (C_3A), prior to its eventual change to sulfoaluminate.

The anhydrous minerals shown in Table 4.3 react chemically with water by hydrating at various rates to form the cement paste matrix of a concrete or mortar, which is principally composed of calcium silicate hydrate (C-S-H) and calcium hydroxide (portlandite $Ca(OH)_2$) with small amounts of calcium aluminium hydrate and calcium monosulfate.

Table 4.2 Cement types and compositions as specified by Standard: BS EN 197-1/2000 [1]

Cement type	Cement: %	Mineral admixture: %	Comment
CEM I	95–100	0	Portland cement
CEM II	60–94	6–35	Portland–slag cement Six subgroups based on admixture type
CEM III	5–64	36–95	Blastfurnace slag cement Three subgroups: A, B and C
CEM IV	45–89	11–55	Pozzolanic cement Two subgroups: A and B
CEM V	20–64	36–80	Composite cement Two subgroups: A and B

Table 4.3 Mineralogical composition of a typical Portland cement

Mineral/ chemical name	Cement chemical notation	Chemical formula	Typical proportion by % mass	Hydration characteristics
Alite, tricalcium silicate	C_3S	$3CaO \cdot SiO_2$	45–65	Moderate reaction rate forms C-S-H + $Ca(OH)_2$
Belite, dicalcium silicate	C_2S	$2CaO \cdot SiO_2$	10–30	Slow hydration rate forms C-S-H + minor $Ca(OH)_2$
Tricalcium aluminate	C_3A	$3CaO \cdot Al_2O_3$	5–12	Rapid reaction rate, initially forms ettringite then monosulfate
Ferrite, tetra aluminoferrite	C_4AF	$4CaO \cdot Al_2O_3 \cdot Fe_2O_3$	6–12	Generally slow to react, forms iron substituted ettringite then monosulfate

4.3. Types of chemical concrete deterioration

Alan Poole, AB Poole & Associates, UK

As has already been noted, there are two separate distinct chemical mechanisms that can cause concrete to deteriorate. First, extrinsic reactions are a chemical attack from agencies outside the concrete and usually involve deleterious solutions acting from the concrete surface inwards. Consequently, the zone of deterioration usually starts at the surface and progresses inwards. The second, the intrinsic mechanisms, involve interactions between components within the concrete, for example, between the cement paste matrix and the aggregates used. Typically, an investigation of a deteriorated concrete will find several deleterious mechanisms working together, sometimes with intrinsic mechanisms allowing the ingress of damaging external (extrinsic) agencies access to the interior of the concrete.

Thus, in any investigation, it is important that the outer layers of the concrete are retained and examined for damage by extrinsic reactions and that the inner portions are examined for possible intrinsic effects. In most laboratory investigations, concrete cores (typically 100 mm in diameter) are taken from the damaged structure so that both the outer zone and the internal condition of the concrete can be examined. The most important laboratory tools for investigating damage to concrete are the optical microscope, the petrological polarising microscope, the scanning electron microscope (SEM) and microanalysis systems based on SEM and energy-dispersive X-ray spectroscopy (EDX). In addition a variety of analytical chemical, X-ray and standard test techniques may be required to accurately identify the causes of deterioration.

Although this chapter is concerned with the deleterious effects on concrete due to chemical interactions, the first indications of the resulting deterioration are usually physical effects on the surfaces of the concrete structure. These effects can usually be identified by careful inspection. However, it is important to differentiate these from those due entirely to physical causes, for example subsidence or a physical impact.

Except for the damage resulting from external physical causes and from the normal effects of the setting, hardening and ageing of a normal concrete structure, almost all concrete deterioration has a chemical mechanism as its root cause. The damage typically produces physical deterioration and cracking of the structure. Alternatively, the fabric of the concrete may decay, leading to a loss of strength, increased permeability, volume expansion or shrinkage of the concrete element, delamination of surface layers, exudation at the surface and crack development. Consequently, it is convenient, first, to categorise the mechanisms as follows.

Extrinsic chemical mechanisms

- weathering and leaching of the surface and the cementitious matrix
- acid attack, chemical dissolution or alteration
- external sulfate attack
- salt attack (Figure 4.1)

Intrinsic chemical mechanisms and agents

- internal sulfate attack
- unstable aggregate components, for example, pyrite and the mundic problem
- reduction of pore solution pH, allowing corrosion of the steel reinforcement
- reactive aggregate components (Figure 4.2)

4.3.1 Observation of features of damage and their possible causes

Several summaries of features of deterioration mechanisms are available in recent literature, for example in *Lea's Chemistry of Cement and Concrete* [2]. Table 4.4 lists a series of observable damage features and

Figure 4.1 An example of extrinsic salt attack from a high saline water table on a concrete pipe support (left), compared to a new replacement support (right)

Figure 4.2 A concrete retaining wall developing the characteristic map cracking pattern in an unconstrained concrete panel due to an intrinsic reaction between aggregate particles and the alkalis in the cement, the ASR

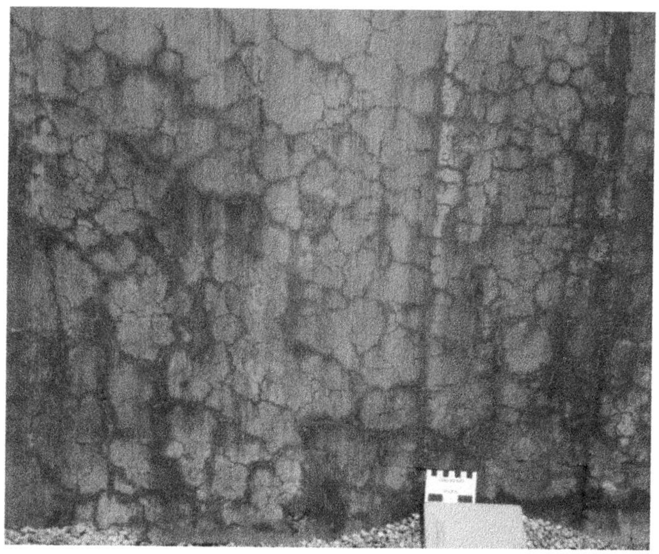

comments on their possible underlying cause. Those features that are possible indicators of some form of degradation due to chemical attack are indicated in italics. Note that certain surface features, such as shrinkage cracks and surface crazing, can be inherent features of concrete as a material, although they are usually indicative of an inappropriate structural design, mix design, curing or placement methods.

A recent approach to laboratory investigation provides a quantitative assessment of deterioration by calculating a damage rating index. Deterioration is identified using microscope observations of a cut and polished surface of a core sample. As shown in Table 4.5, a weighting factor is given to each feature. Scores for features

Table 4.4 Possible damage features on concrete structures and their possible causes[a]

Observed feature	First appearance	Possible causes or faults
Surface cracking		
Plastic settlement	Hours post casting	High w/c ratio, aggregate grading, workmanship
Plastic shrinkage	Hours post casting	Mix design, poor curing workmanship
Drying shrinkage	Days to months	Mix design, long-term shrinkage, poor design
Surface crazing	Days to weeks	High w/c ratio, excess surface trowelling
Thermal cracks (mass concrete)	Days, weeks or years	Inappropriate mix design
Map pattern cracking (as on a political map)	4 to 12 years	*Possible alkali–aggregate reaction (AAR)* *Delayed ettringite formation (DEF)*
Movement or displacement	Months to years	Settlement, seasonal change, *possible AAR*
Perpendicular to maximum load	4 to 12 years	Excess loading, *restrained AAR*
Regular surface pattern	Months to years	Freeze–thaw, reflection of *reinforcement corrosion*
Surface staining		
Surface bleaching	Months to years	*Chemical attack, AAR crack exudations*
Algae development	Months to years	Surface moisture, wet conditions
Iron oxidation staining	Months to years	*Pyrite in aggregate, reinforcement corrosion*
Other discolouration	Months to years	*Chemical solutions, atmospheric grime*
Exudation		
Carbonate (micro-crystalline)	Months to years	*Effect of moisture into or through the matrix*
Sulfate crystalline	Months to years	*Leaching and precipitation by water*
Iron oxides or hydroxides	Months to years	*Steel reinforcement corrosion* *Sulfide minerals leaching from the aggregate*
Gels and fluids	Months to years	*AAR gel, leached soluble salts*
Pop-outs and spalling		
Size, shape, depth and focus	Months to years	*Reinforcement corrosion, AAR, expansive gel at aggregates near the surface*
Surface degradation		
Degraded surface layer	Months to years	Frost attack, *leaching, external sulfate attack*
Porous surface zone	Many months or years	*Acid attack, salt attack*
Friable surface zone	Many months or years	*Extrinsic sulfate or salt attack*
Internal degradation		
Porous and carbonated matrix and a friable crumbling interior structure Multiple subparallel interior fractures	Months to years	*DEF* *Extrinsic thaumasite sulfate attack (TSA)* Freeze–thaw damage
Internal crack networks		
Network of internal cracking on a millimetre or centimetre scale Cracks infilled or empty	Months to years	*Expansion due to gypsum development.* *Gypsum in the aggregate reacting with hydrated cement.* *AAR between cement and aggregate* Drying shrinkage

[a]Items printed in italics indicate a possible chemical deterioration mechanism.

Table 4.5 Damage features used for the damage rating index and their weighting factors relating to AAR

Damage feature	Weighting factor
Cracks in coarse aggregate	0.25
Cracks with gel in coarse aggregate	2.0
Open cracks in coarse aggregate	3.0
Debonding around coarse aggregate	3.0
Reaction rims around aggregates	0.5
Cracks in the cement paste	2.0
Cracks with gel in the cement paste	4.0
Gel in air voids	0.5

are summed and normalised for an area of 100 cm^2. Although devised by Grattan-Bellew and Mitchell [3] to evaluate the severity of the development of the alkali–silica reaction (ASR), with appropriate modification, this approach may be useful in comparative evaluations of a range of chemical deterioration mechanisms.

The development of cracking at the surface of a concrete structure and interpreting the underlying cause have been the subject of many investigations. One of the most comprehensive evaluations of non-structural cracking has been prepared by the Concrete Society in technical report 22 [4]. This report provides tables and figures showing commonly observed cracks in a typical concrete structure and their possible causes.

However, although the physical features of damage often have a physical cause, many are indicative of an underlying chemical deterioration. Table 4.6 identifies some of the most common features of damage visible on close inspection of a concrete surface, highlighting those that may be indicative of deterioration initiated or developed through chemical action.

4.4. Specific concrete damage types and mechanisms

4.4.1 Deterioration resulting from weathering and leaching

Alan Poole, AB Poole & Associates, UK

Many concrete structures are subject to the effects of moisture, ranging from water-retaining structures, concrete water pipes, marine defences and the concrete foundations of structures that are subjected to rain wash, wetting and drying, or moisture condensation. The extent to which moisture penetrates the fabric of a concrete will depend on its quality, in particular its inherent porosity and permeability. Factors influencing the possible deleterious effect on the concrete over time include whether the water is flowing, as in a water pipe, whether there are repeated wetting and drying cycles, and the pH and salt content of the water.

The leaching of soluble components from concrete will, in most cases, cause deterioration with a possible loss of strength and an increase in permeability. One of the most soluble components in a cementitious matrix is the calcium hydroxide formed when the cement hydrates as the concrete sets and hardens. Calcium hydroxide (portlandite) is unusual among salts in that it is more soluble at low temperatures. It dissolves more easily if the water is acidified, as may be the case with acid rain.

A very common observation when water leaches from concrete is the development of calcium carbonate coatings or exudations, sometimes with the development of 'stalactite' growths from overhangs and soffits. This carbonate is the result of the calcium hydroxide carried in solution interacting with atmospheric carbon dioxide as the moisture dries on a surface (Figure 4.3).

Table 4.6 Common surface features visible on a concrete structure indicative of degradation processes

Type	Surface feature	Time to appearance (UK conditions)	Possible cause(s)	Comment
Concrete construction				
Plastic settlement	Cracks and 'draping' over reinforcement	30 min to 3 hours	Wet mix, bleeding settlement	Mix design, deep sections
Plastic shrinkage	Shallow surface crazing, laitance	30 min to 8 hours	Surface over-trowelling, cement-rich mix, bleeding	Wet mix design, typically slabs
Self-desiccation shrinkage	Surface dusting of cement-rich laitance	Hours to days	Over-trowelling, bleeding and rapid surface drying	Wet mix and poor curing
Early thermal contraction	Cracking	Weeks to months	Excess heat generation, restrained sections	Mix design, thick sections
Long-term drying shrinkage	Mid-section open shrinkage cracks	Several months to years	Volume reduction during hardening and ageing	Restrained concrete sections
Surface crazing	Superficial pattern cracking, laitance	1 day to weeks	Shrinkage of the cement-rich surface layer	Rich wet mix, over-trowelling
Reinforcement corrosion	Expansive cracking and rust staining	2 years plus	Reduced passivation, low pH, rusting of the steel	Reflects location of reinforcing steel
Possible chemical damage indicators				
Weathering and leaching	Calcite exudations and porous surfaces	Months to years	Moisture ingress points, air and water carbonation	Poor joints and open cracks, reduced pH
External acid or alkali attack	Friable degraded cement matrix	Months to years	Dissolution of $Ca(OH)_2$ and C-S-H gels, pore fluid with reduced pH	Whitish matrix and exposed aggregate
External sulfate attack	Friable degraded matrix, ettringite	1 to many years	Dissolution of $Ca(OH)_2$ and C-A-H by sulfates from soils	Worst effect at or near ground level
Seawater attack	Surface damage and secondary minerals	1 to many years	$Ca(OH)_2$ or C-S-H altered to brucite, aragonite or ettringite	Worst damage in intertidal zone
TSA	Degraded matrix forms a white mush	1 to several years	SO_4 ions rapidly attack C-S-H and degrade the matrix	Requires Ca^{++} ions and low temperature
Internal sulfate attack	Efflorescence, cracks and spalls	1 to many years	Development of ettringite in cement matrix	SO_4 source, usually from aggregates
DEF	Expansion cracks, high-temperature curing	1 to several years	Ettringite formation in cracks or on aggregates	Expansion of cement matrix
External salt attack	Concrete degradation and salt precipitation	Months to years	High saline water table and capillary rise effects	Worst effects near ground surface
AAR	Pattern cracking, pop-outs and exudation	8 to 12 years	Reactive aggregate, high-alkali pore fluids	Gel product can crack the concrete

Figure 4.3 A common example of leaching of calcium hydroxide from a concrete soffit and its conversion to calcite on exposure to the atmosphere

In some concretes where leaching has occurred, secondary rims of calcite can be observed as rims to otherwise empty voids. They can sometimes completely fill small voids. In some examples, evidence of multiple cycles of solution and precipitation can be inferred from the clear discrete multiple layers of crystals coating the walls of voids, as in Figure 4.4. A secondary consequence of the leaching of calcium hydroxide from the concrete is the lowering of the pH of the pore fluids from 12–13 down to nearer 7–8, which subsequently removes the effect of pacification on the steel reinforcement and allows rusting to initiate. If air and moisture can penetrate the fabric of the concrete, then carbonation (the conversion of calcium hydroxide to calcite) of the cementitious matrix occurs, which reduces the pH. However, carbonation can also block the pores in the matrix, reducing its permeability and so slowing the ingress of air and moisture.

Figure 4.4 Thin petrographic section of a 2-mm elongated air void in concrete showing multiple rims of iron hydroxides, calcite and ettringite. This demonstrates that there have been multiple cycles of solutions permeating through the concrete fabric with subsequent cycles of precipitation on drying

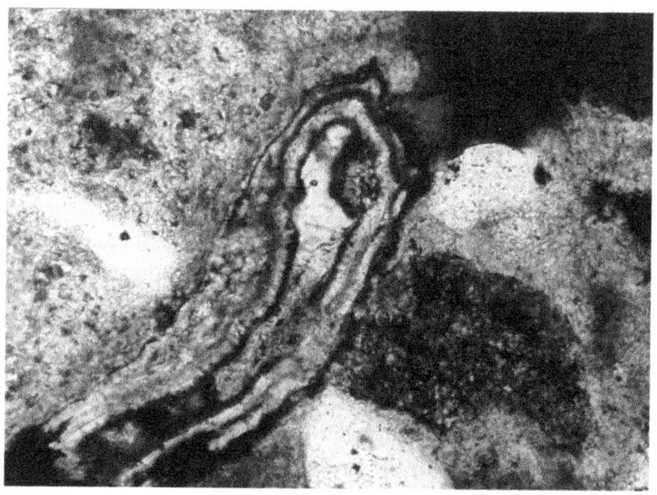

Driving rain and cycles of wetting and drying assist the penetration of moisture into the surface of exposed concrete. The leaching and deterioration that occurs is, in many cases, related to the orientation of the structure with respect to the prevailing wind and rain directions, with weather faces suffering more than sheltered portions. When undertaking an inspection, it is sometimes helpful to mark up diagrams or photographs with the degree of exposure as an aid to understanding the causes of any observed deterioration.

4.4.2 Acid and alkali attacks against concrete

Alan Poole, AB Poole & Associates, UK

The minerals in the cementitious matrix of a concrete are susceptible to rapid attack by acid solutions with a pH below 4. Most notably, calcium hydroxide is degraded by acid solutions, but these also attack C-S-H and in some cases the aggregates. The most important factors in a corrosive acid attack are the pH and the rate of flow of the fluid over the concrete surfaces. BS 8110, now superseded by BS EN 1992-1-1:2004 [5], recommends that unprotected Portland cement concrete should not be used where solutions are consistently below pH 5.5. However, the reaction products can limit the corrosive effect. If one product is insoluble, it can form a protective layer over the concrete surface. An example quoted by Lea [6] relates to an attack on concrete from humic acid in marsh water, which may form an insoluble coating of calcium humate on the concrete.

Rainwater, which may contain carbonic acid, and groundwater percolating though decaying organic matter can range downward to pH 4, although they rarely drop below pH 5. Distilled water is particularly aggressive because of its high solvent capacity, low buffering and an acidity of approximately pH 6 (due to the absorption of carbon dioxide from the air).

Although Portland cement concretes are themselves alkaline, they are attacked by strong alkali solutions if the concentration exceeds 20%. Concretes made from high alumina cement and calcium aluminate cement are more readily attacked by alkali solution, which dissolves the alumina gel and attacks the hydrated calcium aluminates.

4.4.3 Sulfate attack against concrete

Alan Poole, AB Poole & Associates, UK

The sulfate actions that cause concrete to deteriorate are complex. The most common form is an extrinsic attack by sulfate-containing groundwater or soil against concrete foundations, but some decay mechanisms arise from sulfate being available in the matrix pore solutions or sulfate-containing aggregate components. However, for practical purposes, the extent of an attack on the cementitious matrix of a concrete depends on the concentration of sulfate ions in solution, the rate of movement of sulfate-bearing solutions on or within the concrete, the local environmental conditions and the ambient temperature.

4.4.3.1 Extrinsic sulfate attack from groundwater and soil

Evaluating the aggressiveness of ground conditions against buried concrete is covered in BRE Digest 363 (1991 and 1996), although this has been revised and replaced by BRE Special Digest 1 [7], which is a good review of the various types of sulfate actions and the practical procedures for field and laboratory assessments of the aggressiveness.

The most common source of sulfate in soils is gypsum ($CaSO_4 \cdot 2H_2O$), but this has limited solubility, equating to a maximum of approximately 1200 ppm of SO_3. Beyond this concentration, other cations such as Mg^+, Na^+ or K^+ must also be present to balance the sulfate ions in solution. The oxidation of pyrite and the biological decomposition of organic matter in the ground may also increase the sulfate content, provided there is sufficient oxygen.

Concretes with a low permeability have good resistance to sulfate attacks. Most typically the damage affects the concrete just below or above ground level and is usually the result of moisture permeating into the concrete. Sulfate ions typically attack the calcium hydroxide and calcium aluminate components of the cement matrix, leading to the formation of secondary ettringite and sometimes gypsum within the fabric of the hardened concrete. This causes a marked expansion of the concrete. These reactions also degrade the cementitious matrix of the concrete, which becomes friable, mush-like, porous and whitish in colour. Other indications are the development of cracking, generally subparallel to surfaces, exfoliation and spalling. In many examples, salts precipitate onto surfaces (Figure 4.5). Analysing these salts is important when investigating the causes of deterioration in concrete. Lea in 1968 [8] and Nixon in 1978 [9] reported the heaving of concrete floor slabs that had been laid over a fill containing sulfates, such as pyrite (FeS_2). In some of these cases, bacteria appear to be involved in the decay of pyrite. The problem involved the fill material, since there was no evidence of a sulfate attack against the overlying concrete [10,11].

Although normal sewage effluent is alkaline, bacterial action in concrete sewage pipes can produce hydrogen sulfide. This can then suffer from anaerobic bacterial action to produce sulfurous and sulfuric acids, which can attack and dissolve concrete surfaces. A detailed review of this problem was provided by Poole and Sims [12].

4.4.3.2 Seawater attack

An extrinsic attack against concrete resulting from exposure to seawater is most damaging in the intertidal zone because of the cyclic wetting and drying and the abrasive action of wave impacts. In colder regions, freeze–thaw may cause additional physical damage. The deterioration of concrete in marine environments has become of considerable importance because of the proliferation of various offshore installations.

Normal seawater contains approximately 3.5% of soluble salts by mass, comprising 1.1% sodium, 0.04% potassium, 0.1% magnesium, 0.04% calcium, 2% chloride, 0.3% sulfate (SO_4) and trace amounts of other ions. The pH of normal seawater is between 7.8 and 8.3. Depending on temperature, it contains about 10 ppm of dissolved carbonate and 80 ppm of bicarbonate. In restricted sea areas, such as the Arabian Gulf and saline lagoons, the total salt content can be much higher, up to 10% and above, as reported by Purser and Seibold [13].

Figure 4.5 An incipient spall of concrete with a permeable limestone aggregate in a tunnel lining due to sulfate attack. The exfoliation crack runs across the field of view and the white crystals are thenardite (Na_2SO_4). (Courtesy of D St John)

The chemical deterioration arising from this extrinsic action results in a complex series of reactions, which initially affect the concrete surface, then migrate inwards, typically developing a layered sequence of secondary minerals resulting from the attack on the calcium hydroxide and aluminate phases in the cementitious matrix of the concrete. The matrix becomes leached and carbonated. The sequence inwards of secondary minerals may include brucite, then the patchy development of aragonite and finally ettringite as acicular crystals in the matrix and voids. Where organic matter is decomposing on the seabed, for example in sheltered estuaries, the pH can fall to below 7, which inhibits carbonation. Conjeaud [14] showed that SO_4^{2-} and Cl^- diffuse rapidly into mortar, but the development of an almost impermeable aragonite layer forming over the brucite in the concrete surface, due to reactions with bicarbonate and magnesium ions, rapidly slows this diffusion.

4.4.3.3 Thaumasite sulfate attack

The importance of thaumasite sulfate attack (TSA) has been recognised only in recent years. It is a rapid and often destructive cause of damage but is different from the more typical sulfate attack described above. TSA was first recognised in bridge foundations in western England on the M5 motorway in 1998. A first report on the causes of this deterioration was published by the Department for the Environment, Transport and the Regions (DETR) in 1999 [15], which has since been updated by reports published by DETR in 2000 and 2002, and also by Nixon and Longworth [16] and by BRE Digest 1 [7]. Introductory guides on thaumasite include Hartshorn and Sims [17], Benstead [18] and Crammond [19].

A thaumasite attack differs from a conventional sulfate attack in that it forms preferentially at low temperatures below 15°C. The chief reaction product is thaumasite but ettringite and gypsum may also form. The reaction requires both sulfate and calcium ions to be available in the pore solution of the concrete. Unlike ettringite, thaumasite contains a carbonate component, which may come from the aggregate used, from the cement or from the atmosphere. Initially, thaumasite may form in voids without damaging the concrete. It may be a precursor to TSA, in which case it degrades the cement matrix by attacking the calcium silicate rather than the aluminate phases and produces a soft white paste-like mush in the matrix, leading to a loss of strength and eventual disintegration of the concrete. It is possible that ettringite acts as a nucleating centre for the formation of thaumasite, as suggested by Taylor [20]. Examples of thaumasite overgrowths on ettringite have been reported by Erlin and Stark [21]. Thaumasite may develop in concrete tunnels and underground structures, where high humidity and low temperatures prevail. In one example, Hagelia *et al.* [22] reported the extensive development of TSA degradation in concrete and shotcrete tunnel linings in sulfide-bearing alum shales.

In many cases, it is difficult to differentiate between ettringite and thaumasite by petrographic means as both have acicular crystals, although the thaumasite may have a pale yellowish colour when viewed in a thin section. Differentiation may require X-ray diffraction analysis or SEM-EDX microanalysis, where the presence of aluminium differentiates ettringite from thaumasite. A more detailed description of the petrographic features of TSA has been presented in *Concrete Petrography* [12].

4.4.4 Internal sulfate attack

Alan Poole, AB Poole & Associates, UK

An intrinsic sulfate attack against concrete involves the sulfate already present in the concrete and does not require an outside source. Excess sulfate in concrete usually arises from contaminated aggregate materials, which may contain particles of pyrite or pyrrhotite. If oxidised, these can produce sulfate ions. In the simplest case, the degradation of a pyrite aggregate particle near the concrete surface may lead only to an unsightly rust stain, as in Figure 4.6. Usually, the damage is confined to rust staining and pop-outs.

Figure 4.6 Oxidation of pyrite aggregate particles in mass concrete leading to rust staining on the vertical surface

In the presence of oxygen and alkali pore fluids, pyrrhotite (FeS), pyrite and marcasite (both FeS_2), and chalcopyrite ($CuFeS_2$) in concrete can oxidise to produce ferric hydroxide ($Fe(OH)_3$) and a range of other sulfates. These may react in complex ways with the alkali pore solutions to give ferrous hydroxide, alkaline and calcium sulfates, sometimes producing a secondary development of ettringite (Figure 4.7).

In other cases, aggregates may sometimes be contaminated with other sulfate-bearing materials, such as gypsum. In the Middle East, where sources of good quality aggregates are difficult to find, crushed sedimentary material is sometimes contaminated with calcium and magnesium sulfates, as discussed by

Figure 4.7 Thin petrographic section showing secondary ettringite needles in a 0.8-mm void in concrete

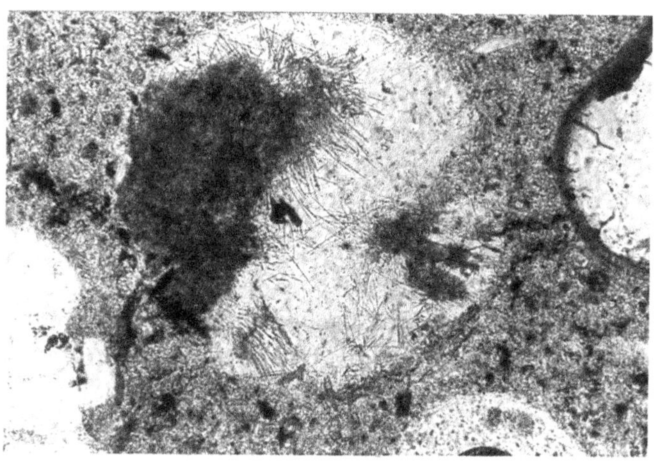

Fookes and Collis [23,24] and Sims [25]. The presence of excess sulfate from such contamination within the concrete can lead to the formation of ettringite in the cement matrix, which subsequently may produce expansive cracking and a general deterioration of the concrete. The dimensions of these ettringite crystals are usually below optical resolution, so that identification requires X-ray diffraction analysis. Where gypsum contamination is present in the concrete mix, it will react with the cement alkalis to form calcium hydroxide, which first develops on the surface of gypsum particles and then grows inwards, converting the gypsum to calcium hydroxide [26]. If fine grained, the gypsum reacts rapidly, leading to efflorescence, spalling, scaling and cracking, which can proceed in a matter of weeks or months. X-ray diffraction analysis showing the presence of gypsum and ettringite provides diagnostic evidence of this type of sulfate attack.

4.4.4.1 The mundic problem

Deterioration of concrete resulting from the oxidation of iron sulfides in the aggregates used is sometimes referred to as the 'mundic problem' (from the Cornish word for pyrite), due to reported cases of concrete deterioration from Cornwall.

A sulfate attack can arise from the in situ oxidation of pyrite and other sulfide minerals in metalliferous mine wastes, wall rocks containing disseminated pyrite and in mineral tailings used as aggregates in Devon and Cornwall. These minerals then decompose. Sulfurous and sulfuric acids form, which attack the concrete. This was referred to as the 'mundic block problem' by Bromley and Pettifer [27]. It was most commonly seen in domestic structures made with concrete blocks in parts of the UK. Similar problems have also been reported from other countries. In some examples, the deterioration had become sufficiently severe to render the concrete unsafe.

Securing loans against such properties became difficult and led to a requirement for petrographic screening of domestic properties made from concrete and built before the mid-1950s. This identified properties containing potentially deleterious aggregates, which were then often considered to be 'not mortgageable' even though the concrete showed no sign of deterioration. The system of classification used was adopted by the Royal Institution of Chartered Surveyors. It was based on a petrographic examination of the concrete [28]. The system is based on four classes. In simple terms, they are as follows

- A: sound concrete with no mundic-related aggregate
- A/B: sound concrete with a minority of mundic-related aggregate
- B: sound concrete composed partly or wholly of mundic-related aggregates
- C: unsound concrete, which may be due to mundic-related aggregate or other causes.

The petrographic assessments are based on transmitted and reflected light microscopy. In some cases, SEM is used to identify very fine-grained secondary products. Poole and Sims [12] point out that even the walls of a simple dwelling are complex systems, with contrasting behaviour with respect to concrete stability. They suggest that foundation concrete is often saturated with groundwater depleted in oxygen, so that the pyrite remains unaltered. The pyrite present in uncarbonated concrete may produce superficial oxidation products that are insoluble in the alkaline pore fluids and passivate further oxidation of the pyrite. However, once the cement matrix is carbonated, the pyrite oxidises rapidly. The observed concrete deterioration of carbonated concrete arises from a direct sulfate attack against the cement, but secondary gypsum, which may develop at cement/aggregate interfaces, may also result in expansion damage, as gypsum replaces calcite. Another degradation mechanism arises from gypsum developing as lenticular veinlets from the oxidation of finely disseminated pyrite in mudstones, which can lead to bulk expansion of the aggregate particles.

4.4.4.2 Delayed ettringite formation

Another important internal sulfate attack is referred to as delayed ettringite formation (DEF). This mechanism was originally considered to be limited to precast concrete elements subjected to a curing regime with an elevated temperature, such as steam curing. However, its occurrence appears to be more widespread and may be encountered in large concrete sections where the heat of hydration produced during curing may be a factor. A temperature in excess of 65°C appears to be necessary to trigger its development. It also appears that DEF and AAR often occur together [29].

In normal hydration, the sulfate present (added to cement as a setting regulator) first forms ettringite and then hydrated sulfoaluminate compounds, which form while the concrete is still in a plastic or semi-plastic state and so are accommodated without localised disruption. Secondary ettringite is commonly found in voids in any concrete subjected to moisture ingress (Figure 4.7). However, DEF is sometimes referred to as a heat-induced internal sulfate attack in which the ettringite formed develops after the concrete has fully hardened, but it too is triggered by moisture ingress. The extent of the expansion due to DEF increases with the concrete curing temperature up to a maximum of about 90°C. Examples of damage due to DEF have been reported in precast concrete bridges by Knights and Baldwin [30], and in cases of cast-in-place concrete by Grantham *et al.* [31], Eden *et al.* [32] and Thomas *et al.* [33].

The mechanism behind DEF expansion remains a matter of debate, but there are three commonly proposed hypotheses.

- Expansion correlates with the growth of ettringite on aggregate surfaces and in microcracks in the cement paste [34].
- The cement paste in voids at aggregate surfaces expands, thus allowing ettringite to form [20,35,36].
- Pre-existing microcracks formed by thermal stresses become infilled with ettringite, leading to expansion [37].

The principal risk factors for the development of DEF-induced expansion appear to be high curing temperatures and the availability of moisture. Other factors include high cement content, high sulfate content and associated AAR.

A common effect of DEF expansion is the cracking of the affected concrete. Since it is an intrinsic reaction, the whole cement paste matrix expands [20] driven by processes within the paste. SEM analysis has been used to detect the development of ettringite within the paste. The typical development of cracking is usually accompanied by ring fractures around aggregate particles, which then infill with ettringite needles orientated perpendicular to the cement margins.

4.4.5 Salt attack against concrete

Alan Poole, AB Poole & Associates, UK

As with the other forms of extrinsic chemical attack against concrete already noted, an attack by salt solutions can cause significant damage and reduction in the durability of concrete. The corrosive effect of sodium chloride on reinforcing steel is well known. The effect of sulfates on concrete was discussed in Section 4.4.3 and seawater attacks, in particular, in Section 4.4.3.2. An attack against concrete by salt solutions depends on several factors: the salt solution composition, the concentration of the attacking solution, the local temperature regime, the length of exposure time and the quality of the concrete, most particularly its permeability. Degradation involves complex multiple reactions, usually affecting the cement paste, but in broad and simplistic terms, it may be considered analogous to the geological process in which rock gradually degrades by weathering to soil. With concrete, the damage results in spalling, cracking and degradation of the cementitious matrix of the concrete. The growth of salt crystals within microcracks induces stresses that assist their propagation.

The most damaging conditions arise in concrete foundations where there is a high saline water table and evaporative climatic conditions. In most cases, capillary rise draws the salt solution from the ground into the concrete and the subsequent evaporation means the solution becomes concentrated so that the salt crystallises on the surface of the concrete. This effect is clearly visible in the concrete fence posts shown in Figure 4.8. The worst damage has occurred just above ground level. This type of damage is most common along the coastal margins of desert countries, such as around the Arabian Gulf. Reviews are given in Fookes and Collis [38] and Walker [39]. A more detailed review of salt weathering processes is given in a book by Goudie and Viles [40].

This type of salt damage in concrete foundations can be avoided by minimising and reducing concrete permeability, by tanking the foundations or by cladding the foundations proper in an additional sacrificial layer of concrete.

4.4.6 Alkali–aggregate reaction

Alan Poole, AB Poole & Associates, UK

This is an intrinsic damage mechanism involving a reaction between alkali pore solution in the cement matrix and certain components in the concrete aggregate. It was first recognised in some Californian highway structures in the late 1930s by Stanton *et al.* [41,42]. The reaction produces an alkali–silica gel, which swells on absorbing water and exerts sufficient force to crack the surrounding concrete. In experimental studies, Diamond [43,44] showed that the expansive force generated by this gel varied considerably depending in part on its composition but could reach values as high as 10 MPa.

Figure 4.8 Severe salt degradation of concrete fence posts in Bahrain due to capillary rise from a high saline water table. Note the most severe damage occurs above ground level where the salt solutions concentrate and dry out through evaporation

This deleterious reaction has since been recognised in many other countries and has come to be recognised as a worldwide problem [45]. Post the Second World War, this recognition led to global collaborative efforts to research, address and mitigate its effects. Commonly, a typical sequence of investigation and research develops, as each country initially recognises ASR and identifies it in their concrete structures, usually with the particular combination of materials, conditions and manifestations specific to that country. Research investigations then follow to determine the mechanisms and controlling factors, which are then followed by the development of test procedures and avoidance specifications, which, again, are specific to the country concerned.

An important aspect of global collaboration to solve the ASR problem has been through the series of international conferences on alkali–aggregate reactions (ICAAR). The first of these was a seminar of experts in Køge, Denmark, in 1974. Since then, these conferences have grown and developed, with contributions presented by researchers and practitioners from all around the world. The *16th ICAAR Conference* in Lisbon was the latest in that series. Today, a general global consensus concerning testing methods and relevant specifications is slowly emerging, largely due to the cooperative spirit embodied in these conferences and the resultant practical work of committees and other working groups under the auspices of RILEM (Réunion Internationale des Laboratoires et Experts des Matériaux systèmes de construction et ouvrages).

4.4.6.1 Types of AAR

The initial conclusions of research into this deleterious reaction in concrete indicate that there may be as many as three separate types of alkali reactivity, although there is more recent compelling evidence that the first two listed below are very likely to be just variants of the third. These are

- alkali–silicate reactivity [45,46]
- alkali–carbonate reactivity [45,47–49]
- alkali–silica reactivity [41,45].

ALKALI–SILICATE REACTION

The alkali–silicate reaction involves argillite, greywacke and other sedimentary and metamorphic aggregates containing disordered forms of cryptocrystalline silica in a concrete containing high-alkali cement. It is slow to initiate and develop the hydrophilic gel, which produces the expansive forces necessary for cracking to occur. Once initiated, it typically appears to continue for many decades. It was originally considered to be a different form of ASR, but the distinction is now thought to be an arbitrary one. Nevertheless, there is evidence to suggest that under certain circumstances, alkali leaches from these aggregates to contribute to the slow development of the reaction [50].

ALKALI–CARBONATE REACTION

The alkali–carbonate reaction is well documented, in so far as it develops distinct de-dolomitised rims at the margins of aggregate particles of fine-grained dolomite (Figure 4.9). According to Katayama *et al.* in a series of publications from 2010 to 2017 [49], the de-dolomitisation involves the formation of a brucite rim with a thin outer zone of hydrotalcite. Magnesium silicate gel forms in the rims as an intermediate product between ASR gel and the brucite. Over time, a secondary calcite replaces the adjacent C-S-H gels of the cement paste forming a halo, while ASR gel develops within the aggregate particles and is the principal cause of expansion. Several types of rim have been identified [51], but whether any of them contribute to the observed expansion of the concrete remains unclear and a matter of research [52]. Certainly, Katayama [49] demonstrated that at least some reactions involving carbonate aggregates are actually expansive due to ASR and coined the term 'alkali–carbonate reaction' [45]. His conclusion was that the alkali–carbonate reaction is a special case of ASR, which appears to be vindicated in that, although it can be very difficult

Figure 4.9 Thin de-dolomitised rims at the margins of coarse aggregate particles of fine-grained dolomite. Ground surface, width of field 40 mm

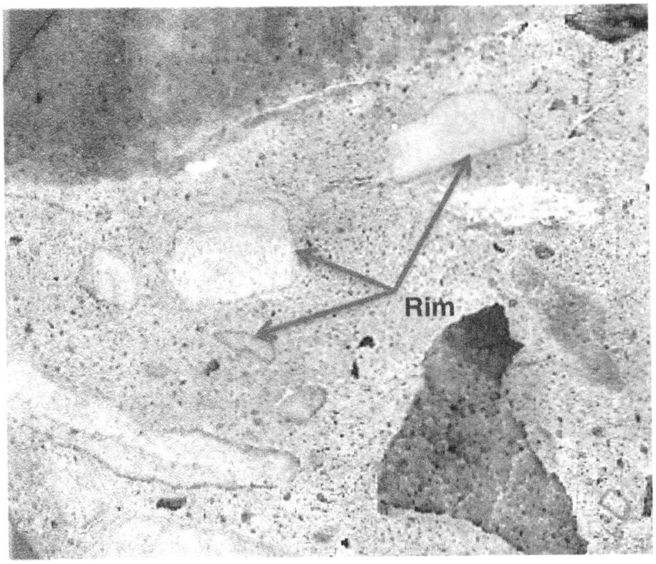

to positively identify the alkali–silica gel reaction product in some cases, it is present. He concludes that the formation of de-dolomitised rims and the formation of brucite [53] are not associated with expansion in themselves.

ALKALI–SILICA REACTION

The role of siliceous components in the aggregates used in concretes affected by ASR arises from a hydrolysis reaction between disordered or poorly crystalline silica in the aggregate (Table 4.1) and the ionised alkali pore fluids in the concrete. If these reactants are present in sufficient concentrations, a reaction takes place, forming an alkali–silica gel product at reaction sites within the concrete. Because the cement pore fluids contain sodium or potassium hydroxides, these metal ions are also present in the gel product. Such gels are hygroscopic and absorb water from the surrounding pore fluids and produce sufficient swelling pressure to crack the concrete. These cracks typically initiate at the reaction sites on or in aggregate particles, leading to a star-shaped crack pattern observable on concrete surfaces. The reactions forming the gel are complicated and multi-staged in detail but in the simplest form can be expressed by the idealised equations [54]:

$$4SiO_2 + 2NaOH = Na_2Si_4O_9 + H_2O$$

$$3SiO_2 + 2NaOH = Na_2Si_3O_7 + H_2O$$

Dent Glasser and Kataoka [55] suggested that the reaction proceeds in at least two stages. In stage 1, there is an acid–base reaction between the silica and hydroxyl, which causes the silica to disintegrate as the siloxane bridges are attacked:

$$H_{0.38}Si_{2.19} + 0.38NaOH = Na_{0.38}SiO_{2.19} + 0.38H_2O$$

Stage 2 overlaps with the first. The hydroscopic gel absorbs water, which in the presence of calcium ions leads to the disruptive expansion:

$$H_{0.38}Si_{2.19} + 1.62NaOH = 2Na^{2+} + H_2SiO_4^{2-}$$

The alkali–silica gel reaction product, if present, can normally be identified using a polarising petrographic microscope, provided the concrete sample has been prepared appropriately and the thin section prepared using an oil-based cutting and grinding lubricant (Figure 4.10). However, the gel is readily leached by water percolating through cracks and the gel itself is readily carbonated by atmospheric carbon dioxide. A detailed account of the methods of petrographic investigation of concrete affected by AAR is given in *Concrete Petrography* [12].

4.4.6.2 Reactive components in ASR-affected concrete

As noted in the introduction, concrete is a mix of aggregates, cement (most commonly Portland cement, CEM I) and water sufficient to hydrate the cementitious minerals. In addition, modern concretes usually contain a proportion of mineral admixtures such as GGBS, pulverised fly ash or silica fume. They are also likely to contain very small amounts of chemical additions, such as plasticisers or water reducers, that improve the rheological properties of the concrete mix and the physical and mechanical properties of the hardened concrete.

Of the three components necessary to initiate and develop alkali–aggregate reactivity in concrete, one is contained within the aggregate. Several mineral components sometimes found within aggregates have the potential to cause ASR, and these have been listed in Table 4.1. In rare examples, the reactive forms of silica, such as cristobalite and tridymite, appear to be derived from silica particles within the raw feed used for cement manufacture. They remain in the feed as high-temperature polymorphs but are otherwise unreacted after firing in the kiln, and so can be found within the anhydrous cement. Such particles are small with large surface areas, and consequently, they react rapidly, typically within a matter of days, to produce gels, which may exude onto the concrete surface.

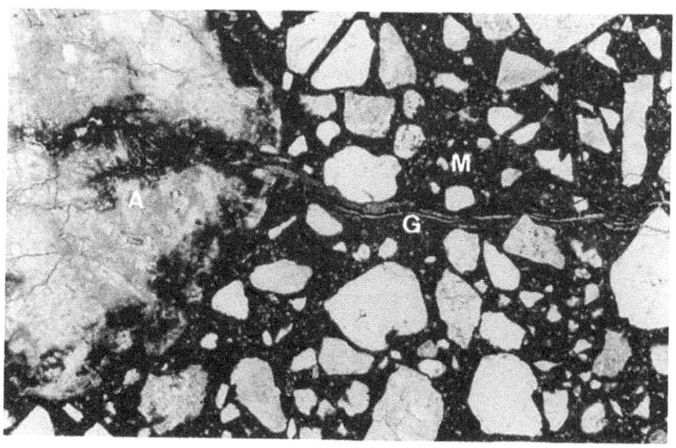

Figure 4.10 Thin section micrograph of concrete showing a gel-filled crack (G) through a cement matrix (M) derived from the aggregate particle (A) on the left. Width of field of view 6 mm

The source of the alkali that is also necessary for ASR is most usually from the cement used. Although, in some rare examples, alkalis derived by degradation of aggregates, such as the leaching from alkali feldspars in geologically weathered rocks, or from extraneous sources, such as de-icing chemicals, have been reported in the literature [52].

Moisture is the third component necessary for ASR to initiate and to provide the swelling pressures within the hydrophilic alkali–silica gel reaction product that forms. Experimental ASR concretes show little evidence of expansion if the relative humidity is kept below 80%, but expansion increases rapidly at higher humidities. In one example [56], alkali-reactive mortar bars stored for 40 days in dry air showed no signs of expansion, but they expanded rapidly when placed in water, suggesting that although ASR had occurred, insufficient water was available for the gel to swell until the bars were wetted. It is also commonly observed that the effects of ASR on concrete structures, such as cracking and deterioration, are most noticeable on surfaces facing the local prevailing wind and rain directions. The cycle of wetting and drying on weather faces is thought to introduce a pumping effect, with moisture moving into and out of the surface layers, thus accelerating the reaction in those zones.

The alkali–silica gel reaction product is variable in composition. Table 4.7 lists typical compositions as means of 20 and general ranges according to Hanson *et al.* [57]. CaO is difficult to assess because of the tendency of the gels to carbonate before or after they have been sampled for analysis. Attempts have been made to relate the potential swelling pressure exerted by the gel to its composition, but this has proved difficult. Calcium ions appear to reduce the swelling pressure, but other variables include the rate of reaction and temperature. Certainly, swelling pressure is primarily dependent on the hygroscopic absorption of water, but the composition of the gel is not a reliable indicator. Viscosity appears greater in silica-rich gels, but viscous silica-rich gels can coexist with fluid alkali-rich sols.

There is evidence that, in some cases, the maximum expansive effects due to ASR are at a particular and often low percentage of the reactive component in the aggregates. In laboratory-prepared concrete or mortar specimens, the expansion produced above and below this pessimum proportion is lower [58–60], as indicated in Figure 4.11.

The diagrammatic representation of the types of 'pessimum effect' shown in Figure 4.11 illustrates three different possible variants. Although it is reasonable to hypothesise that as the proportion of reactive component in the aggregate increases so the expansion resulting from the increase in the gel reaction product will also increase, there are difficulties in explaining why the expansion should decrease as the reactive component increases beyond a critical level of the 'pessimum proportion'. Various hypotheses have been proposed [45] to account for this effect. One suggestion is that if additional reactive centres are available, the gel produced will be more widely distributed, allowing for more rapid carbonation and stabilisation of the gel. Another suggestion is that a greater number of reactive centres will lead to a more rapid reaction

Table 4.7 Alkali–silica gel reaction products

Product	Mean: %	Range: %
SiO_2	48.6	27–75
Na_2O	0.9	0–18
K_2O	6.8	–
CaO	27.4	0–35

Figure 4.11 Curves illustrating the concept of a 'pessimum proportion' for maximum expansion due to ASR. Curve A is typical of some UK aggregates with a pessimum at about 6% reactive component. Curve B is more typical with a broad curve with a pessimum at about 30%. Curve C has no pessimum and is more typical of slowly reacting aggregates

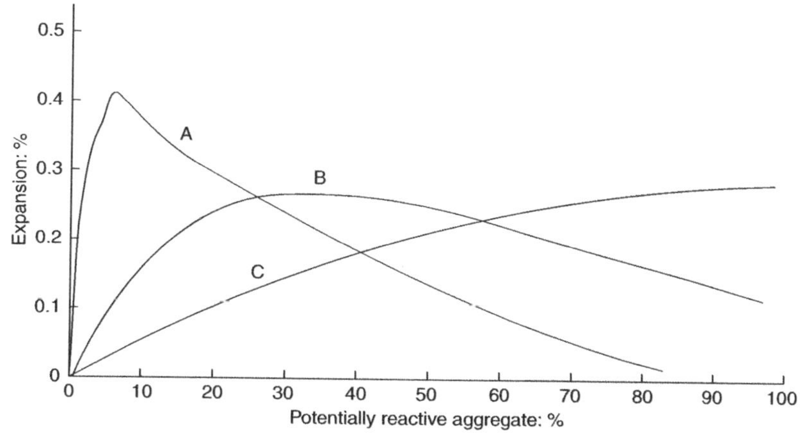

and depletion of the alkali hydroxide reactant and a consequent slowing of the reaction and expansion. The slowly reacting alkali aggregate concrete implies that there is a slow development of the gel reaction product, and this will produce a slow expansion with possibilities for changes (and carbonation) of the gel composition with time, so no pessimum proportion is observed, as illustrated by curve C in Figure 4.11.

4.4.6.3 Effects of ASR on structures

Although damage resulting from AAR in concrete structures is relatively rare, many reports have been published over the years since its first recognition by Stanton in the late 1930s. The effects of this expansive reaction are first recognised in a concrete structure with the appearance of cracking and the exudation of alkali–silica gel on concrete surfaces or broken surfaces. These often exhibit a whitish matt appearance due to carbonation of the gel in the atmosphere. In some cases, reaction centres on aggregate particles close to the concrete surface may produce an expansive gel, which can spall off a roughly conical area of concrete a few centimetres across, which is referred to as a 'pop-out'. The misalignment of structural elements in a concrete structure that are sometimes seen during inspections can be due to uneven ASR expansion of adjacent concrete elements and can seriously compromise the structure.

Inspections of concrete structures should, of course, be followed by a more detailed laboratory investigation involving petrography and other relevant methods, including test procedures, to confirm the cause of any observation of deterioration. If an initial inspection provides clear evidence of ASR damage, that should be confirmed by a laboratory investigation. However, in many examples, the ASR can open pathways into the concrete, allowing other deterioration mechanisms to develop, so that the ASR is only one of the causes of the observed damage.

One typical feature is the carbonation of ASR gel. Figure 4.12 shows a well-advanced attack on concrete. Another very common feature is a pattern of so-called map cracking (like the boundaries on a political map), which develop on unconstrained concrete slabs, as illustrated in Figures 4.2 and 4.13. The latter also illustrates the tendency for the cracks to have stained or bleached margins, which are thought to be due to the alkali in the exuding alkali gel bleaching or staining the concrete surface on each side of the

Figure 4.12 Broken piece of concrete showing extensive carbonation of the ASR gel. This example is from a concrete jetty in Cyprus

Figure 4.13 Typical ASR cracking on a concrete surface. Note the staining associated with the cracks. This example is from Lake Neuchâtel, Switzerland

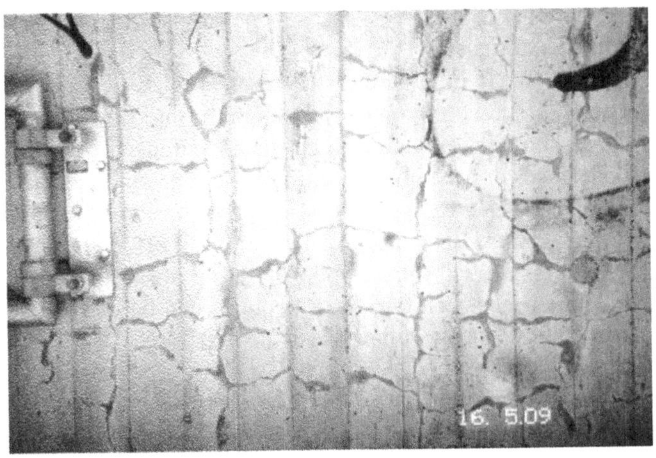

crack and destroying any organic fungal growth. These examples of map cracking occur on unconstrained ASR-affected concrete surfaces. If the concrete is subjected to loading, such as the bridge support column in Figure 4.14, then the cracking can be partly aligned parallel to the direction of maximum stress. In the example shown in Figure 4.14, an attempt has been made to prevent vertical splitting of the column and its consequent failure by fixing steel hoops every half metre.

The misalignment of concrete elements in a structure referred to above is sometimes obvious, such as the bending of the parapet rail on the concrete walkway of a small dam in Norway, as illustrated in Figure 4.15.

Figure 4.14 Vertical ASR cracks in a bridge support column of a motorway overbridge in Devon. Note the DEMEC measuring points, which are fixed to allow movement monitoring across the cracks

Figure 4.15 The bend in the parapet rail indicates misalignment between adjacent concrete structural elements of a bridge in Norway

Misalignments can readily be quantified and monitored by careful surveying, but in some cases, the uneven expansion of different elements can lead to the seizing of mechanical moving components in the structure, such as moveable sluice gates.

A follow-up laboratory investigation usually takes the form of a detailed petrographic study of cores taken from the structure. If possible, these should be 100 mm or larger in diameter and should be taken from both damaged and undamaged parts of the structure for comparison. The petrographer will be able to identify ASR gel if present (Figure 4.10) and probably the mineral components in the aggregate that form the seat of the reaction. They will also be able to identify any other deleterious mechanisms that may be present, for example, DEF. In some cases it may be necessary to use more sophisticated investigative tools, such as X-ray diffraction, chemical techniques and SEM to augment or confirm the initial work.

An example of alkali–silica damage is illustrated by Figures 4.16 and 4.17. The ASR damage is to bridge abutments that carry a main road bridge into the city of Exeter in Devon. Figure 4.16 shows the ASR cracking on the weather face of the west abutment. The pattern is partially aligned with the steel reinforcement system. It is clear where sample cores have been extracted. The holes have been made good. The series of demountable mechanical strain gauge (DEMEC) measuring points can be seen arranged in triangular patterns. These allow crack movements to be monitored. Figure 4.17 shows a remedial approach carried out on the eastern abutment. The cracked and damaged concrete abutment has been pinned back to the soil embankment behind the concrete facing using long soil anchors.

Figure 4.16 North side of the south-west abutment showing that the ASR cracks are partially aligned with the underlying steel reinforcement. Sample 100-mm core locations and triangular arrangements of DEMEC gauge monitoring points can also be seen

Figure 4.17 Close-up view of the cracked north-east abutment bridge deck support showing the ends of the soil anchors, which are tied back into the soil embankment behind the concrete abutment

4.4.6.4 Testing and specification for avoidance of ASR

Effective specifications to avoid damage due to ASR typically rely on the removal of one or more of the three essential requirements for the reaction to initiate and proceed, namely

- a sufficiently alkaline pore solution in the concrete
- a critical concentration of a reactive silica component in the aggregate
- the availability of sufficient water in the concrete.

Consequently, numerous tests have been devised to predict the likelihood of damaging expansion occurring in concrete. These embrace chemical determination of the alkali present in the cement or extractable from the aggregate, examination and testing of the aggregates for alkali-reactive components and the testing of specimens of the proposed concrete for expansive behaviour. This last is usually undertaken under laboratory conditions and extrapolated to predict the behaviour of the real concrete structure. Many countries around the world have developed specifications for the materials to be used for concrete to avoid the likelihood of damage due to ASR specific to their particular materials, climate and other circumstances. These have been summarised in *Alkali–Aggregate Reaction in Concrete, A World Review* [45].

The principal source of alkali is from the cement used, so most countries have mandatory limits on alkali content (in many, this is set at 0.6% by weight of cement). The use of mineral admixtures, such as GGBS, is frequently used to reduce the effective alkali content and to modify the physical and mechanical properties of the concrete. Again, many countries have developed lists of those aggregates with mineral components susceptible to ASR and restrictions or embargos have been placed on the use of such materials.

Alkali determination utilises well-established chemical, spectrometric or X-ray fluorescent techniques. There are standardised national analytical procedures, for example BS 4550-2:1970 and ASTM C114-07 (2007). The identification of potentially reactive mineral components usually involves a petrographic analysis of the aggregate. Again, there are standard procedures, such as BS 812-104:1994 or for aggregates within concrete, BS 1881-211:2016 and ASTM C856 (2013).

If after an initial evaluation the potential for expansive AAR is considered a possibility, then further laboratory testing should be undertaken, most commonly by preparing prism specimens of the concrete or of a mortar utilising the crushed aggregate materials. Such prism specimens are kept in humid conditions or in alkali solution and, in some cases, at elevated temperature. They are then monitored for length changes with time. Numerous tests of this kind are in use in countries around the world (45). Some examples are ASTM C227 (2078b), BS 812-123 (2011), BS DD 249 (1999), CSA A23.2-14A (2014), and RILEM AAR-2 and AAR-3 (2017).

The range and variety of tests in use arise from the different materials, environmental circumstances and manifestations of ASR that arise in different countries in different parts of the world. In recent times, an attempt to evaluate the ASR risks to concrete have been developed globally by an international group of experts under the auspices of RILEM, who published their recommendations in AAR-7.1 [61]. These recommendations determine the level of risk and precautions necessary under the following general headings

- degree of reactivity of aggregates used (divided into five levels based on test results)
- local environmental conditions (divided into three levels of severity)
- level of precaution required (divided into four levels based on acceptable damage)
- precautionary measures necessary (divided into four levels of increasing avoidance measures).

Most of the criteria are based on results obtained from existing test procedures. The values determined for each of the four headings above can be shown as a matrix and used to assess the acceptable risk and required precautions that should be associated with any particular concrete structure. The details of this methodological approach are explained in *Alkali–Aggregate Reaction in Concrete: A World Review* [45]. This approach is now gaining wider acceptance globally.

4.4.6.5 Repair and remediation of ASR-affected structures

Although several methods for arresting the progress of ASR damage and examples of remedial measures that can be taken are described in the published literature, at present, there is no entirely satisfactory long-term solution. This is largely because the damaging expansive reaction works throughout the body of the concrete and it is very difficult to reduce the alkali level within the interior pore solution of the concrete or to entirely exclude moisture from permeating the fabric of the concrete.

When dealing with ASR-affected concrete, it is necessary to take account of the practical requirements of the particular structure. Some of the more important considerations can be listed as follows

- What is the expected design life, its serviceability and the relevant safety requirements?
- How critical is the structure within the local infrastructure?
- Can the structure be taken out of service for a period?
- What are the construction details of the structure and what modifications are feasible?
- What are the access and operational constraints?
- What is the budget for remediation work?

As already noted, several possible repair methods have been tried and reported.

- The use of additional reinforcement to strengthen the structure using a 'cadge' containing rings (Figure 4.14) or by an internal strengthening or supporting system, such as post-tensioning cables or rods (Figure 4.17).
- Additional support for the structure using external props. These were used, one on each side of the central carriageway support column (whose foundations were compromised by ASR), on the Marsh Mills Viaducts, which carry the major A38 into Plymouth, Devon, but were discarded in favour of a complete replacement after a few years.
- If expansion of adjacent elements is causing difficulties, for example, by jamming moving gates on a sluice or on turbine support platforms, then slots cut into the structure can release the stress between segments of the structure. This technique was used in the Beauharnois hydroelectric plant on the St Lawrence Seaway, Canada.
- Surface protection to prevent water ingress has been used quite widely, although the long-term success of methods such as surface cladding and impermeable coatings appear limited.
- Sealing cracks by injecting polymers, which are usually based on silanes, is another possibility, but the life expectancy and durability of penetrative sealers is relatively short compared to the life of a structure, so it may be necessary to renew the treatment every 5 years or so [62].
- Chemical or electrochemical methods have been used to modify the chemistry of the concrete pore fluid and so arrest further reactivity. Lithium salts have been used quite widely, since the gels they form are non-expansive, but ensuring the effective penetration of lithium into the concrete is very difficult, and consequently, in many examples, it has been ineffective [63].

As noted above, because the reaction proceeds in the heart of the concrete, any remedial actions will be of limited benefit unless they affect the centre of the concrete. However, in conclusion, a reduction in the availability of water by using coatings or other measures will slow the reaction. Also, the use of additional reinforcement to constrain the expansive movement of the concrete can make an ASR-affected structure more robust and extend its expected design life.

4.4.7 Corrosion of the steel reinforcement

John Broomfield, Broomfield Consultants, UK

4.4.7.1 How concrete protects reinforcing steel from corrosion

This section discusses the corrosion of steel in concrete. This is one of the most important deterioration mechanisms of reinforced concrete and has major financial implications around the world. Although concrete is a very durable material, corrosion of the reinforcement is a major cause of deterioration of the built infrastructure, costing most major economies billions of pounds for damage to highway bridges, marine structures, buildings and industrial plants.

CHEMICAL AND PHYSICAL STRUCTURE OF CONCRETE AND IMPLICATIONS FOR ITS DURABILITY

When considering why steel corrodes in concrete, the most important question is the converse: why does steel not corrode in concrete in most situations? Concrete is a porous material, typically with 10–20% porosity, which is due to fine gel pores, coarser capillary pores and the porosity of the coarse aggregate. Water and oxygen diffuse easily into concrete and can reach the reinforcing steel, regardless of the concrete cover over the steel. Depending on the pore sizes and the amount of water in the environment, the pores may be relatively empty of water, lined with water or filled with water. However, despite this porosity, most concrete exposed to the outdoor environment does not corrode. This is because of the chemical rather than the physical composition or barrier properties of concrete.

Steel does not corrode in concrete because of the alkalinity of the cement paste. The pore water in hardened Portland cement usually has a pH in the range 12.5–13.5. This is due to large reserves of calcium hydroxide $Ca(OH)_2$, which is pH 12.4, supplemented by far smaller amounts of sodium and potassium oxides or hydroxides, with sodium hydroxide having a pH of 14. In this pH range, steel does not corrode due to passivation, as discussed in the following section.

PASSIVATION OF STEEL IN CONCRETE

There are many oxides of iron and steel, which form in different conditions, from the friable rust we see on unpainted ironwork to dense protective oxides formed at high temperatures in boilers and other industrial plants. Iron can corrode at a rapid linear rate, a controlled parabolic rate or a very slow logarithmic rate, depending on the properties and quality of the oxide formed and the environment in which it forms.

In atmospheric conditions, several factors influence the nature of the corrosion products and, hence, the rate of corrosion. One of the most important is the pH of the steel surface. If the pH is higher than about 10.5, a passive oxide layer will form on the steel surface. This passive layer is generally agreed to be a dense film of γFeOOH of the order of 1 nm thick. However, in concrete, the situation is complicated by the formation of a dense mineral layer many micrometres thick on the steel surface, which may supplement, or substitute for, the action of a 'true' passive layer. This passive layer is dynamic. It breaks down and reforms and will last indefinitely as long as the correct conditions persist.

Concrete is, of course, vulnerable to attack. For instance, acids with a pH of less than 7 will attack concrete. Soft peaty water can also attack it. This form of attack is obvious, and the embedded reinforcing steel is vulnerable to corrosion once it is exposed and no longer protected by a pH of greater than 10.5. However, there are two chemical species that can break down the passive layer without destroying the concrete first. They are the chloride ion and the carbon dioxide molecule, as will be discussed in the next two subsections.

4.4.7.2 Carbonation attack

The atmosphere typically contains between 0.03 and 1% carbon dioxide (CO_2). Like most gases at room temperature, carbon dioxide is an acid gas. It forms carbonic acid when dissolved in water. This weak acid will react with the calcium, potassium and sodium hydroxides in the pore water, reducing the pH to below the critical 10.5 threshold for steel corrosion.

THE CARBONATION MECHANISM

Chemically, the reaction is a simple one and familiar from high-school chemistry. In practice, gaseous carbon dioxide enters the pores in the concrete and dissolves in the pore water to form carbonic acid, which then reacts with dissolved calcium hydroxide to form insoluble calcium carbonate, which precipitates out of solution and lines the pores. There are also some minor reactions with other constituents and some shrinkage of the cement matrix. However, there is no significant deleterious effect due to carbonation on the concrete itself. The damage is the reduction of pH, which allows the steel to depassivate and start to corrode actively.

CARBONATION RATES

All concrete will carbonate at some rate or other. The carbonation rate depends on the amount of moisture present. If there are cycles of wetting and drying, then carbon dioxide will penetrate the concrete when it is dry and then dissolve to form carbonic acid when it is wet. The carbonation rate is also governed by the porosity of the concrete and the amount of alkaline reserves in it. The rate of carbonation follows a roughly parabolic relationship:

$$x = kt^{1/2}$$

where x is the carbonation depth in mm, k is the carbonation coefficient in mm/year$^{1/2}$ and t is time in years.

This equation is an approximation, as the power law exponent can vary from about 0.4 to 0.6, depending upon exposure conditions and the concrete [64].

Carbonation depths can be measured by a range of techniques. The most popular is the phenolphthalein indicator test, as described in Section 4.4.7.5. In an investigation of several field surveys, Trend 2000 [65] found the following range of k values.

- For 11 buildings 8–24 years old, the constant k ranged from 1.2 to 6.7 mm/year$^{1/2}$, with an average of 3 mm/year$^{1/2}$.
- For seven car parks 14–41 years old, the constant k ranged from 2.2 to 7.6 mm/year$^{1/2}$, with an average 4.27 mm/year$^{1/2}$.
- For a 10-year-old jetty and a 90-year-old bridge, k was 1.8 mm/year$^{1/2}$ for the jetty and 1.6 mm/year$^{1/2}$ for the bridge.

It is quite common to find office buildings constructed in the 1960s and 1970s with severe carbonation-induced corrosion, whereas bridges of the same era will show negligible carbonation depths and no carbonation-induced corrosion. This is due to bridges being built with a higher cement content, with a higher and better-controlled cover to the reinforcement and with better compaction of the concrete, ensuring dense, low-porosity concrete around the steel.

The carbonation advances as a 'front' into the concrete. This can be somewhat uneven and irregular, but it is useful, as it is possible to define a maximum, minimum and average carbonation depth and compare it with the cover depth. Using the simple $x = kt^{1/2}$ equation, it is possible to determine a range for the carbonation coefficient k and to estimate the time to corrosion.

4.4.7.3 Chloride attack

As stated in Section 4.4.7.1, another chemical agent that can initiate corrosion without attacking the concrete first is the chloride ion. Although almost all concrete exposed to the environment will eventually carbonate, only concrete exposed to chlorides will suffer from a chloride attack. However, our ability to protect our structures from chloride ingress has been very poor in the past, and therefore, the present cost of repairing and maintaining bridges and marine structures is very high due to chloride ingress and subsequent reinforcement corrosion.

SOURCES OF CHLORIDE

The chloride ion is, of course, 50% of the constituents of common salt, NaCl (60% by weight). It is used in de-icing salt and is present at a concentration of about 3.5% in seawater. Therefore, any reinforced-concrete structure exposed to seawater, de-icing salts or any other source of chloride is vulnerable to chloride-induced corrosion.

Chlorides will migrate into concrete by several mechanisms, as discussed below, unless adequate protection is put in place during construction. However, there are also problems with chlorides being cast into concrete due to contamination of the mix constituents. Up until the 1960s, it was thought by many engineers that the chloride ions were bound chemically to tricalcium aluminate (C_3A) to form chloroaluminates. While this undoubtedly happens, it is not sufficient to prevent chlorides from attacking the passive layer when the concrete is made with seawater or where calcium chloride was used as a set accelerator for cold weather concreting, which was banned in the 1970s. In addition, carbonation breaks down chloroaluminates, freeing up chloride ions and leading to corrosion, after perhaps several decades of good service.

TRANSPORT OF CHLORIDE THROUGH CONCRETE

Like carbonation, the rate of chloride ingress is often approximated with Fick's law of diffusion. There are further complications here. The initial mechanism appears to be suction, namely capillary action, especially when the surface is dry. Salt water is rapidly absorbed by dry concrete. There is then some capillary movement of the salt-laden water through the pores followed by 'true' diffusion. There are other opposing mechanisms that slow the chlorides down. These include a chemical reaction to form chloroaluminates and absorption onto the pore surfaces. The detailed transport mechanisms of chloride ions into concrete are discussed by Kropp and Hilsdorf [66]. Classical diffusion modelling is widely used to predict chloride transport into concrete. Fick's second law of diffusion can be applied, and there is a range of models used, as discussed by Broomfield [67].

CHLORIDE THRESHOLDS

In principle, there is a chloride threshold for corrosion of steel in concrete. Hausmann [68] showed, both theoretically and in practice, that for calcium hydroxide solutions, above a chloride-to-hydroxide ion ratio of 0.6, the probability of corrosion rises rapidly. However, in real concrete, this threshold is harder to pin down [69] for the following reasons.

- Concrete pH varies with the type of cement and other mix constituents. Since pH is a logarithmic scale, a pH change from 12 to 13 is a factor of 10 change in the hydroxide ion concentration.
- Chlorides can be bound chemically by aluminates and physically by absorption to pore surfaces. This can temporarily or permanently remove chlorides from the pore water.

- Carbonation will free chemically bound chlorides, so that there is a 'synergistic' effect, leading to the onset of corrosion in chloride-contaminated concrete due to the progression of chlorides into the concrete, initiating depassivation.
- There must be sufficient moisture for corrosion to progress, so even depassivated steel does not always corrode.
- Corrosion proceeds by the formation of anodes and cathodes, as described in Section 4.4.7.4. Therefore, some areas of steel will not corrode as they are cathodic sites, which must sustain a non-corroding reaction for the corrosion reaction to occur at the anodic sites. This is despite the presence of chlorides at or near anodic sites.
- Corrosion initiates at voids on the steel surface where the alkaline cement paste/pore water is not in contact with the steel, so chlorides can more easily initiate corrosion. More voids mean a lower threshold; fewer voids mean a higher threshold.

It is generally accepted that corrosion can occur at chloride levels from 0.1–1% chloride by mass of cement. A commonly used threshold is 0.4%, although National Highways (UK) prefers 0.3%, and others consider 0.2% to be the best value.

4.4.7.4 Corrosion process after depassivation

Regardless of the source of the depassivation of the steel, once it has occurred, the mechanism is the same. Iron dissolves at the anodic site of depassivation, initially producing soluble ferrous ions (Fe^{2+}) and electrons (e^-). The ferrous ions react with hydroxide ions, then water and oxygen, to ultimately form rust. At an adjacent cathodic site, the electron reacts with water and oxygen to form hydroxide ions. This is illustrated in Figure 4.18. As can be seen from the figure, there is no initial requirement for oxygen at the anode, only at the cathode. Water is required to dissolve the Fe^{2+} ion but not for the anodic reaction. However, water and oxygen are required to form rust. This corrosion product has a volume of about 6.5 times that of the steel consumed.

EXPANSIVE OXIDE FORMATION

The formation of the expansive oxide leads to the cracking and spalling of concrete. This will initiate in corners and areas with low cover and may occur in sheets or delamination, as cracks propagate between

Figure 4.18 Anodic, cathodic, oxidation and hydration reactions of corroding steel in concrete. (Reproduced with permission from Broomfield [67] © 2023 Taylor & Francis)

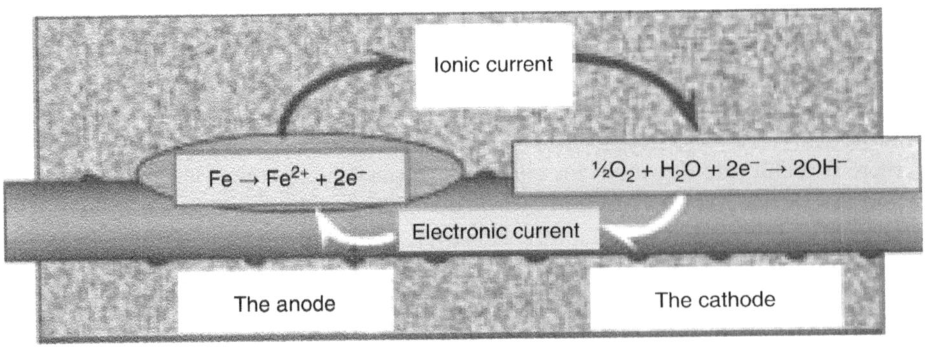

$$Fe^{2+} + 2OH^- \rightarrow Fe(OH)_2 \quad \text{Ferrous hydroxide}$$
$$4Fe(OH)_2 + O_2 + 2H_2O \rightarrow 4Fe(OH)_3 \quad \text{Ferric hydroxide}$$
$$2Fe(OH)_3 \rightarrow Fe_2O_3 \cdot H_2O + 2H_2O \quad \text{Hydrated ferric oxide (rust)}$$

Figure 4.19 Schematic illustration of corner spall and delamination of concrete due to corrosion

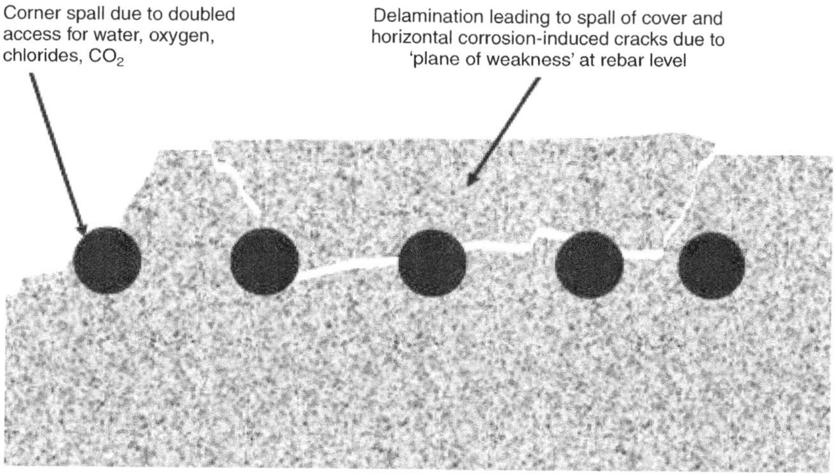

adjacent reinforcing bars, as shown schematically in Figure 4.19. In laboratory tests, Rodriguez *et al.* [70] found that 15–40 μm of section loss generated enough corrosion product to crack the concrete on bars with a cover-to-bar diameter ratio of between 2 and 4.

NON-EXPANSIVE OXIDE FORMATION

From the anodic reaction and subsequent oxide-forming reactions in Figure 4.18, we can see that there must be sufficient moisture to initially dissolve the ferrous ion (Fe^{2+}) in the presence of sufficient oxygen and water to allow the subsequent reactions to occur. If the concrete is saturated or close to saturation, but with an oxygenated cathode close by and a low electrical resistance path through both the steel and the concrete connecting them, then the iron can dissolve and stay in solution. This form of corrosion, often called 'black rust' or 'anaerobic corrosion', can be very rapid. Broomfield [67] gives examples of section losses of greater than 0.5 mm/year for a steel reinforcing bar in a bridge deck in the UK under a damaged waterproofing membrane, which trapped salt-laden water. A more extreme example was a precast post-tensioned marine bridge pile in Florida, where tendon failures showed corrosion rates of 3 mm/year.

TYPICAL CORROSION RATES

Methods of measuring corrosion rates for steel in concrete will be discussed more fully in Section 4.4.7.5. It is important to understand the ranges of corrosion rates found in concrete. Rates of section loss of less than 1 μm per year are generally considered to reflect passive, non-corroding conditions. As stated above, a steel section loss of 15–40 μm was found to lead to cracking and spalling of the cover concrete when the cover-to-bar diameter ratio was between 2 and 4. Generally, a section loss of greater than 10 μm/year is considered to be active corrosion and will ultimately lead to cracking and spalling.

4.4.7.5 Testing for corrosion

There is a wide range of tests for assessing the condition of deteriorated concrete. This section will concentrate on the more technically sophisticated tests directly related to corrosion. In any systematic survey of a deteriorating reinforced-concrete structure, a detailed visual survey should be conducted. If there is corrosion damage, the cracks, spalled and delaminated concrete, and exposed rebars will be noted, along with a 'make safe' survey to remove any loose concrete that could cause injury or damage. Further guidance on testing and surveying is provided by Broomfield [67], Concrete Society Technical Report 60 [71], the technical guide by the Concrete Bridge Development Group [72] and NACE/AMPP SP 0308 [73].

MEASURING THE CHLORIDE CONTENT IN HARDENED CONCRETE

If there is a source of external chlorides or any chance that chlorides were cast into the concrete, then the chloride content of the concrete should be measured in all three dimensions: horizontally, vertically and with depth. The method is destructive, requiring the collection of drilled powder from the structure. Cores may also be taken but will need to be crushed to powder for analysis. Samples may be collected at progressive incremental depths into the concrete. Three increments are the minimum to be useful, but sampling at five depths or more will give sufficient points on a graph to allow the fitting of a curve for modelling chloride ingress rates and back-calculating the effective diffusion coefficient.

In Europe, the standard BS EN 14629 [74] is used by UKAS (the UK Accreditation Service) and other comparable accredited test houses when evaluating acid-soluble chlorides in powder samples. Approximately 25 g of sample is required for an accurate analysis. This can be obtained from drilling a pair of 25-mm-diameter holes 15–25 mm deep in the concrete. Depth measurements at 15 mm increments should be collected with care to minimise cross-contamination of successive samples.

In North America, two different approaches are used for analysis. There are several ASTM standards for analysing chlorides, which fall into two separate groups. One group is for acid-soluble chlorides, like BS EN 14629 [74]. The other is for water-soluble chlorides. This latter test is less reproducible as it requires a solid sample, such as a core, to be broken up into chunks and then boiled to dissolve out the chlorides. However, the results can vary with the size grading of the chunks and powder. This test is important for certain concrete-containing aggregates that contain well-bound chlorides that show up on an acid-soluble analysis. These aggregates are found in certain parts of Canada and, to a lesser extent, in the USA.

There are field kits for measuring chloride contents, which use chloride test strips or a selective chloride ion electrode on site. These require considerable experience and expertise as well as an understanding of their limitations. In the UK, the speed of turnaround and the modest cost of laboratory chloride analysis mean that field analysis is rarely used.

MEASURING CARBONATION DEPTHS

Carbonation depths are measured using chemical indicators that change colour in the range pH 7–9. The most popular and effective method is based on phenolphthalein. The indicator turns pink above pH 9.5, which is below pH 10.5 at which steel starts to corrode. The whole carbonation front will be some 10–15 mm wide, with the pH falling from 12.5 to 8, so there is some inaccuracy in the measurement, but given the variability in the front itself, the measurement is useful. BS EN 14630 [75] describes a method of preparing a phenolphthalein solution, spraying it onto a freshly broken surface and measuring the carbonation depth.

HALF-CELL OR REFERENCE ELECTRODE POTENTIAL SURVEY

A half-cell or reference electrode is a pure metal in a solution of its own ions, such as copper in copper sulfate or silver in silver chloride. When coupled to another metal surrounded by its own ions, such as a corroding reinforcing bar in damp concrete, a small electrochemical cell is formed so that a voltage difference exists between the reinforcing bar and the reference electrode. The reference electrode is moved across the surface of the concrete and touched down on a regular grid using a damp sponge for contact. The higher the voltage measured, the more active is the corrosion cell. The anodes and cathodes (Figure 4.18) can, therefore, be mapped by plotting lines of constant potential on the concrete surface. The basic arrangement is shown schematically in Figure 4.20. The methodology of conducting and interpreting a reference electrode potential survey is described in ASTM C876 [76] and in Concrete Society Technical Report 60 [71].

Figure 4.20 Schematic illustration of a reference electrode set-up for a potential survey

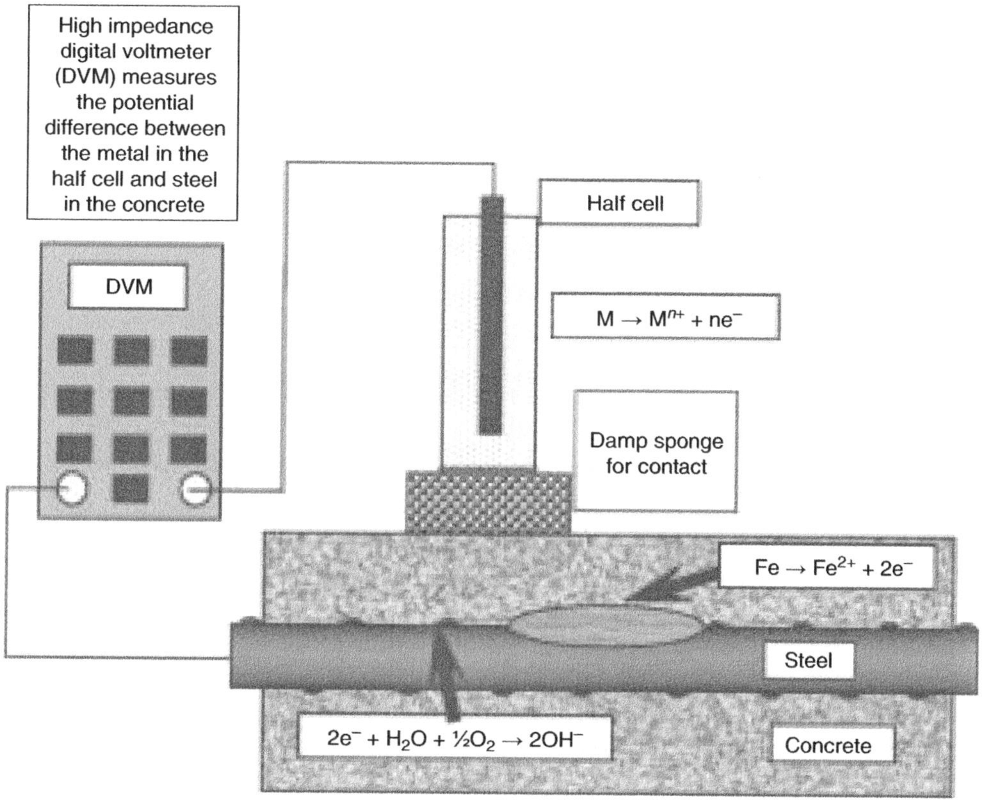

CORROSION RATE MEASUREMENT

Corrosion rates are measured using the polarisation resistance or linear polarisation (LPR) technique. A simple explanation of the LPR corrosion rate-measuring technique for steel in concrete is that the equipment consists of a reference electrode or half-cell and an auxiliary electrode. These are combined in a single head and placed in contact with the concrete surface via a conductive sponge. A schematic illustration of the arrangement is shown in Figure 4.21.

After storing the value of the electrical potential (voltage) between the steel and reference electrode, a small current is applied from the LPR meter. This changes the potential reading in the meter. The higher the corrosion rate, the more current is needed to shift the potential by a fixed amount (up to 20 mV). The corrosion rate is proportional to the applied current divided by the voltage change.

The system records the corrosion current in microamperes. However, that corrosion current is from a certain area of the reinforcing steel. The area is determined from the bar diameter, the size of the head and any system for confining the measurement to specific areas. The current can, therefore, be turned into a current density in μA/cm. Knowing that the corrosion reaction is $Fe \rightarrow Fe^{2+} + 2e^-$, we can use Faraday's law of electrochemical equivalents to convert the current density into a section loss. The conversion tells us that 1 μA/cm^2 is equivalent to a section loss of 11.5 μm/year. The corrosion rate values discussed in Section 4.4.7.4 can then be used to interpret the results. Corrosion rate measurements and interpretation are discussed more fully by Broomfield [67] and in RILEM TC 154 [77] and Concrete Society Technical Report 60 [71,78,79].

Figure 4.21 Schematic illustration of a typical LPR probe arrangement

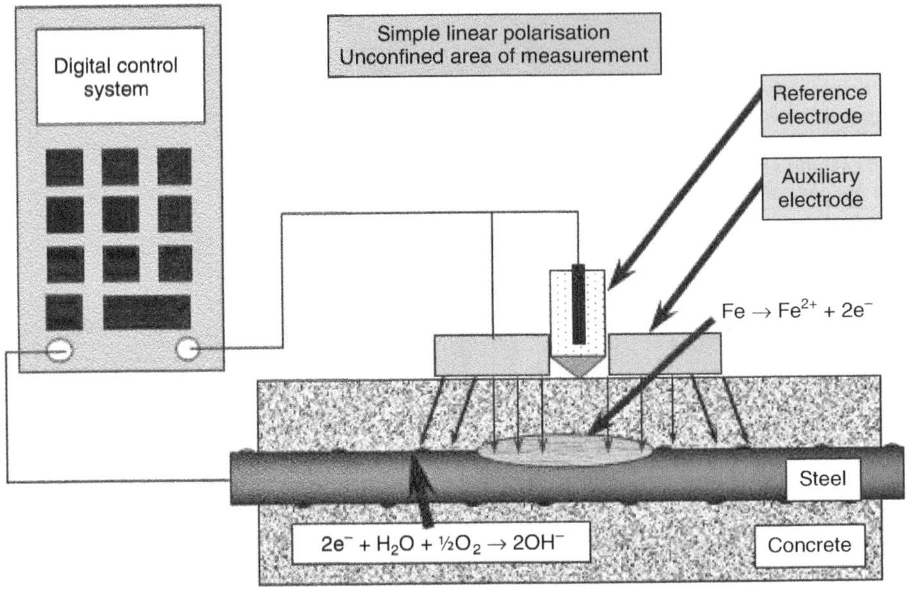

CONCRETE RESISTIVITY

Since corrosion is an electrochemical process, the electrical resistivity of the concrete affects the corrosion rate of the concrete, as an ionic current (electric current in the form of a flow of charged ions in the pore water) must pass from the anodes to the cathodes for corrosion to occur (Figure 4.18).

The four-probe resistivity meter or Wenner probe was developed for measuring soil resistivity (ASTM G57-06) [80]. Specialised modifications of the Wenner probe are frequently used for measuring concrete resistivity on site. The measurement can be used to indicate the possible corrosion activity if steel is depassivated. A proprietary version of the system is shown in Figure 4.22. The system uses four probes. The outer probes pass a current through the concrete while the inner probes detect the voltage difference. This approach eliminates any effects due to surface contact resistance.

Interpreting the resistivity results is empirical. Table 4.8 gives an interpretation of resistivity measurements from the Wenner four-probe system for depassivated steel.

Further information on measuring resistivity is provided by Broomfield [67], Concrete Society Technical Report 60 [71] and RILEM TC 154 [81].

COVER-DEPTH TESTING

Although the cover depth is not strictly a corrosion-related measurement, the cover is critical in assessing the likelihood and severity of corrosion. Correlating the cover depth with the carbonation depth and the chloride content at typical cover depths will inform the engineer of the extent of existing corrosion and allow a prediction of the time to corrosion. Then, using modelling techniques, as discussed in Sections 4.4.7.2 and 4.4.7.3, and looking at the available repair options, as discussed by Broomfield [67], the engineer can identify the most cost-effective repair options. For example, the engineer can decide whether it is worthwhile trying to keep chlorides out with a coating or other barrier or whether the chloride level is so high at the reinforcement depth that corrosion has started and, therefore, active corrosion control is necessary.

Figure 4.22 Combined four probe resistivity meter with reference electrode and temperature sensor showing logger, computer, reinforcement connection. (Courtesy of CITech GMbH)

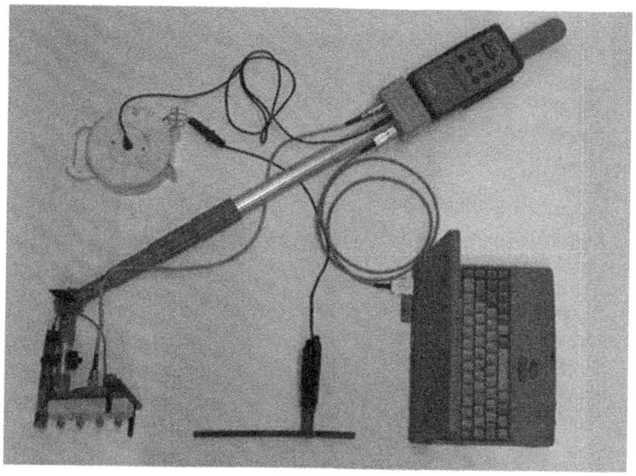

Table 4.8 Interpreting resistivity measurements

Resistivity: kΩ cm	Corrosion rate
>20	Low
10–20	Low to moderate
5–10	High
<5	Very high

Electromagnetic cover meters are widely used in the construction industry. These are now digital devices that can log results and measure bar diameters as well as cover depths. Scanning devices that plot the bar locations and depths are now available. These are discussed by Broomfield [67] and the Concrete Bridge Development Group [72]. The only standard for applying a cover meter is BS 1881 [82].

OTHER TESTING TECHNIQUES

There is a huge range of other test methods, especially for locating flaws, cracks and delamination in concrete. These include ground-penetrating radar, infrared thermography, impact echo, radiography and ultrasonic pulse velocity. None are as widely used or easily applied as the techniques discussed above, although they may yield a wider range of information about the properties of the concrete and what is embedded in it.

There are also many tests that will inform the engineer about the properties of concrete that impact its ability to resist ingress or corrosion. These include permeability and surface absorption [66]. Finally, petrographic analysis is an important part of any investigation of a concrete structure; see Eden [83] and ASTM C856 [84].

4.4.7.6 Corrosion monitoring

The electrical equipment for measuring the corrosion rate, potential and concrete resistivity described in the previous sections is portable and battery operated. The measurement probes or heads are placed on the concrete surface. It is also possible to install these probes permanently in the concrete, either during construction or after problems are identified. They can then be linked, either to a monitoring location where the meters can be connected to take repeat readings or to a logging system that takes measurements automatically at fixed intervals.

CORROSION-MONITORING PROBES

The following permanent monitoring probes can be embedded in concrete

- corrosion rate (LPR) probes
- concrete resistivity probes
- reference electrodes
- temperature probes
- moisture (relative humidity) probes.

Figure 4.23 Permanent probe for monitoring the corrosion rate fixed into a reinforcement cage prior to casting the concrete

Figure 4.24 Ladder probes for measuring the galvanic current prior to casting concrete. (Courtesy of Dr-Ing M Raupach)

These can be linked to other probes, such as strain gauges, for long-term assessment of a structure. Figure 4.23 shows a permanent monitoring probe consisting of a reference electrode, an LPR probe and a temperature sensor. It can also measure concrete resistance. Figure 4.24 shows a 'ladder probe', which measures the development of the galvanic current between carbon steel 'rungs' and a stainless steel cathode to monitor the progress of the corrosion front into the concrete.

CORROSION-MONITORING SYSTEMS

Corrosion-monitoring systems have been installed during construction in several major bridge and tunnel projects in Europe, including the Dartford Tunnel Redecking project and the Great Belt Bridge/Tunnel project. Data from the Dartford Tunnel project are shown and discussed by Broomfield *et al.* [79]. These data were collected from over 60 probes since 2001. An overview of monitoring systems for reinforced-concrete structures can be found in Buenfeld *et al.* [85].

REFERENCES

1. BSI (2000) EN 197-1:2000. Cement – Part 1: Specification, performance and conformity criteria for common cements. BSI, London, UK.
2. Hewlett PC and Liska M (eds) (2019) *Lea's Chemistry of Cement and Concrete*, 5th edn (Hewlett PC and Liska M (eds)). Butterworth-Heinemann, Oxford, UK.
3. Grattan-Bellew PE and Mitchell LD (2006) Quantitative petrographic analysis of concrete – the damage rating index (DRI) method. A review. *Proceedings of the Marc-André Bérubé Symposium on Alkali–Aggregate Reactivity in Concrete, Montreal, Canada* (Fournier B (ed.)), pp. 321–334.
4. Concrete Society (1992 & 2010) *Non-Structural Cracks in Concrete.* Concrete Society, London, UK, Technical report 22.
5. BSI (1997) EN 1992-1-1:2004. Eurocode 2. Design of concrete structures. Part 1.1. General rules and rules for buildings. BSI, London, UK.
6. Lea FM (1970) *The Chemistry of Cement and Concrete*, 3rd edn. Edward Arnold, London, UK (see also [2] Hewlett for the 4th edn).
7. BRE (Building Research Establishment) (1991) *Sulfate and Acid Resistance in Ground.* BRE, Watford, UK, Digest 363 (superseded by Special digest 1).
8. Lea FM (1968) Some studies on the performance of concrete structures in sulphate-bearing environments. *Symposium in Honour of Thorbergur Thorvaldson – Performance of Concrete: Resistance of Concrete to Sulphate and Other Environmental Conditions* (Swenson EG (ed.)). University Press, Toronto, ON, Canada, pp. 56–65.
9. Nixon PJ (1978) Floor heave in buildings due to the use of pyritic shales as fill material. *Chemistry and Industry* **4**: 160–164.
10. Ballivy G, Rivard P, Pepin T, Tanguay MG and Dion A (2002) Damages to residential buildings related to pyritic rockfills: Field results of an investigation on the south shore of Montreal, Quebec, Canada. *Canadian Journal of Civil Engineering* **29(2)**: 246–255.
11. Anderson WH (2008) *Foundation Problems and Pyrite Oxidation in the Chattanooga Shale, Estill County, Kentucky.* Kentucky Geological Survey, Lexington, KY, USA, Report of Investigations 18, Series XII.
12. Poole AB and Sims I (2016) *Concrete Petrography. A Handbook of Investigative Techniques*, 2nd edn. CRC Press, Taylor & Francis, p. 794.
13. Purser BH and Seibold E (1973) The principal environmental factors influencing Holocene sedimentation and diagenesis in the Persian Gulf. In *The Persian Gulf. Holocene Carbonate Sedimentation and Diagenesis in a Shallow Epicontinental Sea* (Purser BH (ed.)). Springer-Verlag, Berlin, Heidelberg, New York, pp. 1–11.

14. Conjeaud ML (1980) Mechanism of sea water attack on cement mortar. In *Performance of Concrete in Marine Environment*. American Concrete Institute, Detroit, MI, Report ACI SP-65, pp. 31–61.
15. Thaumasite Expert Group (1999) *The Thaumasite Form of Sulphate Attack: Risks, Diagnosis, Remedial Works and Guidance on New Construction*. Department for the Environment, Transport and the Regions, London, UK, Report (updated in 2000 and 2002), www.saftey.odpmgov.uk/bregs/tegreports/index.htm.
16. Nixon PJ and Longworth I (2003) Thaumasite Expert Group report: Review after three years' experience. *Concrete* **37(4)**: 22–23.
17. Hartshorn SA and Sims I (1998) Thaumasite – A brief guide for engineers. *Concrete* **32(8)**: 24–27.
18. Bensted J (1999) Thaumasite – Background and nature in deterioration of cements, mortars and concretes. *Cement and Concrete Composites* **21(2)**: 117–121.
19. Crammond NJ (2002) The thaumasite form of sulfate attack in the UK. *Proceedings of the 1st International Conference on Thaumasite in Cementitious Materials*. Building Research Establishment, Watford, UK.
20. Taylor HWF (1997) *Cement Chemistry*, 2nd edn. Thomas Telford, London, UK.
21. Erlin B and Stark DC (1965/1966) Identification and occurrence of thaumasite in concrete. *Symposium on Aggressive Fluids, Highway Research Record 113*. Highway Research Board, pp. 108–113.
22. Hagelia P, Sibbick RG, Crammond NJ, Gronhaug A and Larsen CK (2001) Thaumasite and subsequent secondary calcite deposition in sprayed concrete in contact with sulphide bearing alum shale, Oslo, Norway. *Proceedings of the 8th Euroseminar on Microscopy Applied to Building Materials, Athens, Greece* (Stamatakis M, Georgali B, Fragoulis D and Toumbakari EE (eds)), pp. 131–138.
23. Fookes PG and Collis L (1975) Problems in the Middle East. *Concrete* **9(7)**: 12–19.
24. Fookes PG and Collis L (1975) Aggregates and the Middle East. *Concrete* **9(11)**: 14–19.
25. Sims I (2006) Selection of materials for concrete in a hot and aggressive climate. *Proceedings of the 8th International Conference on Concrete in Hot and Aggressive Environments, Manama, Bahrain*. Bahrain Society of Engineers.
26. Crammond NJ (1984) Examination of mortar bars containing varying percentages of coarsely crystalline gypsum as aggregate. *Cement and Concrete Research* **14(2)**: 225–230.
27. Bromley AV and Pettifer K (1997) *Sulfide-Related Degradation of Concrete in Southwest England*. Building Research Establishment/CRC, Watford, UK, Report 325.
28. Stimson CC (1997) *The Mundic Problem. A Guidance Note. Recommended Sampling. Examination and Classification Procedure for Suspect Concrete Building Materials in Cornwall and Parts of Devon*, 2nd edn. Royal Institution of Chartered Surveyors, London, UK (3rd edn, 2015).
29. Blackwell BQ and Pettifer K (1992) Alkali reactivity of greywacke aggregates in Maentwrog Dam (North Wales). *Magazine of Concrete Research* **44(161)**: 255–264.
30. Knights JC and Baldwin NJR (2010) Diagnosis of delayed ettringite formation in precast concrete blocks. *Proceedings of the 13th International Conference and Exhibition, Structural Faults and Repair, Edinburgh, UK*, pp. 10–17.
31. Grantham MG, Gray MJ and Eden MA (1999) Delayed ettringite formation in foundation bases – A case study. *Proceedings of Structural Faults and Repair, Edinburgh, UK*. Edinburgh University Press.
32. Eden MA, White PS and Wimpenny DE (2007) A laboratory investigation of concrete with suspected delayed ettringite formation – A case study from a bridge in Malaysia. *Proceedings of the 11th Euroseminar on Microscopy Applied to Building Materials, Porto, Portugal*.
33. Thomas MDA, Folliard KJ, Drimalas T and Ramlochan T. (2007) Diagnosing delayed ettringite in concrete structures. *11th Euroseminar on Microscopy Applied to Building Materials, Porto, Portugal*.
34. Diamond S (1996) Delayed ettringite formation – Processes and problems. *Cement and Concrete Composites* **18(3)**: 205–215.

35. Scrivener KL and Taylor HFW (1993) Delayed ettringite formation: A microstructural and microanalytical study. *Advances in Cement Research* **5(20)**: 139–146.
36. Glasser FP, Damidot D and Atkins M (1995) Phase development in cement in relation to the secondary ettringite problem. *Advances in Cement Research* **7(26)**: 57–68.
37. Collepardi M (2003) A state of the art review on delayed ettringite attack on concrete. *Cement and Concrete Composites* **25(4–5)**: 401–407.
38. Fookes PG and Collis L (1977) *Problems in the Middle East.* Cement and Concrete Association, Slough, UK, Viewpoint Publication, pp. 1–7 (originally published in 1975 in *Concrete* **9(7)**: 12–19).
39. Walker MJ (ed.) (2002) *Guide to the Construction of Reinforced Concrete in the Arabian Peninsula, Working Party Report.* Published jointly by CIRIA and the Concrete Society, Crowthorne & London, UK, CIRIA Publication C577 and the Concrete Society Special Publication CS136.
40. Goudie A and Viles H (1997) *Salt Weathering Hazards.* Wiley, Chichester, UK, p. 241.
41. Stanton TE (1940) Expansion of concrete through reaction between cement and aggregate. *Proceedings of the American Society of Civil Engineers* **66(10)**: 1781–1811.
42. Stanton TE, Porter OJ, Meder LC and Nicol A (1942) California experience with the expansion of concrete through reaction between cement and aggregate. *Journal American Concrete Institute* **13(3)**: 209–236 (and discussion in Supplement in November 1942).
43. Diamond S (1975) A review of alkali–silica reaction and expansion mechanisms. I: Alkalies in cements and in concrete pore solutions. *Cement and Concrete Research* **5(4)**: 329–345.
44. Diamond S (1989) ASR – Another look at mechanisms. *Proceedings of the 8th International Conference on Alkali–Aggregate Reaction, Okada, Kyoto, Japan* (Nishibayashi S and Kawamura M (eds)), pp. 83–94.
45. Sims I and Poole AB (eds) (2017) *Alkali–Aggregate Reaction in Concrete: A World Review.* CRC Press, Taylor & Francis, Boca Raton, London, New York, Leiden.
46. Gilliot JE (1975) Practical implications of the mechanisms of alkali–aggregate reactions *Symposium on Alkali–Aggregate Reaction, Preventative Measures, Reykjavik, Iceland*, pp. 213–230.
47. Swenson EG (1957) *A Reactive Aggregate Undetected by ASTM Tests.* ASTM, West Conshohocken, PA, USA, Bulletin 226, pp. 48–51.
48. Milanesi CA, Marfil SA, Batic OR and Maiza PJ (1996) The alkali–carbonate reaction and its reaction products – An experience with Argentinean dolomitic rocks. *Cement and Concrete Research*, **26(10)**: 1579–1571.
49. Grattan-Bellew PE and Katayama T (2017) So-called alkali–carbonate reaction (ACR). In *Alkali–Aggregate Reaction in Concrete, A World Review* (Sims I and Poole AB (eds)). CRC Press, Taylor & Francis, Boca Raton, FL, London, New York, Leiden, pp. 63–88.
50. Stark D and Bhatty MSY (1986) Alkali silica reactivity: Effect of alkali in aggregate on expansion. In *Alkalis in Concrete* (Dodson VH (ed.)). American Society for Testing and Materials, Philadelphia, PA, USA, Report ASTM STP 930, pp. 16–33.
51. Poole AB (1981) Alkali carbonate reaction in concrete. *Proceedings of the 6th International Conference on Alkali–Aggregate Reaction, Cape Town, South Africa, S252/13*, pp. 1–13.
52. Lagerblad B (2012) Alkali release from silicate minerals and alkali-silica reaction in concrete. *Proceedings of the 14th International Conference on Alkali–Aggregate Reaction in Concrete, Austin, TX, USA* (Drimalas T, Ideker JH and Fournier B (eds)), 052411-LAGE, pp. 1–6.
53. Tang M and Deng M (2004) Progress on the studies of alkali–carbonate reaction. *Proceedings of the 12th International Conference on Alkali–Reaction in Concrete* (Tang M and Deng M (eds)). International Academic Publishers/World Publishing Beijing, China, vol. 1, pp. 51–59.
54. Hansen WC (1944) Studies relating to the mechanism by alkali–aggregate produces expansion in concrete. *Journal of the American Concrete Institute* **15**: 213–217.
55. Dent Glasser LS and Kataoka N (1981) Chemistry of the alkali–aggregate reaction. *Cement and Concrete Research* **11(1)**: 191–200.

56. Vivian HE (1950) The reaction product of alkalis and opal. In *Studies in Cement-Aggregate Reaction.* Commonwealth Scientific and Industrial Research Organisation, Australia, CSIRO Bulletin 256, pp. 60–81.
57. Hanson KF, Van Dam TJ, Peterson KR and Sutter LL (2003) Effect of sample preparation on the chemical composition and morphology of alkali–silica reaction products. *TRB Annual Meeting.*
58. Hobbs DW (1981) The alkali–silica reaction – a model for predicting expansion in mortar. *Magazine of Concrete Research* **33(117)**: 208–220.
59. Hobbs DW (1988) *Alkali–Silica Reaction in Concrete.* Thomas Telford, London, UK, pp. 1–183.
60. Ozal MA (1975) The pessimum proportion as a reference point in modulating alkali–silica reaction. *Symposium on Alkali–Aggregate Reaction, Preventative Measures, Reykjavik, Iceland.* Icelandic Building Research Institute, pp. 113–129.
61. Nixon PJ and Sims I (eds) (2016) *RILEM Recommendations for the Prevention of Damage by Alkali–aggregate Reaction in New Concrete Structures.* Springer, Dordrecht, Paris, State-of-the-art report 17, pp. 1–168.
62. Fournier B, Bérubé MA, Folliard KJ and Thomas MDA (2010) *Report on the Diagnosis, Prognosis and Mitigation of Alkali–Silica Reaction (ASR) in Transportation Structures.* Federal Highway Administration, Washington, DC, USA.
63. Folliard KJ, Thomas MDA, Fournier B, *et al.* (2012) Evaluation of mitigation measures applied to ASR-affected concrete elements. Preliminary findings from Austin TX exposure site. *Proceedings of the 14th International Conference on Alkali–Aggregate Reaction in Concrete, Austin, TX, USA.*
64. Parrott LJ (1987) *A Review of Carbonation in Reinforced Concrete.* Cement and Concrete Association and BRE, Watford, UK.
65. Trend 2000 (2001) *Evaluation of Life Performance and Modelling. Corrosion of Steel in Concrete.* BRE, J Broomfield Consultants and Risk Review Ltd, London, UK, Report D29; DTI DMI 5.1 (R 2).
66. Kropp J and Hilsdorf HJ (eds) (1995) *Performance Criteria for Concrete Durability.* E&FN Spon, London, UK, RILEM Report 12.
67. Broomfield JP (2023) *Corrosion of Steel in Concrete: Understanding, Investigation and Repair,* 3rd edn. CRC Press, Taylor & Francis, London, UK.
68. Hausmann DA (1967) Steel corrosion in concrete: How does it occur? *Materials Protection* **6(11)**: 19–23.
69. Glass GK, Roberts AC and Davison N (2004) Achieving high chloride threshold levels on steel in concrete. *Proceedings of Corrosion 2004.* NACE International, Houston, TX, USA.
70. Rodriguez J, Ortega LM and García AM (1994) Assessment of structural elements with corroded reinforcement. In *Corrosion and Corrosion Protection of Steel in Concrete* (Swamy RN (ed.)). Sheffield Academic Press, Sheffield, UK, pp. 171–185.
71. Concrete Society (2004) *Electrochemical Tests for Reinforcement Corrosion.* Concrete Society, Camberley, UK, Technical report 60.
72. Concrete Bridge Development Group (2002) *Guide to Testing and Monitoring the Durability of Concrete Structures.* Concrete Society, Camberley, UK, Technical guide 2.
73. Association of Materials Protection and Performance (2018) *Inspection Methods for Corrosion Evaluation of Conventionally Reinforced Concrete Structures.* Association of Materials Protection and Performance, Houston TX, USA, Report NACE/AMPP SP 0308-2018.
74. BSI (2007) EN 14629. Products and systems for the protection and repair of concrete structures. Test methods, determination of chloride content in hardened concrete. BSI, London, UK.
75. BSI (2006) EN 14630. Products and systems for the protection and repair of concrete structures. Test methods, determination of carbonation depth in hardened concrete by the phenolphthalein method. BSI, London, UK.

76. ASTM (1991) C876. Standard test method for half-cell potentials of uncoated reinforcing steel in concrete. ASTM, West Conshohocken, PA, USA.
77. RILEM TC 154-EMC (2004) Electrochemical techniques for measuring metallic corrosion: Test methods for on-site corrosion rate measurement of steel reinforcement in concrete by means of the polarization resistance method. *Materials and Structures* **37**: 623–643.
78. Langford P and Broomfield J (1987) Monitoring the corrosion of reinforcing steel. *Construction Repair* **1(2)**: 32–36.
79. Broomfield J, Davies K, Hladky K and Noyce PA (2003) Monitoring of reinforcement corrosion in concrete structures in the field. *Proceedings of the 1st International Conference on Concrete Repair, St Malo* (Grantham M, *et al.* (eds)). BRE Press, Watford, UK.
80. ASTM (2006) G57-06. Standard test method for field measurement of soil resistivity using the Wenner four-electrode method. ASTM, West Conshohocken, PA, USA.
81. RILEM TC 154-EMC (2000) Electrochemical techniques for measuring metallic corrosion: Test methods for on-site measurement of resistivity of concrete. *Materials and Structures* **33**: 603–611.
82. BSI (1988) BS 1881:204. Testing Concrete. Recommendations for the use of electromagnetic cover meters. BSI, London, UK.
83. Applied Petrography Group (2010) *A Code of Practice for the Petrographic Examination of Concrete* (Eden MA (ed.)). Geological Society of London, London, UK, APG Report SR 2, http://www.appliedpetrographygroup.com/reports.htm.
84. ASTM (2020) C856/C856M-20. Standard practice for petrographic examination of hardened concrete. ASTM, West Conshohocken, PA, USA.
85. Buenfeld NR, *et al.* (2008) *Intelligent Monitoring of Concrete Structures*. CIRIA, London, UK.

Soutsos M
ISBN 978-0-7277-6375-4
https://doi.org/10.1680/icehcd.63754.083

Chapter 5
Material selection to improve durability

5.1. Introduction

Marios Soutsos, Queen's University Belfast, UK

Construction materials that can be used to improve the durability of reinforced concrete (RC) structures include

- chemical admixtures
- mineral admixtures
  - fly ash (PFA)
  - ground granulated blastfurnace slag (GGBS)
  - silica fume (microsilica)
  - metakaolin (MK)
- alternative reinforcing materials to conventional steel reinforcement (black bars)
  - epoxy-coated steel reinforcement
  - galvanised steel reinforcement
  - stainless steel reinforcement
  - glass-fibre-reinforced polymer (GFRP).

The next section of this chapter deals with chemical admixtures, their composition, the beneficial properties they impart on fresh and hardened concrete and their mode of action in concrete.

The third section deals with supplementary cementitious materials, otherwise known as mineral admixtures or pozzolans, which can be used to improve the durability of RC structures. Their method of manufacture, chemical composition and their mode of action in concrete (fresh and early-age properties and hardened properties, especially with respect to improving the resistance of concrete to degradation mechanisms) are discussed.

The final section deals with four alternative reinforcing materials that can be used in concrete structures in place of conventional steel reinforcement (black bars). Their method of manufacture, how they improve durability, the properties that affect design procedures that have been developed for conventional reinforcement and the reputation that they have based on past applications and performance in real structures are discussed.

5.2. Chemical admixtures

Robert Viles, Fosroc International Limited, UK (†)

5.2.1 Introduction

The quality of concrete depends substantially upon using an appropriate mix design [1,2]. Admixtures have a potentially profound effect on both the durability and sustainability of concrete structures. Admixtures can, by various mechanisms, reduce the water-to-cement (w/c) ratio, reduce permeability, improve workability and ensure compaction, which are all important determinants of concrete quality. Moreover, they can be used to reduce the cement content or allow greater use of cement replacements, thus reducing the carbon footprint and, often, the cost of the concrete.

5.2.2 Water-reducing admixtures

5.2.2.1 Effect of plasticisers on water reduction

Water-reducing admixtures, commonly known as 'plasticisers', are usually natural or synthetic polymer compositions that act as dispersants and workability aids for cementitious compositions. The earliest forms of plasticiser were the sodium and calcium salts of lignosulfonate, which have been available since the early 1950s (Figure 5.1).

Lignosulfonates are natural polymer by-products, occurring as a result of sulfite bleaching of paper in manufacture. They are water-soluble anionic polyelectrolytes having surface active properties. They are a cost-effective way to improve and extend the workability of concrete as they can enable a similar slump or flow to be achieved while using less mixing water. This results in a reduction of porosity and permeability. Consequently, either or both of the following can occur

- an increase in strength and a reduction in permeability
- a reduction in cement content.

Figure 5.1 Part of a lignosulfonate molecule

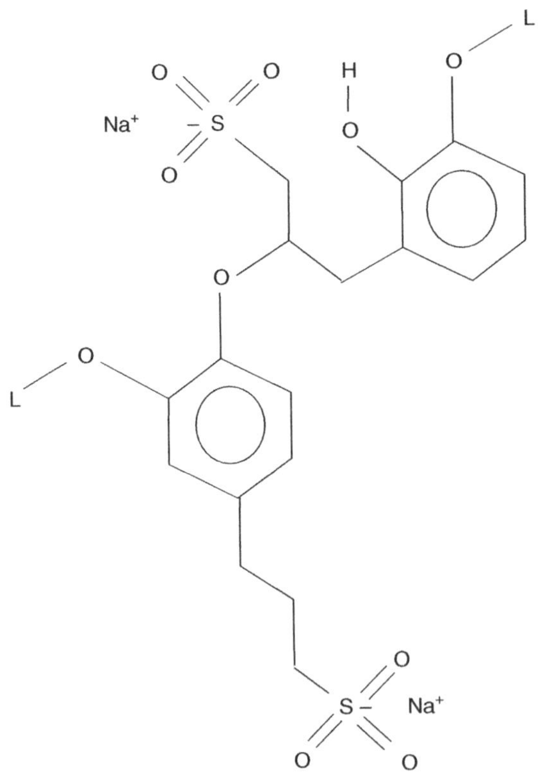

The performance of a lignosulfonate plasticiser depends on its origin, processing and refinement. The natural sugar or saccharide content notably influences early setting and strength by retarding hydration. The sugar level can be controlled by processing, and additional accelerators can be incorporated [3,4]. Adding other components can improve performance according to requirements. The water reduction achieved by using lignosulfonates may be about 10% relative to that of a control mix without an admixture, which is not high performance but can be sufficient and cost effective.

Improved synthetic water-reducing polyelectrolyte polymer admixtures were developed and introduced as cement dispersants from the 1960s and 1970s. Two similar types are sulfonated naphthalene formaldehyde (SNF) condensates and sulfonated melamine formaldehyde (SMF) condensates (Figures 5.2 and 5.3). Both are referred to as superplasticisers since they have a relatively high water-reducing capability.

Like most polymer admixtures, the chemistry of SMF and SNF was known well before there was an uptake of use in concrete [5]. The organic part of such solutions is negatively charged, the overall balance being kept by the cation in solution (usually sodium or calcium). Dispersion is considered to be due principally to electrostatic repulsion, either because the anion is adsorbed on the cement particle making it negatively charged or because cations are adsorbed onto the cement, which are then associated with the anions. In both cases, like charges repel thus keeping the particles apart (Figure 5.4). Cement has a complex variable chemistry, not a simple model. Thus, material and dispersion effects are described in more detail elsewhere [3].

The performance of SMF and SNF depends on molecular weight, active matter contents, dosage rates and other constituents that may be added or present from the process. Crude SNF, in particular, can contain significant levels of sodium sulfate, which not only reduces the active matter but can influence dispersion, as it can affect the adsorption of the plasticizer [6]. SMF and SNF have similar potential water-reduction capacities (10–25% according to slump and cement content required) but vary in the intrinsic air content entrained in the concrete. SMF and SNF are compatible with each other and with lignosulfonates in formulated admixtures.

The next significant generation of molecules was polycarboxylates. Common derivatives used in concrete are polycarboxylate ethers (PCEs) (Figure 5.5).

Figure 5.2 Sulfonated naphthalene formaldehyde (SNF)

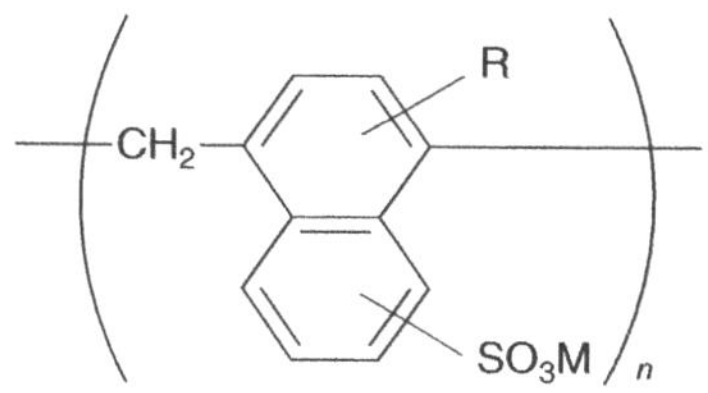

Figure 5.3 Sulfonated melamine formaldehyde (SMF)

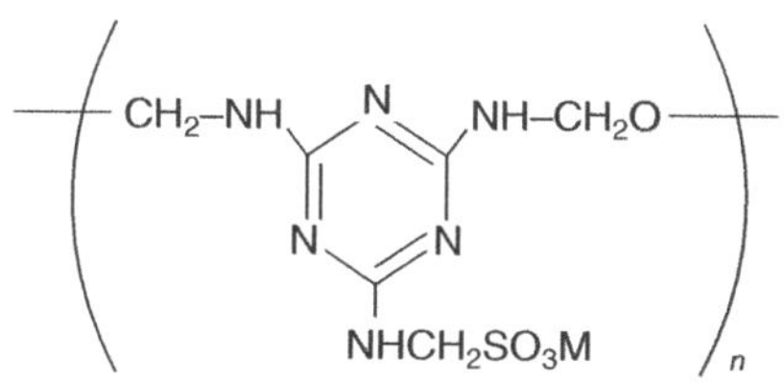

Figure 5.4 Electrostatic dispersion of cement particles

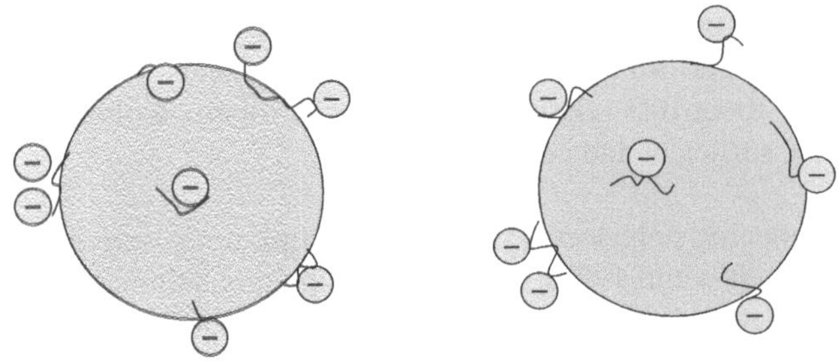

Basic polycarboxylates were already known to be dispersants and used in mineral pigment and ink dispersions before being used in cement. Some have even been used together with dispersants such as SMF [7]. After their very high dispersion power was realised, the chemistry was designed to make them work more effectively with cements and various concrete applications. There are several chemical routes for making these products commercially, and developments continue apace. The chemistry was summarised well by Plank [8]. Two basic effects of PCEs on cement and concrete can be manipulated by design, namely

- initial workability
- workability retention.

Broadly, an increase in carboxylate functionality on the backbone of the polymer will increase adsorption onto the cement particle and tends to increase the retention of workability. The attached side chain (often a polyether) has a marked effect on initial dispersion power and potential initial water reduction. In simple terms, the side chains get in each other's way physically and they tend to avoid 'entanglement' (called steric hindrance). This keeps the attached particles of cement apart, as shown in the simple schematic in Figure 5.6. Powerful dispersants tend to have long and even branched side chains, for example, polyethylene oxide $(\text{-}CH_2CH_2O\text{-})_p$ or $(EO)_p$, as shown, but often mixtures or hybrid molecules are used to balance workability and water-reducing capability with workability retention. Some molecules can also lead to an undesirable rheological sensitivity or adverse filling or finishing effects, such as stickiness. Others deal better with contamination, such as the clay minerals in aggregates. Changing the molecular design or formulation can, therefore, produce a wide range of properties. Cement chemistry, for example, that of soluble sulfate, is an important determinant for how different cements behave with such admixtures [9]. Note that admixtures can be formulated to meet specific requirements.

The benefit of strength improvement is seen in many instances [10]. Strength is, by far, the correct parameter for determining durability. Moreover, a higher strength is broadly correlated with lower permeability. A reduction in permeability reduces the rate of ingress of aggressive elements, such as chloride and carbon dioxide.

Figure 5.5 Basic polycarboxylate molecule (EO, ethylene oxide)

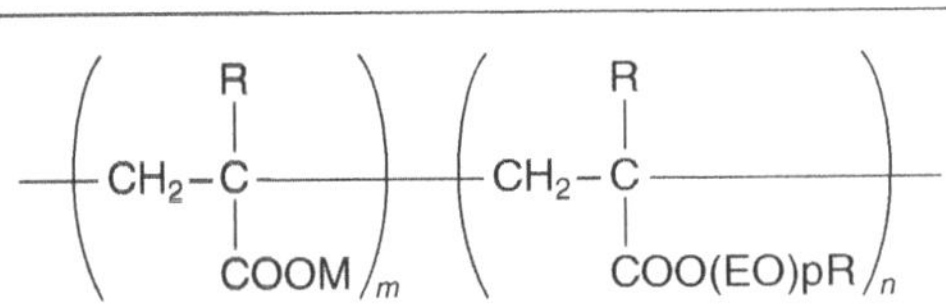

Figure 5.6 Separation of cement particles facilitated by steric hindrance

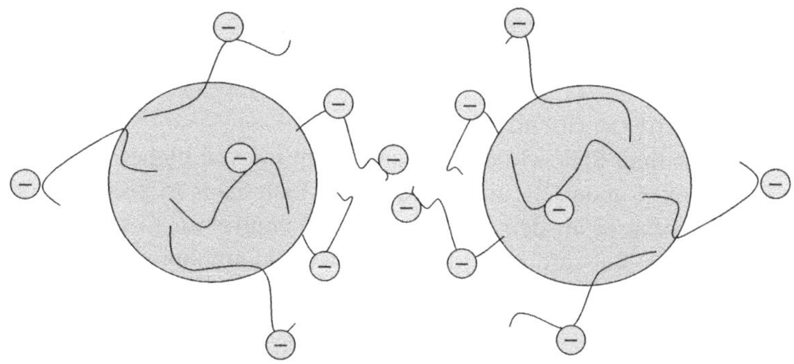

The effects on permeability are plotted in Figure 5.7 for a range of plasticisers. Note that models of ingress that just look at simple diffusion can be misleading, so that test data should be carefully interpreted.

5.2.2.2 Cement content

As described above, admixtures are important elements in controlling cement content. Reducing the proportion of cement reduces costs but has other effects too. As a contribution to sustainability, it has been argued that making full use of admixtures could result in a global reduction of the consumption of cement by up to 20%. The exact figure is arguably higher, notably if PCEs are used, but the order of magnitude is probably correct [11,12]. Note that most aggregates have a lower permeability than hardened cement paste. In some cases, therefore, simply specifying a *minimum* cement content can be contrary to durability and sustainability. Moreover, a higher cement content will lead to higher exothermicity and a greater chance of subsequent thermal contraction and increased shrinkage cracking [13].

Figure 5.7 Effect of plasticisers on water permeability

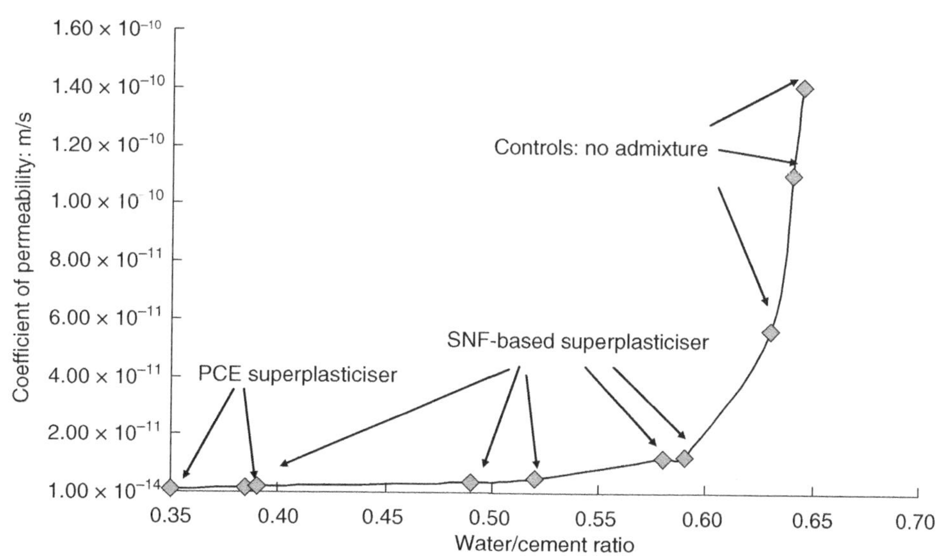

The extent of cement reduction is limited since some binder is needed for cohesion and to fill the space between aggregates. Good grading of aggregates to minimise cement content and maximise workability is possibly the most important aspect of concrete technology. There are also counter effects, such as a higher permeability of the interfacial zone between the cement paste and aggregate [14].

5.2.2.3 Effect of air content on durability

SMF tends to include less air than SNF while PCEs can result in even higher air contents. Most concrete technologists aim at a content of about 2% air in concrete, which leads to good workability and practical rheology. The use of antifoaming or air-detraining agents can control air to some extent. These are often included in formulated products, but care is needed on choice and amount since effective agents are often near to compatibility limits with the liquid admixture.

Air-entraining agents are sometimes added to confer beneficial freeze–thaw resistance. Depending on its form, however, air can be deleterious to durability. Strength is reduced by about 5% for each 1% of air by volume. Air also increases the permeability.

In addition, it is often not appreciated that air can affect parameters such as the chloride threshold for initiating corrosion of the steel reinforcement (rebar). At a 2% air level, the generally recognised 'threshold' for corrosion is typically about 0.3% chloride by weight of cement. As air is eliminated and the solid alkali level at the rebar is increased (one is a partial consequence of the other), then that threshold may be dramatically increased [15] (Figure 5.8). This explains why voids, especially large voids, are usually the initiation sites of corrosion, which causes physical damage and reduces the durability of RC. Using self-compacting concrete and eliminating voids help to increase the chloride threshold.

Another issue is the reserve of solid alkali at the rebar surface. Note that *excessive* use of pozzolans can actually reduce the level of corrosion protection [16]. Other work, however, indicates that in an environment with very high alkalinity and a high level of ordinary Portland cement (OPC), certain pozzolans do not seem to increase corrosion [17]. Therefore, while the beneficial effects of pozzolans are increased use of waste,

Figure 5.8 Chloride threshold and voids. OPC, Ordinary Portland cement, SRPC, sulfate-resistant PC, PFA, pulverised fly ash, GGBS, ground granulated blastfurnace slag (Data courtesy of G Glass, Imperial College London)

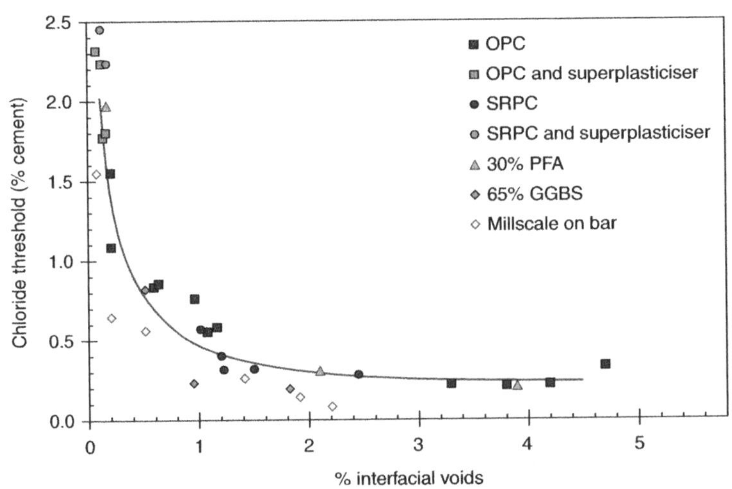

reduction of cement content, reduction of alkali–aggregate reaction (AAR) with reactive aggregates and, at moderate levels, reduced permeability, there may be adverse effects, possibly due to the loss of alkalinity.

5.2.3 Waterproofing admixtures

A porous structure such as concrete cannot properly be described as totally 'waterproof' since an impermeable membrane is required for that, but there are classes of admixtures that can contribute to lowering the permeability, lowering water absorption and slowing the ingress of aggressive elements other than by reducing the w/c ratio.

Sometimes these are also called 'corrosion inhibitors', but truly, they may sometimes have no real effect upon the chloride threshold for corrosion and only delay ingress (as can be determined by ASTM G109).

There are different proprietary approaches [18]. One uses aliphatic carboxylic acids as alkanolamine salts to provide a partial pore lining (and blocking) with increased hydrophobicity. Such formulations can also be combined with a powerful superplasticiser for full effect (Figure 5.9). Another approach is to effect crystallisation within the pore structure, the solid crystal material acting as a pore blocker [19].

5.2.4 Shrinkage-reducing admixtures

Shrinkage occurs because of several features of concrete. First, water can be lost from the system. If that occurs in the early (plastic) stage, it can lead to what is known as 'plastic shrinkage', often accompanied by characteristic random cracking. Note that 'map' cracking could be due to other problems, such as an alkali–aggregate attack. For cracking to occur, it is simply necessary for the tensile stress to exceed the tensile strength.

Second, the volume of the hydrated cement is less than that of the unhydrated product and added water, although there can be early expansion reactions. Hydration shrinkage is sometimes called 'autogenous shrinkage', although that also includes the effect of capillary surface tension, which increases with the development of the capillary pore structure. Researchers are continuing to investigate the mechanism in detail [20–22]. Admixtures to address shrinkage include polyhydroxy compounds that reportedly reduce the driving surface tension force [23,24]. In addition to a reduction in surface tension, there are probably slight

Figure 5.9 Effectiveness of a typical waterproofing admixture

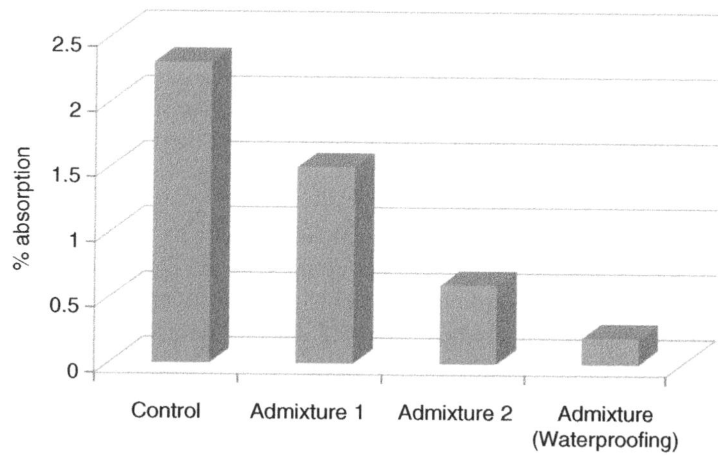

changes in hydrate morphology and a lower tendency to lose water. Such admixtures can be deployed to offset high shrinkage in high-performance concrete [25] or to simply reduce joints. Best control is effected with expansive agents, which include graded calcium oxide and proprietary compounds comprising calcium sulfoaluminate or calcium aluminate. Reducing the cement content and w/c ratio with a good superplasticiser can usually contribute to lowering the shrinkage.

Rarely is shrinkage in concrete fully compensated for due to the cost of incorporating additions at effective levels. Thus, most concrete structures have at least some cracks. Cracking can be exacerbated by high exothermicity, which causes additional stress on cooling. Modern OPCs are designed to give high early and maximum strength up to 28 days. As such, without care, that also can lead to deleterious thermal cracking, which, contrary to popular belief, will not necessarily heal due to the subsequent hydration of the unhydrated cement. That may have been the case when cements had higher content of dicalcium silicate, which hydrates slowly, and were coarser ground. Controlled curing, limiting concrete thickness, using cement replacements and adequate rebar density will also assist in reducing cracking. Reducing the amount of water needed using superplasticisers tends to reduce cracking, although there are limits and water does have a high heat capacity. Admixtures can allow greater cement replacement but different interactions occur.

Using pozzolans to reduce the exothermicity does not always prevent thermal cracking. High addition rates of silica fume will increase shrinkage [26], whereas the hydration of silica fume and blastfurnace slag under certain conditions can be rapid, raising the exothermicity peak. Moreover, low-calcium pulverised fuel ash (PFA) can be slow and more difficult to cure, particularly near the surface. This often necessitates the use of curing compounds, internally or externally, to avoid surface cracking. Cracks of width 300 microns are very porous to chloride ions, which is not surprising given that the hydrated radius of a chloride ion is about 3.2 Å or 3.2×10^{-4} microns. Research indicates that controlling cracks to less than about 30 microns is required to prevent ready chloride passage [27]. Such cracks are difficult to see on dry concrete.

5.2.5 Crack-healing additives

Strategies and additives designed to effect self-healing of concrete cracks have been reviewed [28]. Solutions that are potentially reactive when a crack occurs vary widely from encapsulated polymers to bacteria, which can produce calcium carbonate. Note that the global environmental impact of cement and concrete production has been highlighted [29]. That report also discusses solutions and arguments for prolonging the life of structures. These are a new class of admixtures, which may have a significant effect on the lifetime, durability and sustainability of concrete.

5.2.6 Corrosion inhibitors

Corrosion inhibitors can raise the concentration threshold for a chloride attack or reduce corrosion rates. The best corrosion inhibitor is alkalinity, as described earlier and illustrated to some extent by a Pourbaix diagram (Figure 5.10). After initiation, the pH in a corrosion pit can drop considerably compared with the bulk composition.

One long-proven inhibitor for cast-in-place concrete is calcium nitrite [30]. Nitrite will inhibit corrosion due to chloride ingress until the molar concentration of nitrite is exceeded by chloride. Practically, that means that about 30 litres of a 35–40% aqueous solution per cubic metre of concrete is required to give good protection. Lower doses can be employed, but when corrosion does initiate, the ensuing corrosion rates can be very high. That can partially be accounted for by a large cathode driving a small anode, but there is also often a slight reduction in resistivity too. Despite claims to the contrary, the author has seen no evidence that calcium nitrate is a corrosion inhibitor. Both calcium nitrite and calcium nitrate are accelerators. Lithium nitrite has also been proposed as an inhibitor, but this is a potential hazardous material being explosive

Figure 5.10 Pourbaix diagram showing potential plotted against pH for iron in water

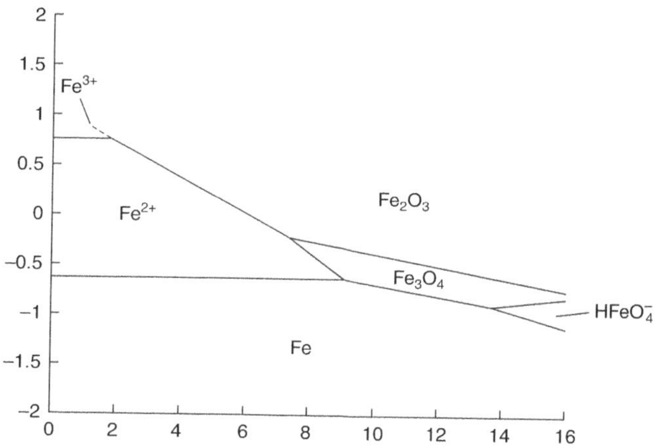

when dry. Nitrites are toxic and require careful handling but the evidence points to their long-term retention in concrete rather than being readily leached [31,32].

Alkanolamines are another group of corrosion inhibitors, claimed to be effective at addition rates as low as a few litres per cubic metre of concrete [33]. There is good evidence that corrosion rates can be reduced, but effectiveness is often greater at addition levels higher than recommended, although much of the work has been with in vitro tests. There is poor evidence that alkanolamines can migrate through concrete to cause any significant reduction in corrosion rates when applied topically to a formed concrete surface. Penetration levels can be detected and are low (several ppm through 25 mm of concrete). Some inhibitors may, at best, be pore or crack blockers. New naturally occurring and thus highly sustainable inhibitors have been described [34].

Using corrosion inhibitors in mortars for patch repairs is not always successful and may accelerate corrosion in untreated areas [35]. Electrochemical means, such as cathodic protection using impressed or galvanic means, are more reliable. That comes at a cost but is probably well worthwhile for critical elements.

5.2.7 Polymers and polymer dispersions

While many water-reducing admixtures are polymers, polymer admixtures have other uses too.

5.2.7.1 Thickeners and stabilisers

Polymers with a high molecular weight are used when placing concrete underwater. A very good solution comprises small additions of polyethylene oxide as anti-washout aids. Some cellulose ethers may also be suitable at slightly higher addition rates. The latter class also helps with water retention and with thickening and slowing down segregation. Polymers can also act as pumping aids. Polyacrylamides have very high activity but must be chosen carefully since they also tend to flocculate depending on the chemistry and molecular weight. They are usually added as powders.

Other often formulated products based on polymers, such as Welan gum, may be added to ensure mix stability and concrete cohesion, thus avoiding water-bleeding and the segregation of aggregates [36].

5.2.7.2 Water-dispersible powder polymers and polymer dispersions

Dispersions, such as acrylic, styrene butadiene rubber, vinyl acetate/ethylene and other co- and terpolymers, have a beneficial effect in reducing permeability, reducing brittleness, improving freeze–thaw resistance, improving the tensile and flexural strength and improving wear resistance. Additions to most concrete mixes is limited by cost, except for high-performance screeds and flooring. A reduction in modulus and an increase in creep also have implications for structural concrete. Use in mortars, tile adhesives and grouts and surface treatments is more common where application factors (adhesion and cohesion) and robustness are important. Additions may be dispersible powders or liquid dispersions. Although it is argued that some polymers are susceptible to chemical attack, hydrolysis or environmental exposure breakdown, research has shown them to be quite durable [37]. Some 'polymer concretes' are not cementitious but are based upon epoxy or polyester binders and the like, and are not admixtures for concrete.

5.2.8 Air-entrainment agents

The inclusion of air increases workability but there are some potential drawbacks, as described earlier. Many plasticisers tend to increase the air content due to a reduction in surface tension. Air-entrainment agents can be designed for the following reasons

- to increase workability, for example of mortar or 'semi-dry precast' mixes
- to entrain air to provide freeze–thaw resistance
- to produce foams for filling voids or for lightweight products.

It is important to stabilise the air bubbles entrained. Many surfactants can entrain air but few provide stability until set. Synthetic surfactants are used, but some of the most stable formulations contain natural products such as Vinsol® resin (fine air for freeze–thaw resistance) and even keratin for foams.

Lightweight products provide enhanced insulation for buildings. In addition, they require a minimal use of primary raw materials and make high use of wastes.

Void fill materials have minimal strength and are used to stabilise the ground. For example, high strength is undesirable for trench fill. In mines and tunnels, vast voids can be filled economically.

The improvement to freeze–thaw resistance is a result of providing space into which pore water can expand, thus mitigating any disruptive expansion. The mechanism and other details have been described in several papers [38,39].

5.2.9 Retarders and accelerators

Retardation can result in improved compaction and a denser pore structure, both of which improve durability. For bulk concrete, however, it is wrong to assume that the heat output is reduced by using a retarder, as it may simply delay the output of heat and may not necessarily reduce the peak temperature. Excessive retardation can lead to segregation and even surface water bleeding, leading to poor quality surfaces.

Accelerators are often used to improve the build and cohesion of shotcrete, notably in tunnels (Figure 5.11). Early strength is improved, which may be critically important for ground stabilisation, for example, but often later hydration, dimensional stability and strength may be reduced, particularly for some alkaline accelerators such as sodium aluminate [38], which is highly alkaline and causes skin burns. For that reason, accelerators based on non-alkaline materials are used, such as modified aluminium sulfate.

Accelerators (both setting and hydration) are often beneficial for cold weather working and hastening shuttering removal, but care should be taken with both addition rate and type of admixture. Accelerators comprise various salts, including nitrates, nitrites, thiocyanates or formates. Alkanolamines can also accelerate setting

Figure 5.11 Applying shotcrete. (Courtesy of Fosroc International Ltd)

and increase early or late strength, depending upon the concentration deployed and the structure. Chloride is a very effective hydration accelerator but products containing chloride should never be used in RC.

5.2.10 Future trends for admixtures

The development of polycarboxylates will continue with an emphasis on local adaptations and formulations. Admixtures, generally, will begin to use a greater amount of biomass as raw materials to improve sustainability. They may have to be tailored for different, low carbon footprint, cement chemistry. There may be a trend towards using solid or concentrated products to minimise water transportation.

Most of the current solutions have been developed for calcareous hydrating cements, such as calcium silicates and aluminates. Future developments may necessitate quite different chemistry and innovation. An important lesson that the author has learned about concrete is that what is 'good' under a selected set of parameters can be 'bad' when one or more of the parameters is changed, such as the concentration. There are many variables.

Finally, to return to the opening remarks, no admixture can fully compensate for poor mix design. Thus, the starting point should be to ensure that there is good grading of aggregates and sands.

REFERENCES

1. Sakai K (ed.) (1996) *Integrated Design and Environmental Issues in Concrete Technology*. Taylor & Francis, London, UK.
2. Day KW, Aldred J and Hudson B (2017) *Concrete Mix Design, Quality Control and Specification*, 4th edn. CRC Press, Boca Raton, FL, USA.
3. Ramachandran VS (1995) *Concrete Admixtures Handbook: Properties, Science and Technology*, 2nd edn. Noyes Publications, Park Ridge, NJ, USA.
4. Taylor HFW (1997) *Cement Chemistry*, 2nd edn. Thomas Telford, London, UK.
5. Auten RW and Rainey JL (1946) *Resinous Compositions and Methods of Making the Same*. US Patent 2,407,599.
6. Kim B-G, Jiang S, Jolicoeur C and Aïtcin P-C (2000) The adsorption behavior of PNS superplasticizer and its relation to fluidity of cement paste. *Cement and Concrete Research* **30(6)**: 887–893.

7. Akao K, *et al.* (1988) *Dispersant for Calcium Carbonate*. Japanese Patent JP 1457558 (C).
8. Plank J (2017) Current status of concrete admixture technology. *9th International Symposium on Cement and Concrete, Wuhan, China.*
9. Magarotto R, *et al.* (2003) Effect of alkaline sulphates on the performance of superplasticizers. *11th International Congress on the Chemistry of Cement, Durban, South Africa*, pp. 569–580.
10. Sobelov K and Soboleva S (2003) The effect of complex admixtures on strength of ultra high strength cement. *11th International Congress on the Chemistry of Cement, Durban, South Africa*, pp. 465–474.
11. Aïtcin PC (2005) Admixtures and sustainability. *Proceedings of 6th International Conference on Admixtures – Enhancing Concrete Performance, Dundee*, pp. 122–128.
12. Cheung J, *et al.* (2018) Admixtures and sustainability. *Cement and Concrete Research* **114**: 79–89.
13. Concrete Society (1982) *Non-structural Cracks in Concrete*. Concrete Society, Camberley, UK, Technical report 22, pp. 18–19.
14. Yang CC and Cho SW (2005) Approximate migration coefficient of percolated interfacial transition zone by using accelerated chloride diffusion test. *Cement and Concrete Research* **35(2)**: 344–350.
15. Buenfeld NR, *et al.* (2001) *Process for the Protection of Reinforcement in Reinforced Concrete*. International Patent WO 2001/055056.
16. Manera M, *et al.* (2006) Effect of silica fume on the initiation of chloride induced corrosion in reinforced concrete. *8th Canmet/ACI International Conference on Superplasticizers and Other Chemical Admixtures in Concrete, Sorrento*, pp. 187–200.
17. da Silva FG and Liborio JBL (2006) A study of steel bar reinforcement corrosion in concretes with SF and SRH using electrochemical impedance spectroscopy. *Materials Research* **9(2)**: 209–215.
18. Aldykiewiez AJ, *et al.* (2006) Use of concrete admixtures to produce 'waterproof' concrete. *8th Canmet/ACI International Conference on Super-plasticisers and Other Chemical Admixtures in Concrete, Sorrento*, pp. 53–61.
19. Iob A, Saricimen H, Narasimhan S and Abbas NM (1993) Spectroscopic and microscopic studies of a commercial concrete waterproofing material. *Cement and Concrete Research* **23(5)**: 1085–1094.
20. Barcelo L, Moranville M and Clavaud B (2005) Autogenous shrinkage of concrete: A balance between autogenous swelling and self-desiccation. *Cement and Concrete Research* **35(1)**: 177–183.
21. Mora-Ruacho J, Gettu R and Aguado A (2009) Influence of shrinkage-reducing admixtures on the reduction of plastic shrinkage cracking in concrete. *Cement and Concrete Research* **39(3)**: 141–146.
22. Rajabipour F, Sant G and Weiss J, *et al.* (2008) Interactions between shrinkage reducing admixtures (SRA) and cement paste's pore solution. *Cement and Concrete Research* **38(5)**: 606–615.
23. Shlonimskaya N, *et al.* (2014) An application of computer-aided molecular design (CAMD) using the signature molecular descriptor – Part 2: Evaluating newly identified surface tension-reducing substances for potential use as shrinkage-reducing admixtures. *Journal of the American Ceramic Society* **97(2)**: 378–385.
24. Bettencourt Ribiero A, Carrajola A and Gonçalves A (2006) Behavior of mortars with different dosages of shrinkage reducing agents. *8th Canmet/ACI International Conference on Superplasticizers and Other Chemical Admixtures in Concrete, SP-239-6, Sorrento*, pp. 77–92.
25. Sugamata T, *et al.* (2006) Characteristics of concrete containing a shrinkage-reducing superplasticizer for ultra-high strength concrete. *8th Canmet/ACI International Conference on Superplasticizers and Other Chemical Admixtures in Concrete, SP-239-4, Sorrento*, pp. 51–66.
26. Bentz DP (2008) A review of early-age properties of cement-based materials. *Cement and Concrete Research* **38(2)**: 196–204.
27. Ismail M, Toumi A, François R and Gagné R (2008) Effect of crack opening on the local diffusion of chloride in cracked mortar samples. *Cement and Concrete Research* **38(8–9)**: 1106–1111.
28. Van Tittelboom K and De Belie N (2013) Self-healing in cementitious materials: A review. *Materials* **6(6)**: 2182–2217.

29. van Bruegel K (2017) Circularity: A new avenue for mitigating the footprint of ageing infrastructure. *9th International Symposium on Cement and Concrete (ISCC 2017)*.
30. Berke NS and Hicks MC (2004) Predicting long-term durability of steel reinforced concrete with calcium nitrite corrosion inhibitor. *Cement and Concrete Composites* **26(3)**: 191–198.
31. Balonis M and Glasser FP (2011) Calcium nitrite corrosion inhibitor in Portland cement: Influence of nitrite on chloride binding and mineralogy. *Journal of the American Ceramic Society* **94**: 2230–2241.
32. Balonis M (2010) *The Influence of Inorganic Chemical Accelerators and Corrosion Inhibitors on the Mineralogy of Hydrated Portland Cement Systems*. PhD thesis, University of Aberdeen, Aberdeen, UK. http://citeseerx.ist.psu.edu/viewdoc/download?doi=10.1.1.458.2386&rep=rep1&type=pdf.
33. Munteanu VF and Kinney FD (2000) Corrosion inhibition properties of a complex inhibitor – Mechanism of inhibition. *2000 Canmet/ACI Conference on Durability of Concrete*, pp. 255–269.
34. Tang SW, *et al.* (2015) Recent durability studies on concrete structure. *Cement and Concrete Research* **78**: 147
35. Sprinkel MM (2003) *Evaluation of Corrosion Inhibitors for Concrete Bridge Deck Patches and Overlays*. Virginia Transportation Research Council, USA, Report VTRC 03-R14.
36. Üzer E and Plank J (2016) Impact of welan gum stabilizer on the dispersing performance of polycarboxylate superplasticizers. *Cement and Concrete Research* **82**: 100–106.
37. Schulze J and Killermann O (2001) Long term performance of redispersible powders in mortars. *Cement and Concrete Research* **31(3)**: 357–362.
38. Chatterji S (2003) Freezing of air-entrained cement-based materials and specific actions of air-entraining agents. *Cement and Concrete Composites* **25(7)**: 759–765.
39. Du L and Folliard KJ (2003) Mechanisms of air entrainment in concrete. *Cement and Concrete Research* **35(8)**: 1463–1471.

5.3. Supplementary cementitious materials
5.3.1 Introduction

Marios Soutsos, Queen's University Belfast, UK

Supplementary cementitious materials (SCMs) are also referred to as

- mineral admixtures
- pozzolans
- cement replacement materials
- additions.

A *pozzolan* is defined as 'a siliceous or siliceous and aluminous material which in itself possesses little or no cementitious value but which will, in a finely divided form and in the presence of moisture, chemically react with calcium hydroxide at ordinary temperature to form compounds possessing cementitious properties' [1].

BS EN 206 [2] defines additions as 'finely divided materials used in concrete in order to improve certain properties or to achieve special properties' and recognises two broad divisions: type 1, which are nearly inert, and type 2, which are pozzolanic or latent hydraulic additions.

The main type 2 additions used in concrete are

- fly ash
- GGBS
- silica fume (microsilica)
- MK.

GGBS is, strictly speaking, not a pozzolan but a hydraulic latent binder. The calcium oxide content is high, and if it is mixed with water, it will set and harden. As such, an alternative term to 'pozzolan' has been sought to group all the above together. The one that appears to be the most preferred is SCMs. MK is the only one from this group of materials that is not a by-product of another industry. For SCMs, the method of manufacture, chemical composition and the mode of action in concrete (fresh and early-age properties and hardened properties, especially with respect to improving the resistance of concrete to degradation mechanisms) are discussed in this chapter.

REFERENCES

1. ACI Committee 201 on Durability of Concrete (2008) *Guide to Durable Concrete*. American Concrete Institute, Farmington Hills, MI, USA, Report 201.2R-08.
2. BSI (2013) EN 206-1. Concrete. Specification, performance, production and conformity. BSI, London, UK.

5.3.2 Fly ash for enhanced concrete durability

Kosmas Sideris, Democritus University of Thrace, Greece

5.3.2.1 Introduction

Fly ash has been used as a cementitious material in concrete production for more than 50 years. It is an industrial by-product of burning coal or lignite in plants that generate electricity. The chemical and mineralogical composition of fly ash depends mainly on the chemical and mineralogical composition of the raw material used (coal or lignite) as well as on the type of the furnace used in the electricity generating plant. Fly ashes are mainly characterised as siliceous or calcareous with the accepted difference being the proportion of reactive calcium oxide, which is $\leq$10% for siliceous ash and >10% for calcareous fly ash. Siliceous fly ash is the predominant material used in concrete and cement production. Calcareous fly ash has both pozzolanic and hydraulic properties. Depending on its composition and fineness, fly ash may have beneficial effects on the fresh and hardened properties of concrete mixtures. This chapter describes how fly ash can be used to enhance the durability of concrete mixtures.

5.3.2.2 Transport properties

Portland cement concrete is a porous material. Its porosity allows the ingress of water and salt solutions. Fly ash reacts with the calcium hydroxide originating from the hydration of cement to produce calcium silicate and aluminate hydrates of the same or similar types as are formed in the normal hydration of cement [1–3]. The secondary hydrates formed due to the pozzolanic reaction fill the capillaries and reduce the diffusion of aggressive fluids. Bigger pores transform into smaller ones [3,4], and the permeability of concrete mixtures is, therefore, much reduced.

Manmohan and Mehta [5] examined the porosity of cement pastes at the ages of 28 days and 1 year. Mixtures were produced with a water-to-cementitious materials (w/cm) ratio of 0.50. The reference mixture was produced with Portland cement. In the alternative mixture, 30% of the Portland cement was replaced with fly ash. The researchers concluded that alternative mixture had a higher porosity and greater volume of macropores at 28 days. Both mixtures had a lower porosity at 1 year, but the difference was more significant in the fly ash mixture. The capillary-sized pores were sharply reduced between 28 days and 1 year in this mixture and there was a significant shift from macroporosity to mesoporosity, thus reducing the number of water-permeable pores in the mixture. This finding emphasises the importance of long-term curing for mixtures produced with slowly reacting fly ashes.

Figure 5.12 Total porosity (1 102 940 Å < D < 75Å) and water-permeable porosity (1 102 940 Å < D < 1000 Å) of cement pastes produced with HCFA. (Data taken from Sideris [4])

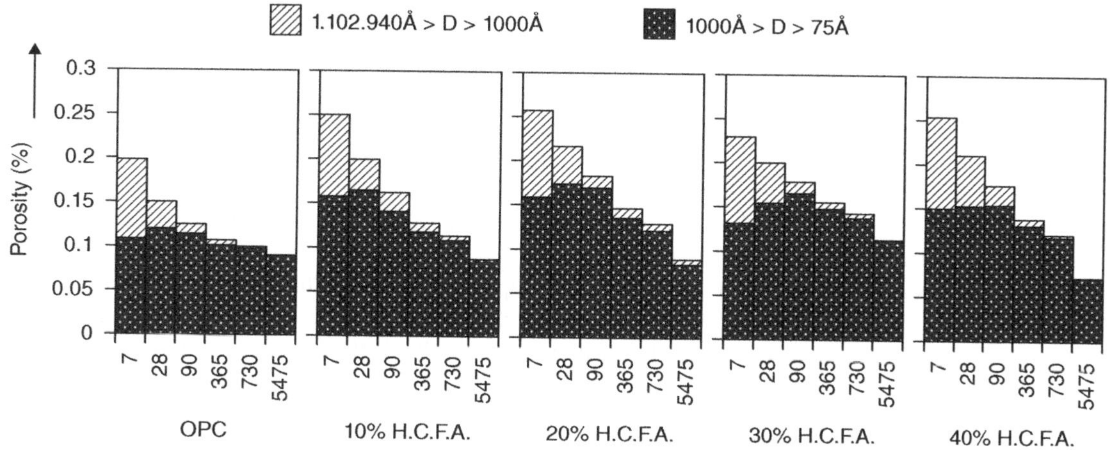

Papayianni-Papadopoulou [1] concluded that high-calcium fly ashes (HCFAs) exhibit both pozzolanic and hydraulic behaviour. She also reported that when mixed with water, some HCFAs harden due to the hydration of crystalline calcium silicate, the reaction between free lime and some of the glass, and the formation of ettringite. Taylor [6] also reported that the hydration products in these mixtures are almost the same as those that form in Portland cement mixes with the addition of strätlingite and C_2AH_8.

The porosity of mixtures produced with HCFA at different ages is plotted in Figure 5.12 [4]. HCFA replaced Portland cement at different levels up to 40%. All cement pastes produced had a w/c ratio of 0.50. Porosity measurements were taken up to the age of 730 days (2 years). The porosity of cement pastes at 5475 days (15 years) was calculated using the cement hydration model [7,8]. The total porosity of fly ash cement pastes slightly increased at all ages compared with Portland cement paste. However, the water-permeable porosity decreased or remained the same as measured for Portland cement pastes at ages greater than 90 days for replacement levels of 20% to 40%.

One can, therefore, conclude that fly ash mixtures become impermeable due to the pore refinement at later ages. Among the first researchers to study the permeability of the concrete used for pipes and the effects of replacing 30% or 50% of the cement with class F fly ash was Davis *et al.* [9]. Concrete produced with fly ash was more permeable at 28 days, especially at higher replacement levels. This trend was different at 180 days, since the fly ash concretes were less permeable.

5.3.2.3 Carbonation

The carbonation resistance of specimens produced with different types of cements and fly ash after 7 days of moist curing was studied by Kasai *et al.* [10]. These researchers concluded the following.

- The carbonation of mortars containing fly ash progressed rapidly up to the age of 3 months and subsequently slowed down.
- The carbonation depth of specimens increased with the coefficient of permeability. The carbonation depth of mortar specimens produced with fly ash cements was higher than that measured for reference specimens produced with OPC.

Moreover, Nagataki and Ohga [11] concluded that the carbonation depth of mortars increased when fly ash was used. They also reported the following.

- The carbonation rate was reduced when the curing period was prolonged.
- Specimens produced with higher amounts of fly ash were more sensitive to the curing period.

These results are consistent with the slower hydration and development of a discontinuous pore system in fly ash cement pastes.

Sideris *et al.* [12] produced nine different cement mortars using HCFAs with lignite in replacement levels up to 60%. Different w/cm ratios were used, since the aim was to produce mixtures with the same fluidity. Carbonation measurements were taken at different ages up to 18 months. These authors reported that the carbonation depths of mixtures produced with fly ash were greater than those measured for the reference mixture produced with OPC only. However, the carbonation rate, that is the increase in carbonation depth at later ages compared with the initial measurement (6 months), was greatest for the plain cement mixture. The carbonation rate generally reduced as the fly ash content increased. The susceptibility of fly ash mixtures to carbonation was also reported by Papadakis *et al.* [13]. All the above researchers attributed this finding to pore refinement.

Sideris and Savva [14] produced five different concrete mixtures. Two natural pozzolans and two lignite fly ashes replaced Portland cement at a constant level of 20%. The concretes were produced with a w/cm ratio of 0.65. The content of cementitious materials was 350 kg/m^3 for all concrete mixtures. Concrete specimens were cured in a curing chamber for 28 days. After this age, specimens were exposed in two different environments: (*a*) to a coastal environment and left unprotected from rain or (*b*) kept in indoor conditions. Carbonation measurements were performed at different ages up to 4 years. All concretes produced with the addition of natural pozzolans and fly ashes had an increased carbonation depth in both curing environments. Among all the blended cement mixtures tested, specimens with pretreated high-calcium lignite fly ash had the lowest carbonation rate, mainly due to the high calcium oxide content of the ash (Figure 5.13).

Papadakis [15] estimated the carbonation resistance of mortars produced with two different fly ashes: a low-calcium fly ash and a high-calcium lignite fly ash. Both ashes were used to replace cement or aggregates at percentages of 10% and 30%. All concrete specimens were water-cured for 28 days. After this age, they were inserted in a controlled carbonation chamber (3% content of carbon dioxide, temperature of 25°C and relative humidity of 61%), where they remained for 100 days. He reported that the carbonation depth decreased as the amount of aggregate replaced by fly ash increased and increased as the amount of cement replaced by fly ash increased. The lowest carbonation depth was observed for HCFA. The combined pozzolanic and hydraulic activity of this type of ash resulted in a more refined pore system in the mortar.

The influence of fly ash on carbonation depth of concrete mixtures depends on many parameters, such as

- chemical and mineralogical composition of the ash
- fineness of the ash
- w/cm ratio
- level of replacement and the type of replacement (replacing cement or aggregates)
- type of OPC used (chemical composition and fineness) and the curing period of the specimens.

Figure 5.13 Carbonation depth of concrete mixtures produced with OPC, natural pozzolans (SkE and ME) and lignite fly ashes (PtFA and MFA). (Data taken from Sideris and Savva [14])

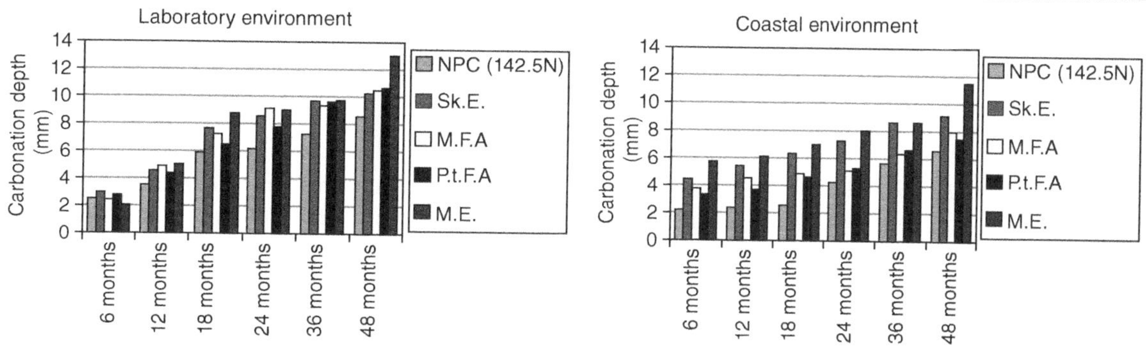

Because of the number of parameters, there is much conflicting data in the literature, so one must be very careful when examining the influence of a specific fly ash on the resistance of a concrete mixture against carbonation.

5.3.2.4 Resistance to chloride ingress

There are many research results in the literature on the influence of fly ash on the ingress of chlorides in concrete. Most of the researchers have concluded that fly ash improves the resistance of concrete to chloride ion penetration. The extent of the improvement depends on the following factors [14]

- chemical and mineralogical composition
- replacement level and fineness of the ash
- w/c ratio, and to a lesser extent, the composition and fineness of the Portland cement used
- maturity of the concrete
- exposure conditions (mainly temperature and carbonation of the concrete)
- extension of the initial curing period of the mixtures.

Papadakis [15] used the rapid chloride permeability test [16] to assess the resistance to chloride ingress of different mortars produced with low-calcium fly ash or high-calcium lignite fly ash. Both ashes were used to replace cement or aggregates at percentages of 10% to 30%. The control specimen produced with OPC passed an electrical charge higher than 4000 coulombs. This quite high value was attributed to the high cement paste volume and the absence of coarse aggregate. All specimens incorporating fly ashes, whether the ash substituted for aggregate or cement, passes a lower electrical charge.

Note that most published research focuses on the chloride permeability of fly ash mixtures measured according to the standards at 28 days. Irrespective of the test set-up and the standard used, they do not take advantage of the pore refinement in the mixture at later ages due to the slower rate of the pozzolanic reaction of the fly ash. This is why both the total porosity and the permeability of concrete containing fly ash is significantly reduced at later ages, as mentioned above. ASTM C1202 [16] notes that the curing period should be extended up to 56 days before concretes containing SCMs are tested.

The chloride diffusion coefficient of different concrete blocks exposed to the south-eastern coast of the United Kingdom was measured by Thomas and Bamforth [17]. Concrete blocks were produced with three different mixtures. The first one was produced with OPC only, while fly ash and slag were added to the other two mixtures at percentages of 30% and 70%, respectively. The compressive strength of all mixtures measured at 28 days was the same. The w/cm ratios were 0.66, 0.54 and 0.44, respectively. Cores were taken from concrete blocks and tested at 6 months or 1, 2, 3, 6 and 8 years. These researchers concluded that the rate of reduction of diffusivity was far greater for concrete containing fly ash or slag than for plain Portland cement concrete. They also noted that diffusion coefficients for fly ash or slag concretes are likely to decrease to 1 or 2 orders of magnitude less than that of similar grade Portland cement concrete during a 100-year service life.

This reduction in the chloride diffusion coefficient of fly ash concrete mixtures has a significant impact on the long-term performance of concrete in chloride environments and on the interpretation of data from tests based on estimating the chloride diffusion coefficient (D_{28}). This long-term reduction in the chloride diffusion coefficient is known as the *ageing factor*. The chloride diffusion coefficient at a particular time t during the service life of a structure is calculated using the following equation:

$$D_{\mathrm{app}(t)} = D_{\mathrm{app}(t0)}\left(\frac{t_0}{t}\right)^a$$

where $D_{\mathrm{app}(t0)}$ is the apparent diffusion coefficient measured at a reference time of t_0 (usually 28 days) and a is an ageing factor giving the decrease over time of the apparent diffusion coefficient.

According to fib Bulletin 34 [18], the ageing factor $a = 0.30$ for OPC and 0.60 for fly ash cements with a fly ash replacement higher than 20%. However, the bulletin suggested that to estimate accurately the ageing factor a in concretes containing pozzolanic materials, long-term laboratory tests of at least 1 year should be performed.

5.3.2.5 Sulfate resistance

The principal method for limiting sulfate attacks is to use a sulfate-resistant cement (such as type II or type V Portland cements instead of type I Portland cement according to ASTM [19] or EN197-1 [20]) or by ensuring that the concrete has a low w/cm. Among the sulfate-resistant cements described in EN197-1 [20], sulfate-resistant pozzolanic cements CEM IV/A-SR and CEM IV/B-SR should contain combinations of natural pozzolanas and siliceous fly ash only up to a content of 55% per weight of cement. In this case, the C_3A content of the cement should be lower than 9%, almost double the upper limit for sulfate-resistant Portland cement CEM I-SR5. Another alternative is the use of additives, such as natural pozzolanas, fly ashes or silica fume. Pozzolanic materials have an important role in the long-term durability of mixtures exposed to sulfate environments because they contribute chemically and physically to the properties of the concrete. Their effect is thought to be explained by two primary mechanisms [21]. The 'dilution effect' is the decrease of the C_3A content of the binder due to the replacement of a portion of the Portland cement. The 'pozzolanic effect' is the formation of secondary C-S-H. The latter results in a less permeable mortar, and the excess calcium hydroxide has been consumed and rendered unavailable for the expansive formation of ettringite or gypsum compounds [1,22,23].

Several attempts have been made to predict the sulfate resistance of a fly ash based on its chemical composition. Among the first is that of Dunstan [24]. He introduced the resistance value:

$$R = \frac{[\mathrm{CaO}] - 5}{[\mathrm{Fe_2O_3}]}$$

According to Dunstan, a fly ash will improve the sulfate resistance of the mixture if $R < 1.5$, whereas the sulfate resistance decreases if R for the ash is higher than 3. In later research, Dunstan [25] reported that the sulfate resistance of mortars tested according to ASTM C1012 is improved if fly ash with $R \leq 3.5$ is used.

In research by Tikalsky and Carrasquillo [26], the CaO/SiO_2 ratio of the fly ash was introduced as an indicator of their contribution to the sulfate resistance of mixtures. According to theses researchers, fly ash with a CaO/SiO_2 ratio of less than 0.25 contained no detectable C_3A and provided a level of resistance to sulfate attack equal to or greater than that of cement type II alone. When fly ash had a CaO/SiO_2 ratio greater than 0.60, the sulfate resistance of the mixtures was lower. Erdogan *et al.* [27] reported that the (Free CaO)/Fe_2O_3 and (C_3A + Free alumina)/(Free CaO) ratios of cement and high-lime fly ash mixtures can be used to determine their sulfate susceptibility. According to these authors, there is a clear relation between these ratios and the sulfate resistance of the mixtures: sulfate resistance increases when the (Free CaO)/Fe_2O_3 ratio decreases and the (C_3A + free alumina)/(Free CaO) ratio increases.

Mehta [28] observed that the sulfate resistance of a concrete is not improved if the fly ash is high in reactive aluminate phases. He found that sulfate resistance depends on the type of aluminate phases at the time of sulfate exposure. Upon immersion in sulfate solution, pastes containing monosulfoaluminate or calcium aluminate hydrate suffered from strength loss due to ettringite formation. In contrast, blended cement pastes containing fly ash that promoted the formation of ettringite prior to immersion into sulfate solution showed superior resistance. The amount of free silicon dioxide, the amount of free calcium oxide and the fineness of the material are also of great importance [2,29].

Fay and Pierce [30] studied air-entrained fly ash concrete using three fly ashes from the United States with a range of calcium oxide contents from 11% to 28.8%. They used 10% to 100% fly ash by weight of the total cementitious material. To accelerate the testing, specimens were cured for 14 days in a 100% humidity room and for another 14 days in a 50% humidity room before they were immersed in 10% Na_2SO_4 solution or subjected to cyclic soaking and drying phases in 2.1% Na_2SO_4 solution. These authors concluded that both low-calcium class C and class F fly ashes can be effective replacements for cement in controlling sulfate expansion. However, class F fly ash at a replacement level of 30% significantly improved the sulfate durability for the cementitious levels and appeared to be optimum for both test conditions. To achieve the best sulfate durability for low-calcium class C fly ash, they recommended replacement levels of greater than 30%. For high-calcium class C fly ashes, the corresponding level was greater than 75%.

Sideris *et al.* [12] investigated the sulfate resistance of eight different cement mortars prepared using two high-calcium lignite fly ashes (modified fly ash (MFA) and PFA) in replacement levels of up to 60% (Table 5.1). Different w/cm ratios were used to keep the fluidity of the mixtures constant. All cements were produced in the laboratory by mixing clinker, pozzolanic material and gypsum ($CaSO_4{\cdot}2H_2O$). The sulfur trioxide content of the pozzolanic material was taken into consideration and the amount of gypsum added was adjusted to keep the total sulfur trioxide content of the binder (clinker + pozzolan) equal to 3.5%. This was not achieved for the mixtures with a high PFA content. The high sulfur trioxide of PFA increased the total amount of sulfur trioxide available in the mix (clinker + gypsum), although no gypsum was added. The sulfate resistance of binders was evaluated using expansion measurements of mortar prisms immersed in a 5% Na_2SO_4 solution, according to ASTM C1012 [19], up to an age of 2 years. These authors concluded that lignite fly ashes with a high calcium oxide and sulfur trioxide content may have a positive effect on the sulfate resistance of blended cements. Specimens of blended cements containing PFA had the smallest expansion among all cement mixtures tested in this research and developed high compressive strength from early ages. These specimens also gave the least reduction in compressive and flexural strengths due to sulfate attack. On the other hand, specimens containing cements blended with MFA had the worst performance,

Table 5.1 Compressive and flexural strength loss, 1-year and final expansion, deterioration age and age when first crack observed for all mixtures. (Data from Sideris *et al.* [12])

Mixture	Compressive strength loss: %	1-year expansion: %	Final expansion (at 24 months): %	Deterioration age: months
CEM I	30	1.10	1.10	6–8
10% MFA	–[a]	0.544	0.544	3–4
20% MFA	–[a]	0.428	0.428	3–4
30% MFA	30	0.408	0.408	4–6
30% PFA	19	0.033	0.049	12–16
40% PFA	15	0.033	0.055	16–20
50% PFA	−0.8[b]	0.016	0.022	No deterioration
60% PFA	−8[b]	0.012	0.016	No deterioration

[a]Totally deteriorated so not measured.
[b]Strength increased

since they deteriorated entirely very quickly. The *R*-values for the two ashes used were $R_{MFA} = 1.86$ and $R_{PFA} = 1.21$. Even though in both cases the *R*-value was much less than 3.5, the contributions of the fly ashes to the sulfate resistance of the mixtures were wholly different. MFA had a deleterious effect, whereas PFA had a beneficial one. The two lignite fly ashes had similar chemical compositions but the sulfate resistance of the mixtures was very different, indicating that the mineralogical composition is more important.

5.3.2.6 Freezing and thawing resistance

Much research has been published on the freeze and thawing resistance of fly ash concretes [31–41]. Most researchers have concluded that the air content has the greatest influence on the freeze–thaw resistance of concrete and that adding fly ash had no influence on the freeze–thaw resistance of concrete if the air content of the mixtures is kept constant. However, if the fly ash contains some residual unburnt carbon, higher dosages of air entrainers may be used to achieve the target air content. Sometimes unstable air may occur.

Adding fly ash changes the pore size distribution due to pozzolanic reactions [42]. Pore refinement due to pozzolanic reactions leads to the breaking of the continuity of the capillary pore structure. Malhotra *et al.* [43] reported that pozzolanic reactions alter the pore structure of the cement paste and, thus, densify the transition zone between the paste and aggregates.

5.3.2.7 Alkali–aggregate reactions

Replacing cement by SCMs generally reduces the expansion due to AAR [41]. This has, in general, been thought to be caused by a reduction in pH (i.e. the hydroxyl concentration) of pore water and a lower content of calcium hydroxide. In the author's opinion, a reduction in pH occurs if the alkalis are busy dissolving SCMs catalytically, but the implication is that this may cause only a delay to the AAR until the SCM has fully reacted.

According to Thomas [44], the level of cement replacement required to control AAR expansion increases as (*a*) the silicon dioxide content of SCM decreases, (*b*) the calcium oxide content of SCM increases, (*c*) the alkali content of SCM increases, (*d*) the alkali availability in the concrete increases or (*e*) the reactivity of the aggregate increases. Shehata and Thomas [45] showed that low-calcium fly ashes (ASTM class F) are much more effective in reducing AAR expansion (typical cement replacement level >20%) than HCFAs (ASTM class C).

Saha *et al.* [46] support this conclusion. They also reported that fly ashes with higher percentages of silica and alumina and lower percentages of alkali and calcium oxides are the most effective in reducing the alkali–silica reaction (ASR). They attributed this conclusion to the fact that fly ash prevents aggregate dissolution due to alkali attack by providing a surface area with sacrificial amorphous silicate area. Moreover, the fly ash also modifies the chemical composition of the ASR gel and reduces its swelling properties so that the gel can move freely into the voids between the aggregate and the binder paste.

REFERENCES

1. Papayianni-Papadopoulou J (1981) *Research for the Possibility of Using Ptolemaida's Fly Ash for Concrete Production.* PhD thesis, Aristotle University of Thessaloniki, Greece. (In Greek.)
2. Mehta PK and Monteiro PJM (2014) *Concrete: Microstructure, Properties and Materials.* McGraw Hill Education, London, UK.
3. Siddique R and Khan MI (2011) *Supplementary Cementing Materials.* Springer.
4. Sideris KK (1996) *Influence of Natural Pozzolanas and Fly Ash on the Compressive Strength and Porosity of Cement Mortars and Concretes* PhD thesis, Democritus University of Thrace, Greece. (In Greek.)
5. Manmohan D and Mehta PK (1981) Influence of pozzolanic, slag, and chemical admixtures on pore size distribution and permeability of hardened cement pastes. *Cement Concrete and Aggregates* **3(1)**: 63–67.
6. Taylor HFW (1997) *Cement Chemistry*, 2nd edn. Thomas Telford, London, UK.
7. Sideris K and Sideris KK (1997) The cement hydration equation and its application to several hydration criteria according to the literature. *Proceedings of the 10th International Congress on the Chemistry of Cement, Gothenburg, Sweden* (Justnes H (ed.)), Vol.2, Paper 2ii061.
8. Sideris K and Sideris KK (2003) *Ten Years of the Cement Hydration Equation and Its Application to Chemistry and Physics of Cement Paste, Mortar and Concrete.* Kyriakidis Bros Publications, Thessaloniki, Greece.
9. Davis RE, Carlson RW, Kelly JW and Davis HE (1937) Properties of cements and concretes containing fly ash. *Journal of American Concrete Institute* **33(5)**: 577–612.
10. Kasai Y, Matsui I, Fukushima U and Kamohara H (1983) Air permeability of blended cement mortars. *Proceedings of the 1st International Conference on the Use of Fly Ash, Silica Fume, Slag and Other Mineral By-products in Concrete, ACI SP 79*, pp. 435–451.
11. Nagataki S and Ohga H (1992) Combined effect of carbonation and chloride on corrosion of reinforcement in fly ash concrete. *Proceedings of the 4th International Conference on the Use of Fly Ash, Silica Fume, Slag, and Natural Pozzolans in Concrete, Istanbul, Turkey, ACI SP132*, pp. 227–244.
12. Sideris KK, Savva A and Papayianni J (2006) Sulfate resistance and carbonation of plain and blended cements. *Cement and Concrete Composites* **28(1)**: 47–56.
13. Papadakis VG, Fardis MN and Vayenas CG (1992) Hydration and carbonation of pozzolanic cements. *ACI Materials Journal* **89(2)**: 119–130.
14. Sideris KK and Savva A (2001) Resistance of fly ash and natural pozzolans blended cement mortars and concrete to carbonation, sulfate attack and chloride ion penetration. *Proceedings of the 7th CANMET/ACI International Conference on Fly Ash, Silica Fume, Slag and Natural Pozzolans in Concrete, Madras, India*, vol. II, pp. 275–293.
15. Papadakis VG (2000) Effect of supplementary cementing materials on concrete resistance against carbonation and chloride ingress. *Cement and Concrete Research* **30(2)**: 291–299.
16. ASTM (2022) C1202. Standard test method for electrical indication of concrete's ability to resist chloride ion penetration. ASTM, West Conshohocken, PA, USA.

17. Thomas MDA and Bamforth PB (1999) Modelling chloride diffusion in concrete: Effect of fly ash and slag. *Cement and Concrete Research* **29(4)**: 487–495.
18. fib (2006) *Model Code for Service Life Design*. Bulletin 34.
19. ASTM (2021) ASTM C1012. Standard test method for length change of hydraulic-cement mortars exposed to a sulfate solution. ASTM, West Conshohocken, PA, USA.
20. CEN (European Committee for Standardization) (2011) EN 197-1. Cement. Part 1. Composition, specifications and conformity criteria for common cements. CEN, Brussels, Belgium.
21. Tikalsky PJ, Carrasquillo RL and Snow PG (1990) Sulfate resistance of concrete containing fly ash. *Proceedings of the G. M. Idorn International Symposium on Durability of Concrete, ACI SP-131*, pp. 255–265.
22. Mehta PK (1981) Studies on blended Portland cements containing Santorinian earth. *Cement and Concrete Research* **11(4)**: 507–518.
23. Al-Amoudi OSB (1999) Mechanisms of sulfate attack in plain and blended cements: A review. *Proceedings of the International Conference Extending Performance of Concrete Structures, Dundee*, pp. 247–260.
24. Dunstan ER Jr (1980) A possible method for identifying fly ashes that will improve sulfate resistance of concretes. *Cement Concrete and Aggregates* **2(1)**: 20–30.
25. Dunstan, ER (1987) Sulfate resistance of fly ash concretes: The *R*-value. *Proceedings of the Katharine and Bryant Mather International Conference on Concrete Durability*, pp. 2027–2040.
26. Tikalsky PJ and Carrasquillo RL (1992) Influence of fly ash on the sulfate resistance of concrete. *ACI Materials Journal* **89(1)**: 69–75.
27. Erdogan TY, Tokyay M and Ramyar K (1992) Investigations on the sulfate resistance of high-lime fly ash incorporating PC-FA Mortars. *Proceedings of the 4th CANMET/ACI International Conference on Fly Ash, Silica Fume, Slag and Natural Pozzolans in Concrete, Istanbul, Turkey, ACI SP-132*, pp. 271–280.
28. Mehta PK (1986) Effect of fly ash composition on sulfate resistance of cements. *ACI Materials Journal* **83(6)**: 994–1000.
29. Stamatakis M, Fragulis D and Papageorgiou A (1997) Quality of Greek fly ash and its influence on blended cement production. *Proceedings of the Conference on the Use of Fly Ash on Construction, Greece*, pp. 213–228. (In Greek.)
30. Fay KFV and Pierce JS (1989) Sulfate resistance of concretes with various fly ashes. *ASTM Standardization News* **17(12)**: 32–37.
31. Perenchio WF and Klieger P (1976) *Further Laboratory Studies of Portland-Pozzolan Cements*. Portland Cement Association, Skokie, IL, USA, Bulletin RD041.01T.
32. Virtanen J (1983) Freeze–thaw resistance of concrete containing blast furnace slag, fly ash or condensed silica fume. *Proceedings of the 1st International Conference on the Use of Fly Ash, Silica Fume, Slag and Other Mineral By-products, ACI SP-79*, pp. 923–942.
33. Carette GG and Malhotra VM (1984) *Characterization of Canadian Fly Ashes and Their Performance in Concrete*. CANMET, Energy, Mines and Resources Canada, Report MRP/MSL 84–137.
34. Malhotra VM, Carette GG, Bilodeau A and Sivasundram V (1990) *Some Aspects of Durability of High Volume ASTM Class F (Low-calcium) Fly Ash Concrete*. Mineral Sciences Laboratories, Report MSL-90-20 (OP & J).
35. Schmidt M (1992) Cement with inter-ground additives – Capabilities and environmental relief. Part 2. *ZKG International, Edition B* **45(6)**: 296–301.
36. Joshi RC and Lohtia RP (1993) Effects of premature freezing temperatures on compressive strength, elasticity and microstructure of high volume fly ash concrete. *Proceedings of 3rd Canadian Symposium on Cement and Concrete, Ottawa, Canada*.
37. Thomas MDA (1997) Laboratory and field studies of salt scaling in fly ash concrete. *Proceedings of the RILEM International Workshop on Frost Resistance of Concrete with and without Deicing Chemicals, Essen, Germany* (Setzer MJ and Auberg R (eds)).

38. Naik TR, Ramme BR, Kraus RN and Siddique R (2003) Long-term performance of high-volume fly ash concrete pavements. *ACI Materials Journal* **100(2)**: 150–155.
39. Knutsson A (2010) *Freeze/Thaw Durability of Concrete with Fly Ash*. MSc thesis, Chalmers University of Technology, Gothenburg, Sweden.
40. Majko RM and Pistilli MF (1984) Optimizing the amount of Class C fly ash in concrete mixtures. *Cement, Concrete and Aggregates* **6(2)**: 105–119.
41. Thomas MDA (2011) The effect of supplementary cementing materials on alkali silica reaction: A review. *Cement and Concrete Research* **41(12)**: 1224–1231.
42. Schiessl P and Härdtl R (1994) Relationship between durability and pore structure properties of concretes containing fly ash. *Proceedings of PK Mehta Symposium on Durability of Concrete, Nice, France* (Khayat KH and Aïtcin P-C (eds)), pp. 99–118.
43. Malhotra VM, Carette GG and Bremmer TW (1982) *Durability of Concrete Containing Granulated Blast Furnace Slag or Fly Ash or Both in Marine Environment*. CANMET, Energy, Mines and Resources Canada, Report 80-18E.
44. Thomas M (2013) *Supplementary Cementing Materials in Concrete*. CRC Press, Taylor & Francis.
45. Shehata MH and Thomas MDA (2000) The effect of fly ash composition on the expansion of concrete due to alkali silica reaction. *Cement and Concrete Research* **30(7)**: 1063–1072.
46. Saha AK, Khan MNN, Sarker PK, Shaikh FA and Pramanik A (2018) The ASR mechanism of reactive aggregates in concrete and its mitigation by fly ash: A critical review. *Construction and Building Materials* **171**: 743–758.

5.3.3 Ground granulated blastfurnace slag

John Reddy, Ecocem Materials Ltd, Republic of Ireland (and MD Connell, first edition)

5.3.3.1 Introduction

Ground granulated blastfurnace slag (GGBS) is one of several materials that are described as additions, which are increasingly being used in combination with CEM I to produce combination cements. These additions fall into two main groups: latent hydraulic materials and pozzolans. GGBS is a latent hydraulic material, which when combined with water forms hydrate that remains stable when submerged in water. Pozzolans are mentioned in Section 5.3.1. Pozzolans do not react with water and require activation with another material.

The reason for the increased use of GGBS and other additions was initially driven by improved durability and cost savings. Both drivers remain, but increasingly its environmental credentials are becoming the key driver. In the UK and other parts of this world, the materials are separately added in powder form at the concrete mixer. The process for certification is detailed in BS 8500 [1]. In much of Europe there is significant resistance to this practice, and it is virtually prohibited in some countries. In others, there are regulations to prevent this practice. This is even though factory combinations are widely used throughout Europe.

5.3.3.2 History of use

Emil Langen first discovered the potential hydraulic properties of GGBS in 1862, and production of a cement containing GGBS commenced in 1892 in Germany. Manufacture in the UK is believed to have started soon afterwards in the Teesside area of the north-east of England. After 1914, GGBS was also produced in Scotland, although the blastfurnaces in Scotland are now closed. The impetus for these developments is the availability of large quantities of blastfurnace slag from local ironmaking works.

The use of these cements was encouraged in 1923 by the publication of BS 146 [2]. In the following years, GGBS-based cements were used on a diverse range of structures, such as sea defence walls and canal works. Later, during the 1950s and 1960s, GGBS-based cements with slag proportions of around 60% were used for many of hydroelectric dams in the north of Scotland. Recognition of the lower heat development of these cements was enhanced with the publication of BS 4246 [3] in 1968.

GGBS has been commercially available in the UK since 1969, and an Agreement Board Certificate was issued in 1972. BS 6699 [4], the first British Standard for GGBS, was published in 1986, and last updated in 1992. This standard has now been reviewed and published as European standard EN 15167 [5].

Utilisation in the UK has grown from relatively modest levels of 200 tonnes per annum in the early 1970s to over 2 million tonnes manufactured at five locations and supplied via several distribution depots. In Western Europe, approximately 18 million tonnes of granulated blastfurnace slag are used annually as a cement constituent, 80% of which is used either as a factory blend or interground. The remaining 20% is used in the UK as well as Ireland, the Netherlands and Belgium as a direct addition to the concrete mixer. In North America, approximately 95% of granulated blastfurnace slag is used as GGBS as a direct addition to the concrete mixer. Worldwide in 2004, 200 million tonnes of blastfurnace slag were produced, 51% of which was granulated for use in cementitious applications. The proportion of slag converted by granulation for use as a cement component continues to increase, particularly as a separate addition.

5.3.3.3 Granulated slag production and furnace operation

High-grade iron ores from around the world are used to produce iron. These are first blended with limestone and coke by mechanical means in huge ore blending beds. The blended mixture is then partly burnt to form sinter cake. This process, known as sintering, removes moisture and some sulfur, and causes the formation of nodules, which are fundamental to efficient blastfurnace operation.

Precise mixtures of sinter, coke, rubble ore and pelletised concentrate are continually fed into the top of the blast furnace (Figure 5.14). Other minor additions, such as silica and lime sources, can also be introduced to fine-tune and correct the composition of the slag. The temperature is raised progressively to around 2000°C by hot blasts and the combustion of fuel. Some additional fuel is also injected through the tuyeres with the hot blast. Molten iron and slag are drawn off at regular intervals throughout the process from tap holes at the base of the blast furnace. A blast furnace is a high-temperature, high-pressure reaction vessel that operates continuously. All gases produced are collected and processed for reuse. The production of high-quality iron combined with efficient furnace operation is largely controlled by the consistency and composition of the slag.

After being tapped from the furnace, the slag is separated from the iron and cooled. The cooling rate of the molten material determines its physical and chemical characteristics. Two principal methods of cooling are used. One is to pour the liquid slag into pits and leave the material to gradually air-cool to ambient temperature over many hours. Such air-cooled material is of no use as a cementitious material and is generally used as road stone.

The other method is to cool the blastfurnace slag rapidly by quenching with water. There are two methods of achieving this: granulation (Figure 5.15) and pelletisation. Both methods use water. Granulation relies on large volumes of water, whereas pelletisation uses water sprays through which pellets of slag are propelled. Both methods result in a glassy homogeneous non-crystalline material that has latent hydraulic properties.

5.3.3.4 Production of GGBS

The granulated slag is selected, stored and allowed to drain prior to processing. GGBS is produced by drying and reducing the practical size of the slag down to a similar size to that of Portland cement (CEM I).

Figure 5.14 Section through a typical blast furnace

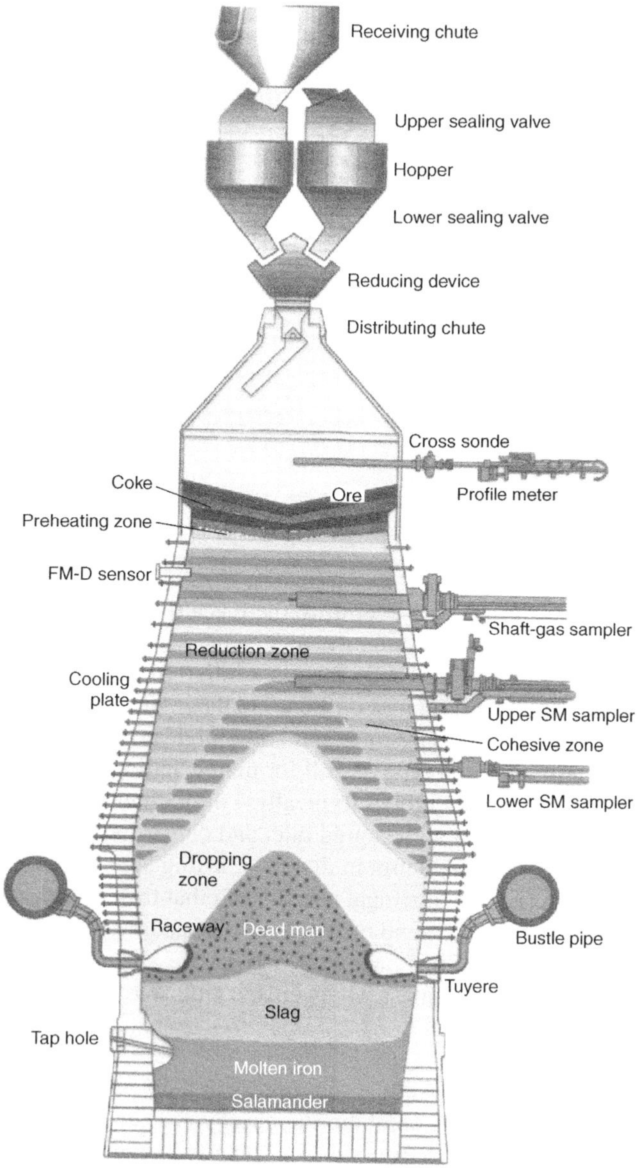

TYPES OF MILL

In the UK, much work has gone into refining the process for reducing the particle size, as this is highly energy intensive. Three main types of equipment are in use throughout the cement industry.

- *High-pressure rollers:* There are two large-diameter rollers, one fixed, the other moving and held against the fixed roller under high pressure. Material is passed between the two rollers in a recirculation system. Material of the right size is extracted by separators and a baghouse. This process requires a high level of maintenance and has low availability.

Figure 5.15 Granulation in progress

- *Vertical mills:* Similar in basic design to a flour mill, there is a rotating table and several rollers that apply high pressure. The material passes under the rollers and is crushed between them and the table. The material is recirculated and separated as is done in a roller mill. The advantage of this system is that the table rollers and separator are housed in one chamber so that the material is recirculated within the chamber. Like the high-pressure rollers, this is a high-maintenance low-availability process.
- *Ball mills:* These are the most common type of mill for final grinding in the cement industry. A ball mill consists of large-diameter steel tubes (typically 4 m in diameter), which may have one or more chambers. The tubes are capped with slotted inlet and outlet diaphragms. Each tube contains steel balls of various sizes, from 50 to 20 mm in diameter, filling approximately 30% of its volume. A tube is rotated at 95% of a critical centrifugal speed, such that the balls cascade inside the tube. The material circulates through the tube and milled by the cascading balls. The final product is extracted by a separator and collected in a baghouse. This process is more energy intensive than the other processes but has a large thermal mass which aids drying, and it has high availability and low-maintenance requirements.

Hybrid or combi systems are also used. In these, a roller press in the feed circuit pre-crushes the feed prior to sending it to a ball mill. These systems combine the energy efficiency of a roller press with the high availability of a ball mill. Such systems are considered by some to be the most effective method of milling.

INFLUENCE OF FINENESS AND PARTICLE SIZE DISTRIBUTION

GGBS is milled to a fine particle size, typically 400–500 m^2/kg compared with 300–400 m^2/kg for CEM I. Generally, the finer a material is, the more reactive it is likely to be, although fine materials can have an adverse effect on water demand. This influence of fineness on reactivity is less significant than the influence of chemical composition. Much research has been conducted into using the particle size distribution as a control tool, but it has yet to be adopted. This distribution may affect the water demand and, as such, the performance of GGBS in concrete. However, GGBS is used in combination with CEM I, typically in a 50% blend, and it is the overall particle size distribution that is influential, as the combination tends to dilute the influence of any individual powder.

CHEMICAL COMPOSITION AND GLASS CONTENT

The chemical composition in combination with the glass content has a significant influence on the performance of GGBS. The likely reactivity of a particular slag can be assessed by a number of chemical formulae. The two most common are:

V ratio $CaO + MgO/SiO_2$

Ih $CaO + MgO/Al_2O_3/SiO_2$

In both cases, the higher values indicate increased reactivity. Table 5.2 gives the range of compositions that are available in the UK.

The glass content is a measure of the granulation efficiency: higher values are indicative of improved reactivity for a given composition. A minimum specified limit of glass content is not required by EN 15167-1 [5], but glass content may be requested by an auditing body or end user.

ADDITIVES AND MINOR ADDITIONS

Additives can be included to a maximum of 1%, of which 0.5% can be organic. Grinding aids are the most common additive, although other materials are used. These materials incorporated into the grinding process aid either grinding efficiencies or other aspects of performance but must not promote corrosion of the reinforcement or impair other properties of the cement or the concrete and mortar made from the cement. The Portland cement industry uses these materials extensively, and they are used to some extent by the GGBS industry elsewhere in the world. Minor additions are inorganic materials that can be included to a maximum of 5% but do not contribute to performance or grinding. These materials are used in the manufacture of factory-blended cements with GGBS to EN 197-1 [6] but are not permitted in the production of GGBS to EN 15167-1 [5].

APPEARANCE

GGBS is off-white in colour and is often specified by architects due to the lighter tone of the resulting concrete. Because of its finer particle size, smoother more uniform finishes are achievable.

Table 5.2 Chemical composition of base blastfurnace slags used in the UK

Parameter	Typical range: %	
CaO content	42–38	
Al_2O_3 content	14–12	
SiO_2 content	36–34	
MgO content	9–7	
Na_2Oeq	0.7–0.4	
V ratio	1.3–1.4	
Ih	1.5–1.8	
Glass granulate content	Granules: 90–98%	Pellets: 80–90%[a]

[a]Generally only the fine fraction passing 20 mm is used for GGBS. Larger sizes can be used, but the glass content will be lower.

5.3.3.5 Certification and standards

EN 15167

Published in 2006, EN 15167 [5] is a harmonised European norm with CE marking. The standard specifies the chemical composition with limiting values for loss on ignition, magnesia, chloride, manganese and sulfur expressed as both sulfide and sulfate. Several of the physical and mechanical properties are like those specified in the CEM I standards. The strength performance has limiting values for activity index and the compressive strength of mortar containing 50% GGBS expressed as a proportion of a CEM I control. Initial setting times for mortar containing 50% GGBS must be no more than twice the setting time of CEM I. Furthermore, the harmonised EN 206 [7] presents three methodologies that govern the use of additions to a combination in a mixer: (*a*) equivalent performance of combinations, (*b*) equivalent concrete performance, and (*c*) the *k*-factor. The national annexes of EN 206 include the methodology to be applied for the use of GGBS as a standalone mixer addition in a specific country.

EN 197

EN 197 [6] is the European standard for cement and includes CEM I and combinations of CEM I with GGBS, pozzolans, fly ash, silica fume and limestone fillers. The standard refers only to factory-produced cements and combinations. CEM II/A-S may contain 6–20% of GGBS. CEM III/A may contain 36–65% of GGBS. CEM III/B may contain 66–80% of GGBS. CEM III/C may contain 80–95% of GGBS. Specific suitability for the use of these cement types or combinations in concrete is set out in the national annexes of EN 206.

BS 8500

BS 8500 [1] is the UK annex to EN 206 [7]. It includes a procedure for establishing the conformity of combinations that are made in a mixer. GGBS has been tested with individual grades of CEM I. There is a maximum replacement level that satisfies the minimum strength requirements for the different strength grades.

5.3.3.6 Influence on plastic properties

When tested as a paste, the stiffening rate of combinations of GGBS and CEM I was slower than that of a similar paste containing CEM I alone. The difference increases with increasing proportions of GGBS (Figure 5.16). The initial setting time is influenced by the curing temperature as well as the characteristics of the Portland cement component. This means that concrete containing increased amounts of GGBS will remain workable longer. This may be desirable in mass pour situations or hot weather conditions by preventing cold joints.

Figure 5.16 Typical setting times for combinations of GGBS and CEM I

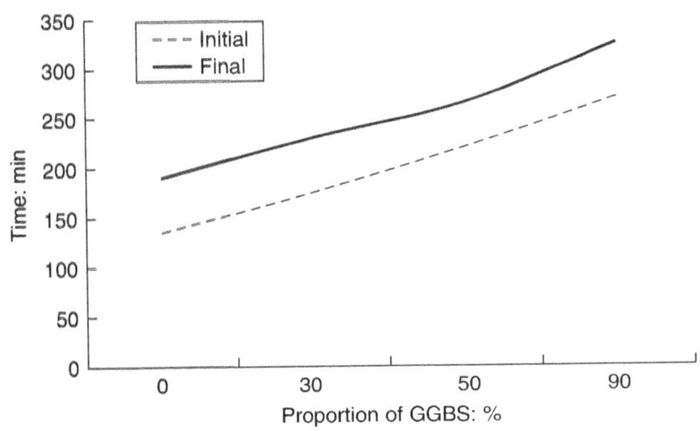

AMBIENT TEMPERATURE

All cement combinations containing GGBS are influenced by ambient conditions. Because GGBS is a latent hydraulic material, this influence is magnified depending on the proportion of GGBS. A high proportion of 70% or above in thin sections can have a significant influence, and special precautions may be required. Generally, large sections of at least 600 mm have sufficient thermal capacity not to be influenced by the ambient temperature, even with high replacement levels.

PLACING AND FINISHING

As indicated above, the time that concrete remains open is extended in line with increasing proportions of GGBS. This is further influenced by the ambient temperature. In thin sections, such as flat slabs, this influence can be significant in cold conditions, and GGBS proportions are often reduced to 20% or not used at all. Because the concrete remains plastic for longer, formwork pressures are likely to be increased, which may affect the pour rate. However, work in Dundee [8] has indicated that these increases are not significant and are within the factor of safety in the formwork design code.

BLEEDING, PLASTIC SHRINKAGE AND SETTLEMENT CRACKING

As a further result of the extended plastic time, the potential bleed capacity may be increased when high percentages of GGBS are used. Bleeding may result in plastic settlement cracking if restraint is present, but this can be rectified by re-vibration. Plastic shrinkage occurs when the evaporation rate exceeds the bleed rate. As concrete with GGBS remains plastic for longer, it may be more vulnerable to this form of cracking. However, suitable curing systems will prevent this problem.

5.3.3.7 Influence on mix design and hardened properties

RATE OF STRENGTH DEVELOPMENT

As GGBS hydrates slower than CEM I, the rate of strength development will be slower. Increasing the proportion of GGBS increases this effect. As a result, the relation between the w/c ratio and strength changes. Figure 5.17 gives typical values when normal addition rates of up to 50% are in use. The biggest difference is seen at early ages. Typically, 7-day strengths are 50–65% of the characteristic 28-day strength in comparison to the usual 70–80% of a similar CEM I concrete. For normal addition rates of up to 50%, 28-day strengths are like those of a comparable CEM I concrete. As the hydration process for GGBS is largely pozzolanic, significant strength development is achieved at later ages. This enhanced late strength development of GGBS beyond 28 days is an added performance benefit. Data are available that demonstrate continued development up to 5 years.

Figure 5.17 Typical strength development in concrete containing different proportions of GGBS. Total cement content: 350 kg/m³. (Reproduced with permission of Trent Valley Aggregates)

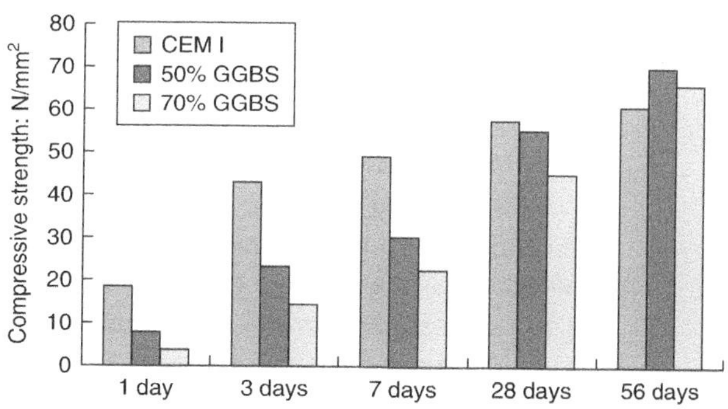

SULFIDE BLUE

GGBS contains small amounts of sulfides, which on hydration form complex polysulfide products that are blue-green in colour. This colouration is sometimes evident on the surface of concrete and is indicative of the presence of GGBS in concrete when broken surfaces are exposed. Upon contact with air, these products oxidise, and the colours fade to the normal overall off-white appearance. The oxidation has no effect on the properties of the concrete other than its colour. The process of oxidation and the time taken to achieve a uniform colour are influenced by several interrelated factors, which are summarised in Table 5.3.

5.3.3.8 Influence on thermal properties

THICK SECTIONS

GGBS combinations have long been recognised as a material with low-heat-generating properties, and high proportions of GGBS are often specified to minimise the temperature rise in thick sections. Increasing the GGBS content reduces the rate at which heat is generated as well as the rate of heat evolution, which results in lower peak temperatures and shallower thermal gradients. Several documents give advice, such as CIRIA reports 91 [9], 135 [10] and C660 [11], all of which refer to cement type as one of the means of influencing temperature rise. Figure 5.18 illustrates the influence of increasing GGBS content.

THIN SECTIONS

In thin sections, the lower rate of heat evolution affects the rate of in situ strength development, particularly at low ambient temperatures. Depending on the proportion of GGBS, the mix design and the formwork type (in terms of insulating capacity and the strength required to release the formwork), the formwork striking times may be extended. In these conditions when high proportions of GGBS are used, extended curing times are required. Concrete maturation methods and temperature-matched curing provide accurate in situ strength estimates that allow for optimal decision-making on striking the formwork.

Table 5.3 Factors influencing sulfide colour and its rate of oxidation

Influencing factor	Influence on rate of colour change
Proportion of GGBS	Higher proportions of GGBS give a deeper colour and take longer to oxidise.
Concrete quality	High-performance concrete with a low w/c ratio and high cement content takes longer to oxidise.
Types of formwork	Low-permeability surfaces, such as treated plywood and polished steel, are likely to produce a greater colour intensity
Release agents	Some types leave an impervious surface on the concrete, which inhibits oxidation.
Curing and curing membranes	In essence, effective curing, such as underwater or in close contact with polythene, slows down the rate of oxidation. Sprayed-on membranes have a similar influence.
Ambient conditions	Cold damp conditions slow down the rate of oxidation. Warm dry weather has the opposite effect
Surface treatments	Lacquer treatments seal in the colour. An acid treatment can expose the colour and seal the colour in.

Figure 5.18 Influence of GGBS percentage on the temperature rise in concrete elements of different thicknesses [12]

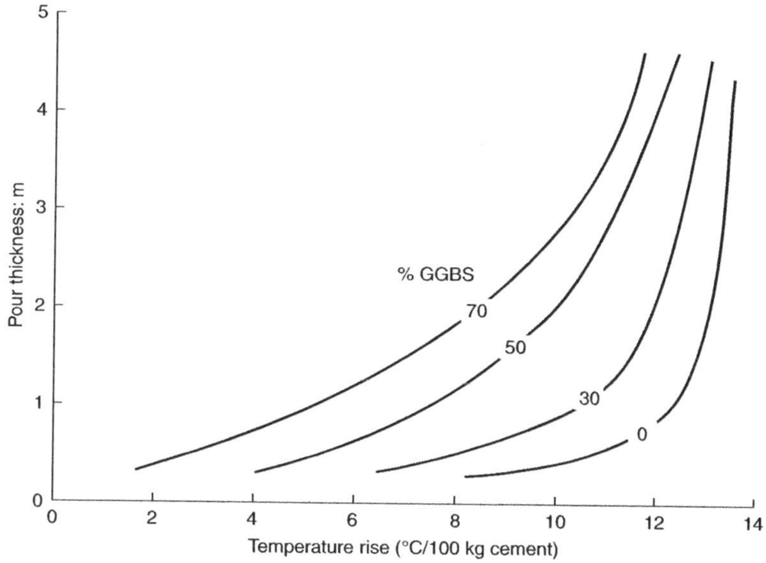

5.3.3.9 Influence on durability

The general term 'durability' is a measure of the ability of a concrete structure to survive in a given environment. GGBS concrete is inherently different in character to Portland cement concrete in that GGBS concrete has low permeability, a different hydrate structure and different cement particle composition. Concrete is subject to many different types of potential attack, which are considered separately in this section with respect to the advantages of using GGBS in place of Portland cement.

SULFATE ATTACK

Sulfate attacks are a significant mechanism for failure in environments containing high-sulfate groundwater or in marine environments. Partial replacement of CEM I by GGBS generally increases the resistance of concrete to a sulfate attack, and this is acknowledged in all major European codes of practice. The sulfate resistance of GGBS concrete depends on several parameters, principally the ratio of GGBS to Portland cement, the C_3A (tricalcium aluminate) content of the CEM I, the alumina content of the slag and the w/c ratio. Generally, the resistance increases with increasing replacement by GGBS, less C_3A in the CEM I, less alumina in the slag and a lower w/c ratio. When the GGBS content is 65% or greater, most combinations of GGBS and CEM I provide equivalent or better sulfate resistance than sulfate-resistant Portland cement (SRPC). This is confirmed by BRE Special Digest 1 [13], whose recommendations have been incorporated into BS 8500 [1]. These recommendations indicate that CIIIA cement (36–65% GGBS) is equivalent to SRPC whereas CIIIB (66–80% GGBS) is superior, as the recommendations allow lower cement contents and high w/c ratios for these cements compared with SRPC. A special designation '+SR' maybe declared for CIIIA and CIIIB when the alumina (Al_2O_3) content of the GGBS used <14%.

The mechanisms associated with sulfate attacks are discussed in detail in Chapter 4. The influence of GGBS on the mechanisms is debated by researchers. The principal process by which GGBS improves performance in sulfate conditions is by dilution of the vulnerable hydrates, such as C_3A, produced by CEM I. This is further enhanced by the more refined pore structure, which is less permeable to ingress by sulfate solutions. In situ concrete foundations in sulfate-bearing ground are also less likely to suffer from thermal cracking when GGBS is used, further reducing possible paths for the ingress of sulfate-bearing solutions.

THAUMASITE ATTACK

Thaumasite requires a source of carbonate and sulfate to form, and these may be provided in many ways, as Table 5.4 indicates. The formation takes place rapidly in wet conditions at temperatures below 15°C but can occur at higher temperatures. Thaumasite forms when sulfates react with calcium carbonate and calcium silicates. This is distinct from a conventional sulfate attack, in which the calcium aluminate phases are attacked. At a Building Research Establishment (BRE) conference [14], it was reported that after the initial diagnosis on the M5 bridges, a significant number of structures had been identified.

A 7-year research programme conducted at BRE indicated that concrete containing 70% GGBS was particularly resistant to this form of attack, and the following recommendations were developed for the following ground conditions as determined by BRE SD1 [13] or BS 8500 [1].

- Design chemical class 3 (DC-3): Use a cement type containing at least 70% GGBS and adopt a minimum cement content of 380 kg/m^3 and a maximum w/c ratio of 0.45.
- Design chemical class 4 (DC-4): Use a cement type containing at least 70% GGBS and adopt a minimum cement content of 400 kg/m^3 and a maximum w/c ratio of 0.40.

ACID ATTACK

An acid attack may occur in industrial environments and in low-pH soil environments (e.g. peat). In general, GGBS concrete is more resistant to dilute acids than that made with Portland cement or SRPC. Many European standards recommend the use of GGBS concrete for acid conditions. However, note that no concrete is completely immune to attack in acidic conditions, and the quality of the concrete (in particular the w/c ratio and compaction) can be more important than the type of cement. EN 206 [7] offers guidance on the mitigation of concrete to acid attack at pH > 4. Below this level, additional preventative measures may be required. There are novel cement types, such as supersulfated cement and alkali-activated slag, that have enhanced acid resistance in comparison to EN 197 cements and combinations.

FREEZE–THAW RESISTANCE

The freeze–thaw resistances of CEM I concrete and GGBS concrete of similar strength and air content are essentially the same. There are no published reports of frost damage occurring in actual structures containing GGBS that are directly attributed to its presence. However, because of its slower rate of strength gain, GGBS concrete may be more susceptible to frost attack at an early stage. There is some debate on the relation between the indicated performance from laboratory test methods and field performance. Some tests that use freeze cycles down to 20°C can produce high degrees of scaling, but this is not considered to

Table 5.4 Potential sources of carbonate and sulfate

Carbonate	Sulfate
Limestone aggregate	Lower Lias clay (disturbed and used as a backfill)
Limestone filler	Kimmeridge clay
Limestone fines	Rhaetic mudstone
Limestone-filled cements	Railway ash spoil from coal and oil shale
Groundwater high in carbon ions	Seawater
	De-icing salts
	Other sulfide-bearing soils
	Sulfate-bearing bricks, gypsum plasters

be representative of field performance. There are limiting values in the national annexes of EN 206 that are prescriptive in the amount of GGBS that may be used in XF4.

CHLORIDE ATTACK

A chloride attack is a significant mechanism for failure in marine environments and structures exposed to de-icing salt. Concrete containing GGBS provides high resistance to the penetration of chloride. Tables 5.5 [15] and 5.6 [16] show typical diffusivities for various types of concrete. For 65% GGBS, the diffusivity is reduced by an order of magnitude compared with normal CEM I. This reduction in diffusivity has been confirmed by many investigators and would appear to be due to two mechanisms. First, the incorporation of GGBS reduces the permeability of the concrete; second, evidence suggests that the hardened paste of slag cement immobilises the chloride ions, thus removing the concentration gradient needed for diffusion. A report by Bamforth [17] summarises the research, concluding that significant benefits are achieved at addition levels in excess of 40%, with increasing resistance up to 65%.

In practice, it is common for chloride ingress in the near-surface zone to be rapid, but for further migration to occur very slowly with little change in the initial profile over many years (Figure 5.19). In addition to substantially reducing the rate of chloride ingress by a combination of enhanced pore structure and chloride binding, GGBS also influences the chemistry of the pore solution. Although opinion is divided, some research has suggested that GGBS may reduce the chloride threshold [18,19] by reducing the level of free chloride to a lesser extent than the reduction in hydroxide levels.

RESISTANCE TO CARBONATION

GGBS concretes will carbonate more rapidly than CEM I concrete. For concretes of a reasonable strength grade (C25/30) or greater and moderate levels of GGBS (50%), the difference is marginal when concretes of equal grade are compared. Greater differences are reported for lower-strength grades with higher GGBS

Table 5.5 Diffusion of chloride ions at 25°C in cement pastes with w/c ratio = 0.5 [15]

Type of cement	Diffusivity: 10^{-8} cm²/S
SRPC	100.0
OPC	44.7
70% OPC/30% PFA	14.7
35% OPC/65% GGBS	4.1

Table 5.6 Diffusion of coefficients in OPC and Portland blastfurnace cement [16]

Diffused ion	Ionic diffusivity: 10^{-8} cm²/S	
	OPC	Portland blastfurnace cement
Sodium	2.38	0.10
Potassium	3.58	0.21
Chloride	4.47	0.41

Figure 5.19 Chloride profiles recorded over 10 years of splash zone exposure for 70% GGBS (w/c ratio 0.48) [17]

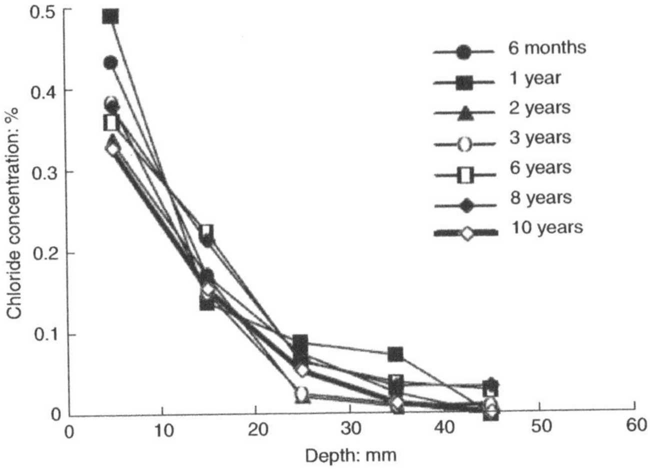

additions (<70%). As a general rule for predictive purposes, a buffering capacity of 0.6 can be assumed [17]. The national annexes of EN 206 give guidance on the mix design requirements and cover required to ensure carbonation resistance when using high percentages of GGBS.

ALKALI–SILICA REACTION

ASR and the associated but rare alkali–carbonate reaction can occur in concrete, depending on the character and proportion of the reactive aggregate and the availability of mobile metal alkalis, details of which are discussed in Chapter 4.

It is accepted worldwide that the inclusion of GGBS in concrete reduces the risk of damage due to ASR. There have been many investigations, and without exception, these have confirmed the ability of GGBS to reduce significantly the deleterious expansion caused by ASR. Expansion due to ASR reduces with increasing replacement of CEM I by GGBS. In the UK, it is generally agreed that a minimum of 50% GGBS can be a suitable preventative measure, provided that the total alkali content of the GGBS/CEM I blend does not exceed 1.1% Na_2O equivalent. At lower replacement levels, Concrete Society TR30 [20] and BRE Digest 330 [21] recommend that it should be assumed that GGBS contributes half of the total alkali content. In a report that attempted to clarify this conflict, Higgins and Connell [22] found that at low replacement levels (25% GGBS) the half value was appropriate, but at higher replacement levels (50% GGBS) there was zero contribution of alkalis from the GGBS. In practice, the alkali content of GGBS produced in the UK is such that concrete mixes can readily be designed that satisfy the requirements of both approaches. BS 8500 [1] has incorporated these views and recommends that the advice given in BRE Digest 330 [21] be adopted.

There is some debate over why GGBS is effective in reducing the deleterious expansion due to alkali reactions. The alkalis present in GGBS are predominantly acid-soluble, with a very small proportion water-soluble in comparison with those present in CEM I, which are predominantly water-soluble. Water-soluble alkali becomes available as soon as hydration commences and can move to reaction sites. As such, GGBS would tend to act as a diluting influence on the water and total alkali available; however, as hydration progresses, the acid-soluble alkali in GGBS is likely to be released into solution. Also, GGBS remains effective at controlling expansion even when excess alkali of up to 8 kg/m^3 Na_2Oeq is included in the concrete [22]. It has been reported that GGBS can be activated by alkalis, and it is thought that some of the released alkali is consumed in reactions with the GGBS. Additionally, as hydration progresses, the pore structure becomes increasingly refined, so that diffusion coefficients reduce dramatically [16], thus reducing the ability for alkali and water to move to reaction sites.

5.3.3.10 Environmental aspects

Environmental aspects are becoming increasingly important in construction, and the impact of construction materials are now considered by clients, architects and designers. The manufacture of Portland cement is thought to contribute 5% of the world's carbon dioxide emissions. Additions can significantly reduce several environmental aspects and make a positive impact on sustainability. GGBS is a by-product that does not require any additional mineral extraction. Using it is a very low energy and carbon dioxide intensive process compared to using CEM I. Environmental product declarations give the full life-cycle assessment of products. These are being adapted by the construction industry and are being incorporated in green building specifications such as BREEAM and LEED. Table 5.7 gives values for some of these environmental aspects, and Table 5.8 shows the impact of similar grades of concrete with different cement types.

5.3.3.11 Specified application

Driven by the improved durability obtained with GGBS, it is often specified for major contracts, particularly when an extended design life is required or particularly aggressive environments are an issue. The following are some typical examples where GGBS has been used for these reasons, both internationally and in the UK.

- Humber Bridge: For many years, this structure was the longest single-span crossing in the world. GGBS was used at 70% replacement as an addition in the anchor blocks and foundations and 40% in the slipformed towers.

Table 5.7 Environmental profiles

Issue	Measured as	Impact	
		Manufacture of 1 tonne of GGBS[a]	Manufacture of 1 tonne of Portland cement (CEM I)
Climate change	CO_2 equivalent	0.07 tonne[b]	0.88 tonne
Energy use	Primary energy[c]	1300 MJ	7000 MJ
Mineral extraction	Weight quarried	0	1.5 tonnes
Waste disposal	Weight to tip	1 tonne saved[d]	0.02 tonne

[a]The profile for GGBS is the impacts involved in processing granulated blastfurnace slag to produce GGBS. No account has been taken of the impacts of ironmaking because the slag evolves, irrespective of whether or not it can be used.
[b]Extracted from the BRE environmental profile.
[c]Includes losses involved in the generation and distribution of electricity.
[d]The use of slag for the manufacture of GGBS potentially saves it from having to be disposed in a tip.

Table 5.8 Calculated environmental impacts for 1 tonne of concrete

Impact	Reference concrete: Portland cement (CEM I)	GGBS concrete	PFA concrete
Greenhouse gas	142 kg (100%)	85.4 kg (60%)	118 kg (83%)
Primary energy use	1070 MJ (100%)	760 MJ (71%)	925 MJ (86%)
Mineral extraction	1048 kg (100%)	965 kg (92%)	1007 kg (96%)

- Bahrain Causeway: All concrete elements were cast with blended cement imported to the Gulf region. This produced a concrete capable of resisting the extremely aggressive conditions in the Gulf area.
- Queen Elizabeth II Bridge across the Thames in London: All concrete was specified to contain 70% GGBS, mainly to give an extended design life but also for the light colour of the concrete.
- Blackpool sea defence work constructed over 10 years up to 2008: GGBS was specified in much of the concrete, including precast step units to improve durability.
- Berwick-upon-Tweed sea defence cast-in-place wave wall: 70% GGBS was specified and used despite cold weather conditions, steel formwork and relatively thin sections.
- The new Tyne Tunnel, expanded in 2011: This submerged tube construction was specified to contain 70% GGBS to ensure a watertight durable construction.
- Tsing Ma Bridge, Hong Kong: The concrete for the towers of this suspension bridge was specified to have an extremely low coefficient of permeability, and GGBS was exported from the UK to achieve the required permeability.

5.3.3.12 Special applications

GGBS is used mainly to produce cast-in-place concrete, either by ready-mix concrete companies or, on larger projects, on site by bespoke batching plants. Over the many years of use in the UK, several special applications have developed, partly driven by economy compared with CEM I but also by the specific properties of GGBS.

SECANT PILING

A high proportion of GGBS, 80% and above, is used to produce concrete with very low early strength for cutting the secant but still retaining long-term durability, low permeability, and strength. BRE IP17/05 [23] gives details of the process.

GROUT CUT-OFF WALLS

Several specialist piling contractors have developed a method of producing cut-off walls to satisfy the demand for impermeable structures around existing and new waste disposal sites. GGBS in combination with bentonite is used to produce slurry that is pumped into trenches during excavation. These are typically 600 mm wide and deep enough to toe into a lower impermeable soil layer. The slurry provides support for the sides of the trench during excavation and subsequently hardens to form a continuous wall that is stiff and impermeable to leachate and gas but still remains semi-plastic and able to take ground movement.

SOIL STABILISATION AND HYDRAULICALLY BOUND MATERIALS

Partly driven by economic construction demands and the environmental aspects of construction, the use of soil stabilisation and hydraulically bound materials has grown considerably. A wide range of processes are now in common use for roads and airfields but also for building construction, reinstatement of brownfield sites and stabilising contaminated ground. Several codes of practice, advice documents and papers describe the performance of GGBS in these applications [24–27].

PRECAST CONCRETE APPLICATIONS

Traditionally, the requirement for early formwork striking, lifting of units and transfer of prestressed loads has restricted the use of GGBS in precast concrete applications. However, driven by environmental aspects, durability requirements and the colour of finished products, precast operators have developed mix designs and curing systems that have resulted in significant use of GGBS in many applications, such as hollow core flooring, pigmented precast products and structural sea defence units.

5.3.3.13 Conclusion

Cement manufacturers worldwide are increasingly examining the use of alternative materials, partly to reduce emissions but also to reduce manufacturing costs. In the UK, the most common cement type is one containing an addition such as GGBS, fly ash or a limestone. The use of GGBS, most commonly as an addition at the mixer but also as factory-produced CEM II or III, has resulted in significant reductions in the various environmental aspects of cement manufacture, including carbon dioxide emissions. This use has also resulted in significant potential benefits to durability, leading to increases in the design life and reductions in the whole-life cost of structures.

REFERENCES

1. BSI (2006) BS 8500-1:2006. Concrete – Complementary British Standard to BS EN 206-1. Part 1. Method of specifying and guidance for the specifier. BSI, London, UK.
2. BSI (1996) BS 146:1996. Specification for Portland blastfurnace cements. BSI, London, UK.
3. BSI (1996) BS 4246:1996. Specification for high slag blastfurnace cement. BSI, London, UK.
4. BSI (1992) BS 6699:1992. Specification for ground granulated blastfurnace slag for use with Portland cement. BSI, London, UK.
5. BSI (2006) BS EN 15167-1:2006. Ground granulated blastfurnace slag for use in concrete mortar and grout. Part 1: Definitions, specifications and conformity criteria. BSI, London, UK.
6. BSI (2000) BS EN 197-1:2000. Cement. Composition, specifications and conformity criteria for common cements. BSI, London, UK.
7. BSI (2000) BS EN 206-1:2000. Concrete. Part 1. Specification performance composition and conformity. BSI, London, UK.
8. McCarthy MJ, Caliskan S and Dhir RK (2004) Update on developing guidance for design formwork pressures. *Concrete Journal* **38(5)**: 45–46.
9. Harrison T (1992) *Early-age Thermal Crack Control in Concrete*, revised edn. CIRIA, London, UK, Report 91.
10. Bamforth PB and Price WF (1995) *Concreting Deep Lifts and Large Volume Pours*. CIRIA, London, UK, Report 135.
11. Bamforth PB (2007) *Early-age Thermal Crack Control in Concrete*. CIRIA, London, UK, Report C660.
12. Bamforth PB (1982) *Early-age Thermal Cracking in Concrete*. Institute of Concrete Technology, Camberley, UK, Technical note TN/2.
13. BRE (Building Research Establishment) (2005) *Concrete in Aggressive Ground*. BRE, Watford, UK, Special digest 1.
14. BRE (2002) *Proceedings of the 1st International Conference on Thaumasite in Cementitious Materials*. BRE, Watford, UK.
15. Page CL, Short NR and El Tarras A (1981) Diffusion of chloride ions into hardened cement pastes. *Cement and Concrete Research* **11(3)**: 395–406.
16. Bakker RFM (1981) About the cause of the resistance of blastfurnace cement to the alkali silica reaction. *5th International Conference on Alkali Aggregate Reactions in Concrete, Cape Town*, Paper S252/29.
17. Bamforth PB (2004) *Enhancing Reinforced Concrete Durability*. Concrete Society, Camberley, UK, Technical report 61.
18. Bamforth PB (1998) *Guide to the Prevention of Corrosion in Reinforced Concrete Exposed to Salt, Part 1. Performance of RC Blocks Exposed for 10 years in Marine Splash Zone*. Taywood Engineering, Watford, UK, Report 1303/98/10248.
19. Tuutti K (1982) *Corrosion of Steel in Concrete*. PhD thesis (monograph), Division of Building Materials, Swedish Cement and Concrete Research Institute, Stockholm, Sweden, Report 4.82.

20. Concrete Society (1987) *Alkali–silica Reaction: Minimising the Risk of Damage to Concrete*. Concrete Society, Camberley, UK, Technical report 30.
21. BRE (2004) *Alkali Aggregate Reactions in Concrete, Part 2*. BRE, Watford, UK, BRE Digest 330.
22. Higgins D and Connell MD (1992) Effectiveness of granulated blastfurnace slag in preventing alkali–silica reaction. *9th International Conference on Alkali–Aggregate Reaction, London*, pp. 175–183.
23. BRE (2005) *Concretes with High GGBS for Use in Hard/Firm Secant Piling*. BRE, Watford, UK, BRE Information paper 17/05.
24. BSI (2004) BS EN 14227. Hydraulically bound mixtures, Parts 1–5. Specifications. BSI, London, UK.
25. Britpave (2005) *The Immediate Trafficking of Cement-bound Materials*. Britpave, Camberley, UK, Technical Report BP/14.
26. Britpave (2005) *Stabilisation of Sulfate-Bearing soils*. Britpave, Camberley, UK, Technical guidelines BP/16.
27. Kennedy J (2006) *Hydraulically Bound Mixtures for Pavements*. Concrete Centre, Camberley, UK, Report CCIP-009.

5.3.4 Silica fume for concrete

Robert C Lewis, Ferroglobe Plc, UK

5.3.4.1 The material

The terms 'condensed silica fume', 'microsilica', 'silica fume' and 'volatilised silica' are often used to describe the by-products extracted from the exhaust gases of silicon, ferrosilicon and other metal alloy smelting furnaces. However, the terms 'microsilica' and 'silica fume' are used to describe condensed silica fumes that are of high quality for use in the cement and concrete industry. The latter term, 'silica fume', is used in the European standard EN 13263.05.

Silica fume was first 'obtained' in Norway in 1947, when environmental concerns made it compulsory to filter the exhaust gases from furnaces. The major portion of these fumes was a very fine powder composed of a high percentage of silicon dioxide. As the pozzolanic reactivity of silicon dioxide was well known, extensive research was undertaken, principally at the Norwegian Institute of Technology. Over 7000 papers describe the application of silica fume and silica fume concrete.

Large-scale filtering began in the 1970s, and the first standard, NS 3050, for the use of silica fume in a factory-produced cement, was granted in 1976.

PRODUCTION AND EXTRACTION

Silica fume is produced during the high-temperature reduction of quartz in an electric arc furnace. The main product is silicon or ferrosilicon (Figure 5.20). Due to the vast amounts of electricity needed, these furnaces are in countries with abundant electrical capacity: China, Europe including Scandinavia, the Americas including Canada, South Africa and Australia, among others.

High-purity quartz is heated to 2000°C with coal, coke or wood chips as fuel, and an electric arc separates out the metal. As the quartz is reduced, it releases silicon monoxide gas. This mixes with oxygen in the upper parts of the furnace, where it oxidises and condenses into microspheres of amorphous silicon dioxide.

The fumes are drawn out of the furnace through a pre-collector and a cyclone, which remove the larger coarse particles of unburnt wood or carbon. The fumes are then blown into a series of special filter bags.

Figure 5.20 Typical plant layout (left to right): furnace, main stack, pre-collector, cooling pipes and fan, and baghouse

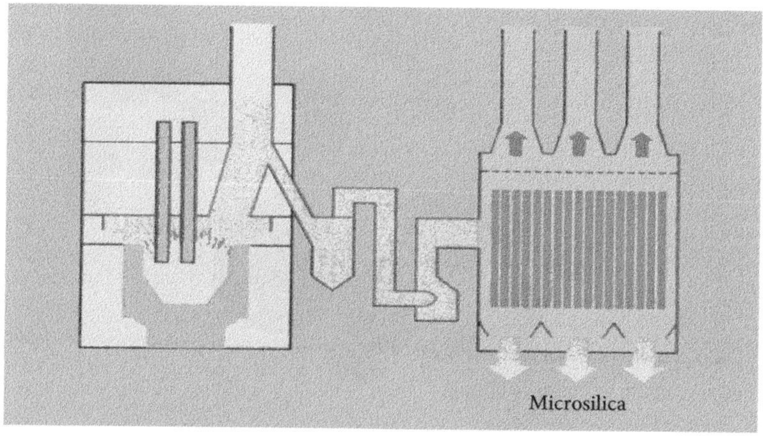

The chemistry of the process is very complex and temperature dependent. The silicon carbide formed initially plays an important intermediate role, as does the unstable silicon monoxide gas, which forms the silica fume. The following summarises the process:

$$t > 1520°\mathrm{C}$$
$$\mathrm{SiO_2 + 3C = SiC + 2CO}$$
$$\downarrow$$
$$t > 1800°\mathrm{C}$$
$$\mathrm{3SiO_2 + 2SiC = Si + 4SiO + 2CO}$$

The unstable gas travels up in the furnace, where it reacts with oxygen to form silicon dioxide (Figure 5.21):

$$\mathrm{4SiO + 2O_2 = 4SiO_2}$$

The silica fume is not collected as purely this material. Other particles and chemicals make up the powder that is collected in the filter bags (Table 5.9).

CHARACTERISTICS

When collected, silica fume is an ultrafine powder having the following basic properties

- at least 85% silicon dioxide content
- mean particle size between 0.1 and 0.2 μm
- minimum specific surface area of 15 000 m^2/kg
- spherical particle shape.

The powder is normally grey, but this can vary according to the source. Operation of the furnace, the raw materials and the quality of metal produced all affect the colour of the powder. This variation can show up in the material from different furnaces; thus, it is advisable to ensure consistency by getting a supply from a single source.

Figure 5.21 Undensified silica fume (scanning electron micrograph, SEM)

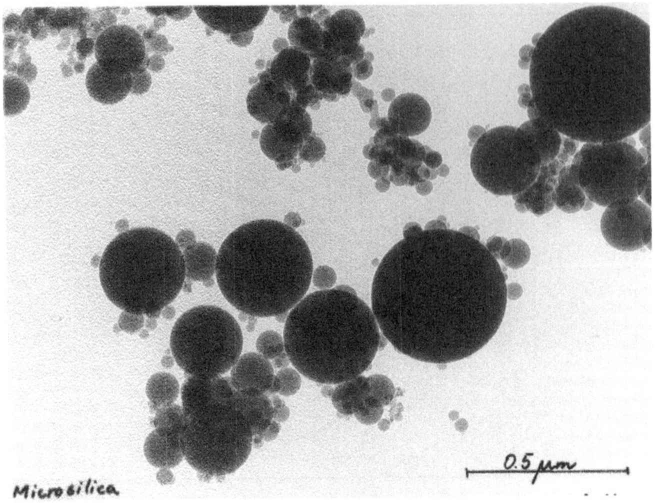

Table 5.9 Comparison of cementitious materials available in the UK

Property	Portland cement	Fly ash	GGBS	Silica fume
Physical data for cementitious materials				
Surface area: m^2/kg	350–500	300–600	300–500	15 000–20 000
Bulk density: kg/m^3	1300–1400	1000	1000–1200	200–300
Specific gravity	3.12	2.30	2.90	2.20
Chemical data for cementitious materials				
SiO_2	20	50	38	92
Fe_2O_3	3.5	10.4	0.3	1.2
Al_2O_3	5.0	28	11	0.7
CaO	65	3	40	0.2
MgO	0.1	2	7.5	0.2
$Na_2O + K_2O$	0.8	3.2	1.2	2.0

With the ongoing focus on the environment, research continues to improve the efficiency of the furnaces producing the silicon and ferrosilicon alloys. This may include the use of 're-burners' to remove organic materials from the fume, new arc furnace operations or different raw materials. As such, some characteristics of the silica fume produced may vary, and to maintain quality, it is advisable to study trend analyses. These can be requested from any plant that conforms to EN 13263 or ASTM C1240 and are compulsory for such compliance.

HEALTH AND SAFETY

As with all fine powders, there are potential health risks, particularly since the inhalation of silicon dioxide causes the lung disease silicosis. Studies of the material and of workers in the ferrosilicon industry [1,2] have found that it is the crystalline form of silicon dioxide that causes silicosis. Silica fume, produced as described, is amorphous and, so, non-hazardous.

The limits for the respirable dust have been set at 3 mg/m^3 (low) and 5 mg/m^3 (high), and exposure to concentrations above the high limit are not advised. Use of the slurry form will minimise the dust potential. It is always recommended that workers handling the material in either form use appropriate protective equipment (gloves, dust masks, protective glasses and such) and use procedures that minimise the generation of dust. Although spillages of either powder or slurry can be treated as 'common mess' and dealt with accordingly, an exception is when there is the potential for mixing with hydrofluoric acid. This will produce silicon tetrafluoride, which is a toxic gas.

The CAS number for both powder and slurry containing amorphous silicon dioxide is 7631-86-9. For full health and safety information, it is always advisable to contact the producers.

FORMS OF SILICA FUME

As the powder is 100 times finer than ordinary cement, there are transportation, storage and dispensing considerations to be taken into account. To accommodate some of the difficulties, the material is commercially available in various forms. The differences between these forms are in the shape and size of the particles, which do not greatly affect the chemical make-up of the material or how it reacts. These differences do, however, influence how it is used, and careful thought should be given to the type of silica fume chosen for a specific application. The main forms are as follows.

- *Undensified* with a bulk density of 200–350 kg/m^3. Due to the very low bulk density, this form is considered impractical to use for normal concrete production. The main uses are in refractory products and formulated bagged materials, such as grouts, mortars, concrete repair systems and protective coatings.
- *Densified* with a bulk density of 500–650 kg/m^3. During densification, the ultrafine particles become loosely agglomerated, making the particulate size larger. This makes the powder easier to handle, with less dust, than the undensified form. This material is used in processes with high-shear mixing facilities, such as precast works, concrete roof tile works and ready-mixed concrete plants with 'wet' mixing units.
- *Micropelletised* with a bulk density of 700–1000 kg/m^3. Micropelletisation involves forming the powder into small spheres about 0.5–1 mm in diameter. The material in this form does not readily break down in conventional concrete mixing and is best suited to intergrinding with cement clinker to produce a composite cement. Icelandic cement is made in this fashion and contains 7.5% silica fume to combat the potential ASR from the local aggregates.
- *Slurry* with a specific gravity of 1400 kg/m^3. This material is produced by mixing the undensified powder and water in equal proportions, by weight, to produce a stable slurry.

Mixing and maintaining a stable slurry require expensive hi-tech equipment and cannot be done easily. Therefore, all slurries should be obtained from one specific supplier to maintain quality. In this form, the material is easily introduced into the concrete mix. It is also the most practical form for dispensing by weight or volume. It can be used for virtually all forms of concrete, from semi-dry mixes for precast products to self-compacting concrete and is ideally suited to the ready-mixed industry.

Some major global projects have used 'instant slurry' produced on site. Densified powder is mixed with water in a high-shear, purpose-built mixer tank to form a slurry, which is then stored in a larger tank with low circulation. The powder can be conveniently shipped without the 50% water. This slurry can then be rapidly dispersed in the concrete mixing process. However, this instant slurry cannot be stored, unlike factory-produced slurry. It will settle out within a few days if not kept circulating. A nominal time for such a slurry to stay 'live' is 3 days.

5.3.4.2 Inclusion in concrete

STANDARDS AND SPECIFICATIONS

Various national standards, codes of practice and recommendations have been published since the use of silica fume has become globally accepted. The ones in major use are as follows

- USA — ASTM C1240 'Silica fume for use in concrete'
- Europe (CEN) — EN 13263.05 'Silica fume for concrete'
- Canada — CSA A3001-13 'Cementitious materials for use in concrete'
- Japan — JIS A 6207 2000 'Silica fume for use in concrete'
- Australia — AS 3582 'Supplementary cementing materials (1992)'
- China — GB/T 18736 'Mineral admixtures for high strength and high performance concrete'
- India — IS 15388 'Silica fume – specification'.

Check with the issuing body – or a supplier – to ensure that the latest version of a standard is being used.

The main standards accepted worldwide are EN 13263 parts 1 and 2 and ASTM C1240. EN 13263 requires a minimum of 6 months testing before compliance can be shown, and the quality control has to be 'advised to the customer on request' in the form of a 6-month trend analysis. Table 5.10 gives the mandatory chemical and physical requirements. Several of the standards also contain optional requirements. Where there are blanks in the table, no mandatory requirements exist. Note that the Norwegian and French standards have now been superseded by the European standard, but they have been left in for comparison.

EFFECTS ON FRESH CONCRETE

Due to the nature and size of the particles in silica fume, a small addition to a concrete mix will produce marked changes in both the physical and chemical properties.

At the typical dosage of 8–10% by cement weight, the primary physical effect is due to adding between 50 000 and 100 000 microspheres per grain of cement. This means that the mix will be suffused with fine material, thus increasing the cohesiveness of the concrete. When using a powder form of silica fume, to maintain mixability and workability, more water is required. Therefore, such powders are most often used with plasticisers or superplasticisers.

Regarding workability, note that a fresh silica fume concrete will have a lower slump than a similar ordinary concrete, due to the greater cohesion. When the mix is supplied with energy, as in pumping, vibrating or tamping, the silica fume particles, being spherical, will act as ball bearings and lubricate the mix, giving it a greater mobility than a similar ordinary concrete. Silica fume concrete is often considered to be thixotropic. Thus, when measuring the slump of a silica fume concrete, note that the figure will indicate only the consistency of the concrete and will not relate to its workability. The most favourable test for such concrete is the DIN-standard flow table, or similar, which describes how the concrete reacts to an energy input and, thus, gives a better visual appraisal of the workability of the mix.

In mixes using a slurry as an addition, there can be a slight increase in water demand to maintain a given slump. This demand is normally offset by using a standard plasticiser. Another way of negating the effect is to modify the fines in the mix, namely reducing the sand content by a small percentage.

In the ready-mix industry, a slurry is nearly always enhanced by a small amount of a plasticiser, which ensures full dispersion and maintains the w/c ratio. As the concrete is more cohesive, it is less susceptible to segregation, even at very high workabilities, such as with a flowing or self-compacting concrete. This lack of segregation also makes slurries ideal for incorporation into a high-fluidity grout.

Table 5.10 Comparison of standards – chemical and physical requirements[a]

	USA ASTM C1240	Norway[b] NS 3045	Canada CSA A3001-13	Australia AS 3582	Europe EN 13263-1	France[b] NF P 18-502 1992	Japan JIS A6207 2000
SiO_2: %	85.0	85	85	85	85	85	85
SO_3: %			1.0	3	2.0	2.5	3.0
Cl: %		Report if >0.10		Report	0.3	0.2	0.1
CaO: %		2			1.0	1.2	1.0
MgO							5.0
Si (free): %					0.4		0.4
Total alkalis						4	
Free carbon						4	
Moisture content: %	3.0			2		1	3.0
Loss on ignition: %	6.0	5	6.0	6	4.0		5.0
Specific surface: m^2/g		>12			15–35	20–35	>15
Bulk density	Report			Report			
Pozzolanic activity index:	85% after 7 days of accelerated curing with variable w/cm	95% after 28 days of normal curing with w/cm = 0.5	85% after 7 days of accelerated curing	Report	100% after 28 days of normal curing with water-to-powder ratio = 0.5		95% after 7 days or 105% after 28 days with w/cm = 0.5
Retained on 45 μm sieve: %	10		10				
Density: kg/m^3						2100–2300	Report
Autoclave expansion: %			0.2				
Canadian foaming test			No visible foam				
Notes		Characteristic values			Characteristic values. Not an official standard, in the approval process	Material type A	

[a]Several of the standards also contain optional requirements. Where there are blanks in the table, no mandatory requirements exist.
[b]Norwegian and French standards have now been superseded by the European standard, but they have been left in for comparison.

Because the particles are ultrafine, there is a much greater contact surface area, which improves the bonding between the fresh concrete and the substrate or reinforcement.

Another aspect of this non-segregation and the filling in of the major voids in fresh concrete is that a silica fume concrete will produce virtually no bleed water. The concrete must, therefore, be cured, in accordance with good site practice, as soon as it has been placed, compacted and finished. The lack of bleed water means that processes such as power-floating can be commenced much sooner than with ordinary concretes.

However, the lower slump, lack of bleed water and 'gelling' (stiffening when not agitated) does not indicate that it sets rapidly. Silica fume is a pozzolana and, therefore, calcium hydroxide is required to activate it. Calcium hydroxide is produced when the cement hydrates, and thus, the silica fume can work only after the cement has started reacting. The setting times for silica fume concretes are generally like those of ordinary concretes, except when they are specifically designed for ultra-high strength or low heat.

MIX DESIGN CRITERIA

Most producers of silica fume consider that the powder forms are best suited to specialised production or mixing facilities or those precasters and ready-mix producers using high-power, forced-action mixers. This is inherently due to the water demand of the powders and the subsequent difficulty in achieving dispersion through a drier mix.

In most ready-mix or precast operations, it is normal for a slurry to be added at the same time as the normal water addition. Although a slurry disperses more readily, care is needed to ensure the mix is uniform before applying the concrete.

There are variations of dosage for given types of application, which serve as guidelines for initial trial mixes. Note that, even with the precedent of past work, a trial mix should always be prepared before acceptance of a mix design.

As silica fume is most often used to produce high-performance concrete, the dosage is always given as an addition to the total cementitious content of the mix. The following are often used as starting points

- pumping aid: 2–4%
- normal concretes: 4–7%
- high strength: 8–10% (higher values may be used in special circumstances)
- high chemical resistance: 10–12% (higher values may be used in special circumstances)
- underwater concretes: 10–15%.

The effect of having a fine filler due to adding a material with a bulk density one-fifth of that of the cement will nearly always require a modification to the mix constituents, particularly the coarse-to-fine aggregate ratio, to achieve the optimum rheology. Particle packing software can find the best balance of coarse-to-fine products when blending the mix, thus leading to the optimum mix rheology. In nearly all ready-mix production, a plasticiser – or, more often, a superplasticiser – is used to give optimum dispersion while maintaining the w/c ratio.

In some countries, silica fume is added as a large-scale cement replacement across the board in ready-mix production, but this practice is decreasing as quality control restrictions become tighter.

Mixing times need to be adjusted to allow for the maximum dispersion of the silica fume. This is most important when using any of the powder forms to prevent agglomeration within the mix. When used in a ready-mix environment, the slurry form does not need excessive mixing times, and the 'mass action' of a truck mixer has been known to produce better results than laboratory trials for the same mix.

Like most cementitious materials, silica fume functions more efficiently with some types of chemical admixtures than with others, and trial work is advised. Most of the producers of silica fume have basic mix designs for various situations and aggregate types.

EFFECTS ON THE SETTING AND HARDENING OF CONCRETE

As the concrete sets and hardens, the pozzolanic action of the silica fume becomes more important than the physical effects. The silica fume reacts with the liberated calcium hydroxide to produce calcium silicate hydrates. These both increase the strength and reduce the permeability by densifying the matrix of the concrete.

Silica fume, having a greater surface area and higher silicon dioxide content, is much more reactive than fly ash or GGBS [3]. This increased reactivity initially appears to push the hydration of the C_3S fraction of the cement [4], thus creating more calcium hydroxide, but settles down to more normal rates beyond 2 days.

The high reactivity and consumption of the calcium hydroxide has prompted questions relating to the pH levels of the concrete and the related effects on steel passivity and carbonation rates. Studies have shown that the effect on the carbonation rate depends greatly on the quality of the mix produced. Good-quality well-proportioned silica fume concrete does not exhibit any greater carbonation than an OPC concrete. The reduction of pH in a concrete mix is usually from around 14.0 to 13.0–12.0, which is well above the level for steel passivity. It has been established that a 25% addition of silica fume [4] is required to use up all the calcium hydroxide produced in a concrete. Studies have shown that the pH still does not drop below 12.0. In normal practice, the highest dosage advised for concrete is 15%, which should have no deleterious effects.

This reduction in the alkalinity and the binding of the K^+ and Na^+ ions in the pore solution [5] is one of the ways in which the addition of silica fume decreases the risk of ASR in concrete [6].

As the silica fume reacts and produces calcium silicate hydrates, the voids and pores within the concrete become filled because the crystals that form bridge the gaps between cement grains and aggregate particles. Coupled with the physical filling effect, the matrix of the concrete becomes very homogeneous and dense, improving strength and impermeability [7]. The relatively porous section that surrounds the aggregate grains in normal concrete – the transition zone – is virtually absent in high-quality silica fume concrete.

The measure of reactivity of silica fume, namely its cementing efficiency, is between four and five times that of OPC [8]. This implies that a concrete would still achieve the required strength if a large amount of Portland cement were to be replaced by a small dosage of silica fume. Although this is possible, it is not considered to be 'ethical'. Reducing the cementitious content to very low levels, while still achieving the required strength, will have adverse effects on the durability of the concrete, despite the benefits imparted by incorporating the silica fume.

Even though the rate of reaction is very high in the initial stages, not all the silica fume will be used up, as studies have shown that unreacted material is still present at later ages [9].

In general, the heat evolution of a silica fume mix depends on the design. If high early strength is needed, addition of silica fume to a high-cement content mix will produce a higher heat of hydration in the initial stages, lessening as the reactivity slows. Additions to 'normal' concretes do not usually produce significant changes in heat evolution.

Silica fume concrete is very susceptible to temperature variations while hardening. The rate of strength gain can be reduced with low temperatures, below the 20°C optimum, and accelerated with increased temperatures [10]. This relates to concrete and not to some of the specialised repair materials available, which can be used at very low temperatures.

BLENDED CEMENT MIXES

While silica fume is compatible with both fly ash and GGBS, it is a pozzolanic material and, hence, will give differing results depending on the mix designs used. High fly ash percentages are affected by the ability of the silica fume to rapidly take up calcium hydroxide. This may give good early strengths but a reduced rate of long-term gain. GGBS blends are less affected because of the latent hydraulic nature of the GGBS. When high replacement levels of GGBS are used (50–70%), silica fume can be added to improve the early-age strength, as it reacts faster in the first 3 days, or to improve the consistency of the fresh concrete. High levels of GGBS can cause problems with high water contents, leading to segregation and bleeding, not only on the surface of the concrete but also within the matrix itself. Silica fume can virtually eliminate this bleeding and, hence, maintain the integrity of the concrete. With more standard levels of fly ash and GGBS (25–40%), silica fume can be added to give enhanced performance. In such cases where there would be a minor reduction in strength due to using these materials, this can be offset by the silica fume, so that high early and ultimate strengths can be achieved without an excessive increase in the cost of the concrete. These triple-blend cements employ the beneficial characteristics of both pozzolanic materials in producing an optimally durable concrete. This type of concrete is being specified where concrete structures are expected to last for upwards of 100 years, such as the Storebælt in Denmark, the Tsing Ma Bridge in Hong Kong and the Burj Khalifa (the world's tallest building in 2023). As previously stated, caution must be exercised and full trials conducted. Although silica fume can enhance concretes with high replacement levels of fly ash or GGBS, none of these materials can react properly without there being sufficient cement in the mix to produce free lime.

There is always a maximum level when replacing cement with pozzolanas, and the target should always be the ultimate quality of the concrete, not just the required compressive strength.

5.3.4.3 Hardened concrete: Mechanical properties

COMPRESSIVE STRENGTH

Generally, the first property associated with silica fume concrete is its high compressive strength. Many reports [11–13] show that adding silica fume to a concrete mix increases the strength of that mix by between 30% and 100%, depending on the type of mix, the type of cement, the amount of silica fume, the use of plasticisers, aggregate types and curing regimes. In general, to achieve the same strength, the w/c ratio must be increased. For a set w/c ratio, the strength is higher [13] (Figure 5.22).

Silica fume concrete undergoes a marked reduction in strength gain when subjected to early-age drying [14], which can be as much as 20%. Combinations of silica fume and fly ash do not seem to be subject to this reduction [15].

With correct design, ultra-high-strength concrete can be produced using normal ready-mix facilities. In the USA, it is common to use 100–130 MPa concrete in tall buildings, and 83 MPa silica fume concrete was used to build the 79-storey office block at 311 South Wacker Drive in Chicago. Silica fume concrete of 100 MPa was used in the Petronas Towers in Kuala Lumpur. Concrete of 100 MPa was used in a triple-blend format for the Burj Khalifa, and 150 MPa has now been used in various towers in Japan.

Figure 5.22 Strength at 28 days plotted against w/c ratio

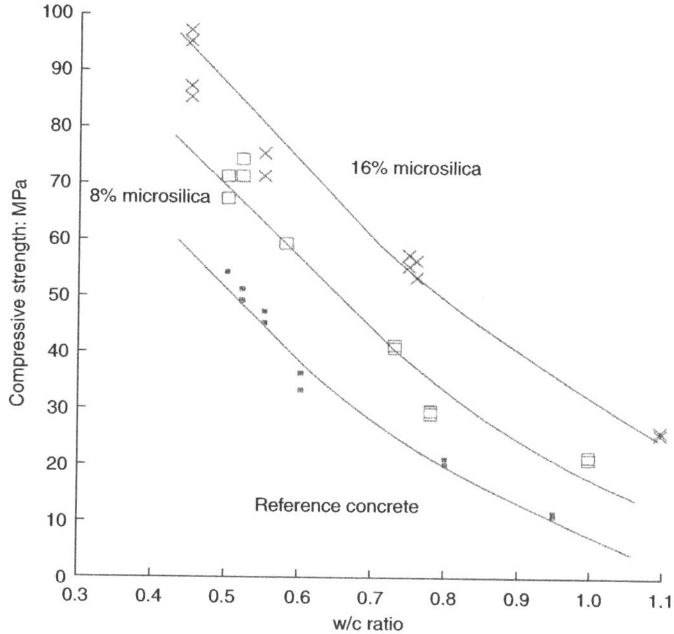

TENSILE AND FLEXURAL STRENGTHS

The relation between tensile, flexural and compressive strengths in silica fume concrete is very much like that of ordinary concrete. An increase in the compressive strength using silica fume results in a similar relative increase in the tensile and flexural strengths. This is important when silica fume concrete is used in flooring, bridging or roadway projects. The increased tensile strength allows for a possible reduction in the slab thickness while maintaining the high compressive strength, and thus, a reduction in the overall slab weight and cost. Curing has a very high influence on these particular qualities. Dry curing at early ages significantly reduces the tensile strength [14].

BRITTLENESS AND ELASTIC MODULUS

The stronger a concrete is, the more brittle it becomes, and silica fume concrete is no exception to this rule. However, the strengths concerned are the high to ultra-high values, 150–250 MPa and over, and not the 50–100 MPa concrete used for most projects [16,17]. The elastic modulus does not follow the same trend as the tensile strength and shows only a slight increase in comparison to the compressive strength. Therefore, high- and ultra-high-strength concretes can be used for tall structures without loss of ductility [18].

BONDING

Silica fume concrete has a much finer paste phase, and so bonds better with substrates, old concrete, reinforcement, fibres and aggregates. Investigations have shown [19,20] that the aggregate/cement interface is altered when silica fume is present, and pull-out tests [21,22] show improved strength. Bonding to fibres is greatly improved [22–24]. This is particularly beneficial in steel fibre/silica fume-modified shotcrete. It removes the need for mesh reinforcement, which would have to be fixed to the substrate. The high reactivity and extreme cohesiveness of silica fume shotcrete and gunite also reduces the need for an accelerator. Such use of silica fume has resulted in rebound figures of less than 5%.

SHRINKAGE

Cement pastes shrink more when they contain silica fume [25]. So, when paste, mortar or grout systems are used, a compensator is usually added.

In concretes, however, the shrinkage is related to the aggregate volume and aggregate quality. Many reports [25,26] show that the addition of silica fume reduces shrinkage in concrete when close control is exercised over the mix design and aggregate selection. Ensuring that the grout matrix is distributed evenly across and around the aggregates reduces shrinkage, as there are no large pockets of grout that shrink. Most concretes are made with too little aggregate and too much paste. Curing is, again, stressed in these papers, due to the marked effect that curing can have on this type of concrete.

CREEP

There is little information about the effect of silica fume on the creep of concrete. What is available refers to high-strength high-dosage mixes. In general, it appears that silica fume concrete exhibits less creep in comparison with ordinary concrete [26,27].

FIRE RESISTANCE

Fire resistance is not often considered for normal applications of concrete, though it is still of great importance. High-strength concrete can explode when exposed to fire. Several tests [28–30] have shown that under normal fire conditions, silica fume concrete does not behave any differently to normal concretes. Ultra-high-strength silica fume concrete may be susceptible to this type of failure due to increased brittleness, and consideration should be given to the mix design (i.e. minimising the free water content). High-performance silica fume concrete, designed with optimum particle packing and the inclusion of poly-fibres, has been shown to be very effective at resisting damage, even in high-carbon fires, such as those with burning petrol or similar fuel in a tunnel or storage facility. Report 216 published by the American Concrete Institute (ACI) is a good source for data on this aspect of silica fume concrete.

ABRASION AND EROSION

Low w/c ratio high-strength silica fume concrete has a very high resistance to abrasion and erosion. A large amount of silica fume concrete has been produced to specifically utilise this quality. A major repair project on the Kinzua Dam, USA, has been studied for over 40 years, and results show good performance of the concrete used [31]. A high number of hydropower projects in India have utilised silica fume concrete for this performance.

5.3.4.4 Hardened concrete: Durability properties

Adding silica fume to concrete, with its potential for greater strengths, both compressive and tensile, its more refined pore structure and its lower permeability, can increase durability, leading to a longer working lifespan than a conventional concrete in the same environment.

PERMEABILITY

The two main methodologies of measuring permeability are either static, such as allowing a concrete to dry out and noting the weight loss, or active, by subjecting the material to a liquid or gas under pressure, and measuring the depth of penetration.

In studies of drying methods [16,32], the efficiency factor for silica fume concrete was between 6 and 8. This indicates that the physical size and high reactivity of the particles in silica fume have more influence on the permeability than on the compressive strength.

In active tests of the permeability with water under pressure, early tests [32] showed that silica fume concrete ranged from having very little permeability to being impermeable. Since these tests in the 1960s, several comprehensive evaluations [33–35] have confirmed the previous results. Examinations of mature specimens [36] showed that they had reduced permeability. A microscopic study revealed that they had a very dense microstructure and that the weak layer normally surrounding the aggregate grains was virtually absent. In all these tests, close attention was paid to the curing regimes, finding that the curing time had a very marked influence on the results.

Permeability should not be confused with porosity, as the pore structure is modified but not decreased [16].

The permeability of concretes, particularly to chemicals such as chlorides and sulfates, is a great concern around the world regarding durability. Research has been ongoing, and results are frequently added to the literature.

SULFATE RESISTANCE

A major study was initiated in Oslo when silica fume concretes were first being tested. Various specimens were buried in acidic sulfate-rich ground in Oslo. Results after 20 years of burial are available [37,38]. These results indicate that the silica fume concretes performed as well as sulfate-resistant ones, which was confirmed in the 40-year report and further laboratory tests [39,40].

When used with GGBS or PFA [41], silica fume mixes had a greater resistance to sulfate attacks than special sulfate-resistant cements. Thus, silica fume concrete has been specified in places such as the Arabian Gulf to prevent severe deterioration of the concrete [42].

This sulfate resistance of the concrete can be attributed to

- the refined pore structure and, thus, the reduced passage of harmful ions [43]
- the increased amount of aluminium incorporated in the silica fume, thus reducing the amount of alumina available for ettringite formation
- the lower calcium hydroxide content.

CHLORIDE RESISTANCE

For chloride attacks against the reinforcement, it is not always necessary for chlorides to penetrate the concrete, as some chlorides may already be present in the aggregates or even in the mixing water. Thus, the ability of concrete to withstand chloride penetration, from seawater or de-icing salts, is important, as is the reactivity of the mix in binding up the aggressive fraction of the chlorides that are present in the pore water. Studies [5,44–46] discuss the varying effects of the lower permeability and the reduction of pH in the pore water and how this relates to the presence of chlorides and the passivity of the embedded steel.

It is considered that the lower the pH, then the lower the threshold for depassivation of the steel by chlorides. Research has shown that the chloride-penetration coefficient is much lower in silica fume concretes [47], thus negating the effect of the lower threshold. In general, for equivalent strengths, the initiation of chloride attack will be delayed in a concrete containing silica fume. The initial curing and correct amount of cover are still important, as reducing the penetration rate is not effective if the distance to be travelled is also reduced.

In the USA, a special rapid test – ASTM C1202 – has been applied to compare the chloride penetration of high-strength silica fume concrete with latex-modified and low w/c ratio concrete [48]. The test is based on the resistivity of the concrete to the passage of an electrical charge. Properly designed, produced and finished silica fume concrete normally achieves a very low rating.

Continuing developments in testing chloride penetration – such as the rapid migration test and NT Build 443 'Chloride penetration' – have given a variety of options for looking at this durability factor. Some specifications for high-durability concretes often require a series of such tests for water and chloride resistance. In all these tests, properly designed and produced silica fume concrete has a very high resistance to attack.

CARBONATION

Results for tests on carbonation and carbonation rates are somewhat varied and contradictory depending on the viewpoint taken when analysing the findings. A study into the effect of silica fume on carbonation and the transport of oxygen [49] found that adding up to 20% silica fume caused a slight reduction in both of these in water-saturated concrete. In essence, the reports [14,49,50] conclude that for equal strengths of any concrete below 40 MPa, carbonation is higher in silica fume concrete. Concretes above 40 MPa had a lower carbonation rate, and it is only these concretes that are deemed susceptible to attack and damage if there is reinforcement. As silica fume concrete normally has an average strength above 40 MPa, it is a moot point as to whether carbonation is a serious risk. Correct curing procedures are essential to ensure optimum performance of silica fume concrete.

LEACHING AND EFFLORESCENCE

Leaching and efflorescence mainly occur when one surface of a concrete structure is subjected to either continuous water contact or intermittent wetting and drying. The excess calcium hydroxide is leached through the concrete to the surface, where it carbonates to leave a white powdery deposit. Although a problem of aesthetics more than anything, if sufficient leaching occurs, porosity and permeability will be increased, leading ultimately to a weaker concrete. Adding silica fume can reduce leaching [51] because of the refined pore structure and the increased consumption of calcium hydroxide. Research indicates that the more efficient the curing and the longer curing time before exposure, the more resistant the concrete becomes.

FROST RESISTANCE

As the major producers and users in the early years were the Scandinavian countries, the frost resistance of silica fume concrete has been well scrutinised. Investigators have used the silica fume on its own, with superplasticisers, with varying dosages of air entrainers and with different aggregates and using various curing regimes.

Air is entrained into concrete to increase the frost resistance. However, it is necessary to entrain the correct amount of air with the right dispersion of bubbles so that the mix is stable enough to withstand compaction and will reach the required strength.

The effect of silica fume addition has been compared for many different concretes [52,53]. It is difficult to entrain air into a silica fume mix without using a plasticiser. However, by increasing the dosage of air entrainer and adding a plasticiser, it is easy to achieve the desired level of air. There is some speculation as to the reason for the increased dosage of air entrainer, the most likely being that the air and the silica fume compete for the same space in the mix. If the silica fume is fully dispersed in the mix, the entrained air has a greater stability than in control mixes without the air entrainer. The variations in air content for given dosages, as sometimes happens when using fly ash, were not noticed for silica fume concretes.

The combination of silica fume and air entrainment seems to be the best option for frost protection. The silica fume maintains good stability and there is uniform spacing of the air bubbles. For a mix design guideline for a concrete of 30–50 MPa, then 8% silica fume and 5% air entrainment are optimum. In all cases, the concrete should be cured for the longest allowable time before exposure to the working environment.

ALKALI–SILICA REACTION

To proceed into a lengthy discourse on this particular problem, its causes and its prevalence would be inappropriate here, and the subject is well covered in Section 4.4.6 of this book as well as in the general literature. To understand the effect of silica fume on this form of attack, it is necessary to remember the three main factors that are required for a potential reaction

- a high alkali content in the mix
- reactive aggregates
- available water.

Silica fume reacts with the liberated calcium hydroxide to form calcium silicate hydrates, and this reduction leads to a lowering of the pH and a lower risk of reaction due to the high alkali content. In the formation of the calcium silicate hydrates, the K^+ and Na^+ ions are bound into the matrix and cannot react with any potentially siliceous aggregates.

Moreover, the minute size and the pozzolanic reactivity of the silica fume greatly refine the pore structure of the concrete, reducing the permeability, which means that less water can enter the concrete.

Thus, a normal dosage of silica fume, 10% by cement weight, can negate the main factors that lead to ASR in concrete. Many reports [54–58] document numerous studies on this.

5.3.4.5 Summary

The addition of silica fume to concrete produces significant changes in the structure of the matrix, giving a densified, refined pore system and greater strength through both physical effects and the pozzolanic reaction. In most cases, it is the refinement of the pore system, which reduces the permeability, that has the greatest effect on the performance of the concrete, rather than the increased strength. Usage can be made of these advantages when designing a concrete to comply with onerous requirements or to provide greater resistance to certain hostile environments. Note that silica fume should be considered as an addition to a mix rather than a replacement for cementitious content. A sensible mix design is essential.

Silica fume concrete is susceptible to poor curing conditions, the effects of which are more pronounced than in ordinary concrete. Close attention to curing methods and times is important to ensure optimum performance.

Tailoring the design of a silica fume concrete to the specific requirements should always be a matter of consultation between the client, the contractor, the ready-mix or precast supplier and the technical section of the silica fume supplier. Reference should be made to any previous project or use, and trial work is 'compulsory' to ensure correct usage of the material and an appreciation of the characteristics of this type of concrete.

Acknowledgements

Background information for this chapter, as well as that released by Elkem Materials and Ferroglobe Plc, was obtained from the following sources.

- FIP (1988) *Condensed Silica Fume in Concrete*. Thomas Telford, London.
- Concrete Society (1993) *Microsilica in Concrete*. Concrete Society, Camberley, UK, Technical report 41.
- Fidjestøl P and Lewis R (2019) Microsilica as an addition. In *Lea's Chemistry of Cement and Concrete*, 4th edn (P. Hewlett (ed.)). Butterworth-Heinemann, Oxford, Ch. 12.

FURTHER READING

- ACI (2006) *Guide for the Use of Silica Fume.* American Concrete Institute, Farmington Hills, MI, USA, Report 234R-06.
- Elkem Materials, www.elkem.com.
- Ferroglobe Plc, www.ferroglobe.com.
- Silica Fume Association, www.silicafume.org.
- Hewlett PC and Liska M (eds) (2019) *Lea's Chemistry of Cement and Concrete*, 5th edn. Butterworth-Heinemann, Elsevier.
- Christensen P (1982) *Afprovningaf kvalitetskarakterisering af beton tilsat silica.* Teknologisk Institut, Byggeteknik, Copenhagen, Denmark. (In Danish.)
- Carette GG and Malhotra VM (1983) Mechanical properties, durability and drying shrinkage of Portland cement concrete incorporating silica fume. *Cement, Concrete and Aggregates* **5(1)**: 3–13.
- Huang C-y and Feldman RF (1985) Dependence of frost resistance on the pore structure of mortar containing silica fume. *Journal of the American Concrete Institute* **82(5)**: 740–743.
- Feldman RF and Huang C-y (1984) Microstructural properties of blended cement mortars and their relation to durability. *RILEM Seminar on Durability of Concrete Structures under Normal Outdoor Exposure, Hanover.*
- Virtanen J (1985) *Mineral By-products and Freeze–Thaw Resistance of Concrete.* Dansk Betonforening, Copenhagen, Denmark, Publication 22:85.
- Hooton RD (1987) Some aspects of durability with condensed silica fume in pastes, mortars and concretes. *International Workshop on Condensed Silica Fume in Concrete, Montreal.*
- Malhotra VM, Painter KA and Bilodeau A (1987) Mechanical properties and freezing and thawing resistance of high strength concrete incorporating silica fume. *International Workshop on Condensed Silica Fume in Concrete, Montreal.*

REFERENCES

1. Aïtcin P-C and Regourd, M. (1985) The use of condensed silica fume to control alkali silica reaction – A field case study. *Cement and Concrete Research* **15(4)**: 711–719.
2. Jahr J (1980) *Possible Health Hazards from Different Types of Amorphous Silicas – Suggested Threshold Limit Values.* Institute of Occupational Health, Oslo, Norway, Revised report HD806/79.
3. Regourd M (1983) Pozzolanic reactivity of condensed silica fume. In *Condensed Silica Fume* (Aïtcin P-C (ed.)). Editions de l' Université de Sherbrooke, Quebec, Canada, pp. 20–24.
4. Andrija D (1986) Hydration of alite and C_3A and changes of some structural characteristics of cement paste by the addition of silica fume. *Proceedings of the 8th International Congress on Chemistry of Cement, Rio de Janeiro*, vol. 4, pp. 279–285.
5. Page CL and Vennesland O (1983) Pore solution composition and chloride binding capacity of silica fume cement pastes. *Materials and Structures* **16(1)**: 19–25.
6. Parker DG (1986) *Alkali Aggregate Reactivity and Condensed Silica Fume.* Elkem Materials, Kristiansand, Norway, Report SA 854/3C.
7. Diamond S (1986) The microstructures of cement paste in concrete. *Proceedings of the 8th International Congress on Chemistry of Cement, Rio de Janeiro*, vol. 1, pp. 122–147.
8. Page CL (1983) *Influence of Microsilica on Compressive Strength of Concrete Made from British Cement and Aggregates.* Elkem Materials, Kristiansand, Norway.
9. Li S, Roy DM and Kumar A (1985) Quantitative determination of pozzolanas in hydrated systems of cement or $Ca(OH)_2$ with fly ash or silica fume. *Cement and Concrete Research* **15(6)**: 1079–1086.
10. Sandvik M (1981) *Fasthetsutvikling for silicabetong ved ulike tempera-turniva.* Norwegian Institute of Technology, Trondheim, Norway, Report STF65 F81016. (In Norwegian.)
11. Loland KE (1983) Fasthets-og deformasjonsegenskaper i herdnet tilstand – herdebetingelser. *Bruk av silika i betong, Norsk Sivilingeniorers Forening* (seminar), *Oslo.* (In Norwegian.)

12. Loland KE and Hustad T (1981) *Report 2: Mechanical Properties*. Norwegian Institute of Technology, Trondheim, Norway, Report STF65 A81031.
13. Sellevold EJ and Radjy FF (1983) *Condensed Silica Fume (Microsilica) in Concrete: Water Demand and Strength Development*, American Concrete Institute, Farmington Hills, MI, USA, Report SP-79.
14. Johansen R (1981) *Report 6: Long Term Effects*. Norwegian Institute of Technology, Trondheim, Norway, Report STF65 A81031.
15. Maage M and Hammer TA (1985) *Modifisert Portlandsement. Delrapport 3, Fasthetsutvikling og E-modul*. Norwegian Institute of Technology, Trondheim, Oslo, Report STF65 A85041. (In Norwegian.)
16. Sellevold EJ, Bager DH, Klitgaaxd Jensen E and Knudsen T (1982) *Silica Fume-Cement Pastes: Hydration and Pore Structure*. Norwegian Institute of Technology, Trondheim, Norway, Report BML 82.610.
17. Justesen CF (1981) *Performance of Density Injection Grout for Prestressing Tendons*. Technical Advisory Service, Aalborg Portland, Aalborg, Denmark.
18. de Larrard F, Boulay C and Rossi P (1987) Fracture toughness of high strength concrete. *Proceedings of the Symposium on the Utilisation of High Strength Concrete, Stavanger*, pp. 215–223.
19. Carles-Gibergues A, Grandet J and Ollivier JP (1982) Contact zone between cement paste and aggregate. *Proceedings of the International Conference on Bond in Concrete* (Bartos P (ed.)). Applied Science, London, UK, pp. 24–33.
20. Gjorv OE, Monteiro PJM and Mehta PK (1986) *Effect of Condensed Silica Fume on the Steel–Concrete Bond*. Norwegian Institute of Technology, Trondheim, Norway, Report BML 86.201.
21. Monteiro PJ, Gjorv OE and Mehta PK (1986) *Effect of Condensed Silica Fume on the Steel–Cement Paste Transition Zone*. Norwegian Institute of Technology, Trondheim, Norway, Report BML 86.205.
22. Bache HH (1981) Densified cement/ultrafine particle-based materials. *2nd International Conference on Superplasticisers in Concrete, Ottawa*.
23. Krenchel H and Shah S (1985) Applications of polypropylene fibres in Scandinavia. *Concrete International: Design and Construction* **7(3)**: 32–34.
24. Ramakrishnan V and Srinivasan V (1983) Performance characteristics of fibre reinforced condensed silica fume concrete. *American Concrete Institute Special Publication* **79**: 797–812.
25. Traetteberg A and Alstad R (1981) *Volumstabilitet i blandingssementer med rajernslagg og silikastov*. Norwegian Institute of Technology, Trondheim, Norway, Report STF65 A81034. (In Norwegian.)
26. Johansen R (1979) *Silicastov i fabrikksbetong. Langtidseffekter*. Norwegian Institute of Technology, Trondheim, Norway, Report STF65 F79019. (In Norwegian.)
27. Buil M and Acker P (1985) Creep of silica fume concrete. *Cement and Concrete Research* **15**: 463–466.
28. Wolsiefer J (1982) Ultra high strength field placeable concrete in the range 10,000 to 18,000 psi (69 to 124 MPa). *American Concrete Institute Annual Conference, Atlanta*.
29. Maage M and Rueslatten H (1987) *Trykkfasthet og blaeredannelse pa brannpakjent hoyfastbetong*. Norwegian Institute of Technology, Trondheim, Norway, Report STF65 A87006. (In Norwegian.)
30. Shirley ST, Burg RG and Fiorato AE (1988) Fire endurance of high strength concrete slabs. *ACI Materials Journal* **85**: 102–108.
31. Holland TC (1983) *Abrasion–Erosion Evaluation of Concrete Admixtures for Stilling Basin Repairs, Kinzua Dam, Pennsylvania*. US Army Engineer Waterways Experiment Station, Structures Laboratory, Vicksburg, MS, USA, Miscellaneous paper SL-83-16.
32. Sorensen EV (1982) *Concrete with Condensed Silica Fume. A Preliminary Study of Strength and Permeability*. Norwegian Institute of Technology, Trondheim, Norway, Report BML 82.610. (In Danish.)
33. Markestad SA (1977) *An Investigation of Concrete in Regard to Permeability Problems and Factors Influencing the Results of Permeability Tests*. Norwegian Institute of Technology, Trondheim, Norway, Report STF65 A77027.

34. Hustad T and Loland KE (1981) *Report 4: Permeability*. Norwegian Institute of Technology, Trondheim, Norway, Report STF65 A81031.
35. Sandvik M (1983) *Silicabetong: herdevarme, egenkapsutvikling*. Norwegian Institute of Technology, Trondheim, Norway, Report STF65 A83063. (In Norwegian.)
36. Skurdal S (1982) *Egenskapsutvikling for silicabetong ved forskjellige herdetemperaturer*. Norwegian Institute of Technology, Trondheim, Norway, Report BML 82.416. (In Norwegian.)
37. Maage M (1984) *Effect of Microsilica on the Durability of Concrete Structures*. Norwegian Institute of Technology, Trondheim, Norway, Report STF65 A84019.
38. Fiskaa O, *et al.* (1971) *Betong i Alunskifier*. Norwegian Geotechnical Institute, Oslo, Norway, Publication 86. (In Norwegian.)
39. Fiskaa OM (1973) *Betong i Alkunskifier*. Norwegian Geotechnical Institute, Oslo, Norway, Publication 101. (In Norwegian.)
40. Mather K (1980) Factors affecting the sulphate resistance of mortars. *Proceedings of the 7th International Conference on Chemistry of Cements, Paris*, vol. 4, pp. 580–585.
41. Carlsen R and Vennesland O (1982) *Sementers sulfat- og sjovannsbestan-dighet*. Norwegian Institute of Technology, Trondheim, Norway, Report STF65 F82010. (In Norwegian.)
42. Dakhil FH, Al-Gahtani AS, Al-Saadoun SS, Bader MA and Medallah KY (1987) Proposal for a code of practice to ensure durability of concrete construction in the Arabian Gulf environment. *Proceedings of the 2nd International Conference on Deterioration and Repair of Reinforced Concrete in the Arabian Gulf, Bahrain.*
43. Popovic K, Ukraincik V and Djurekovic A (1984) Improvement of mortar and concrete durability by the use of condensed silica fume. *Durability of Building Materials* **2(2)**: 171–186.
44. Mehta PK (1981) Sulphate resistance of blended Portland cements containing pozzolans and granulated blastfurnace slag. *Proceedings of the 5th International Symposium on Concrete Technology, Monterey, CA.*
45. Page CL and Havdahl J (1985) Electrochemical monitoring of corrosion of steel in microsilica cement pastes. *Materials and Structures* **18(103)**: 41–47.
46. Monteiro PJM, Gjorv OE and Mehta PK (1985) Microstructure of the steel–cement paste interface in the presence of chloride. *Cement and Concrete Research* **15(5)**: 781–784.
47. Fischer KP, *et al.* (1982) *Corrosion of Steel in Concrete: Some Fundamental Aspects of Concrete with Added Silica*. Norwegian Geotechnical Institute, Oslo, Norway, Report 51304–06.
48. Christensen DW, Sorensen EV and Radjy FF (1984) Rockbond: A new microsilica concrete bridge deck overlay material. *Proceedings of the International Bridge Conference, Pittsburgh, PA*, pp. 151–160.
49. Vennesland O and Gjorv OE (1983) *Silica concrete – Protection against corrosion of embedded steel.* American Concrete Institute, Farmington Hills, MI, USA, Report SP-79.
50. Vennesland O (1981) *Report 3: Corrosion Properties*. Norwegian Institute of Technology, Trondheim, Norway, Report STF65 A81031.
51. Samuelsson P (1982) *The Influence of Silica Fume on the Risk of Efflorescence on Concrete Surfaces.* Norwegian Institute of Technology, Trondheim, Norway, Report BML 82.610. (In Swedish.)
52. Okkenhaug K and Gjorv OE (1982) *Influence of Condensed Silica Fume on the Air-Void System in Concrete*. Norwegian Institute of Technology, Trondheim, Norway, Report STF65 A82044.
53. Okkenhaug K (1983) *Silikastovets innvirkning pa luftens stabilitet i betong med L-stoff og med L-stoff i kombinasjon med P-stoff.* Swedish Cement and Concrete Research Institute at the Institute of Technology, Stockholm, Sweden, CBI Report 2:83. (In Norwegian.)
54. Asgeirsson H and Gudmundsson G (1979) Pozzolanic activity of silica dust. *Cement and Concrete Research* **9(2)**: 249–252.
55. Parker DG (1986) *Microsilica Concrete. Part 2: In Use*. Concrete Society, Camberley, UK, Current practice sheet 110.
56. Perry C and Gillott JE (1985) The feasibility of using silica fume to control concrete expansion due to alkali aggregate reaction. *Durability of Building Materials* **3**: 133–146.

57. Glasser FP and Marr J (1984) The effect of mineral additives on the composition of cement pore fluids. *Proceedings of the British Ceramic Society* **35**: 419–429.
58. Sellevold EJ, Bager DH, Jensen EK and Knudsen T (1982) *Silica Fume–Cement Pastes: Hydration and Pore Structure*. Norwegian Institute of Technology, Trondheim, Norway, Report BML 82.610.

5.3.5 Cement–MK materials

Asia Shvarzman, Antex Western, Winnipeg, Canada
Alexander Mezhov, Technion – Israel Institute of Technology, Israel
Konstantin Kovler, Technion – Israel Institute of Technology, Israel

5.3.5.1 Introduction

The use of calcined clays as pozzolanic admixtures for enhancing the mechanical properties and durability of cementitious materials has been known since the time of the Romans. Metakaolin (MK), $Al_2Si_2O_7$, is an amorphous product of kaolinite dehydration, which exhibits strong pozzolanic activity. MK is not a by-product, unlike the most often used mineral admixtures such as silica fume and fly ash.

This section is an updated version of the review chapter in the previous version of this book [1]. During the decade or so since first publication, interest in MK as a highly active and effective pozzolana for partial replacement of the cement in concrete has grown and the number of publications on cement–MK materials has increased significantly.

5.3.5.2 Production of MK

The raw material used in the production of MK is kaolinite clay. Kaolinite, $Al_2Si_2O_5(OH)_4$, is a dioctahedral 1:1 layer silicate mineral with well-formed six-sided flakes, frequently with a prominent elongation in only one direction (Figure 5.23). Kaolinite is derived from the weathering of aluminosilicate minerals and named after Kao-ling in China.

Figure 5.23 High-resolution SEM micrograph of kaolinite

Different methods have been tried to enhance the potential reactivity of natural pozzolanas (clay minerals and zeolites) and to increase the strength and durability of pozzolana-containing cement and concrete. Among these methods are calcination, acid treatment, mechano-chemical treatment, hydrothermal treatment, elevated temperature curing of pozzolana-containing cementitious materials, and the addition of alkalis to pozzolana-containing cement paste and concrete [2–4].

THERMAL METHOD

Clay minerals are mostly crystalline aluminosilicates and include some accessory minerals. They have limited pozzolanic activity in their original state, but their properties can be enhanced by thermal treatment. The thermal properties of clay minerals are largely determined by the crystal structure and chemical composition of their essential constituents and by the nature and amount of accessory, associated minerals. The exact selection of calcination temperature is essential, as the reactivity of the calcined product increases with increasing calcination temperature to an optimum and declines again as the temperature increases further. Under heating, three main processes – dehydration, dehydroxylation and recrystallisation – occur, depending on the temperature

- low temperatures, below ~400°C: dehydration
- intermediate temperatures of 400–750°C: dehydroxylation
- high-temperature reactions above 750°C: resulting mostly in the recrystallisation of new phases.

The thermal reactions of kaolinite are as follows

$$\text{Kaolinite} \rightarrow \text{MK}(\sim 750^\circ\text{C}) \rightarrow \text{Spinel}(\sim 900^\circ\text{C}) \rightarrow \text{Mullite}(\sim 1000^\circ\text{C})$$

The optimum calcination temperature for maximising the strength of portlandite–MK paste varies slightly. For example, Ambroise *et al.* [5] determined the effect of kaolinite calcining temperature (600–800°C) on the strength development of 1:1 portlandite–MK pastes with a water-to-solid ratio of 0.72 for different curing times and found that the optimal calcination temperature was 700°C [6–8] . Fialips *et al.* [3] and Odler [4] found that the optimum calcination temperature of kaolinite was 650°C.

On heating, kaolinite undergoes both chemical and structural transformations. The thermal transformations of kaolinite have been the subject of many investigations, which have showed that heating temperature, rate and time as well as the cooling parameters at the end of the production cycle significantly influence dehydroxylation. Heating can have both positive and negative effects on the reactivity of pozzolana [9,10].

The major quantitative criterion for evaluating the effectiveness of the heat treatment of kaolinite is the degree of dehydroxylation. The dehydroxylation of kaolinites in normal atmospheres results in a mass loss of 13.95%, which corresponds to the transition of $SiO_2{\cdot}2Al_2O_3{\cdot}2H_2O$ to $SiO_2{\cdot}2Al_2O_3$. Typically for most kaolinites, the reaction is strongly endothermic in the range from 450 to 600°C (the exact temperature interval depends on the crystallinity and particle size). The well-defined endothermic peak (Figure 5.24) corresponds to the formation of a new disordered phase, MK ($Al_2Si_2O_7$), via dehydroxylation followed by the outflow of water, as shown by mass spectrometry (Figure 5.24). The dehydroxylation of kaolinite is preceded by a 'pre-dehydroxylation' stage in the range 80–150°C, where the evolution of a small amount of water is observed. Between 900°C and 1100°C, the exothermal formation of crystalline phases, like spinel and mullite ($Al_6Si_2O_{13}$), where the silica is totally or partly segregated, is characterised by an exothermic peak on the differential thermal analysis (DTA) curve and insignificant weight loss on the thermogravimetric analysis (TGA) curve. According to Shvarzman *et al.* [11], dehydroxylation is accompanied by amorphisation of the kaolinite, which affects the pozzolanic activity of MK.

Figure 5.24 (a) Mass spectroscopic analysis of water released by pure kaolinite during calcination. (b) DTA and TGA curves for pure kaolinite [11]

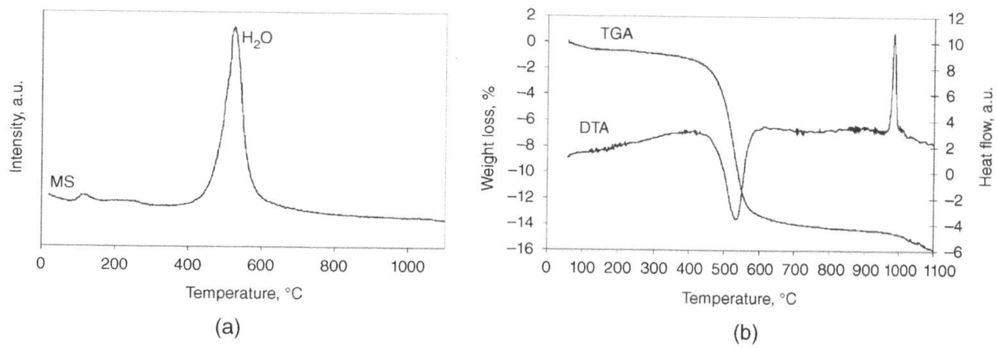

MECHANO-CHEMICAL METHOD

Kaolinite can be also activated by mechano-chemical methods [12–18], which are usually based on combined grinding (either dry or wet) and intercalating kaolinite with molecules that otherwise would not readily insert between the kaolinite layers. The mechano-chemical treatment of kaolinite results in partial dehydroxylation of the kaolinite [2]. New additional phases of kaolinite are produced. Modification of the hydroxyl surfaces can be extensive, even causing mild heating. The mechano-chemical modification of kaolinite through grinding results in a material with a high surface area. Intense local heating can occur, resulting in the formation of new surfaces. Mechano-chemical activation causes significant changes to the structure of kaolinite (by increasing the number of lattice defects), the surface energy and the chemical reactivity. Makó *et al.* [13] concluded that destruction of the crystal structure of kaolinite (by the rupture of the O–H, Al–OH, Al–O–Si and Si–O bonds) was significantly influenced by the kaolinite-to-quartz ratio and by the grinding time.

5.3.5.3 Hydration of cement–MK systems

PORTLANDITE–MK REACTIONS

Many researchers have studied the kinetics of portlandite–MK reactions, detailing the nature of the hydration products formed. The hydrates formed are essentially C-S-H, C_2ASH_8, C_3AH_6 and C_4AH_{13}. The MK content, curing conditions and presence of chemical activators are usually the main influencing factors on the hydration reaction and the new-formed hydrated products [19–24].

Murat [25] studied MK–lime mixtures. For a mass ratio of 1:1, hydration products such as calcium silicate hydrate (C-S-H) and gehlenite hydrate C_2ASH_8 were observed.

De Silva and Glasser [26] reported that in a MK–lime mixture of the same composition at 55°C, C-S-H developed after 24 hours of hydration. After 90 days, the hydration products were C_2ASH_8, C_4AH_{13}, hydrogarnet and C-S-H. They indicated that C_2ASH_8 and C_4AH_{13} are probably present as metastable phases, which after long curing periods convert to hydrogarnet with variable composition, $C_3AS_zH_{6-2z}$. This phenomenon may lead to a negative influence on the performance of MK-blended matrixes, mainly on durability.

Cai *et al.* [27] showed that 15 wt% of MK replacement is considered to be optimal for increasing the hydration rate of a blended binder. Kalpokaitė-Dičkuvienė *et al.* [28] examined the influence of MK in the hydration of a ternary paste that also contained cement and fly ash. They found that the hydration rate increased with an increase of the content of amorphous fly ash.

Cabrera and Rojas [29] investigated the reaction kinetics of an MK–lime mixture in water at 60°C. They reported that the first product of hydration was C-S-H, which was detected clearly at 6 hours. At 12 hours of hydration, C-S-H, C_2ASH_8 and C_4AH_{13} were clearly detected. After 21 hour of hydration, hydrogarnet appeared. According to Rojas and Cabrera [19], hydrogarnet is not formed from a transformation reaction but as a result of a direct reaction between MK and lime.

Wild and Khatib [30] suggested that in Portland cement (PC)–MK–water systems at ambient temperature, the most likely phase to coexist with C-S-H at short curing times and low MK levels is C_4AH_{13}, whereas at long curing times and high levels of MK, it is C_2ASH_8.

Frías and Cabrera [31] related the absence of C_4AH_{13} in the reaction kinetics of a PC-MK system to the presence of SO_4^{-2} ions in the matrix.

Shi and Day [21] concluded that the increase in the curing temperature accelerated the reaction rate but did not affect the hydration products in the pastes.

According to Murat [25] and Wild *et al.* [32], the hydration of MK may occur in the following form:

$$AS_2 + 7CH + 9H \rightarrow C_4AH_{13} + C_3S_2H_3 \quad (5.1)$$

$$AS_2 + 5CH + 3H \rightarrow C_3AH_6 + 2CSH \quad (5.2)$$

$$AS_2 + 3.5CH + 6H \rightarrow C_2ASH_8 + 0.5C_3S_2H_3 \quad (5.3)$$

MICROSTRUCTURE OF PORTLANDITE–MK PASTES

Figure 5.25 is an SEM image of hydrated portlandite–MK paste aged 90 days. Thermally treated (under 600°C) standard pure kaolin was used to prepare the portlandite–MK paste. The paste with the treated standard kaolin is a dense, well-formed crystalline structure containing a great amount of gehlenite hydrate (fibrous plates) and some nodular spongy areas, corresponding to a C-S-H microstructure. The Ca:Si and

Figure 5.25 SEM-micrograph of portlandite–MK paste at 90 days [33]

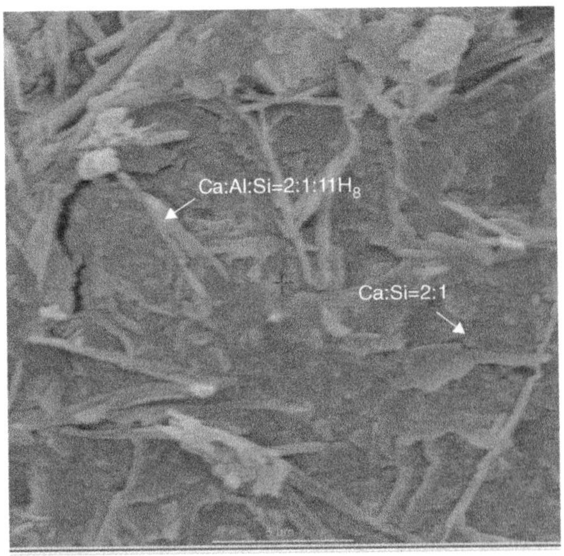

Ca:Si:Al ratios were measured by energy-dispersive X-ray spectroscopy (EDX) as the average of three measurements taken in the same zone. According to the EDX, the Ca:Si ratio was 2 for C-S-H, whereas the Ca:Al:Si ratio was 2:1:1 for gehlenite hydrate.

CHEMICAL ACTIVITY OF MK

The reactivity of a calcined clay such as MK with calcium hydroxide is often determined by a so-called Chappelle test, in which a dilute slurry of the calcined clay is reacted with excess calcium hydroxide at 95°C for 18 hours. At the end of this period, the amount of unreacted calcium hydroxide is determined by titration. The results are expressed as the weight of $Ca(OH)_2$ in milligrams absorbed by 1 g of calcined clay, for example, MK [34].

Compressive strength tests are another way to determine the chemical activity of MK. The compressive strength of pastes with different molar ratios of lime–MK [$CaO:(Al_2O_3{\cdot}2SiO_2)$] is presented in Figure 5.26. The molar ratio influences the development of compressive strength. There is an optimal $CaO:(Al_2O_3{\cdot}2SiO_2)$ molar ratio of 2.2–4.1, which corresponds to an MK content of 42–55%, from the viewpoint of compressive strength (chemical activity) [35]. In the molar ratio range 0.2–4.1, a drastic increase in compressive strength was observed. As can be seen from Figure 5.26, such intensive growth in strength was observed within a very short curing time (within the first 3 weeks). However, for molar ratios >4.1, there was a severe decrease in compressive strength at all ages tested.

EFFECT OF PORTLANDITE-TO-MK RATIO ON PHASE FORMATION AND VOLUMETRIC QUANTITIES

DTA and TGA curves tested at 90 days for portlandite–MK paste with different molar ratios are shown in Figures 5.27 and 5.28, respectively. Figure 5.29 shows the weight loss associated with decomposition of hydration products as a function of the portlandite-to-MK molar ratio. Figure 5.30 shows the content of hydrated products as a function of the portlandite-to-MK molar ratio at 90 days.

As can be seen from Figure 5.27, the main phases formed by hydration are C_2ASH_8 (gehlenite hydrate) and C-S-H. The following is apparent from Figures 5.27 and 5.28.

- The dehydration of calcium silicate hydrates, which is manifested by endothermicity at ~115–170°C, increased with increasing molar ratio. The low endothermicity below 100°C, which occurred for all samples, is most likely due to moisture adsorbed when preparing the samples for the thermal analysis.
- The endothermicity centred at ~200°C, which is due to the dehydration of gehlenite hydrate, C_2ASH_8, increased with increasing molar ratio.

Figure 5.26 Compressive strength plotted against the molar ratio of $CaO/(Al_2O_3{\cdot}2SiO_2)$ [33]

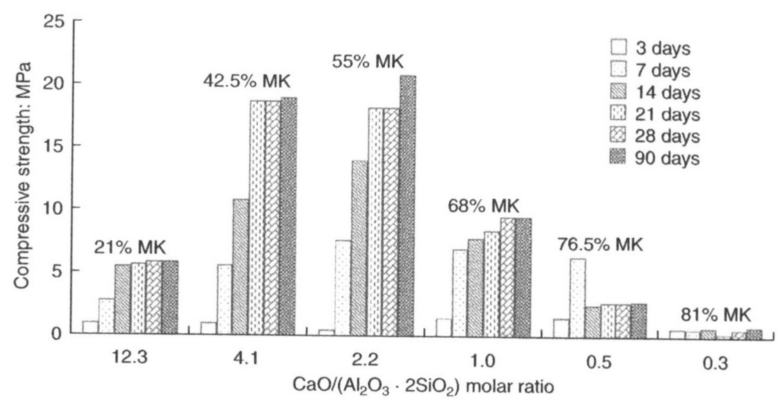

Figure 5.27 DTA curves for portlandite–MK paste with different $CaO:(Al_2O_3{\cdot}2SiO_2)$ molar ratios [33]

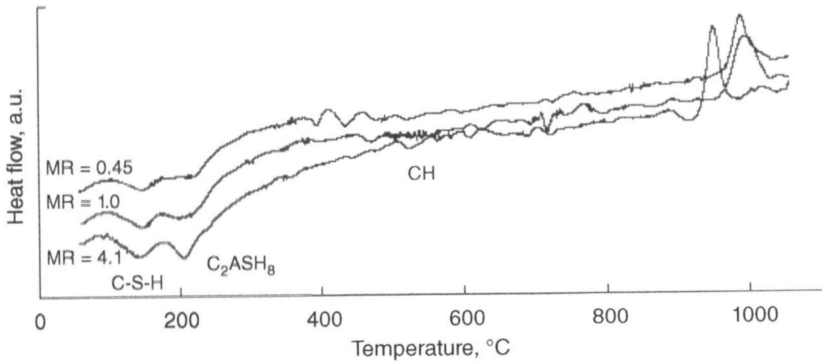

Figure 5.28 TGA curves for portlandite–MK paste with different $CaO:(Al_2O_3{\cdot}2SiO_2)$ molar ratios [33]

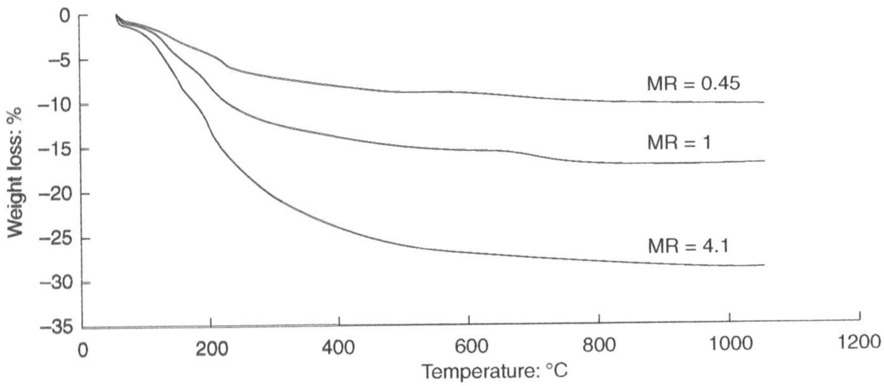

Figure 5.29 Influence of $CaO:(Al_2O_3{\cdot}2SiO_2)$ molar ratio on weight loss and compressive strength at 90 days [33]

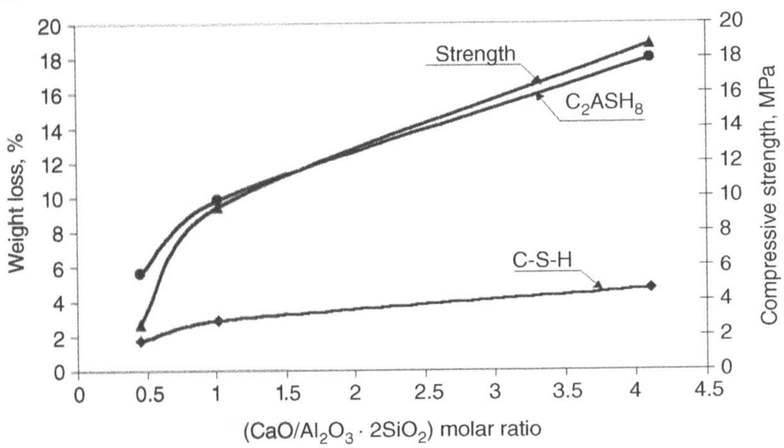

Figure 5.30 Influence of CaO:(Al_2O_3·$2SiO_2$) molar ratio on content of hydrated products at 90 days [33]

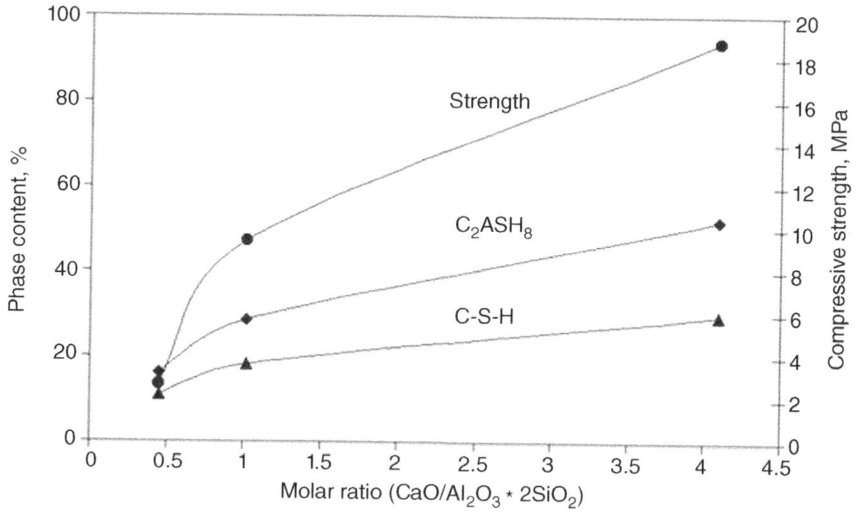

5.3.5.4 Properties of cement–MK systems at early ages

The properties of MK-based cementitious systems at early ages often have a primary influence on their durability. Thus, it is important to discuss at least some of these properties before discussing durability.

WATER DEMAND

The mean diameter of commercial MK particles is about 1–5 μm. The small sizes of the MK particles mean that the material has a large specific surface area of about 12 000–18 000 m^2/kg. The high specific surface area of MK increases the water demand, which should be considered in concrete mix design.

WORKABILITY

MK adversely affects the workability of concrete [36–38]. This is the consequence of high water demand, one of the disadvantages of MK. Bai *et al.* [39] reported a decrease in both slump and compacting factor and an increase in Vebe time as the content of MK in concrete mixtures increased from 0% to 15%. Rols *et al.* [40] reported that using MK required higher quantities of superplasticiser compared to a control concrete. Wild *et al.* [41] found it necessary to employ up to 3% superplasticiser to produce moderate slumps (75 mm) in MK concrete (10% MK) with a water-to-binder (w/b) ratio of 0.45.

HEAT OF HYDRATION

Recent work indicates that PC-MK blends produce more heat of hydration and have a higher temperature rise in concrete and mortar than does Portland cement alone. For example, Zhang and Malhotra [42] reported a temperature rise 7°C greater than Portland cement concrete when 10% of Portland cement was replaced with MK, whereas 10% silica fume produced a temperature rise only slightly above (0.5°C) that of the Portland cement concrete. Ambroise *et al.* [5], in similar work on PC-MK mortar using semi-adiabatic calorimetry, reported temperature rises above that of the control Portland cement mortar by 8, 6, and 1°C, at Portland cement replacement levels by MK of, respectively, 10%, 20% and 30%. This enhanced temperature rise could lead to severe thermal stress and cracking, particularly in a large volume of mass concrete.

EARLY-AGE STRENGTH

The effect of MK on compressive strength development has been the subject of several studies. Recent works have shown that when MK is used as a partial cement replacement in concrete, it significantly enhances early strength and increases long-term strength. This is because, by acting as a filler, it accelerates initial cement hydration. Moreover, in the early stages of curing, it rapidly consumes portlandite, which is produced by cement hydration and hardly contributes to the strength.

In an investigation into the properties of PC-MK pastes at w/b ratios between 0.25 and 0.54 and MK contents of 0%, 10%, 20%, 30%, 40% or 50%, Ambroise *et al.* [5] reported that the maximum strength occurred at 10% replacement with MK. A similar influence of MK on the strength of paste and mortars has been reported by several authors [43,44].

CHEMICAL SHRINKAGE

Wild *et al.* [32] investigated the autogenous and chemical shrinkage and swelling of PC-MK pastes with different contents of MK (5–25%) and a w/b ratio of 0.55. The specimens were monitored for 1 to 45 days. Chemical shrinkage was found to reach a maximum for between 10% and 15% MK and then decreased sharply for higher MK contents. Li *et al.* [45] examined the chemical shrinkage of an MK-based geopolymer. Their results showed that, unlike conventional cements, the chemical deformation of the MK-based geopolymer can be divided into three stages: shrinkage, expansion and final shrinkage.

AUTOGENOUS SHRINKAGE

Autogenous shrinkage can be a major problem for high-performance concrete. Certain pozzolanas, such as silica fume and fly ash, may result in increased autogenous shrinkage because of the refinement of the pore structure.

MK affects the autogenous shrinkage of cement pastes for four reasons: (*a*) cement dilution by MK, because if there is less cement there is less shrinkage, (*b*) the heterogeneous nucleation of hydrates on the surface of MK particles, which accelerates cement hydration and, consequently, increases shrinkage, (*c*) the pozzolanic reaction of MK with CH, and (*d*) an increase in capillary tension due to the refinement of the pore size distribution, leading to an increase in autogenous shrinkage.

Using MK as a partial cement replacement may lead to reduced autogenous shrinkage. For example, Rols *et al.* [40] measured autogenous shrinkage in a concrete containing 9% MK. At all times, the autogenous shrinkage was lower than that in control specimens.

Brooks and Megat Johari [46] found that at very young ages from the initial setting to up to 20 hours, the autogenous shrinkage of concrete with a low w/b ratio of 0.28 decreased with an increase of the MK content.

Wild *et al.* [32] showed that, for cement–MK pastes (w/b = 0.55), autogenous shrinkage increased with MK content up to a maximum of 10%. For MK contents above 15%, it appeared to be comparable to that of the control cement paste. Expansion continued for up to 14 days for all compositions (0–25% MK), except for 10% MK. For cement–MK pastes (w/b = 0.50), Kinuthia *et al.* [47] observed that 5% and 10% MK increased the autogenous shrinkage, whereas at 15% and 20% MK, there was a significant decrease.

According to Gleize *et al.* [48], a significant part of the reduction in long-term autogenous shrinkage in cement–MK pastes is due to the pozzolanic activity of MK. A long-term autogenous shrinkage of cement–MK paste with w/b ratios of 0.3 or 0.5 decreased as the cement replacement level by MK increased (5%, 10%, 15% and 20%).

Yang *et al.* [49] found that the partial substitution of fly ash by MK in a blended binder increased the autogenous shrinkage due to densification of the binding gel. Moreover, the introduction of MK limited water evaporation, which in turn reduced drying shrinkage.

Shen *et al.* [50] investigated the autogenous shrinkage of high-performance concrete cured at 80°C with 10% replacement of MK. The shrinkage–expansion process was followed by a temperature regime of curing. Overall, concrete prepared with MK had higher autogenous shrinkage than that of a reference. However, drying shrinkage measured between 1 and 56 days was lower for the concrete with MK, which agrees with other results in the literature [51,52].

5.3.5.5 Properties of hardened cement–MK systems

POROSITY

In most cases, mortar- and concrete-containing materials with pozzolanic characteristics have porosities equal or superior to that of OPC concrete. The porosity depends on the characteristics of the pozzolanic materials, such as fineness, specific surface area, mineralogical and chemical composition, content and quality of amorphous phases and, importantly, the w/b ratio. However, it is well known that the critical factor affecting the performance and durability of a concrete structure is the pore size distribution, rather than the total porosity.

The main physical effect of MK on the microstructure of hardened concrete is a refinement of the pore structure. Khatib and Wild [44] reported a refinement of the pore structure and an increase of the total intruded pore volume between 14 and 28 days for pastes with 5%, 10% or 15% of MK. Shen *et al.* [50] observed a decrease of the total porosity from 14.4% to 11.3% for a high-strength concrete prepared with 10% MK.

Rojas and Cabrera [31] studied the effect of MK content on pore size distribution and the degree of hydration of cement–MK paste. The total porosity decreased after up to 28–56 days of curing time. Above this age, the total porosity of the MK mixed was higher compared to a Portland cement paste. The capillary porosity for MK mixes was lower compared to the Portland cement paste. The strongest effect of MK on the refining of the pore structure was detected in pores with a radius smaller than 100 Å. Between 7 and 90 days, the gel porosity of MK mixes increased, whereas the porosity in an OPC mix was practically constant.

The curing conditions influence the porosity of a MK system [53]. The pore volumes of an autoclaved plain concrete and an MK concrete were less than those of a moist plain concrete and an MK concrete.

The improvement in pore structure is reflected in a higher compressive strength, and it also significantly affects the permeability of the material. Another important factor is the improvement in the quality of the interfacial transition zone. The partial replacement of cement with ultra-fine particles of MK densifies the interfacial transition zone because of the microfiller effect due to the relatively fine particles of MK [54].

COMPRESSIVE STRENGTH

The development of strength in concretes containing MK is similar but not identical to that observed in mortars and pastes. Factors such as the content of amorphous phases, type of cement, type and content of chemical admixture, aggregate properties and type of mix influence the strength development of MK concrete. Caldarone *et al.* [55], Wild *et al.* [41], and Shvarzman [33] reported that concretes made from MK and Portland cement had strengths that were slightly greater than those of concretes made from silica fume and Portland cement at the same level of cement replacement by pozzolanas, as shown in Figure 5.31.

Wild *et al.* [41] and Sabir *et al.* [37] studied the influence of two factors, namely the filler effect and curing temperature, on the compressive strength. The filler effect, which is the immediate cause of the acceleration of hydration during the first 24 hours and the pozzolanic reaction, has its maximum effect within the first 7–14 days for all MK levels between 5% and 30%. The degree to which strength is enhanced declines beyond 14 days, although strength gains relative to a control still occur after 90 days. Little strength advantage is gained for MK levels of more than 15%. Similar findings were observed by Curcio *et al.* [56] in mortars containing 15% MK.

Figure 5.31 Development of compressive strength of concretes made with silica fume (SF) or MK (w/b = 0.40) [33]

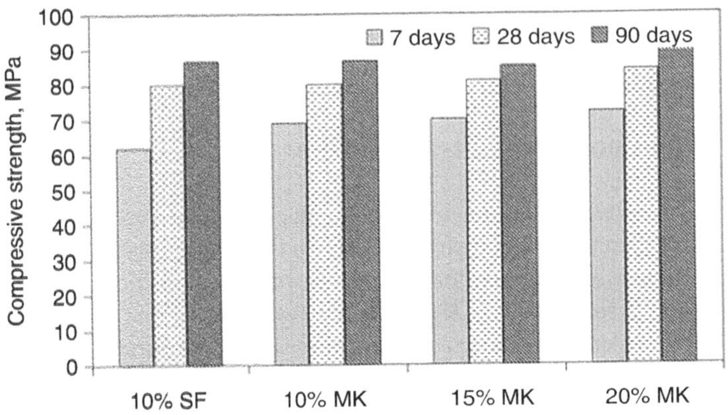

The influence of curing temperature on strength development in concretes containing up to 15% MK was studied by Sabir *et al.* [37]. They reported increased early strength (7 days) of concrete at a curing temperature of 50°C compared to specimens cured at 20°C. The acceleration in strength development due to the high temperature diminishes in the long term.

Khatib and Hibbert [57] investigated the influence of incorporating GGBS and MK on the strength of concrete. The incorporation of MK caused an increase in compressive strength, especially during early curing. The decrease in compressive strength during the early curing of GGBS concrete was compensated for by the inclusion of MK. Adding MK to concrete containing GGBS increases the flexural strength. The effect was particularly beneficial at 20% MK and 60% GGBS, as the flexural strength exceeded that of a control after 90 days of curing.

Muduli and Mukharjee [36] examined the influence of MK on the mechanical properties of concrete containing recycled or natural aggregates. Incorporating MK increased the material properties of a concrete prepared with either natural or recycled aggregates. Moreover, concrete with 15 wt% addition of MK had similar mechanical properties as concrete with natural aggregates but without MK addition.

Antoni *et al.* [58], investigated the mechanical properties of a ternary blended binder. Samples comprising 30 wt% of MK and 15 wt% of limestone had a higher compressive strength at 7 and 28 days than a reference containing 100 wt% cement. This was due to reactions between aluminates from MK and calcium carbonate, which resulted in the formation of additional monosulfate.

The development of the hydration strength of ternary blended binders prepared with cement, limestone, fly ash or MK was investigated by Medjigbodo *et al.* [59]. Those researchers focused on the influence of the particle size of limestone on the properties of the blended binder. Limestone with the smallest particles predictably increased the 1-day strength and accelerated the hydration kinetics.

da Silva Andrade *et al.* [60] examined a ternary cement–MK paste containing nanosilica. Mortar with 15% of MK and 3% of nanosilica had a 44% increase in compressive strength in comparison to a reference. However, the simultaneous incorporation of MK and nanosilica significantly decreased the rheological properties.

Overall, to avoid a reduction in the mechanical properties of a concrete, most of these researchers found that the maximum amount of MK in a binary binder was 30 wt% [37,41,57–59].

FLEXURAL STRENGTH

The relation between flexural strength and compressive strength in MK concrete is like that of ordinary concrete. Shvarzman [33] found that the flexural strength of MK concrete is higher than that of silica fume concrete (Figure 5.32).

ELASTICITY

The modulus of elasticity of concrete at 28 days, as reported by Shvarzman [33], is illustrated in Figure 5.33. At 28 days, the concrete with silica fume had a higher modulus of elasticity compared with MK concrete. The modulus of elasticity increased with an increase in MK content.

Figure 5.32 Flexural strength of concretes made with silica fume (SF) or MK (w/b = 0.40) [33]

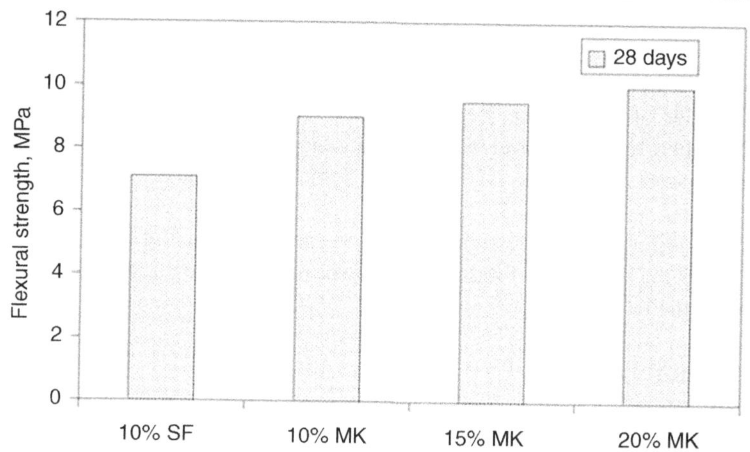

Figure 5.33 Modulus of elasticity of concretes made with silica fume (SF) or MK (w/b = 0.40) [33]

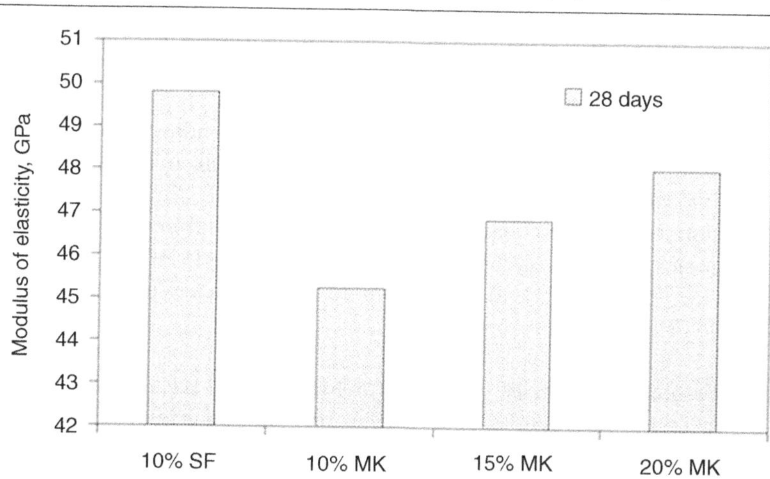

Figure 5.34 Drying shrinkage of concrete made with 10% silica fume or 10% MK (w/b = 0.40) [33]

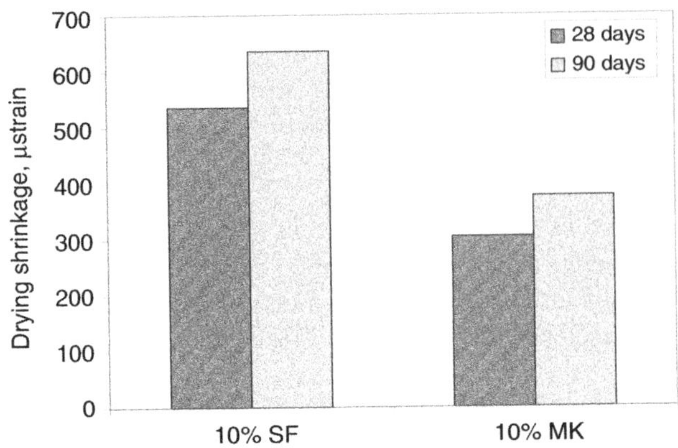

DRYING SHRINKAGE

Shvarzman [33] found that MK concrete had significantly lower drying shrinkage compared with concrete containing the same dosage of silica fume. Figure 5.34 shows the drying shrinkage of concretes made with 10% silica fume or MK measured at 28 and 90 days. For example, the MK concrete had a drying shrinkage of only 376×10^{-6} compared with 636×10^{-6} for silica fume concrete at 90 days.

Medjigbodo *et al.* [59] continued the investigation started earlier [58] using the same high level of cement substitution (up to 45 wt%). The ternary binder had a lower ultimate shrinkage, but the shrinkage rate was higher in comparison to the reference.

5.3.5.6 Aspects of durability

PORTLANDITE CONSUMPTION

The amount of MK required for complete elimination of CH depends on several factors, mainly purity of MK, Portland cement composition, w/b ratio and curing condition. The reduction in CH content results in superior strength and durability [30,37]. Oriol and Pera [61] determined the level of portlandite consumption in PC-MK pastes under microwave curing conditions. They reported that the total elimination of CH was achievable in binders containing 15% MK. They also reported that between 30% and 40% MK is required to remove all the CH in PC-MK pastes when cured in lime-saturated water at room temperature for 28 days.

RESISTANCE TO CHLORIDE ION PENETRATION

The ingress of chloride ions into reinforced concrete and the corresponding corrosion is a major problem in the cement and building materials industry. The ability of concrete to bind chloride ions and retard their penetration is an important factor in determining the service life of concrete structures affected by chloride-induced corrosion. In hardened cement, chloride may be bound in the C-S-H gel or as a result of the formation of the complex calcium oxychloride, Friedel's salt ($Ca_2Al(OH)_6Cl{\cdot}2H_2O$). PC-MK paste can bind a considerable amount of the chloride dissolved in pore water with a relatively low reduction in pH due to the formation of stable Friedel's salt.

The increase in chloride-binding capacity observed for MK-blended paste samples could be attributed to the participation of calcium aluminate species in the formation of Friedel's salt, which would otherwise be engaged in the formation of C_2ASH_8 and C_4AH_{13} [62]. The mechanism for how chloride ions are bound in lime–MK mixtures was studied by Zibara *et al.* [63] and Saikia *et al.* [64]. The influence of the

Figure 5.35 Resistance of concretes made with silica fume (SF) or MK to chloride ion penetration (w/b = 0.40) [33]

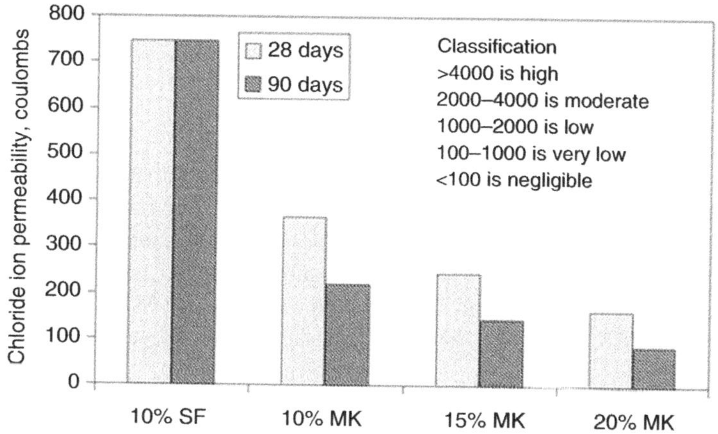

C/S and C/A ratios of hydration products on the chloride-ion-binding capacity of lime–MK systems was investigated. The alumina content of the mixture has an important influence on the chloride-ion-binding capacity. At a high C/A ratio, the formation of monocarboaluminate is favoured, which has a high ability to form Friedel's salt and does so at low chloride concentrations (less than 0.1 M). With a low C/A ratio, the formation of strätlingite is favoured, with little formation of monocarboaluminate. Strätlingite does convert to Friedel's salt, but at a much lower rate and much of the strätlingite persists at higher chloride concentrations.

Boddy *et al.* [65] studied the chloride-penetration resistance of concrete containing MK. A higher MK content and lower w/b ratio decreased diffusion, permeability and conductivity and increased resistivity. Resistance to chloride migration increased with increasing MK content and decreasing w/b ratio. Similar results have been obtained by other authors [33,66,67].

Shvarzman [33] found that the resistance of MK concrete to chloride ion penetration was significantly higher than that of a silica fume concrete (Figure 5.35). At 28 days, the charge passing through the MK concrete (365 coulombs) was very low in comparison with silica fume concrete (750 coulombs). The charge passed through the MK concrete decreased with increased curing time. At 90 days, the charge passing through the 10% MK concrete was only 160 coulombs and 86 for concrete with 20% MK. According to ASTM C1202, if the charge passing through concrete is below 100 coulombs, then the concrete is inert and can be considered as impermeable to chlorides.

RESISTANCE TO FREEZING AND THAWING

Both the scaling resistance of concrete surfaces exposed to freezing and thawing cycles in the presence of de-icing chemicals (ASTM C672) and the direct freeze–thaw test under water-saturated conditions (ASTM C666) have shown that cement–MK mixes exhibited high freeze–thaw durability, like that of silica fume concrete [55].

CARBONATION

The carbonation of concretes mixes with MK is characterised by a significant increase of the carbonation depth with an increase of the replacement ratio [33,68,69]. This may be because the replacement of cement by MK decreases the content of portlandite in hydrate products due to pozzolanic reactions.

SULFATE RESISTANCE

The reported properties of MK suggest that it should be, in general, highly effective in promoting sulfate resistance [70,71]. Sulfate attacks against concrete are primarily attributed to sodium, magnesium and calcium salts. However, the samples exposed to sodium sulfate and magnesium sulfate environments show different behaviour, which indicates that the mechanisms of deterioration in these environments are different.

Khatib and Wild [72] have investigated the effect of cement type with different C_3A contents on the resistance of MK mortar to sulfate attack under moist curing conditions. The results for sulfate expansion demonstrated that sulfate resistance increased as the replacement level of cement with MK increased, up to at least 25% replacement. After prolonged periods of exposure to Na_2SO_4 solution, there was a significant strength loss for Portland cement mortars and mortars with a low level of MK (5% and 10%). In contrast, for mortars with a high level of MK (15–25%), there was consistent strength gain.

Al-Akhras [53] studied the effect of MK replacement of cement on the resistance of concrete to sodium sulfate attack under different parameters, including MK replacement level, w/b ratio, curing type, initial moist curing and air content. The effect of w/b ratio for plain and MK concrete after 18 months of sulfate exposure on the reduction of compressive strength and on expansion is shown in Figures 5.36 and 5.37, respectively. The sulfate resistance of MK concrete increased with increasing MK replacement level. Autoclaved MK concrete showed superior sulfate resistance compared to moist-cured MK concrete. Increasing the air content from 1.5% to 5% improved the sulfate resistance of MK concrete.

Limestone cement mortar is susceptible to the thaumasite kind of sulfate attack at low temperatures. Incorporating MK improved the resistance of limestone cements against sulfate attack [73]. MK can be effective in reducing the thaumasite type of sulfate attack in concrete containing limestone filler [74].

Lee *et al.* [75] found that MK negatively influences the performance of mortar or paste specimens when exposed to magnesium sulfate solution. This is partially attributed to the formation of gypsum, but not ettringite or thaumasite. Another possible reason may be the decalcification of primary and secondary C-S-H gel following the formation of M-S-H (magnesium silicate hydrate) gel.

Figure 5.36 Effect of w/b ratio on the reduction of the compressive strength of MK concrete [53]

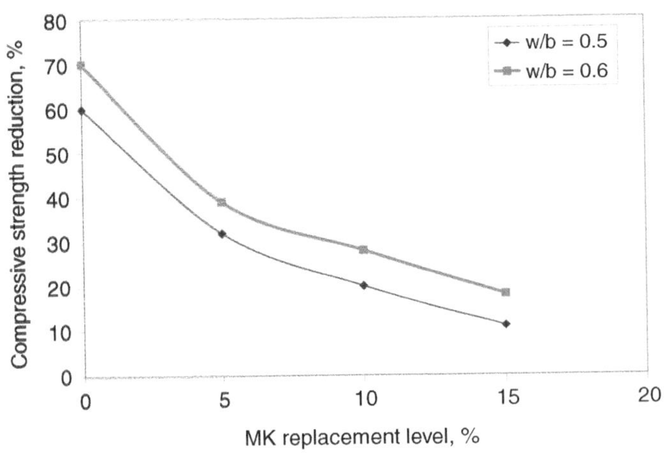

Figure 5.37 Effect of w/b ratio on the 18-month sulfate expansion of MK concrete [53]

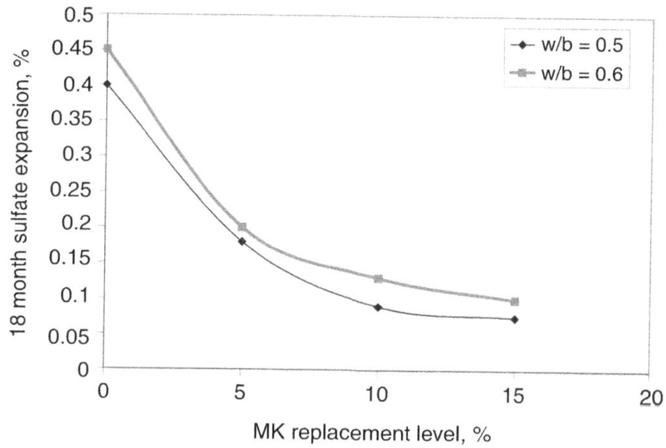

Shi *et al.* [76] studied the sulfate resistance of ternary binder containing 35% substitution of cement by limestone (L) and MK (calcinated clay, CC). Mortars prepared with the following ratio $CC/(CC + L) \geq 0.5$ had excellent sulfate resistance.

PERFORMANCE AT ELEVATED TEMPERATURES

The fire resistance of concrete is highly dependent on its constituent materials, particularly the pozzolanic admixture. Poon *et al.* [77] studied the performance of MK concrete (0–20% MK) at elevated temperatures up to 800°C. The MK concrete showed a distinct pattern of strength gain and loss at elevated temperatures. After gaining an increase in compressive strength at 200°C, it maintained higher strengths compared to a corresponding silica fume, fly ash and reference concretes up to 400°C. Within the range 400–800°C, the MK concretes suffered more loss and had lower residual strengths than the other concretes. The MK concretes had a high increase in the loss of impermeability and durability after heating compared to the other tested concretes. However, MK concrete with 5% cement replacement showed better performance than the corresponding OPC concrete and silica fume concrete due to the reduced CH content, which initiates internal cracking during heating and disintegration on cooling.

PERFORMANCE OF FIBRE-REINFORCED MORTAR

The incorporation of fibres in cementitious materials has gained importance, in part because of the reduction of shrinkage cracking. Thus, fibres can improve the durability of the materials. In general, the compressive strength of a material reduces and its permeability increases because of the embedding of some types of fibres into the matrix. However, the loss of resistance due to fibres embedded in the mortar is compensated for by the increase in strength caused by MK to the mortar [78].

Although the addition of pozzolana to cementitious materials reinforced with glass fibres is well known, not all pozzolanas have a beneficial effect on the mechanical properties and durability. MK is a pozzolanic material that significantly enhances both of these [79].

ALKALI–SILICA REACTIONS

Pozzolanas containing MK and manufactured at several sites across the United States have been shown to reduce ASR in concrete when tested according to ASTM standards C227, C441, C1293, and C1260 [80,81]. The pozzolanas may have had different but uniform amounts of MK depending upon their geographic location and geologic formation. MK lowers the alkalinity of the pore solution by consuming much of the free lime. This decreases the silica solubility and the propensity for alkali–silica reactivity. Shi *et al.* [82] tested ASR in concrete consisting of alkali-activated slag, fly ash and MK. They found that ASR can be mitigated by increasing the MK content.

EFFLORESCENCE CONTROL

Dissolved calcium hydroxide can migrate to the surface of a structure and form unattractive efflorescent deposits. The progressive consumption of $Ca(OH)_2$ by MK lowers the amount of calcium hydroxide in the system and decreases efflorescence [55].

5.3.5.7 Applications of MK

High-reactivity MK can be considered to be a highly pozzolanic material. It appears to have excellent potential as an SCM for high-performance cementitious materials [83]. In addition, attempts have been made to implement MK in geopolymers, blended cements and hydraulic lime. Here is a short list of recent applications of cement–MK concrete reported by BASF.

- High-reactivity MK was specified for the new US embassy in Beijing. The 50 000 m^2 building cost USD 434 million and was opened on 8 August 2008. The dedication was timed to coincide with the commencement of the 2008 Olympic Games.
- MK with a high-range water-reducing admixture has been used to produce very high-strength concrete. For example, New York's Department of Transportation approved MK for use in high-performance concrete jobs. As a result, the concrete admixture was applied in high-visibility work on the Erie Canal, Interstate 87 bridges over the Hudson River, and other projects around the state. For the Erie Canal work, Cranesville Block of Amsterdam, NY, is supplying high-performance MK concrete for refacing the lock walls and abutments and for extending a dam apron.
- Departments of transportation from Florida to Wyoming, the Port Authority of New York and New Jersey, and the Pennsylvania Turnpike Commission have approved high-reactivity MK for use in high-performance concrete.
- MK was used in a 23-storey condominium constructed in a seismically active zone (Figures 5.38 and 5.39). The high early strength of the MK-enriched concrete prevented thermal cracking despite concerns about the heat of hydration in the thick columns.
- The resistance MK provides to the penetration of chloride ions is also attracting interest for marine applications. MK concrete was used for a wharf at Shakespeare Bay in Queen Charlotte Sound, New Zealand. The wharf was built on concrete piles up to 30 m long and has a 200-m-long deck.
- Underwater cellular MK concrete was used to repair and protect piles and pile caps on a 183-m-long, 50-year-old dock on the East River, New York City.
- High-reactivity MK has been successfully used in concrete mix design in the construction of a super tall, upscale condominium building at 432 Park Avenue, New York City. It is in the heart of Manhattan, close to Central Park, the Rockefeller Center and Broadway (Figures 5.40 and 5.41). This is the tallest residential structural concrete building in the western hemisphere, standing at 420 m in height. It is also remarkable because of its small, square footprint, 28 m. The mixes included local coarse and fine aggregates, slag, small amounts of silica fume and high-reactivity MK, which improved the strength and durability of the concrete and lightened the colour when combined with white Portland cement.

5.3.5.8 Summary

- MK can be considered to be a highly pozzolanic material and appears to have excellent potential as an SCM for high-performance cementitious materials.
- The main phases that form in an hydrated portlandite–MK system are C_2ASH_8 and C-S-H.

Figure 5.38 Late afternoon sunlight colours the lower floors of the structure of this Los Angeles condominium. The large columns and beams of the moment frame, exposed at the centre of the east and west facades, is in counterpoint to the minimal columns required around the rest of the building. (Courtesy of Michael Chusid, www.chusid.com)

Figure 5.39 The condominium believed to be the first high-rise structure to use high-reactivity MK in concrete, which has a strength of 80 MPa (12 000 psi). (Courtesy of Michael Chusid, www.chusid.com)

Figure 5.40 Upscale condominium building under construction at 432 Park Avenue, New York City. (Photograph Joe Nasvik)

Figure 5.41 Close view of a condominium building at 432 Park Avenue, New York City. (Photograph Joe Nasvik)

- The hydration of a portlandite–MK system is very intensive at early age. The rates of both strength development and the formation of hydration products are a maximum at 2 weeks.
- The portlandite-to-MK molar ratio influences the hydration rate, type, content and volume of the hydration phases and, consequently, strength development.
- The type, content and volume of hydrated products as well as their morphology significantly influence the development of compressive strength.
- The content of MK influences the compressive strength of high-performance concrete.
- The optimum level of cement replacement seems to be around 10–20%, as this maximizes the strength. The compressive strength of concrete made with 10–20% of MK was 20–25% higher than that of a reference concrete at 90 days.
- MK concrete has lower drying shrinkage compared with silica fume concrete.
- Cement–MK mixes have high freeze–thaw durability, like silica fume concrete.
- The addition of MK reduces the ingress of chloride by improving the microstructure and chloride-binding behaviour. The resistance of MK concrete to the penetration of chloride ions is higher than that of silica fume concrete.
- MK enhances both the mechanical properties and the durability of glass-fibre-reinforced cementitious materials and reduces ASR and efflorescence.
- MK has a beneficial effect on the resistance to sodium sulfate attack, reduces the thaumasite type of sulfate attack, but negatively influences the resistance to magnesium sulfate attack.

REFERENCES

1. Shvarzman A and Kovler K (2009) Cement-metakaolin materials. In *Concrete Durability* (Soutsos M (ed.)). Thomas Telford, London, UK, Chapter 7.2.4, pp. 231–258.
2. Mendelovici E (1997) Comparative study of the effects of thermal and mechanical treatments on the structures of clay minerals. *Journal of Thermal Analysis* **49**: 1385–1397, 10.1007/BF01983697.
3. Fialips C, Petit S and Decarreau A (2000) Hydrothermal formation of kaolinite from various metakaolins. *Clay Minerals* **35**: 559–572, 10.1180/000985500547025.
4. Odler I (2000) *Special Inorganic Cements*. E&FN Spon, London, UK.
5. Ambroise J, Maximilien S and Pera J (1994) Properties of metakaolin blended cements. *Advanced Cement Based Materials* **1**: 161–168, 10.1016/1065-7355(94)90007-8.
6. Kenne Diffo BB, Elimbi A, Cyr M, Dika Manga J and Tchakoute Kouamo H (2015) Effect of the rate of calcination of kaolin on the properties of metakaolin-based geopolymers. *Journal of the Asian Ceramic Society* **3**: 130–138, 10.1016/J.JASCER.2014.12.003.
7. Claverie M, Martin F, Tardy JP, *et al.* (2015) Structural and chemical changes in kaolinite caused by flash calcination: Formation of spherical particles. *Applied Clay Science* **114**: 247–255, 10.1016/J.CLAY.2015.05.031.
8. Kovářík T, Bělský P, Novotný P, *et al.* (2015) Structural and physical changes of re-calcined metakaolin regarding its reactivity. *Construction and Building Materials* **80**: 98–104, 10.1016/J.CONBUILDMAT.2014.12.062.
9. Akolekar D, Chaffee A and Howe RF (1997) The transformation of kaolin to low-silica X zeolite. *Zeolites* **19**: 359–365, 10.1016/S0144-2449(97)00132-2.
10. He C, Osbaeck B and Makovicky E (1995) Pozzolanic reactions of six principal clay minerals: Activation, reactivity assessments and technological effects. *Cement and Concrete Research* **25**: 1691–1702, 10.1016/0008-8846(95)00165-4.
11. Shvarzman A, Kovler K, Grader G and Shter G (2003) The effect of dehydroxylation/amorphization degree on pozzolanic activity of kaolinite. *Cement and Concrete Research* **33**: 405–416, 10.1016/S0008-8846(02)00975-4.

12. Torres Sánchez R, Basaldella E and Marco J (1999) The effect of thermal and mechanical treatments on kaolinite: Characterization by XPS and IEP measurements. *Journal of Colloid and Interface Science* **215**: 339–344, 10.1006/JCIS.1999.6241.
13. Makó É, Frost RL, Kristóf J and Horváth E (2001) The effect of quartz content on the mechanochemical activation of kaolinite. *Journal of Colloid and Interface Science* **244**: 359–364, 10.1006/jcis.2001.7953.
14. Frost RL, Makó É, Kristóf J, Horváth E and Kloprogge JT (2001) Mechanochemical treatment of kaolinite. *Journal of Colloid and Interface Science* **239**: 458–466, 10.1006/JCIS.2001.7591.
15. Souri A, Kazemi-Kamyab H, Snellings R, *et al.* (2015) Pozzolanic activity of mechanochemically and thermally activated kaolins in cement. *Cement and Concrete Research* **77**: 47–59, 10.1016/J.CEMCONRES.2015.04.017.
16. Baccour A, Sahnoun RD and Bouaziz J (2014) Effects of mechanochemical treatment on the properties of kaolin and phosphate–kaolin materials. *Powder Technology* **264**: 477–483, 10.1016/J.POWTEC.2014.05.034.
17. Vizcayno C, de Gutiérrez RM, Castello R, Rodriguez E and Guerrero CE (2010) Pozzolan obtained by mechanochemical and thermal treatments of kaolin. *Applied Clay Science* **49**: 405–413, 10.1016/J.CLAY.2009.09.008.
18. Balczár I, Korim T, Kovács A and Makó É (2016) Mechanochemical and thermal activation of kaolin for manufacturing geopolymer mortars – Comparative study. *Ceramics International* **42**: 15367–15375, 10.1016/J.CERAMINT.2016.06.182.
19. Rojas MF and Cabrera J (2002) The effect of temperature on the hydration rate and stability of the hydration phases of metakaolin–lime–water systems. *Cement and Concrete Research* **32**: 133–138, 10.1016/S0008-8846(01)00642-1.
20. Coleman NJ and McWhinnie WR (2000) The solid state chemistry of metakaolin-blended ordinary Portland cement. *Journal of Materials Science* **35**: 2701–2710, 10.1023/A:1004753926277.
21. Shi C and Day RL (2000) Pozzolanic reaction in the presence of chemical activators: Part II – Reaction products and mechanism. *Cement and Concrete Research* **30**: 607–613, 10.1016/S0008-8846(00)00214-3.
22. Love CA, Richardson IG and Brough AR (2007) Composition and structure of C–S–H in white Portland cement–20% metakaolin pastes hydrated at 25°C. *Cement and Concrete Research* **37**: 109–117, 10.1016/J.CEMCONRES.2006.11.012.
23. Rojas MF (2006) Study of hydrated phases present in a MK–lime system cured at 60°C and 60 months of reaction. *Cement and Concrete Research* **36**: 827–831, 10.1016/J.CEMCONRES.2006.01.001.
24. Lagier F and Kurtis K (2007) Influence of Portland cement composition on early age reactions with metakaolin. *Cement and Concrete Research* **37**: 1411–1417, 10.1016/J.CEMCONRES.2007.07.002.
25. Murat M (1983) Hydration reaction and hardening of calcined clays and related minerals. I. Preliminary investigation on metakaolinite. *Cement and Concrete Research* **13**: 259–266, 10.1016/0008-8846(83)90109-6.
26. De Silva PS and Glasser FP (1992) Pozzolanic activation of metakaolin. *Advances in Cement Research* **4**: 167–178, 10.1680/adcr.1992.4.6.167.
27. Cai R, He Z, Tang S, Wu T and Chen E (2018) The early hydration of metakaolin blended cements by non-contact impedance measurement. *Cement and Concrete Composites* **92**: 70–81, 10.1016/J.CEMCONCOMP.2018.06.001.
28. Kalpokaitė-Dičkuvienė R, Baltušnikas A, Levinskas R and Čėsnienė J (2019) Incinerator residual ash – Metakaolin blended cements: Effect on cement hydration and properties, *Construction and Building Materials* **206**: 297–306, 10.1016/J.CONBUILDMAT.2019.02.060.
29. Cabrera J and Rojas MF (2001) Mechanism of hydration of the metakaolin–lime–water system. *Cement and Concrete Research* **31**: 177–182, 10.1016/S0008-8846(00)00456-7.

30. Wild S and Khatib JM (1997) Portlandite consumption in metakaolin cement pastes and mortars. *Cement and Concrete Research* **27**: 137–146, 10.1016/S0008-8846(96)00187-1.
31. Frías M and Cabrera J (2001) Influence of MK on the reaction kinetics in MK/lime and MK-blended cement systems at 20°C. *Cement and Concrete Research* **31**: 519–527, 10.1016/S0008-8846(00)00465-8.
32. Wild S, Khatib JM and Roose LJ (1998) Chemical shrinkage and autogenous shrinkage of Portland cement–metakaolin pastes. *Advances in Cement Research* **10**: 109–119, 10.1680/adcr.1998.10.3.109.
33. Shvarzman A (2003) *Properties and Microstructure of High Performance Cementitious Systems with Metakaolin*. PhD thesis, Technion – Israel Institute of Technology, Israel.
34. Largent R (1971) Estimation de l'activite pouzzolanique – recherche d'un essai. *Bulletin de liaison des laboratoires des ponts et chaussées*. https://trid.trb.org/view/1059001 (accessed 22/04/2019).
35. Shvarzman A, Kovler K, Grader G and Shter GE (2003) Hydration of portlandite-metakaolin and Portland cement-metakaolin systems. *Proceedings of the 11th International Congress on Cement Chemistry, Durban, South Africa*, pp. 843–852.
36. Muduli R and Mukharjee BB (2019) Effect of incorporation of metakaolin and recycled coarse aggregate on properties of concrete. *Journal of Cleaner Production* **209**: 398–414, 10.1016/J.JCLEPRO.2018.10.221.
37. Sabir B, Wild S and Bai J (2001) Metakaolin and calcined clays as pozzolans for concrete: A review. *Cement and Concrete Composites* **23**: 441–454, 10.1016/S0958-9465(00)00092-5.
38. Nazário Santos F, Gomes de Sousa SR, Faria Bombard AJ and Lopes Vieira S (2017) Rheological study of cement paste with metakaolin and/or limestone filler using mixture design of experiments. *Construction and Building Materials* **143**: 92–103, 10.1016/J.CONBUILDMAT.2017.03.001.
39. Bai J, Sabir BB, Wild S and Kinuthia JM (2000) Strength development in concrete incorporating PFA and metakaolin. *Magazine of Concrete Research* **52**: 153–162, 10.1680/macr.2000.52.3.153.
40. Rols S, Mbessa M, Ambroise J and Pera J (1999) Influence of ultra-fine particle-type on properties of very high-strength concrete. *ACI Special Publication* **186**: 671–686, 10.14359/5584.
41. Wild S, Khatib JM and Jones A (1996) Relative strength, pozzolanic activity and cement hydration in superplasticised metakaolin concrete. *Cement and Concrete Research* **26**: 1537–1544, 10.1016/0008-8846(96)00148-2.
42. Zhang MH and Malhotra VM (1995) Characteristics of a thermally activated alumino-silicate pozzolanic material and its use in concrete. *Cement and Concrete Research* **25**: 1713–1725, 10.1016/0008-8846(95)00167-0.
43. Kim H-S, Lee S-H and Moon H-Y (2007) Strength properties and durability aspects of high strength concrete using Korean metakaolin. *Construction and Building Materials* **21**: 1229–1237, 10.1016/J.CONBUILDMAT.2006.05.007.
44. Khatib JM and Wild S (1996) Pore size distribution of metakaolin paste. *Cement and Concrete Research* **26**: 1545–1553, 10.1016/0008-8846(96)00147-0.
45. Li Z, Zhang S, Zuo Y, Chen W and Ye G (2019) Chemical deformation of metakaolin based geopolymer. *Cement and Concrete Research* **120**: 108–118, 10.1016/J.CEMCONRES.2019.03.017.
46. Brooks JJ and Megat Johari M (2001) Effect of metakaolin on creep and shrinkage of concrete. *Cement and Concrete Composites* **23**: 495–502, 10.1016/S0958-9465(00)00095-0.
47. Kinuthia JM, Wild S, Sabir BB and Bai J (2000) Self-compensating autogenous shrinkage in Portland cement–metakaolin–fly ash pastes. *Advances in Cement Research* **12**: 35–43, 10.1680/adcr.2000.12.1.35.
48. Gleize PJP, Cyr M and Escadeillas G (2007) Effects of metakaolin on autogenous shrinkage of cement pastes. *Cement and Concrete Composites* **29**: 80–87, 10.1016/J.CEMCONCOMP.2006.09.005.
49. Yang T, Zhu H and Zhang Z (2017) Influence of fly ash on the pore structure and shrinkage characteristics of metakaolin-based geopolymer pastes and mortars. *Construction and Building Materials* **153**: 284–293, 10.1016/J.CONBUILDMAT.2017.05.067.

50. Shen P, Lu L, Chen W, Wang F and Hu S (2017) Efficiency of metakaolin in steam cured high strength concrete. *Construction and Building Materials* **152**: 357–366, 10.1016/J.CONBUILDMAT.2017.07.006.
51. Wang G, Kong Y and Shui Z (2015) Influence of surface treated metakaolin with coupling agent on the properties of concrete. *Materials and Structures* **48**: 261–267, 10.1617/s11527-013-0181-7.
52. Güneyisi E, Gesoğlu M and Özbay E (2010) Strength and drying shrinkage properties of self-compacting concretes incorporating multi-system blended mineral admixtures. *Construction and Building Materials* **24**: 1878–1887, 10.1016/J.CONBUILDMAT.2010.04.015.
53. Al-Akhras NM (2006) Durability of metakaolin concrete to sulfate attack. *Cement and Concrete Research* **36**: 1727–1734, 10.1016/J.CEMCONRES.2006.03.026.
54. Asbridge A, Page C and Page M (2002) Effects of metakaolin, water/binder ratio and interfacial transition zones on the microhardness of cement mortars. *Cement and Concrete Research* **32**: 1365–1369, 10.1016/S0008-8846(02)00798-6.
55. Caldarone MA, Gruber KA and Burg RG (1994) High reactivity metakaolin (HRM): A new generation mineral admixture for high performance concrete. *Concrete International* **16(11)**: 37–41.
56. Curcio F, DeAngelis B and Pagliolico S (1998) Metakaolin as a pozzolanic microfiller for high-performance mortars. *Cement and Concrete Research* **28**: 803–809, 10.1016/S0008-8846(98)00045-3.
57. Khatib JM and Hibbert JJ (2005) Selected engineering properties of concrete incorporating slag and metakaolin. *Construction and Building Materials* **19**: 460–472, 10.1016/J.CONBUILDMAT.2004.07.017.
58. Antoni M, Rossen J, Martirena F and Scrivener K (2012) Cement substitution by a combination of metakaolin and limestone. *Cement and Concrete Research* **42**: 1579–1589, 10.1016/J.CEMCONRES.2012.09.006.
59. Medjigbodo G, Rozière E, Charrier K, Izoret L and Loukili A (2018) Hydration, shrinkage, and durability of ternary binders containing Portland cement, limestone filler and metakaolin. *Construction and Building Materials* **183**: 114–126, 10.1016/J.CONBUILDMAT.2018.06.138.
60. da Silva Andrade D, da Silva Rêgo JH, Cesar Morais P and Frías Rojas M (2018) Chemical and mechanical characterization of ternary cement pastes containing metakaolin and nanosilica. *Construction and Building Materials* **159**: 18–26, 10.1016/J.CONBUILDMAT.2017.10.123.
61. Oriol M and Pera J (1995) Pozzolanic activity of metakaolin under microwave treatment. *Cement and Concrete Research* **25**: 265–270, 10.1016/0008-8846(95)00007-0.
62. Gruber K, Ramlochan T, Boddy A, Hooton R and Thomas MD (2001) Increasing concrete durability with high-reactivity metakaolin. *Cement and Concrete Composites* **23**: 479–484, 10.1016/S0958-9465(00)00097-4.
63. Zibara H, Hooton RD, Thomas MDA and Stanish K (2008) Influence of the C/S and C/A ratios of hydration products on the chloride ion binding capacity of lime-SF and lime-MK mixtures. *Cement and Concrete Research* **38**: 422–426, 10.1016/J.CEMCONRES.2007.08.024.
64. Saikia N, Kato S and Kojima T (2006) Thermogravimetric investigation on the chloride binding behaviour of MK–lime paste. *Thermochimica Acta* **444**: 16–25, 10.1016/J.TCA.2006.02.012.
65. Boddy A, Hooton R and Gruber K (2001) Long-term testing of the chloride-penetration resistance of concrete containing high-reactivity metakaolin. *Cement and Concrete Research* **31**: 759–765, 10.1016/S0008-8846(01)00492-6.
66. Asbridge A, Chadbourn G and Page C (2001) Effects of metakaolin and the interfacial transition zone on the diffusion of chloride ions through cement mortars. *Cement and Concrete Research* **31**: 1567–1572, 10.1016/S0008-8846(01)00598-1.

67. Basheer PA, Gilleece PR, Long A and McCarter W (2002) Monitoring electrical resistance of concretes containing alternative cementitious materials to assess their resistance to chloride penetration. *Cement and Concrete Composites* **24**: 437–449, 10.1016/S0958-9465(01)00075-0.
68. Bucher R, Diederich P, Escadeillas G and Cyr M (2017) Service life of metakaolin-based concrete exposed to carbonation: Comparison with blended cement containing fly ash, blast furnace slag and limestone filler. *Cement and Concrete Research* **99**: 18–29, 10.1016/J.CEMCONRES.2017.04.013.
69. Mikhailenko P, Cassagnabère F, Emam A and Lachemi M (2018) Influence of physico-chemical characteristics on the carbonation of cement paste at high replacement rates of metakaolin. *Construction and Building Materials* **158**: 164–172, 10.1016/J.CONBUILDMAT.2017.10.021.
70. Vu D, Stroeven P and Bui V (2001) Strength and durability aspects of calcined kaolin-blended Portland cement mortar and concrete. *Cement and Concrete Composites* **23**: 471–478, 10.1016/S0958-9465(00)00091-3.
71. Courard L, Darimont A, Schouterden M, *et al.* (2003) Durability of mortars modified with metakaolin. *Cement and Concrete Research* **33**: 1473–1479, 10.1016/S0008-8846(03)00090-5.
72. Khatib J and Wild S (1998) Sulphate resistance of metakaolin mortar. *Cement and Concrete Research* **28**: 83–92, 10.1016/S0008-8846(97)00210-X.
73. Tsivilis S, Kakali G, Skaropoulou A, Sharp JH and Swamy RN (2003) Use of mineral admixtures to prevent thaumasite formation in limestone cement mortar. *Cement and Concrete Composites* **25**: 969–976, 10.1016/S0958-9465(03)00153-7.
74. Smallwood I, Wild S and Morgan E (2003) The resistance of metakaolin (MK)–Portland cement (PC) concrete to the thaumasite-type of sulfate attack (TSA) – Programme of research and preliminary results. *Cement and Concrete Composites* **25**: 931–938, 10.1016/S0958-9465(03)00150-1.
75. Lee ST, Moon HY, Hooton RD and Kim JP (2005) Effect of solution concentrations and replacement levels of metakaolin on the resistance of mortars exposed to magnesium sulfate solutions. *Cement and Concrete Research* **35**: 1314–1323, 10.1016/J.CEMCONRES.2004.10.035.
76. Shi Z, Ferreiro S, Lothenbach B, *et al.* (2019) Sulfate resistance of calcined clay – Limestone – Portland cements. *Cement and Concrete Research* **116**: 238–251, 10.1016/J.CEMCONRES.2018.11.003.
77. Poon C-S, Azhar S, Anson M and Wong Y-L (2003) Performance of metakaolin concrete at elevated temperatures. *Cement and Concrete Composites* **25**: 83–89, 10.1016/S0958-9465(01)00061-0.
78. de Gutiérrez RM, Díaz LN and Delvasto S (2005) Effect of pozzolans on the performance of fiber-reinforced mortars. *Cement and Concrete Composites* **27**: 593–598, 10.1016/J.CEMCONCOMP.2004.09.010.
79. Thiery J and Soukatchoff P (1989) *Method for Selecting a Pozzolan Intended to be Incorporated into a Composite Material Comprising Cement and Glass*. US Patent 4,994,114. https://patents.google.com/patent/US4994114A/en (accessed 22/04/2019).
80. Barger G, Hansen E, Wood M, *et al.* (2001) Production and use of calcined natural pozzolans in concrete. *Cement, Concrete and Aggregates* **23**: 73–80, 10.1520/CCA10478J.
81. Ramlochan T, Thomas M and Gruber KA (2000) The effect of metakaolin on alkali–silica reaction in concrete. *Cement and Concrete Research* **30**: 339–344, 10.1016/S0008-8846(99)00261-6.
82. Shi Z, Shi C, Zhang J, *et al.* (2018) Alkali–silica reaction in waterglass-activated slag mortars incorporating fly ash and metakaolin. *Cement and Concrete Research* **108**: 10–19, 10.1016/J.CEMCONRES.2018.03.002.
83. Wang F, Kovler K, Provis JL, *et al.* (2018) Metakaolin. In *Properties of Fresh and Hardened Concrete Containing Supplementary Cementitious Materials* (De Belie N, Soutsos M and Gruyaert E (eds)). Springer, Cham, Switzerland, pp. 153–179, 10.1007/978-3-319-70606-1_5.

5.4. Alternative reinforcing materials to conventional steel reinforcement or black bars

5.4.1 Introduction

Marios Soutsos, Queen's University Belfast, UK

Concrete, if properly designed, will offer adequate protection for steel reinforcement in most environments, but that assumes that an appropriate cover to the reinforcement has been provided. There are environments in which, however, improving the concrete mix may not be sufficient to protect the steel reinforcement. Alternative reinforcing materials to conventional steel reinforcement (black bars) may, therefore, be considered to ensure that an RC structure is durable

- epoxy-coated steel reinforcement
- galvanised steel reinforcement
- stainless steel reinforcement
- fibre-reinforced polymer (FRP) reinforcement.

The first two of these are still black bars, but they have a protective coating, whereas the third offers improved protection due to a change in the composition of the metal. The fourth is the only one that is not metallic.

There do not seem to have been any major developments in the use of epoxy-coated steel reinforcement and stainless steel reinforcement since the first edition of this book. Readers can access the relevant sections of the first edition from the link below:

https://www.icevirtuallibrary.com/doi/book/10.1680/icehcd.63754

Thus, this section discusses manufacturing methods for galvanised steel reinforcement and FRP reinforcement, how they improve durability, their properties that may affect the design procedures developed for conventional reinforcement, and their reputation based on past applications and performance in real structures.

5.4.2 Galvanised steel reinforcement

Desmond Makepeace, Galvanizers Association, UK
Iqbal Johal, Galvanizers Association, UK

5.4.2.1 Introduction

Galvanised steel reinforcement has been in use for a considerable period, its first application being in the 1930s, although it has been used extensively only since the 1960s and 1970s [1–4]. To fully understand the benefits of galvanised reinforcement, it is important to understand a little about the coating process and the properties of the coating.

5.4.2.2 Galvanising: The process

Hot-dip galvanising is a batch process carried out under factory-controlled conditions. Pieces are immersed in a series of tanks, as shown in Figure 5.42. It is, however, important that the steel is free from heavy oil or grease contamination, paint and welding slag, which are not removed during normal pretreatment.

Typically, steel is degreased and pickled in hydrochloric acid to remove rust, mill scale and light oils. This is followed by immersion in a flux solution, which forms a coating on the steel surface that will aid the reaction between the iron and the molten zinc by increasing the surface wettability. Finally, after drying, the steel is immersed in a bath containing molten zinc, which is typically at a temperature of about 450°C [5–7].

Figure 5.42 Schematic layout of a galvanising plant [5]

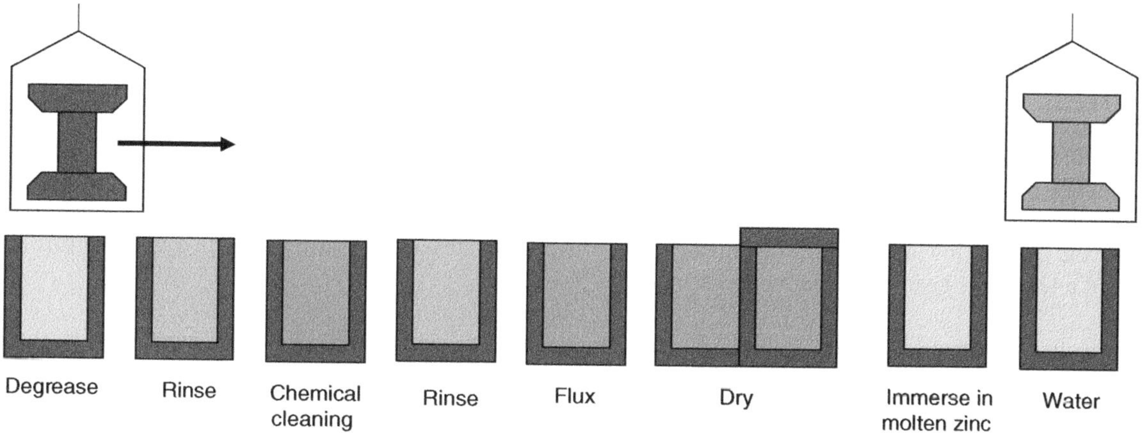

During immersion, the iron reacts with the molten zinc, and a zinc–iron alloy coating is formed, which is metallurgically bonded to the base steel. The alloy layer continues to grow during immersion, gradually becoming richer in zinc, until upon withdrawal from the galvanising bath, some free zinc is pulled out on the surface of the steel reinforcement and solidifies to form an outer layer of pure zinc.

5.4.2.3 Coating properties

The different layers within a galvanised coating offer a unique combination of mechanical properties [5,6,8]. The pure zinc outer layer is very soft and acts as a cushion against impacts, thus protecting the coating from damage during handling and installation. Conversely, the alloy layer beneath is very hard and wear-resistant, typically being harder than the base steel itself and is not easily damaged by abrasion. As a result of these properties, a galvanised coating is robust and far more difficult to damage during normal handling than other corrosion-protection systems. Once formed, a galvanised coating corrodes at a very slow and predictable rate.

5.4.2.4 Protecting steel reinforcement by hot-dip galvanising

A protective layer forms on uncoated reinforcement due to passivation of the steel surface at the high pH encountered in fresh concrete. As a result, an RC structure might be expected to last for a long time [9]. However, this strongly depends upon the performance of the reinforcement, which is affected by a variety of factors, including concrete composition, the w/c ratio, the level of compaction, the amount of concrete coverage, the density and porosity of the concrete, and the service environment in which the structure will operate [10].

It is fair to say that a structure cast from good-quality concrete that provides adequate coverage to the reinforcement will achieve a long life. However, in practice, variables such as concrete quality and accurate positioning of the reinforcement to ensure adequate coverage cannot be guaranteed and strongly depend upon the operatives conducting the work. The steel reinforcement in concrete can corrode due to carbonation or chloride attack [9].

In carbonation, carbon dioxide dissolves into any moisture in the concrete to form an acidic solution. This then reacts with the calcium hydroxide in the concrete and reduces the pH. Carbonation takes place over an extended period, and its effects are not usually seen in the short term, as the carbonation front must reach the steel reinforcement for it to have an effect. However, as it penetrates further into the concrete and reaches the reinforcement, problems may occur, as once the pH at the reinforcement/concrete interface falls below about 11.5, uncoated reinforcement will depassivate [4,11,12] and corrosion may occur, resulting in delamination of the reinforcement and, ultimately, the formation of cracks and spalling of the concrete.

Although a galvanised coating will react with both acids and alkalis, it normally gives good performance in the pH range 5.5–12.5 [4,14] (Figure 5.43). It is, therefore, far more tolerant of alkalis. However, when concrete is first cast, it will typically have a pH of about 13, leading to a short period of attack during which about 10 microns of the coating may be lost [15–17]. However, as the concrete dries and cures, its pH falls. Once the pH is below about 12.5, the galvanised coating passivates against further attack due to the formation of a tenacious layer of calcium hydroxyzincates [2,18,19].

Due to its tolerance of alkalis, galvanised reinforcement withstands carbonation far more readily than uncoated reinforcement [15,17]. Although the coating will normally operate well within a pH range of 5.5–12.5, a more conservative maximum pH of around 9.5 is normally specified for galvanised reinforcement, below which passivation of the galvanised reinforcement ceases [3]. Given the amount of calcium hydroxide present in concrete, a pH of 9.5 cannot be easily reached, so the use of galvanised reinforcement virtually eliminates the risk of carbonation.

The chloride content of concrete also affects the performance of the reinforcement. There are several potential sources of chloride, including calcium chloride, which is added to concrete as a hardening agent, estuary-sourced aggregates, sea spray and de-icing salts. The last two can potentially be encountered once the structure is in service, depending upon its location and the climatic conditions.

Uncoated reinforcement has a relatively low chloride threshold. The American Concrete Institute suggests that above 0.15% of chloride by mass, or 0.4 kg/m^3, corrosion of the reinforcement will likely initiate. Galvanised reinforcement is far more tolerant of chlorides [1–4,6,9,11,12,15–17,20,21] and will significantly delay the onset of corrosion due to chloride attack, as shown in Figure 5.44.

Figure 5.43 Corrosion rate of zinc plotted against the pH of the service environment [13]

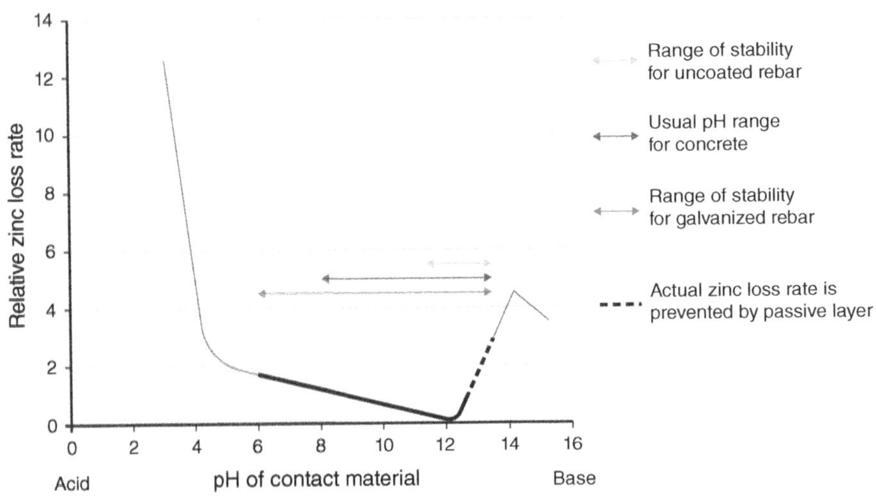

Figure 5.44 Diagram showing the much-increased chloride threshold for galvanised steel reinforcement [2]

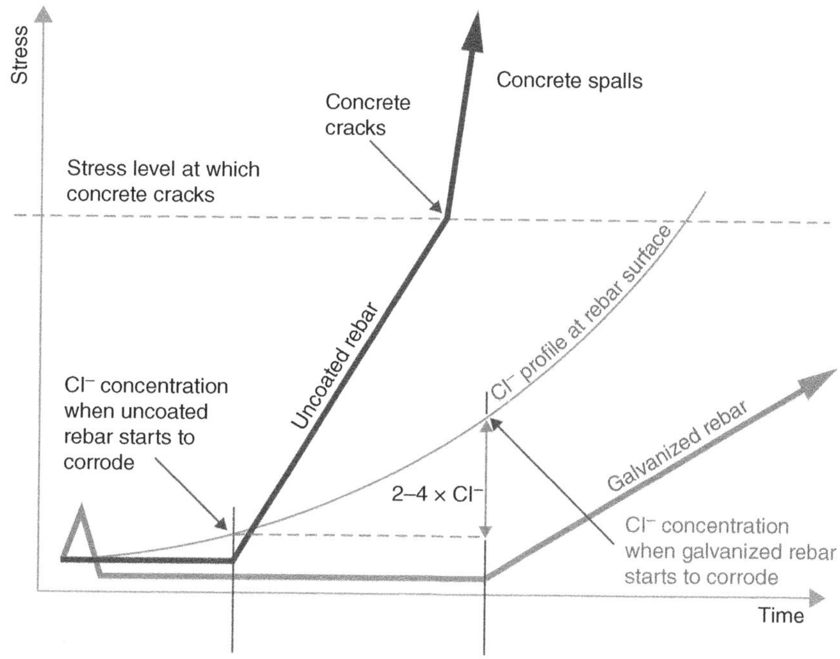

Typically, the threshold quoted for galvanised reinforcement is between two and four times greater than that for uncoated reinforcement. However, in practice, galvanised reinforcement has been shown to offer excellent protection with little or no deterioration of the coating at six to eight times the chloride content for uncoated steel reinforcement. Indeed, in a single reported case, the galvanised reinforcement was performing well despite the chloride content being ten times greater than the accepted threshold for uncoated steel reinforcement [2].

Yeomans [15] calculated the time to onset of corrosion due to chloride attack based upon conservative field data for galvanised reinforcement. There was an increase from 14 years for uncoated reinforcement to 44 years for galvanised reinforcement.

In addition to withstanding carbonation and higher chloride levels, a galvanised coating also offers the unique property of sacrificial protection. This means that if a small area of the coating becomes damaged, the surrounding coating will sacrifice itself to prevent rusting of the exposed steel surface and also prevent undercutting of the remaining sound coating [4–6]. This is of particular importance, as the reinforcement may well be cut to length on site, thus exposing steel at the cut ends.

Even where corrosion of the galvanised coating takes place, the products produced have a smaller volume than rust [3] and diffuse [3,16,17] within the concrete, so filling any porosity that may be present. The net effect is that zinc corrosion products do not produce the same high tensile stresses as rust and, therefore, do not result in disbondment of the reinforcement and cracking and spalling of the concrete.

5.4.2.5 Galvanised reinforcement bond strength

One of the other key advantages of galvanised reinforcement is the more consistent bond strength of the concrete [1,2,6,14,18,20–24]. Tests conducted at the Building Research Establishment indicate that galvanised

Figure 5.45 Variation in bond strength for different reinforcement conditions, as measured at the University of California [18]

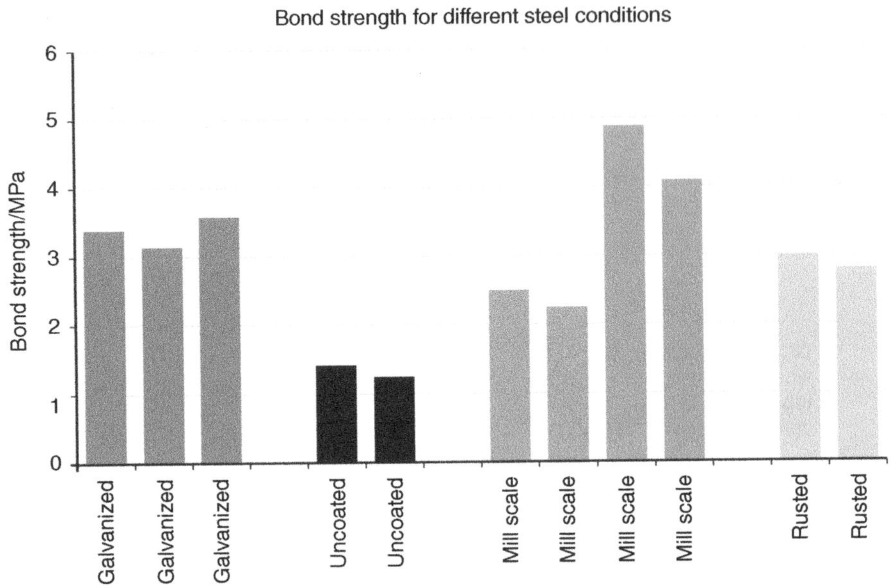

reinforcement has a bond strength of 3.3–3.6 MPa, which compares favourably to the 1.3–4.8 MPa for uncoated reinforcement. Although there are some high results for uncoated reinforcement, there is a much wider spread. Additional tests conducted at the University of California according to the ACI standard also show the consistent nature of results for galvanised reinforcement, as shown in Figure 5.45.

5.4.2.6 Advantages of galvanised steel reinforcement

Galvanising, therefore, offers a variety of advantages, as summarised below.

- Galvanised reinforcement corrodes slowly in an atmospheric environment and, therefore, does not corrode prior to being embedded in the concrete.
- The coating is robust and does not require any special handling.
- The galvanised reinforcement forms a strong bond with the concrete.
- Once embedded in the concrete, galvanised reinforcement withstands carbonation better than uncoated reinforcement and has a higher chloride threshold.
- If passivation breaks down, the galvanised coating corrodes slowly. The corrosion products migrate into the concrete away from the reinforcement, thus filling any porosity.
- The corrosion products formed do not produce the same tensile stresses as rust.
- Due to the good performance of a galvanised coating, variations in concrete quality and level of coverage are less critical.
- Galvanised reinforcement allows a reduction in concrete cover of up to 10 mm, enabling thinner designs with better material efficiency without compromising resistance to carbonation.

5.4.2.7 Cost of galvanised steel reinforcement

Although hot-dip galvanising offers many technical advantages over uncoated reinforcement, there is a cost implication. In recent data provided by a major steel reinforcement supplier and major galvanising groups, the costs for different forms of steel reinforcement are given in the form of an index, as shown in Table 5.11

Table 5.11 Cost of different types of steel reinforcement

Reinforcement	Indexed cost
Uncoated	100
Galvanised (straight bar)	165
Galvanised (bent bar)	200
Fusion-bonded epoxy coated	220
Stainless steel	510

(Hyten, 2008, personal communication; Beech T and Wedge BE, 2008, personal communication; and Catley S, 2008, personal communication). Note that while these figures were accurate at the start of 2008, variations in the price of steel, stainless steel, zinc, etc., may lead to changes in the relative indexes.

Calculations have shown that galvanising would be expected to add 6–10% to the overall cost of the steel reinforcement. However, the cost of the structural frame and skin is typically only 25–30% of the total build cost, so the cost of galvanising all the reinforcement would add only 1.5–3.0% to the total build cost, a figure that could be further reduced by the selective use of galvanised steel reinforcement [4,6].

For a modest increase in initial cost, it is, therefore, possible to effectively eliminate any possibility of carbonation-related issues and very significantly extend the maintenance-free life of the structure at places where there is likely to be a chloride build-up in the concrete.

5.4.2.8 Case histories and recent applications

The performance of galvanised steel reinforcement has been assessed in several structures, most notably bridges in North America, where its use is commonplace in concrete structures in marine environments or locations that are likely to be subject to high levels of de-icing salts during the winter months [2,25,26]. The results of these assessments are summarised in Table 5.12. The percentage chloride figures are compared to the 0.15% threshold for chloride by mass.

The first five bridges still exist and the galvanised reinforcement is in pristine condition, the most notable feature being the high chloride levels measured in four of the cases, which exceeds the typically stated threshold of two to four times that for uncoated reinforcement.

Table 5.12 Summary of the performance of the galvanised reinforcement used in various bridges [2,26]

Bridge	Location	Erected	Last inspection	% chloride (×threshold)	Galvanised coating thickness: μm
Boca Chica	Florida Keys, USA	1972	1999	0.38–0.74 (2.5–4.9)	170
Tioga Bridge	Pennsylvania, USA	1974	2001	0.35 (2.3)	198
Curtis Road Bridge	Michigan, USA	1976	2002	0.14–0.96 (0.9–6.4)	155
Spring Street Bridge	Vermont, USA	1971	2002	0.43–1.14 (2.9–7.6)	191
Evanston Interchange	Wyoming, USA	1975	2002	0.85–1.55 (5.7–10.3)	236
Algoa Bay Footbridge	South Africa	1964	2005	0.15–1.26 (1.0–8.4)	240

Galvanised reinforcement was placed on the seaward side of the Algoa Bay Footbridge (now demolished), where again there was a high build-up of chloride. The other point of interest, though, is that carbonation had progressed to a depth of up to 23 mm, which was beyond the first layer of reinforcement, which had only 15 mm of coverage. The good condition of the galvanised reinforcement illustrated the non-effect of carbonation on its performance.

The Saint-Nazaire Bridge in France was built in the 1970s and is still the longest bridge in the country (Figure 5.46). Most of its concrete reinforcement was not protected against corrosion, and this has required an extensive programme of maintenance and repairs. Around 63 tonnes of galvanised steel reinforcement were used in very specific locations, namely the flanges of the T-beam ends and the connection slabs cast-in-place at the height of the flanges. No repairs have been reported in the specific locations where the galvanised reinforcement was used.

Over 27 000 tonnes of galvanised steel reinforcement ensure the long-term durability of the main span towers, approach span piers, abutments and road deck panels of the Mario M Cuomo Bridge in the USA (Figures 5.47 and 5.48). The new bridge replaced the previous Tappan Zee Bridge over the Hudson River north of New York City. The structure is a 5-km-long, eight-lane, twin-span, cable-stayed bridge designed

Figure 5.46 Saint-Nazaire Bridge, France, built in the 1970s. There has been an extensive programme of maintenance and repairs for the unprotected reinforcement. No repairs have been reported in the specific locations where galvanised reinforcement was used. (© EGGA)

Figure 5.47 The Mario M Cuomo Bridge is a 5-km-long, eight-lane, twin-span, cable-stayed bridge designed for a minimum 100-year life using galvanised reinforcement. (© New York State Thruway Authority)

Figure 5.48 Prior to selecting galvanised reinforcing steel for the Mario M Cuomo Bridge, the New York State Thruway Authority developed a corrosion protection plan to identify exposure, degradation mechanisms, design and construction strategies as well as life-cycle costs. (© New York State Thruway Authority)

for a minimum 100-year life using galvanised reinforcement. The bridge deck comprises nearly 6000 pre-cast panels, each 3.66 m long and 6.7–13.7 m wide, for the approach spans and 963 main span deck panels. The panels were joined by pouring concrete around the protruding galvanised steel reinforcement.

5.4.2.9 Specifications for galvanised reinforcing steel

Galvanised reinforcing steel is specified according to EN 10348-2:2019 'Steel for the reinforcement of concrete – Galvanized reinforcing steel – Part 2: Galvanized reinforcing steel products'. EN 10348-2 gives requirements for galvanised reinforcing steel products that have been manufactured from bars that meet the requirements of EN 10080. Requirements for galvanising the bars are in accordance with the international standard for batch galvanising, EN ISO 1461, with the exception that coating thickness requirements for reinforcing steel products are specifically defined, as listed in Table 5.13.

EN 10348-2 also ensures that the rib geometry (rib height or indentation depth) is satisfactorily retained after galvanising. To avoid any possible consequences for the mechanical properties, the standard sets minimum bend diameters for bars that are bent before galvanising.

5.4.2.10 Summary

Hot-dip galvanised reinforcement has been used extensively in structures around the world. The increased cost of using galvanised reinforcement is minimal compared with the costs of potential future maintenance work, which might otherwise be required. Its use can also facilitate designs with reduced concrete cover, in particular to prevent against carbonation.

Table 5.13 Coating thickness requirements of EN 10348-2:2019

Steel diameter: mm	Coating mass: g/m^2	Coating thickness: microns
>6	610	85
≤ 6	505	70

REFERENCES

1. Yeomans SR (1991) *ILZRO Project ZE341. Comparative Studies of Galvanized and Epoxy Coated Steel Reinforcement in Concrete.* University of New South Wales, Australia, Research report R103.
2. International Zinc Association (2006) *Hot Dip Galvanized Reinforcing Steel. A Concrete Investment.* International Zinc Association, Brussels, Belgium.
3. Yeomans SR (1993) *Considerations of the Characteristics and Use of Coated Steel Reinforcement in Concrete.* University of New South Wales, Australia, Report NISTIR 5211.
4. Yeomans SR (2005) Corrosion protection for steel in concrete. The case for galvanizing. *6th Asian Pacific General Galvanizing Conference, Cairns, Australia.*
5. Galvanizers Association (2016) *Engineers and Architects' Guide: Hot Dip Galvanizing.* Galvanizers Association, Sutton Coldfield, UK.
6. Galvanizers Association of Australia (1999) *After Fabrication Hot Dip Galvanizing.* Galvanizers Association of Australia, Melbourne, Australia.
7. Galvanizers Association (2015) *The Galvanizing Process.* Galvanizers Association, Sutton Coldfield, UK, HDG datasheet 2.
8. Galvanizers Association. *Galvafact – Toughness & Abrasion Resistance.* Galvanizers Association, Sutton Coldfield, UK.
9. Yeomans SR (2004) *Galvanized Steel Reinforcement in Concrete.* Elsevier Science.
10. Davies H (1990) Performance of concrete reinforcement products. *Transactions of the Institute of Metal Finishing* **60(3)**: 103–108.
11. Yeomans SR (2006) Corrosion of steel reinforcement in concrete. *CFC Rebar Workshop, Singapore.*
12. Yeomans SR (1994) A conceptual model for the corrosion of galvanized steel reinforcement in concrete. *International Conference on Corrosion and Corrosion Protection of Steel in Concrete, Sheffield, UK.*
13. European General Galvanizers Association (2021) *Galvanized Reinforcement in Concrete Structures, An Introduction for Engineers and Designers.* European General Galvanizers Association, Caterham, UK.
14. Porter F (1991) *Zinc Handbook: Properties, Processing and Use in Design.* Marcel Dekker, New York, NY, USA.
15. Yeomans SR (2006) Processing and characteristics of galvanized steel for use in concrete. *CFC Rebar Workshop, Singapore.*
16. Nordic Galvanizers (2007) *Galvanizing Handbook.* Nordic Galvanizers, Stockholm, Sweden.
17. Fratesi R (2002) Galvanized reinforcing steel bars in concrete. *COST 521 Workshop, Luxembourg.*
18. European General Galvanizers Association (2021) *Galvanized Reinforcement in Concrete Structures: An Introduction for Engineers and Designers.* European General Galvanizers Association, Caterham, UK.
19. White R, Goodwin FE and Rourke D (2007) Galvanized reinforcement for concrete: Research update. *7th Asian Pacific General Galvanizing Conference, Beijing, China.*
20. Yeomans SR (1998) Corrosion of the zinc alloy coating in galvanized reinforced concrete. *Corrosion 98 Conference, San Diego,* paper 653.
21. American Galvanizers Association (1996) *Galvanizing for Corrosion Protection: A Specifiers Guide to Reinforcing Steel.* American Galvanizers Association, Centennial, CO, USA.
22. American Galvanizers Association (2006) *Hot Dip Galvanizing for Corrosion Protection: A Specifiers Guide.* American Galvanizers Association, Centennial, CO, USA.
23. Kayyali OA and Yeomans SR (1995) Bond and slip of coated reinforcement in concrete. *Construction and Building Materials* **9(4)**: 219–226.
24. Swamy RN (1994) *Corrosion and Corrosion Protection of Steel in Concrete.* Sheffield Academic Press, Sheffield, UK, vols 1 and 2.
25. Galvanized Rebar Resource Center. *Case Histories.* www.galvanizedrebar.com.
26. Galvanizers Association of Southern Africa (2005) Case history 04/2005. Hot dip galvanized steel reinforcement in concrete. *Hot Dip Galvanizing Today* **2(3)**: 51.

5.4.3 Fibre-reinforced polymer reinforcement

Marco Rossini, University of Miami, USA
John Myers, Missouri S&T, USA
Antonio Nanni, University of Miami, USA

5.4.3.1 Applications and uses of fibre-reinforced polymer reinforcement

One of the primary causes of premature failure in RC and prestressed concrete is corrosion of the steel reinforcement. When steel corrodes, its effective cross-section reduces, and, therefore, so does its load-carrying capacity. Furthermore, corrosion products have an expansive action that causes the adjacent concrete to crack and spall, promoting the penetration of aggressive chemicals. In addition, a chemical electrolytic cell often develops within the structure, causing a battery effect that accelerates corrosion of the steel reinforcement. The corrosion rate is accelerated in chloride-rich environments, such as a bridge deck subjected to de-icing agents or a marine structure exposed to salt water. Other sensitive applications include RC or prestressed concrete elements subjected to low-pH solutions, such as containment vessels for waste-water treatment or chemical storage.

One traditional solution for corrosion consists of protecting the reinforcement by reducing the porosity of the concrete. This can be achieved by increasing the amount of cement or using concrete admixtures. Another traditional approach is to increase the clear cover of concrete to retard aggressive agents from reaching the reinforcement layer. The use of ferrous corrosion-resistant solutions has also been tried with varying degree of effectiveness. Epoxy-coated and galvanised steel bars and strands have proved to be ineffective in highly corrosive marine environments, leading some governing agencies in the US, including the Florida Department of Transportation, to ban their use. Stainless steel bars are readily available and have good corrosion resistance. However, their prohibitive cost prevents large-scale deployment.

Being a non-ferrous material, fibre-reinforced polymer (FRP) is non-corrosive and impervious to chloride attack. Therefore, using FRP reinforcement can extend the service life of RC and prestressed concrete structures. Typical applications for FRP include in highly corrosive chloride-contaminated or low-pH environments or, simply, for concrete elements that are required to be thin and so have inadequate cover for traditional steel reinforcement. Other applications take advantage of the electromagnetic transparency of FRP, such as when concrete members are adjacent to high voltages, magnetic fields, stray currents, or radio or high-frequency signals. Typical applications include the reinforcement of magnetic resonance imaging units in hospitals, high-voltage substations, transformer pads, cable ducting, and inductance loops in highways and tollbooths. In addition to being non-corrosive and electrically transparent, FRP reinforcement is also thermally non-conductive. Some applications take advantage of this property, for example, by surrounding thermal shunts with FRP, such as for cast balcony units and connections in sandwich walls.

Unlike steel, FRP reinforcement is highly anisotropic. Thus, it has good mechanical properties in the direction of the fibres but low transverse strength and stiffness. The anisotropy makes cutting and demolition easier, and some temporary applications take full advantage of this property. An example are soft-eyes that make it easier for tunnel-boring machines to penetrate concrete when launching an excavation. In sequential excavation tunnelling, FRP reinforcement is used to stabilise the soil temporarily and is then consumed by the tunnelling and mining equipment. Similarly, FRP soil nails are used in passive removable applications.

FRP reinforcement has been used in concrete structures for more than 40 years. To date, over 270 bridges have been built using FRP reinforcement in the US and Canada, according to statistics collected by the American Composites Manufacturers Association [1]. There are also several applications in China, Russia and the Middle East, plus some use in Japan, Europe, Oceania and South America. The Halls River Bridge (Figure 5.49) is a successful example of a structure built entirely using corrosion-resistant solutions, mostly

Figure 5.49 Aerial view of the Halls River Bridge, Florida

FRP reinforcement. All the FRP-reinforced bridges that have been built to date are meeting their design requirements and weathering the challenge of time without signs of degradation.

5.4.3.2 Properties of FRP reinforcement

FRP reinforcement is made from continuous fibres, typically glass, basalt, carbon or aramid. Fibres are impregnated with a polymeric thermosetting resin, typically vinyl ester, epoxy, polyester or acrylic. The fibres provide tensile strength, and the resin acts as a binder. The resin has an important role in transferring stresses from fibre to fibre as well as in protecting the fibres from environmental exposure. Figure 5.50 shows glass-fibre-reinforced-polymer (GFRP) bars of different sizes and surface preparations produced by various manufacturers.

Figure 5.50 GFRP bars from various manufacturers

Unlike steel, which exhibits isotropic ductile behaviour, FRP reinforcement is linear elastic up to failure. Furthermore, it has anisotropic and asymmetric behaviour with good tensile strength and stiffness, poor transverse properties and a tendency to fibre-buckling, which limits its compressive properties. Furthermore, FRP reinforcement exhibits a size effect known as shear lag. As the diameter of an FRP bar increases, loads cannot be efficiently transferred from the external layers to the core of the bar. Thus, the tensile strength in terms of stresses of an FRP bar reduces with increasing diameter, as shown in Table 5.14. FRP reinforcement is typically stronger but less stiff than steel. These are appealing features for prestressed concrete applications, where losses are directly proportional to the elastic modulus of the reinforcement and the initial pull is limited by the elastic capacity of the material. Prestressing of FRP reinforcement is possible, if adequate anchoring is provided at pull.

The mechanical properties of FRP reinforcement are mostly dictated by the choice of fibre. GFRP is the most common and the least expensive alternative. It is typically deployed in RC applications in the shape of solid bars. The ultimate tensile strength of GFRP reinforcement ranges from 500 to 1500 MPa, depending on the size, shape, constituents and manufacturing processes. The elastic modulus ranges from 45 to 50 GPa. Glass fibres are manufactured by large multinational companies, such as Owens Corning, Pittsburgh Plate Glass Company, Vetrotex CertainTeed, Saint-Gobain and others. Basalt FRP (BFRP) has recently been developed as an alternative to GFRP. It is undergoing standardisation to ensure adequate quality control of the basalt fibres and their composites. Carbon FRP (CFRP) is the most expensive and the best performing among commercially available FRP reinforcement solutions. It is typically deployed in prestressed concrete applications in the shape of multi-wire strands. The ultimate tensile strength of CFRP reinforcement ranges from 2000 to 3000 MPa, depending on the size, shape, constituents and manufacturing processes. Its elastic modulus is approximately 155 GPa. Aramid fibres are organic fibres widely used in the ballistics industry, and aramid FRP has been deployed in prestressed concrete applications. However, its poor durability and relevant material cost have limited its development.

Even if FRP reinforcement does not corrode, it can degrade in certain environments. In general, the degradation rates are orders of magnitude lower than those of steel and the processes do not entail loss of the effective cross-section, which can result in the complete dissipation of steel reinforcement. The degradation of FRP is predictable and accounted for at the design stage by simply introducing an environmental

Table 5.14 Mechanical properties of GFRP bars [2]

GFRP bar designation	Nominal diameter: mm	Nominal area: mm^2	Measured area: mm^2	Minimum guaranteed tensile strength:	
				kN	MPa
M6	6.3	32	30–55	27	844
M10	9.5	71	67–104	59	831
M13	12.7	129	119–169	96	744
M16	15.9	199	186–251	130	653
M19	19.1	284	268–347	182	641
M22	22.2	387	365–460	241	623
M25	25.4	510	476–589	297	582
M29	28.7	645	603–733	365	566
M32	32.3	819	744–894	437	534

knock-down factor to the material strength. No further actions are required to ensure the performance is met through the service life of the structure. Conversely, the degradation of steel reinforcement is difficult to predict and cannot be accounted for at the design stage. Therefore, constant monitoring and repairs are required to ensure that steel-reinforced structures meet their design performance through their service life. The durability of GFRP bars after up to 15 years of service in bridge decks and superstructures has been validated through an extensive monitoring campaign in the US and Canada [3,4].

The physical properties and durability of FRP reinforcement are dictated by the choice of both the fibre and the resin. Different types of glass fibres have different resistances to alkaline and corrosive environments. Electrical-grade glass was used in the past but has been replaced by a corrosion-resistant alternative to meet the durability requirements for more than 100 years of service life. Considering other types of fibres, carbon has a very good chemical resistance. The performance of basalt fibres is being validated. The issues related to the durability of aramid fibres are well known and related to their organic nature. Concerning the choice of resin, unsaturated polyester was commonly applied in the past, but its poor durability in moist environments is limiting its use. In the US, only vinyl ester and epoxy resins are allowed, given their superior performance. These resins are produced by multinational companies such as Ashland Chemical, Dow Chemical, Reichhold Chemical, DSM and others.

The role of an FRP reinforcement manufacturer is to combine fibres and resin into a pultruded composite rod. During pultrusion, fibres are impregnated with resin and drawn through a heated die from which they emerge as a semi-final product. Then, various surface preparation techniques can be employed to enhance the bonding of the FRP bar to the concrete. These include sand-coating, surface deformations like those of traditional steel bars and carving. In addition, several small-diameter composite rods can be twisted into a single strand for prestressed concrete applications. Several variables affect the quality of the final product, mainly temperature, pultrusion speed, resin formulation and fibre size. Nevertheless, once defined and controlled, pultrusion is a stable and reliable process.

The emphasis in commercial FRP manufacturing is currently on GFRP bars for RC. There are several manufacturers of GFRP bars worldwide, whereas there are few CFRP strand manufacturers and prestressed concrete commercial applications represent a small portion of their market. Recently, some manufacturers have begun production of BFRP bars alongside GFRP bars [5]. Conversely, although commercial aramid FRP solutions were widely developed through the 1980s and 1990s, currently there is limited commercialisation.

5.4.3.3 Material specifications and testing

The standardisation of FRP reinforcement began in the early 1990s and is eventually reaching maturity. In the US, standard specifications for GFRP bars are covered by ASTM D7957 [2]. Specifications for other FRP products are variously covered by documents published by ACI, the International Code Council and the American Association of State Highway and Transportation Officials (AASHTO) [6–8]. These documents will be withdrawn or harmonised once relevant ASTM standards are published. In Canada, standard specifications for FRP bars are covered by CSA S807 [9]. In Europe, while a consensus standard is under preparation, test methods are standardised by ISO 10406 [10]. In Russia, standard specifications for FRP bars are covered by GOST 31938 [11]. In China, standard specifications for FRP bars are covered by GB/T 26743 [12].

To ensure that commercial FRP products meet standard requirements, testing is performed as detailed by relevant specifications. Historically, testing of FRP reinforcement has been performed based on the test methods outlined in ACI 440.3R [13]. The document was developed to facilitate the preparation of design guidelines and for a transition to ASTM standards. In the US, most individual test methods have already been published as ASTM standards, such as ASTM D7205 [14], which provides a standard test method to measure the tensile properties of FRP bars. In Europe, ISO 10406 [10] is typically preferred to its ASTM counterparts. In Canada, Russia and China, national standards [9,11,12] provide specific test methods in addition to material specifications. Harmonisation is being promoted among different national standards where possible; thus, test methods and minimum requirements are typically similar among different standards.

Specifications must standardise FRP reinforcement without limiting the potential for innovation nor penalising virtuous manufacturers. Thus, taking GFRP bars standardised by ASTM D7957 [2] as an example, manufacturers are free to refine constituents, optimise the pultrusion process and select the surface preparation of their choice if the bar is round and meets the minimum requirements in terms of mechanical performance and durability. Some variations on the shape and size of bars produced by different manufacturers do exist. To address the issue, nominal bar sizes are defined to be equivalent to traditional steel bar sizes. A range is defined for the measured area of a GFRP bar to meet the ASTM D7957 [2] specification, as shown in Table 5.14. Mechanical properties for certification are measured on the nominal area, which is also the reference for design.

Standardised tests enable a consistent determination of the relevant engineering properties and give the designer, contractor and owner confidence that the FRP reinforcement deployed will meet the properties that were assumed in the design calculations. Standardisation enables independent quality control and allows the prequalification of certified manufacturers that are then allowed to bid on public projects.

5.4.3.4 Design guidelines

Based on the experience gained from past projects and a voluminous amount of academic research, several design guidelines have been published internationally. In the US, guidelines for the design of buildings reinforced with FRP bars are provided by ACI 440.1R [15], whereas guidelines for the design of buildings prestressed with FRP reinforcement are provided by ACI 440.4R [16]. A document written in mandatory language is under preparation by ACI Committee 440 and covers building code specifications for buildings reinforced with GFRP bars. For infrastructural applications, the design of passive GFRP reinforcement is regulated by the second edition of *AASHTO LRFD Bridge Design Guide Specifications for GFRP-Reinforced Concrete* [17], whereas the design of active CFRP reinforcement is regulated by the first edition of AASHTO's *Guide Specifications for the Design of Concrete Bridge Beams Prestressed with CFRP Systems* [8]. The two documents have been recently developed by two separate task forces of researchers, practitioners and government officials [18,19]. In Canada, two separate documents have been published: CSA S806 – 'Design and construction of building structures with fibre-reinforced polymers' [20] and, CSA S6 – 'Canadian highway bridge design code' [21], which includes provisions for the use of FRP bars and strands. In Europe, guidelines for the design of FRP reinforcement are included in *fib Model Code 2010* [22] and discussion is undergoing for the inclusion of FRP reinforcement in 'Eurocode 2' [23]. In Russia, guidelines for the design of FRP reinforcement are provided by SP 295.1325800 – 'Concrete structures reinforced with fibre-reinforced polymer bars: Design rules' [24]. Similarly, in China, design guidelines are provided by GB 50608 – 'Technical code for infrastructure application of FRP composites' [25]. These documents provide designers with authoritative guidance using a traditional regulatory format.

Design with FRP reinforcement follows the same principles established for RC and prestressed concrete members reinforced or prestressed with conventional steel bars and strands. Strain compatibility and stress equilibrium hold valid under traditional Bernoulli assumptions. The main difference is in the constitutive model used to describe the behaviour of the reinforcement and the failure criterion. The brittle nature of FRP reinforcement implies the possibility that over-reinforced flexural members may fail because of

Figure 5.51 Stress–strain curves for steel and GFRP bars [18]

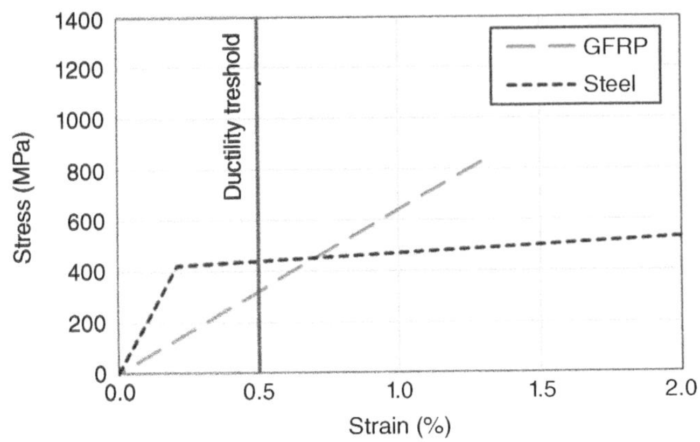

concrete collapse in the compression zone or that under-reinforced flexural members may fail because of reinforcement rupture in the tension zone. Whereas under-reinforcement is typically preferred when designing for traditional steel bars and strands because it ensures a ductile failure, over-reinforcement is sometimes preferred when designing for FRP. Irrespective of the failure mode, thanks to its relatively low elastic modulus, an FRP-reinforced member will typically experience large-enough deformations that an upcoming collapse can be foreseen before it occurs. GFRP bars have ultimate strains well beyond the ductile threshold traditionally set at 5000 microstrain for steel reinforcement [18], as shown in Figure 5.51. This pseudo-ductile behaviour is adequate for static applications. However, the deployment of FRP reinforcement in earthquake-resistant members, where high ductility is required, is not allowed.

Taking the AASHTO approach to GFRP-RC design [17] as an example, resistance factors vary between the two failure modes. Resistance factors are applied a posteriori to the flexural strength of a member. A very conservative safety factor of 0.55 is applied to the flexural strength of under-reinforced members, whereas over-reinforced members have a safety factor of 0.75 in line with traditional steel reinforcement. Further conservativeness is added by defining the guaranteed strength of FRP reinforcement at the 99.9th strength percentile (average minus three standard deviations), despite a characteristic strength at the 95th percentile (average minus 1.64 standard deviations) as done for traditional construction materials. The design strength of FRP reinforcement is computed by multiplying the guaranteed strength by an already mentioned environmental knock-down factor ranging from 0.7 to 0.8 for GFRP, depending on the exposure conditions.

The design of GFRP-RC is rarely governed by strength considerations. Given the relatively low elastic modulus of GFRP, serviceability considerations typically govern. These include limiting crack width and deflection. Furthermore, the stresses induced in the reinforcement by sustained and cyclic loads must be limited within allowable thresholds to avoid the occurrence of creep ruptures, cyclic fatigue failures or a combination of the two.

For shear design, the second edition of *AASHTO LRFD Bridge Design Guide Specifications for GFRP-Reinforced Concrete* [17] adopts the modified compression field theory, which is the standard approach in the eighth edition of *AASHTO LRFD Bridge Design Specifications* [26]. The sectional shear strength is computed as a sum of the contribution of the concrete to the contribution of the FRP shear reinforcement, if provided. The relatively low elastic modulus of GFRP longitudinal bars results in a reduced dowel action, affecting the concrete contribution. Furthermore, where GFRP stirrups are present, their exploitability is reduced to a portion of their strength to limit the opening of transverse cracks and avoid failures at the bend, which has reduced strength because the fibres would have buckled during bending.

Furthermore, the second edition of *AASHTO LRFD Bridge Design Guide Specifications for GFRP-Reinforced Concrete* [17] is the first guideline covering the design of torqued and compressed GFRP-RC elements, thus allowing the design of any structural component of a bridge, from foundations to traffic barriers. The approach to torque is consistent with the eighth edition of *AASHTO LRFD Bridge Design Specifications* [26] and includes adjustments like those discussed for shear design. The design of compressed elements must account for the reduced strength and stiffness of GFRP bars under compression as opposed to their good tensile properties. Thus, the contribution of GFRP in compression is neglected.

These additional safety layers are being refined, and reduction factors are being progressively calibrated toward a more efficient design as experimental evidence becomes available and experience is gathered from real applications. With the publication of the second edition of *AASHTO LRFD Bridge Design Guide Specifications for GFRP-Reinforced Concrete* [17], the amount of FRP bars required to reinforce a standard bridge deck has been reduced from a factor of 2.0 to a factor of 1.5 with respect to a traditional steel reinforcement design. This corresponds to an increment of 10% in the initial investment for an estimated 25% reduction in total costs over the service life of the structure, and without the need for extraordinary maintenance or repair operations [27].

Further optimisation can be achieved by adopting an empirical design method, also included in the second edition of *AASHTO LRFD Bridge Design Guide Specifications for GFRP-Reinforced Concrete* [17]. The empirical method means GFRP-RC bridge decks can be designed by mostly relying on the arching effect that develops inside the concrete and using a limited amount of reinforcement. This approach is like that prescribed by the 'Canadian highway bridge design code' [21].

5.4.3.5 Construction

Generally, installing FRP bars is mainly the same as installing traditional steel bars. Conversely, prestressing with FRP strands entails some added complexity and requires the use of specific anchors to avoid damaging the strands when pulling them. In the US, construction specifications for buildings reinforced with FRP bars are provided in mandatory language by ACI 440.5 [28]. For infrastructural applications, construction specifications are provided alongside AASHTO design specifications for GFRP-RC and CFRP-PC [8,17]. In Canada, construction specifications for buildings reinforced or prestressed with FRP are provided by CSA S806 [20]. Other national standards outside the US incorporate some handling and construction considerations alongside material or design specifications [9,11,21,24,25].

During construction, FRP reinforcement has two fundamental differences with respect to steel bars and strands. Being a brittle material, FRP cannot be bent on site to form hooks and stirrups but must be shaped during manufacture before the resin sets. Furthermore, being flammable, FRP cannot be cut using a torch, so a grinder should be used instead. Recently, thermoplastic resins have been proposed as an alternative to traditional thermosetting solutions. Theoretically, a thermoplastic bar could be heated and bent to shape at any time, making it easier to manufacture complex shapes and, potentially, transforming the supply chain. However, the quality control required for bending would require the operation to be carried out in a controlled facility environment and not at a construction site.

There are other minor differences, including the relatively low modulus of elasticity of FRP, which must be accounted for during construction. Typically, FRP bars need to be supported with chairs at a spacing of two-thirds of that needed for steel bars. Furthermore, FRP is lighter than concrete, so that concerns have been expressed that it may float when pouring the concrete. Therefore, it is recommended that at least 50% of bar intersections are tied using plastic-coated wire or zip ties. As an additional precaution, cages may be tied down in some locations. These provisions are conservative, and no issues have been recorded during field construction. The potential abrasion of FRP reinforcement during construction should be considered. Small scratches can be visually assessed and coated. Wider damage will require the bar to be rejected. Experience has proved that FRP reinforcement can withstand typical construction loads and handling without damage and without the need for particular precautions. As an example, Figure 5.52 shows the installation of an FRP cage for an auger-cast pile. Finally, FRP reinforcement is sensitive to prolonged UV exposure, which may negatively affect the mechanical performance of the resin. This has to be considered during storage. Typically, FRP reinforcement should not be exposed to sunlight for more than 4 months during construction.

Recent experience and rigorous investigations have proven that using FRP reinforcement in construction is as efficient as using traditional steel bars and, in some cases, more convenient [29]. Thanks to its light weight at one-quarter the weight of steel, ease of transportation and reduced need for machinery, construction with FRP reinforcement may be faster, cheaper and require less energy.

5.4.3.6 Successful projects

Over the last 40 years, many successful projects using FRP reinforcement have been completed worldwide, making it challenging to provide a comprehensive summary. Metrics collected by [1] count over 270 bridges built using FRP reinforcement in the US and Canada. The Canadian Network of Centres of Excellence on Intelligent Sensing for Innovative Structures has published some key metrics on the performance of bridge

Figure 5.52 Installation of an FRP pile during the construction of the Innovation Bridge, University of Miami

structures after 5 to 8 years of service [30]. More recently, ACI conducted a similar investigation on structures that have been in service for more than 15 years. Analyses show that there has been no appreciable degradation of the GFRP bars, in line with experimental predictions. Some relevant projects are discussed hereinafter.

The Floodway Bridge over the Red River near Winnipeg (Figure 5.53) was one of the largest deployments of FRP bars in a single project. There are two bridge structures, each with eight spans giving a total length of 348 m. GFRP bars were used in the barrier walls and in the deck slab that was designed according to the empirical approach. The project required over 150 tonnes of GFRP bars, equivalent to approximately 500 tonnes of steel, accounting for the difference in weight. FRP reinforcement was chosen based on purely economic considerations after a full life-cycle analysis was performed. The Floodway Bridge was among the first GFRP-RC bridge decks to be constructed outside of a research project. This is important to note, as there were no external factors influencing the decision beyond a cost–benefit analysis.

Built on the Coral Gables campus of the University of Miami, the Innovation Bridge was one of the first structures to be built using solely FRP reinforcement [31]. Different material systems were deployed to exploit the specific features of each type of fibre. CFRP strands were used to pre-tension the two 20-m double-T girders constituting the main structure of the bridge (Figure 5.54). GFRP and BFRP bars were used to reinforce the deck, abutments and auger-cast piles. The installation of a pile is illustrated in Figure 5.52. The structure was one of the first deployments of BFRP in a bridge project, and one of the first deployment of FRP in a bridge substructure. The reinforcement had a variety of peculiar shapes, including closed-loop filament-wound stirrups to ensure that there was ideal bonding to the concrete. Pre-assembled pile cages and deck meshes significantly sped up construction on site and allowed for better quality control.

The Florida Department of Transportation is one of the most proactive agencies in promoting the deployment of FRP reinforcement in bridges and other transportation infrastructure. Built in Homosassa, Florida, the Halls River Bridge (Figure 5.49) is a five-span vehicular bridge entirely built using corrosion-resistant solutions and mostly FRP reinforcement. The bridge was a conceptual prototype for future FRP deployments in the state of Florida. The structure includes CFRP-prestressed concrete (PC) bearing piles, CFRP-PC/GFRP-RC sheet piles, high-strength carbon steel-PC/GFRP-RC sheet piles, GFRP-RC pile bent caps and bulkhead caps, a GFRP-RC bridge deck, GFRP-RC traffic railings, GFRP-RC approach slabs and a GFRP-RC gravity wall. The unprecedented variety and completeness of the material and structural solutions deployed make the Halls River Bridge an invaluable source of data. Thus, monitoring protocols

Figure 5.53 Floodway Bridge over the Red River near Winnipeg

Figure 5.54 Prestressed CFRP girder for the Innovation Bridge

were implemented in the design and construction stages, which will continue throughout the service life of the structure. A life-cycle cost analysis alongside a complete life-cycle assessment analysis found that this complete FRP-RC/PC design had the least impact, both economic and environmental perspectives [29].

5.4.3.7 Ongoing research

A voluminous amount of research has addressed the fundamental questions related to the deployment of FRP reinforcement in structural applications. These include durability, mechanical characterisation, design algorithms and construction procedures. Nevertheless, research is undergoing to facilitate transfer to current practice and develop more efficient and rational design and construction approaches. Furthermore, new research areas are exploring the exploitation of some untapped technology potential.

Prestressing CFRP and aramid FRP reinforcement has historically been a technological challenge. Carbon and aramid fibres have a high tensile strength but also a high material cost. To justify their deployment from an economic perspective, extreme pulling is required to exploit as much tensile strength as possible, which requires complex anchor systems due to the limited shear strength of FRP. Moreover, there is the risk of failure during pulling. Limiting the initial pull would simplify pre-tensioning operations and enable steel-like constructability with traditional anchors. However, this solution requires a material system that is mechanically more efficient and is cheap enough. Thanks to its relatively low elastic modulus, which reduces prestress losses, and relatively low material cost, GFRP strands (Figure 5.55) are being developed for prestressed concrete applications within the MILDGLASS project funded by NCHRP-IDEA [32]. The project is focusing on those prestressed concrete members for which mild levels of prestress are enough but where corrosion of the steel reinforcement is a concern. These include bridge substructures – piles and sheet piles – exposed to seawater but also precast deck panels in cold-weather regions.

Historically, thermosetting resins like vinyl ester and epoxy have been preferred to thermoplastic alternatives because of their chemical stability and higher glass transition temperature. However, the inability to modify the shape of an FRP bar once the resin has cured restricts the development of an efficient logistics chain. In the steel industry, a steel mill is not responsible for shaping and cutting bars to order. Instead, a bar fabricator stocks a large quantity of coiled straight bars that are readily shaped and cut once orders are received from a construction site. In the FRP industry, the bar manufacturer must bend and shape the FRP bars and stirrups when the bar is pultruded. Therefore, field operations may be forced to plan their schedules

Figure 5.55 Prototype GFRP strand

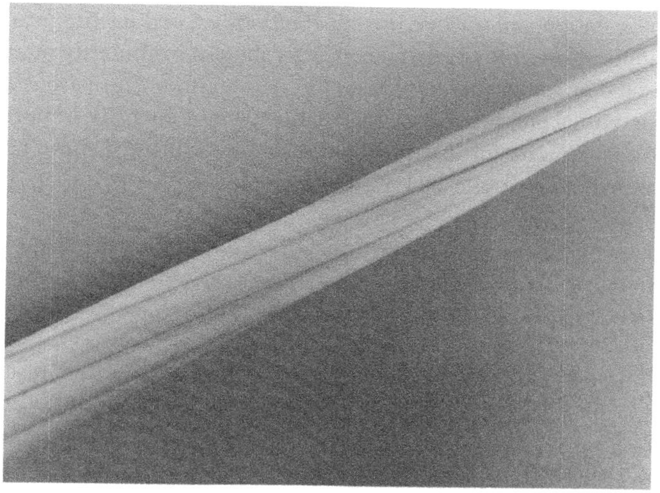

according to the speed of a pultrusion line. Any adjustments during construction may represent a challenge for bar suppliers. Using a thermoplastic resin would allow the production of stocks of coiled FRP bars that could subsequently be heated, shaped and cut to length as orders are received. This solution would sensibly speed up the delivery of complex shapes and stirrups to construction sites. A new generation of thermoplastic resins is being developed with mechanical and physical properties comparable to state-of-the-practice thermosetting solutions. The development of thermoplastic FRP reinforcement solutions is also being investigated within the MILDGLASS project funded by NCHRP-IDEA [33].

A traditional solution for addressing corrosion of the reinforcement entails designing concrete mixes with limited permeability and superior durability. This approach lays the durability burden on cement manufacturers and ready-mix producers. It typically involves cement-rich concrete mixes and exacerbates the concrete cost and the environmental impact of the industry. The use of non-corrosive reinforcement relieves the need for low-permeability mixes and even allows the use of chloride-contaminated aggregates, recycled materials, bypass dust and seawater for mixing. The performance of chloride-contaminated concrete mixes coupled with FRP reinforcement has been investigated through the SEACON project co-funded by the European Commission and the Federal Highway Administration through the Infravation platform [34]. The Halls River Bridge served as a demonstrator for the project along with a sister structure built in Italy. Some portions of the bulkhead cap and the retaining wall were cast using seawater concrete, recycled concrete and recycled asphalt aggregates, and these are being monitored to assess their performance. The use of seawater and recycled material would reduce the environmental impact of the concrete industry. Similarly, alternative binders are being extensively investigated to see if they could reduce carbon emissions and energy consumption due to cement production [35]. Calcium sulfoaluminate cement is a promising solution with good mechanical performance and workability. However, the resulting concrete mixes are characterised by lower alkalinity and higher carbonation rates compared to Portland cement; therefore, they provide less protection against corrosion of the traditional steel reinforcement. Coupling with GFRP may be ideal, given the optimal durability of glass in less alkaline environments. The issue is under investigation.

5.4.3.8 Conclusions

After 40 years of research and development, the deployment of FRP reinforcement has finally moved from state-of-the-art exploratory applications to industrial scale. The technology has undergone standardisation and is being implemented in ordinary bidding procedures in important markets including the United States, Canada, Russia and China. FRP reinforcement is delivering on the expectation of good durability with minimal maintenance intervention and service life expectations exceeding 100 years. The increasing commercial interest in FRP applications follows rising concerns about the durability of an ageing infrastructural network, an issue highlighted by increasing maintenance costs and some disastrous collapses. Furthermore, sea level rises, high temperatures and the increasing frequency and intensity of hurricanes are pushing the demand for resilient and durable coastal infrastructures able to withstand these new challenges. Specific features of FRP reinforcement are summarised in the following.

- FRP is a composite material. Bars are produced by pultrusion of long resin-impregnated fibres. Surface preparations, like sand-coating, ensure that they bond well to the concrete. Before curing, straight bars can be bent to form stirrups and hooks. Several rods can be twisted together to form a prestressing strand.
- FRP is orthotropic. It has linear elastic behaviour up to failure, good tensile properties, low transverse resistance and tendency for fibre-buckling in compression. Typically, FRP is stronger but less stiff than steel under tensile loads.
- FRP is non-corrosive. Laboratory testing and field assessments confirm expected service lives exceeding 100 years in chloride-contaminated, carbonated or alkaline concrete.
- Various fibres are available. Carbon and sometimes aramid fibres are used in prestressing applications. Glass and lately basalt fibres are used in passive reinforcement. GFRP strands are being developed for mild prestressing applications.
- Various resins are available. Polyester was once widely used but has been dismissed because of poor durability. Vinyl ester and epoxy are the current standard. Thermoplastic resins are being investigated and may transform the FRP industry.
- The properties of FRP bars and strands are regulated by material specifications published in the US and internationally. Similarly, test methods have been standardised. This allows for pre-certification of manufacturers and their participation in public bidding.
- The design and construction of FRP-reinforced or prestressed concrete structures is regulated by design codes and guidelines published in the US and internationally. There are some differences with respect to traditional steel reinforcement.
- The design of FRP reinforcement is typically governed by service considerations due to the relatively low elastic modulus. The potential for tensile failure of the reinforcement must be accounted for in designing flexural members. The use of FRP reinforcement is not allowed in earthquake-resistant members.
- Construction with FRP is less demanding due to the light weight of the reinforcement. However, being brittle, FRP cannot be bent on site, and being flammable, it cannot be cut or welded using torches.
- Life-cycle costs and life-cycle assessment analyses have proven FRP to be the least impacting solution, both economically and environmentally, due to the extended service life and the minimal need for maintenance.
- Several successful FRP projects have been completed. More than 270 bridges have been built using FRP reinforcement in the United States and Canada to date.

REFERENCES

1. American Composites Manufacturers Association (2016) *List of GFRP Reinforced Bridges in North America*. ACMA, Arlington, VA, USA.
2. ASTM (2017) D7957/D7957M–17. Standard specification for solid round glass fiber reinforced polymer bars for concrete reinforcement. ASTM, West Conshohocken, PA, USA.

3. Gooranorimi O and Nanni A (2017) GFRP Reinforcement in Concrete after 15 Years of Service. *Journal of Composites for Construction* **21(5)**: 04017024. 10.1061/(ASCE)CC.1943-5614.0000806.
4. Benmokrane B, Nazair C, Loranger M-A and Manalo A (2018) Field durability study of vinyl-ester–based GFRP rebars in concrete bridge barriers. *Journal of Bridge Engineering* **23(12)**: 04018094. 10.1061/(ASCE)BE.1943-5592.0001315.
5. Emparanza AR, Kampmann R and De Caso y Basalo F (2018) State-of-the-practice of global manufacturing of FRP rebar and specifications. *ACI Special Publication* **327(45)**: 1–14.
6. ACI Committee 440 (2008) ACI 440.6-08. Specification for carbon and glass fiber-reinforced polymer bar materials for concrete reinforcement. ACI, Farmington Hills, MI, USA.
7. International Code Council (2016) AC454. Acceptance criteria for fiber-reinforced polymer (FRP) bars for internal reinforcement of concrete members. ICC, Brea, CA, USA.
8. American Association of State Highway and Transportation Officials (2018) *Guide Specifications for the Design of Concrete Bridge Beams Prestressed with CFRP Systems*, 1st edn. AASHTO, Washington, DC, USA.
9. CSA (Canadian Standards Association) (2010) S807-10. Specification for fibre-reinforced polymers. CSA, Mississauga, ON, Canada.
10. ISO (2015) 10406-1:2015. Fibre-reinforced polymer (FRP) reinforcement of concrete – Test methods – Part 1: FRP bars and grids. ISO, Geneva, Switzerland.
11. Euro-Asian Council for Standardization, Metrology and Certification (2012) GOST 31938-2012. Fibre-reinforced polymer bar for concrete reinforcement. EASC, Minsk, Belarus.
12. Standardization Administration of China (2011) GB/T 26743-2011. Fiber reinforced composite bars for civil engineering. SAC, Beijing, China.
13. ACI Committee 440 (2012) ACI 440.3R-12. Guide test methods for fiber-reinforced polymer (FRP) composites for reinforcing or strengthening concrete and masonry structures. ACI, Farmington Hills, MI, USA.
14. ASTM (2006) D7205/D7205M-06. Standard test method for tensile properties of fiber reinforced polymer matrix composite bars. ASTM, West Conshohocken, PA, USA.
15. ACI Committee 440 (2015) ACI 440.1R-15. Guide for the design and construction of structural concrete reinforced with fiber-reinforced polymer (FRP) bars. ACI, Farmington Hills, MI, USA.
16. ACI Committee 440 (2004) ACI 440.4R-04. Prestressing concrete structures with FRP tendons. ACI, Farmington Hills, MI, USA.
17. AASHTO (American Association of State Highway and Transportation Officials) (2018) *AASHTO LRFD Bridge Design Guide Specifications for GFRP-Reinforced Concrete*, 2nd edn. AASHTO, Washington, DC, USA.
18. Rossini M, Matta F, Nolan S, Potter W and Nanni A (2018) AASHTO Design Specifications for GFRP-RC Bridges. *Proceedings of Italian Concrete Days*. Springer International Publishing, pp. 432–444.
19. Belarbi A (2017) Guide specifications for the design of concrete bridge beams prestressed with CFRP systems. NCHRP 12-97. *Proceedings of the Highway Subcommittee on Bridges and Structures Annual Meeting*. American Association of State Highway and Transportation Officials, Washington, DC, USA.
20. CSA (2012) S806. Design and construction of building structures with fibre-reinforced polymers. CSA, Mississauga, ON, Canada.
21. CSA (2014) CAN/CSA-S6-14. Canadian highway bridge design code. CSA, Mississauga, ON, Canada.
22. fib (2013) *Model Code for Concrete Structures 2010*. Wilhelm Ernst & Sohn, Berlin, Germany.
23. CEN (European Committee for Standardization) (2004) Eurocode 2. Design of concrete structures. Part 1-1. General rules and rules for buildings. CEN, Brussels, Belgium.

24. Ministry of Construction Industry, Housing and Utilities Sector (2018) SP 295.1325800.2017. Concrete structures reinforced with fibre-reinforced polymer bars, design rules. Minstroy, Moscow, Russia.
25. Standardization Administration of China (2010) GB 50608-2010. Technical code for infrastructure application of FRP composites. SAC, Beijing, China.
26. AASHTO (2017) *AASHTO LRFD Bridge Design Specifications*, 8th edn. AASHTO, Washington, DC, USA.
27. Rossini, M, Nanni, A, Matta, F, *et al.* (2019) Overview of AASHTO design specifications for GFRP-RC bridges: Toledo Bridge as case study. *IABSE Symposium 2019: Towards a Resilient Built Environment – Risk and Asset Management.* International Association for Bridge and Structural Engineering, Guimarães, Portugal.
28. ACI Committee 440 (2008) ACI 440.5-08. Specification for construction with fiber-reinforced polymer reinforcing bars. ACI, Farmington Hills, MI, USA.
29. Cadenazzi T, Rossini M, Nolan S, *et al.* (2018) Resilience and economical sustainability of a FRP reinforced concrete bridge in Florida: LCC analysis at the design stage. *Life Cycle Analysis and Assessment in Civil Engineering: Towards an Integrated Vision: Proceedings of the 6th International Symposium on Life-Cycle Civil Engineering, Ghent, Belgium* (Caspeele R, Taerwe L and Frangopol DM (eds)). CRC Press, London, UK, pp. 2913–2920.
30. Mufti A, Onofrei M, Benmokrane B, *et al.* (2005) *Studies of Concrete Reinforced with GFRP Specimens from Field Demonstration Projects.* Canadian Network of Centres of Excellence on Intelligent Sensing for Innovative Structures, Winnipeg, Manitoba, Canada, Technical report.
31. Spadea S, Rossini M and Nanni A (2018) Design analysis and experimental behavior of precast concrete double-tee girders prestressed with carbon-fiber-reinforced polymer strands. *PCI Journal* **63(1)**: 72–84.
32. Rossini M, Jalles P, Balconi G and Nanni A (2018) GFRP strand prototype: Experimental investigation and technology readiness. *9th International Conference on Fibre-Reinforced Polymer (FRP) Composites in Civil Engineering.* International Institute for FRP in Construction, Paris, France.
33. Nanni A, Rossini M, Dotelli G and Spadea S (2019) MILDGLASS: GFRP strands for resilient mild prestressed concrete. In *NCHRP-IDEA Annual Progress Report: New IDEAS for Highway Systems.* Transportation Research Board, Washington, DC, USA, pp. 246–248.
34. Bertola F, Gastaldi D, Canonico F and Nanni A (2017) SEACON Project: Sustainable concrete using seawater, salt-contaminated aggregates, and non-corrosive reinforcement. *Proceedings of the 14th International Conference on Durability of Building Materials and Components* (De Schutter G, De Belie N, Janssens A and Van Den Bossche N (eds)). RILEM, Ghent, Belgium, pp. 353–354.
35. Aranda MAG and De la Torre AG (2013) Sulfoaluminate cement. In *Eco-efficient Concrete* (Pachego-Torgal F, Jalali S, Labrincha J and John VM (eds)). Woodhead Publishing, Cambridge, UK, pp. 488–522.

Soutsos M
ISBN 978-0-7277-6375-4
https://doi.org/10.1680/icehcd.63754.183

Chapter 6
Construction processes to improve durability

Dave Cullen (†)
Jacobs, UK

Jon Knights
Jon Knights Materials Consulting, UK

Don Wimpenny
Jacobs, UK

6.1. Construction processes
6.1.1 Introduction

For concrete to be durable, the designer's intentions must be translated into the actual structure. The concrete must be produced to give a consistent final in situ product, allowing for any variations in the constituent materials, production conditions and site practices. It is not sufficient for test specimens to give consistent results, although this is certainly a necessary first step. The concrete must be transported, placed and fully compacted without segregation, honeycombing or excessive bleeding. It must not develop cracks (beyond the design intent) or other faults, either before or after hardening, and must be initially maintained in an environment that will allow it to develop its intended properties for the duration of its projected service life.

It must not be forgotten that most structural concrete is reinforced with corrodible steel. There is a huge difference between the durability of unreinforced concrete and that of reinforced concrete. Many ancient structures survive after millennia for no other reason than that they were not reinforced. Clearly, durability significantly depends on the properties of the concrete between the reinforcing steel and the elements to which the structure is exposed.

For the design intent to be manifest in the actual structure, the designer, after consulting with the client and the contractor, needs to address issues with the design intent in relation to the constructability and consequent durability as early as possible. Delays and cost increases may occur on a project either if there is a realisation that a design cannot be practically built or if problems occur on site during construction that potentially affect the serviceability of the structure.

At the early (concept) design stage, it is desirable to engage experts in the durability of concrete structures to ensure that all aspects of design and construction dovetail with the durability required in the design intent. This may also mean occasionally challenging the specification requirements to facilitate the most efficient and cost-effective design and construction.

A reality check is sometimes needed to adjust the specified requirements for concrete durability to reflect the other constraints during construction. Contractually, this can sometimes be problematic but often worthwhile to allow an efficient design intent that is in line with construction processes.

It is essential, therefore, that the design process remains iterative at the early stages to allow for reviews of constructability issues and concerns. It is also essential for the designer to communicate fully the design intent to the contractor to ensure their complete understanding.

Some examples that can affect the performance of a structure if design, construction and durability are not considered holistically are

- effective choices between the use of precast concrete versus cast-in-place concrete
- specifications and method statements that do not fully address the project requirements for construction in extreme conditions (e.g. hot or cold weather, tidal areas and restricted access)
- designer cover requirements not in line with durability requirements for the climatic region
- overly conservative reinforcement design and detailing, causing reinforcement congestion (i.e. difficulties in fabrication and fixing and concrete placement problems)
- use of concrete with excessive binder contents, which increases the risk of early thermal cracking, shrinkage and internal chemical attacks (i.e. alkali–aggregate reaction and delayed ettringite formation)
- inadequate design for abrasion resistance or aggressive chemical attack
- excessive bay sizes or wall lengths to improve the construction programme or poor construction sequencing (e.g. for slabs and walls), both of which increase the risk of cracking
- proprietary concrete products with lower cover than prescriptive codes require (e.g. concrete pipes in aggressive ground and permanent formwork in high-chloride-bearing environments)
- local aggregates that do not give sufficient control of early thermal cracking
- jump-form or slipform construction without due consideration of the concrete mix design requirements
- inadequate curing standards for critical concrete mixes and elements.

Value engineering design options and construction processes are sometimes worthwhile. Careful consideration of the cost benefit of certain design and construction processes can optimise the programme, improve durability and reduce the environmental impact. Cost, therefore, should, where possible, remain secondary to the production of a practical, buildable and efficient design.

Examples of this are

- options for reducing the cover to the reinforcement to increase design efficiency and constructability (e.g. high-performance self-compacting concrete, steel/synthetic fibre concrete and precast concrete design and fabrication)
- optioneering construction processes and materials (e.g. jump-form versus slipform, off-site fabrication versus on-site construction, and sprayed concrete)
- use of maturity systems to allow optimisation of mould removal or formwork striking
- detailing of elements to control water paths and accumulation.

This chapter examines the way concrete production and construction processes influence the quality and durability of the as-constructed works.

Documentation has a key role in communicating requirements, and this is discussed in the following sections.

6.1.2 The specification

The specification is the key document for communicating to the construction team the materials and processes to be used. A failure to produce a clear and comprehensive set of requirements is likely to lead to disputes or as-constructed works that do not reflect the intended design.

The concrete specification typically occurs as a section within a series of sections making up the overall construction specification. It may utilise industry specifications, such as the UK's *Civil Engineering Specification for the Water Industry* (CESWI) [1] or ACI 318 – 'Building code requirements for structural concrete' [2]. It is important that the specification chosen is broadly relevant to the type of construction. For example, codes for the water industry like CESWI and general specifications like ACI 318 are not the best basis for specifying highway structures or tunnel linings.

The first part of a specification typically will comprise clauses describing the concreting materials and reinforcement, including their testing and storage. The second part will have clauses dealing with workmanship, such as concrete production (e.g. batching and mixing), concrete handling (e.g. finishing and curing) and inspection and testing (e.g. strength monitoring and dimensional tolerances). At the heart of the specification is a tabulated schedule of mixes to be used in different applications summarising the essential requirements for the concrete.

In Europe, standard EN 206 for concrete specification, performance and production has been agreed between participating countries, but its adoption permits each country to modify its provisions to some extent. In the UK, EN 206 has been issued as BS EN 206 [3] along with a complementary standard BS 8500. BS 8500-1 [4] provides an indication of the mixes (i.e. limiting proportions of binder type, content and maximum water-to-binder ratio) and the minimum cover values (excluding deviation for construction tolerance) to be used for different exposure conditions and design working lives. BS 8500-2 [5] provides methods of specifying mixes, as listed in Table 6.1. Product conformity certification is provided by independent bodies, such as the Quality Scheme for Ready Mixed Concrete and the British Standards Kitemark™ Scheme.

Selecting the specification method is important as it alters the information to be provided to or by the producer and the responsibilities of the various parties. For example, a designed mix must normally meet performance requirements for consistence and strength as well as restrictions on the type of materials (e.g. binder type) and proportions (e.g. maximum water-to-binder ratio). In contrast, a standardised prescribed mix has a restricted range of permissible constituents and specific batch weights for different slump classes, against which compliance is assessed. Designated and standardised prescribed mixes can be viewed as narrower versions of designed and prescribed mixes with preset options that provide a simplified approach appropriate to some types of construction. Proprietary mixes are typically specialised mixes (e.g. self-compacting concrete) developed by the producer, who may withhold information on the composition as intellectual property and, instead, declare conformance with the specification.

If durability modelling has been used in the design, it may be appropriate to run tests to confirm whether the parameter values assumed in the design have been achieved, for example, the inverse carbonation resistance and chloride diffusion coefficient. In theory, it is possible to rely totally on meeting the performance requirements without needing to constrain the producer. In practice, a performance-based approach is rarely adopted because prescribed specifications are adequate for most applications and there is uncertainty in whether performance testing will provide a reliable and representative means of assessing compliance and potential performance in service. Nevertheless, the performance-based approach allows innovation and potential cost savings if well planned and executed. It is also the intention of the European Standards Committee (CEN) to enable performance-based specifications within the existing prescriptive framework.

Table 6.1 Different specified mix types

Specified mix type	Compliance
Designed	Constituent materials: binder type, aggregate type and size, etc. Basic concrete requirement: alkali–silica reaction (ASR), chloride content, temperature, volumetric yield Limiting proportions: minimum binder content, maximum water-to-binder ratio Performance: consistence, strength Additional specified requirements: density, air content
Designated	Product conformity certification and control Constituent materials: restricted binder type, aggregate type and size, etc. Basic concrete requirement: ASR, chloride content, temperature, volumetric yield Limiting proportions: minimum binder content, maximum water-to-binder ratio Performance: consistence, strength, air content (PAV1 and PAV2)
Prescribed	Constituent materials: binder type, aggregate type and size, etc. Basic concrete requirement: chloride content, temperature Batching to batch masses within allowable tolerances
Standardised prescribed	Constituent materials: restricted binder type, aggregate type and size, etc. Basic concrete requirement: ASR, chloride content, temperature, volumetric yield Batching to batch masses within allowable tolerances Performance: consistence
Proprietary	Constituent materials: binder type, aggregate type and size, etc. Basic concrete requirement: ASR, chloride content, temperature, volumetric yield Other requirements selected in consultation with the producer

The difficulties in identifying appropriate performance tests for quality and durability are reflected in the range of practices around the world. In Australia, the volume of permeable voids or the NordTest Build chloride migration tests are specified by the transport authorities in Victoria and Queensland, respectively (Chapter 8). In the USA, a movement called P2P (prescription to performance) is promoting the adoption of durability performance tests [6]. The rapid chloride-permeability test (RCPT) is favoured, but it has significant shortcomings, both as a compliance test and as an indicator of durability. It is becoming increasingly common in the Middle East to witness the use of chloride migration testing as a specification requirement.

A study of different durability design and quality assurance approaches found that final documentation on construction quality and compliance with the specified durability, such as the in situ chloride migration coefficient, is key to any rational approach for better control and increased durability [7]. The value of such documentation in encouraging good practice may be enhanced by supplementing self-certification by independent inspection and testing of critical parameters, such as concrete cover depth.

6.1.3 Method statements

Method statements provide important feedback to the designer from the contractor as to whether the requirements of the specification, especially in terms of the construction processes, have been fully appreciated and properly allowed for in planning the work. It is a safe assumption that if the method statement is silent on a specified requirement, then it will not be addressed.

A method statement typically comprises general method statements utilised in previous projects, supplemented by amended or new procedures specific to the project. The method statements and their contents may well be outlined in the specification and should cover

- pre-construction trials
- construction and striking of formwork and falsework
- transportation, handling and compaction of the concrete
- curing and protection
- cold weather concreting (where applicable)
- hot weather concreting (where applicable)
- large or complex pours (where applicable)
- temperature and early-thermal crack control measures
- joints
- repairs.

A method statement should contain adequate details of the methods of working, such as the intended plant and equipment (including any standby provision), hold points and responses to non-conformance or an emergency. In addition, the method statement must address the health and safety risks and manage the environmental impact of the work.

The temperature and early-thermal crack control plans need to be developed at an early stage, as these may necessitate the casting of mock-ups or insulated blocks, thermal modelling or changes to the batching plant, concrete mix, and the curing and protection measures. Early-age thermal cracking and high peak temperatures with an associated risk of delayed ettringite formation may result in long delays due to resin injection works and expansion testing of core samples.

Investigations of construction disputes, defects and durability problems have identified tight construction programmes in conjunction with inadequate planning and poor communication as common root causes of many problems. Method statements are key documents for preventing the latter two issues.

6.1.4 Inspection and test plans

Primarily, inspection and test plans (ITPs) for concrete construction are useful in providing the following

- a checklist of necessary actions, including identifying the roles and responsibilities of each of the parties
- a record of assurance that the materials and processes comply with the specified requirements (including where this may not be the case).

Developing ITPs for a specific application can also be a useful planning tool. Some aspects will be clear at the outset, such as the main form of construction, geometry etc., but to be of maximum relevance and benefit, an ITP should be refined as late in the process as practicable, namely when the details of the final mix design, construction methodology and sequence have largely been defined and finalised.

The content of an ITP will depend heavily on the design basis and the method and level of detail in the specification. In some cases, specific tests are required to confirm the assumptions made in the durability design, such as limits (e.g. chloride diffusion and water permeability). Testing has an important role in the overall quality assurance and is intrinsic to ensuring the durability for any structure. As well as forming a relatively simple checklist of necessary actions, ITPs provide evidence of compliance with the specified requirements and demonstrate that there are appropriate processes to minimise the risk of any defects that are likely to lead to early durability failures.

As with method statements, if there is no reference to a particular check or witness hold point, for example, then there is a good chance that it will be forgotten, overlooked or not recorded.

Typically, and most usefully, actions are set out chronologically, with the content and terminology closely aligned with the method statements for the site operations. However, in some cases, they can include a wider use, covering aspects of planning, for example, cross-referencing the design and drawing sign-off or including a follow-up response to non-conformance. Provision of contingency in case of emergency is sometimes overlooked, such as the preparation of a construction joint if there is a need to terminate a large pour.

The level of detail included is an important consideration. ITPs can be too generic, not providing enough detail so that important items may not be addressed. Conversely, there may be excessive detail that makes the process cumbersome and problematical to use. There is anecdotal evidence of over-elaborate ITPs, for example, nuclear outage maintenance works, for which sequencing of the workflow is critical, the requirements are fixed and ITPs are agreed way in advance of the works. The result can be a document unusable on site so that it is completed before or after the works as a tick box exercise. It is generally prudent to start with more detail and eliminate where not necessary, rather than being tempted to adopt overgeneralisations such as ‘check documentation’ or ‘make cubes’ without adequate cross references to where the specifics are detailed.

A clear specification provides useful information on what is needed to ensure the design is appropriate; good construction documents provide evidence that the design has or has not been incorporated into the works.

Typically, the format will include numbered sections within main headings. A good ITP will include clear summary information for each line item within each section, with clear references to supporting documents. For example

Main section headings

- Pre-construction approvals, for example, binder certification, delivery dockets and aggregate testing confirmations
- Pre-construction activities, for example, equipment set-up and checks on rebar, formwork, batcher and other plant
- During construction, for example, checking concrete delivery records, water and admixtures added on site, sampling and testing workability
- Finishing, curing and transporting test specimens
- Post-pour and follow-up actions, for example, thermal control measures, and falsework and formwork stripping

The information needed for each specific action or task includes

- unique inspection reference
- description
- reference to the control document or clause
- frequency, for example, once per shift or every 100 m^3
- acceptance criteria, for example, consistence limits for acceptance
- description of verification record, for example, current certificate and record of a proforma reference

- witness or sign-off confirmations for the various parties, for example, contractor, supervisor, engineer, site agent, client representative, project manager and third parties. Roles and responsibilities should be clear, with nominated individuals where appropriate and an indication of who should take corrective action.

ITPs should be relatively short and easy to use for construction teams. In many cases, the main ITP could be supported by supplementary records. For example there could be a single line item for up-to-date certifications of aggregate tests, with a clear reference to a document that sets out the detail.

6.2. Concrete production control

6.2.1 Constituent materials

6.2.1.1 Introduction

Most specifications provide limiting values for proportions and performance, including durability, with the choice of specific constituents and target proportions left to the contractor. It is important in finalising the mix design that the form of the construction and the principal handling and placement methods are considered early enough in the process. For example, an economically designed mix for placement with skips and chutes may not perform well if there is a late decision to pump the concrete. This could be confirmed based on what is established and readily available in the supply chain or trials may be needed (Sections 6.2.5 and 6.3.1).

6.2.1.2 Selection and control of binders

A specification may give limiting values for cement types and combinations (usually including options for ground granulated blastfurnace slag (GGBS) or fly ash), limits on replacement levels, and minimum or maximum total binder content. In most cases, a specification will indicate a cement combination, or type, within which there is a range or replacement levels. In Europe, cement combinations are defined by composition in BS EN 206 [3] and BS EN 197 [8]. Possibly the most commonly specified is Portland cement with 36 to 65% GGBS, designated CEM III/A if factory blended or CIIIA if batched as separate powders. In broad terms, the equivalent for durability performance is Portland cement with 21% to 35% fly ash, CEM II/B-V or CIIB-V. However, this is becoming increasingly less common as reliable sources of fly ash become scarcer. Sometimes a specification will provide options of two or more cement combinations, so that several often conflicting factors need to be considered carefully in the final selection, including fresh concrete performance, heat of hydration, risk of thermal cracking, construction sequence, and formwork stripping and demoulding expectations. If available, mixer blended combinations provide more flexibility, which can be of benefit if thermal considerations are paramount.

For most significant concrete production, binders will be delivered in purpose-built bulk tankers and stored in weatherproof silos, with the silo, silo charging pipe and couplers clearly marked with the cementitious material type.

Silica fume is many times finer than other cementitious components, and when processed or used in undensified form there can be health and safety risks related to respiratory illness. The material is commonly supplied as a slurry, with typically 50% solids, which should be kept agitated. Densified powder is available and commonly used if slurry is impractical or not possible, for example preparatory 'dry batch' sprayed concrete. However, caution is needed to ensure full dispersion of the material during mixing to ensure that there are no excessive agglomerations of silica that may negatively affect performance or increase the risk of ASR.

Commonly, a minimum total binder content is included for durability, with no upper limit. As well as a commercial benefit, there can be technical advantages in keeping the total binder to the minimum needed for placement. Optimising the combined grading as far as possible is useful in allowing the lowest binder content to be used [9].

All too often, binder contents that are higher than necessary are adopted to overcome shortcomings elsewhere, possibly aggregate grading or shape, a lack of understanding or to achieve compliant strength at 28 days. Issues with locally available aggregates may mean the former is unavoidable, but too many producers maintain that the primary route to greater strength or workability is by increasing the binder content without due consideration of other options that may be available. Where possible, producers should be encouraged to consider refining a combined aggregate grading and using different admixtures dosages and combinations, for example. There is a tendency for modern cements to be ground finer to promote early-age strength gain or achieve compliance with the specified 28-day strength. However, if commercial or contractual considerations allow, there can be significant benefits in relaxing strength compliance to 56 or 90 days [10]. This is particularly the case if higher levels of replacement are specified or where thermal cracking is a potential issue, as later-age compliance can significantly reduce the amount of binder and produce a cheaper mix that is more resistant to cracking.

Accurate batching of cementitious materials is essential, but quality testing of the cement by the concrete producer is not frequently done. In any case, test results on already received cement are often too late for mix adjustments to be made. Another complicating factor is that cement flow in a silo is often not a simple matter of first in at the top, first out at the bottom. Cement from the top can often funnel down the centre of the silo, leaving substantial amounts of the previous delivery around the sides.

6.2.1.3 Selection and control of aggregates

Specifications will vary in their requirements for aggregate properties, for example where abrasion-resistant concrete is needed. Except for certain architectural applications, there will normally be a considerable degree of latitude for the concrete producer to select the most appropriate materials. Due to the volumes needed and the relatively high costs of transportation, locally won, naturally occurring materials are normally used. Exceptions include the Arabian Peninsula, where a lack of local coarse aggregates suitable for high-performance concrete has led to the importation of higher-quality basalt aggregates from the Indian subcontinent. The cost of transportation is offset by economies of scale and the technical and commercial advantages in achieving the required performance.

Other types of aggregate include manufactured (e.g. air-cooled blastfurnace slag) or crushed concrete 'recycled' aggregates. However, care is needed in considering the latter. Some areas impose limits on recycled aggregate in structural concrete due to the inherent variability and the potential for inclusion of potentially weak or deleterious materials, such as wood, bituminous residues and other contaminants [3,11].

Some codes, for example BS EN 12620 [12], allow for a wide range of physical and chemical properties, some of which are unsuitable for moderate- or high-performance applications. The onus is on the specifier to select the appropriate requirements. The importance of this should not be underestimated, particularly if there is a focus on durability.

Overall aggregate grading and particle texture and shape have a significant effect on the properties of fresh concrete, especially water demand and workability or consistence. Crushed rocks are generally more angular with a greater surface area to be wetted. There can be problems in achieving the required rheological characteristics when using coarse particles with an irregular shape (flakiness). Conversely, relatively rounded particles can help with achieving workability at lower cement contents and can provide benefits in pumping etc., but smooth rounded coarse particles, such as river or wadi gravels, may produce relatively poor bonds with the binding matrix, especially when coated with silt, effectively limiting the ultimate strengths that are achievable.

If resistance to fire spalling is a consideration, commonly in buildings and transportation tunnels, some quartz-based aggregates, such as flint or chert, may be unsuitable, as they typically exhibit explosive spalling at around 350°C, compared to approximately 500°C for granite and limestone [13].

Limestone or dolomitic aggregates can also be beneficial for concrete subject to acid attack. As the aggregate particles have attached acid species, the overall level of attack against the concrete is reduced by a factor of approximately four compared to siliceous or granitic rock types [14]. Also, limestone typically has a significantly lower coefficient of thermal expansion than rock types with a higher silica content, which is an advantage in reducing the risk of thermal cracking. Some manufactured lightweight aggregates provide even greater advantages in this regard, but it is rarely economically viable to include them for this reason alone.

Optimising the combined grading as far as possible within commercial and other constraints is useful for producing a concrete with the minimum paste, as the cement matrix is fundamentally less stable than the aggregate. Other things being equal, optimising the packing density to maximise the aggregate content will improve the thermal performance, shrinkage and durability. For some concrete applications, such as sprayed and tremie concrete, the adoption of combined aggregate grading curves is recommended [15].

During a major infrastructure project in the UK, particle shape and grading issues with the locally available aggregates presented problems in achieving the flow characteristics needed for underwater placement. The addition of a small amount of intermediate-sized coarse aggregate provided significant improvements in the fresh properties, resulting in reduced thermal cracking issues, substantial savings in construction and programme (reduced number of tremies and lifts), and client confidence in the durability of a critical nuclear-safety-related foundation structure.

A good specification should identify specific measures to reduce the risk of built-in deterioration mechanisms, with limits on alkali reactivity, sulfate content and chloride content. Limits on other properties, like water absorption, levels of other harmful impurities, soundness and drying shrinkage, can be general indicators of aggregate quality, and the potential variability in the properties of fresh and hardened concrete. Even a very small proportion of heavily weathered coarse aggregate particles, for example, basalt, can effectively limit the ultimate strength achievable in high-performance concretes, as the weaker particles crush preferentially under compressive loading.

Natural aggregates are inherently variable, so some variation in the properties and performance of the concrete is inevitable. Before selecting a source for extensive use, some research into the quantity and quality available is prudent. Some major projects with a long construction period will have specific requirements for a quarry report, which should demonstrate that the quarry can consistently produce the quantities and quality of aggregate materials required. One specification for nuclear-related structures calls for a detailed quarry report for each of the proposed aggregate sources, which must be accepted by a geologist in the client organisation as part of the prequalification.

Marine dredged aggregate, particularly sand, is a significant source of aggregate for concrete in some areas of the UK. The chloride distribution in stockpiles may be very variable, and although regular testing is required due to this important durability consideration, it is not always clear that sampling and testing regimes recognise this variability. On large or prestigious projects, it may be prudent to consider additional sampling and testing of the aggregates, or other assurances above those imposed by national or regional codes, to reduce any risk of there being an unacceptably high level of built-in chlorides in the concrete. However, it can be difficult to include 'special' requirements in established supply chain processes, particularly where precedent and commercial imperative have led to very large volumes being routinely supplied.

6.2.1.4 Fine aggregates

Usually, the most significant feature of aggregates causing batch-to-batch variation in water content at a given slump is the grading and silt content (usually defined as particles passing the 45 or 63 micron sieve) of the fine aggregate. Relatively small changes in fines content can have a significant effect on the water demand or workability of the mix. Failing to adjust the added water to take account of the variation in the moisture content of fine aggregate will also lead to a variation in workability. Moisture checks on representative samples using definitive oven or microwave drying can be useful to confirm the performance of in-line moisture probes.

It is not simple, or rarely convenient, to check the sand grade at the point at which it enters the batching process. ITPs should include checks for grading and excessive fines or clay content if concrete properties or water demand are suspect.

The most obvious effect of changes in grading and fines content is probably changes in the workability achieved. Other potential issues with high silt contents are a retardation in setting, delayed strength development, potentially lower ultimate strength (particularly if additional water was added to offset the effect of increased fines) and the colour of architectural concrete.

The effects of variation in a sand tend to be relatively immediately reflected in consistence, strength tests or setting time and should be managed by the routine control systems, although special interventions may occasionally be needed.

6.2.1.5 Coarse aggregates

The characteristics of coarse aggregate that can have an immediate effect on the concrete include dust content, particle shape, surface coatings that reduce bonding, and the inclusion of weak and unstable particles. Some coarse aggregates, such as basalt, can undergo excessive volume changes in response to moisture, leading to potentially disruptive swelling or shrinkage over time. Others may be affected by the alkali–aggregate reaction, which causes disruptive expansion over an extended period. Control of such features is a matter of thorough investigation of new sources of aggregate by geologists, rather than day-to-day control by concrete technologists.

6.2.2 Batching

Modern batching plants usually have hardware and software to record the actual batch quantities of every mix. However, this is reliant on the data inputs, and the assumed water content of the aggregate may not adequately reflect any variation. The information generated is usually too extensive for a primary inspection to be useful, but software can automatically detect variations from the target design and can incorporate batch quantity information in the analysis of test data.

For consistent concrete, batching accuracy must be properly controlled. Acceptable batching tolerances from BS EN 206 [3] are given in Table 6.2.

There are also requirements for the producer to operate a calibration programme and review the production control system at least every 2 years to ensure its suitability and effectiveness. In the UK, the calibration of

Table 6.2 BS EN 206 batching tolerances

Constituent	Tolerance[a]
Cement	±3% of required quantity
Water	±3% of required quantity
Total aggregates	±3% of required quantity
Additions and fibres used at 5% by mass of cement	±3% of required quantity
Admixtures, additions and fibres used at ≤5% by mass of cement	±5% of required quantity

[a]The tolerance is the difference between the target value and the measured value

weighing systems is required at least annually by the Quality Scheme for Ready Mixed Concrete [16], but shorter intervals are often adopted by producers. Recalibration is required after modification or breakdown of equipment. Calibration checks should be performed if there is any doubt.

The regulation of water content in concrete is a significant ongoing problem, being responsible for much of the variability experienced. Although it is easy to control the amount of water added to concrete during batching, it is more difficult to allow for the moisture in the aggregates. It is important that concrete has the required workability, so water addition is often regulated by assessing this. Consequently, the total water content in concrete is usually not the target mix design value and it may not be accurately reflected by the batch records.

If cement combinations are blended at the batcher, fundamental errors are theoretically possible due to confusion of Portland cement, GGBS or fly ash. These can usually be detected at an early stage, so there are limited risks for the long-term durability, although they can be expensive and time-consuming to rectify. Some facilities that have experienced such problems have moved from potentially fallible labelled or colour-coded silos to more robust systems, such as using different couplings on silos and delivery vehicles to render simple mistakes in delivery essentially impossible.

Fibres should be dispensed using purpose-made equipment that disperses the fibres in the mixer without forming clumps. Polypropylene microfibres, which are incorporated for fire resistance or to prevent plastic shrinkage cracking, are supplied in bags or cut from a strand. They are typically dispersed into the mixer by air blowers. Steel fibres are commonly collated with a water-soluble glue, and it is important for the mixing to be adequate to allow the glue to break down. Ideally, the batching and mixing plant will be specifically designed for fibre dosing, as problems with inadequate dispersion and balling of steel fibres are more common when dosing equipment is retrofitted.

6.2.3 Mixing

For all concrete, it is important that it is fully mixed. If in doubt, checks in line with the principles of ASTM C94 [17] can be useful. Central mixing is usually regarded as likely to result in better control, and a forced action mixer is the most reliable way of achieving thorough mixing,

Concrete consistence can be assessed from the power consumption of the pan or truck mixer. In addition, moisture sensors in the mixer pan can be used to assess the uniformity of mixing and enhance the control of concrete consistence [18].

Whether pan or truck mixed, it is important that the mixing blades should not be excessively worn. If steel fibre-reinforced concrete is routinely mixed, then rates of wear are likely to be greater and checks should be more frequent. An abrasive heavyweight aggregate (magnetite) in gravity structures can lead to very high rates of wear in the central batcher and delivery vehicles.

The order of adding materials can also be important. For most new and established batching set-ups, this is allowed for, but for some relatively unusual material combinations, more thought is needed. For example, for some lightweight or very low water-to-binder ratio mixes, the timing of water and admixture addition may need to be adjusted. There needs to be sufficient water addition and mixing before admixtures are added, so that they are not directly in contact with dry cementitious material or 'dry' aggregates with potentially very high absorption. If not properly considered, the result can be an ineffective admixture or variable performance.

Documentation from plant suppliers tends to include production rates per hour or per day based on relatively standard production, but with minimal mixing time. Mixing times for high-performance or special mixes may need to be relatively extended, and it is prudent to allow for this in planning and procurement.

6.2.4 Monitoring

6.2.4.1 Early monitoring

Monitoring the constituent materials, autographic records and consistence are important because they provide an early indication of potential problems. Cement production alters seasonally, leading to a fall in specific surface area in summer. The quality of fly ash can also vary depending on the coal source and how the power plant is operated. The contamination of aggregate with clay minerals can result in an increased water demand and reduced strength and durability.

Control charts of parameters, such as the specific surface area of the cement, aggregate grade and concrete consistence, can be used to forecast adverse trends in the compressive strength. Some specifications require binders and admixtures to be sampled and tested to assess their interaction [19]. This is especially important for sprayed concrete, as early strength development is critical. Calorimetry can provide useful additional information on changes in the hydration reaction.

Accurately monitoring water demand can assist in identifying such problems, but an accurate and complete record of the water content of the mix may not be readily accessible from batching plant records, as added water is not always automatically recorded and does not include any water added on site.

Assessing batching using a rapid analysis machine or microwave sensors requires some knowledge of the mix, including the silt content of the aggregate when assessing the binder content [20] and the moisture content and absorption of the aggregate when assessing the water content [21]. Microwave sensors can assess the moisture content of the aggregate and fresh concrete, which has the potential for enhanced control. It has been claimed, for example, that such equipment allowed the flow range to be controlled to ±20 mm compared to a normal range of ±50 mm [22].

The workability or consistence of concrete is important in terms of its placing and compaction and also as an indication of its underlying quality and potential long-term properties. In this respect, a slump test is quick and simple but not very informative, especially at slump values higher than 210 mm. Two-point tests, which measure the resistance of concrete to shear at two shear rates, can provide additional information about the stability of fresh concrete, but these are best performed in a laboratory [23].

Finally, if there is doubt about the correct batching of a mix, then a plastic density test will help confirm the volumetric yield of the concrete and any gross deviation (>5%) from the theoretical density.

6.2.4.2 Monitoring compressive strength

Measuring the compressive strength is the most common method of monitoring concrete quality. Various systems of increasing effectiveness have been devised. First, there were Shewhart charts, which are standard time-sequence plots of individual results with statistically calculated 'warning' and 'action' limits. It was later found that more reliable action could be based on plotting the running means of three or four consecutive results. A further improvement was the 'cusum' analysis. A cusum is the cumulative sum of differences between successive results and a target value. A change of slope on a cusum graph is about three times as effective in detecting change as a Shewhart chart.

In the UK, compressive strength at 7 days is used to predict the 28-day strength, which is then subject to three analyses

- cusum M to monitor the mean strength
- cusum R to monitor the standard deviation
- cusum C to monitor the correlation between the 7-day and 28-day strength values.

The deviation of a value from the target is calculated, and the cumulative value is plotted graphically. A transparent mask in the shape of a truncated V (Figure 6.1) is used to assess whether the progressive

Figure 6.1 Chart and mask showing the decrease in the cusum mean with a change point in the 3 N/mm^2 standard deviation

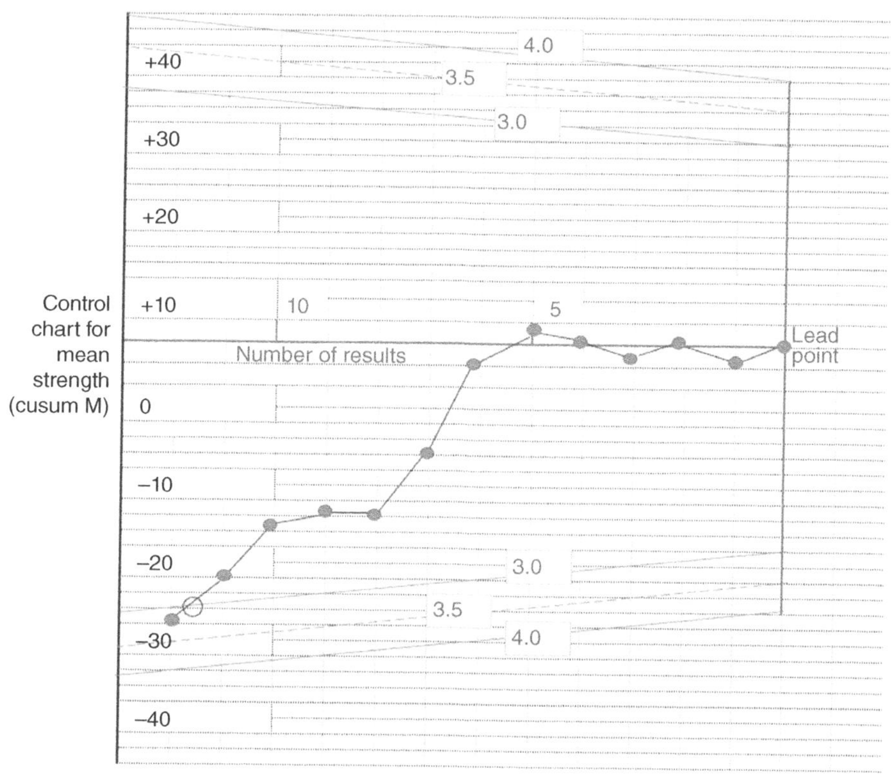

deviation from the target, as represented by the slope of the cusum plot, requires action. Usually, this is due to a change in the binder content, the assumed standard deviation or the 7-day to 28-day strength correlation.

For a concrete plant to generate data more quickly, a 'family' of mixes can be formed from mixes with similar constituents and properties. One mix in the family becomes the control mix and the strength data from other mixes is transformed so that it can be analysed as though it originated from the control grade. This means that the concrete on a particular project may not be monitored by the producer and may not be tested unless the specification requires identity testing to supplement the producer's conformity testing.

In the previous edition of this book, Ken Day proposed that in addition to compressive strength, other parameters, such as density, slump and temperature, should also be subject to a cusum analysis, so that changes requiring action can be independently verified, thus allowing prompter action.

Note that interpreting strength data relies on the correct sampling, manufacture, curing and testing of the specimens. When investigating low strength values, it is prudent to check the validity of the test results. For example, the early storage of specimens at temperatures 5–10°C higher than those within cooled insulated boxes during hot weather resulted in a 5–10% decrease in the 7-day and 28-day strengths [24]. A low hardened density of specimens can indicate gross compaction problems, or, if it correlates with high workability, excessive water content or possible segregation. The basic point is that any such explanations may point to a need to amend procedures or train staff, rather than adjust the mix.

Predicting strength development from ages earlier than 7 days is feasible. This is important if achieving an early strength limit is required, such as for striking the formwork or stopping curing. For example, 70% of the characteristic strength is required for curing class 4 to BS EN 13670 [25]. The strength development may be estimated from the curing history by using a relation between strength and maturity or equivalent age (Arrhenius or Sadgrove are recommended). Note that the relation between strength and the logarithm of the equivalent age is commonly bilinear if GGBS and fly ash are incorporated into the concrete. This break in the relation at 3–7 days must be taken into consideration when predicting the strength from values at early ages [26].

6.2.4.3 Monitoring durability

Monitoring strength can provide an indirect indication of quality and durability in combination with other information, such as density and consistence. However, there is a need for robust, simple, rapid and reliable direct tests for durability that can be applied to samples in the laboratory as part of production control and, ideally, to in situ concrete.

Table 6.3 lists a typical set of durability test parameters for concrete in the Arabian Peninsula. However, note that these are mostly suitable only at the mix development and initial acceptance of a mix [27]. Long-term tests, such as for bulk chloride diffusion, will not yield results in a timescale that would allow concrete production to be altered or prevent further construction, leading to a risk of rejection.

A monthly water absorption test is required for some precast products, such as pipes and tunnel segments [33,34].

The RCPT is routinely used to monitor concrete in America and parts of the Arabian Peninsula. The test has significant limitations but is sensitive to silica fume content. A project involving the construction of approximately 3000 bored piles in aggressive ground experienced an issue with the concrete in the form of non-compliant coulomb values, high excess voids content, high water-to-binder ratio measured by microwave tests of fresh concrete and high strength variability. A root cause analysis identified a failure to disperse the

Table 6.3 Typical set of durability test parameters for concrete in the Arabian Peninsula

Test	Method	Age: days	Limit on average of two determinations
Water penetration test	BS EN 12390-8 [28]	28	≤40 mm unreinforced ≤20 mm reinforced
Water absorption (30 minutes)	BS1881-122 [29]	28	≤2.25% unreinforced ≤1.75% reinforced
Initial surface absorption	BS1881-208 [30]	28	≤0.15 ml/(m^2 s) (30 minutes)
RCPT	ASTM C1202 [31]	28	≤1000 coulombs
Chloride migration coefficient	NTB 492 [32]	56	≤2.5 × 10^{-12} m^2/s (dependent on durability design)

silica fume in combination with a loss of workability, leading to compaction issues. In this case, the RCPT in conjunction with other tests indicated a problem, but the results could not be directly used to quantify the impact on durability. Parameters such as the chloride diffusion coefficient were estimated and used to predict deterioration in service, which was then used in a structural assessment of the piles.

Initial surface absorption tests and Figg tests provide some indication of permeability, albeit dependent on the moisture content of the concrete. Autoclam and PERMIT provide sophisticated tests for a range of durability parameters, including air and water permeability, sorptivity and chloride migration coefficient. However, the cost, duration and complexity of these tests does not make them well suited to application on site. In contrast, electrical resistivity has been used to monitor concrete quality on large projects and has the advantage of being relatively simple and quick [7].

It is possible that a future solution for monitoring durability will involve a suite of parameters. Batch records and rheological measurements in the truck mixer made available in real time could be combined with electrical resistivity measurement at early ages. However, these may cover only the most common mixes and causes of variability and durability problems, such as water content control. They should be supplemented by indicator tests specific to the mix and the expected service conditions.

In addition to monitoring the concrete, parameters associated with durability, such as cover depth, compaction, cracking and curing, should also be measured (Section 6.3).

6.2.5 Laboratory trials

For many construction projects of a certain size or prestige, it may be necessary to undertake development trials of the concrete mix to determine the properties of and control measures for the concrete mixtures if these are not available from the suite of tried and tested mixes from the concrete supply chain. This may be appropriate if locally available constituents are new or unknown, or in more remote site-batched situations or where specific project requirements require bespoke or novel concrete products or casting methods (e.g. high- or low-density, underwater, self-compacting or ultra-high-performance fibre-reinforced concrete or concrete using alkali-activated cementitious materials).

The aim of concrete trials is, therefore, to establish the fresh and hardened properties desired for the required application. Numerous tests can be carried out on concretes, depending on the project requirements. Table 6.4 provides a (non-exhaustive) selection of the most common.

Table 6.4 Examples of testing and development of concrete mixtures during laboratory trials

Test description	Remarks
Preliminary trials	Often carried out with a minimum of three mix designs to establish the initial parameters
Maximum water-to-binder ratio	To establish the maximum to achieve the design strength and consistence
Minimum cement content	To establish the minimum needed to achieve the design strength and consistence
Cement type	Required for durability or early-age thermal crack control
Maximum aggregate size	Dependent on section size and reinforcement congestion
Target workability or consistency	Requirements for placeability and compaction
Compressive strength and hardened density	At a range of ages (e.g. 1, 2, 3, 7, 14, 28, 56 and 90 days)
Water absorption	A general measure of quality and durability (often a specification requirement)
Bleed (rate and capacity)	Important property for many concrete applications (i.e. slabs and thick or deep sections). Good control of the bleed can reduce the risk of plastic shrinkage. Poor control can lead to bleed channels and sand runs.
Secondary trials	Carried out after the initial parameters have been defined
Pressure bleed (e.g. Bauer filtration)	Pressure bleed may often be useful for high-pressure pump mixes or mixes for deep foundations
Slump or slump flow target and retention with time	Retention of adequate workability is often required to be extended (e.g. 450 mm slump flow after 5 hours)
Setting time	Necessary for large volume pours or concrete requiring extensive transportation (e.g. to remote locations or in long tunnels)
Water permeability	Durability test often required for aggressive exposure conditions
Chloride diffusion or migration coefficient	Used in aggressive exposure conditions with chlorides (e.g. marine or highway structures)
Carbonation resistance coefficient	Used in aggressive conditions where high levels of carbon dioxide may be a risk (e.g. tunnels or industrial use)

One aspect that is key to the successful development and testing of concrete mixtures is understanding the role of water and how to control it. First, it is essential to determine the absorption characteristics of the aggregates in use, so that the actual free (or effective) water content in the mix can be accounted for.

In addition, it is also often useful to establish the relations among the fresh properties, such as consistence (workability) and its retention over time, for the proposed admixtures to be used. New generations of powerful superplasticising admixtures (such as polycarboxylate ethers and derivatives) can have a significant effect on the rheology of concretes but with only small changes in the free water content. Moreover, concretes containing new generation admixtures can sometimes lose desirable flow properties quickly when combined with certain cement types. It is, therefore, essential to map these relations as early as possible to enable changes or combinations of admixture formulations that give the desired properties for successful placement and compaction.

It is also important in many circumstances to undertake sensitivity trials on the candidate mix designs. This involves the trialling and testing of mixes following small changes in constituent proportions to determine their effect on concrete properties. This can be valuable in understanding the changes in a concrete that may be observed during batching and serve as an aid to production control. It also leads to a clearer understanding of the role of the free water content of the mix, its water-to-binder ratio and the effect on rheology, again aiding control during production and ensuring that the specification limits are not breached.

After the initial trials, the sensitivity of the concrete formula to changes in the constituent proportions can be assessed.

- Trial 1: Target batch masses defined in initial trials.
- Trial 2: Free water-to-binder ratio increased by increasing the water and decreasing the binder content based on the permissible batching tolerance (e.g. 3% by mass).
- Trial 3: Free water-to-binder ratio decreased by decreasing the water and increasing the binder content based on the permissible batching tolerance (e.g. 3% by mass).
- Trial 4: Target batch masses defined in initial trials but with additional admixture based on the permissible batching tolerance (e.g. 5% by volume).
- Trial 5: Target batch masses defined in initial trials but with reduced admixture based on the permissible batching tolerance (e.g. 5% by volume).
- Trial 6: Target batch masses defined in initial trials but with the addition of fine aggregates and a reduction in coarse aggregates based on the permissible batching tolerance (e.g. 3% by mass).
- Trial 7: Target batch masses defined in initial trials but with a reduction of fine aggregates and the addition of coarse aggregates based on the permissible batching tolerance (e.g. 3% by mass).

The outcomes of any trials should be written up in the form of a technical submission, with accompanying test data, to demonstrate that the desired properties can be achieved. This typically forms part of the materials submission and compliance record for the project. It should also allow the generation of method statements applicable to the production, control, placement, finishing and curing of the developed mixes.

Clear record-keeping is a key aspect of design, not only for diligence in project recording but also in the event of any disputes that may arise during or after the project.

6.3. Concrete control on site

6.3.1 Field trials

Specifications typically call for three full-scale trial batches on separate days. If placement is relatively straightforward and the concrete producer and construction teams are established and suitably experienced, there may be no need for trials, or they could be very limited. For example, trials may just confirm the workability, which is, typically, slightly higher for a production plant compared with laboratory-scale batches for the same mix proportions, with historic records being relied upon for compressive strength.

The first days of precast production could be treated as field trials, with units being rejected as appropriate and the process being optimised before quality is acceptable and production commences. For precast tunnel segments, there will usually be a requirement for at least two rings to be erected and these should be used to confirm the acceptability of the finish (especially the blowholes around the gasket) and the dimensions [35].

For some large or prestigious projects, or where there are complex placement methods or other issues, it may be prudent to undertake larger-scale pre-production trials to confirm that the proposed mixes are compatible with the construction methods and equipment and thus, to provide confidence that the specified performance and durability requirements can be reliably achieved.

For some forms of construction, for example, sprayed concrete, robust trials are an inherent part of the process. For example, in participating countries, EN 14487-1 [36] sets out requirements for pre-construction trials, based on the application (repair, free-standing structure or primary ground support), and lists three inspection categories for low, moderate and high durability requirements and risk.

Mock-ups or insulated 'hot block' trials are often employed to determine the temperature rise and cooling characteristics of a concrete mix, which inform the thermal modelling and control measures for construction (Section 6.3.8). It is important to trial the actual mix proposed for construction, with the binder type and content and coarse aggregate type being key factors. The careful positioning and fixing of thermocouples are important, especially near surfaces. Duplicate thermocouples and short cable runs (less than 10 m) should be used to reduce errors and the risk of damage from site activities.

Several demanding and often conflicting requirements need to be considered for tremie concrete in deep foundations. Furthermore, defects, for example, segregation, can be very difficult to rectify. Understanding the rheology of the mix is important, especially with modern concretes and due to the increasing use of high-performance chemical admixtures. Full-scale trials are recommended to assess the performance of fresh concrete [37].

Other examples where pre-construction trials can be a critical factor in successful construction include underwater concreting and in situ tunnel construction with high flow and lengthy placement times. For each application, the scope of testing should encompass the expected construction methods and conditions. The extent of the trial works should be proportionate to the size, complexity and risks of the project.

The process, plant, materials and personnel should be representative of those to be used during construction. This can present practical difficulties. For example, if a full suite of sprayed concrete trials is undertaken early in the programme, the actual plant and personnel, such as the nozzle operators, may be unavailable. In addition, the ambient conditions could differ from those during construction. Trials should use similar plant (same make and model if possible) and a certified nozzle operator. Subsequently, the information can be supplemented after limited confirmation trials as part of site start-up.

The factors listed below should be considered when planning trials (e.g. the sensitivity of the mix to expected batching tolerances). If it is not practical to vary a factor (e.g. the ambient temperature), then accurate recording during the trials supplemented by close observation in the construction phase can mitigate the residual risks

- temperature
- reinforcement density
- minimum dimensions
- mixing regime
- transportation method and duration
- placement method and duration
- finishing and curing methods
- batching tolerances for water content.

The outcome of testing should feed into the contractor's construction programme, the detailed specifications and the ITP requirements. Consideration should be given to additional testing if, for example, there are significant changes to the process, plant or materials, the sequence and timings, or the placement conditions, or if there are placement problems or an unacceptable level of defects.

The greatest benefit can be obtained from larger-scale trials, if it is possible to plan these with enough time for any necessary changes or perceived improvements to be incorporated into the works. However, this is often not practical, as procurement processes may mean the supply chain is not engaged sufficiently early in the programme.

A key consideration during field trials is to address construction issues that may lead to defects in the works, as even well-designed and executed concrete repairs can promote early durability failures.

6.3.2 Placing

Concrete is commonly placed by direct delivery from the back of a mixer truck using the chute or, if appropriate, by dumper, skip, kibble, pump, conveyor or tremie or elephant trunking. More specialised methods are occasionally required and include concrete hoists, crane-mounted conveyors and aerial cableways. The objective is to place the concrete efficiently and safely such that it retains homogeneity and its required properties.

Essentially, most concrete construction involves repeating a series of similar or identical procedures. However, increases in the complexity or size of a concrete pour or significant changes in the conditions (e.g. temperature or tide conditions) increase the risk of things going wrong. Apart from getting the fresh properties of the mix right and confirming this by trials for the selected placement method (if appropriate), the most important factors for successful concrete placing are proper planning and resourcing.

CIRIA 135 [38] includes a useful checklist for planning deep lifts and complex pours. The key considerations can be summarised as follows

- concrete supply
- site set-up and communications
- concreting sequence, including finishing
- plant and equipment
- resources and labour
- pour geometry and access
- ambient conditions
- any special considerations (e.g. thermal) or other constraints
- backup and contingency.

Appropriate measures should be implemented for hot and cold weather. ACI 305 [39], ACI 306R-16 [40] and the Good Concrete Guide 8 [41] provide useful guidance and reference additional information. Placing concrete in inclement weather should be avoided as far as possible. For example, heavy rain can damage finishes and reduce the effective cover if the pour cannot be protected.

Adequate consideration should be given to the welfare of site labour. Concreting work can be physically demanding, so providing rest periods and drinking water reduces the health and safety risks and the potential for mistakes due to fatigue. Casting at night under floodlights is common practice in the Arabian Peninsula to avoid the hottest part of the day.

Regardless of planning, things can go wrong, especially with demanding or prolonged pours. Probably, the most likely are delays in placing, leading to the potential for cold joints and other defects, such as honeycombing. It is essential to have contingency plans, for example

- additional placing equipment or sources of concrete supply
- procedures to reduce the thickness of layers in deep lifts

- approved procedures to add water or admixture on site to restore concrete workability (retempering)
- resources to install a vertical stop end and prepare a horizontal construction joint.

Special consideration should be given to pours carrying significant safety, commercial, technical or programme risks, for example, architectural concrete or diaphragm wall pours.

The addition of water on site should be under the control of the concrete producer and the volume recorded. Truck water meters should be calibrated in advance, and the plant batching records checked to ensure the specified maximum water-to-binder ratio is not exceeded. The procedures should ideally include a table showing the water addition needed to adjust a measured workability to the target value. Admixtures can similarly be dosed on site by a technician, using calibrated equipment, for example, a measuring cylinder. If water or an admixture is added on site, the concrete should be fully mixed before sampling, retesting or placing. It is prudent to take samples for compressive strength testing whenever water or admixture has been added on site.

There are many reasons for concrete supply problems, including plant breakdown, variable concrete quality leading to rejected loads, the availability of mixer trucks, traffic problems and high winds, which affect the use of cranes. A risk assessment-based approach that considers the likelihood and consequences of problems can be a useful tool in deciding what practicable mitigations are appropriate.

Improper placement techniques can adversely affect the concrete. Care should be taken to prevent the concrete from free-falling through the reinforcement. The loss of paste due to the 'sieving' effect of the reinforcement can lead to honeycombing at the bottom of the lift. The paste coating the top surface of the reinforcing bars may weaken the bonding (Figure 6.2). With some deep lifts, it is worth temporarily moving aside some of the top reinforcement to allow access points for placing by elephant trunking (comprising a hopper and tremie pipe) or by pump line. When concreting underwater using a tremie, the pipe should remain at least 1 m below the surface of the placed concrete to exclude water.

To facilitate placement and monitoring of concrete, hatches and lighting should be installed in tunnel forms and the formwork for deep lifts [38].

Figure 6.2 Loss of fines and poor compaction leading to honeycombing at the bottom of a deep pour. (Courtesy of D Wimpenny)

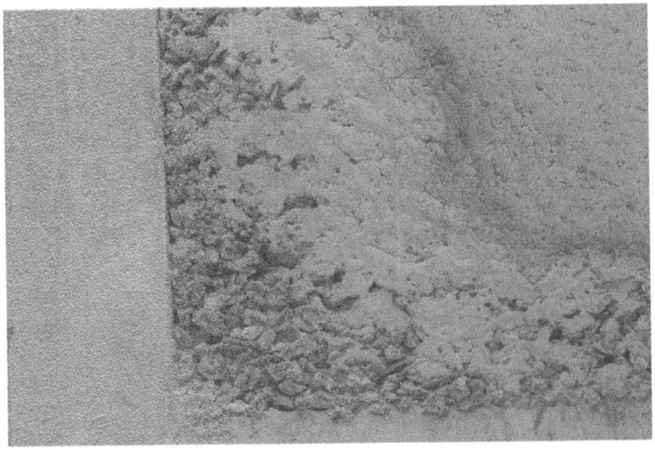

Figure 6.3 Poker vibration for full compaction

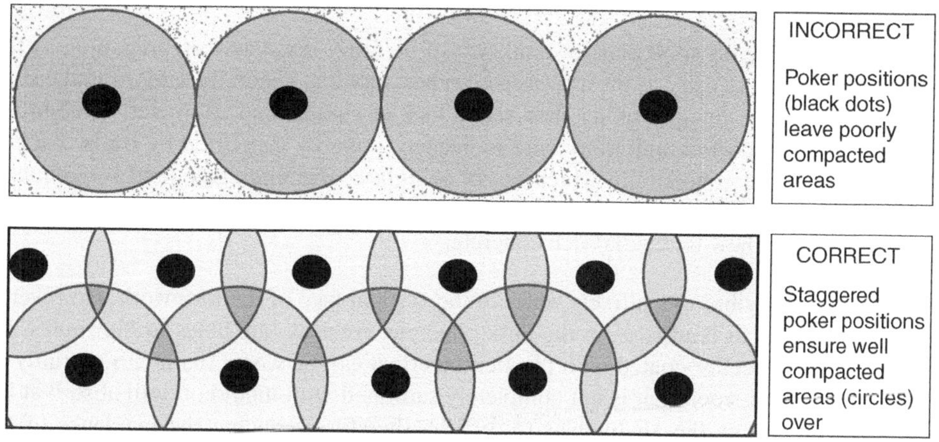

6.3.3 Compaction

Most concrete is placed using internal poker vibrators. Other methods include vibrating tables or forms for precast units, such as tunnel segments, and beam vibrators for slabs up to 150 mm in thickness [42]. The vertical and horizontal zone of influence of any vibrator will depend on its size, type, vibration frequency and magnitude. For conventional vibrating pokers to be effective, it is important that they are inserted at appropriate centres, depending on the diameter of the head, so that the radius of action encompasses the whole section (Figure 6.3 and Table 6.5).

The concrete should be observed to flow under vibration, which should be continued for several seconds in each location until no more air is expelled. A poker vibrator is inserted vertically and withdrawn slowly, leaving no hole. Insertion at the top of a heap is likely to cause a lack of compaction at the foot of the heap as liquefied concrete flows out over concrete unaffected by the vibration, entrapping air voids. Concrete should ideally be placed in consecutive layers, usually around 500 mm thick depending on the equipment used. It is important for the poker to extend through the entire thickness of a new layer and into the layer below so that the two are fully compacted into a homogenous mass.

The aims are to achieve full compaction by expelling all the air, full encapsulation of the bars and other embedded items, and full filling of the form. Excessive vibration can cause segregation, in which coarse aggregate particles sink and a mortar-rich layer rises to the surface. This surface layer can have lower abrasion resistance and increased shrinkage, cracking and permeability to aggressive agents. These effects are substantially exacerbated if the concrete has a higher water content than the correct mix design. Attempting

Table 6.5 Poker vibrator sizes [42]

Diameter of head: mm	Radius of action: mm	Approximate rate of compaction: m^3/hour	Use
20–30	80–150	Up to 2	Very thin or congested sections
35–40	130–250	2–4	Thin walls or columns
50–75	180–350	3–8	General work

to move concrete along the formwork by making it flow is a common cause of over-vibration and, depending on the mix design, presents a segregation risk.

Most well-designed mixes with good plastic stability will be fairly resistant to segregation, as the cement matrix has around the same density as most aggregate types, and it is generally safer to make sure that the concrete is vibrated for long enough, rather than risk a lack of compaction. However, some mixes can be more sensitive to over-compaction and more care is needed, possibly supported by trials. An example of the latter is density-adjusted concrete, in which heavy or light coarse aggregate (and sometimes also fine aggregate particles) readily sink or float, leading to segregation. Field trials are recommended to familiarise the construction team with how to handle such concrete.

Vibration, especially form vibration, attracts water to the vertical face of the formwork. However, air voids may not rise to the surface as readily as in the bulk concrete, creating bug holes or blowholes. If the face is impermeable, made of steel or coated with plastic, the effect can be worse than with partially absorbent plywood, for example. New wood that is not completely saturated with mould oil will absorb at least some of the excess water and allow the air bubbles to diffuse, therefore reducing the incidence of blemishes. Controlled-permeability formwork (CPF), as described below, is a further exploitation of such absorbency. Blowholes can be reduced but not eliminated (e.g. by changing the release agent or vibration or using defoaming agents in the mix). Excessive blowholes may be an issue for fair-faced architectural finishes, surfaces that will receive a coating or a membrane, and for tunnel segments if they are immediately adjacent to a gasket, but they are rarely of direct significance for durability.

If plastic shrinkage cracking is identified in good time, re-vibration of the concrete down to just beyond the top mat of reinforcing and refinishing can be very effective in closing the cracks, but it is critical that no vibration is applied to concrete that has begun to set.

Special provision needs to be made for awkward or unusual geometries, particularly dense reinforcement and hard-to-reach corners (Figure 6.4). Air released during compaction will rise, along with some water. Therefore, with horizontal or shallow sloping top shutters, there will inevitably be a greater incidence of

Figure 6.4 Uncompacted area in a hard-to-reach corner (left) and large 'blowholes' left by a horizontal top shutter (right)

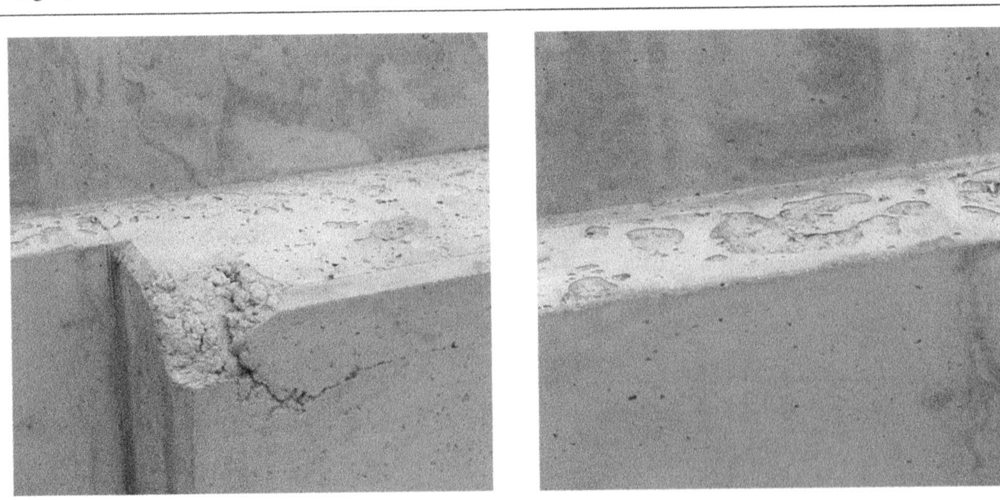

blowholes. This may be addressed using special measures, such as CPF, or in some cases it may be acceptable to include additional cover, for example, an extrados for tunnel segments.

The exposure of workers to the risks of hand-arm vibration syndrome and noise can be reduced by using self-compacting concrete, which needs no vibration.

6.3.4 Finishing

6.3.4.1 Introduction

The finishing process and the final type and quality of the finish for formed (cast against formwork) and unformed surfaces have an important influence on durability. For example, the correct finishing of unformed surfaces improves the density and hardness of the surface, thus increasing its abrasion and impact resistance (Figure 6.5). This is particularly the case for floors (Section 6.4.2).

Finishing is an important opportunity for detecting and taking action in the event of excessive bleeding or plastic cracking. Correct finishing can help reduce the effect of water gain by bleeding, but poor finishing can exacerbate problems or hide them until they become evident in service. An example of the latter was a water-retaining structure, in which cracking was noticed prior to commissioning. Close inspection of the surface suggested that power-floating the slab had generated crazing and, perhaps, also obscured the underlying plastic cracking (Figure 6.6).

Finishing is highly dependent on the training, skill and diligence of site staff. Shortcuts during finishing, like sprinkling water onto surfaces to aid finishing or working an evaporation retarder into the surface, can reduce the durability of the surface. Such practices need to be eliminated.

Some special finishes are specifically chosen to enhance the aesthetics or the chemical or physical durability, including

- power-floating (Section 6.4.2)
- dry-shaking (Section 6.4.2)
- CPF (Section 6.3.6.4)
- sealers and anti-graffiti coatings (Section 6.4.1).

Figure 6.5 Effect of finishing on abrasion resistance [43]

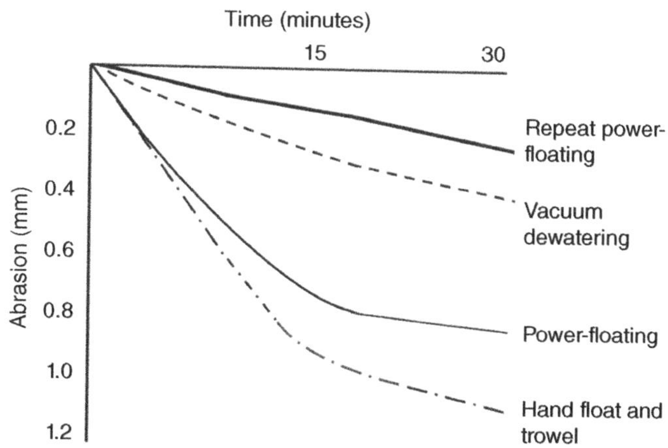

Figure 6.6 Crazing and plastic shrinkage of a power-floated concrete floor slab. (Courtesy of D Wimpenny)

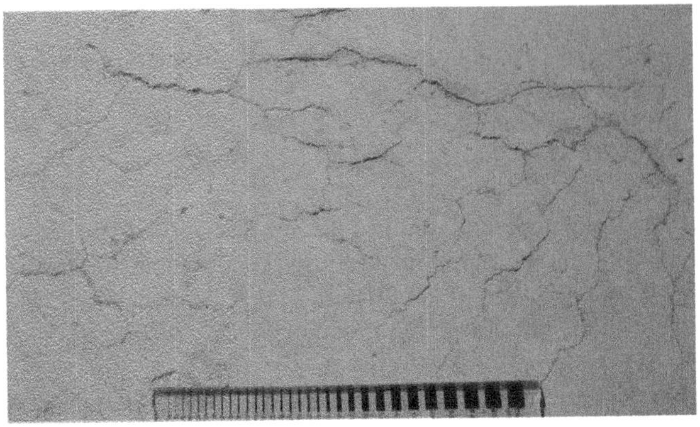

Some specific problems that occur in plastic concrete and can be detected and influenced during finishing are described in the following section.

6.3.4.2 Bleeding, settlement and plastic cracking

Even for concrete that does not have excessive water content, a certain amount of bleeding is normal. Essentially, the solids in concrete, including the cementitious material, are denser than water and so tend to settle as the placed concrete comes to rest. The displaced water rises in the paste and tends to form voids under anything blocking its upwards passage. This includes coarse aggregate, steel reinforcement, and plumbing and electrical fittings. These voids are a substantial factor in the permeability of the concrete in a horizontal direction. They also partly explain the reduced bonding of the top steel. However, this is exacerbated by the general settlement of the concrete away from the lower surface of the bars as the bleeding takes place.

Bleeding can lead to water gain on the surface and a layer of porous friable laitance as well as sand runs on vertical elements (Figure 6.7). Excessive bleeding can be reduced by changing the mix to reduce the

Figure 6.7 Sand runs at an in situ stitch between precast panels. (Courtesy of D Wimpenny)

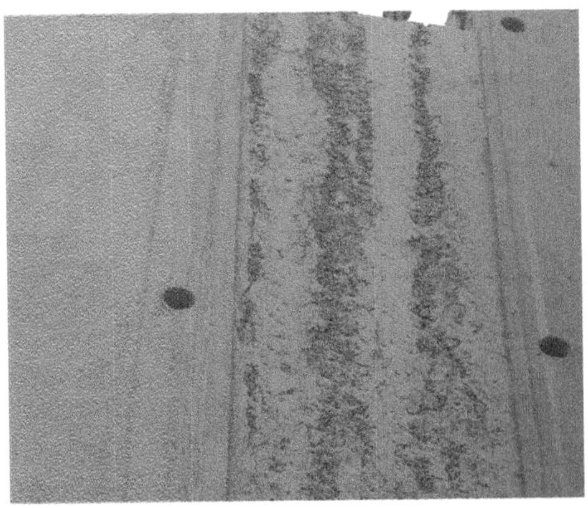

water content or to include finer particles (e.g. fly ash) that slow the rate of rise. A concrete squeegee or more sophisticated techniques, such as vacuum dewatering, can also be used to remove bleed water prior to finishing [44].

Two kinds of cracking can occur in concrete prior to setting: plastic settlement and plastic shrinkage. Plastic settlement cracking is caused by concrete 'breaking its back' over obstructions to bleeding settlement and typically occurs within 10 minutes to 3 hours of casting [45]. These obstructions often include the top reinforcement and changes in depth of the concrete section, for example, trough and waffle floors. Such cracking is diagnosed by the alignment of the cracking over the reinforcement or because it occurs in areas where section changes occur. Placing a straightedge on a deep beam can identify shallow depressions in the concrete between the lines of cracking. These cracks can be closed by re-vibration just before initial setting occurs. They may appear to be closed by surface trowelling, but, since this only closes them at the surface, they will often subsequently open later under drying shrinkage. This type of cracking can have a significant impact on durability, as it if often wide and extends down to and aligns with the reinforcement. Voids beneath the reinforcement can also encourage the initiation of corrosion.

Plastic shrinkage cracking typically occurs 30 minutes to 6 hours after casting when water evaporates from the surface layer of the concrete, reducing its volume and initiating tension. This risk is increased in hot, windy weather and when the concrete has a low rate of bleeding, for example, due to the inclusion of silica fume. ACI 305 [39] provides a nomogram for assessing the approximate evaporation rate. The rate of evaporation can be moderated with shading, windbreaks, aliphatic alcohol evaporation retarder and early curing using mist sprays and polythene sheeting prior to spraying on a curing membrane. It is important that evaporation retarders are not worked into the surface during finishing. Polypropylene microfibres can also be used to reduce the risk of cracking. A dosage of at least 200 m^2 of surface area of fibres per cubic metre is recommended [46].

6.3.5 Reinforcement

The fixing, bending and placement of the reinforcement impacts important aspects relating to the durability, like the cover (Section 6.3.9), the control of cracking and the ability to place and compact the concrete.

Highly congested reinforcement may require a concrete with a nominal maximum aggregate size of 14 or 10 mm, which may, in turn, require a higher binder content. It may be necessary to temporarily leave some bars loose in the top layer of the steel reinforcement so that these can be moved to provide access for placing the concrete. For piling and diaphragm walling, the spacing of the vertical and horizontal bars should be at least 100 and 150 mm, respectively, to allow the concrete to flow [47].

The reinforcement should be stored on racks above ground and protected from contamination by sea spray and salt-laden dust. A failure to prevent chloride contamination can lead to premature corrosion. If the reinforcement has been contaminated, it should be cleaned by water-jetting with clean water or abrasive-blasting with approved media. The effectiveness of the cleaning can be established on site by measurements before and after treatment using potassium thiocyanate strips or chloride test strips (www.hach.com).

Corrosion of the reinforcement in the absence of contamination is not normally an issue for structural performance or durability provided the rust is uniform and well adhered. However, if there is break in construction of more than a week, then the starter bars should be protected using plastic sheaths or cement grout slurry to prevent rust from staining the concrete below.

The reinforcement should not be rebent on site, particularly at lower ambient temperatures, as this could damage it. Welding the reinforcement on site should also be avoided. Cutting stainless steel reinforcement

Figure 6.8 Reinforcement fixing in progress. (Courtesy of J Knights)

with a disc cutter can impair its durability if it becomes contaminated by mild steel or thermal oxides [48]. The fusion-bonded epoxy coating of rebar can be damaged during fabrication and handling. Corrosion concentrates at points of damage and work on site should closely follow ASTM D3963 [49].

The bimetallic corrosion of carbon steel reinforcement is theoretically possible if it is connected to a metal that is lower or less reactive in the electrochemical series, such as stainless steel [50]. The risk of significant corrosion depends on the environment and the relative area of the two metals, and this should be assessed in accordance with the guidance in PD 6484 or, for stainless steel, advisory desk note 339 [51].

The reinforcement must be firmly secured to prevent movement during concreting to maintain the correct cover (Section 6.3.9). When concrete is sprayed, the reinforcement must be held sufficiently rigid to prevent vibration during spraying, but ideally, small diameter well-spaced bars should be used to reduce the risk of shadowing.

Figure 6.8 is an example of reinforcement fixing for a deep pour, showing the challenges in correctly fixing and thoroughly inspecting the reinforcement.

6.3.6 Formwork

6.3.6.1 General

Formwork allows concrete to fulfil its role as a universal building material. It has an important influence on durability through its effect on the characteristics of the concrete surface and the concrete covering the reinforcement. There is a clear distinction between formwork and falsework. The definition of formwork, often also referred to as shuttering, is 'the mould to contain the fresh concrete. It consists of the face contact material and the members to support it'. Distinct from formwork, the definition of falsework is 'any temporary structure used to support a permanent structure until it is self-supporting'. Examples of some different formwork types are shown in Figures 6.9 and 6.10.

Figure 6.9 Panel formwork system for a high wall. (Courtesy of J Knights)

The cost of formwork and falsework is a significant element in the total cost of concrete construction, ranging from 35% to 40% for a typical concrete frame and up to 95% when working over water [52].

Formwork should be designed only by competent personnel who have relevant design and practical experience. Formwork should be constructed to ensure that the concrete is formed to the required shape, size and position, as well as providing the required surface finish on the concrete. The supporting system, that is, tie rods, horizontal walings and vertical supports (soldiers), and bracings for slab formwork should be installed in accordance with the formwork drawings and the supplier's recommendations.

Figure 6.10 Large construction project showing jump-form systems for core wall construction. (Courtesy of J Knights)

It is important to ensure that formwork design and construction provides the following attributes

- sufficient strength to take the pressure or weight of fresh concrete during placing and compacting
- ease of erection and striking
- sufficient stability in adverse weather conditions
- ease and safety in handling
- provision of suitable access for handling, placing and compacting concrete.

The design and construction should follow appropriate safety regulations relating to access, working areas, platforms, toe boards and guard rails. Joints between members must be tight to prevent loss of grout, which can cause honeycombing and similar faults that can affect the concrete finish and the durability of the completed structure.

There are four main uses for formwork: walls, columns, beams and slabs. Formwork for the underside of slabs is termed soffit formwork. BS 5975 [53] – 'Code of practice for temporary works procedures and the permissible stress design of falsework' should be used in conjunction with formwork design guides, particularly when designing slab soffit formwork. Further guidance on the design of temporary works is given in PAS 8812 [54].

A useful review of construction techniques to maximise speed and economy is provided in BRE 495 [55]. One way of eliminating the need to strike formwork is to use permanent formwork designs. These systems can also enhance the durability of structures. Further guidance is provided in CIRIA C558 [56].

A wide range of formwork materials are available, from backing members to the contact surface (Table 6.6).

6.3.6.2 Release agents

The contact surfaces of formwork require treatment with release agents so that the formwork can easily be removed without the contact surface adhering to the surface of the concrete and causing damage. Timber faces can often be absorbent if not pretreated, so it is especially important to treat these surfaces with at least one coat of release agent prior to use. However, care should be taken to apply the correct amount of release agent, as it is common for it to be applied too liberally, which can cause staining and retardation of the concrete surface. Release agents should always be applied in accordance with the manufacturer's recommendations.

Common types of release agent are

- neat oils with surfactants – commonly used on steel surfaces
- mould cream emulsions – good general purpose release agent for timber and plywood
- chemical release agents – applied by spray for high-quality work
- vegetable oil-based release agents – more environmentally friendly and biodegradable water-based products.

6.3.6.3 Sample panels and trial pours

The specification may necessitate the production of sample panels or trial pours that incorporate significant features in the design of the structures to allow acceptance of the mix, method statement and finish. Figure 6.11 illustrates a hypothetical example of a trial panel designed to incorporate several features that will affect the appearance and durability of the structure.

Table 6.6 Different formwork materials

Material	Characteristics
Timber	Most common for backing members. This traditional material is easily cut and assembled. Grade by strength class.
Aluminium	Strong and light. Typically used in proprietary systems for backing members (e.g. beams and joists), although may deflect more than steel or timber.
Steel	Proprietary and purpose made. Used in panels (contact surfaces), frames and moulds. Used for high-precision and low-tolerance works. Very high levels of reuse possible. Its high-surface heat transfer coefficient may need to be accounted for in designs to control early thermal cracking.
Plywood	Traditionally used for contact surfaces. Can be backed with polymers to provided fine finishes. Lower levels of reuse before replacement required. Lower surface heat transfer (more insulating).
Plastics	Glass-reinforced plastics are commonly used for complex shapes. High levels of reuse. Proprietary systems available.
Expanded polystyrene	Single-use contact surface that can be sculpted to form complex shapes and void formers. Can be used as permanent insulating material. May also be waterproof.
Rubber	Typically used as form liners to produce patterned concrete surfaces. There are bespoke designs and 'pour-your-own' liquid resins.
Fabrics	Used directly to contain concrete as a permanent formwork. Commonly used in underwater applications (scour protection, stop ends, etc.)
Expanded metal	Commonly used as stop-end materials to form a joint requiring little or no preparation. Also reduces formwork pressures in tall structure construction. Single use.
Cardboard	Single use, wax coated. Typically used for column formers and void formers.
CPF	Special type of contact surface liner. It enhances the concrete surface properties (Section 6.3.6.4)

6.3.6.4 Controlled-permeability formwork

Experience with exploiting the beneficial effects of slightly absorbent formwork led to the development of techniques aimed at deliberately dewatering the outer layer of the concrete. These include inserting permeable layers inside the conventional formwork to permit bleed water to escape. This has several benefits.

- It produces a more uniform and attractive appearance, reducing the presence of blowholes.
- It reduces the water-to-binder ratio of the concrete cover to the steel reinforcement, thus giving a higher strength and lower permeability.
- While in place, it improves curing.

The impact of CPF on the occurrence of blowholes is illustrated in Figure 6.12.

A clear distinction should be made between CPF and other open textured materials such as geotextiles, sacking and absorbent liners, which may remove water from the concrete surface, but in an uncontrolled and variable manner. The basic elements of a CPF system are

- a filter that allows passage of excess air and water while retaining fine particles (cement)
- a drain that allows air and water to pass out through the formwork
- a support system to resist the pressure of the concrete (i.e. the formwork).

Figure 6.11 Composite sample or trial panel [52]

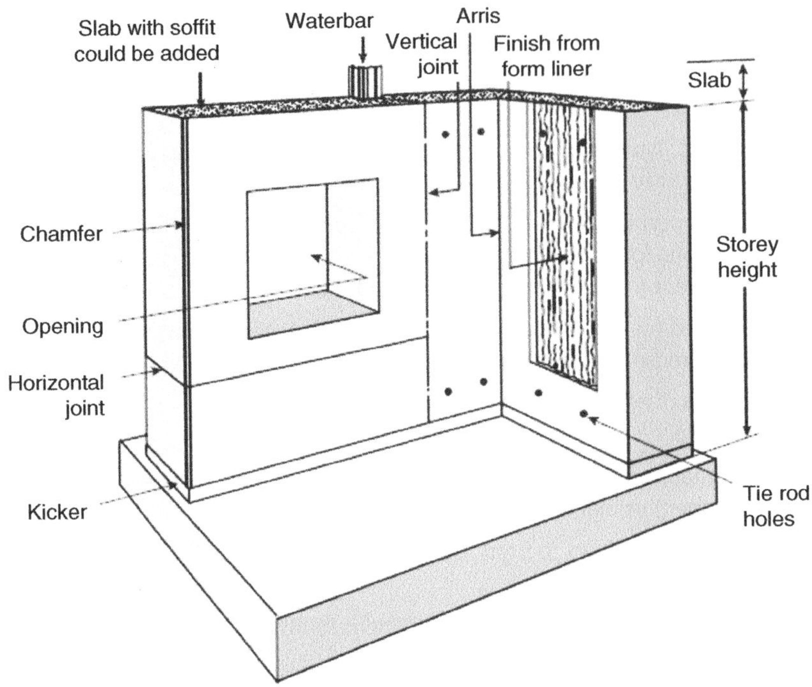

With CPF attached to the internal face of formwork, the concreting process is undertaken in the normal way. Guidance on CPF is given in CIRIA report C511 [57] and by McKenna [58]. CPF enhances the properties of concrete to a depth of approximately 15 mm. It provides approximately a 45% reduction in the chloride diffusion coefficient and a 10% reduction in the surface chloride level [59]. This is typically equivalent to a 10 mm increase in cover.

Figure 6.12 Trial section of a half-round tunnel construction. Greater incidence of blowholes at the invert, which are significantly mitigated by the use of CPF to the left of the structure. (Courtesy of D Cullen)

There are typically three types of CPF liner systems commercially available, as well as more ad hoc systems that function in a similar way to proprietary CPF systems (i.e. softboard or geotextile grids). The general classification is as follows [57].

- Type I: Two-layer filter fabric fixed and tensioned in situ, combining the filter and drainage function. Woven fabric with drainage holes. Can be reused with careful cleaning (e.g. Textile Form and Silk Form).
- Type II: Single-layer non-woven filter fabric fixed over structural support and tensioned in situ, generally single-use products (e.g. Zemdrain and Formtex/Fibretex).
- Type III: Two-layer non-woven system combining a filter fabric bonded to a backing grid. Pretensioned system maintained by the backing grid. Multiple reuse (e.g. Zemdrain MD).

Type II and III systems are generally the most commonly used. Geotextiles and softboard help drain the surface but are less efficient in enhancing the surface properties of the concrete. It is important that the CPF is appropriately tensioned to avoid ripples.

6.3.6.5 Defects

Poor design or use of formwork can directly or indirectly result in defects affecting the strength, durability or appearance of the concrete section (Table 6.7).

6.3.7 Curing and protection

Curing is the action taken to maintain moisture and temperature conditions to allow hydration of cementitious materials, both hydraulic cements and pozzolanic materials. The objectives of curing are to prevent the loss of moisture from concrete and to maintain a favourable concrete temperature for a relatively short period of time so that the concrete will be sufficiently strong and impermeable. Most of the hydration

Table 6.7 Defects associated with formwork

Defect	Cause
Blowholes	Form faces orientated approximately horizontally, trapping air Choice of release agent
Chipping and spalling	Premature or clumsy removal of formwork Formwork not designed for easy removal
Cover depth deviates from target	Excessive deflection of form face Rippling in CPF (Figure 6.13)
Cracking, scaling and deflection	Inadequate or premature removal of support
Dusty and weak surface	Overapplication or choice of release agent Retardation from sugars in timber formwork
Early-age thermal cracking	Formwork type or time of removal does not match early-age thermal cracking design
Honeycombing Steps in surface	Grout loss from joints (Figure 6.14)
Plucking to surface	Premature removal of formwork combined with cold weather and choice of release agent
Variable colour	Large variation in absorbency of form face

Figure 6.13 Effect on concrete surface and cover of rippling to CPF. (Courtesy of D Wimpenny)

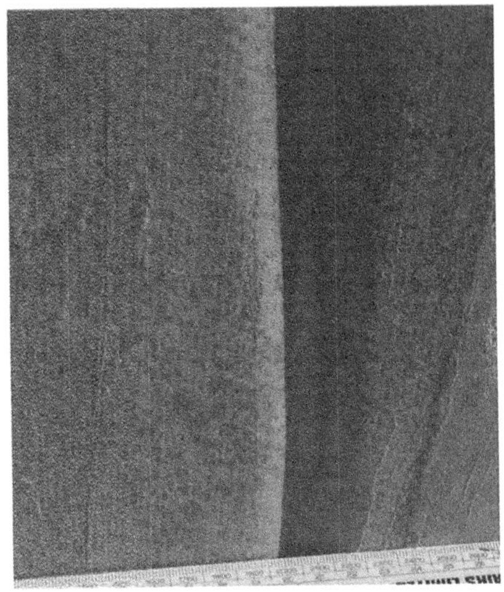

Figure 6.14 Grout loss leading to honeycombing on column. (Courtesy of J Knights)

of concrete takes place within the first month, in ideal conditions, but will continue slowly with time. Continuous increases in strength have been recorded exceeding 30 years [60].

Both at depth and at the near surface, curing has a beneficial influence on the properties of hardened concrete, such as strength, permeability, abrasion resistance, volume stability, early-age thermal cracking, frost resistance and resistance to chloride ingress.

Poor curing practices adversely affect the desirable properties of concrete. Proper curing is essential for obtaining the maximum durability of the concrete covering the reinforcement, especially if the concrete is exposed to severe environments.

Curing is even more important than ever before [61].

- Today's cements gain strength earlier and allow contractors to remove formwork soon after concrete placement. This encourages curing operations to be discontinued prematurely.
- The lower water-to-cement ratios used with modern concretes tend to cause self-desiccation. Ingress of water from proper curing is necessary to control this.
- Many modern concrete mixtures contain mineral admixtures, such as fly ash and GGBS, that have slower reaction rates. Curing over longer periods of time is needed for the proper development of the properties of these mixtures.

Worldwide, it is common to witness contractors either taking shortcuts with curing requirements or ignoring them almost completely. The concrete industry must do a better job in educating contractors, engineers and asset owners on the importance of good curing for protecting and enhancing their structures. Cather [62] has suggested that one way to highlight the importance of curing is to make it a separately billed item in the schedule of prices for a project.

Figure 6.15 illustrates the effect of moist curing on strength development, and Figure 6.16 shows the decrease in permeability of cement mortars with increasing moist curing periods for a range of water-to-binder ratios.

Moist curing of concrete is important, in general, for the following reasons.

- The development of both strength and impermeability is due to hydration processes and stops when the concrete is dry. In most cases, durability is much more dependent on impermeability than on strength.

Figure 6.15 Effect of moist curing time on the strength gain of concrete [63]

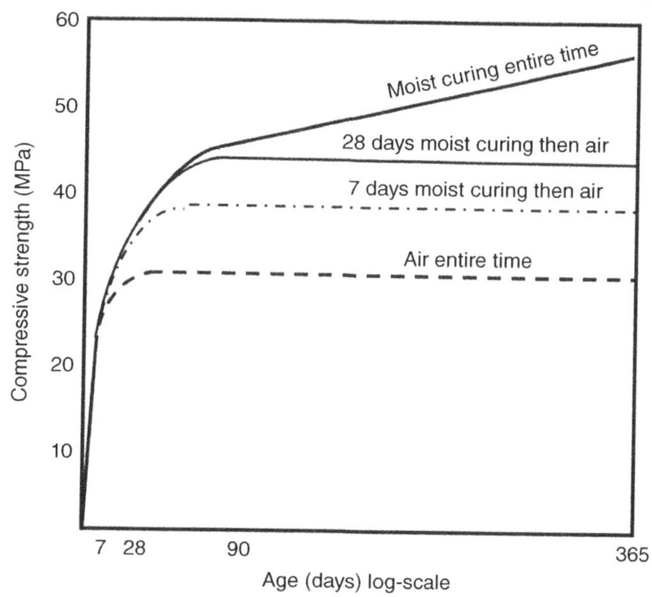

Figure 6.16 Effect of water-to-binder (w/b) ratio and curing duration on the permeability of mortar [63]

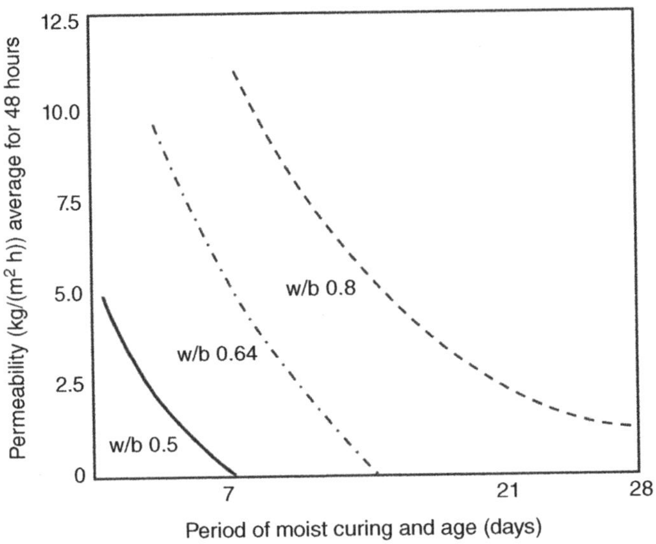

- The development of impermeability is even more strongly dependent on the water-to-binder ratio than is strength. At water-to-binder ratios below 0.45, it may take place in a few days, but at values of more than 0.50, it can take weeks of curing. Above 0.60, it can require months.
- It is important that concrete dries as uniformly as possible because drying of surface layers causes differential stresses, potentially leading to cracking and curling in floor slabs.
- Shrinkage is mostly drying shrinkage, commencing as concrete dries. However, high-strength concretes with a low water-to-binder ratio can be subject to autogenous shrinkage, in which all the free water is consumed by the hydration reaction at a much earlier stage than with drying. Since such concrete rapidly becomes substantially impermeable, it is difficult to introduce further water. Thus, it is essential that none of the original water escapes. Note that concretes that are wet cured for 7 days have less shrinkage due to drying and autogenous effects than concrete with no water curing (Figure 6.17).

Figure 6.17 Length change of concrete samples exposed to different curing regimes [63]

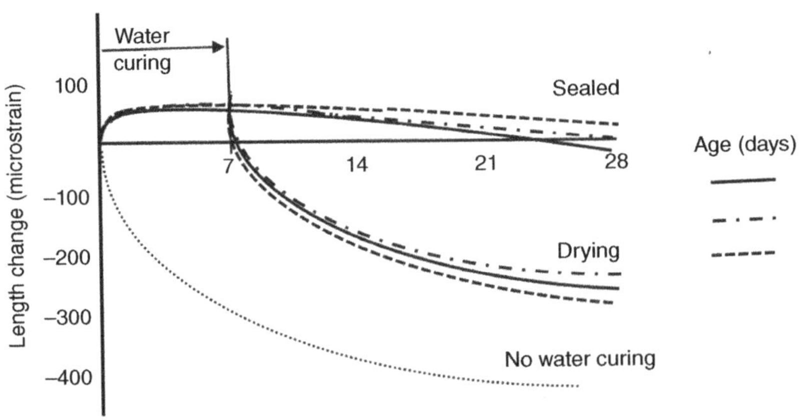

Concrete can be kept moist by three main curing methods, namely

- supplying additional moisture, for example, ponding, immersion, spraying or fogging, and saturated wet coverings
- sealing in the mix water, for example, covering concrete with impervious sheeting or membrane-forming curing compounds
- accelerated curing, for example, steam curing, heating coils and pads.

Curing should be started immediately following the placement and finishing operations to prevent rapid surface evaporation.

Since all the desirable properties of concrete are improved by curing, the curing period should be as long as practicable. ACI 301 [64] recommends a minimum of 7 days for structural concrete, reduced to 3 days for high-early-strength concretes.

6.3.8 Temperature control and early thermal cracking

The hydration reactions that take place between cement and water are exothermic, generating heat. The rate of heat generated depends on the type of cement used and its fineness. The peak temperatures (total heat) produced in concrete depend on the type and amount of cement in the concrete mixture, the initial placement temperature, the element geometry, the type of formwork and the insulation.

Thus, to control cracking and mitigate the risk, it is important to choose the right type and quantity of cement in the concrete mixture. Other aspects, such as aggregate type, reinforcement density, pour sequencing and restraint reduction all reduce the risk, but ultimately, it is the quantity and type of cement that are the key aspects of design.

As the concrete increases in temperature, it expands. In some cases, compressive stresses are generated if there are restraints, for example, existing structures or elements. Once the peak temperature has been reached, the concrete will start to cool and contract. Restraining this thermal contraction results in the development of tensile stresses. As young concrete is still relatively weak but with increasing stiffness (i.e. elastic modulus), the stresses due to contraction may cause cracking. The magnitudes of expansion and contraction depend on the type of aggregates in the concrete mixture, and as such, this may be a key parameter during design.

Mass concrete (without reinforcement) may crack in an uncontrolled manner. However, provided that sufficient mitigation measures are in place, this is not normally a concern for durability. Moreover, early thermal cracking is rarely a structural issue. Typically, thick sections of mass concrete are susceptible to cracking due to the temperature differential between the centre of the element and the surface, as the surface layer cools more quickly than the centre causing internal tensile stresses within the section. Surface cracks form, which often close with further cooling.

Controlling early-age thermal cracking in structural concrete relies on the use of reinforcement to control crack widths. The need to control crack widths ranges from being purely an aesthetic concern to the exacting requirement for waterproofing of liquid-retaining and containment structures. To ensure durability, the Eurocodes require a limit of 0.3 mm for crack widths. This limiting crack width is reduced for liquid-retaining structures, depending on the hydraulic gradient at the outer surface of the structure (i.e. the ratio of the hydraulic head to the thickness of the retaining element), as plotted in Figure 6.18. This requirement assumes that concrete will self-heal (autogenous healing) according to the relation between crack width and hydraulic gradient. That is, the higher the gradient, the smaller the limiting crack width will be.

Figure 6.18 Limiting crack widths for self-healing plotted against the pressure gradient across the section [65]

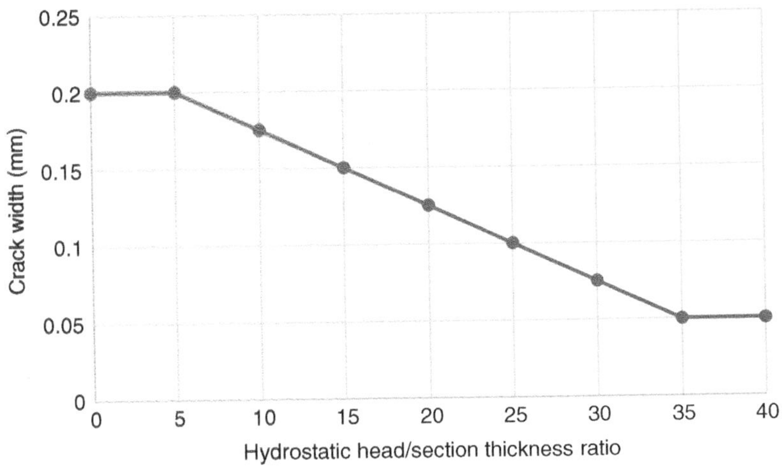

Standards on the provisions for design and for controlling crack widths are presented in EN 1992-1-1 [66] and EN 1992-3 [67]. Comprehensive guidance is provided in CIRIA C660 [65] and its successor CIRIA C766 [68]. Further useful information and guidance may also be found in ACI 224R [69] and ACI 207.2R [70].

Ideally, controlling temperature-dependent cracking should be considered during the design phase for a structure. The specification should consider the binder type and content required for each concrete mix, particularly for sections with a minimum dimension over 0.5 m, to mitigate excessive cracking. The specification should also state the limits on peak temperatures and differential temperatures between the concrete surface and the core. The reinforcement design has to be calculated specifically to control crack widths.

Aside from the risk to the durability of the reinforcement due to cracking, it is important to assign peak temperature limits for thick concrete sections to avoid long-term deterioration, such as delayed ettringite formation. Typically, limits of around 65°C are suggested for concretes containing Portland cement only, increasing to 75–80°C for concretes containing higher replacement levels of fly ash and GGBS. Excellent guidance on the mitigation of delayed ettringite formation is given in BRE information paper 11/01 [71]. American guidance on limiting peak temperatures is provided in ACI 308R-16 [72].

In structurally critical pours, or where specified, monitoring peak temperatures and the differentials between centre and surface temperatures may be a compliance requirement. Embedded thermocouples are commonly used to measure temperature, which can then be recorded at frequent intervals once the concrete is placed. Automated logging devices that record the evolution of temperature are available, often transmitting the information via a modem over the internet. CIRIA C660 [65] also provides further guidance on the monitoring of concrete temperatures.

If appropriate, this information, along with a little pre-construction testing and development, can be used to derive maturity curves for the in situ concrete, which can be used to estimate the in situ strength. This allows optimisation of the striking of the formwork and precast moulds and the post-tensioning operations.

Avoiding excessive peak temperatures may require additional measures. The temperature of fresh concrete can be reduced with chilled water, flaked ice or even liquid nitrogen, thus reducing the peak in situ temperature rise. Simple calculations can be used to estimate the effect of cooling concrete and its constituents (ACI 207.4R [73]). Cooling pipes, although not common in the UK, may also be necessary to reduce the peak temperature at the core of the concrete. Specialist advice should be sought if considering this measure.

Insulating or heating the concrete surface can also be successful in controlling excessive differential temperatures. Often, only a limited thickness of insulation is required, for example, proprietary 10-mm frost blankets. Such blankets can protect the concrete from extreme temperature differentials but are also thin enough to allow adequate cooling of the element. If the concrete is too well insulated, it will take longer to cool, which means the insulation will need to be retained for longer, potentially impacting the construction programme.

In summary, the judicious selection of materials, carefully designing the reinforcement to control cracks, practical measures for in situ temperature control and insulating exposed surfaces will mean that, in many cases, early thermal cracking can be successfully moderated.

6.3.9 Cover

The most frequently encountered cause of significant deterioration in concrete structures is corrosion of the reinforcing steel due to inadequate protection by the cover concrete. The cover depth on construction drawings is usually the target or nominal cover value and the actual depth will deviate from this value depending on the care taken when bending, fixing and placing the reinforcement. Commonly assumed cover deviations based on Eurocode 2 (BS EN 1992-1-1 [66]) are listed in Table 6.8 for different casting conditions.

The assumed cover deviation is added to the minimum cover for durability to give the target or nominal cover. However, it is important to stress that the actual cover deviation should be established on site by cover meter surveys, ideally during trials. There can be a high variability in measurements of cover depth on site (Figure 6.19). The type, frequency and correct use of spacer blocks and chairs have a significant influence in maintaining cover during concrete placement. These aspects are covered by BS 7973 Parts 1 and 2 [74,75]. Spacers can be crushed, dislodged or even used wrongly, resulting in incorrect cover. The cover to the steel, the fixing of the blocks and the general rigidity of the cage should be checked in the forms. This is a hold point prior to concreting. A cover meter survey should be undertaken after concreting to ensure that any movement of the reinforcement has not compromised the minimum cover for durability. When interpreting the cover meter results, the effect of measurement accuracy and laps should be considered. A simple specification requirement that the minimum cover for durability should be met by 95% of the reinforcement and the absolute minimum should not be less than 85% of the nominal value has been proposed [76].

It is not desirable to provide an excessive thickness of cover because, apart from the extra cost and weight, this concrete is essentially unreinforced and can develop substantial cracking through shrinkage or stresses

Table 6.8 Cover deviations for different casting conditions

Casting conditions	Cover deviation: mm
Precasting under factory production control	5
Casting against formwork	10
Casting against blinding	15
Casting against a temporary pile casing	25
Casting directly against the ground	50[a]

[a]Arguably, this includes cover deviation and an allowance for a reduction in the quality of the concrete.

Figure 6.19 Histogram of 900 cover meter values from a motorway overbridge (nominal cover 30 mm)

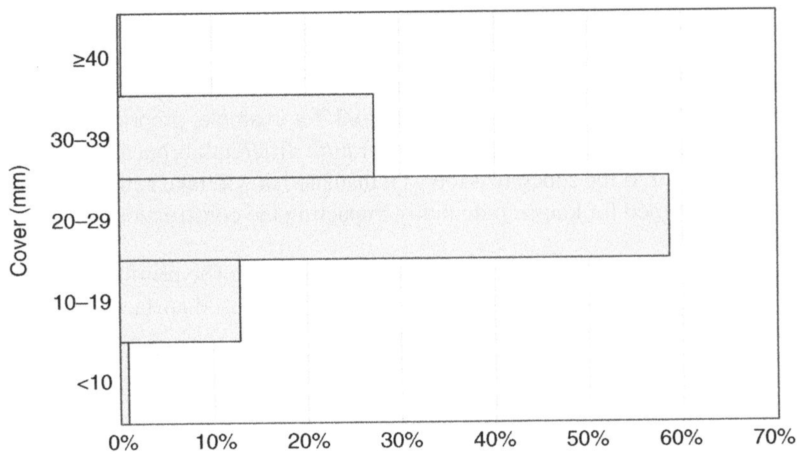

due to loading, such as bursting stresses in precast tunnel segments. For a structural fire design to Eurocode 2 (BS EN 1992-1-2 [77]), if the distance between the surface exposed to fire and the axis of the reinforcement is 70 mm or more, then surface reinforcement with a spacing of not greater than 100 mm and a diameter of not less than 4 mm should be provided, unless tests have shown that surface spalling does not occur. To ensure durability, this surface reinforcement should be non-ferrous.

The protection provided to the steel also depends on the properties of the concrete. The type of cementitious material, the maximum water-to-binder ratio and the cover depth are key parameters in a durability design. The design can be undermined if the mix does not match that specified, if curing is inadequate or if the concrete cover is penetrated by defects, such as voids, cracks, bleed channels or inclusions. Reinforcement spacers can be a weak point in the cover zone and, ideally, should be manufactured from concrete with comparable properties as and good bonding to the concrete in the element [78]. Steel chairs should not be used unless protected by a robust plastic coating.

6.3.10 Making good and repairs

Concrete repairs fall into two broad categories: making construction defects good and addressing deterioration to extend the life of an asset or structure. Neither subject is covered here in detail, although the key points affecting durability are mentioned. The focus is on making new-build construction good. Asset management planning for historical structures is touched on where relevant, but this is a wide-ranging subject in itself. BRE information paper 9/07 [79] has information on reactive and proactive approaches to maintaining structures.

Major repairs during construction resulting in the need for significant redesign, strengthening or reinstatement with structural concrete are not addressed and can be regarded as new construction rather than repair.

Often, the need for repairs is urgent, and construction teams and asset owners can be under pressure to opt for what could be described as a quick fix. It is understandable that new-build defects are dealt with as quickly and economically as possible, but there needs to be adequate consideration of the durability of the repairs in the contract specification. Owners of ageing assets may recognise the need for repair only when deterioration impacts safety, commercial value or functionality. The result is often a solution that addresses the immediate problem but without proper consideration of longer-term implications, including the need for repeat interventions.

The best way to ensure that new construction results in a durable structure is to avoid the need for repairs. Occasionally, the measures proposed to avoid the need for later repairs are considered too onerous and the decision to repair retrospectively is taken instead, for example, using injections for early thermal cracking. In general, this approach is appropriate only in low-risk cases, for example, structures in a benign environment with a relatively short design life. The construction benefits should be assessed against the implications for the asset owner or operator in terms of cost, safety and disruption due to increased intervention in service.

For the best long-term performance, repairs should be properly designed, although this has not always been the case historically. Understanding the nature of the defect or damage and why a repair is needed is of key importance. For example, crack injection may be ineffective if the cracking is subject to long-term movement or is the result of an ongoing expansive reaction, such as ASR. In the latter case, crack injection may encourage new cracking by preventing the stresses and the silica gel from being released. Patch repairs of corrosion-related damage are likely to lead to the formation of an incipient anode and corrosion in areas adjacent to the repair in a relatively short time.

In the UK, common repair principles and methods are outlined in BS EN 1504-9 [80]. Requirements for testing and certification of repair compounds and associated quality assurance procedures are set out in the parts of BS EN 1504 relevant for each repair method. Concrete Society Technical Report 69 [81] provides useful guidance on the application of BS EN 1504. Even with appropriate design and material selection, poorly executed repairs will not be durable, so trials are strongly recommended. If the need for repairs can be reasonably foreseen, for example, to the exposed surfaces of diaphragm walls, it is appropriate to establish in advance agreed repair methods to avoid the need for delays.

One of the most common repair methods is a patch repair using modified cementitious or epoxy resin-based repair compounds, hand-applied or poured within shutters. The list below includes some common points to consider for all types of patch repair.

- Percussive breakout methods can damage the reinforcement. Where practicable, high-pressure water-jetting (at least 110 MPa or 15 000 psi) is useful.
- The damaged areas that need to be broken out can be significantly greater than expected. It is, therefore, important to consider the structural implications.
- It is important to excavate with a sufficient clearance behind the reinforcement, for example, 25 mm, to provide a mechanical key for the patch repair.
- Feathered edges can fail relatively quickly and should be avoided by saw cutting the perimeter at least 10 mm deep and perpendicular to the substrate surface.
- The manufacturer's guidance on the maximum and minimum application thicknesses should be strictly followed and if necessary, the repair completed in several layers.
- Most important of all, adequate substrate preparation should be undertaken, with loose friable material and fractured aggregate removed back to clean sound concrete.

6.4. Issues with specialised concretes

6.4.1 Architectural

Architectural concrete usually requires special care to achieve a specified colour or surface texture. Getting a uniform appearance requires consistency with the concrete mix, formwork face and curing regime. For example, differences in absorption by the formwork will result in variations in the water content and shade of the surface. Water flowing across the surface of shiny impermeable formwork can expose black cement grains, leading to a dark colouration.

Achieving the required finish can necessitate

- particular binder types, such as high GGBS levels, to produce a paler surface
- avoiding pyrite and lignite contamination of the aggregate
- stainless steel tie wire
- pigments, stains, sealers and coatings
- carefully planned tie locations with rebates to hide construction joints
- shaped polystyrene and patterned elastomeric formwork liner
- washing and brushing the surface to expose the aggregate
- abrasive blasting or tooling the hardened concrete to form needle-gun, bush-hammered, point-tooled or hammered ribbed finishes [82]
- trials to establish an acceptable finish
- very careful making good to match the colour and texture.

In relation to the above, it is important that if a surface retarder is used to facilitate the washing and brushing process, it must be applied evenly at the manufacturer's dosage rate and must not contaminate the reinforcement. The depth of textured surfaces, tooling and joint rebates must be allowed for in the nominal cover to ensure the minimum cover for durability is maintained. Staining due to pyrite, lignite or tie wire can be very unsightly and warrant special attention. If stainless steel tie wire is not used, then soffit formwork must be cleared of all discarded tie wire. The fine aggregate sources should not have a history of contamination, unless additional precautions are applied.

In addition to the durability of the concrete to deterioration, the impact of ageing and weathering on appearance is also important. Finishes can be rapidly marred by efflorescence, biological growth, pollution or poorly planned water drainage [83]. Efflorescence is due to the deposition of calcium carbonate and salts when a surface dries following an extended period of dampness [84]. It is especially a problem for concrete that is darkly pigmented. Mould and lichen grow most rapidly on surfaces exposed to warm, humid conditions and sunlight. Bridge piers with a high proportion of GGBS developed unsightly discolouration after 2–3 years of service in Hong Kong [85]. This was treated by pressure-washing with a biocide followed by the application of silane. Water draining across a surface can lead to staining and exacerbate the impact of pollution and biological growth. A common example of this is water running off balustrades or windowsills, leading to vertical stripes down the concrete below [86].

Positive steps should be taken to avoid damage to architectural concrete due to ongoing construction, including rust staining from exposed starter bars and impact damage from construction traffic. Skill is required when matching colour and texture to make good blemishes and damage. Differences in water absorption by the repair and parent concrete may become evident on surfaces exposed to dampness in service. Anti-graffiti coatings and sealers are necessary in some public spaces. The maintenance of these needs to be planned to ensure a durable appearance. For example, anti-graffiti coatings can be either 'permanent' or sacrificial, and each has a different impact on appearance and the maintenance regime. Integral pigments will fade (especially darker shades and more vivid colours) and staining applied to a surface will be lost by abrasion. Expectations over the durability of appearance and its maintenance need to be managed.

6.4.2 Floors

Finishing is more important for floor slabs than for basic concrete elements. The floor is a key element of many facilities, such as workshops and warehouses. Floors are often specified with requirements for surface regularity and abrasion and slip resistance. Non-compliance with these requirements or premature deterioration of the surface can impact the safe use of a facility, often necessitating disruptive remedial work, such as grinding.

The modern practice for floors is to use steel fibre reinforcement instead of bars or mesh and casting in large areas rather than in narrow bays. This requires superplasticised high-workability concrete combined with laser-levelling and power-floating. Using these techniques with this type of concrete needs care, both in ensuring the correct design and the workmanship.

Occasionally, dry-shake finishes or monolithic screeds are used to enhance abrasion and slip resistance. These are not a replacement for good concrete practice. Monolithic screeds are very prone to curling and delamination due to drying so are best avoided.

After striking off the surface to the required level, it is essential that it is not worked until it is sufficiently stiff to permit foot traffic without leaving any marks. This is especially the case if the concrete bleeds. The surface must not dry before this stage or evaporation cracking will occur. If necessary, an evaporation retardant or a fine mist spray can be applied. Equally, any remaining bleed water must be removed by a squeegee or other means before the surface is worked. Working the surface before bleeding is complete can create a layer of dense concrete at the surface, which traps the bleed water below, leading to blistering and delamination. The risk of delamination can be exacerbated by the presence of entrained air and mixes. Thus, air-entraining admixtures are not recommended if the surface is to be power-floated. Working the surface in the presence of water from bleeding or other sources will result in a surface with reduced strength and abrasion resistance. If power-floating is carried out too early, then it will spread a layer of laitance rather than densifying the surface. This is a particular problem for concrete containing fly ash or GGBS in cold conditions due to the extended setting time.

Curing and protecting floors is very important to reduce the risk of cracking and to ensure good abrasion resistance. Mixes with silica fume can have a very low bleed rate, so the early commencement of curing is necessary to reduce plastic shrinkage cracking, unless synthetic microfibres are incorporated into the mix (Section 6.3.4). The abrasion resistance of concrete mixes containing GGBS and fly ash is particularly sensitive to poor curing, which can result in increased abrasion and dusting of the surface, especially when combined with excessive bleeding.

6.4.3 Fibre-reinforced concrete

Reinforcement with steel and synthetic macrofibres has increasingly been adopted for sprayed concrete and precast linings in tunnel and mining applications as well as for floors. Concrete Society Technical Reports 63 and 65 provide useful guidance on steel and synthetic fibres [87,88].

Fibre-reinforced concrete has significant durability advantages over concrete with conventional reinforcing bar, including

- enhanced impact and abrasion resistance (although this is contested) [89]
- high chemical and corrosion resistance with synthetic fibres [90]
- no corrosion-related spalling or corrosion risk from stray currents due to the discrete and random orientation of the steel fibres [91]
- high resistance to carbonation due to the formation of a portlandite-rich layer on steel fibres
- increased chloride threshold level for steel fibres [92].

The chloride threshold of steel fibres may be 2.5–4.7% by mass of binder compared to 0.4–0.6% for conventional reinforcement [93]. This may be due to a modification of the surface of cold-drawn wire leading to reduced roughness and increased homogeneity [92].

There are a range of opinions as to the durability design of steel-fibre-reinforced concrete. Some designers rely on field evidence that corrosion is limited to the fibres at the surface of the concrete, although others include a depth of loss of fibres due to corrosion based on predictive modelling [90]. It is also important to consider the durability of fibres, which suffer from cross cracks. The limiting crack width to control the corrosion of steel fibres appears to be between 0.1 and 0.25 mm [94–96].

In some cases, the speckling of surfaces caused by the rusting of exposed fibres will be aesthetically unacceptable and, in these cases, galvanised or stainless fibres may be required. Galvanising the fibres should not be relied upon if they are exposed to chlorides.

Polypropylene fibres have good durability [97]. For a mortar reinforced with polypropylene-based synthetic fibres, a modest reduction in tensile strength of up to 15% was found after 18 years [98]. Information from the manufacturers of the polypropylene used in some fibres indicates that sodium chloride has a negligible effect but that the fibres are vulnerable oxidising agents and chlorine (gas) [99]. There is also some evidence that microbes derived from marine-dredged aggregate may grow into the interface between synthetic fibres and the concrete matrix, thus accelerating degradation [100].

The batching and mixing of fibres need to be undertaken with care to prevent the fibres from forming clumps or balls. Steel fibres are sometimes collated with water-soluble glue to reduce this risk. The fibre manufacturer's recommendations on the method of adding fibres should be followed. Some fibre manufacturers provide additional support to ensure correct batching and mixing, including non-destructive methods to monitor the homogeneity of the fibre content during discharge (Dramix eyeD® www.bekaert.com). Trials and visual monitoring during concrete discharge are essential for avoiding significant problems. For example, fibre balling has been identified in precast segments (Figure 6.20). In that case, no reliable non-destructive method of finding or treating fibre balls was identified so that a structural and durability assessment led to a requirement for long-term monitoring in service.

6.4.4 Self-compacting concrete

Self-compacting concrete has advantages over concrete that must be vibrated in terms of reduced noise and labour during placing. However, the concrete mix design and the planning of the pour must allow for the specific pour characteristics [101,102]. Trials and testing are essential to ensure that the proposed mix, methods and personnel produce the intended design. Formwork must be grout-tight.

The primary risk is segregation of the concrete, so the concrete must not be subject to extraneous vibration, movement or working. The risk that the cement paste and aggregate will separate is increased by

Figure 6.20 Fibre ball revealed within a precast segment by saw cutting. (Courtesy of D Wimpenny)

washout if the concrete is placed underwater. In such cases, viscosity modifiers or anti-washout admixtures are recommended.

Testing should include assessments of the flow, passing ability and stability of fresh concrete in accordance with BS EN 206 [3]. Consideration should also be given to performing a rapid segregation test, such as ASTM C1712 [103]. A summary of the testing undertaken on a transfer beam constructed using self-compacting concrete is given in Table 6.9 [101].

Self-compacting concrete mixes typically have a higher powder content than normal concrete mixes and, as such, may be subject to a higher temperature rise during hydration and increased plastic and long-term shrinkage, which may increase the risk of cracking and delayed ettringite formation.

Self-compacting concrete does not have the same surface characteristics as a concrete subject to normal internal vibration or intense external vibration during precast production. The finish of formed faces can be excellent, but blowholes may form and necessitate very gentle rodding of the concrete adjacent to a face [102].

A failure to properly design or use self-compacting concrete can result in thick layers of grout and areas of poor compaction and honeycombing where the cement paste is depleted, for example, due to grout loss from formwork. These defects can undermine the intended durability.

6.4.5 Sprayed concrete

Sprayed concrete is commonly used for repairing and strengthening existing structures, for constructing free-standing structures and for ground strengthening and support [36]. Durability is particularly important for permanent tunnel linings. Sprayed concrete typically has a 6–10 mm maximum aggregate size and is reinforced with steel fabric or mesh or with fibres. The concrete mixes commonly utilise silica fume to improve bonding and early strength development. The shrinkage of sprayed concrete is higher than with normal mixes, which can lead to cracking and opening of joints and, thus, water penetration.

Sprayed concrete can be produced as a wet mix. The fresh concrete is pumped to a nozzle where a setting accelerator is added. It can also be produced as a dry mix. The dry powder is pumped to a nozzle and both water and accelerator are added. The latter process does not provide an accurate measurable water content

Table 6.9 Test methods for self-compacting concrete

Test	Method	Limit
Flow tests		
Slump flow	BS EN 12350-8 [104] ASTM C1611 [105]	550–800 mm
T500	ASTM C1611 [105]	2–5 seconds
Visual stability index	ASTM C1611 [105]	<2
Passing ability tests		
J-ring	ASTM C1621 [106] BS EN12350-12 [107]	Δ flow 25–50 mm Δ height ≤15 mm
Stability tests		
Stability penetration	ASTM C1712 [103]	≤10 mm depth
Sieve segregation test	BS EN 12350-11 [108]	≤15% on sieve

for the mix, so that dry mixes should not be used for structures requiring durability over long service lives, such as permanent tunnel linings.

To assess its permeability, sprayed concrete for watertight applications is sometimes subject to the water penetration tests described in the chapter on durability performance tests in BS EN 12390-8 [28]. The European specification for sprayed concrete and BS EN 14487-1 [36] impose a water penetration limit of 50 mm.

Accelerators for sprayed concrete are based on aluminium sulfate. This creates a difficulty in meeting sulfate limits of 3.5% by mass of cement and raises questions about the impact of the sulfate on durability. Although there does not appear to be evidence of delayed ettringite formation or internal sulfate attack connected with the high sulfate content, sulfate attack has been witnessed in the sprayed concrete used to repair erosion of inclined seawalls. An unusual form of sulfate attack was reported within the sprayed concrete of marine tunnels, which was due in part to the formation of Mn-Fe biofilms [109]. This suggests that biofilms may have a wider role in the deterioration of concrete in marine environments than is currently appreciated.

Sprayed concrete should be compacted well by the spraying process, but the quality is highly dependent on the skill of the nozzle operator, who should have appropriate certification from a recognised industry body, for example, the Nozzleman Certification Scheme run by EFNARC. Commons problems are sand pockets, voids due to the shadowing effect of bars and the incorporation of rebound material. The latter leads to a porous friable surface and a reduction in bonding to subsequent layers. Pulsation in the pumping equipment can lead to bands of weak porous material and a reduction in strength and durability. Hammer soundness testing is helpful in identifying loss of bonding between layers, and this should be accompanied by routine bond testing and cores for assessing strength and compaction.

6.4.6 Slipformed concrete

Slipforming is a specialist method for constructing vertical concrete structures and horizontal paving structures (e.g. slabs and barriers). It is one of the fastest methods of concrete construction, often operated on a 24-hour basis. Effectively, the concrete is extruded from the moving formwork during the process.

Horizontal slipforming is typically used for paving, kerbs and barriers. The equipment for horizontal slipforming is normally self-propelled and supported plant machinery, which may undertake one or more roles (spreading, placing and finishing), depending on the size of the project. Kerbing or barrier machines have moulds specially designed for the desired shape of the structure. Levels are kept by string line or laser.

For vertical structures, the process utilises a self-climbing formwork system with two or three decks and shuttering that straddle the structural element and are supported by a yoke lifted by a series of mechanical jacks. The jacks are supported by rods or tubes positioned within the wall or structure, and they are laterally braced by the rising concrete structure. Rates of rise of up to 450 mm/hour can be achieved with good practice and management. Levels and verticality are typically kept by laser line, particularly for high-rise structures.

As slipforming is often a continuous process, it should be treated as a factory production procedure, with detailed planning and careful control of the flow of materials onto and into the structure, including the reinforcement, boxouts, embedded items and, ultimately, the concrete, as the form travels along or upwards. The flow and movement of the labour force are also key for ensuring timely fixing of the reinforcement, the accuracy and control of the structure, the concrete placement and the finishing operations, often in 24-hour working conditions. Good management and supervision are essential elements of a successful slipform process.

Moreover, to ensure timely concrete placement and good finishing, concrete production and, as importantly, the concrete setting rate must be monitored constantly to take account of the climb rate of the form, the complexity of the structure and the climatic conditions (temperature, humidity, wind speed, etc.). Even and level placement of the concrete around the form is essential to avoid problems as the form rises.

Minor defects that affect most concrete construction operations can also affect slipform concrete (e.g. poor compaction, honeycombing, bleeding and segregation). More serious defects principally depend on the ration of the rate of concrete set and the rate of climb, namely

- areas of concrete slumping and the possible falling out of sections of the wall (concrete not set)
- areas of the concrete surface tearing or the lifting of sections of the wall (concrete too stiff).

The concrete can be remediated either after the slipform process is complete or undertaken during the climb (or 'slide'). If too serious, the slide may need to be halted to allow for repairs to be completed while access is available. Not identifying and addressing a problem sufficiently early can result in expensive remedial works or even rejection of the works [110].

Further guidance can be found in reports published by the Concrete Society (2008) [111] and the American Concrete Institute [112,113].

6.4.7 Piling and diaphragm walling concrete

The placing of concrete for piling and diaphragm walling is usually undertaken by a specialist subcontractor. It typically involves casting underwater or using a temporary support fluid or casing. The cage is either placed into the open excavation or can be inserted into the plastic concrete before any temporary casing is withdrawn.

Table 6.10 Tests for piling and diaphragm wall mixes

Test	Standard	Acceptance
Consistency tests		
Slump	BS EN 12350-2 [114]	150 ± 30 mm
Flow table[a]	BS EN 12350-5 [115]	560 ± 50 mm for a tremie underwater[b] 600 ± 50 mm for a tremie under a support fluid[b]
L-box	Modified BS EN 12350-10 [116] to measure time to end of box	Time <12 seconds
Consistency retention		As specified
Stability tests		
Visual stability	BS EN 12350-5 [115] ASTM C1611 [105]	No signs of segregation or bleed Visual stability index = 0
Bleeding	Modified ASTM C232 [117] to include bleed rate	<0.1 ml/min after 2 hours
Column segregation	Modified ASTM C1610 [118] with 2-hour standing time	<5%
Static segregation	AASHTO PP 58-12 [119]	Hardened visual stability index = 0 or 1
Pressure bleed	Bauer filtration (CIA, 2012) [120]	15 litres/m^3 or 22 ml direct filtration

[a]The slump-flow test in BS EN 12350-8 [104] is an alternative to the flow table test.
[b]Tolerance for the initial discharge (±40 mm for a composite sample).

The tremie must be kept embedded at least 1 m below the concrete to avoid entry of water or support fluid. The concrete mix design and consistence must be controlled to ensure adequate flow and self-weight compaction throughout an extended casting period while avoiding excessive bleeding or segregation. The timing of the insertion of the cage and withdrawal of the temporary casing is critical to avoid distortion to the cage and inclusion of soil.

Unsurprisingly, given their underlying complexity, piles and diaphragm walls are common sources of problems, including excessive segregation and bleeding. The ICE *Specification for Piling and Embedded Retaining Walls* provides some useful guidance on the concrete mix and methodology. Relevant tests are listed in Table 6.10 [47].

REFERENCES

1. UK Water Industry Research (2011) *Civil Engineering Specification for the Water Industry*, 7th edn. UK Water Industry Research, Swindon, UK.
2. ACI (American Concrete Institute) (2014) 318-14. Building code requirements for structural concrete. ACI 318R-14. Commentary. ACI, Farmington Hills, MI, USA.
3. BSI (2013) BS EN 206. Concrete. Specification, performance, production and conformity. BSI, London, UK.
4. BSI (2015) BS 8500-1. Concrete. Complementary British Standard to BS EN 206. Part 1. Method of specifying and guidance for the specifier. BSI, London, UK.
5. BSI (2015) BS 8500-2. Concrete. Complementary British Standard to BS EN 206. Part 2: Specification for constituent materials and concrete. BSI, London, UK.
6. Bickley J, Hooton RD and Hover KC (2006) *Preparation of a Performance Based Specification for Cast-in Place Concrete*. RMC Research Foundation, Silver Spring, MD, USA.
7. Gjørv OE (2013) Durability design and quality assurance of major concrete infrastructure. *Advances in Concrete Construction* **1(1)**: 45–63.
8. BSI (2011) BS EN 197. Cement. Composition, specifications and conformity criteria for common cements. BSI, London, UK.
9. Dewar JD (1999) *Computer Modelling of Concrete Mixtures*. E&FN Spon, London, UK.
10. Concrete Society (2011) *Cementitious Materials*. concrete Society, Camberley, UK, Technical Report 74, pp. 23–25.
11. BRE (Building Research Establishment) (1998) *Recycled Aggregates*. BRE, Watford, UK, Digest 433.
12. BSI (2002) BS EN 12620. Aggregates for concrete. BSI, London, UK.
13. Khoury GA (2000) Effect of fire on concrete and concrete structures. *Progress in Structural Engineering and Materials* **2(4)**: 429–447.
14. Pomeroy RD (1992) *The Problem of Hydrogen Sulphide in Sewers*, 2nd edn (Boon AG (ed.)). Clay Pipe Development Association, http://www.cpda.co.uk/publications (accessed 22/09/2022).
15. European Federation of National Associations Representing for Concrete (1996) *The European Specification for Sprayed Concrete*. EFNARC, www.EFNARC.org/publications.
16. Quality Scheme for Ready Mixed Concrete (2003) *Quality and Product Conformity Regulations*. QSRMC, Hampton, Middlesex, UK.
17. ASTM (2017) C94. Standard specification for ready-mixed concrete. ASTM, West Conshohocken, PA, USA.
18. Abel P and Cudley C (2018) Concreting at Hinkley Point. *Institute of Concrete Technology 46th Annual Convention Symposium, University of Derby*.
19. Crossrail (2012) *Materials and Workmanship Specification, Sprayed Concrete Linings*. Crossrail, Section KT20.
20. Wimpenny D (1991) *A Critical Appraisal of the RAM for Concrete*. Pioneer Concrete, UK, Report ITR5.

21. Nagi M and Whiting D (1994) Determination of water content of fresh concrete using a microwave oven. *Cement, Concrete and Aggregates* **16(2)**: 125–131.
22. Cameron S and Chudley S (2018) Team Concrete – a collaborative approach to new nuclear. *Institute of Concrete Technology 46th Annual Convention Symposium, University of Derby.*
23. Ellis C and Wimpenny DE (1990) The assessment of mix stability using the two-point test. *Proceedings of the International Conference on the Rheology of Fresh Cement and Concrete, University of Liverpool.* British Society of Rheology, pp. 281–291.
24. Wimpenny D (1990) *The Effect of Pre-hardening Temperature on Concrete Strength.* Pioneer Concrete, UK, Report ITR1.
25. BSI (2009) BS EN 13670. Execution of concrete structures. BSI, London, UK.
26. CIRIA (Construction Industry Research and Information Association) (1995) *Formwork Striking Times: Criteria, Prediction and Methods of Assessment.* CIRIA, London, UK, Report 136.
27. Concrete Society (2008) *Guide to the Design of Concrete Structures in the Arabian Peninsula.* Concrete Society, Camberley, UK, Report CS163.
28. BSI (2000) BS EN 12390-8. Testing hardened concrete. Part 8. Depth of penetration of water under pressure. BSI, London, UK.
29. BSI (2011) BS 1881-122. Testing concrete – Method for determination of water absorption. BSI, London, UK.
30. BSI (2008) BS 1881-208. Testing concrete – Recommendations for the determination of the initial surface absorption of concrete. BSI, London, UK.
31. ASTM (2017) C1202. Standard test method for electrical indication of concrete's ability to resist chloride ion penetration. ASTM, West Conshohocken, PA, USA.
32. NordTest Build (1999) NT Build 492. Concrete, mortar and cement-based repair materials. Chloride migration coefficients from non-steady state migration experiments. NordTest Build, Espoo, Finland.
33. BRE (2005) *Concrete in Aggressive Ground.* BRE, Watford, UK, Special digest 1.
34. BSI (2002) BS EN 1916. Concrete pipes and fittings, unreinforced, steel fibre and reinforced. BSI, London, UK.
35. ICE (2010) *Specification for Tunnelling*, 3rd edn. British Tunnelling Society/Institution of Civil Engineers, Thomas Telford.
36. BSI (2005) BS EN 14487-1. Sprayed concrete. Definitions, specifications and conformity. BSI, London, UK.
37. European Federation of Foundation Contractors (2017) *Best Practice Guide to Tremie Concrete for Deep Foundations.* EFFI/DFI, www.effc.org.
38. CIRIA (1995) *Concreting Deep Lifts and Large Volume Pours.* CIRIA, London, UK, Report 135.
39. ACI (2010) *Guide to Hot Weather Concreting.* ACI, Farmington Hills, MI, USA, Report 305.
40. ACI (2016) *Guide to Cold Weather Concreting.* ACI, Farmington Hills, MI, USA, Report 306.
41. Concrete Society (2016) *Concrete Practice: Guidance on the Practical Aspects of Concreting.* Concrete Society, Camberley, UK, Good Concrete Guide 8, pp. 23–25.
42. British Cement Association (1993) *Placing and Compacting, Concrete on Site 5.* British Cement Association, Camberley, UK.
43. Vassou V, Kettle RJ and Sadegzadeh MA (2002) Summary of the development in the abrasion resistance testing of industrial concrete floors. *Proceedings of the International Seminar Concrete Floors and Slabs, University of Dundee, Scotland, UK*, pp. 245–259.
44. Garber G (2006) *Design and Construction of Concrete Floors*, 2nd edn. Butterworth-Heinemann, Oxford, UK, p. 121.
45. Concrete Society (2010) *Non-structural Cracks in Concrete.* Concrete Society, Camberley, UK, Technical Report 22.
46. Davies D, Clarke A and Male P (1992) Microsilica, foam and fibres. *CIOB/Concrete Society Meeting, Halcrow, UK.*

47. ICE (2017) *ICE Specification for Piling and Embedded Retaining Walls*, 3rd edn. Thomas Telford, pp. 124–133 and pp. 211–213.
48. Concrete Society (1998) *Guidance on the Use of Stainless Steel Reinforcement.* Concrete Society, Camberley, UK, Technical report 51.
49. ASTM (2015) D3963. Standard specification for fabrication and jobsite handling of epoxy-coated steel reinforcing bars. ASTM, West Conshohocken, PA, USA.
50. BSI (1979) BSPD 6484. Commentary on corrosion at bimetallic contacts and its alleviation. BSI, London, UK.
51. Steel Construction Institute (2009) *Stainless Steel in Contact with Other Metals.* Steel Construction Institute, Ascot, UK, Advisory desk note 339.
52. Concrete Society (2013) *Formwork: A Guide to Good Practice*, 3rd edn. Concrete Society, Camberley, UK, Good formwork guide CS030.
53. BSI (2008) BS 5975. Code of practice for temporary works procedures and the permissible stress design of falsework. BSI, London, UK.
54. BSI (2016) PAS 8812. Temporary works. Application of European standards in design. BSI, London, UK.
55. BRE (2007) *Formwork for Modern, Efficient Concrete Construction.* BRE, Watford, UK, Report 495.
56. CIRIA (2001) *Permanent Formwork in Construction.* CIRIA, London, UK, Report C558.
57. CIRIA (2000) *Controlled Permeability Formwork.* CIRIA/Concrete Society, London, UK, Joint report C511.
58. McKenna P (2005) Controlled permeability formwork, Current Practice Sheet 144. *Concrete* **39(10)**: 28–31.
59. Bamforth PB (2004) *Enhancing Reinforced Concrete Durability. Guidance on Selecting Measures for Minimising the Risk of Corrosion of Reinforcement in Concrete.* Concrete Society, Camberley, UK, Technical report 61, pp. 65–68.
60. Washa GW and Wendt KF (1975) Fifty year properties of concrete. *Journal of the American Concrete Institute* **72(1)**: 20–28.
61. Neville AM (1996) *Properties of Concrete*, 4th edn. Wiley, London, UK.
62. Cather B (1994) Curing: The true story? *Magazine of Concrete Research* **46(168)**: 157–161.
63. Kosmatka S and Wilson M (2011) *Design and Control of Concrete Mixtures: The Guide to Applications, Methods and Materials.* Portland Cement Association, www.cement.org.
64. ACI (2010) *Specifications for Structural Concrete.* ACI, Farmington Hills, MI, USA, Report 301.
65. CIRIA (2007) *Early-age Thermal Crack Control in Concrete.* CIRIA, London, UK, Report C660.
66. BSI (2008) BS EN 1992-1-1. Eurocode 2. Design of concrete structures. Part 1-1. General rules and rules for buildings. BSI, London, UK.
67. BSI (2006) BS EN 1992-3. Eurocode 2. Design of concrete structures. Part 3. Liquid retaining and containment structures. BSI, London, UK.
68. CIRIA (2018) *Control of Cracking Caused by Restrained Deformation in Concrete.* CIRIA, London, UK, Report C766.
69. ACI (2001) *Control of Cracking in Concrete Structures.* ACI, Farmington Hills, MI, USA, Report 224R.
70. ACI (2007) *Report on Thermal and Volume Change Effects on Cracking of Mass Concrete.* ACI, Farmington Hills, MI, USA, Report 207.2R.
71. BRE (2001) *Delayed Ettringite Formation: In-situ Concrete.* BRE, Watford, UK, Information paper 11/01.
72. ACI (2011) *Specification for Curing Concrete.* ACI, Farmington Hills, MI, USA, Report 308.1M.
73. ACI (2012) *Cooling and Insulating Systems for Mass Concrete.* ACI, Farmington Hills, MI, USA, Report 207.4R.
74. BSI (2001) BS 7973-1. Spacers and chairs for steel reinforcement and their specification. Product performance requirements. BSI, London, UK.

75. BSI (2001) BS 7973-2. Spacers and chairs for steel reinforcement and their specification. Fixing and application of spacers and chairs and tying of reinforcement. BSI, London, UK.
76. Sharp BN (1997) Specifications and acceptance criteria for cover: A 'black hole'. *Concrete* **31(6)**: 34–38.
77. BSI (2004) BS EN 1992-1-2. Eurocode 2. Design of concrete structures. General rules. Structural fire design. BSI, London, UK.
78. Alzyoud S, Wong HS and Buenfeld NR (2014) Effect of spacers in concrete transport properties and chloride penetration. In *Year Book 2014–2015*. Institute of Concrete Technology, pp. 95–97.
79. BRE (2007) *Performance-based Intervention for Durable Concrete Repairs*. BRE, Watford, UK, Information paper 9/07.
80. BSI (2008) BS EN 1504-9. Products and systems for the protection and repair of concrete structures. Definitions, requirements, quality control and evaluation of conformity. Part 9. General principles for use of products and systems. BSI, London, UK.
81. Concrete Society (2009) *Repair of Concrete Structures with Reference to BS EN 1504*. Concrete Society, Camberley, UK, Technical report 69.
82. Concrete Society (2013) *Visual Concrete – Finishes*. Concrete Society, Camberley, UK, Report CS170.
83. Comité Euro-International du Béton (1992) Weathering. In *Durable Concrete Structures*. Thomas Telford, CEB design guide No 183, pp. 58–65.
84. Concrete Society (2013) *Visual Concrete – Control of Blemishes*. Concrete Society, Camberley, UK, Report CS172.
85. Kay E and Hunter C (1999) Cleaning off mould. *Concrete Engineering International* June/July, pp. 44–46.
86. Concrete Society (2013) *Visual Concrete – Weathering, Stains and Efflorescence*. Concrete Society, Camberley, UK, Report CS173.
87. Concrete Society (2007) *Guidance for the Design of Steel-fibre-reinforced Concrete*. Concrete Society, Camberley, UK, Technical report 63.
88. Concrete Society (2007) *Guidance on the Use of Macro-synthetic-fibre-reinforced Concrete*. Concrete Society, Camberley, UK, Technical report 65.
89. Cabrion J-L (1999) Anti-abrasion concrete. *Concrete Engineering International*, March, pp. 64–66.
90. Wimpenny DE, Angerer W, Cooper A and Bernard S (2009) The use of steel and synthetic fibres in concrete under extreme conditions. *Proceedings of the 24th Biennial Conference of the Concrete Institute of Australia, Sydney*.
91. Solgaard A, Carsana M, Geiker M, Kuter A and Bertolini L (2013) Experimental observations of stray current effects on steel fibers embedded in mortar. *Corrosion Science* **74**: 1–12.
92. Dauberschmidt C (2012) Corrosion mechanism of steel fibres in concrete – What can we learn for RC structures? *Rilem International Workshop on Present and Future Challenges for RC Structures, ETH Zurich, Switzerland*.
93. Dauberschmidt C and Raupach M (2005) *Passivity and Depassivation of Steel Fibres and Cold-drawn Steel Wire in Chloride Containing Pore Solution and Concrete*. European Federation of Corrosion.
94. ACI (2010) *Report on the Physical Properties and Durability of Fiber-Reinforced Concrete*. ACI, Farmington Hills, MI, USA, Report 544.5R.
95. Nordström E (2005) *Durability of Sprayed Concrete, Steel Fibre Corrosion in Cracks*. PhD thesis, Luleå University of Technology, Sweden.
96. Mangat PS and Gurusamy K (1988) Long-term properties under marine exposure of steel fibre reinforced concrete containing PFA. *Materials and Structures* **21(5)**: 352–358.
97. Bernard ES (2004) Durability of cracked fibre reinforced shotcrete. *Proceedings of 2nd International Conference on Engineering Developments in Shotcrete, Cairns, Australia*. CRC Press, Taylor & Francis, London, UK, pp. 67–80.

98. Hannant DJ (1999) The effects of age up to 18 years under various exposure conditions on the tensile properties of polypropylene fibre reinforced cement composite. *Materials and Structures* **32(216)**: 83–88.
99. Basell (2002) *Profax and Moplen Polypropylene Chemical Resistance.* Basell Polyolefins, Data Sheet 08/02.
100. Hughes P, Fairhurst D, Sherrington I, *et al.* (2013) Microscopic examination of a new mechanism for accelerated degradation of synthetic fibre reinforced marine concrete. *Construction and Building Materials* **41**: 498–504.
101. Wieloch M, Bui V, Madrio H and Wimpenny D (2013) Self-compacting concrete use in bridge transfer beams. *Understanding Concrete, Proceedings of the 26th Biennial Conference of the Concrete Institute of Australia, Gold Coast, Queensland, Australia.*
102. Knights J and Wimpenny DE (2002) Some aspects of the design and control of self-compacting concrete for an underwater application. *Innovations and Developments in Concrete Materials and Construction, Proceedings of the International Conference, University of Dundee*, pp. 443–453.
103. ASTM (2017) C1712. Standard test method for rapid assessment of static segregation resistance of self-consolidating concrete using penetration test. ASTM, West Conshohocken, PA, USA.
104. BSI (2010) BS EN 12350-8. Testing fresh concrete. Self-compacting concrete. Slump-flow test. BSI, London, UK.
105. ASTM (2014) C1611. Standard test method for slump flow of self-consolidating concrete. ASTM, West Conshohocken, PA, USA.
106. ASTM (2009) C1621. Standard test method for passing ability of self-consolidating concrete by J-ring. ASTM, West Conshohocken, PA, USA.
107. BSI (2010) BS EN 12350-12. Testing fresh concrete. Part 12. Self-compacting concrete. J-ring test. BSI, London, UK.
108. BSI (2010) BS EN 12350-11. Testing fresh concrete. Part 11. Self-compacting concrete. Sieve segregation test. BSI, London, UK.
109. Hagelia P (2011) *Deterioration Mechanisms and Durability of Sprayed Concrete for Rock Support in Tunnels.* PhD thesis, Technical University of Delft, Netherlands.
110. Wimpenny DE, Knights JC and Slater D (2008) Temperature simulation by FEA as a tool in forensic investigation. *4th International Conference on Forensic Engineering, From Failure to Understanding.* Institution of Civil Engineers, London, UK.
111. Concrete Society (2008) *Slipforming of Vertical Structures.* Concrete Society, Camberley, UK, Report CS162.
112. ACI (2014) *Guide to Formwork for Concrete.* ACI, Farmington Hills, MI, USA, Report 347R.
113. ACI (2015) *Guide for Construction of Concrete Pavements.* ACI, Farmington Hills, MI, USA, Report 325.9R.
114. BSI (2009) BS EN 12350-2. Testing fresh concrete. Part 2. Slump-test. BSI, London, UK.
115. BSI (2009) BS EN 12350-5. Testing fresh concrete. Part 5. Flow table test. BSI, London, UK.
116. BSI (2010) BS EN 12350. Testing fresh concrete. Part 10. Self-compacting concrete – L box test. BSI, London, UK.
117. ASTM (2014) C232. Standard test method for bleeding of concrete. ASTM, West Conshohocken, PA, USA.
118. ASTM (2017) C1610. Standard test method for static segregation of self-consolidating concrete using column technique. ASTM, West Conshohocken, PA, USA.
119. American Association of State Highway and Transportation Officials (2012) PP 58-12. Standard specification for the static segregation of self-consolidating concrete (SCC) cylinders. AASHTO, Washington, DC, USA.
120. Concrete Institute of Australia (2012) *Tremie Concrete for Deep Foundations. Recommended Practice.* CIA, Australia, Report Z17.

Soutsos M
ISBN 978-0-7277-6375-4
https://doi.org/10.1680/icehcd.63754.233

Chapter 7
Design aspects that can reduce risks from deterioration mechanisms

Bryan Marsh and Fragkoulis Kanavaris
Arup, UK

7.1. Introduction

The design form of a structure can be fundamental to its durability. This is illustrated by many examples from history, such as gargoyles channelling water away from vulnerable building fabric. Nevertheless, much of the detailed guidance on achieving durable concrete tends to concentrate on the materials. Although undeniably of major importance, this is often not the complete story and the design of the element itself can play a vital role. Indeed, there are circumstances where an appropriate design can reduce the demand for technologically challenging materials solutions and reduce the dependence on high standards of skilled workmanship. The lack of attention paid to the durability aspect of design has been held responsible for the premature loss of serviceability of many highway structures [1].

This chapter is intended as an introduction for readers who are unfamiliar with how design aspects can be used to minimise the risk of deterioration of concrete structures in various exposure environments. It does not simply give standard solutions but rather provides an extra tool for the designer's 'toolkit'. Moreover, nothing in this chapter is new or original. The need to consider durability at the design stage has long been pointed out within concrete design standards [2].

In most 'normal' applications, adequate durability can be achieved by the simple combination of good quality concrete, sufficient cover to the reinforcement and good workmanship. A more considered approach, however, may be needed for any of a number of reasons, including

- particularly aggressive exposure conditions
- unavailability or prohibitive cost of suitable materials
- very long intended working life
- unavailability of skilled or trained labour
- unusually high consequences of failure, for example, in safety-critical structures.

It is difficult to define any dividing line between the design and materials aspects of durability, if, indeed, there is one. In many cases, the optimum solution is likely to involve a combination of both design and materials measures. It is useful, therefore, for designers to have some knowledge of the deterioration mechanisms of reinforced concrete and the properties and limitations of the range of available materials. Likewise, it is useful for materials advisers to be aware of where a simple or effective solution to a potential durability problem may be achieved through appropriate design measures. In this chapter, design measures are considered to be anything beyond the selection of the composition of the concrete (e.g. cement type and content, free water-to-cement (w/c) ratio and admixtures) and depth of cover to the reinforcement, although the latter has obvious consequences for the structural design and even material use.

To set the scene, one can do little better than to quote, somewhat extensively, relevant parts of the European design standard, Eurocode 0 [3]. This standard sets down a very clear basic requirement that

> A structure shall be designed and executed in such a way that it will, during its intended life, with appropriate degrees of reliability and in an economic way:
>
> - sustain all actions and influences likely to occur during execution and use, and
> - remain fit for the use for which it is required.

It adds that a structure …

> … shall be designed to have adequate:
> - structural resistance
> - serviceability, and
> - durability

and that these basic requirements …

> … should be met:
> - by the choice of suitable materials,
> - by appropriate design and detailing, and
> - by specifying control procedures for design, production, execution, and use relevant to the particular project.

There should be no doubt after reading this that durability is as much a part of the responsibility of the designer as is structural loading, and not simply an 'add-on' to be dealt with by the concrete technologist much further down the line when the structural form and details have already been decided.

7.2. The decision process

Durability must be considered right at the onset of design, even at the concept stage. It is necessary at this stage to identify the performance requirements and any particular threats to durability that could affect the proposed structure or element, and also to identify any parts that are vital to the function of the structure, namely the durability-critical areas. Once these have been identified, a durability strategy and plan can be formulated. Again quoting Eurocode 0 [3]

> In order to achieve an adequately durable structure, the following should be taken into account:
>
> - the intended or foreseeable use of the structure
> - the required design criteria
> - the expected environmental conditions
> - the composition, properties and performance of the materials and products
> - the properties of the soil
> - the choice of structural system
> - the shape of members and the structural detailing
> - the quality of workmanship, and level of control
> - the particular protective measures
> - the intended maintenance during the design working life.

Table 7.1 Design life categories from Eurocode 0 [3]

Category	Indicative design working life: years	Examples
1	10	Temporary structures
2	10–25	Replaceable structural parts (e.g. bearings)
3	15–30	Agricultural and similar structures
4	50	Building and other common structures
5	100	Monumental building structures, bridges and other civil engineering structures

7.2.1 Performance requirements

Before informed decisions can be made about a durability strategy, it is essential to know the performance required for the structure or element under consideration. Most deterioration mechanisms are time-dependent, so the intended working life should be defined, at least in broad terms. Clearly, a public monument will generally require a much longer service life than the hardstanding for temporary site buildings associated with the construction of that monument. Eurocode 0 [3] recommends that the intended or design working life for a structure should be stated, and it defines several categories, as listed in Table 7.1.

Specific design measures are most likely to be required for the higher categories, although, for exceptional circumstances with extreme exposure conditions or if there are very high consequences of failure, they may be required even for the lowest category.

7.2.2 Acceptable or possible maintenance

Although there is little that can be done directly to maintain concrete itself, other than perhaps the reapplication of any protective coating that may have been applied, the overall maintenance of a structure can have a large influence on durability. Examples include ensuring the continued efficacy of joint sealants and drainage channels and repairing any cracks that may develop throughout the service life of the structure. Thus, an essential element to consider within a durability design is the acceptability of regular maintenance to the owner or operator of the structure and, perhaps more importantly, the practicality and cost of such maintenance. A motorway bridge, for example, that requires frequent carriageway closures to facilitate maintenance of joints could hardly be said to be a good or efficient design. Moreover, the nature of some structures may make maintenance effectively impossible, such as hazardous waste containment structures or deep foundations. The inspection and maintenance regime can normally be agreed among the designer, contractor and asset owner during the design stage of the project. Furthermore, durability and structural health monitoring can also provide useful information about the potential maintenance needs of a structure and the condition of the concrete. This may become particularly relevant for novel low-carbon concrete technologies with a limited application history. A corresponding instrumentation and monitoring plan could be derived as part of the durability and structural design of the structure.

7.2.3 Threats to durability

Once the intended working life is known, it is necessary to determine the threats to the achievement of the required durability. These will generally be in the exposure environment and may include the following

- chlorides originating from de-icing salts or seawater
- sulfates or other aggressive chemicals in the ground

- freezing of water-saturated surfaces
- aggressive chemicals from industrial processes
- deleterious chemical reactions within the concrete itself.

Most, if not all, of these threats depend upon the presence of water so this, although not inherently deleterious in itself, should perhaps also appear in this list. Likewise, temperature might appear because increased ambient temperature will accelerate most chemical reactions as well as permeation rates and the rate of corrosion of steel. There also are other physical deterioration mechanisms, such as those related to the wearing of surfaces and abrasion damage; however, these are less related to environmental conditions.

7.2.4 Durability-critical areas

Certain elements or parts of a structure will be more crucial to its function than others and the consequences of a durability failure will be greater. If failure of a certain element or part of a structure seriously inhibit the ability of the structure to perform its required function, this part can be classed as a durability-critical area. Such areas clearly warrant special consideration with respect to durability. The criticality of particular elements and structures is recognised by Eurocode 0 [3] through the definition of consequence class (CC), which ranges from CC1 for least critical to CC3 for the most critical elements and structures.

7.2.5 Durability strategies and plans

How the required durability can be provided depends on the particular requirements (e.g. intended working life, acceptability and ease of maintenance) and the durability threats provided by the exposure environment. Generally, a durability strategy will focus on either deterioration avoidance or deterioration management. Deterioration avoidance is achieved through measures that effectively eliminate the risk of deterioration over the required working life by preventing degradation mechanisms or their deleterious effects. Deterioration management does not prevent a degradation mechanism or its deleterious effects but controls the rate of deterioration such that its effect does not become critical over the required working life. The normal materials measure of the combination of suitable concrete quality and adequate cover to the reinforcement in external environments is an example of deterioration management. As few solutions are entirely foolproof, it may be argued that deterioration avoidance is often a special case of deterioration management but where the expected effective life of the measure is usually well in excess of the required working life. Some examples are given below.

7.2.5.1 Deterioration avoidance

- Remove the degradation factor by changing the exposure environment (e.g. over-cladding, extending the building envelope to bring the element into the indoor environment, waterproofing, and anti-carbonation and anti-chloride coatings).
- Make the structure immune to the degradation factor (e.g. non-corroding reinforcement, cathodic prevention or unreinforced concrete).

7.2.5.2 Deterioration management

- Use corrosion-resistant reinforcement to prolong the propagation time from corrosion initiation to manifestation of damage.
- Use controlled-permeability formwork to create a dense low-penetrability layer at the surface to slow the ingress of any aggressive media.
- Use coatings or impregnants to slow the ingress of any aggressive media.

- Use additional cover or enhanced concrete quality to slow the rate of progress of aggressive media towards the reinforcement.
- Control water movement to reduce the mobility of aggressive waterborne media or to divert them away from the concrete surface.

7.3. Deterioration mechanisms and appropriate durability strategies

The main mechanisms of deterioration are covered in depth in other parts of this book (Chapters 3 and 4) and elsewhere. Nevertheless, an understanding of the basic mechanisms by which concrete can deteriorate in different exposure conditions is essential when selecting appropriate design measures. For example, it is important to understand whether the main threat to durability is corrosion of the reinforcement or whether an attack against the concrete itself is more likely. Eurocode 0 [3] states that the 'environmental conditions shall be identified at the design stage so that their significance can be assessed in relation to durability and adequate provisions can be made for protection of the materials used in the structure'. The following sections review each of the major deterioration mechanisms affecting concrete and its reinforcement and suggest appropriate durability strategies. The actual choice of strategy depends on many factors, not least of which is cost. An assessment of cost-effectiveness, however, should be based on the expected benefit over the whole required working life of the structure rather than just the initial upfront cost. When performing a cost–benefit analysis, consideration should be given to the possible costs of disruption caused by a durability failure in service. The reasons behind decisions to provide specific measures beyond what might be considered normal should be clearly recorded and communicated along the construction chain to ensure that vital measures are not removed at a later stage under the guise of otherwise well-intentioned 'value engineering'. An example, best not identified, involved the specification of stainless steel reinforcement in a vulnerable detail, as it was felt that it might be difficult to attain the required cover during construction. Post-construction checks revealed the pessimistically predicted lack of cover, but unfortunately the specification of stainless steel reinforcement had been lost somewhere along the way.

7.3.1 Corrosion of the reinforcement

Corrosion can initiate when the protective action of the concrete cover is compromised, usually either by reduced alkalinity (carbonation) or by a build-up of chlorides above the threshold for corrosion. Nevertheless, corrosion is normally expected to occur if there is a sufficient supply of both oxygen and moisture. Design measures to prevent reinforcement corrosion should, therefore, address at least one of these factors

- maintaining the alkalinity
- preventing the excessive chloride build-up, including raising the chloride threshold
- reducing the moisture content
- reducing the oxygen availability
- reducing the susceptibility of the reinforcement to corrosion (e.g. using epoxy-coated, galvanised, stainless steel or non-ferrous reinforcement).

Alternatively, deal with corrosion once it starts by installing a cathodic protection system during construction (cathodic prevention).

7.3.1.1 Carbonation-induced corrosion

Carbonation-induced corrosion is readily dealt with for normal design lives (50–120 years) by providing an adequate depth of good quality concrete in accordance with design code recommendations. Specific design measures are necessary only when there is a requirement for an exceptionally long design life or situations where increased temperature and carbon dioxide are anticipated. Measures to reduce oxygen availability are likely to be largely impractical, but moisture content can be reduced by enclosing any otherwise externally exposed surfaces. If this is impractical or undesirable, the susceptibility of the reinforcement to corrosion can be reduced with epoxy-coated, galvanised or stainless steel reinforcement. Anticarbonation coatings can assist in maintaining the alkalinity but have a relatively short life before reapplication is necessary, so are best suited to remedial applications rather than new construction.

7.3.1.2 Chloride-induced corrosion

Chloride-induced corrosion can be a greater problem than carbonation-induced corrosion and can be more difficult to deal with by materials measures alone, particularly where a long service life is required or where exposure conditions are severe, such as in the Middle East. Materials measures, such as cement type and free w/c ratio combined with an adequate cover to the reinforcement and, sometimes, a hydrophobic surface impregnant, can be adequate for many conditions. If such conditions are likely to prove inadequate or if, for example, it is not desirable to provide deep cover, it may be necessary to employ design measures. Such measures may involve changing the exposure environment, for example, by using permanent formwork, cladding, waterproofing or hydrophobic and pore-blocking admixtures. Alternatively, protection against deterioration could be realised by, for example, using stainless steel reinforcement whose chloride threshold is greater than the level of chloride that can be expected to ingress to the depth of the outer bars within the intended working life of the structure.

7.3.2 External chemical attacks

Certain chemicals in natural groundwater, industrial processes or contaminated land can attack concrete. The severity of an attack depends on many factors, including the aggressiveness of the chemical, its concentration, the groundwater mobility, the temperature and, of course, the chemical resistance of the concrete. Most naturally occurring ground conditions can be resisted adequately by careful selection of the cement type, content and free w/c ratio. More severe conditions or very long design lives, for example, may require specific design measures. These may involve preventing the concrete surface from making contact with the aggressive environment or by providing extra concrete cover to act as a sacrificial layer and thus, permit some degree of attack to occur but without threatening the structural integrity. Reducing the movement of groundwater in concrete by using drainage can greatly reduce the rate of attack and bring it down to a manageable level.

7.3.3 Internal chemical attacks

Internal chemical attacks, including the alkali–silica reaction and delayed ettringite formation, can usually be avoided by materials measures, such as choice of cement type and minimising the cement content. Nevertheless, this type of attack depends upon the ready availability of water, so design measures that prevent the concrete from making contact with water can be effective in avoiding problems. An example is the provision of effective waterproofing to underground concrete. Delayed ettringite formation depends also on a high temperature developing within the concrete during hydration or accelerated curing, so design measures that reduce the temperature to an acceptable level can be effective (< 65–80°C depending on cement type). An example of such a measure would be the incorporation of cooling pipes within a massive pour, which may also be employed to avoid cracking due to early-age thermal contraction.

7.3.4 Salt recrystallisation

Salt recrystallisation can be a major problem in marine structures in hot climates or in land-based structures in hot dry climates where the ground contains high concentrations of salts, such as chlorides or sulfates, as is the case in much of the Middle East. Damage can occur when salt-laden water is absorbed into the concrete below ground level but is left behind in the pore structure when the water evaporates to the atmosphere above ground after having risen through the concrete by wicking. Concrete is inherently porous, so this form of potential deterioration can be avoided only by preventing water from entering the pore structure. A possible materials solution is to incorporate hydrophobic and pore-blocking admixtures into the concrete mix. A design solution is to prevent ingress of salt-laden water into the concrete surface by effective tanking of the concrete below ground. Special details may be required to cope with piles that cannot practically be tanked. These may comprise coating the top of the pile with a material, such as a polymer-modified cementitious layer, that is impermeable to groundwater but which still allows the transfer of forces between pile cap and pile.

7.3.5 Freezing and thawing

So-called freezing and thawing damage, or frost damage, can occur when water-saturated concrete is allowed to freeze. When water changes to ice it expands, which results in tensile stresses if there is no empty space into which it can expand. If these stresses exceed the strength of the concrete, then damage will occur. Materials solutions involve either using a certain minimum strength of concrete or, more effectively, ensuring that there is empty space in the form of a void system of entrained air. The necessary conditions for damage are saturation of the concrete surface and, of course, freezing temperatures. Design measures should, therefore, address either of these factors, although in most cases the materials solutions should be adequate. A surface can be prevented from becoming saturated by some form of waterproofing or surface impregnation, or even by a simple shelter. Alternatively, freezing can be prevented again by a shelter or by integral heating, such as has been done with some bridge decks to avoid the use of de-icing chemicals. The extent of damage to be anticipated from freezing and thawing action depends on the expected number of freeze–thaw cycles, and therefore, consideration could also be given to the location of the structure and to the duration that sub-zero temperatures are sustained throughout the service life of the structure.

7.3.6 Cracking and crack widths

The influence of cracking and crack widths on the corrosion of the reinforcement has been the subject of much debate. Codes of practice have long provided limits for acceptable crack widths related to the perceived severity of the exposure environment, although the most recently developed code, Eurocode 2 [4], has bucked this trend, to some extent, to reflect the findings of much research and experience that show there is no general correlation between the reinforcement corrosion rate and the crack width for cracks that intersect the reinforcement bars. Control of cracking within defined limits may, however, be desirable for reasons such as appearance and watertightness and for durability in environments with high chemical aggressiveness or severe wave action, particularly if there are sand particles.

7.4. Specific design and additional protective measures

Although the term 'additional protective measure' came to prominence through BRE Special Digest 1 [5] for the design of concrete in aggressive ground, it is a logical term for any measure deliberately applied to reduce the dependence on the concrete alone for durability.

Measures introduced into BRE Special Digest 1 to provide durability for buried concrete under conditions where there is a high risk of the thaumasite form of sulfate attack (TSA) are as follows.

- *Drainage:* The risk of a serious chemical attack against concrete surfaces is much reduced if there is not a ready supply of mobile groundwater to constantly replenish the aggressive ions as they are consumed in reactions with the concrete. Good drainage can be employed to interrupt the groundwater flow towards the structure.
- *Protective coating:* As described in Section 7.4.3, a waterproof coating will effectively change the exposure environment of the actual concrete surface from aggressive to benign by preventing the aggressive groundwater from actually making contact with the concrete surface. Some waterproofing measures, however, are not always fully effective and their protective properties may decline with time. For these reasons, BRE Special Digest 1 still requires a good quality concrete to be used in conjunction with the waterproofing, and the concrete acts as a second line of defence if necessary.

- *Sacrificial cover:* TSA can transform the outer layer of concrete into a weak incohesive material reminiscent of toothpaste. If the surface of the concrete is not important structurally, the cover to the reinforcement can be increased by a thickness of concrete intended to be sacrificial should such degradation occur. Provided the thickness of the sacrificial layer is chosen correctly, the deleterious reaction can be tolerated over the intended working life of the structure without fear of distress. Based on early observations of structures severely affected by TSA, Special Digest 1 suggests that the thickness of a sacrificial layer should be 50 mm. This measure is not suitable for piles that rely on skin friction, as the friction would be seriously diminished by loss or weakening of the concrete surface.
- *Permeable formwork:* Controlled-permeability formwork uses a fine mesh type of lining, like a geotextile, usually fixed to the inside of conventional formwork. The permeable liner allows excess water and air to pass through and escape but retains the cement, aggregates and sufficient water for hydration. This can result in a significant reduction in the effective free w/c ratio over a few millimetres depth at the surface of the concrete. The actual depth of the effect will depend on the concrete composition and the efficiency of the formwork liner. The lower free w/c ratio at the surface increases the impenetrability of the surface and, hence, reduces the extent of any chemical attack. The concrete surface should also provide enhanced resistance to chloride ingress.
- *Increased concrete quality:* Resistance against chemical attacks increases with the quality of the concrete, as determined by the free w/c ratio and cement type. BRE Special Digest 1 [5] lists four qualities of concrete in design chemical classes from DC-1 (which has no specific limitation on composition) to DC-4 (which permits only a limited range of cement types and generally at very low w/c ratios depending on the actual cement type). One means of providing enhanced protection is, thus, to use a concrete of a higher design chemical class, although this is strictly more of a materials measure than a design measure.

7.4.1 Unreinforced concrete

In some circumstances, it may be possible to design a structure without any reinforcement, hence removing any risk of damage from corrosion. The highways authorities in the UK encourage this approach if it can be achieved in elements of highway structures, such as abutments, wing walls and retaining walls. Clearly, this is an option for only a very limited number of structures, but it is one that is likely to be particularly effective.

7.4.2 Provision of adequate falls and drainage routes

As noted in Section 7.2.3, most of the deterioration mechanisms affecting concrete and its reinforcement depend upon the presence of water. Failure to manage water by moving it away from the vulnerable parts of structures has been the cause of much premature deterioration. This is particularly the case if the water contains chlorides or aggressive chemicals.

7.4.3 Waterproofing (including integral) and tanking

Preventing contact between the concrete and water can be an effective means of avoiding deterioration in an element or structure, provided the waterproofing remains effective over the whole required working life. Following the example of the UK highways authorities, BS 8500-1 [6] allows the exposure class for concrete protected by waterproofing to be reduced from an onerous classification such as XD3 (where the concrete is subject, for example, to direct splashing with water laden with de-icing salt) to XC3/4, which is the normal exposure class for externally exposed superstructure elements remote from any sources of chloride. It may be argued that XC3/4 is not technically the correct exposure classification for concrete protected by waterproofing, but it provides the quality of concrete felt to be appropriate. Of course, if the waterproofing is fully effective over the required working life of the structure or element, the quality of concrete need not be particularly high, as there is no significant degradation factor directly affecting it. Nevertheless, such an approach would generally be held to be unwise, as it allows little by way of a safety factor should the waterproofing prove not to be 100% effective over the whole area or for the full required working life. BS 8500 allows for this by suggesting that XD1 might be more appropriate in this circumstance.

7.4.4 Corrosion-resistant reinforcement

7.4.4.1 Stainless steel reinforcement

The chemical composition of stainless steel can give it a much greater chloride threshold at which corrosion is initiated than normal carbon steel reinforcement. This means it can tolerate a much greater degree of chloride ingress before damage results. Nevertheless, stainless steel is not completely immune to corrosion should the chloride reach very high levels, and the design must, therefore, ensure that this threshold is not reached during the intended working life. Moreover, the chloride threshold will be significantly reduced should the surrounding concrete fall to a near-neutral pH through carbonation. For these reasons, it is recommended that the minimum cover should not be reduced below 40 mm [7], and clearly, the concrete should be of low penetrability to restrict chloride ingress and carbonation.

A barrier to the use of stainless steel reinforcement has traditionally been its greater initial cost than carbon steel. Nevertheless, whole-life costing can mean that the selective use of stainless steel is worthwhile, particularly in very aggressive chloride environments and where a long life is required with a high degree of reliability. Generally, stainless steel bars need to be used only in the most vulnerable locations. There is little risk of problems due to bimetallic corrosion between the two types of steel.

The term 'stainless steel' covers a wide range of materials with varying compositions, properties and suitability for use as reinforcement. The most suitable types for concrete are austenitic (grade 1.4436) and duplex (grade 1.4462) stainless steels, as covered by European standard EN 10088-1 [8]. Martensitic and ferritic stainless steels are generally not suitable.

7.4.4.2 Epoxy-coated reinforcement

Fusion-bonded epoxy-coated reinforcement improves the corrosion resistance of normal carbon steel reinforcement, as there is an epoxy resin barrier between the concrete and the reinforcement. Its success, therefore, relies on the impermeability of the epoxy coating, which protects the steel surface from chlorides, oxygen and moisture. The quality of the coating is, thus, an important factor in the effectiveness of epoxy-coated reinforcement, and it has gained a poor reputation after several incidences of failure, particularly in bridge decks in the USA. The bond between the reinforcement and the concrete may be lower than that with conventional carbon steel. Moreover, the bars cannot be bent once the epoxy coating has been applied. These factors should be considered in the structural design.

7.4.4.3 Galvanised reinforcement

Despite its promise, the performance of galvanised steel reinforcement is such that it is generally regarded as unsuitable for use in concrete in chloride environments. It can, however, be useful if there is a risk of carbonation-induced corrosion that cannot readily be dealt with by the normal method of providing an adequate thickness of good quality cover. An example could be where a particularly slender element is required to have a long working life.

7.4.4.4 Non-ferrous reinforcement

Reinforcement made from fibre-reinforced polymer is immune from corrosion in both carbonation and chloride environments. Its properties as a structural reinforcement are, however, quite different from conventional carbon steel reinforcement, which means it cannot simply be substituted. Possibly because of this, its use has largely been restricted to specialist applications where advantage has been made of some of its other properties, such as being non-magnetic and non-conducting.

7.4.5 Cathodic protection and prevention

Cathodic protection has a long record of protecting structures already suffering from reinforcement corrosion from further deterioration through the application of a small electrical current that counteracts the electrochemical corrosion process. However, the reinforcement must be electrically continuous, which often requires intrusive works to make connections between bars. Moreover, any concrete already damaged by cracking and spalling must be repaired. An anode system has to be installed in a drilled hole or by adding an overlay. Cathodic prevention is a special form of cathodic protection in which all the necessary features, such as the electrical continuity of the reinforcement and a suitable anode system, are installed during construction, thus avoiding much of the expense of a retrofitted cathodic protection system.

7.4.6 Coatings and impregnants

The application of a hydrophobic pore-lining impregnant to an exposed concrete surface can significantly reduce the rate of ingress of waterborne chlorides impinging on that surface, for example, by splashing. Indeed, such treatment has been required for all highway structures in the UK adjacent to carriageways [9]. At the time of writing, this requirement has been suspended due to concerns over the effectiveness of test methods but is expected to be reinstated. The impregnant was originally specified to be a silane, but because of health and safety concerns alternative materials, such as siloxanes, may be preferable. By lining the walls of the pores for just a small depth, water molecules are repelled and the normal uptake of water through capillary absorption is greatly reduced. This has the effect of also greatly reducing the rate of ingress of chlorides, as this depends on water acting as a carrier. This type of surface impregnation is only really effective against splashing or spray and is not suitable for use on parts of structures in direct contact with water, such as in the tidal or submerged zones of marine structures.

7.4.7 Integral bridges

The UK Highways Agency design standard BD 57/01 [10] requires that, unless inappropriate for the particular circumstances, bridges up to 60 m in length and with skews of up to 30° should be designed as integral bridges. An integral bridge is one where the deck is connected directly to the abutments without any movement joints (Figure 7.1). Experience has shown that failed or imperfect joints have been a major

Figure 7.1 Concrete bridge with integral abutments and piers

cause of deterioration of bridges in the UK because they have allowed the passage of salt-laden water onto the unprotected concrete surfaces of piers and abutments.

7.5. Examples of the application of design measures for durability

7.5.1 Stonecutters Bridge

The Stonecutters Bridge in Hong Kong was required to have a design life of 120 years in an aggressive marine environment of high heat and high humidity (Figure 7.2). Informed use of predictive modelling showed that certain corner reinforcement bars were much more vulnerable to corrosion than the rest of the outer layer of reinforcement. This was largely because of the two-dimensional ingress of chlorides possible at corner locations. The concrete materials specification was already very high, and it was felt undesirable to further increase the cover to these bars. Since the consequences of these bars corroding were unacceptable, it was decided to use stainless steel reinforcement in these durability-critical areas. In other areas, the normal carbon steel reinforcement was felt to be adequately protected by the combination of cover depth and concrete quality. A similar approach was adopted for the Queensferry Crossing in Scotland, as only stainless steel reinforcement was used in the outer 100 mm within the splash, spray and tidal zones.

7.5.2 Subarctic concrete gravity-based structure

The design brief for a concrete gravity platform in the Russian subarctic in an area of very high annual ice movement (approximately 5000 km/year) called for a 30-year working life with no opportunity for maintenance or repair to the concrete surfaces. A study of the available information published on ice abrasion of concrete revealed that there was no reliable way to predict the likely loss of cover to the reinforcement over that period. Moreover, even when using concrete designed to provide the greatest resistance to ice abrasion, it was not possible to be absolutely confident that even, say, 100 mm of sacrificial cover would be adequate over 30 years with no opportunity for maintenance or repair. The design solution was to provide a sacrificial steel jacket together with a generous cover depth of highly abrasion-resistant concrete.

Figure 7.2 Stonecutters Bridge, Hong Kong, incorporating the selective use of stainless steel reinforcement

7.5.3 Underground structures in the Middle East

Structures in the Middle East with continuous concrete structures extending from the groundwater level to above ground are at risk of damage due to salt recrystallisation. If possible, concrete below ground must be fully tanked to prevent water uptake. Integral permeability-reducing admixtures may also be an appropriate alternative to tanking where the aggressivity of the ground permits. If tanking is not possible, such as for piles, the continuity of the water path can be broken by a waterproof seal (e.g. a polymer-modified cementitious layer) at the top of the pile where it enters the pile cap [11].

7.5.4 Railway tunnels in the UK

As part of a major infrastructure project in the UK, several sprayed concrete and cast-in-place concrete tunnels have been designed with a requirement for a service life of 120 years. Water management and deterioration avoidance were imperative for the long-term functioning of this safety-critical infrastructure. The design solution involved the combination of a waterproofing membrane at the extrados of the tunnel sections, low-permeability concrete with a w/c ratio of less than 0.45 and even an allowance for a 0.20-mm crack width depending on the crack formation characteristics (flexural or tensile crack).

7.5.5 Remedially applied design measure

A design solution was post-applied to a ground-bearing slab during the construction of a school as an insufficiently sulfate-resistant concrete had been either specified or supplied in error. Drainage trenches were installed beyond the boundary of the slab to interrupt the flow of groundwater and effectively change the exposure environment to one of lower aggressivity through removal of mobile groundwater.

7.5.6 An example of where design has contributed to deterioration

An attack against the concrete in the foundations of bridges on the M5 motorway in the UK by TSA led to a major revision of guidance on concrete in aggressive ground, culminating in the publication of BRE Special Digest 1 [5]. Prior to this time, TSA had seldom been identified as a deterioration mechanism in concrete structures and there were no specific avoidance measures within the UK guidance at the time. The severity of the attack was surprising but would probably have been much less had the design of the drainage been better. Instead of taking aggressive groundwater away from the concrete bridge foundations, the drainage actually created a sump where water collected, which contributed to an otherwise largely unforeseeable deterioration mechanism. It is possible that the deterioration may not have occurred or would have been much less severe had the drainage design been effective in moving water away from the concrete surfaces.

7.6. Summary

There is rarely a single means of achieving durability in a concrete structure. Consideration should, therefore, be given at the design stage to the protection that can be achieved both by suitable design aspects and through appropriate materials measures. Design aspects may be effective in reducing the reliance on expensive or technically difficult materials measures or the need for a high quality of workmanship where it may not readily be available. This chapter has attempted to provide an insight into the types of design measures that are available and how they should be targeted at the specific deterioration mechanisms identified in the exposure environment.

REFERENCES

1. HA (Highways Agency) (2001) Design for durability, BA 57/01. In *Design Manual for Roads and Bridges*. HA, London, UK, vol. 1, section 3, part 8.
2. BSI (British Standards Institution) (1985) BS 8110-1. Structural use of concrete. Code of practice for design and construction. BSI, London, UK.

3. CEN (European Committee for Standardization) (2005) EN 1990:2002 + A1:2005. Eurocode 0. Basis of structural design. CEN, Brussels, Belgium.
4. CEN (2014) EN 1992-1-1:2004 + A1:2014. Eurocode 2. Design of concrete structures. Part 1-1. General rules and rules for buildings. CEN, Brussels, Belgium.
5. Building Research Establishment (2005) *Concrete in Aggressive Ground*. BRE, Watford, UK, Special digest 1.
6. BSI (2016) BS 8500-1:2015 + A1:2016. Concrete. Production, performance and conformity criteria. Part 1. Method of specifying and guidance for the specifier. BSI, London, UK.
7. Concrete Society (1998) *Guidance on the Use of Stainless Steel Reinforcement*. Concrete Society, Camberley, UK, Technical report 51.
8. CEN (2014) EN 10088-1. Stainless steels. Part 1. List of stainless steels. CEN, Brussels, Belgium.
9. HA (2003) The impregnation of reinforced and prestressed concrete highway structures using hydrophobic pore-lining impregnants, BD 43/03. In *Design Manual for Roads and Bridges*. HA, London, UK, vol. 2, section 4, part 2.
10. HA (2001) Design for durability, BD 57/01. In *Design Manual for Roads and Bridges*. HA, London, UK, vol. 1, section 3, part 7.
11. Concrete Society (2008) *Guide to the Design of Concrete Structures in the Arabian Peninsula*. Concrete Society, Camberley, UK. Report CS 163.

Soutsos M
ISBN 978-0-7277-6375-4
https://doi.org/10.1680/icehcd.63754.247

Chapter 8
Durability performance tests

Peter A Claisse (†)
Coventry University, UK

Alan Richardson
Northumbria University, UK

Don Wimpenny
Jacobs, UK

8.1. Introduction

The tests described here cover two main areas, namely physical and chemical material tests. It will help the designer and specifier to choose the correct test to determine durability for a given service life. From the tests used, a reasonable evaluation of life expectancy can be determined, which is related to the expected overall durability of the structure. Durability is defined within Eurocode 2 [1]: 'A durable structure shall meet the requirements of serviceability, strength and stability throughout its intended working life, without significant loss of utility or excessive maintenance'. To achieve this design requirement, concrete must be tested to permit a reasoned choice within the design parameters required or specified.

Concrete can deteriorate for a variety of reasons, and concrete damage is often the result of a combination of factors, starting with the initiation of cracks, which leads to processes that involve deleterious chemical reactions and hydrostatic breakdown. The rate of crack propagation is controlled by ionic and molecular transport systems. The result is microstructural changes that degrade the physical properties of the concrete.

The main causes of degradation are environmental factors, such as seasonal temperature variations, cyclical freezing and thawing, rainfall and relative humidity changes, and the concentration of deleterious chemicals in the atmosphere or water that are in contact with the concrete. Geographical location is also an important consideration for durability.

In essence, the long-term performance of a concrete depends on its interactions with the service environment, and the penetration of deleterious substances is highly significant. The latter is controlled by the permeability of the near-surface concrete. Essentially, durability tests are all intended to measure the pore structure of the concrete [2]. In general, 'the single most important parameter that leads to premature deterioration of concrete, is the ingress of moisture into the concrete' [3].

Applications for explicit durability design are outlined in a discussion document published by the Concrete Society [4], which outlines the criteria for durable concrete design to achieve the desired length of service (long or short) and the performance requirements for 'normal' structures.

The specifier may prefer to evaluate concrete performance at the point of placement. This is logically sound but logistically difficult for several reasons. Although there is some guidance in various documents, there is no standard method for sampling concrete at the point of placement, so that methods in use vary widely. Questions remain about how to secure a representative sample, how to do so in a safe manner, whether sampling should be done immediately prior to impact at the point of placement, after impact, or before or after consolidation, and whether to retrieve and then re-compact a sample that has been placed, consolidated and, perhaps, finished.

Requirements for point-of-placement sampling and testing should, therefore, be accompanied by a specific sampling protocol. Further, it is to be expected that concrete properties, especially air content and density, will be affected by the placing, consolidation, finishing and curing operations [5]. Clearly, the on-site testing of concrete in its plastic state is subject to many vagaries.

Regarding durability testing, 'there is no universally agreed definition for durability performance test methods ... a test that directly assesses the resistance to a standardised deterioration process e.g. the freeze–thaw test, or a test that directly assesses the resistance to one of the phases of a deterioration process, e.g. carbonation tests, or a test that directly assesses a performance related parameter e.g. measurement of cover' [4]. The location or point of testing involves two considerations: where to test and when to test?

The following sections describe a range of suitable tests to assess aspects affecting the durability of concrete.

8.2. Tests for assessing durability

A performance test is typically chosen because of its apparent ability to differentiate between concrete with different levels of durability in a practical timescale. However, to provide rapid results, the testing conditions are often unrepresentative of those in service. For instance, the beneficial effect of secondary cementitious materials, such as fly ash, may not be apparent in accelerated tests conducted at early ages relative to the intended service life. Some tests are best suited for qualification or initial type approval of a mix, whereas others are suitable for routine production control.

There are three approaches to durability design that have a direct relevance to the scope and role of performance testing

- deemed to satisfy
- durability modelling
- design by testing.

In the deemed to satisfy design approach, concrete is produced to meet specified proportions and performance requirements, which in combination, mean that the concrete is deemed to satisfy the durability requirements based on experience. For example, in BS 8500-1 [6], which is a complementary standard to BS EN 206 [7], the maximum water-to-cement (w/c) ratio and minimum cement content are specified, together with the characteristic cylinder/cube compressive strength class for different cementitious binder types. Depending on the exposure and application, further requirements will apply, such as the minimum air content for freeze–thaw resistance and initial surface absorption for precast tunnel segments for sulfate resistance [8]. For more onerous exposure conditions and critical applications, such as nuclear installations, additional performance testing may be required. These may be imposed through a regional code, an application-related specification or a specification particular to the project.

In the durability modelling approach, performance testing often has a direct role in confirming the assumptions in the design. For example, testing accelerated carbonation or chloride migration in the laboratory can be used to verify the parameters input into the durability models for carbonation and chloride reinforcement corrosion. However, interpreting such tests requires care, not least because the standard compacted and cured samples are unlikely to replicate the structural element. For accelerated carbonation testing, the design method mandates the adjustment required when using test results [9]. For durability modelling, the testing will normally be required to determine a value for each of the basic physical properties of the concrete, such as the coefficients of permeability and diffusion and the capillary suction pressure, which in turn depend on the pore sizes.

In the design by testing approach, examples of the structural member, typically precast elements, are subject to testing to establish performance under set conditions related to a specific design case. The testing is more usually applied to structural design but also has potential application in relation to durability. For example, precast tunnel segments may be subject to heating to match temperature curves simulating a specific fire type, such as a hydrocarbon fire, to assess the fire spalling resistance.

Introductions to performance specifications and testing concrete for durability are given in Concrete Society Technical Reports 109 [10] and 31 [11].

8.3. Physical material testing

Without proper concrete design and production control, deteriorating concrete can compromise structural integrity, pose serious liability issues and create significant problems throughout a structure. There are various tests for durability because concrete is a complex material with many aspects worthy of testing. All aspects of durability rely upon the transport of liquids or gases into the concrete matrix, where they may cause physical damage or chemical or electrical instability. The main ways by which moisture moves within concrete are capillary suction, absorption, and water and gas permeability. This movement ultimately affects durability. As noted in the introduction, the long-term performance of a concrete depends on its interactions with the service environment, and the penetration of deleterious substances is highly significant.

8.3.1 Water absorption testing

The tests in this section determine the rate of absorption of water into concrete. This rate will depend on the capillary suction pressure and the permeability.

Absorption testing to BS 1881-122 [12] requires a minimum of three concrete cores to be cut from the concrete to be tested or cast with a diameter of 75 mm and 75 mm long. Note that cast specimens have a lower absorption due to the surface laitance covering. Longer or shorter specimens can be used but the results obtained must be modified by a correction factor, as absorption is effective only in the first 50 mm from the surface that is in contact with the water. For example, core samples 150 mm long (Figure 8.1) are subject to a correction factor of 1.2 due to their length. The specimen size is critical for obtaining reasonable saturation.

The apparatus required to carry out this test is a balance capable of weighing specimens up to 5 kg with an accuracy of 0.001 kg and a ventilated drying oven to maintain a temperature of 105°C ± 5°C. A diamond-tipped core cutter with an accuracy of ±3 mm may also be needed. After the samples have been dried for 72 hours in the oven, they are fully submersed in water in a curing tank, which is then placed inside an airtight container for 30 minutes.

The test is quick, cheap and reliable, and outliers are rare. It gives an insight into the internal absorption qualities of the concrete. The mass of the sample is measured before and after submersion. Water absorption is the change in mass as a percentage of the initial dry mass of the sample

Figure 8.1 Core samples 150 mm long

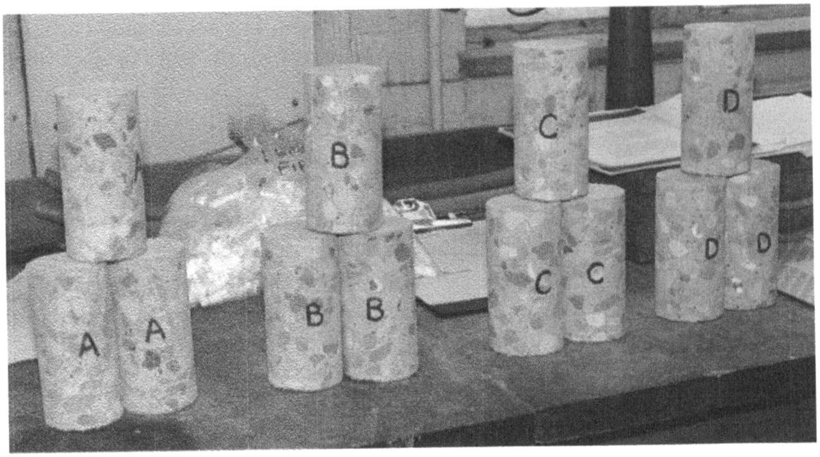

Dill [13] states

> General experience has shown that typical values for water absorption of concrete, as established by BS 1881, 30 minute immersion procedure are:
>
> - Low absorption < 3%
> - Average absorption 3–4%.

The unit weight of water compared to the unit weight of concrete is lower by a factor of approximately 2.3, which is key when interpreting the results. Small changes in weight represent significant amounts of absorbed water. So, a 1% difference in absorption by weight is a significant change in absorption, for example, when comparing plain and monofilament concrete regarding enhanced durability.

Another absorption test is ASTM C1585 – 'Standard test method for measurement of rate of absorption of water by hydraulic-cement concretes' [14]. It uses only two cores or cylinders of concrete as the test material and water absorption occurs in just one direction. It is a useful test when a single direction of absorption is needed, and rate of water absorption is required.

8.3.2 Capillary absorption

Several test methods have been published, including ASTM C1585 [14] and RTA T362 [15]. These involve drying a sample and placing one surface in contact with a source of moisture for a prescribed period. The height of capillary rise or the increase in mass is recorded. The extent of drying prior to the exposure to water is important. Commonly, samples are conditioned at elevated temperature, typically at or above 105°C until a steady mass is achieved. Alternatively, a sample is placed in an oven or desiccator under controlled temperature and humidity conditions for a fixed duration.

Capillary absorption testing can provide a useful indication of the rapid mass water movement that occurs when concrete is subject to cyclic wetting and drying in service, which has an impact on concrete quality and curing [16]. Typical limits for capillary absorption expressed as maximum rise in the RTA T362 test are 8–25 mm [17].

Capillary absorption has been extensively studied by Hall [18]. The sorptivity summarises the capillary absorption characteristics of porous construction materials:

$$I = S\sqrt{t}$$

where S is the sorptivity (m/s$^{0.5}$), I is the depth of infiltration and t is time.

Capillary absorption can also be assessed by measuring the mass of water absorbed, which can be expressed as the volume of water absorbed per unit of area in unit time. Figure 8.2 indicates the effect of the proportion of ground granulated blastfurnace slag (GGBS), water-to-binder (w/b) ratio and curing method (100% and 60% relative humidity) on 75-mm-diameter cores subject to a 2-hour test at an age of approximately 91 days.

8.3.3 Porosity

Test methods for porosity accessible to water are given in AFNOR NF P18-459 [19] and by RILEM [20]. The sample can be a moulded specimen or core with a minimum dimension at least three times the maximum aggregate size and a volume of 0.4 to 1.6 litres. At least three samples should be tested.

The test is like the water absorption test in BS 1881-122 (Section 8.3.1), except that the sample is conditioned in a vacuum at 25 mbar for 4 hours prior to testing to remove air, which could otherwise be entrapped. The sample is gradually submerged under water while the vacuum is maintained. After 44 hours, the mass of the sample in water M_w is measured. After removal of surface water using a cloth or sponge, the mass in air M_{air} is measured. The sample is then dried to constant mass at 105 ± 5°C (less than 0.05% change over 24 hours). After cooling for at least 3 hours, the mass M_{dry} is measured.

The porosity accessible to liquid is given by:

$$\text{Porosity} = \frac{(M_{air} - M_{dry})}{(M_{air} - M_w)}$$

The maximum porosity is typically limited to 10–15%, depending on the application [21].

Figure 8.2 Capillary absorption

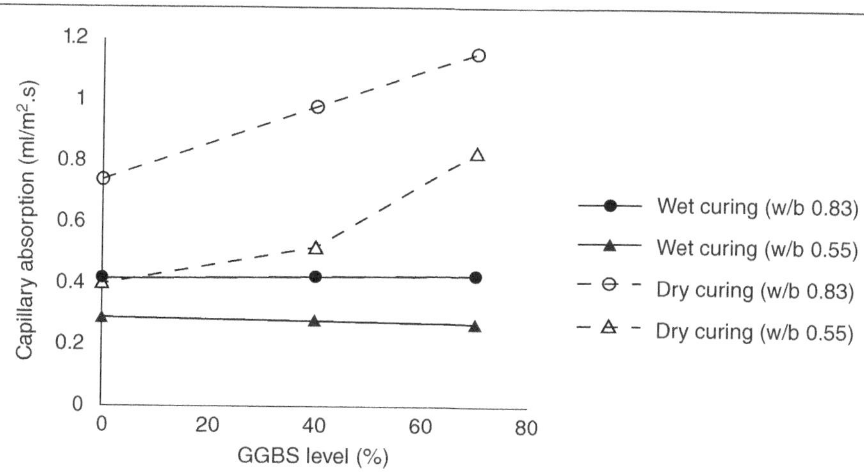

8.3.4 Surface absorption

The surface properties of concrete differ from the core in that laitance is present, which generally has a different composition due to fines being drawn to the surface during trowelling or compaction. The surface of the concrete is the interface between it and its environment, and it can be subject to potential threats to the durability of the material. It is, therefore, a very important boundary to assess with regards to the resistance of the material to stop or slow the ingress of deleterious materials.

In an initial surface absorption test (ISAT), surface absorption is measured in a uniaxial direction, as described using BS 1881-208 [22]. The test arrangement is shown in Figure 8.3. A watertight cap is placed on the surface of the test specimen and the concrete is subject to a head of 200 mm of water pressure over an area of 5000 mm^2. The movement of the meniscus in the capillary is measured when the tap is closed. 'The method measures the rate at which water is absorbed into the surface and must not therefore be thought of as a test to measure the bulk permeability of concrete' [23].

A theoretical analysis of the test by Claisse [24] gives the flow rate F_v in m^3/s as

$$F_\mathrm{V} = A\left(\frac{Ks\varepsilon}{re}\right)^{1/2} t^{-1/2}$$

where A is the area of the cap (m^2), K is the intrinsic permeability of the concrete (m^2), s is the surface tension (N/m), ε is the porosity, r is the radius of the pores (m), e is the viscosity of the water (Pa s) and t is time (s)

Figure 8.3 Schematic arrangement of ISAT test [24]

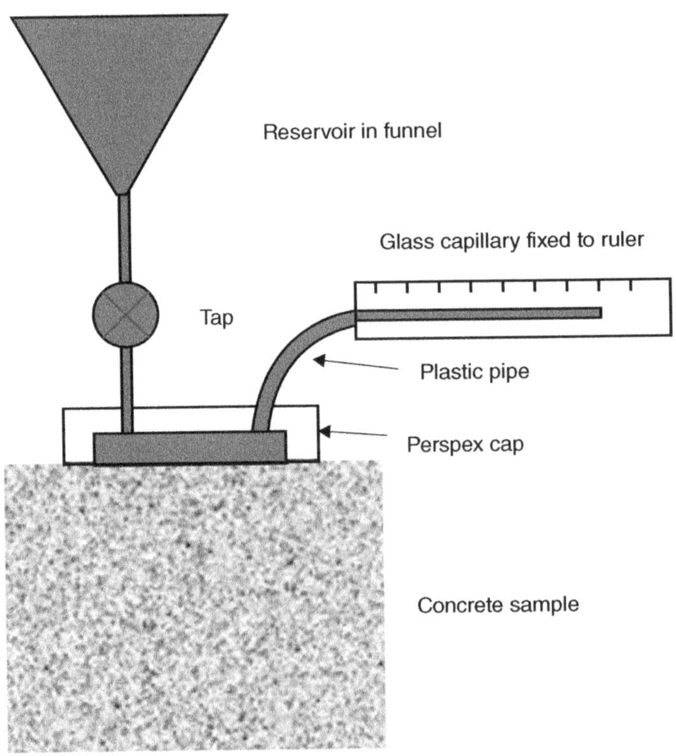

measured from when the concrete is first wetted. Note that the height of the reservoir above the cap (i.e. the pressure head) does not appear in this equation because the capillary suction pressure is much more significant.

8.3.5 Boiled absorption or volume of permeable voids test

This test is in ASTM C642 [25] and has been historically applied to sprayed concrete [26]. The test is also given in AS 1012.21 [27] and specified by VicRoads in Australia for structural concrete [28]. The test involves drying a concrete sample in an oven at 110°C to constant mass A (less than 0.5% change in 24 hours), immersing it in water at 21°C for at least 48 hours to constant mass B and then boiling for 5 hours followed by gradual cooling to 20–25°C and measurement of mass C. Finally, the sample is suspended in water and its apparent mass D is measured.

The percentage volume of permeable voids (VPV) can be calculated as:

$$\mathrm{VPV} = 100 \times \left\{1 - \left(\frac{A - D}{C - D}\right)\right\}$$

Limits on VPV values for test cores from sprayed concrete and structural concrete are in the range 14–17% [26, 28]. A high VPV does not necessarily mean poor quality or low durability, particularly when mass water movement is not the prevalent transport mechanism and Portland cement has been replaced by other cementitious materials. This test is reasonably simple in principle and apparently effective in assessing concrete performance [29]. However, in practice the test is time-consuming and has limitations.

8.3.6 Water penetration under pressure (permeability)

The penetration of water under pressure is pertinent to the design of water-retaining and water-resisting structures, such as shaft and tunnel linings. Water penetration tests include DIN 1048-5 (withdrawn) [30] and BS EN 12390-8 [31].

In the BS EN 12390-8 test, the sample is at least 100 m long and requires a flat surface for testing with a minimum dimension of 150 mm. The surface of the sample to be subject to pressure should be wire-brushed immediately after demoulding the sample, presumably to open pores. This surface is then subject to a water pressure of 500 kPa (5 bar equivalent to 50-m head). After 72 hours, the sample is broken open by axial splitting and the depth of penetration is recorded. Note that splitting a sample of fibre-reinforced concrete can take some time, so that the penetration front may be indistinct at the time of measurement. The penetration front undulates due to local features, such as cracks, fibres and coarse aggregate particles. Therefore, the water penetration should be measured at increments across the width of the broken surface, and the average and maximum calculated. Moreover, this test is reported to have a high coefficient of variation [32]. It is recommended that compliance should be based on at least three tests.

The non-steady-state partial penetration of concrete by liquids can be described by Valenta's equation for uniaxial penetration [33]:

$$d = \left(\frac{2ktP}{v}\right)^{0.5}$$

where d is the depth of penetration (m), k is the coefficient of water permeability (m/s), P is the pressure head (m), t is time (s) and v is the porosity. This equation is strictly applicable only to completely dry concrete in which the pores are empty, whereas the samples in the tests above are not tested in this condition.

Steady-state water flow is described by the Darcy equation, but testing this is difficult due to the low permeability of high-quality concrete and the extended timescale for steady-state conditions to be established.

The maximum water penetration is typically limited to 10–20 mm, equivalent to a Valenta coefficient of water permeability of 0.5–1 × 10^{-12} m/s, depending on the porosity.

8.3.7 Air and water permeability

Air permeability can provide a useful indication of the quality and durability of concrete, particularly in relation to deterioration processes that involve the penetration of gases, such as carbonation. It is highly affected by the moisture condition of the concrete.

Figg [34] developed a test for air permeability. A 50-mm-deep hole is drilled into the concrete surface, thus avoiding the problem with the surface laitance. Figg's air permeability test method is based on reducing the pressure in the drilled hole in the concrete through a hypodermic needle using a vacuum pump. The rate at which the pressure subsequently returns to atmospheric gives a measure of the permeability of the concrete through which the air then flows.

Figure 8.4 Schematic representation of a Figg test [35]

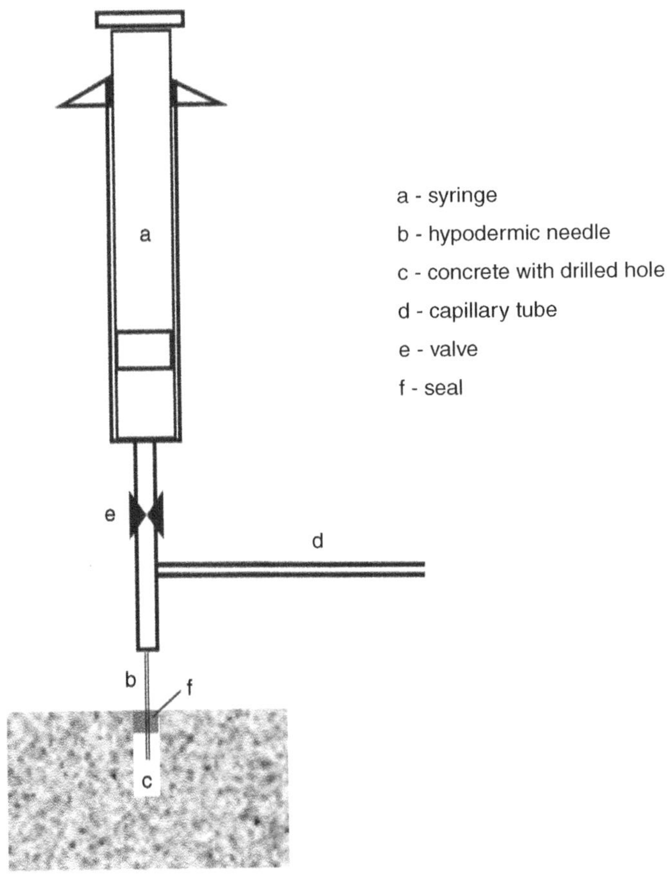

In a modified version of the test, a 10-mm-diameter hole is drilled to a depth of 40 mm (Figure 8.4). The hole must be carefully formed to avoid cracking of the surrounding concrete and thoroughly cleaned of dust. A silicone plug is used to seal the top of the hole and a vacuum pump is used to reduce the pressure in the enclosed space to −55 kPa. The time taken for the pressure to rise to −50 kPa is recorded. The test has the advantage of reducing the influence of the condition of the concrete surface, but the volume of the concrete tested is relatively small and the sample locations must be repaired.

Commercial versions of the test are available.

Alternative test methods include the Autoclam. The permeability is expressed in terms of an empirical index with a value of >300 s indicative of good quality concrete [36].

8.3.8 Absorption and permeability tests for in situ concrete

The two tests described in Sections 8.3.4 and 8.3.7 – the ISAT and Figg test – can be used for in situ concrete testing. Two other in situ tests have been developed from the ISAT and the Figg tests. The ISAT apparatus can be used to carry out an air permeability test by evacuating the cap and using the procedure from the Figg test. Similarly, a water absorption measurement can be made in a drilled hole.

Claisse [24] presents a full description, test results and theoretical analysis of the in situ tests and concludes as follows.

- Any water present in the pores of the sample can severely affect the gas migration measurements. This problem may be overcome by introducing some indicating silica gel desiccant and applying a vacuum until the colour of the indicator changes. This will indicate that the vacuum has dried the sample sufficiently to get a good reading.
- The tests measure different properties, in that the water tests measure capillary suction as well as permeability.
- A comparison was made of the ability of the tests to detect the difference between different sample types. These results did not indicate a very clear preference for either surface or drilled hole tests. However, the analysis of the different tests concluded that the water tests give more consistent results than the gas tests. If possible, a site test programme should use both tests.
- It was found that a full analytical solution to determine permeability values from the vacuum tests was not possible because the distance over which the vacuum decays is not known. This means that a permeability value cannot be obtained from the test results of the gas tests. A solution to this problem that involved drilling additional holes was developed.

8.4. Chloride migration and electrical resistivity testing

8.4.1 Rapid chloride test

Chloride permeability tested using ASTM C1202 [37] expresses the results as coulombs of charge that have passed through a 50-mm-thick 100-mm-diameter sample with sodium chloride on one side, sodium hydroxide on the other and an applied potential of 60 V for 6 hours. There is no simple conversion to give a chloride diffusion coefficient. To ensure the results are useful, at least three test cylinders should be tested from each sample. A high chloride permeability would be 4000 coulombs whereas a low permeability would be 2000 coulombs. In some project specifications, a maximum charge passed of 1000 coulombs is specified, and the mix design is left to the contractor with no further checks for durability.

The significant transport processes that take place during the test are diffusion and electromigration. The combined effect of the two processes is given by the Nernst–Planck equation. Analytical solutions to this equation are available, but they assume that chloride ions are the only species of charged particle in the system.

Figure 8.5 Schematic representation of the progress of a chloride ion [24]

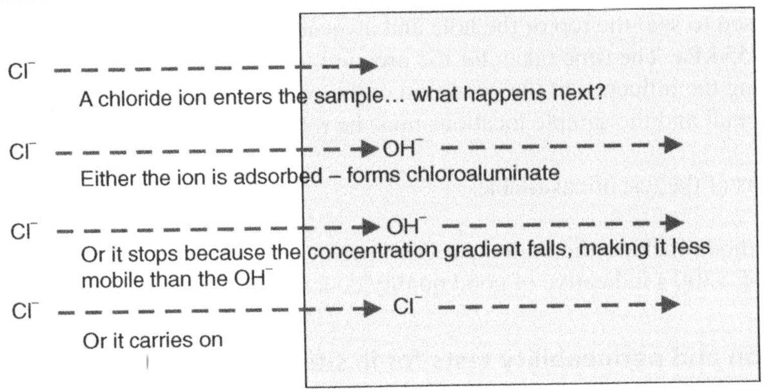

However, Kirchhoff's law states: The current density into any point will equal the current out of it. The effect of Kirchhoff's law is shown in Figure 8.5. When a chloride ion enters the sample, it cannot just stop because this would violate the law. Something must carry the charge away. Thus, the result of the test (measured as charge passed) depends significantly on the ability of other ions, such as OH^-, to carry the charge. Indeed, at the start of the test, there are no chloride ions in the sample, and often, they never penetrate right through in the 6 hours. This leads to highly misleading results when pozzolanic cement replacements are used because the pozzolanic reaction depletes the OH^- ions.

Figure 8.6 shows results for two similar mixes, one with silica fume cement replacement and one without. The chloride values from a simple diffusion test like NT Build 443 [38], which is described below, are shown on the *x*-axis plotted against the charge passed on the *y*-axis from the voltage-driven

Figure 8.6 Comparison of charge passed in a simple diffusion test for mixes with w/c = 0.3 and with 0% and 20% silica fume replacement [24]

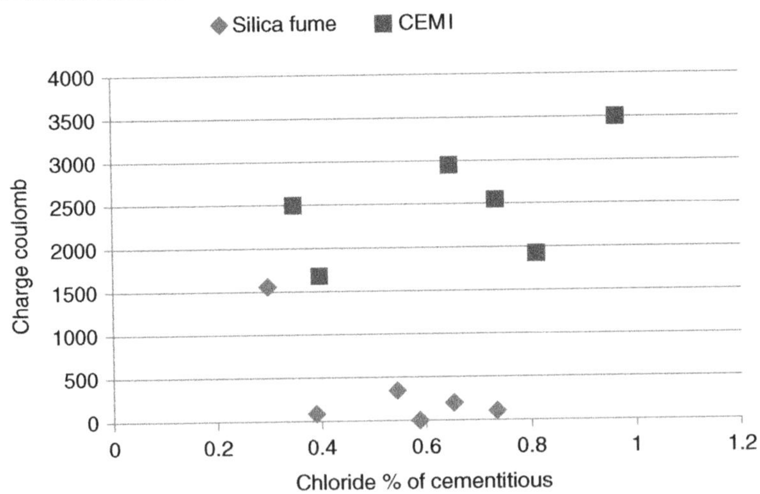

ASTM C1202 migration test. The pozzolanic mixes are generally below 1000 coulombs and the CEM I mixes above (the only silica fume mix above 1000 coulombs was from 28 days of cold curing, so the pozzolanic reaction would not have progressed). These data indicate that the ASTM C1202 test is not suitable for comparing pozzolanic mixes with other mixes. Adding silica fume (or other pozzolans) to almost any mix can significantly reduce the charge passed without necessarily improving strength or durability.

8.4.2 NT Build 443 bulk diffusion test

NT Build 443 [38] is a bulk diffusion test. It involves submerging a portion of a cylinder or core in sodium chloride solution for at least 35 days. This long test duration (which may be even longer for high-strength concrete) is why the voltage-driven tests in ASTM C1202 and NT Build 492 [39], which can be completed in a day, are very commonly used, despite their limitations. The test portion is the 10 to 60 mm depth increment taken from the outermost end of a core or cylinder to represent the cover concrete. The samples must have a minimum diameter of 75 mm (or three times the maximum aggregate size) and all surfaces of the test portion, except the 10-mm-deep face, are sealed with an epoxy or polyurethane coating. At the end of the test, the exposed surface is ground in at least eight increments (the first being 1 mm deep). These dust samples are subject to chemical analysis to generate a profile of the acid-soluble chloride content with depth. The initial chloride content of the sample is found by crushing a 20-mm-thick slice.

An apparent chloride diffusion value can be estimated from the following equation based on Fick's second law of diffusion:

$$C(x,t) = C_s - \left[(C_s - C_i)\mathrm{erf}\left(\frac{x}{\sqrt{4D_e t}}\right)\right]$$

where $C(x,t)$ is the chloride content at depth x at time t, C_s is the surface chloride content, C_i is the initial chloride content, x is the depth below the exposed surface to the middle of the increment, D_e is the effective chloride migration coefficient (m^2/s), t is the exposure time and erf is the error function.

8.4.3 NT Build 492 accelerated migration test

NT Build 492 [39] is like the ASTM C1202 test in that a voltage is used to drive chlorides into the sample (Figure 8.7). Cylindrical specimens with a diameter of 100 mm and thickness of 50 mm (sliced from cylinders or cores) are conditioned for 18 hours by placing them in saturated calcium hydroxide solution under 10–50 mbar vacuum. The test duration following this preconditioning is 6 to 96 hours depending on the initial current passed. The voltage used is moderated to reduce sample heating, and instead of measuring the electrical charge passed, there is a more direct measurement of chloride penetration using a 0.1M silver nitrate solution. The average penetration depth from seven measurements on a split surface can be converted into a chloride diffusion value using a development of the Nernst–Einstein equation [40]. At least three specimens should be used. The test is reported to have a coefficient of variation of reproducibility of 24% for concrete using slag cement.

It would be misleading to assume that, just because it does not depend on an electrical measurement, this test overcomes all the problems with the ASTM C1202 test. No account is taken of the ion–ion interactions. The depletion of charge carriers in the pozzolanic reaction will reduce the transport of chloride ions in the test without necessarily improving the durability. The presence of steel or conductive ionic species, such as calcium nitrite, could also reduce the chloride penetration [41].

Figure 8.7 Schematic representation of an NT Build 492 test

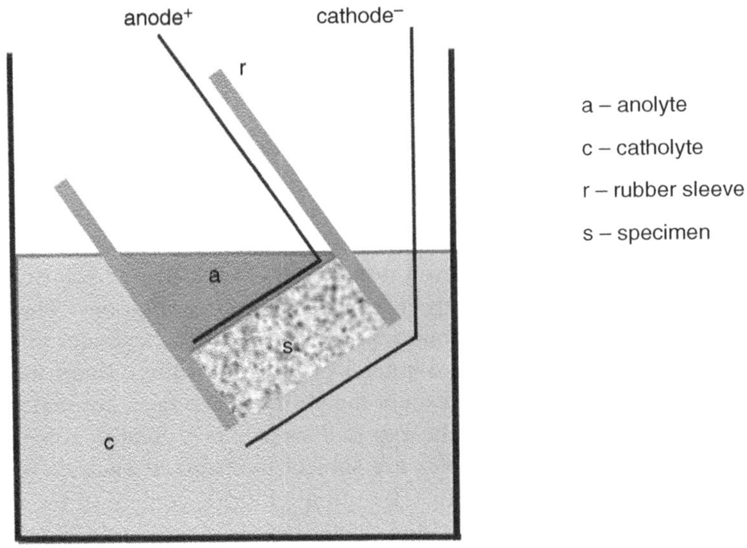

The NordTests or their equivalent are increasingly popular in Europe, Australia and North America [42]. Maximum chloride migration values are typically 4–8 × 10^{-12} m^2/s for NT Build 492 at 28 days and 2–4 × 10^{-12} m^2/s for NT Build 443 at 56 days [18].

8.4.4 Electrical resistivity

Electrical resistivity measurements can be a useful and rapid means of assessing concrete durability and quality, as well as being an important parameter for cathodic protection and earthing and bonding design. It is strongly influenced by the moisture condition of the concrete and the presence and mobility of conductive ions, which may, for example, be significantly depleted by the presence of pozzolanic cement replacements. The resistivity in conjunction with the electrical potential and chloride content can be used to assess the risk of reinforcement corrosion [43]. Note that the resistivity must always be measured with an alternating current. A direct current in concrete will result in ion migration so that the apparent resistivity will change with time.

Andrade has reported that the corrosion rate and the effective chloride diffusion coefficient are inversely proportional to the resistivity for saturated concrete [21,44]:

$$I_{corr} = \frac{26}{\rho_{28}} \text{ and } D_{eff} = \frac{20}{\rho_{28}}$$

where I_{corr} is the corrosion rate ($\mu A/cm^2$), ρ_{28} is the resistivity at 28 days (kΩ cm) and D_{eff} is the effective chloride diffusion coefficient (10^{-12} m^2/s). This is in agreement with the Nernst–Einstein equation [24].

The diffusion coefficient is given by:

$$D = \frac{RT\sigma}{z^2 F_a^{\,2} C}$$

Figure 8.8 Schematic representation of resistivity tests

where D is the diffusion coefficient (m^2/s), the gas constant $R = 8.31$ (J/(mol K)), T is the temperature (K), σ is the electrical conductivity $(\Omega\ m)^{-1}$, which is the inverse of resistivity, z is the valence of the ions conducting the charge (typically Cl^- or OH^-), the Faraday constant $F_a = 9.65 \times 10^4$ coulomb/mol and C is the concentration of ions in the solution (kg/m^3).

Resistivity can be measured in the field using a four-probe Wenner array placed on the exposed surface of a concrete element or in the laboratory with the two-electrode method (TEM) using electrodes placed on opposite surfaces of a moulded or cored specimen (Figure 8.8).

Proprietary systems are available based on the four-probe Wenner array. Alternatively, a soil resistivity meter can be attached to carbon steel rods at 50-mm spacing mounted on a non-conductive bar. Shallow holes are drilled at the test locations to remove surface concrete, which may be affected by leaching and carbonation. The resistivity is calculated as:

$$\rho = \frac{2\pi\delta E}{I}$$

where ρ is the resistivity (Ω cm), δ is the probe spacing (cm), E is the potential drop (volts) across the inner probes and I is the current (amperes) flowing between the outer probes [45].

TEM uses a specimen the same size as the rapid chloride permeability test (51-mm-long slices from 102-mm-diameter cores or cylinders). The specimen is subject to an alternating current of <1.5 mA (frequency 108 Hz) passed through stainless steel plates attached to the flat faces [46]. The resistivity is then calculated as:

$$\rho_{\text{TEM}} = R_{\text{TEM}}\left(\frac{A}{L}\right)$$

where ρ_{TEM} is the electrolytic resistivity (Ω m), R_{TEM} is the absolute resistivity of the sample (Ω), A is the cross-sectional area (m^2) and L is the length (m) of the specimen.

The resistivity indicative of a low corrosion rate is in the range 10–20 kΩ cm [43]. The resistivity limit for monitoring the quality of concrete should be established for the required chloride diffusion coefficient for the specific concrete.

Measurement of conductivity using the ASTM C1760 test [47] is an alternative method of assessing resistivity and chloride diffusion properties.

8.5. Temperature performance

8.5.1 Freeze–thaw tests

There is more than one test available to determine the ability of a cementitious material to withstand the effects of the internal expansion of water caused by freezing conditions, which can initiate microcracking. A single freezing of concrete may not be a significant event with regards to durability, as one event will cause little damage in fully cured concrete. However, repeated cycles of freezing and thawing allow microcracks to propagate, thus increasing the width of the cracks. Eventually, this process will cause spalling of the concrete.

From work carried out by the Concrete Society's Specialist Materials Group [48], conclusions were drawn by Haynes as to the postulated risk of damage, namely

- moisture state and distribution in the pore structure
- availability during the freezing cycle of non-frozen water to form additional ice
- lowest freezing temperature
- length of the freezing cycle if external water is available.

Hobbs [49] suggests that freeze–thaw attacks are one of three main causes of concrete degradation.

8.5.1.1 ASTM C666/C666M-15 – 'Standard test method for resistance of concrete to rapid freezing and thawing'

ASTM C666-15 [50] allows comparative statements to be made about test results when the concrete is made from one or two batches, if the standard deviation for the durability factor is within the acceptable range. To obtain the relative dynamic modulus as required by the ASTM test, a Pundit test for BS EN 12504-4 [51] should be carried out within the test. There are two procedures: A being with the air thawed and B being with the water thawed. As concrete needs to be almost fully saturated before freeze–thaw damage can take place, procedure B is recommended. Expansion tests may be carried out as an optional procedure [50]. However, final strain measurements can be used to evaluate internal cracking and subsequent expansion.

The following information is taken prior to commencement of a freeze–thaw test

- all nominal external dimensions
- saturated weight
- Pundit test, reading for transverse frequency time and relative elastic modulus
- air content of the fresh concrete (BS EN 12350-7, Clause 7) [52]
- pore and void spacing content [53].

8.5.1.2 Freeze–thaw procedure B

In this procedure, the sample undergoes freezing in air. It is thawed in water until the centre of the sample changes from −19.4°C to 6°C and the transition period between the freezing and thawing phases of the cycle does not exceed 10 minutes. ASTM recommends that number of cycles is within the range 100 to 300, whereby the 'indexes of precision are valid at least over these ranges'.

The final report will examine the following data

- consistency limits per batch (slump)
- characteristics of the test specimen if found to deviate from the norm and surface condition

- original weight, ultrasonic pulse velocity and nominal size
- final weight, ultrasonic pulse velocity and size
- final number of freeze–thaw cycles to failure per concrete type
- durability factor for all samples
- paired comparison test on the control batches (*t*-distribution tables) [54]
- relative comparison test showing the average difference in performance, modulus of elasticity and standard deviation per concrete type.

The durability factor provides a final percentage evaluation against the concrete samples prior to the start of the freeze–thaw cycles. The ASTM C666 test gathers a great deal of useful information about durability over a significant number of freeze–thaw cycles compared to other standard tests.

8.5.2 Other freeze–thaw tests

ASTM C672 – 'Standard test method for scaling resistance of concrete surfaces exposed to deicing chemicals' [55]: This provides a visual scale to determine surface scaling of concrete over 50 cycles. This test depends on the operator's judgement, making the results relative.

ASTM C1645-16 – 'Standard test method for freeze–thaw and de-icing salt durability of solid concrete interlocking paving units' [56]: This test is useful when considering the thermal shock effects of using de-icing salts.

BS PD, CEN/TR 15177 – 'Testing the freeze–thaw resistance of concrete. Internal structural damage' [57]: This measures the internal structural damage of concrete using the relative dynamic modulus of elasticity and length change over 56 freeze–thaw cycles to determine internal expansion due to the freezing of water within the concrete. In the authors' experience, 56 cycles are too few to provide meaningful data.

HETEK adaptation of the Boras Method SS 137245 [58]: This measures the mass of material lost by scaling when the concrete is subject to temperature cycles from approximately −20°C to +20°C in combination with 3% sodium chloride solution to represent exposure to de-icing salts.

8.5.3 Fire tests

The performance of the structural frame, whether it is concrete or steel, determines the degree of safety afforded to the occupants. Concrete can be used as a fire protection or as a structural member with cover to the steel providing fire protection. The performance of this material is critical to the structural integrity of the structure.

Explosive spalling occurs during the early part of a fire, usually within the first 30 min or so of a furnace test. It can occur at an early stage just above 150°C. It is characterised by pieces of concrete being violently expelled from the surface, which may occur just once or at intervals and even from the previously spalled parts. Multiple spalling layers are more likely in high-strength than ordinary-strength concrete due to its low permeability. Spalling is often restricted to the unreinforced part of the section and usually does not proceed beyond a reinforcing layer.

A reduction in permeability has been found to be detrimental to performance in fire. This is because steam is produced in concrete at high temperatures. Unless there is an escape route for the steam, internal pressures are generated, which, in conjunction with other stresses, can exceed the tensile strength of the concrete. In addition, the stress–strain characteristics depend on the concrete grade, so that a higher grade concrete is stiffer and more brittle. Thermal stresses may therefore be higher and the concrete less accommodating to internal pressures [59].

The rate at which the temperature rises inside concrete when heated plays a significant role in causing explosive spalling, because if the temperature rises quickly, then the water content inside the concrete will try to escape more quickly, so that more steam is produced, which leads to the explosive spalling of the concrete. However, if the temperature rises slowly, then the steam will be produced at a slower rate, thus allowing the concrete time to release the steam.

Fire performance tests are used to assess damage to the concrete and reinforcement in tunnel linings during a fire event and are described in international standards, such as Eurocode 1 and ISO 834, as well as national regulations, such as the Dutch RWS (Rijkswaterstaat) and German RABT/ZTV curves [60]. Various fire curves prescribing a temperature–time profile are used, depending on the type of fire expected, which in turn is influenced by the intended use of the tunnel [61]. Figure 8.9 shows fire curves for cellulose and hydrocarbon fires. Fire testing is time-consuming and expensive and may be avoided if there is historic evidence of previous acceptable fire test performance with a very similar mix (same fibre geometry and dosage and coarse aggregate type).

Explosive concrete spalling is a particular concern, and this can be addressed by incorporating approximately 2 kg/m^3 of synthetic microfibres in the concrete [62]. Certain cement types have increased resistance to fire, including calcium aluminate cement and geopolymer cement. Figure 8.10 shows the effect of polypropylene fibres and geopolymer cement on fire spalling resistance of tunnel segments [63]. The moisture condition of the concrete at the time of testing is also very important, and concrete that performs satisfactorily after a period of drying may not do so if saturated.

Richardson and Dave [64] cast 150-mm concrete cubes with various dosages of polypropylene fibres. The cubes consisted of medium-strength concrete with type 1 fibre additions, except for a batch with fibre 0.9 + 6 kg/m^3 concrete as the 0.9 value was supplemented with 6 kg/m^3 of type 2 polypropylene fibres. Following the 2-hour heat test, the cubes were left to cool and examined to evaluate the degree of spalling.

The final average compressive strength following the furnace test was 50% lower compared with control cubes. The time to initial spalling is shown below

- plain concrete cubes: 45 minutes to first spall
- fibre 2 kg/m^3 concrete cubes: 75 minutes to first spall

Figure 8.9 Standard fire curves

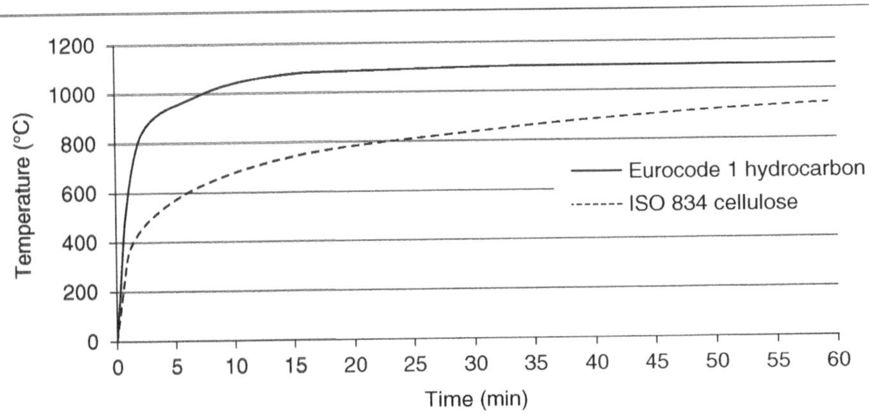

Figure 8.10 Tunnel segments after being exposed to fire at 1100°C for 1 hour

GEL-8SY Synthetic macrofibre reinforced concrete No spalling	**PC-S1** Steel fibre reinforced concrete Explosive spalling 40% of area
PCM-S1 Steel fibre reinforced concrete and synthetic microfibres No spalling	**GNH-FR1** Steel fibre reinforced geopolymer concrete Non-explosive spalling 10% of area

- fibre 0.9 kg/m³ concrete cubes: 105 minutes to first spall
- fibre 0.9 + 6 kg/m³ concrete cubes: 105 minutes first spall.

The time to the first spall is significant for a short fire. Thus, concrete with polypropylene fibre additions will receive greater protection against spalling compared to plain concrete. The final conditions of the spalled cubes are described Table 8.1, namely the number of spalled areas and mass lost, based on the dry weight of the test samples. The results demonstrate the positive effects of polypropylene fibres when used for fire protection.

Technical Report 33, *Assessment and Repair of Fire-damaged Concrete Structures* [65], Eurocode 1 [66] and ISO 834-11 – 'Fire resistance tests. Elements of building construction' [67] are recommended for determining the condition of concrete following exposure to fire and the temperatures the concrete has been exposed to following an unplanned fire event, as well as general guidance.

8.6. Experiments on self-healing and sealing of cementitious materials

The test methods described below form part of a STAR report by the RILEM TC 263 MCI team. One of the authors was a member of the team and contributed to the development of the report. The test described here was carried out at Northumbria University to assess the self-healing capability of cracks in concrete [68]. The aim was to use bacteria and an encapsulated food source to create a sealing/healing agent composed of a calcite precipitation. *Bacillus sphaericus* (LMG 22257) was used to illustrate its efficacy in the repair of concrete work. This strain precipitates calcium carbonate ($CaCO_3$) onto its cell constituents and in its micro-environment by decomposition of urea ($CO(NH_2)_2$) into ammonium (NH^{4+}) and carbonate (CO_3^{2-}).

Table 8.1 Average number of spalled areas per cube and mass lost [64]

Cube type	Mean number of spalled areas per cube	Mean % mass lost
Plain	10	0.58
Fibre 0.9 kg/m³	8	0.49
Fibre 2 kg/m³	2	0.099
Fibre 0.9 + 6 kg/m³	2	0.11

The latter subsequently promotes the microbial deposition of calcium carbonate in a calcium-rich environment. Through this process, the bacterial cell becomes coated with a layer of calcium carbonate. This bio-calcium carbonate has been explored [69] for protecting cementitious materials and natural stone and thus to improve their durability. The bio-deposition treatment consisted of two laboratory steps: creation of a bacterial growth culture and the addition of bacteria to a nutrient broth medium with precipitation precursors. Bio-deposition treatment resulted in the presence of a layer of calcium carbonate crystals.

8.6.1 Test programme

The test programme adopted in this study aimed to demonstrate evidence of crack healing for concrete mixes containing bacteria. Control samples were compared with those in which bacteria were encouraged to grow. A comparison of water volumes flowing through a cracked concrete prism, tested prior to and after bacterial healing, provided evidence of concrete repair. Further, scanning electron microscopy (SEM) and three-dimensional fracture modelling supported the analysis of the repaired concrete at a microscopic level.

8.6.2 Test sample: Mix specification and preparation

Three mixes were designed for test prisms of dimensions length $l = 160 \pm 1$ mm, width $b = 40.0 \pm 0.1$ mm and depth $h = 40.0 \pm 0.1$ mm. These dimensions are in accordance with BS EN 1015-11 and are reflective of samples tested by other researchers [69].

The basic mix design of each concrete sample assumed a mortar mix manufactured in accordance with EN 196-1. Each basic mix utilised a pervious expanded clay aggregate to ensure consistency. The voids within this aggregate afforded potential sites into which bacterial spores could be absorbed. The control mix, containing no bacteria, was labelled sample N. Samples labelled R were produced with a calcium lactate food source and no bacterial spores, whereas samples labelled BAC consisted of the basic mix, the food source and bacterial spores.

The expanded clay aggregate samples (1–4 mm) were prepared by Delft University. The aggregate was placed in a vacuum and then flooded with bacterial spores in a water carrier to impregnate the material. The healing agent solution contained calcium lactate (200 g/l), yeast extract (4 g/l) and bacterial spores of the bacillus genus (10^8 spores per litre) [70].

8.6.3 Water flow test

This test was developed with the RILEM TC 253-MCI team (Figures 8.11 and 8.12). Tests were carried on each prism after a 300-micron crack was induced and again after 56 days to permit self-healing. A constant head water supply (500 mm) was connected to the 5-mm-diameter hole running through the centre of a concrete prism. The other end of the hole was sealed with epoxy glue. A measuring cylinder underneath the prism collected the residual water flowing through the crack. The test time was 5 minutes from the first drop of water entering the measuring cylinder. The contents of the measuring cylinder were then weighed. A sealing efficiency calculation was carried out using the change in flow divided by the original flow and the results expressed as a fraction of the flow for a completely sealed crack.

8.6.4 Limitations of self-healing

Note that while some aggregates will rupture on crack propagation, others will remain intact as the crack propagates around the capsule. If fewer aggregates rupture, there are less bacteria available to react with the catalyst to create calcium carbonate.

8.7. Strength, stiffness and resilience

The tests described here are not single measures of durability. However, they do provide the specifier with values of physical parameters that affect the general durability. The tests should be used appropriately for concrete in situations where these parameters are a key requirement.

Figure 8.11 Water flow test apparatus

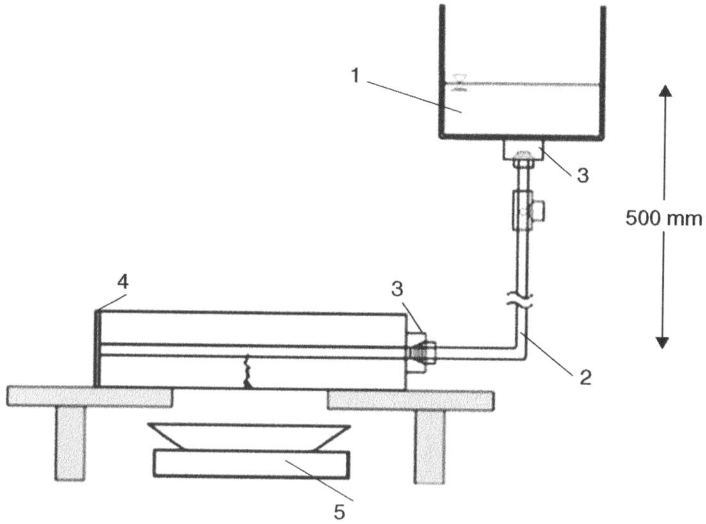

Nomenclature:
1 – Constant head water container
2 – Plastic feed tube
3 – Connector
4 – Resin seal
5 – Electronic scale

Figure 8.12 Test specimen with wire reinforcement and a 5-mm-diameter central hole for the water test

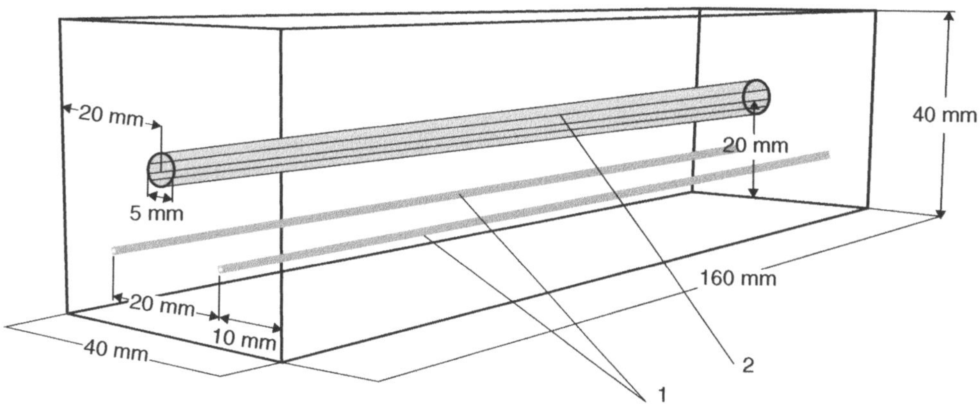

8.7.1 Compressive strength

Compressive strength is widely used as the key indicator of concrete quality. Therefore, it needs to be determined accurately. BS EN 12390-3 is a simple and readily available test for evaluating the compressive strength of concrete test cubes [71]. This test is widely used and accepted by industry. The load at failure is recorded, and this is divided by the surface area of the cube in contact with the platens through which force is applied to the concrete cube. Units used are MPa or N/mm^2.

A failing of this test is that it is carried out retrospectively as samples are taken from the point of placing. Thus, defective concrete is often found only after the work has been carried out and further work may have been added to the defective concrete. Examples are described in Chapter 1.

8.7.2 Modulus of elasticity

The modulus of elasticity may be specified for structures or structural elements for which vertical or lateral deformation, overall stiffness, or vibration control are critical to performance, such as electronics production facilities. ASTM C469 can be used to evaluate this property at 28 days [72]. When deciding whether to specify and test for modulus of elasticity, the specifier should recognise the approximate nature of the relation between elastic modulus and compressive strength [73]. Modulus values can be obtained from compressive strength tests that include strain measurements or pulse velocity tests.

8.7.3 Abrasion resistance

Abrasion resistance may be specified for structural elements where surface wear is a concern for functional performance, such as for industrial floor slabs. The specifier should identify those elements of the structure to which abrasion resistance requirements apply. The specifier has the option to require a maximum abrasion loss and state the associated test method and to state the age at which such values are to be measured. There are multiple abrasion tests, but each varies in the type of wear pattern induced and, thus, each may have varying relevance to the wear anticipated in service for the subject structure.

BS EN 13892-4 [74] describes an abrasion test for horizontal wearing surfaces of concrete using three hardened steel wheels traversing 2850 times around a circular path under a vertical load. The maximum depth of wear is used to define the abrasion class specified to BS 8204-2 [75] for the application and anticipated service conditions. The abrasion class ranges from AR0.5 (0.05 mm) from severe abrasion and impact from steel or plastic wheels to AR4 (0.4 mm) for rubber-tyred traffic.

Other tests for abrasion are ASTM C418 [76], ASTM C779 [77], ASTM C944 [78], ASTM C1138 [79] and the attrition test [80] proposed by the Italian National Research Council. In the latter test, three submerged 100-mm cubes or 100-mm-diameter cores are subject to a 2.5-bar-pressure water jet containing 2-mm standard sand impinging at 45° to one of their flat surfaces. To control variability due to sand particle shape and grading, a reference glass plate is also tested. After 75 minutes, the material lost from the surface is measured by the volume of mercury that fills the eroded depression, and the ratio of this volume to the equivalent loss in the glass control plate is expressed as an index. The abrasion index ranges from 1 to 4 for high to poor resistance, respectively.

8.8. Chemical material testing

8.8.1 Accelerated carbonation testing

The accelerated carbonation test involves subjecting concrete to an atmosphere with an artificially raised level of carbon dioxide and measuring the carbonation depth. A test method is described in fib Bulletin 34 [9]. This depth can be used to calculate an inverse carbonation resistance value.

A moulded specimen with a 100-mm-square surface and 500 mm long is wet cured for 7 days in tap water at 20°C. After curing, the sample is conditioned at 65% relative humidity for 21 days. The sample is then subject to a carbon dioxide concentration of 2% by volume (current atmospheric carbon dioxide levels are approximately 0.04%) for 28 days. The carbonation depth is measured on a fresh split surface using a phenolphthalein solution, which is used to calculate the inverse effective carbonation resistance:

$$R_{ACC,0}^{-1} = \left\{\frac{x_c}{\tau}\right\}^2$$

where $R^{-1}_{ACC,0}$ is the inverse effective carbonation resistance of concrete ((m^2/s)(kg/m^3)), τ is the time constant 420 (s/kg/m^3)$^{0.5}$ for the specific test conditions and x_c is the carbonation depth (m). Carbonation under the accelerated test conditions is supressed by the relatively slow rate of drying of the concrete. Therefore, after adjusting for the effect of the increased carbon dioxide level, a factor of approximately 1.25 is applied to estimate the effective carbonation resistance under natural exposure conditions.

8.8.2 Acid resistance

Testing for acid resistance is problematical, not least because the conditions accompanying the acid exposure and the method by which deterioration is assessed can have an important impact on the results. Exposing a sample to acid followed by abrasion to remove softened surface deposits and then measuring the depth of loss will not replicate the situation where the water around the concrete is static and calcareous material may buffer its effect. The type and concentration of the acid is critically important, and pH alone is inadequate for describing the severity of exposure [81].

Acid resistance is usually assessed using non-standard laboratory tests in which concrete is exposed to concentrated acid. However, a test designed to simulate exposure to silage has been developed by ADAS on behalf of the Environmental Protection Policy Division of the UK Ministry of Agriculture, Fisheries and Food. Sixteen concrete blocks are fixed to a steel wheel rotating at 0.5 rpm and exposed to simulated silage effluent and wear from farm tools, as illustrated in Figure 8.13. The effluent is maintained at a pH of 3.3 and temperature of 23°C. The mean profile and surface roughness of each block is measured every hour using an optical profile gauge for a period of 10–21 days (with 1 day of exposure intended to simulate 1 year of service).

8.8.3 Sulfate resistance

Sulfate resistance has been historically assessed in the UK by a combination of non-standard laboratory tests and long-term field exposure trials [82]. A significant challenge is the requirement to yield a measurable effect within a practical time frame, and for this reason, laboratory test methods often use mortar samples with a relatively high surface area to volume ratio.

A standard test to assess the sulfate resistance of different cementitious binders is given in ASTM C1012 [83]. Mortar prisms 25 mm × 25 mm × 250 mm with a cement-to-sand ratio of 2.75 and w/c ratio of approximately 0.48 are exposed to a 50 g/l sodium sulfate solution. The axial expansion is measured over 15 weeks. The strength of the mortar, assessed from standard cured cubes, should be approximately 20 MPa at the commencement of the test.

The limits on expansion given in ASTM C1157 [84] are as follows

- high resistance: 6-month expansion $\leq$ 0.05%
- moderate resistance: 6-month expansion $\leq$ 0.10%
- low resistance: 6-month expansion $>$ 0.10%.

Concrete using a binder with a high sulfate resistance is required in conjunction with a maximum w/c ratio of 0.45 if the concentration of sulfate (SO_4) in the groundwater is 1.5–10 g/l [85].

Figure 8.13 Agricultural concrete acid exposure test

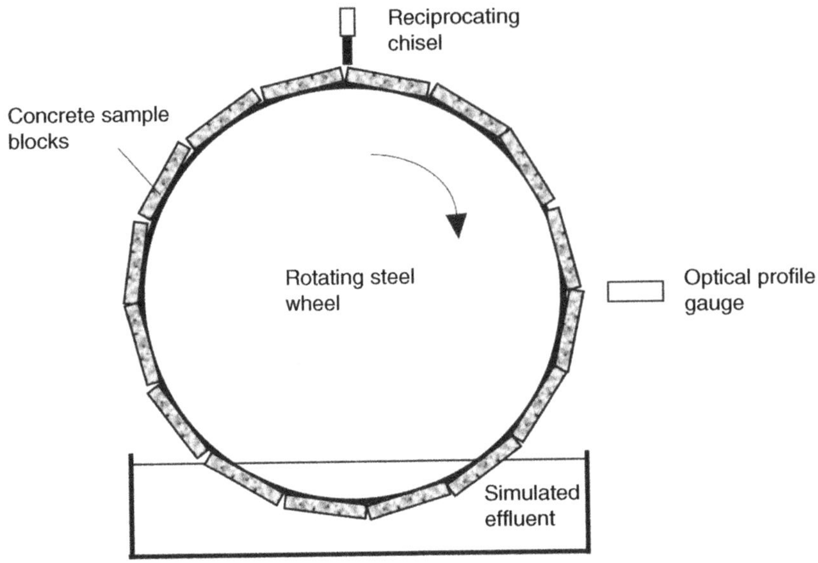

8.9. Conclusion

The tests reviewed here are not meant to be an exhaustive list of every test available of parameters that affect resilience or durability. Bespoke tests are constantly being developed to assess quantitative values that are important to researchers and practitioners alike. The tests have been selected with regard to their practical application and direct usefulness in establishing the parameters surrounding aspects of durability. There are various tests for durability because concrete is a complex material with many aspects worthy of testing. A resilience or durability model may ultimately be developed based upon the test methods described here.

REFERENCES

1. BSI (British Standards Institution) (2004) BS EN 1992-1-1. *Eurocode 2. Design of concrete structures.* General rules and rules for buildings. BSI, London, UK.
2. Pocock D, Corrans J and Al Naboodah (2007) Concrete durability testing in Middle East construction, regional report – Middle East. *Concrete Engineering* **11(2)**: 52–54.
3. Long AE, Henderson GD, Montgomery FR (2001) Why assess the properties of near-surface concrete? *Construction and Building Materials* **15(2–3)**: 65–79.
4. Concrete Society (1996) *Discussion Document – Developments in Durability Design and Performance-Based Specification of Concrete. Prepared by a Working Party of the Materials Group.* Concrete Society, Camberley, UK, Special publication CS109, pp. 1–63.
5. Bickley JA, Hooton RD and Hover KC (2006) *Preparation of a Performance-based Specification for Cast-in-place Concrete.* RMC Research Foundation.
6. BSI (2015) BS 8500-1. Concrete. Complementary British Standard to BS EN 206. Method of specifying and guidance for the specifier. BSI, London, UK.
7. BSI (2013) BS EN 206. Concrete. Specification, performance, production and conformity. BSI, London, UK.
8. BRE (Building Research Establishment) (2005) *Concrete in Aggressive Ground.* BRE, Watford, UK, Special digest 1, 3rd edn.

9. International Federation for Structural Concrete (2006) *Model Code for Service Life Design*. fib, Bulletin 34.
10. Concrete Society (1996) *Developments in Durability Design and Performance-based Specification of Concrete*. Concrete Society, Camberley, UK, Technical report 109 (currently being revised).
11. Concrete Society (2008) *Permeability Testing of Site Concrete*. Concrete Society, Camberley, UK, Technical report 31.
12. BSI (2011) BS 1881-122. Method for determination of water absorption. BSI, London, UK.
13. Dill MJ (2000) *A Review of Testing for Moisture in Building Element*. Construction Industry Research and Information Association, London, UK, Publication C538.
14. ASTM (2013) C1585. Standard test method for measurement of rate of absorption of water by hydraulic-cement concretes. ASTM, West Conshohocken, PA, USA.
15. RMS (Roads and Maritime Services) (2012) *RTA Test Method T362. Interim Test for the Verification of Curing Regime – Sorptivity*. RMS, NSW, Australia.
16. Senbetta E and Scholer CF (1984) A new approach for testing concrete curing efficiency. *ACI Journal* **81(1)**: 82–86.
17. RMS (2013) *QA Specification B80. Concrete Work for Bridges*. RMS, NSW, Australia.
18. Hall C (1989) Water sorptivity of mortars and concretes: A review. *Magazine of Concrete Research* **41(147)**: 51–61.
19. AFNOR (Association Française de Normalisation) (2010) NF P18-459. Concrete. Testing hardened concrete. Testing porosity and density. AFNOR, France.
20. RILEM (1984) CPC 11.3 Absorption of water by immersion under vacuum. *Recommendations for the Testing and Use of Construction Materials*. E&FN Spon, pp. 36–37.
21. Association Française de Génie Civil (2007) *Concrete Design for a Given Structure Service Life*. AFGC, France.
22. BSI (1996) BS 1881-208. Recommendations for the determination of the initial surface absorption of concrete (ISAT). BSI, London, UK.
23. Concrete Society (2008) *Permeability Testing of Site Concrete*. Concrete Society Camberley, UK, Technical report No. **31**.
24. Claisse PA (2014) *Transport Properties of Concrete: Measurement and Applications*. Woodhead Publishing, Cambridge, UK.
25. ASTM (2013) C642. Standard test method for density, absorption, and voids in hardened concrete. ASTM, West Conshohocken, PA, USA.
26. Bolduc L-S and Jolin M (2010) Shotcrete boiled water absorption. *Shotcrete Magazine Winter* **12(1)**: 12–17.
27. Standards Australia (1999) AS 1012.21-1999. Methods of testing concrete. Determination of water absorption and apparent volume of permeable voids in hardened concrete. Standards Australia, Australia.
28. VicRoads (2017) *Standards Specification Section 610, Structural Concrete*. VicRoads, VIC, Australia.
29. Andrews-Phaedonos F (1997) Recommended durability classifications for structural concrete based on the measurement of volume of permeable voids (VPV). *Proceedings of the 4th CANMET/ACI International Conference on Durability of Concrete Conference, Sydney, Australia.*
30. Deutsches Institut für Normung (1991) DIN 1048-5. Testing concrete. Testing of hardened concrete (specimens prepared in mould) (withdrawn). DIN, Germany.
31. BSI (2000) BS EN 12390-8. Testing hardened concrete. Part 8. Depth of penetration of water under pressure. BSI, London, UK.
32. Pocock D and Corrans J (2006) Experience and limitations of concrete durability testing for selection and monitoring concrete quality in Middle East construction. *Proceedings of the 8th International Conference on Concrete in Hot and Aggressive Environments*. Bahrain Society of Engineers, Bahrain.
33. Bamforth PB (1987) The relationship between permeability coefficients for concrete obtained using liquids and gas. *Magazine of Concrete Research* **39(138)**: 3–11.

34. Figg J (1973) Methods of measuring the air and water permeability of concrete. *Magazine of Concrete Research* **25(85)**: 213–219.
35. Concrete Society (2008) *Permeability Testing of Site Concrete*. Concrete Society, Camberley, UK, Technical report No. 31, pp. 20–21.
36. Kreijger PC (1984) The skin of concrete: composition and properties. *Matériaux et Constructions* **17(4)**: 275–283.
37. ASTM (2019) C1202. Standard test method for electrical indication of concrete's ability to resist chloride ion penetration. ASTM, West Conshohocken, PA, USA.
38. NordTest Build (1995) NT Build 443. Concrete, hardened: Accelerated chloride penetration. NordTest Build, Espoo, Finland.
39. NordTest Build (1999) NT Build 492. Concrete, mortar and cement-based repair materials: Chloride migration coefficients from non-steady state migration experiments. NordTest Build, Espoo, Finland.
40. Tang L and Nilsson L (1992) *Chloride Diffusivity in High Strength Concrete at Different Ages*. Nordic Concrete Research, Publication 11, pp 162–170.
41. Hooton RD, Thomas MA and Stanish K (2000) *Prediction of Chloride Penetration. Final Report to the Federal Highway Administration*. University of Toronto, Report DFTH-97-R-00022.
42. American Association of State Highway and Transportation Officials (2015) T537. Standard method of test for predicting chloride penetration of hydraulic cement concrete by the rapid migration procedure. AASHTO, Washington, DC, USA.
43. Concrete Society (2000) *Diagnosis of Deterioration in Concrete Structures*. Concrete Society, Camberley, UK, Technical report 54, pp. 46–47.
44. Andrade C (2014) Resistivity test criteria for durability design and quality control of concrete in chloride exposure. *Concrete in Australia* **40(4)**: 57–64.
45. Kay EA (2002) *Assessment and Renovation of Concrete Structures*. Longman Scientific & Technical, pp 91–92.
46. DuraCrete (1999) *Compliance Testing for Probabilistic Design Purposes*. European Union, Brite EuRam III, Document BE95-1347/R8.
47. ASTM (2012) C1760. Standard test method for bulk electrical conductivity of hardened concrete. ASTM, West Conshohocken, PA, USA.
48. Concrete Society (1996) *Discussion Document – Developments in Durability Design and Performance – Based Specification of Concrete,* Prepared by a Working Party of the Materials Group. Concrete Society Camberley, UK, Special publication CS109, pp. 1–44.
49. Hobbs DW (2002) The durability of concrete. In *Yearbook 2002–2003*. Institute of Concrete Technology, UK, pp. 55–69.
50. ASTM (2015) C666. Standard test method for resistance of concrete to rapid freezing and thawing. ASTM, West Conshohocken, PA, USA.
51. BSI (2004) BS EN 12504-4. Testing concrete. Determination of ultrasonic pulse velocity. BSI, London, UK.
52. BSI (2009) BS EN 12350-7. Testing fresh concrete. Air content. Pressure methods. BSI, London, UK.
53. Concrete Society (1989) *Analysis of Hardened Concrete. A Guide to Test Procedures and Interpretation of Results*. Joint Working Party of the Concrete Society and the Society of the Chemical Industries, UK, Report TR32.
54. Davies OL and Goldsmith PL (eds) (1972) *Statistical Methods in Research and Production*, 4th edn. Longman Group, London, UK.
55. ASTM (2012) C672. Standard test method for scaling resistance of concrete surfaces exposed to deicing chemicals. ASTM, West Conshohocken, PA, USA.
56. ASTM (2016) C1645. Standard test method for freeze–thaw and de-icing salt durability of solid concrete interlocking paving units. ASTM, West Conshohocken, PA, USA.

57. BSI (2006) PD CEN/TR 15177. Testing the freeze–thaw resistance of concrete. Internal structural damage. BSI, London, UK.
58. Henrichsen A, Laugesen P, Geiker M, Pedersen EJ and Thaulow N (1997) *HETEK Method for Test of the Frost Resistance of High Performance Concrete. Summary, Conclusions and Recommendations.* Road Directorate, Danish Ministry of Transport, Copenhagen, Denmark, Report 97.
59. Clayton N and Lennon T (2000) *Fire Safety of Concrete Structures.* BRE, Watford, UK, Digest 395.
60. Maraveas C and Vrakas A (2014) Design of concrete tunnel linings for fire safety. *Structural Engineering International* **24(3)**: 319–329.
61. Technology Committee (2016) *Guidance for Precast Fibre Reinforced Concrete Segments. Vol 1: Design Aspects.* International Tunneling and Underground Space Association, Report 7.
62. Shuttleworth P (2002) Fire protection of concrete tunnel linings. *Proceedings of the International Conference on Innovations and Developments in Concrete Materials and Construction, University of Dundee.*
63. Wimpenny D and Chappell M (2013) FRGC – An innovative material for tunnel segments. *Tunnelling Journal Dec* 2013/Jan 2014, pp 35–39.
64. Richardson AE and Dave UV (2008) The effect of polypropylene fibres within concrete with regard to fire performance in structures. *Structural Survey* **26(5)**: 435–444.
65. Concrete Society (1990) *Assessment and Repair of Fire-damaged Concrete Structures.* Concrete Society, Camberley, UK, Technical report 33.
66. BSI (2002) BS EN 1991-1-2. Eurocode 1. Actions on structures. General actions. Actions on structures exposed to fire. BSI, London, UK.
67. ISO (2014) 834-11. Fire resistance tests. Elements of building construction. Part 11. Specific requirements for the assessment of fire protection to structural steel elements. ISO, Geneva, Switzerland.
68. Richardson A, Amess L, Neville S and Walton C (2017) Crack healing utilising bacterial spores in concrete. *Journal of Green Building* **12(3)**: 103–114.
69. Wiktor V and Jonkers HM (2011) Quantification of crack-healing in novel bacteria-based self-healing concrete. *Cement and Concrete Composites* **33(7)**: 763–770, 10.1016/j.cemconcomp.2011.03.012.
70. Tzivilogou E, Wicktor V, Jonkers HM and Schlangen E (2015) Development of bio-based self healing concrete to increase durability of structures. *Proceedings of the 8th International Conference on Construction in the 21st century, Changing the field: Recent developments for future engineering and construction, Thessalonica, Greece*, pp. 330–337.
71. BSI (2009) BS EN 12390-3. Testing hardened concrete. Compressive strength of test specimens. BSI, London, UK.
72. ASTM (2014) C469/C469M. Standard test method for static modulus of elasticity and Poisson's ratio of concrete in compression. ASTM, West Conshohocken, PA, USA.
73. Hooton RD, Khrapko M, Otieno M and Ismail MA (2016) Responsibilities. In *Performance-Based Specifications and Control of Concrete Durability* (Beushausen H and Fernandez Luco L (eds)). Springer, Dordrecht, Netherlands, State-of-the-report RILEM TC 230-PSC, pp. 179–196.
74. BSI (2002) BS EN 13892-4. Methods of test for screed materials. Part 4. Determination of wear resistance-BCA. BSI, London, UK.
75. BSI (2003) BS 8204-2. Screeds, bases and in situ floorings. Concrete wearing surfaces. Code of practice. BSI, London, UK.
76. ASTM (2012) C418. Standard test method for abrasion resistance of concrete by sandblasting. ASTM, West Conshohocken, PA, USA.
77. ASTM (2012) C779/C779M. Standard test method for abrasion resistance of horizontal concrete surfaces. ASTM, West Conshohocken, PA, USA.
78. ASTM (2012) C944/C944M. Standard test method for abrasion resistance of concrete or mortar surfaces by the rotating-cutter method. ASTM, West Conshohocken, PA, USA.

79. ASTM (1997) C1138. Standard test method for abrasion resistance of concrete (underwater method). ASTM, West Conshohocken, PA, USA.
80. CIRIA (Construction Industry Research and Information Association) (2010) *The Use of Concrete in Maritime Engineering. A Good Practice Guide*. CIRIA, London, UK, Report C674, pp. 217–218.
81. Eglinton MS (1987) *Concrete and Its Chemical Behaviour*. Thomas Telford, London, UK, pp. 69–70.
82. Harrison WH (1992) *Assessing the Risk of Sulphate Attack to Concrete in the Ground*. BRE, Watford, UK, Information paper 15/92.
83. ASTM (2015) C1012. Standard test method for length change of hydraulic-cement mortars exposed to a sulfate solution. ASTM, West Conshohocken, PA, USA.
84. ASTM (2011) C1157/C1157M. Standard performance specification for hydraulic cement. ASTM, West Conshohocken, PA, USA.
85. ACI (American Concrete Institute) (2019) Code 318. Building code requirements for structural concrete. ACI, Farmington Hills, MI, USA, pp. 318–323.

Soutsos M
ISBN 978-0-7277-6375-4
https://doi.org/10.1680/icehcd.63754.273

Chapter 9
Performance-based specifications for concrete

R Doug Hooton
University of Toronto, Canada

Kenneth C Hover
Cornell University, USA

John Bickley (†)
JA Bickley Associates, Canada

9.1. Introduction

As part of its initiative to facilitate the change in the construction industry from prescription specifications to performance specifications (P2P), the US National Ready Mixed Concrete Association (NRMCA) commissioned a review of the international state of the art. In 2006, Bickley *et al.* [1] conducted a literature review, from which a draft performance specification was produced in 2008 [2]. In addition, related papers have been published [3,4]. This draft specification was then used as one of the inputs to a special American Concrete Institute (ACI) task group report on performance specifications [5], which then led to the formation of ACI Committee 329 on Performance Specifications for Concrete, resulting in the creation of a committee report [6]. Also, in Canada, since 2004, the Canadian Standards Association (CSA) A23.1/A23.2 – 'Concrete materials and methods of concrete construction/test methods and standard practices for concrete' [7] has included performance options. The requirements have been refined in more recent editions, including the assignment of roles and responsibilities of owners, specifiers, concrete producers, contractors and testing agencies. Although this chapter reviews the international literature, it mainly deals with performance specifications from the North American perspective. However, the reader should be aware of RILEM technical committee TC230-PSC on Performance Specifications for Concrete, which has produced conference proceedings [8] and a state-of-the-art report on performance specifications [9], which include contributions from the present work.

A major difference between the scope of ACI 329R-14 [6] and CSA A23.1-19 [7] is that the ACI report deals only with performance up to delivery of the concrete to a construction site (i.e. up to the transfer of ownership from the supplier to the contractor), whereas the CSA standard covers performance that also includes the owner's and contractor's requirements.

Although there have been some recent advances, from the review of standards and the literature undertaken at the time of the previous edition of this chapter, it became clear that, while there was an almost universal interest in performance, primarily for durability, few specifications contained any pure performance criteria. Most specifications defined exposure conditions that pertained to each country or region and then tabulated concrete mixture contents and limits that had been shown by previous studies to likely result in the desired durability. These include maximum limits for the water-to-cement (w/c) ratio or the water-to-cementitious materials (w/cm) ratio, minimum cement contents and an acceptable range of air contents. In addition to different levels of interground limestone in cements, there is an almost universal use of supplementary cementitious materials, such as fly ash, granulated ground blastfurnace slag, silica fume and raw or calcined natural pozzolans, either as separate additions or in blended cements. All the specification documents have assumed that statistical quality control will ensure the consistent conformity and reliability of strength.

It also became clear that the term 'performance specification' means many things to many people. This is not necessarily because of any misinterpretation but because there is such a wide array of options, perspectives and valid interpretations, making it imperative that the term be carefully defined in any given context. Parties could agree in principle to execute work under the performance specification umbrella and yet have widely differing views about mutual expectations.

A lack of reliable, consistent and standardised test procedures for evaluating concrete performance is frequently cited as a major barrier to the adoption of performance specifications. Some of the available tests can be expensive, take a long time to run or may not be as precise as desired. Short bid times and quick construction starts create a difficult situation for a concrete supplier faced with the need to develop a performance mixture and to perform prequalification testing. In several jurisdictions, such as state highway departments, some advanced tests have been proven on site and then specified in subsequent years for pay items in contracts.

On the other hand, in the face of an international mindset that says that testing technology has not yet caught up with performance philosophy, a wide range of tests are available today and have been used successfully on important concrete projects. These tests methods can be called into action to support performance-based specifications. Although some may complain that current tests are not ideal or are insufficiently accurate or precise, which of our everyday concrete quality tests are ideal? If a new test only has to be as accurate, as precise or as meaningful as the slump test, there may be many new developments to choose from. Although, perhaps, not used for acceptance, even a non-standard test can be useful if it provides timely information for construction decisions and the site engineer has experience with interpreting the results. In addition, there has been recent and widespread adoption of wireless embedded sensors in concrete by contractors. These allow real-time monitoring of concrete temperatures and maturity, resulting in more rapid construction scheduling.

The advent of performance specifications could significantly change the distribution and sharing of responsibility among owner, contractor and concrete supplier. It would be up to the owner (through design professionals) to clearly specify performance requirements together with the test procedures used for acceptance. For true end-result specifications based on the properties of hardened in-place concrete, meeting these requirements would be the joint responsibility of the contractor and the concrete supplier. Both parties would assume the risk involved and would have to work closely to determine the appropriate concrete mixture. Quality management programmes would also be required by both, since the successful installation of a concrete mixture would be imperative to achieving acceptance by the owner.

One of the increasingly important potential benefits of using performance standards is that when producers propose more sustainable alternative mixtures for concrete (i.e. with lower embodied carbon and lower embodied energy), they can clearly demonstrate that the owner's performance requirements can be met.

The transition to performance specifications as another and complementary way of doing business will require a dedicated educational effort, and the advantages and disadvantages will have to be made concrete, so to speak. The motivation will have to come from demonstrating clear benefits that can be shared at many levels of the industry and not just because it is time for a change.

9.2. What is a performance specification?

It is useful to begin the discussion by sampling a range of definitions of the term 'performance specification'.

9.2.1 Definitions

As documented on its website, the NRMCA [10] describes a performance specification as follows: 'A performance specification is a set of instructions that outlines the functional requirements for hardened concrete depending on the application. The instructions should be clear, achievable, measurable and enforceable.' For example, the performance criteria for interior columns in a building might be compressive strength and weight, since durability is not a concern. Conversely, the performance criteria for a bridge deck might include strength, permeability, scaling, cracking and other criteria related to durability since the concrete will be subjected to a harsh environment. Performance specifications should also clearly specify the test methods and acceptance criteria that will be used to enforce the requirements. Some testing may be required for prequalification and some for job-site acceptance. The specifications should give the contractor and producer flexibility to provide a mix that meets the performance criteria in the way that they choose. The contractor and producer will also work together to develop a mix design for the plastic concrete that meets additional requirements for placing and finishing, such as flow and setting time, while ensuring that the performance requirements are not compromised.

Performance specifications should avoid requirements for means and methods and should avoid limitations on the ingredients or proportions of the concrete mixture.

Some of the key components of a workable performance-based specification system should include [1,2] the following

- a qualification or certification system that establishes the requirements for a concrete production facility, the facility's quality control management system and the facility's personnel
- specification provisions that clearly define the functional requirements of the hardened concrete (the submittal would not be a detailed list of mixture ingredients but rather a certification that the mix will meet the specification requirements, including prequalification test results)
- suppliers, producers and contractors that partner to ensure that the right concrete mixture is developed, delivered, placed and finished
- sufficient flexibility to allow the supplier to provide a concrete mixture that meets the performance criteria (including prequalification test results) while satisfying the contractor's requirements for placing and finishing
- requirements for field acceptance tests needed to verify that the in-place concrete meets the performance criteria, as well as a clear set of instructions defining the actions required if those test requirements are not met, that is, non-conformance.

9.2.2 US Federal Highway Administration

Part of the *Performance Specifications Strategic Roadmap* [11] published by the Federal Highway Administration (FHWA) is devoted to a detailed description of performance specifications. It states, 'A performance specification defines the performance characteristics of the final product and links them to construction, materials and other items under contractor control.'

9.2.3 Canadian standard CSA A23.1

In Canadian standard CSA A23.1 [7], a performance specification for concrete is defined as follows

> A performance concrete specification is a method of specifying a construction product in which a final outcome is given in mandatory language, in a manner that the performance requirements can be measured by accepted industry standards and methods. The processes, materials, or activities used by the contractors, subcontractors, manufacturers, and materials suppliers are then left to their discretion. In some cases, performance requirements can be referenced to this Standard, or other commonly used standards and specifications, such as those covering cementing materials, admixtures, aggregates or construction practices.

9.2.4 UK Department for Transport

A report by this agency [12] defines performance in general as meeting key performance indicators (KPIs)

> The Performance Specification also sets certain Requirements. These may be for the Company to develop a strategy or a new, more effective, metric, or gather information on an issue. In many instances, such as developing broad environmental outcome based metrics, the Requirements will help to inform the setting of KPIs for future road periods. We require that progress in delivering these Requirements is reported annually at the very least.

9.2.5 Cement and Concrete Association of New Zealand

The Cement and Concrete Association of New Zealand [13] states that: 'A performance-based specification prescribes the required properties of the concrete but does not say how they are to be achieved.'

9.3. Why have performance specifications become an issue now?

The competing or sometimes complementary philosophies of prescriptive and performance specifications have been around as long as there have been concrete specifications. In 1928, Committee E-1 of the ACI and the Committee on Engineering Practice of the Concrete Reinforcing Steel Institute (CRSI) proposed their *Joint Code: Standard Building Regulations for the Use of Reinforced Concrete* [14] (Table 9.1). This document was the precursor to what many consider to be the 'first' ACI code, published in 1936, ACI 501-36T, *Building Regulations for Reinforced Concrete* [15]. These early documents permitted use of concrete mixtures without 'preliminary tests of the materials to be used', as long as the w/c ratio met the prescriptive requirements.

The 1928 'Joint Code' required tests of at least one specimen per 100 cubic yards of concrete placed. Interestingly, this early set of regulations permitted a performance-based alternative to the prescriptive w/c requirements for concrete strength, by allowing prequalification of a mixture on the basis of test data correlating strength to w/c ratio. Four different w/c values had to be tested with four specimens each (the forerunner of today's three-point curve), and the w/c ratio approved for production was that corresponding to

Table 9.1 Criteria from the 1928 *Joint Code: Building Regulations for Reinforced Concrete* [14]

Assumed compressive strength[a] at 28 days: pounds per square inch (MPa)	W/c ratio: US gallons per sack of cement	W/c ratio (by mass)	Approximate ratio of cement volume to dry total aggregate volume
1500 (10.3)	8¼	0.73	1:7
2000 (13.8)	7½	0.67	1:6
2500 (17.2)	6¾	0.60	1:5¼
3000 (20.7)	6	0.53	1:4½

[a]Strength value assumed in structural design.

a compressive strength 15% higher than specified (the forerunner of today's so-called overdesign strength). Once the proposed mixture was approved, no substitutions in materials were permitted without additional tests. A bias toward the prescriptive specification of the w/c ratio was apparent, however, as the frequency of testing had to increase to one specimen per 50 cubic yards (40 m^3) placed if the w/c ratio was established based on contractor testing. There is no mention of durability or permeability in the 1928 Joint Code.

The state of the technology in these early days of the industry can be gauged from this statement in Lord's *Handbook of Reinforced Concrete Building Design*, published in 1928 [16]: 'Engineers are so accustomed to thinking of concrete for buildings in terms of 2,000-lb [per in^2] strength at 28 days that it may be novel to consider using a 3,000-lb [per in^2] concrete as the basic mix.' However, changes were soon to be in the works.

S.C. Hollister, visionary engineer and educator, was the president of the ACI in 1933/34. In his outgoing address, he predicted the advent of chemical admixtures and high-strength concrete:

> One may grow so accustomed to the surrounding conditions that they are accepted as a sort of status not subject to review. … Who may say, for example, whether it is possible to achieve mobility or workability with an agent other than water. … We see the many varied and intriguing avenues of development that present themselves. … Imagine, for example, the concrete with an available strength of 10,000 pounds per square inch. Smaller columns, thinner and lighter beams and slabs would at once result. Present limitations … would at least double. A new basis of design would be required. The achievement of today was the goal of yesterday.

Almost 90 years after Hollister's predictions, today's concrete has become a complex and truly 'engineered material', and this development has intensified the debate surrounding performance specifications.

Over this same period that concrete transitioned to being an engineered material, many concrete producers have likewise transitioned from being merely 'truckers' who deliver concrete mixed in accordance with a specified recipe to being well informed on concrete materials, including complex aggregate grading, chemical admixtures and a wide range of cementitious materials. Similarly, when the 1928 code [14] was published, the design professionals had responsibility for preparing detailed prescriptive specifications and conducting careful inspections of the mixing process. More recently, fewer specifications require predetermined concrete recipes or materials and production inspections. Likewise, there has been a shift in the responsibility for concrete ingredients and mix proportions toward the concrete producer and away from the design professional. Today's review of concrete mixture submittals 'for general conformance with the contract documents' is a significant evolution from the fully specified mixture proportions of only a few years ago.

Other changes that have swept the industry include recognition that, for many modern concrete applications, strength is no longer the only, or even the most important, issue. Portland cement is not the only cementitious material, water content and aggregate size are not the only factors that influence slump and the w/c ratio is not the only factor influencing permeability. Air content is easily specified and readily measured in the field, but freeze–thaw durability and scaling resistance are more dependent on the size and distribution of air voids in the paste than on the total air volume in the concrete. Chemical and mineral admixtures affect air, workability, setting time, bleeding, rate of strength gain, and early- and later-age strengths. These same admixtures may or may not be mutually compatible. Moreover, it has become more difficult to write a prescriptive specification that can take advantage of these developments and avoid their pitfalls. It has become evident that evaluating the durability of concrete is more difficult than evaluating strength. It is more difficult to predict or ensure the long-term service life of concrete than it is to predict or ensure the short-term load capacity.

Thus, the simultaneously increasing demand for improved concrete durability and the growing complexity of concrete mixture design and proportioning lead us back to the debate about moving from prescription to performance. Interest has been further fuelled by the changes in construction techniques that have accompanied these newer concrete materials. An example of all these factors is a high-performance, high-density and low-permeability concrete. Such mixtures often blend Portland cement with one or more cementitious materials, such as silica fume and fly ash or slag, use up to two performance grades of a water-reducing admixture, incorporate at least three sizes of aggregate and may have set retarders or corrosion inhibitors plus an air-entraining admixture. Proportioning such a mixture requires experience with these specific materials, including recognition that the normal relations between workability and water content, and between strength and w/cm ratio, need to be recalibrated. Handling, placing, finishing and curing such a mix requires experience as well to accommodate rapid surface drying, rapid setting and a high shrinkage potential. So, if the question is, 'Why discuss performance specifications now when prescriptive specifications have been used since the early ACI codes in 1928?', then one answer is that we are now demanding more of the concrete and that it may be difficult to take full advantage of the wide range of material and construction combinations and options under a strictly prescriptive specification.

9.4. The essence of prescription or performance

The NRMCA's P2P initiative [10] has been directed towards attaining a shift in the focus of concrete specifications. A prescriptive specification focuses on the properties of the raw materials, mixture proportions, the batching, mixing and transport of the fresh concrete, and the full range of construction operations from placing to curing. Prescriptive specifications rely on observed or implied relations between the details specified and the final, in-place, or 'end product' or 'end result' performance of the concrete. Under a prescriptive specification, the performance of the end product may or may not be described. In contrast, a pure performance specification 'starts with the end in mind'. It fully describes the required performance characteristics of the end product, leaving materials selection, proportioning and construction means and methods up to the party contractually bound to comply with the specifications. Under a pure performance specification, it is the responsibility of the concrete producer/contractor team to select the materials and conduct construction operations that will produce the required concrete performance. Proponents of prescriptive specifications say, 'Here is how we want you to proportion and install the concrete, and if done in accordance with these instructions, the results will be satisfactory.' Proponents of performance specifications say, 'Just tell me what you want done; don't tell me how to do it.'

Keys to the concept of performance specifications include the following

(a) the ability of the specification writer to discern the performance characteristics appropriate to the owner's intended use of the concrete
(b) the ability of the specification writer to describe these performance characteristics clearly, unambiguously and quantitatively so that performance can be evaluated
(c) the availability of reliable, repeatable test methods that evaluate the required performance characteristics (along with performance compliance limits that take into account the inherent variability of each test method)
(d) the ability of the concrete producer/contractor team to correlate choices of materials, mixtures and construction techniques to the required characteristics so that projects can be planned and bid, risks and costs can be assessed, and materials and construction operations adjusted to comply with performance requirements.

These four keys present at least the following challenges.

(a) Under current, predominantly prescriptive specifications, the performance of the end product is not always comprehensively spelled out at the specification stage. For example, prescriptive specifications may not explicitly include requirements for abrasion resistance, scaling resistance or limitations on concrete cracking. Nevertheless, unsatisfactory performance in any of these categories is often pointed out after the concrete has been installed. The rationale for finding the concrete unsatisfactory may be that these common end-result requirements are generally implied and that the concrete would have been satisfactory if the prescriptive requirements had been met. In contrast, performance specifications require an 'upfront' description of owner expectations. In most cases, this can take significant additional effort and expertise beyond that required for prescriptive specifications by design professionals working on the owner's behalf. In some cases, the mechanisms involved are understood, but a reliable measure of end result performance is not easily obtained. In such cases, it can be necessary to rely on an 'index test' or to retain a prescriptive option that has served well. For example, strength test results might be used to estimate abrasion resistance based on a correlation with a given set of materials.

(b) Some commonly expected (although uncommonly specified) performance characteristics are not readily or clearly definable or quantified. In-place cracking, movements due to shrinkage, scaling, pop-outs, colour variations or local abrasion are easy to spot but more difficult to describe unambiguously.

(c) Despite an explosion of research and development into new concrete test methods, the industry does not yet have a comprehensive suite of test methods or the predictive models that allow them to reliably predict service life in general.

(d) Some contractors and concrete producers will need additional training to be able to select materials and construction operations that will produce the required concrete. Design professionals will also need additional training or special expertise so that they can develop reliable performance requirements.

9.5. Advantages and disadvantages

The primary advantage in specifying the performance of the end product is that a knowledgeable concrete producer/contractor team has the flexibility to develop a unique combination of materials and construction methods that will achieve the stated objectives of the owner or designer. Performance specifications, therefore, work well when the producer/contractor team has the necessary expertise, the owner/designer can clearly articulate the requirements, and appropriate and sufficiently precise test methods are available for documenting the specified performance. A prescriptive specification may work best when there is a reliable connection between the specified materials, means and methods and the desired outcomes. The simpler the concrete mixture and the less restrictive the required outcomes, the more likely it is that a prescriptive specification will be an efficient and reliable way to specify the concrete. As the P2P initiative is implemented, it will be necessary to identify those opportunities for which performance specifications offer the greatest advantage, as well as to identify situations where a more conventional approach is more appropriate. Further comments on the advantages and disadvantages for both prescriptive and performance specifications are listed in Tables 9.2 and 9.3.

9.6. Available options

Given that any method for specifying concrete materials and construction services will have both advantages and disadvantages, the challenge (and the opportunity) is to develop appropriate specifications that maximise the advantages and minimise the disadvantages. This also means having a range of available specification types that are appropriate for different situations. For example, BS 8500-1 [17] defines five approaches to specifying concrete, provided each conforms to BS8500-2

1. Designed concrete: Like current practice under ACI 318 and other standards; also referred to as deemed-to-comply
2. Designated concrete: A specific and certified mix that meets the requirements of designed concrete

Table 9.2 Advantages and disadvantages of prescriptive specifications

Advantages	Disadvantages
Some designers and producers may have more confidence in, and be more comfortable with, a traditional prescriptive approach.	Some designers and producers may not be confident that a prescriptive specification will lead to the desired end performance.
Expertise is required when writing the specification.	Some specification writers may not have such expertise, especially with modern materials combinations.
The value and effectiveness of the product are 'designed-in' by the specifier.	There are limited opportunities for optimising the concrete after the specification has been written.
Newer materials and methods can be used if the specifier has remained technologically current.	There are limited opportunities to take advantage of the producer's unique access to materials, material combinations, plant, equipment, technology, expertise or knowledge of local materials and conditions.
The specification reflects the specification writer's understanding of the relation between the desired properties of the concrete and the specified materials, means and methods.	The relations implied in the prescriptive requirements may not be as reliable or the same as the relations assumed for the specific materials or project conditions.
The specification writer can control any aspect of the process, from concrete materials selection and proportioning to batching, mixing, transporting, placing, consolidating, finishing and curing the concrete. This control is exercised through prescriptive specification requirements.	The interests of all parties may not be represented in the prescribed specification, i.e. raw materials suppliers, concrete producer, concrete placing contractor, concrete finishing contractor, general contractor, construction manager, owner, investor and end user.
Basic tests to ensure compliance with the specification are inexpensive, generally accepted and commonly available.	Basic tests to ensure compliance with the specification may not tell us as much as we would like to know. Test results may be more variable (less reliable) than supposed. Conventional results may reflect the material as delivered and as subsequently cured under standard conditions, rather than as installed and as cured in the actual field conditions. Standard tests may report results at concrete ages other than are critical for ensuring quality. Standard tests may not enable accurate prediction of longer-term concrete performance in the actual environment.
Prescriptive specifications could be interpreted as limiting the concrete producer's responsibility to adhere to the prescribed requirements and their liability for post-chute influences on concrete behaviour.	The producer often ends up being liable for post-chute concrete behaviour anyway, at least until expensive tests demonstrate placing, consolidation, finishing or curing problems.
The concrete producer just needs to batch the required materials in the required manner. Concrete materials and mixture expertise may not be required for typical applications.	Since limited expertise is required, limited expertise is applied. Thus, a lowest common denominator industry emerges. Concrete producers have limited technical, economic or creative control over the product.

Table 9.2 Continued

Advantages	Disadvantages
In the absence of explicitly defined end results, contractual performance requirements can be implied in addition to the explicitly stated prescriptive requirements. The owner/designer may object to subsequent concrete performance, regardless of compliance with the prescriptive specifications.	In the absence of explicitly defined end results and if no desired end results can be implied, the owner/designer may be dissatisfied with the end product but have limited recourse if all the prescriptive requirements have been met.
Prescriptive specifications can be written so that they clearly separate the concrete producer's responsibility from the concrete contractor's responsibility.	Even with prescriptive specifications, lines of responsibility can be blurred, especially when testing is conducted on concrete sampled anywhere beyond the truck chute or any time after the addition of water at the job site.
Prescriptive specifications 'level the playing field', allowing concrete producers with a wide range of levels of expertise to compete.	Prescriptive specifications diminish the incentives for a given concrete producer to optimise a mixture or to exercise quality control beyond the level of its competitors.

3. Prescribed concrete: Fully prescriptive – a 'recipe' specification
4. Standardised prescribed concrete: A 'standard mix', like many public works-type standard mixes
5. Proprietary concrete: Full performance with a performance agreement between the specifier and the producer.

The official definitions of these options may be a bit difficult for a North American audience to understand, but they cover the range from being pure performance and pure prescriptive to calling for a specific, pre-approved, proprietary concrete mixture. As evidence that 'there is nothing new under the sun', Elwyn Seelye's classic civil engineering reference book, *Specifications and Costs* (first published in 1946) [18] has model specifications for both prescriptive or 'fixed ratio' concrete mixtures and for performance-oriented 'controlled concrete'. In the first case, 'Concrete shall be, by dry volume, of those proportions that are shown on the drawings.' In the second case, 'Controlled concrete shall conform to the following requirements,' which is followed by a table showing 'class of concrete', and '28-day compressive strength' [18].

Blends of philosophies within a given specification are frequently encountered. ACI 318 [19], for example, allows for acceptance of strength based on strength test results (performance), but for durability, limits are placed on the maximum w/cm ratio (prescription.) Likewise, the CSA A23.1 [7] 'performance' specification still includes prescriptive w/cm limits. NRMCA has already developed an example of a 'minimally prescriptive' specification that recognises the current prescriptive limits of ACI 318 [19] but allows for maximum flexibility through performance requirements for the end product.

There are further options in the distinction between performance characteristics that are used for acceptance of the concrete and those used to adjust the amount paid. Payment schemes can be (and have been) developed that provide incentives for good performance and exact penalties for marginal performance, if the concrete that remains in place has met minimum requirements. Examples of such arrangements are discussed in Section 9.2. A related issue, however, is that, given the joint responsibility generally inherent in performance specifications, fairly distributing cash bonuses or penalties among the parties that contributed to realising the quality of the concrete can be difficult.

Table 9.3 Advantages and disadvantages of performance specifications

Advantages	Disadvantages
Designers can focus on what is needed rather than how to achieve it (not all designers are familiar with *how* to best achieve the end results).	Specifying how to achieve satisfactory concrete has been a traditional design responsibility. Engineers may be concerned over a perceived reduction in their control.
There are opportunities to focus on the behaviour and characteristics of concrete that really matter.	The specifier may not be sure what those characteristics are nor how to measure them. Reliable tests may not be available to quantify the desired outcomes. Performance tests may be more expensive, more time-consuming or require more specialised expertise compared to conventional tests.
The concrete producer/contractor team has technical, logistical, economic and 'creative control' or influence over the product. The are opportunities to take advantage of unique materials, material combinations, plant, equipment, technology, expertise and knowledge of local materials and conditions.	The properties of the end product are influenced by materials, concrete production, concrete delivery, mix adjustments by the contractor, placing, consolidation, finishing, adjustments to the mix properties at the surface, ambient conditions, moisture control and temperature control. There are many parties involved, and each party has a unique influence on the product. It may be difficult to separate their influences and responsibilities.
Flexibility in mix proportioning can produce a better overall mixture or a more economical or more sustainable mixture that meets all performance requirements, or both.	There may be increased costs in the prequalification stage and durability-related or in-place testing may be more expensive.
A more durable product has lower life-cycle costs.	It may be difficult to take advantage of the life-cycle economic benefit in a low bid (first-cost) contract.
Assumed relations between concrete performance and mix characteristics can be augmented or replaced with tests of concrete properties.	Tests beyond the routine slump, air and cylinders are likely to be more expensive and more complicated, and their precision must be taken into account in the specification.
In those cases where prescriptive specifications are interpreted to give the producer/contractor full responsibility for the behaviour of the final concrete, even when prescriptive requirements have been met, a switch to performance specifications does not necessarily result in any additional responsibility.	In those cases where prescriptive specifications clearly limit the producer's responsibility to comply with instructions, the switch to performance specifications and the accompanying responsibility for the end results is a considerable increase in responsibility.

9.7. Point of performance

Given the multiple stages of concrete production and installation and that 'custody' of the concrete changes hands multiple times before the properties of the end product are fully developed, one question is as follows: 'At what point in the process do we specify and evaluate the concrete?' This is further complicated because concrete performance is, to some extent, in the eye of the beholder. Contractually speaking, performance is in the eye of the party that wrote the specification and, thus, defined the required performance characteristics. To a concrete producer buying raw materials, the performance of the end product when applied to aggregates may be defined by density, aggregate grading or the uniformity of the fineness modulus of the sand or by compliance with an aggregate specification, such as ASTM C33 – 'Standard specification for concrete aggregates' [21], with no further stipulation as to where or how to extract the rock or how to process it. Similar examples could be given for the requirements relating to the performance of cement, cementitious materials or admixtures. A purely performance-minded concrete producer might not ask for cement mill test reports and might not even demand that cement or cementitious materials meet the specific chemical or fineness requirements of EN 197 [20] or ASTM C150 – 'Standard specification for Portland cement' [21], specifying instead that the cement meets requirements for strength, rate of strength gain, soundness, shrinkage, setting time, water demand, uniformity and limited expansion in alkali–silica reaction (ASR) tests. Likewise, a performance-minded buyer of chemical admixtures might say, 'Don't tell me what they are made of, just guarantee that they will perform in the concrete and that they are mutually compatible.' Although some of these performance characteristics for raw materials are easily specified and readily evaluated with long-established test methods and the responsibility for meeting them is rather clear, the situation gets complicated when the raw ingredients are combined.

For a contractor buying concrete from a producer, performance might be defined in terms of concrete quantity delivered per hour, workability, pumpability, finishability, setting time or early-age strength for formwork or shoring removal. If performance is evaluated at the point of discharge from the concrete truck, responsibility is fairly clear, but as soon as the evaluation moves to the point of discharge from the pump, responsibility starts to get fuzzy. If the concrete is not pumpable, does the concrete producer have to redesign the mixture or does the pumping service have to change its equipment? The project owner, on the other hand, may not be concerned with any of these raw materials, fresh concrete or construction issues. To the owner, performance is defined by having sufficient in-place load-carrying capacity to allow the safe operation of the facility and sufficient in-place durability to withstand the service environment for the financially intended life of the facility. The idea of a performance specification can, therefore, imply different things to different players and achieving that required performance can (and must) become the joint responsibility of more than one party. Since the owner's chief interests are the in-place long-term properties of the concrete, meeting such requirements will necessarily be the joint responsibility of the raw materials suppliers, concrete producer, formwork subcontractor, pumping subcontractor, placing and finishing subcontractor, and the general contractor or construction manager managing the entire process. It is, therefore, necessary to be more definitive and to talk about the point of performance, that is, when and where in the multiple processes from making and curing to protecting concrete are we going to define the required performance of the concrete? The related question is this: 'Who bears the responsibility for achieving the specified performance?'

The following example, as summarised in Table 9.4, focuses on specifying and installing a superflat industrial floor and demonstrates that the terms 'concrete performance', 'performance specification' and 'point of performance' have a different scope and meaning for each of the various parties to the overall project. (This example considers only the flatness of the floor and ignores other critical performance characteristics, such as strength, cracking and abrasion resistance.)

As shown in this example, the performance required by the owner is a consequence of design, materials and construction performance. Performance specifications can, thus, be efficient for the owner, as they quickly define the key operational characteristics of the installed concrete, but they become equally efficient for the concrete producer only when the aspects of concrete materials performance that contribute to the owner's required performance have been identified and can be controlled. The contractor must, likewise, control construction operations to achieve the owner's requirements. Satisfying the owner's performance requirement specification means that the various parties influencing performance accept their mutual responsibility.

Table 9.4 Example of multiple responsibilities and multiple performance criteria for a superflat industrial floor

Example of a performance requirement	Party setting the performance requirement	Party obligated to achieve the specified performance	Point of performance	Is performance measurable?
Superflat floor	Owner/specifier	General contractor	Hardened concrete floor, in service	Flatness can be measured at any time. Conventional flatness criteria are intended to be applied prior to shrinkage and curling.
Finishing to proper tolerances	General contractor	Concrete floor placing and finishing contractor	Hard concrete floor, day after placing	Placing and finishing must be evaluated prior to curling.
Consistent rate of concrete delivery	Concrete floor placing and finishing contractor	Concrete producer	At concrete delivery	Observed on site
Consistent timing and rate of concrete placement	Concrete floor placing and finishing contractor	Concrete producer, pumping contractor and the placing and finishing crew	As observed during placing	Observed on site
Consistent concrete finishability	Concrete floor placing and finishing contractor	Concrete producer	As observed during finishing	Somewhat subjective – influenced by crew, equipment and weather.
Consistent concrete bleeding	Concrete floor placing and finishing contractor	Concrete producer	As observed during finishing	Can be but is rarely measured.
Consistent concrete setting characteristics	Concrete floor placing and finishing contractor	Concrete producer	As observed during finishing	A test for concrete setting has not yet been standardised.
Cement with consistent setting behaviour	Concrete producer	Cement manufacturer	As evaluated at cement delivery to concrete batch plant	Is typically reported by cement producer but rarely measured later.
Cementitious materials with consistent setting behaviour	Concrete producer	Cementitious materials supplier	As evaluated at delivery to concrete batch plant	Can be but is rarely measured.
Chemical admixtures with consistent setting behaviour	Concrete producer	Admixture supplier	As evaluated at delivery to concrete batch plant	Can be but is rarely measured.

Aggregate with consistent grading	Concrete producer	Aggregate producer	As evaluated at aggregate delivery to concrete batch plant	Can be measured but rarely done during concrete production.
Concrete with controlled shrinkage (and related curling)	General contractor	Concrete producer	Prequalify materials, as evaluated prior to construction? Sample at time of construction to verify	ASTM C157 [21] for samples taken on site, but no reliable in situ test.
Contribution to shrinkage (and related curling) of the entire range of concrete ingredients from aggregates to admixtures	Concrete producer	Cement, cementitious materials, aggregate and admixture suppliers	Prequalify materials, as evaluated prior to construction? Sample at time of construction to verify	ASTM C157 [21] for samples taken on site, but no reliable in situ test.
Contribution to shrinkage and related curling from timing of finishing and curing, type and duration of curing, and job-site microclimate	General contractor	Concrete floor placing and finishing contractor	After construction but before service loading	ASTM C157 [21] for samples taken on site, but no reliable in situ test.
Influence of subgrade preparation specification and compliance, installation of reinforcing or dowels	Owner/specifier	General contractor and parties other than the concrete placing and finishing contractor or the concrete producer	Fundamental quality is defined prior to concrete placement, but its effect on floor performance is not evident until after construction	Compaction tests and inspection of reinforcing and dowels prior to concrete placement.
Accuracy of anticipated loads, floor thickness, joint spacing and detailing, and reinforcement	Owner	Designer/specifier	Fundamental quality is defined during design, but its effect on floor performance is not evident until after construction	The end results can be measured, but it can be complicated to attribute effects to specific sources.

From the owner's overall project perspective, there is no question that the most meaningful point of performance is the hardened concrete, in place, at an age that is indicative of the service-life capacity and longevity. However, the frequently acceptable state of the practice is to evaluate concrete properties as sampled at the point of discharge from the concrete truck, the point of placement or both. From the results of these tests, the in-place capacity and durability are projected or assumed based on known or implied relations. (In many cases the properties of in-place concrete can be measured using standard tests developed for that purpose, but such test programmes are not necessarily the norm.) Thus, one could consider a less comprehensive but more conventional performance specification that targets the performance of concrete as sampled at the time of casting. For example, the concrete strength value that is most meaningful in determining structural capacity is the in-place strength, as influenced by materials, mixing, consolidation, curing and in-place time–temperature history. However, most frequently this in-place value is inferred using standard cylinders sampled at the time of placement and cured under laboratory conditions. Thus, one could specify the strength of standard laboratory-cured cylinders as a performance criterion. Similarly, air content in fresh concrete (or even air bubble size in fresh concrete using an air-void analyser) may be a viable performance criterion. Even if a performance-minded specifier chose to ignore slump as a performance criterion, the concrete contractor might demand that the concrete arrives on site at a particular slump, thus reintroducing slump as a performance requirement for the concrete producer.

9.8. Potential performance, prequalification and identity testing

It is difficult, time-consuming and expensive to deal with hardened concrete that has been deemed to be unsatisfactory. This reality heavily influences decisions about how to specify and evaluate concrete, and how to control and ensure its quality. In contrast, consider a typical industrial example of manufacturing steel bolts for construction. The bolts can be made and tested in a factory and shipped only if found to be satisfactory. Alternatively, the bolts could be sampled and tested upon delivery on site and used only if proven to be satisfactory. If the bolts are no good, they can be scrapped or shipped back and replaced with a new batch. Even in the worst case, unsatisfactory bolts in place can still be removed and replaced without destroying the structure.

The quality control environment is far more complex for concrete compared to steel bolts. The properties of concrete will not have developed at the time of shipment, and in almost all cases, the concrete must be installed long before its properties can be reliably measured (even though it can be sampled before installation). Further, the installation process itself affects the concrete properties. Further still, if the concrete is found to be unsatisfactory, it is not a simple matter to replace it with a new batch. For this reason, to limit (but not eliminate) the chances of ending up with an unsatisfactory material in place, the industry has developed several intermediate checks, such as reviewing and approving proposed concrete mixtures, developing prescriptive specifications for raw materials and mix proportions, testing the temperature, slump and air content of fresh concrete, and accelerated strength tests.

With a performance specification, the owners and specifiers can minimise the chances of ending up with unsatisfactory, hardened concrete in place by first demanding evidence that the proposed concrete materials, mixture and methods have the potential to meet the specification requirements. It may, therefore, be necessary to 'prequalify' the materials or methods based on historical records of performance or by providing laboratory test data. It makes sense to approve the use of a 'known' successful combination of materials and construction technique, especially when a fresh set of tests to demonstrate performance requires more time than is normally available before the concrete is needed on site.

However, given the inherent variability in concrete materials, batching and mixing, even if a mixture has been prequalified, it will still be necessary to prove that the concrete actually delivered, placed and finished is, in fact, the same material that was demonstrated to have been satisfactory during the prequalification process. 'Identity testing' is the term used in BS 8500 [17] to describe such on-site testing to validate the identity of a mixture. Identity testing seeks to verify some key characteristic of the concrete that relates to the desired performance. It could take the form of typical slump, air and strength tests, a water-content test, or an air-void analysis of fresh concrete, and use a non-destructive or an in-place method.

For example, consider a performance requirement for an in-place value of 1000 coulombs in the ASTM C1202 – 'Electrical indication of concrete's ability to resist chloride ion penetration test' [21] (also known as the rapid chloride permeability test). Assume that a mix had been prequalified based on pre-construction C1202 testing. During actual construction, the challenge is to perform a suite of tests on concrete sampled at the time of placement that can be used to verify that the concrete as delivered is substantially the same as the concrete that had previously been shown to meet the 1000-coulomb requirement. This could be some combination of water content, fresh unit weight, compressive strength or field calorimetry, for example. There may also be periodic C1202 tests on samples as delivered, accelerated strength testing or tests on cores extracted from the structure. Further, differences in sampling, numbers of samples or test conditions can make it appropriate to set different acceptance criteria for in-place or job-site tests compared to controlled laboratory tests used for prequalifying a concrete mixture.

Although prequalification or pre-certification provides evidence to the owner/specifier that the producer/contractor *can* install a product that meets the performance specifications, it does not prove that such has actually been done. It is likely, therefore, that performance specifications will include requirements for pre-construction demonstrations of suitability that could range from the simple submittal of historical data as evidence of past performance all the way to the casting of demonstrations or sample slabs that could be evaluated by in-place methods. However, it is also likely that additional testing during or after construction would supplement such prequalification. This is because of the significant potential for batch-to-batch variation of the concrete due to variations in the raw materials and the fundamental difficulties in precisely controlling the quantity of water and air.

The Concrete Institute of Australia [22] has developed a comprehensive series of durability planning documents based on knowledge gained from large infrastructure projects. The series of seven recommended practices includes

Z7/01 Durability – Planning
Z7/02 Durability – Exposure classes
Z7/03 Durability – Deemed to comply requirements
Z7/04 Durability – Good practice through design, concrete supply and construction
Z7/05 Durability – Modelling
Z7/06 Durability – Cracks and crack control
Z7/07 Durability – Testing.

As indicated by these titles, these standard practices provide guidance for all stages of construction.

9.9. Concrete performance characteristics

One clear advantage of performance specifications is that they focus attention on the concrete properties that are the most important for a given situation. Conventional testing often concentrates on slump, air content and 28-day cylinder strength, even though one or more of these properties may not be relevant to the owner's desired performance. Further, more relevant performance requirements may not be tested at all.

As presented in the *Highway Materials Course Manual* produced by National Highway Institute [23], one way to look at a broader range of concrete properties is shown in Table 9.5, where concrete is evaluated as it transitions from the fresh to the hardened state. Although the properties of fresh concrete are rather conventional and the hardened concrete list is expanded well beyond the typical cylinder break, the transitional properties are frequently not specified but are nevertheless critical to the safe and economical progress of a concrete construction project. Note that within this list (which could be expanded), relatively few properties are commonly specified and tested even though owner satisfaction is commonly based on a far larger set of criteria.

The NRMCA-sponsored P2P contractors' joint task group developed a list of performance characteristics oriented around various applications. For example, concrete for exterior pavements should place easily, be finishable, have no pockmarks, set in a reasonable time, be resistant to freeze–thaw cycles and de-icer salt

Table 9.5 Concrete properties of interest

Fresh concrete	Transition	Hardened state
Workability	Rate of slump loss	Compressive strength (as specified and at ages of interest for construction scheduling)
Slump	Time to initial set	Tensile strength
Response to vibrator	Time to final set	Flexural strength
Pumpability	Rate of strength gain (compression)	Shear strength
Finishability	Rate of strength gain (tension)	Fatigue strength
Segregation	Rate of stiffness gain	Fracture toughness
Bleeding	Time to frost resistance	Elastic properties
Air content	Tolerable rate of evaporation	Shrinkage
Stability of air bubbles	Plastic shrinkage	Creep
Uniformity of mixing	Drying shrinkage	Porosity
Consistency of properties	Temperature changes	Pore size distribution
Temperature		Permeability
Yield		Air-void system
		Frost resistance
		Abrasion resistance
		Sulfate resistance
		Acid resistance
		Alkali resistance
		Thermal volume change
		Heat capacity
		Thermal conductivity
		Electrical conductivity
		Density
		Radiation absorption
		Colour
		Texture
		Cost

scaling, and have low permeability and low shrinkage. For tilt-up concrete construction, early-age flexural strength is critical along with setting time, consistent colour and low shrinkage. Tilt-up finishability was considered important, but not as critical as for a floor. Indoor slabs-on-grade were said to require low to minimum shrinkage, consistent set times, good finishability and good workability (in that order).

9.10. Exposures and exposure classes

One effective way of specifying concrete durability is to require that the concrete remain serviceable for a minimum period of time in a specified environment. Prescriptive specifications usually approach durability using a deemed-to-comply approach by requiring specific ingredients (such as fly ash or air-entraining admixtures), proportions (such as minimum cementitious materials content or maximum w/cm ratio) or construction operations (such as wet curing for a specified duration). Each of these factors is a means to an end where the required 'end' is durable in-place concrete. Conceptually, requiring that the concrete remain serviceable for a given period when exposed to a particular set of environmental conditions specifies the end result itself. (The term 'remain serviceable', however, would require further quantitative definition.) Nevertheless, clearly and unambiguously specifying the service environment that the concrete must endure puts all bidders on an equal footing regarding the expectations for durability, and it challenges each prospective supplier/contractor team to jointly figure out ways to economically blend concrete materials technology with construction practice to achieve the required endurance.

Standardised 'exposure classes' can be developed to serve as descriptions of common environmental exposures. This has been done effectively in many codes, including EN 206 [24], ACI 318 [19], Canadian A23.1 [7], Australian AS 3600 [25] and China's GB/T50476 [as cited in 26]. A review of durability exposure classes and requirements in these standards was provided by Hooton [27]. EN 206 specifically differentiates between exposures leading to reinforcement corrosion and exposures leading to deterioration of the concrete. The exposure classes in EN 206 are listed in Table 9.6.

Some 31 European countries cooperate in the development of CEN standards, and under this system, matters relating to durability remain under the authority of the individual CEN member countries so that provisions on this matter are given in national annexes. The UK durability recommendations are, for example, given in the complementary standard BS 8500-1 [17] for reinforced or prestressed concrete elements employing normal-weight concrete with a maximum aggregate size of 20 mm for an intended working life of at least 50 years and or at least 100 years. Note that there are some striking differences between the national recommendations of the individual CEN member countries. The only explanation is that the different national standardisation bodies have significantly different views on the concrete quality required to resist a particular exposure class for a given minimum cover and intended working life. There are clearly different views on how particular concrete compositions are likely to perform in the specified service environments or exposure classes [28].

Table 9.6 Exposure classes

Class	Description
X0	No risk (likelihood) of corrosion or attack
XC	Corrosion of reinforcement induced by carbonation
XD	Corrosion of reinforcement induced by chlorides other than from seawater
XS	Corrosion of reinforcement induced by chlorides from seawater
XF	Freeze–thaw attack on concrete with or without de-icing agents
XA	Chemical attack on concrete

Table 9.7 Minimum recommendations for 100-year design life for steel in concrete subject to carbonation-induced corrosion (BS 8500, 2020 [17])

Exposure class and description		Minimum strength class	Maximum w/c ratio	Minimum cement content: kg/m³	Minimum cover to steel: mm	Cement type
XC1	Dry or permanently wet	C20/25	0.70	240	15	CEM I
XC2	Wet, rarely dry	C25/30	0.65	260	25	CEM II
XC3	Moderate humidity	C40/50	0.45	340	30	CEM III (with maximum of 80% GGBS[a])
XC4	Cyclic wet and dry					CEM IV (SRPC[b])

[a]Ground granulated blastfurnace slag
[b]Sulfate-resistant Portland cement

The design recommendations in BS 8500 [17] for carbonation-induced corrosion are aimed at ensuring that the depth and quality of concrete cover are sufficient to achieve a sufficiently long initiation period (t_0), that is the depassivation of steel. BS 8500 gives combinations of required concrete quality and cover to steel for various exposure classes or conditions. The minimum values are summarised in Table 9.7.

For chloride-induced corrosion, the BS 8500 standard gives numerous combinations of required concrete quality and cover to steel for various exposure classes or conditions. These give design engineers some flexibility of choice for the combination for each exposure condition. Some of the minimum values are summarised in Tables 9.8 and 9.9. Different exposure classes apply when the corrosion is induced by chlorides from road de-icing salts (Table 9.8) to those that apply when the chlorides come from seawater (Table 9.9), but the requirements for the concrete quality and cover are largely similar for the equivalent exposure class. There are, however, circumstances in which protection against corrosion cannot be guaranteed by the right selection of materials, proportions of the concrete, depth of cover and attention to sound construction practice. These include, for example, marine exposure in extreme climatic conditions and regions in which aggregates containing excess chloride must be used. One or more of the following additional protective measures may then be taken

- adding a corrosion-inhibiting admixture, such as calcium nitrite, to the fresh concrete
- using corrosion-resistant stainless steel reinforcement bars or epoxy-coated conventional bars
- applying a protective coating to the concrete to reduce chloride or oxygen ingress
- applying cathodic protection to the reinforcement, that is, applying a voltage from an external source sufficient to ensure that all the steel remains permanently cathodic.

Some of the recommendations for concrete exposed to freeze–thaw action in the UK (which has a relatively mild freeze–thaw climate compared to many countries) are shown in Table 9.10. For each exposure class, it gives the option of air entrainment or a higher-quality concrete, and attention to the properties of the aggregate is required for the two most severe cases. The use of a de-icing agent, for example on exposed road surfaces, is included in the definition of the exposure class as it can result in distress from thermal shock when applied.

Table 9.8 Some minimum recommendations for 100-year design life for steel in concrete subject to corrosion induced by chlorides from road de-icing salts (BS 8500, 2020 [17])

Exposure class and description		Minimum strength class	Maximum w/c ratio	Minimum cement content: kg/m³	Minimum cover to steel: mm	Cement type
XD1	Moderate humidity	C45/55	0.40	380	30	CEM I CEM II CEM III with maximum 80% GGBS CEM IV SRPC
XD2	Wet, rarely dry	C35/45	0.45	360	40	CEM I
XD3	Cyclic wet and dry	C45/55	0.35	380	55	CEM II with up 20% fly ash or 35% GGBS SRPC
		C40/50	0.35	380	45	CEM II with 21–35% fly ash CEM III with 36–65% GGBS

EN 206 [24] classifies groundwater and soil as slightly, moderately or highly aggressive (denoted as exposure classes XA1, XA2 and XA3, respectively) according to the sulfate (SO_4^{2-}), carbon dioxide (CO_2), ammonium (NH_4^+) and magnesium (Mg^{2+}) contents and pH level, as shown in Table 9.11 for groundwater, thus encompassing sulfate and acid attack. The concrete requirements show that lower sulfate levels do not require any special considerations over and above those for all concrete construction. With increasing sulfate levels, combinations of sulfate resistance of the binder and overall quality of the concrete are required.

Table 9.9 Some minimum recommendations for 100-year design life for steel in concrete subject to corrosion induced by chlorides from seawater (BS 8500, 2020 [17])

Exposure class and description		Minimum strength class (Section 22.1 of BS 8500)	Maximum w/c ratio	Minimum cement content: kg/m³	Minimum cover to steel: mm	Cement type (Sections 13.8 and 15.4 of BS 8500)
XS1	Exposed to airborne salt but not in direct contact with seawater	C45/55	0.35	380	45	CEM I CEM II with up 20% fly ash or 35% GGBS SRPC
XS2	Permanently submerged	C35/45	0.45	360	40	
XS3	Tidal, splash and spray zones	C45/55	0.35	380	60	CEM II with 21–35% fly ash
		C40/50	0.35	380	45	CEM III with 36–65% GGBS

Table 9.10 Recommendations for 100-year design life for concrete exposed to freeze–thaw attack (BS 8500 [17])

Exposure class and description		Minimum strength class (Section 22.1 of BS 8500)	Maximum w/c ratio	Minimum air content: %	Minimum cement content: kg/m³	Aggregates	Cement types (Sections 13.8 and 15.4 of BS 8500)
XF1	Moderate water saturation, without de-icing agent	C25/35 C28/35	0.6 0.6	3.5 –	280 280		CEM I CEM II CEM III with maximum 80% GGBS SRPC
XF2	Moderate water saturation, with de-icing agent	C25/35 C32/40	0.6 0.55	3.5 –	280 300		
XF3	High water saturation, without de-icing agent	C25/35 C40/50	0.6 0.45	3.5 –	280 340	Freeze–thaw-resistant aggregates	
XF4	High water saturation, with de-icing agent or seawater	C28/35 C40/50	0.55 0.45	3.5 –	300 340		

The European standardisation committees for the design of concrete structures and concrete materials have agreed in principle to move to specifying durability by performance in standard tests. Turning this principle into normative requirements is another matter, and it is accepted that in the next revision of the European standards, this objective can be achieved only for resistance to reinforcement corrosion due to a lack of agreed European tests with the status of a full standard for the other aspects of durability. The European concrete committees have also agreed that there is a need to retain a limiting value, for example maximum w/c ratio, and other requirements, for example cement types, for concretes made with well-established constituents.

Table 9.11 BS EN 206 Classification of groundwater for concrete exposure classes

Property	XA1	XA2	XA3
Groundwater composition			
SO_4^{2-}: mg/l	200–600	600–3000	3000–6000
pH	5.5–6.5	4.5–5.5	4.0–4.5
CO_2: mg/l	15–40	40–100	>100
NH_4^+: mg/l	15–30	30–60	60–100
Mg^{2+}: mg/l	300–1000	1000–3000	>3000
Concrete requirements			
Minimum strength class (Section 22.1.1 of BS 8500)	C30/37	C30/37	C35/45
Maximum w/c ratio	0.55	0.50	0.45
Minimum cement content	300	320	360
Cement type	Any	SRPC or CEM I + an addition with the equivalent performance	

If consensus can be reached, this more limited objective will be a major achievement and it will open the door to more sustainable concrete with proven adequate durability. Substantial progress has been made with many aspects of this development, but in a phrase familiar to most Europeans: 'Nothing is agreed until everything is agreed' (personal communication, Thomas Harrison).

Note that until the 2008 revision of ACI 318 [19], many of these exposures were addressed in various small tables rather than in one table of designated exposure classes, which would have made them more prominent. Due to this lack of visibility, it was rare when a specification based on ACI 318-05 deliberately and clearly pointed out the required exposures. In the 2008 revision, based on the recommendation of the authors, exposure classes – as in most other international codes – were finally adopted.

ACI 318 [19] exposure classes and requirements for each of the major types of exposure are included in Tables 9.12 and 9.13 together with those in the EN 206, Canadian A23.1, Australian AS3600 and Chinese GB T50476 standards. Note that prescriptive minimum cement contents exist in EN 206 (and in the complementary BS 8500 as shown in Tables 9.12 to 9.18) and in the Chinese standard. As well, there are maximum limits on the w/c or w/cm ratio in all the standards except AS 3600. Despite widespread global issues with reinforcement corrosion, other than compressive strength, it is surprising that the only performance requirements for chloride and marine exposures are the ASTM C1202 – 'Electrical indication of concrete's ability to resist chloride ion penetration test' [21] as well as coulomb limits in CSA A23.1 [7], NT Build 492 [29] and the Chinese GB T50476 [as cited in 26]. In part, this may be related to the relatively poor precision of potential tests for measuring resistance to fluid ingress, as shown in Table 9.18.

9.11. Prescription and performance elements in the ACI 318 building code

9.11.1 Durability requirements

The ACI 318 code [19] contains elements of both prescriptive and performance specifications. Chapter 19 of that code contains tables of durability requirements in terms of maximum w/cm ratio, total air content, limitations on supplementary cementitious materials (SCMs), limits on chloride content and requirements for cement types, each as a function of each exposure class. Tables listing exposure classes and requirements and explicitly requiring that exposures be defined in the drawings and specifications were included for the first time only in the 2008 edition of ACI 318. Prior to this, there was confusion about whose responsibility it was to determine durability requirements. In some cases, project specifications imply that it is the contractor's responsibility to consult the code and to determine the appropriate concrete mixture requirements. This approach is unambiguous only when the specifier explicitly defines the expected exposure conditions.

While maximum w/cm ratios, minimum air contents, limitations on SCMs, limits on chloride content and requirements for cement types are clearly prescriptive in nature, some flexibility for materials selection and proportioning remains, in that there are no maximum or minimum limits on the total weight of cement or SCMs nor are there maximum or minimum water contents. The code imposes no requirement on aggregate content except that which derives from the limits on the nominal maximum size of coarse aggregate and good mixture proportioning practice. For resistance to freezing and thawing cycles, even air content is given as a function of coarse aggregate size, which is an approximate (if indirect) way to account for the need for more air as paste content increases. (The increased surface area of smaller aggregates demands more paste, thus a higher air content is required for mixtures with smaller aggregates.)

9.11.1.1 W/cm limits

As discussed earlier, prescriptive requirements for the w/c ratio were incorporated in codes as far back as 1928, and a prescriptive option for the w/c ratio based on strength was built into every code until 1986. (Modern versions of the code use instead the w/cm ratio.) The prescriptive association between w/c (w/cm)

Table 9.12 De-icer and chloride soils exposures and requirements [after 27]

Standard	Exposure class	Exposure	Minimum strength (28 days unless noted): MPa	Maximum w / (c + k (SCM)) or w / (c + SCM)	Minimum cement content: kg/m³	Minimum curing period: days	Minimum cover depth: mm	Air content: %	ASTM C1202 maximum: coulombs	NT 492 maximum: 10^{-12} m²/s
EN 206	XD1	Moderate humidity	C30/C37	0.55[b]	300[c]	[d]	35[a]			
	XD2	Wet, rarely dry	C35/C45	0.55[b]	300	[d]	40[d]			
	XD3	Cyclic wet and dry	C35/C45	0.45[b]	320	[d]	45[d]			
ACI 318	C2	All chloride exposures	31	0.40		7	75			
CSA A23.1	CXL	C1 with extended service life	50 at 56 days	0.40		7	75	5–8	1000 at 91 days	
	C1	Structurally reinforced	35 at 56 days	0.40		7	75	5–8	1500 at 91 days	
	C2	Non-structurally reinforced	32	0.45		7	75	5–8		
	C4	Non-structurally reinforced and not exposed to freezing	25	0.55		3	75			
AS 3600	A2	Saline soils with conductivity 4–8 decisiemens/m	25			3	45			
	B1	Saline soils with conductivity 8–16 decisiemens/m	32			7	50			
	B2	Saline soils with conductivity >16 decisiemens/m	40			7	55			
GB/T50476	IV-C	De-icing spray	35	0.50	300	7	35	4–6		–
(China)	IV-D	Chloride groundwater	40	0.45	320	7	45	4–6		10
50-year design life	IV-E	Directly exposed to de-icers	45	0.40	340	7	50	4–6		6

[a]EN 1992-1-1 cover depth.
[b]Varies in national application rules: XD1 0.45–0.60, XD2 0.40–0.55 and XD3 from 0.40–0.50.
[c]Varies in national application rules: XD1 and XD2 200–360 and XD3 200–400.
[d]EN 13670 has four levels of curing of 12 hours or the time to achieve 35, 50 or 70% of the specified characteristic strength.

Table 9.13 Marine exposures and requirements [after 27]

Standard	Exposure class	Exposure	Minimum strength: MPa	Maximum w / (c + k (SCM)) or w / (c + SCM)	Minimum cement content: kg/m³	Maximum C_3A of Portland cement: %	Minimum curing period: days	Minimum cover depth: mm	ASTM C1202 maximum: coulombs	NT Build 492 maximum: 10^{-12} m²/s
EN 206	XS1	Airborne spray	C30/C37 at 28 days	0.50	300		[b]	35[c]		
	XS2	Permanently submerged	C35/C45 at 28 days	0.45	320		[b]	40[c]		
	XS3	Tidal, splash and spray	C35/C45 at 28 days	0.45	340		[b]	45[c]		
ACI 318	C2 + S1	All marine exposures	31 at 28 days	0.40		10	7	75		
CSA A23.1	CXL + S3	Tidal and splash zones with extended service life	50 at 56 days	0.40		Minimum 4 to maximum 10[a]	7	75	1000 at 91 days	
	C1 + S3	Tidal and splash zones	35 at 56 days	0.40		Minimum 4 to maximum 10[a]	7	75	1500 at 91 days	
	C3 + S3	Permanently submerged	30	0.50		Minimum 4 to maximum 10[a]	3	75		
AS3600	B1	Near coastal (1–50 km)	32				7	40		
	B2	Permanently submerged	40				7	45		
	C1	Airborne spray	50				7	50		
	C2	Tidal and splash zones	50				7	65		
GB/T50476 (China) 50-year design life	III-C	Immersion in seawater	35	0.50	300		7	4–6		–
	III-D	Seawater spray	40	0.45	320		7	4–6		10
	III-E	Tidal and splash zone, mild climate	45	0.40	340		7	4–6		6
	III-F	Tidal and splash zone, hot climate	50	0.36	360		7	4–6		6

[a] Does not apply if blended cements or SCMs are used.
[b] EN 13670 has four levels of curing in increments of 12 hours or the time to achieve 35, 50 or 70% of the specified characteristic strength.
[c] EN 1992-1-1: for 50 years, but for 100 working life, increase cover by 10 mm.

Table 9.14 Carbonation exposures and requirements [after 27]

Standard	Exposure class	Exposure	Minimum 28 day strength: MPa	Maximum w / (c + k (SCM)) or w / (c + SCM)	Minimum cement content: kg/m³	Minimum curing period: days	Minimum cover depth: mm
EN 206	XC1	Dry or permanently wet	C20/25	0.65	260	[a]	25[c]
	XC2	Wet, rarely dry	C25/C30	0.60	280	[a]	35[c]
	XC3	Moderate humidity	C30/C37	0.55	280	[a]	35[c]
	XC4	Cyclic wet and dry	C30/C37	0.50	300	[a]	40[c]
ACI 318	Not covered	If also exposed to freezing and unsaturated	27.6	0.55	–	7	50
CSA A23.1	Not covered	If also exposed to freezing and unsaturated	25	0.55	–	3	40[e]
AS3600[b]	A2	Moderate humidity	25	–	–	3	45
	B1	Tropical	32	–	–	7	40
	B1	Cyclic wet and dry	32	–	–	7	40
GB/T50476 (China) 50-year design life	I-A	Indoor, dry <60% relative humidity	25	0.60	260	3	20
	I-B	Sheltered, outdoors or indoors, humid	30	0.55	280	3	20
	I-C	Outdoors, exposed to rain	35	0.50	300	3	30[d]

[a]EN13670 has four levels of curing of 12 hours or the time to achieve 35, 50 or 70% of the specified characteristic strength.
[b]Carbonation is not mentioned directly.
[c]EN 1992-1-1 for structural class S4.
[d]Cover can be reduced by 5–10 mm if strength increased to 40–45 MPa and w/c ratio lowered to 0.45–0.40.
[e]40 mm if exposed to freezing. 60 mm if exposed to chlorides.

and strength was dropped in 1989, but the more performance-oriented approval of mixtures based on strength test results plotted as a function of w/c ratio (the three-point curve) remained in the ACI 318 code in 2005 but has since been moved to ACI 301-20 – 'Specifications for concrete construction' [37]. Although the w/c ratio had been limited by the code for freeze–thaw durability since 1947 (6 gallons of water and 94 lb sack of Portland cement to give w/c = 0.53), the current more comprehensive tables describe maximum w/cm limits and minimum $f'c$ values for permeability control, de-icer salt scaling resistance and corrosion protection. This table first appeared in 1989. Further, the table of 'Requirements for concrete exposed to sulfate-containing solutions' includes maximum limits on the w/cm ratio depending on sulfate exposure.

The concept behind these tables is fundamentally based on the behaviour of Portland cement concrete. The porosity and permeability of hardened cement paste is intimately connected to the volume of mix water, as much of the volume initially occupied by mix water in the fresh paste remains as pore space in the hardened paste. A more porous paste implies a more porous mortar, and a more porous mortar implies a more porous concrete. This was conclusively demonstrated in many tests ranging from pioneering work on pastes by Powers *et al.* [38] to the classic permeability studies on mortars and concretes conducted by the Bureau of

Table 9.15 Freezing and thawing exposures and requirements [after 27]

Standard	Exposure class	Exposure	Minimum strength (28 days unless noted): MPa	Maximum w / (c + k (SCM)) or w / (c + SCM)	Minimum cement content: kg/m³	Minimum curing period: days	Air content: %
EN 206	XF1[b]	Moderate saturation without de-icers	C30/C37	0.55[c]	300[c]	[a]	
	XF2[b]	Moderate saturation with de-icers	C25/C30	0.55	300	[a]	4.0
	XF3[b]	Highly saturated without de-icers	C30/C37	0.45	320	[a]	4.0
	XF4[b]	Highly saturated with de-icers or seawater	C30/C37	0.45	340	[a]	4.0
ACI 318	F1[b]	Freezing cycles with limited wetting	27.6	0.55		7	3.5–6.5[d]
	F2[b]	Freezing cycles with frequent wetting	31	0.45		7	4.5–7.5[d]
	F3[b]	Freezing cycles and de-icer salts	34.5	0.40		7	
CSA A23.1	F1[b]	Freezing cycles while saturated	50 at 56 days	0.50		7	5–8[d]
	F2[b]	Freezing cycles while unsaturated (If exposed to salts, then also a C class)	35 at 56 days	0.55		3	4–7[d]
AS3600		Occasional exposure to freezing cycles (<25 cycles/year)	32			3	4–8[d]
		Frequent exposure to freezing cycles	40			3	4–8[d]
GB/T50476 (China) 50-year design life	II-C	Moderate climate, high saturation	40	0.45	320	[f]	4.5–6
	II-D	Cold region, high saturation[e]	35	0.50	280	[f]	4.5–6
	II-E	Cold region, high saturation + salt	40	0.45	280	[f]	4.5–6

[a]EN 13670 has four levels of curing of 12 hours or the time to achieve 35, 50 or 70% of the specified characteristic strength.
[b]Also requires frost-resistant aggregates.
[c]National application rules only require 280 kg/m³ and 0.6 w/c in Germany, France and UK, but 320 kg/m³ and 0.50 w/c in Italy.
[d]Values shown for 20-mm maximum aggregate.
[e]Cold climate defined as having 1 month with average temperature < -3°C.
[f]After curing, concrete must not be exposed to freezing for 1 month.

Table 9.16 Sulfate exposures and requirements [after 27]

Standard	Exposure class	$[SO_4]$ in groundwater: mg/l	$[SO_4]$ in soil: mg/kg	Minimum strength (28 days unless noted): MPa	Maximum w / (c + k (SCM)) or w / (c + SCM)	Minimum cemert content: kg/m^3	Maximum C_3A of Portland cement: %	ASTM C452 14 day mortar bar expansion[f]: %	ASTM C1012 mortar bar expansion[g]: %	AS2350–14 mortar bar expansion[d]: %	Minimum curing period: days	Minimum cover depth: mm
EN 206	XA1[b]	200–600	2000–3000	C30/C37	0.55	300	5				[e]	[h]
	XA2[b]	600–3000	3000–12 000	C30/C37	0.50	320	3				[e]	[h]
	XA3[b]	3000–6000	12 000–24 000	C35/C45	0.45	360	0				[e]	[h]
ACI 318 2019	S1	150–1500	1000–2000	27.6	0.50		8		0.10 at 6 months		7	75
	S2	1500–10 000	2000–20 000	31	0.45		5	0.040[a]	0.05 at 6 months[c]		7	75
	S3 option 1	>10 000	>20 000	31	0.45		5	0.040[a]	0.10 at 18 months		7	75
	S3 option 2	>10 000	>20 000	34.5	0.40		5	0.040	0.05 at 6 months[c]		7	75
CSA A23.1	S3	150–1500	1000–2000	30 at 56 days	0.50		8	0.050	0.10 at 6 months		3	75
	S2	1500–10 000	2000–20 000	32 at 56 days	0.45		5	0.035	0.05 at 6 months[c]		7	75
	S1	>10 000	>20 000	35 at 56 days	0.40		5	0.035	0.05 at 6 months[c]		7	75

AS3600	A2[b]	<1000	<5000	25				3	30
	B1[b]	1000–3000	5000–10 000	32				7	40
	B2[b]	3000–10 000	10 000–20 000	40			0.075 at 16 weeks	7	45
	C2[b]	>10 000	>20 000	50			0.075 at 16 weeks	7	55
GB/T50476 (China) 50-year design life	V-C	200–1000+	350–1500+	40	0.45	320			
	V-D	1000–4000+	1500–6000+	45	0.40	340			
	V-E	4000–10 000+	6000–15 000+	50	0.36	360			

[a]Optional
[b]Can be relaxed by one class if in low-permeability clayey soils.
[c]Can be exceeded if <0.10% at 12 m.
[d]15 × 40 × 160 mm bars cured for 7 days then in ASTM C1012 sulfate solution.
[e]EN 13670 has four levels of curing of 12 hours or the time to achieve 35, 50 or 70% of the specified characteristic strength.
[f]Required only for Portland cements when used as sole binder.
[g]Required only for testing blended cements or Portland/SCM combinations.
[h]Cover depths not covered by EN 1992-1-1.

Table 9.17 Acid exposures and requirements [after 27]

Standard	Exposure class	Exposure	Minimum strength (28 days unless noted): MPa	Maximum w / (c + k (SCM)) or w / (c + SCM)	Minimum cement content: kg/m³	Minimum curing period: days	Minimum cover depth: mm	Air content[b]: %	ASTM C1202 maximum: coulombs
EN206	XD1	Moderate humidity	C30/C37	0.55	300	[a]	45[c]		
	XD2	Wet, rarely dry	C35/C45	0.55	300	[a]	50[c]		
	XD3	Cyclic wet and dry	C35/C45	0.45	320	[a]	55[c]		
ACI 350		Exposed to water, waste water or corrosive gases	27.6	0.45	333	7	50 for liquids or 75 against ground	4.5–7.5	
		Exposed to corrosive chemicals	31	0.42	333	7	50 or 75	4.5–7.5	
		Tanks exposed to chlorides	34.5	0.40	333	7	50 or 75	4.5–7.5	
CSA A23.1	AXL	A1 with extended service life	50 at 56 days	0.40		7	75	5–8	1000 at 91 days
	A1	Structurally reinforced concrete exposed to severe agricultural and acid wastes with H_2S gas	35 at 56 days	0.40		7	75	5–8	1500 at 91 days
	A2	Structurally reinforced concrete exposed to moderate to severe agricultural and acid wastes	32	0.45		7	75	5–8	
	A3	Structurally reinforced concrete fully submersed in moderate to severe agricultural and acid wastes	30	0.50		3	75		
	A4	Non-structurally reinforced, exposed to moderate agricultural wastes	25	0.55		3	75		

AS3600	A2	pH >5.5	25			3	45
	B1	pH 4.5–5.5	32			7	50
	B2	pH 4.0–4.5	40			7	55
	C2	pH <4.0	50			7	65
GB/ T50476 (China) 50-year design life	V-C	pH >6.5	40	0.45	320		
	V-D	pH 4.5–5.5	45	0.40	340		
	V-E	pH <4.5	50	0.36	360		

[a]EN 13670 provides ranges of curing durations at different temperatures to obtain either 35, 50 or 70% of characteristic strength.
[b]If exposed to freezing. Values given for 20-mm maximum aggregate.
[c]EN 1992-1-1

Table 9.18 Coefficients of variation (COVs) for various standard fluid transport test methods

Test method	Average COV (within laboratory)	Average COV (between laboratories)	Source of data
Electrical charge (coulombs) AASHTO T277 [35], ASTM C1202, CSA A23.2-23C	12%	18%	Reported in ASTM C1202 [21]
Non-steady-state chloride migration, NT Build 492 [29]	15.2%	23.6%	Tang *et al.* [30]
Steady-state migration, NT Build 355 [31] and [32]	22%	76%	[31,32]
Bulk resistivity AASHTO TP 119	11%	25%	Reported in AASHTO TP 119 [34]
CSA A23.2-26C ASTM C1876	4.3% 4.3%	13.2% 13.2%	Reported in ASTM C1876 [21]
Bulk chloride diffusion, NT Build 443, ASTM C1556	Da 20.1% (apparent diffusion) Cs 13.3% (surface concentration)	Da 28.3% Cs 18.1%	Tang *et al.* [30]
AASHTO T358 Wenner probe, surface resistivity	6.3%	12.5%	AASHTO T358 [33]
Air-permeability method, Swiss standard SIA 262/1-E	3.8%	25.7% (based on tests by 5 labs)	Moro and Torrent [36]
Rate of water absorption, ASTM C1585	6.0%	Not yet determined	Reported in ASTM C1585 [21]

Reclamation in association with dam construction in the western US [39]. The conclusion has always been the same: for any given mixture, permeability decreases as the w/c (or w/cm) ratio decreases. However, it is not true that permeability is uniquely or absolutely defined by the w/c or w/cm ratio across all mixtures when aggregate size and content, total paste content, total water content, paste composition (different types of cementitious materials), age or method of test are allowed to vary.

The high-pressure permeability tests conducted by Ruettgers *et al.* in 1935 [40] showed that pastes are far more permeable than mortars or concretes at the same w/c ratio. For concrete at any given w/c ratio, the permeability can vary by a factor of 10 to 100 as other mixture ingredients change. (The scatter was considerably narrower when the results were recomputed as permeability per lb of cement per ft^3.) Nevertheless, these results were based on Portland cement as the only binder and, therefore, do not begin to reflect the differing capacity of various SCMs to affect permeability. Further, if the issue is in-place permeability, the significant influences of consolidation, finishing and curing need also to be considered.

It is clear, then, that the w/cm values in the tables do not define a particular level of *concrete* permeability (although for a pure cement binder, the limiting values may come closer to defining a level of *cement paste* permeability). It may be that these requirements evolve (or devolve, depending on the point of view) along the lines of ACI's older default requirements for a w/c ratio based on a specified strength, which were used only when no other data on strength plotted against w/c ratio were available. Prescriptive w/cm values could be overridden on the basis of permeability index test data, and proposals along these lines are currently before ACI 318 subcommittee 318A. If a permeability index value were specified (or some similar property that relates to the transport of fluids and dissolved solids through hardened concrete), a value of the w/cm ratio could be determined that would meet that specified requirement for a given set of concrete materials.

Allowing for an increase in the limiting w/cm values when supported by test data can be advantageous to all parties. This is because any given mixture requires a basic water content to achieve the necessary workability (pumpability, compactability and finishability) and the requisite total cementitious materials content is determined by dividing water content by the w/cm ratio. For any given level of workability and water demand, the lower the w/cm ratio, the higher the total cementitious material content and the greater the paste content. More paste requires more air and as paste content goes up, total aggregate content must go down. Higher paste and lower aggregate generally lead to increased shrinkage and creep, and the combination of high paste content and a low w/cm ration increases the risk of plastic shrinkage cracking for any given rate of evaporation. Increased paste content also increases the total heat of hydration so that there is a greater temperature rise and a risk of thermal cracking and strength reduction. Controlling these heat effects requires changing the blend of cementitious materials or using other mixture- and construction-related cooling techniques. Thus, requiring a w/cm ratio that is lower than it needs to be to develop the desired permeability can increase the possibility that the concrete will crack, if the paste content is not limited. Specifying a minimum cement content that is higher than needed to meet strength or durability criteria can have the same deleterious effect.

As a final comment on the w/cm ratio, the ACI 318 tables also require a minimum $f'c$ together with each maximum value of the w/cm ratio. As stated in the code commentary, the minimum strength requirements 'will ensure the use of a high quality cement paste' and help to guard against mismatched specifications such as 'Maximum w/cm shall be 0.40, and *f'c shall be 3000 psi* [20.6 MPa].' The strength requirements are a pragmatic recognition that since the w/cm ratio and permeability are not normally evaluated, strength is often the only indicator of concrete quality. Concrete that truly has a w/cm ratio of about 0.40 is likely to have a 28-day compressive strength well in excess of 5000 psi (35 MPa). Exceptions are possible, of course, given the wide range of materials available, but the current code has no specific provisions for accommodating such exceptions. If a clearly defined end performance requirement were available and specified in lieu of the w/cm ratio, perhaps the strength associated with that characteristic could be determined for a specific mixture.

9.11.1.2 Freeze–thaw durability and scaling resistance

The prescriptive requirements for the w/cm ratio when the concrete is to be exposed to freezing and thawing while in moist condition and for de-icer salt scaling were addressed in Section 9.10, and the total air content requirements were introduced in Table 9.15. Note that ACI 318 has limits only for total fresh air content 'as delivered', with no further distinction between so-called entrained and entrapped air. ACI 318 makes no mention of air bubble size in the fresh concrete or air content, air-void size or air-void spacing in the hardened concrete.

If total air content is seen as a durability characteristic by itself, then one could consider the values in the ACI 318 tables to be performance criteria. However, if the total air content is seen as one of the factors leading to freeze–thaw resistance, with the needed air content varying with paste content and air-void size, then the table limits are prescriptive. Of course, if freeze–thaw resistance is the desired end result, then demonstrated performance in a freeze–thaw test would be the most useful measure, and it is entirely possible that fully successful performance could be achieved in some mixtures at total air contents significantly lower than those required by the table. (This can be the case with low paste content and a stable system made up predominantly of microscopically small air voids.) On the other hand, when the air voids are predominantly large, it is possible that the total air content required by the code will not necessarily lead to the desired durability. The code-mandated total air content is, therefore, only part of the story, and either freeze–thaw tests or air-void system analyses are needed to increase confidence in (but still not guarantee) freeze–thaw durability. This is why using an air-void analyser [41] to test fresh concrete and ASTM C457 – 'Standard test method for microscopical determination of parameters of the air-void system in hardened concrete' [21] are of great interest, especially if specifications move from the prescriptive air content to a more performance-oriented criterion. Also note that since freeze–thaw durability and de-icer salt scaling resistance are dependent on both materials and construction procedures (such as finishing and curing), prequalification of a concrete mixture based on freeze–thaw or scaling tests demonstrates only the potential of the material to achieve the required durability [42]. It may be necessary to demonstrate this potential, however, for certain combinations of cementitious materials and admixtures as proof of compatibility, especially when the producer wants to prove that durability can be achieved at air contents or spacing factors that are outside the values recommended for conventional mixtures. However, if standard freeze–thaw or scaling tests are being contemplated for prequalification testing, the multi-month duration of these procedures must be kept in mind.

If in-place freeze–thaw resistance were the objective, there would be little question that the most meaningful point of sampling would be after all of handling, pumping, squeezing, dropping, pressurising, depressurising, consolidating, finishing and curing in place. Although current code provisions can be interpreted to imply testing fresh concrete at the truck chute, sampling at the point of placement is increasingly common, yet there are no clearly required values for total air content at the point of placement (nor are there standard methods for collecting a sample of fresh concrete at that point). Some specifiers automatically invoke the ACI 318 air content values at placement, which can be too conservative when the air bubbles remaining after handling and consolidation are predominantly small. This leads to the need to batch the concrete at higher than normal air contents to accommodate the normal losses that occur during handling. This can, in turn, lead to significant drops in strength. This confusion is avoided if in-place hardened air-void system criteria or freeze–thaw testing are specified as performance requirements in lieu of reliance on total air. Related discussions can be found in [1].

Table 26.4.2.2(b) in ACI 318-19 [19] sets limits on the maximum percentage of total cementitious materials by weight for fly ash, slag, silica fume and other pozzolans for concrete exposed to cycles of freezing and thawing and application of de-icing chemicals (although, in 2022, there was a ballot to potentially remove this table from the next edition). Under a more performance-oriented approach, a concrete producer might be permitted to demonstrate that resistance to freeze–thaw damage and de-icer scaling can be achieved with a particular combination of cementitious materials when finished and cured, as the contractor intends. With an appropriate lead time, this might be an opportunity for effective prequalification with actual freeze–thaw or scaling tests. However, scaling tests need to relate to real-life exposure conditions, and ASTM C672 – 'Standard test method for scaling resistance of concrete surfaces exposed to deicing chemicals' [21] is known to be overly harsh for concretes containing fly ash or slag, partly due to the inadequate time for the concrete to achieve its potential before the initiation of freeze–thaw cycles and in part due to different setting times (which affect the finishing time) [43]. The Canadian scaling test, CSA A23.2-22C – 'Scaling resistance of concrete surfaces exposed to deicing chemicals using mass loss' [7], which eliminated final finishing and also allows the salt solution to be ponded for 7 days prior to initiating freeze–thaw cycles, appears to be better for field performance [44,45].

9.11.1.3 Corrosion protection

The first line of defence against the corrosion of embedded metals is to inhibit the penetration of water, oxygen, carbon dioxide and salts from the concrete surface to the level of the embedded metal. The ACI 318 code [19] sets minimum requirements for the depth of cover with the requirement that in a corrosive environment, the concrete protection shall be 'suitably increased, and denseness and non-porosity of the protecting concrete shall be considered'. This provides for the interesting interplay of a structural design feature (bar cover) and the material property of the concrete. Given that a deeper bar cover is often associated with wider cracks at the concrete surface, overall performance of the completed structure might be improved by coordinating the bar cover and the material properties of the concrete.

The permeability-related properties of concrete have already been addressed regarding w/cm ratio requirements, but it may be reiterated that including SCMs, such as fly ash, slag, silica fume or other pozzolans, can be an effective way of decreasing the permeability of the concrete. It is expected that within a given set of materials and proportions, permeability will decrease with the w/cm ratio. Across a wide range of material combinations and proportions, the w/cm ratio that leads to a particular and desired permeability is expected to vary. Adjusting the code limit on the w/cm ratio based on mixture composition is not currently permitted within the provisions of ACI 318.

To reduce the amount of potentially corrosive chloride in concrete, the code sets limits on the 'maximum water-soluble chloride ion concentration'. As discussed in more detail [1,2], these limits vary with exposure condition and between reinforced and prestressed concrete, but there are no options for differences in concrete composition, permeability or w/cm ratio.

9.11.1.4 Sulfate durability

In addition to the w/cm limits mentioned earlier, minimum strengths and requirements for allowable types of cementing materials are included in the table requirements. These requirements are based on the nature and severity of the sulfate exposure. Limits on w/cm ratio and associated values of $f'c$ are intended to limit permeability to reduce the ingress of sulfates, and the earlier comments about a mix-specific correlation among w/cm ratio, strength and permeability apply. Given the wide range of cementitious materials available and the wide range of types of sulfate exposure and sulfate concentrations, it is difficult to incorporate a one-size-fits-all approach in the code, but it is equally difficult to allow multiple exceptions and adjustments in the absence of performance tests that would demonstrate the sulfate resistance of a given mixture.

9.11.2 Concrete not exposed to aggressive conditions

Importantly, if durability is not a concern, the code imposes no prescriptive limits of any kind on the w/c ratio, the w/cm ratio, air, cement content or percentages of SCMs. Thus, ACI 318 permits performance specifications for concrete that will not be exposed to aggressive conditions. Except for an internal chloride limit of 1.0%, the ACI 318 code imposes no prescriptive limits in the absence of freezing and thawing or risk of corrosion or sulfate attack, unless the finished structure is intended to have a low permeability.

9.11.3 Strength requirements

As is common in any rational quality control system, the code provisions recognise that accepting concrete whose average strength equals the specified strength is unreasonable, since about half of the concrete accepted would have below-average strength and therefore below-specified strength. The code provisions also recognise that rejecting all concrete with a strength lower than a specified 'minimum' is just as unreasonable, as the cost and frustration associated with a zero-tolerance policy would be prohibitive. The code, therefore, requires that concrete mixtures be selected that demonstrate at least a 99% chance of meeting the two principal strength requirements as per section 26.12.3.1(a) of ACI 318-19 [19].

(1) Every average of any three consecutive strength tests must equal or exceed $f'c$.
(2) No strength test can fall below $f'c$ by more than 500 psi if $f'c \leq 5000$ psi or by more than $0.10f'c$ if $f'c > 5000$ psi.

Working backward from these reasonable requirements, any concrete mixture that meets both of the acceptance criteria 99% of the time will have an acceptably predictable average strength that will always be higher than the specified strength, so here the code imposes a strictly performance-oriented requirement. The amount by which the required average concrete strength must be higher than the specified strength depends on the level of precision of the concrete producer/test laboratory team. (This level of precision is indicated by the standard deviation.) The lower the demonstrated variability in cylinder test results (lower standard deviation), the lower is the required difference between the specified and average strengths.

These provisions contain no arbitrary 'safety factors' and merely express the reality of everyday 'normal' variability. Unfortunately, ACI uses the misleading term 'overdesign' to refer to the difference between the specified and required average strengths. Thus, a fully rational procedure that incorporates a tolerance for occasional low breaks, provides the owner with 99% confidence regardless of producer and is self-adjusting across concrete producers with varying levels of sophistication and quality control is made to look like an arbitrary strength requirement above and beyond that which is really needed.

There is a common and perhaps dangerous misconception that the code requirements for concrete strength 'overdesign' are an extra layer of conservatism in addition to other factors of safety applied in structural design. In actuality, the code has an integrated, three-part approach to establishing structural reliability. First, 'load factors' take into account the variability of service loads and the likelihood that the structure will experience an overload during its service life. The values assigned to these load factors take the nature and predictability of various loads and load combinations into account. Second, the code applies a strength-reduction factor, which accounts for the difference between the computed and actual strengths of a member, based on factors such as variable dimensions and rebar placing tolerances, and reflects the nature and consequences of structural failure. Third, the variability in the strength of the concrete itself is accounted for only by the statistical quality control requirements on concrete strength. The load and strength-reduction factors are based on the assumption that the concrete strength will meet the two strength acceptance requirements 99% of the time. Thus, the code provides a comprehensive and interdependent approach to reliability: load factors, strength-reduction factors and the statistical quality control requirements for concrete strength. The load and strength-reduction factors are not intended to make up for a shortfall in concrete quality. Once the structure is designed, any concrete that fails to meet the code requirements for strength will reduce the designed load-carrying capacity. Thus, an investigation is required if concrete strength test results drop below currently specified values.

These strength requirements began to take their current format with the introduction of the 1971 edition of the ACI 318 code, but the bottom line is that for most concrete producers and labs, compliance with the code requirements for compressive strength requires that the average concrete strength will be about 10% to 15% greater than the specified strength. The 1928 ACI code required 15% greater than $f'c$, and this requirement was carried all the way to the 1963 code, which immediately preceded the 1971 code, at which time today's slightly more rigorous statistical approach appeared.

A final comment on strength is that a fully performance-based, in-place strength requirement is embedded in Section 26.12.6.1(e) of ACI 318-19

> 26.12.6.1(e)—Concrete in an area represented by core tests shall be considered structurally adequate if the average of three cores is equal to at least 85 percent of $f'c$ and if no single core is less than 75 percent of $f'c$. Additional testing of cores extracted from locations represented by erratic core strength results shall be permitted.

This code provision is an interesting example of an in-place requirement that is different from the requirements for the corresponding standard laboratory-cured test. In this case, an in-place strength of 85% $f'c$ is acceptable compared to approximately 1.10 to 1.15$f'c$ for the average of laboratory-cured test results (recall that the statistical quality control provisions require an average strength that often turns out to be 10% to 15% greater than $f'c$). However, some portion of this difference has also been attributed to the differences between cores and cylinders and to the conditioning of cores.

9.12. Changing role of testing

A transition to performance specifications literally means a transition to performance testing as well. However, if we can specify only those properties that we can reliably test, our current range of specifiable performance criteria are limited to our current battery of approved, standardised tests.

From the owner's perspective the most meaningful tests are those that evaluate the concrete properties as influenced by materials, proportions, mixing, transport, placing, consolidating, finishing, curing and concrete temperature, that is, in-place testing of the hardened concrete. Although many proven options are available for evaluating in-place strength, fewer alternatives are available for evaluating durability, although ASTM C457 [21] (microscopic analysis) and C1202 [21] (rapid chloride permeability) are viable candidates in most cases. In general, we have only a few ways of reliably predicting durability without testing samples extracted from the hardened structure. It is also likely that since laboratory-cured cylinders have been traditionally considered an acceptable basis for evaluating concrete strength, they will continue to play a significant role in performance testing. If reliable mixture-specific correlations can be demonstrated between various properties of interest and the results of more conventional tests, such as of density, cylinder strength or beam strength, then such common tests might be considered an acceptable surrogate.

Since the fully hardened, in-place properties are the most desirable but also the most difficult to obtain and are often at least 28 days too late, there is an intense need to evaluate the concrete earlier to get an early warning of problems or to gain early confidence that all will be well. The sooner information is obtained about the early-age hardened properties of any given load of concrete, the sooner any adjustments can be made to the materials, proportions or processes for subsequent concrete placements, the sooner remedial measures can be initiated on the concrete already installed, or the sooner construction practices can be altered (i.e. longer form or shoring removal times or extended curing). Early-age or accelerated strength testing is useful in this regard. It is absolutely essential if the consequences of a later-age discovery of unsatisfactory concrete are more expensive. Likewise, the sooner that durability-related test results, such as ASTM test methods [21] C1202 (rapid chloride), C1876 (bulk resistivity) or C457 (microscopic analysis), become available, the sooner remedial actions, such as requiring a sealer on marginal concrete, or the sooner mix proportions or admixtures can be changed.

From a logistical perspective, tests are needed that can be performed on the fresh concrete at the time of delivery (when we can still accept, adjust or reject the product) and will provide data for predicting the likelihood that the required hardened properties will be achieved. However, aside from slump, air and unit weight (fresh density), we have no tests that are fast enough to allow an accept/reject decision on the truck at hand. The challenge of predicting performance from air or unit weight has already been discussed, and predicting performance based on slump may be impossible. The water content test and air-void analysis are examples of 'fresh concrete' tests that have the potential to deliver truly useful information for predicting long-term performance, but this information can at best be used to influence subsequent loads as the concrete evaluated by these tests will already be in the structure by the time the results are available. As a final comment on tests of fresh concrete, given the influence of subsequent construction operations, it remains entirely possible that concrete that is satisfactory at the point of delivery will not yield the desired performance characteristics in place. Alternatively, if concrete is found to be unsatisfactory at the point of delivery, it is unlikely that subsequent construction operations will bring any substantial improvement.

Thus, there remains a need for some type of screening tests for the fresh concrete, if only to serve as an early warning for material that may be unsatisfactory. Screening tests could include air or water content or density (unit weight of fresh concrete). Cylinders are undeniably useful but cannot return immediate data (except for their as-cast weight).

As discussed earlier in this section, given the variables that affect concrete performance and the consequences of installing unacceptable concrete, it makes sense to pre-certify concrete production facilities as having the capability of making the kind of concrete that is required and to prequalify specific mixtures and materials based on their demonstrated potential to meet all performance requirements. Under normal circumstances, this prequalification would only be step one, as it would be followed by screening tests on delivery and a programme of subsequent tests of hardened concrete. However, given that the ability to meet performance requirements starts with appropriate plant, equipment and concrete-making materials and knowledgeable personnel, it also makes sense to develop certification programmes or expand existing ones that can identify those operations that are broadly capable of meeting the specifications. Likewise, personnel certification programmes should be expanded, especially into skills such as mix design with modern materials, the effect of construction operations on concrete quality and advanced quality control and quality assurance testing. Furthermore, the certification of plants and personnel not only elevates knowledge levels but also establishes pride and credibility.

9.13. Risk and responsibility

Starting with the 2004 edition of CSA A23.1 – 'Concrete materials and methods of concrete construction' – currently the 2019 edition [7] is the main Canadian standard for concrete – the owner has two options for specifying concrete: performance or prescription. Each option delineates what the owner should specify and what the contractor and supplier should do. The standard states that the performance requirements apply 'when the owner requires the concrete supplier to assume responsibility for the performance of the concrete as delivered and the contractor to assume responsibility for the concrete in place'. It is, thus, clear that the responsibility of the concrete supplier ends with the discharge of the appropriate concrete mix from the mixer or delivery unit. The contractor, on the other hand, is responsible for placing, compacting and curing the concrete so that it matures to have the strength and durability characteristics required by the owner. The text of the options is given in Table 5 of CSA A23.1, as shown in Table 9.19.

CSA A23.2-24C – 'Standard practice for sampling, testing, and inspection of concrete for qualification and acceptance purposes' [7], referred to in Table 9.19, details the requirements, roles and responsibilities for qualifying concrete mixtures for both the performance and prescriptive options. For acceptance, the standard gives the requirements and defines the roles and responsibilities of the producer, contractor and owner. The ACI *Guide for Responsibility in Concrete Construction*, developed by ACI Committee 132, was published in *Concrete International* [46]. It details the responsibilities of various parties, including the owner, the licensed design professional, specialty engineers, the general contractor, the designer/builder, subcontractors, the concrete materials supplier, the testing/inspection agency and the construction manager.

While the following quotations are no longer in the current guide (ACI 132R-14), in the previous 2005 edition, ACI's Committee on Responsibility in Concrete Construction (RCC) (ACI 132R-05) reported that: 'Construction has now reached a level of complexity that makes design input from constructors and subcontractors desirable and sometimes essential. This input, whether submitted as value engineering proposals, responses to performance requirements or design alternatives, has a legitimate place in concrete construction.' However, for a conversion to performance specifications, taking this 'legitimate place' will entail some redistribution of risk and responsibility. RCC goes on to cite the 'over-riding principle … that responsibility and authority must be congruent'. This suggests that if the concrete producer and contractor

Table 9.19 Responsibilities in prescriptive specifications compared to performance specifications (adapted from CSA A23.1-19 [7])

Alternative	The owner shall specify	The contractor shall	The supplier shall
Performance: The owner requires the concrete supplier to assume responsibility for the performance of the concrete as delivered and the contractor to assume responsibility for the concrete as placed, finished and cured.	(a) required structural criteria including strength at age (b) required durability criteria including class of exposure (c) additional criteria for durability, volume stability, architectural requirements, sustainability and any additional owner performance, prequalification or verification criteria (d) quality management requirements (e) whether the concrete supplier shall meet the requirements of concrete industry certification programmes (f) any other properties that may be required to meet the owner's performance requirements	(a) work with the supplier to establish concrete mix properties that meet the performance criteria for plastic and hardened concrete, considering the contractor's criteria for construction and placement and the owner's performance criteria (b) submit documentation in accordance with CSA A23.2-24C demonstrating that the owner's pre-performance requirements have been met (c) prepare and implement a quality control plan to ensure that the owner's performance criteria will be met and submit documentation in accordance with CSA A23.2-24C demonstrating that the owner's performance requirements have been met	(a) certify that the plant, equipment and all materials to be used in the concrete comply with the requirements of this standard (b) certify that the mix design satisfies the requirements of this standard (c) certify that production and delivery of concrete will meet the requirements of this standard (d) certify that the concrete complies with the performance criteria specified (e) prepare and implement a quality control plan to ensure that the owner's and contractor's performance requirements will be met, if required (f) provide documentation verifying that the concrete supplier meets industry certification requirements, if specified (g) submit documentation in accordance with CSA A23.2-24C demonstrating that the proposed mix design will achieve the required strength, durability and performance requirements
Prescription: The owner assumes responsibility for the concrete.	(a) mix proportions, including the quantities of any or all materials (admixtures, aggregates, cementing materials and water) by mass per cubic metre of concrete (b) the range of air content (c) the slump range (d) use of a concrete quality plan, if required (e) any other requirements	(a) plan the construction methods based on the owner's mix proportions and parameters (b) obtain approval from the owner for any deviation from the specified mix design or parameters (c) inform the owner of any anticipated problems or deficiencies with the mix parameters relevant to construction	(a) provide verification that the plant, equipment and all materials to be used in the concrete comply with the requirements of this standard (b) demonstrate that the concrete complies with the prescriptive criteria supplied by the owner (c) inform the contractor of any anticipated problems or deficiencies with the mix parameters relevant to construction

are to be held ultimately responsible for concrete performance, then it is reasonable to give them freedom of action to provide a product that meets the demanding performance requirements developed by the design professional. In one sense, this means that the producer/contractor will take on additional responsibility in return for the added freedom to select and proportion materials and to plan the means and methods of construction. Viewed another way, in many current projects, the concrete producer and contractor are already heavily responsible for concrete behaviour, even when following prescriptive specifications. For many producers and contractors, the principle of 'congruency of authority and responsibility' under a performance specification might simply mean getting the freedom that comes with the *current* level of responsibility. However, under a prescriptive specification based on sampling at the truck chute, responsibility for final concrete problems frequently tends to gravitate toward the concrete producer. Under a clear performance specification with at least some in-place testing of hardened concrete, the contractor's joint responsibility for in-place concrete performance may become more visible.

Under a performance specification, design professionals will have a clearer responsibility to articulate exposure conditions and performance characteristics. However, committing to a discrete list of explicitly required concrete performance criteria introduces the risk that an unspecified long-term performance problem may develop. For example, if the stated performance criteria included strength, shrinkage, permeability and frost resistance but did not specifically require immunity to ASR, who is responsible if ASR develops sometime later? This is not to suggest, however, that the responsibility for a problem like ASR is necessarily clear under a conventional prescriptive specification that does not include explicitly stated requirements for special ASR-related testing. Meeting the requirements for aggregates in ASTM C33 – 'Standard specification for concrete aggregates' [21] does include a general requirement for being non-deleterious due to ASR, but the specific test methods and limits to be used are not mentioned, so specifications need to state clearly any ASR requirements (note that there is guidance on mitigating ASR in ASTM C1778 – 'Standard guide for reducing the risk of deleterious alkali-aggregate reaction in concrete' [21]).

Another interesting issue is raised in the RCC document, which states that 'it can be appropriate [for the design professional] to delegate certain aspects of engineering design to specialty engineers working for the constructor or subcontractors. When any of this design work involves engineering (as opposed to simply detailing), it should be done under the control of an engineer who is licensed in the state of the project and who takes responsibility for such work.' Many concrete materials engineers would agree that concrete mixture design and proportioning is in fact 'engineering'. The RCC's report also raises the issue of the design professional's responsibility to review mix submittals: 'The design professional should review the mixture proportions and submittals concerning materials, procedures and testing data, but the constructor remains responsible for compliance with the requirements of the contract documents. If approval is required, the contract documents should state so specifically.'

At the top of the performance specification pyramid, the owner gains the opportunity for clarity of expected performance, but along with the design professional, the owner must accept the risk associated with a finite list of those quantifiable objectives. Under current specifications, there is a tendency to seek relief from the contractor for a wide range of longer-term performance problems, some of which may come from unstated exposures or unspecified service conditions. Pinning these down puts most of the cards on the table. In a related issue, the performance criteria selected must be indicative of the concrete's ability to meet the owner's functional needs. When the UK Department for Transport [12] studied the risks in adopting performance specifications, one of its concerns as owners of the highway network and as buyers of construction services was 'inappropriate application of performance measures resulting in a situation whereby suppliers can meet targets without achieving the desired outcome.' A related concern was 'mismatches between contract performance requirements and client objectives'. If the performance criteria are overspecified relative to the owner's needs, the product is unduly expensive, In contrast, if the criteria are underspecified, the result can be 'poor operational performance, excessive maintenance and premature replacement'.

9.14. A few comments about the current general state of practice

When considering new or alternative approaches to specifying concrete construction, it can be helpful to be reminded of the current state of practice. The following outline has been developed for that purpose, serving as a background for further thinking about specifications in general.

1. Qualifications.
   a. There is a broad range of expertise and experience in the marketplace.
   b. It can be difficult to qualify contractors or producers based on their quality record or expertise.
   c. Experience often works against a knowledgeable producer and contractor in a low bid situation, especially when specifications are subject to interpretation.
2. Pre-placement approval for a concrete mixture (prequalification) is based on one or more of the following
   a. historical records for strength performance
   b. laboratory tests that document the strength record
   c. ACI 301 [37] 'three-point-curve' to demonstrate strength as a function of w/c (w/cm) (formerly contained in ACI 318)
   d. detailed ACI 301 statistical method for mix approval based on strength (formerly contained in ACI 318).

   Producers do not frequently prequalify on basis of air, freeze–thaw testing, scaling resistance testing, corrosion protection or permeability (or its surrogate tests). (The lead time for these tests can be excessive and there is no guarantee that the pre-placement mixture is the same as the mixture actually installed.)
3. We often specify the following.
   a. Minimum cement or minimum total cementitious contents.
      i. The code is silent on this.
      ii. The code does set maximum limits for SCMs by percentage mass of total cementitious materials for flatwork exposed to both de-icing salts and freeze–thaw cycles.
      iii. The code dictates cement type, pozzolan or slag for various sulfate exposures.
   b. Maximum aggregate size. ACI 318 limits only the maximum aggregate size based on clear spacing between reinforcing bars and between reinforcing bars and formwork.
   c. W/c (w/cm).
      i. W/c is either directly incorporated into project specifications for special exposure conditions or included by reference to Chapter 19 of ACI 318 for freeze–thaw, de-icers, corrosion and sulfate exposures.
   d. Temperature.
      i. ACI 301 has maximum and minimum limits.
      ii. ACI 305 recommends mix-specific waivers based on tests for maximum temperature.
      iii. There is no specific temperature guidance in ACI 318, but there are caveats for avoiding harmful effects.
   e. Slump and slump flow.
      i. ACI 318 gives no guidance on slump or slump flow, leaving that to the contractor.
      ii. ACI 301 allows the contractor to set a target slump or slump flow as required to place and consolidate the concrete. Then the selected target slump or slump flow is required to stay within a given tolerance according to ACI 117-10 – 'Specification for tolerances for concrete construction and materials'.

f. Air content is either directly incorporated into project specifications for special exposure conditions or included by reference to Chapter 19 of ACI 318. In ACI 318, values are for 'total air, as delivered'. No hardened air values from ASTM C457 [21] are implied. Code values are tied to old ACI 211 assumptions about mortar content based on water content as a function of maximum aggregate size but with no adjustment for admixtures or aggregate grading.
g. Chloride-penetration resistance for corrosion-sensitive structures. ACI 318 gives no guidance regarding ASTM C1202 'coulomb' values, but the code does specify cover depth and allows adjustments to it based on concrete quality.
h. Compressive strength.
   i. Test specimens are wet cured until tested at a standard laboratory temperature.
   ii. Statistical quality control is used.
   iii. Effectively there is an option for in-place testing via cores.

4. We do not often specify the following.
   a. Air-void system parameters: these are most often checked after the fact if there is a problem and then we often act as though ASTM C457 had been specified.
   b. Shrinkage: we check after the fact if we have a problem and then often act as though ASTM C157 had been specified.
   c. Actual concrete material transport properties, such as actual permeability, sorptivity and diffusivity.
   d. Curing measures required to achieve a durable surface (CSA A23.1-19 [7] addressed this by specifying minimum curing requirements for each exposure class).
   e. Measures of curing effectiveness.
5. We often measure the following
   a. temperature of fresh concrete
   b. slump of fresh concrete
   c. air content of fresh concrete
      i. at the truck chute
      ii. sometimes at the point of placement (although there are no standard procedures for doing so and no clear acceptance criteria for air content beyond the point of discharge from the concrete truck)
   d. strength under laboratory-cured conditions
   e. rapid chloride test (ASTM C1202) when specified
      i. a specimen is commonly cut from a 4-inch cylinder, as sampled at the truck chute
      ii. a specimen is sometimes cut from a core extracted from the structure.
6. We do not often inspect or measure the following
   a. w/c (or w/cm) or various quantities of water
      i. water content in the drum when the truck is loaded
      ii. changes in aggregate moisture content during production
      iii. water added on site
      iv. retempering water or water applied to concrete surface or finishing tools
   b. quality of aggregate (including grading) during production
   c. quantity of aggregate during production
   d. quality of cementitious materials during production
   e. quantity of cementitious materials during production
   f. unit weight (density) or yield of fresh concrete
   g. air bubble size in fresh concrete (air-void analysis)
   h. In-place strength of concrete: field-cured cylinders do *not* give the same results as in-place testing; however, the in-place conditions must be clearly provided along with test results, as cases have

occurred where field-cured cylinders were taken into a heated construction office the day after casting in the winter. One site superintendent stated: 'If I leave them on the structure the strengths will be too low, and the structural engineer will not let me remove the forms.' However, the recent adoption of wireless embedded temperature sensors has allowed in situ maturity measurements, which can be calibrated to the strength of a given mixture and used to allow construction operations, such as form removal.

   i. in-place chloride penetration resistance of concrete
   j. In-place temperature of the concrete; however, the use of wireless embedded sensors on larger projects and for mass concrete placements to evaluate maturity and temperature gradients is becoming quite common.
   k. in-place curing effectiveness
   l. in-place degree of consolidation
   m. in-place air-void system parameters
   n. in-place air-void system parameters at the concrete surface.

7. Tests, variability and precision. The values given below are approximate (half the D2s between-laboratory precision limits as defined by ASTM) and simplified. For complete and accurate information on variability and precision, refer to the relevant ASTM standard.
   a. slump
      i. time dependency is not normally taken into account.
      ii. precision: ± 11 mm
   b. air
      i. volume of a pressure air meter pot = 0.1% volume of a 6-m^3 truck
      ii. precision: ± 0.4%
      iii. One core examined by ASTM C457 contains about 1000 of the 10^{12} air voids in an 80 m^3 placement. Due to the cost per test, we cannot afford to realise statistical significance.
      iv. ASTM C457: precision: ± 1.16% of the air content
      v. spacing factor of fresh concrete measured by an air-void analyser: ± 25%
   c. strength
      i Reported cylinder strength is also a function of cylinder maker and cylinder breaker: precision: ± 4%
      ii. cores are sensitive to moisture conditioning: precision: ± 6.5%
   d. ASTM C1202 – 'Rapid indication of chloride penetration resistance'
      i. multi-laboratory coefficient of variation: ± 18.0% (based on testing single specimens)
      ii. Note that multi-laboratory precision of other fluid transport test methods including NT Build 443 [47] and NT Build 492 [29] is often worse (Table 9.18).
   e. ASTM C157 – 'Length change for drying shrinkage'
      i. precision: ± 0.0037% if cured in water and 0.0069% if cured in air
   f. AASHTO microwave oven test for w/c ratio of fresh concrete
      i. ± 0.03 to ± 0.05.

8. We have reasonably reliable relations between
   a. w/c (w/cm) ratio and the strength of a given concrete mixture at a given maturity
   b. w/c ratio and the permeability of a given hardened Portland cement paste (with no SCMs) at a given maturity
   c. strength of a given hardened Portland cement paste and maturity
   d. total paste content and shrinkage
   e. total water content and shrinkage.

9. We do not have reliable, mix-independent relations between
   a. slump and water content

b. slump and w/c or w/cm ratio
c. slump and strength
d. slump and risk of segregation (in the presence of admixtures)
e. slump and pumpability
f. slump and finishability
g. slump and response to vibrator
h air content (alone) and frost resistance
i. air content (alone) and de-icer scaling resistance
j. strength and durability
k. strength and w/c or w/cm ratio
l. w/c or w/cm ratio and permeability over a broad range of mixtures and binders
m. strength as represented by cylinders cured in the laboratory and cores extracted from the structure
n. strength indicated by either cylinders or cores and the strength of the concrete surface exposed to weather and traffic
o. permeability or conductivity of concrete at mid-depth of a core or cylinder compared to at the surface.

9.15. Key aspects of a performance specification and the resulting challenges

9.15.1 Definitions

There are multiple definitions of the term 'performance specification'. The Cement and Concrete Association of New Zealand [13] clearly states: 'A performance-based specification prescribes the required properties of the concrete but does not say how they are to be achieved.' In the USA, NRMCA [10] defines a performance specification as 'a set of instructions that outlines the functional requirements for hardened concrete depending on the application. The instructions should be clear, achievable, measurable and enforceable'. In either of these definitions there is room for interpretation as to whether the terms 'required properties' or 'hardened concrete' refer to hardened properties in the structure, hardened properties as sampled at the point of delivery, hardened properties as cast from prequalification testing or perhaps some combination. (Added complication comes from the likelihood that a contractor setting performance requirements for a concrete producer would include multiple fresh concrete requirements in addition to any hardened concrete requirements.) In the minds of many, the term 'performance specification' automatically refers to the in-place properties of concrete as influenced by materials, proportions, construction operations, and control of temperature and moisture. For others, the term 'performance specification' implies merely the freedom to supply concrete that meets point-of-discharge requirements without the customary qualifications or limitations on ingredients and proportions and without the need to submit documentation concerning materials or proportions. For still another group, the term 'performance specification' implies that concrete is supplied based on historical records or pre-construction test results, with only spot-checking to verify that the mix delivered remains substantially the same as that originally approved. In reality, a performance specification can be any or all of these and users of the convenient phrase must carefully define it to avoid miscommunication. It may not be helpful to use 'performance specifications' in an agreement if the agreeing parties hold differing concepts of what it means. In all cases, however, there is a need to define the responsibility for product control and to allocate the authority to make the decisions about how to carry out that product responsibility.

9.15.2 Key aspects of a performance specification

To recap what was previously detailed in Section 9.4, the keys aspects of a performance specification are

(a) the ability of the specification writer to discern the performance characteristics appropriate to the owner's intended use of the concrete

(b) the ability of the specification writer to describe these performance characteristics clearly, unambiguously and quantitatively so that performance can be evaluated
(c) the availability of reliable, repeatable test methods that evaluate the required performance characteristics (along with performance compliance limits that take into account the inherent variability of each test method)
(d) the ability of the concrete producer/contractor team to choose combinations of materials, mixtures and construction techniques to meet required characteristics so that projects can be planned and bid, risks and costs can be assessed, and materials and construction operations adjusted to comply with performance requirements.

9.15.3 This is not new

The concept of performance specifications is not new. The ability to gain acceptance of concrete based on proven performance requirements has been part of the ACI 318 Building Code since its early days. Nevertheless, conventional concrete specifications include many prescriptive elements that also have their origins in the earliest code, perhaps reflecting an era when designers were masters of the selection and proportioning of concrete materials and concrete producers purchased raw materials and batched in accordance with the specified instructions. Over the last several decades, however, concrete materials have become increasingly complex, with a growing number of combinations and permutations of cements, cementitious materials, admixtures, and aggregate types and grading. This not only means that there is a wider range of options for meeting any given concrete requirement, but also that it is more difficult for the design professional to stay current with concrete materials and construction technology, as it has become a specialised field. At the same time and as stated previously, there has been a shift in the responsibility for concrete ingredients and mix proportions towards the concrete producer and away from the design professional.

9.15.4 Advantages and disadvantages

As stated previously in Section 9.5, there are advantages and disadvantages to the use of either performance or prescriptive specifications. The challenge is to find ways to combine the two types as appropriate for various applications to maximise the advantages and minimise the disadvantages. Performance may be called for when there are clear economic, logistical or scheduling benefits to be gained and those benefits can be shared with the designer and owner.

9.15.5 Combinations of specification methods

Arguments pitting performance against prescriptive specifications may not be productive. At issue is the most effective combination of specification requirements, with a sequence of prequalification, on-site testing at delivery and in-place evaluation of hardened concrete. One consequence of performance specifications is the ability to link payment to demonstrated quality. This results in penalties for quality that falls short of specified requirements but is acceptable [48–53]. In a few cases, bonuses are also paid for achieving or surpassing the specified minimum quality. Ideally, these bonuses or penalties could be connected to anticipated life-cycle costs, but there are no clear examples of this to date. Where bonuses have been paid, concrete has been found to be either more consistent in quality or, in one documented case, cheaper [49]. Options on some projects would include separate sets or types of requirements for acceptance and for pay.

9.15.6 Prequalification

It makes good sense to qualify a mix for field use based on pre-construction testing or historical records. Such prequalification demonstrates that the mixture and concrete producer have the potential to meet project requirements. However, the reality of batch-to-batch and day-to-day variability in some concrete production facilities (especially the variability in air and water content) makes it necessary to demonstrate that the material delivered to the job (and in some cases as-placed, consolidated, finished and cured) lives up to its prequalification expectations. In some cases, the difference between prequalification results obtained on concrete cast and tested in the laboratory compared to performance of actual production concrete needs to be considered when evaluating laboratory test results.

9.15.7 Performance characteristics

On any given job, regardless of type of specification, many concrete performance characteristics may be expected by the designer and owner. On that same job, some of those expectations will have been incorporated in the formal specifications, whereas others will have been assumed to develop if the few specified properties are achieved. Performance specifications focus on specific concrete properties and hold the potential to clarify what is and what is not expected of the concrete.

9.15.8 Durability

The common concern worldwide is to design durable concrete with an extended service life. Common to most specifications are exposure classes that clearly define expectations about the types of exposure that constitute the service environment of the concrete. Even under current and predominantly prescriptive specifications, ambiguity is greatly diminished when the specifier clearly states the severity of the freeze–thaw environment, salt exposure, need for reduced permeability and sulfate exposure. If these are not clear, it can be difficult for the producer/contractor to comply with building code requirements or to ensure owner/designer satisfaction. As previously shown in Tables 9.11 and 9.10, most national standards, including ACI 318-19 [19], contain requirements for concrete in each exposure class.

9.15.9 Code freedom for concrete in benign environments

The ACI 318 building code does not inhibit performance specifications for concrete that does not have special durability requirements. This type of concrete represents a significant segment of the concrete construction market, which could be exploited to allow performance specifications to be more widely adopted. Specifiers should avoid imposing special durability requirements in situations where no harsh service environment exists, as they may, thus, invite other problems in addition to increasing costs. A specification that leads to a high paste content as a result of minimum cement or maximum w/c requirements can result in increased shrinkage cracking as well as higher permeability and an increased carbon footprint.

9.15.10 Code limits for concrete in aggressive environments

For durable concrete, the current ACI 318 building code limitations on the w/cm ratio and percentage of SCMs (for de-icer salt exposure, for example) are restrictive and can lead to undesirable consequences, such as increased shrinkage and cracking, while nevertheless providing the desired durability. However, these requirements might evolve toward performance alternatives, as in the case of w/cm code requirements that used to apply to concrete strength. The statistical quality control features of the code are reasonable and protect the owner and producer, even though ACI chose the unfortunately misleading term of 'overdesign' to describe the difference between specified strength and required average strength. It is likely that any meaningful performance criteria will have a similar statistical basis.

9.15.11 Share and transfer of responsibility

In making the transition to performance specifications, the concrete production industry needs to be clear about the potential for a significant transfer of responsibility. If true performance of the end product is specified, requiring the in-place assessment of hardened concrete, the concrete producer and contractor become jointly responsible for the quality of the finished structure. However, if the point of performance is specified as the point of discharge of the concrete, the concrete producer's share of responsibility with the contractor is not significantly different than with a prescriptive specification. The owner's risk is likely to be minimised by the point of performance being in-place hardened concrete that is the joint responsibility of the producer/contractor team. The concrete producer's risk is likely to be minimised if acceptance were based on the delivery of a prequalified product. Decisions about the point in the construction sequence at which the concrete properties are to be evaluated thus have a critical impact on risk and responsibility. Further, taking advantage of performance specifications mandates greater dialogue between contractors and their concrete suppliers. Concrete suppliers will need to be proactive in determining the concrete mix characteristics needed by the contractor in placing, compacting and finishing as well as in meeting the performance criteria for the hardened concrete. Contractors cannot be relied upon entirely to clarify these needs when asking for prices.

9.16. International perspective and progress

9.16.1 Overall assessment

A study of the current world literature on specifications makes it clear that performance is a hot topic and that performance specifications are the 'holy grail' that many desire and are seeking. There is worldwide concern about the durability of concrete and the effects of the quality of installation on the final in-place quality. In reviewing a large number of documents from around the world, however, few true performance specifications were found, and these contained only pure performance criteria for some properties of concrete. To paraphrase a popular quotation, 'When all is said and done, there is a great deal more said than done on the topic of performance specifications.'

The adoption of performance-based specifications assumes

- that there are appropriate performance test methods in place to evaluate all the essential properties of concrete
- that performance can be measured in time to affect the construction outcome or that concrete mixtures can be prequalified.

Although new and better test methods will become available with time, nearly all the relevant properties of concrete can be measured to a practical acceptable level, provided the implementation is governed by appropriate procedures (such as the location of sampling points, sample frequency, and the handling and storage of test specimens). These procedures need to be incorporated into contract specifications and mutually agreed by all parties. However, although most parties in construction are familiar with the testing of the properties of fresh and hardened concrete, the biggest challenges in this regard relate to evaluating durability.

At the moment, most specifications that address performance delineate exposure conditions that affect the service life of concrete and list parameters to be verified that are assumed to be indicative of the ability of the concrete to achieve the desired service life. These parameters are based on experience or durability indexes derived from research studies and, in some cases, field experience. In many cases, specifications provide tables of mandatory limits to w/cm ratios, minimum cement contents and air entrainment. There is an almost universal use of SCMs, such as fly ash, ground granulated blastfurnace slag and silica fume. Without exception, the levels of quality control required to guarantee a consistent and economical product are based on statistical analyses of test results, at least for strength.

The literature search uncovered few examples of performance specifications where the quality of the end product was stated and the product details left entirely to the concrete supplier. In all instances where this issue was discussed, the consensus was that the test procedures needed before this approach can be extensively implemented have not yet been proven to be reliable or repeatable to the degree that would make them viable from a practical and legal point of view. Thus, it would seem that any performance criteria in specifications for current general use that are to be made the responsibility of the concrete supplier are limited to slump, air content and strength.

Among the most highly developed international codes and specifications, the Eurocodes with their nationalised amendments may be the most comprehensive and the most complex. BS 8500 [17] is particularly complex. It is considered that the approach to determining and ordering a concrete mixture specified in these standards, BS 8500 in particular, is unnecessarily and unattractively complicated [54–56]. Some of the complication has resulted from having to accommodate the needs of the 31 EU countries in one standard. The Australian [25] and New Zealand standards are progressive, enlightened and clear but are not truly performance based. However, the previously mentioned practices recommended by the Concrete Institute of Australia [22] encompass many of the aspects needed for obtaining durable concrete using a performance specification. The Canadian standard CSA A23.1 [7] is also not strictly performance based (as it still includes maximum w/cm limits) but does contain many elements that are adaptable with little or no modification to a performance-oriented specification, particularly for durable concrete.

9.16.2 US Initiatives

Within the US, many states have taken important steps toward performance specifications, both independently and under the leadership of FHWA's roadmap [11]. Virginia [49] has specified shrinkage and rapid chloride permeability. Kansas has been exploring air-void analyses, and Minnesota has been among the states that have encouraged 'contractor-based mix design' and has been controlling water content in certain applications. The Port Authority of New York and New Jersey has had good experience in measuring the water content of fresh concrete (using the microwave test), shrinkage and rapid chloride permeability [50]. The Florida Department of Transportation requires concrete mixtures for exposure to marine environments to be prequalified using ASTM C1202 – 'Standard test method for electrical indication of concrete's ability to resist chloride ion penetration' [21] and has recently transitioned to test surface resistivity on site-cast cylindrical specimens [57]. Riding *et al.* [58] discuss alternative test methods for specifying resistance to chloride ingress, including ASTM C1202, bulk resistivity and the formation factor (bulk resistivity normalised for the resistivity of the pore fluid of the concrete).

9.17. Testing and quality management

9.17.1 Testing

To provide confidence for all parties with the adoption of performance specifications, there is a need for quick, reliable performance tests for concrete properties, including durability, which have to go far beyond current reliance on the 28-day compressive strength as the sole arbiter of concrete quality. The lack of adequate performance-related test methods certainly hinders the move from prescriptive to performance specifications and there is no question that new developments in testing will facilitate the use of performance specifications.

9.17.2 Multi-stage testing

A transition to performance specifications literally means a transition to performance testing. Prequalification testing would be step one, followed by screening tests of fresh concrete on delivery and a programme of subsequent tests of hardened concrete, possibly of samples that had been taken from the truck chute but more comprehensively from the structure itself. When establishing levels of quality assurance, a DuraCrete report [59] rated in-place testing as the highest level. There is already a developing trend to test the concrete in the finished structure, since the owner's primary concern is that concrete.

9.17.3 Screening tests

Screening tests of fresh concrete are likely to continue to be based on slump, temperature and total air content, although slump may not be required by the specifier in lieu of more meaningful tests of the properties of hardened concrete. Performance-oriented contractors may have slump requirements, however. Tests of fresh concrete, such as for density (fresh density), may become more important for identifying mixes, quantifying the batch-to-batch uniformity and providing additional data concerning air.

9.17.4 Other tests

Other tests can be performed on fresh concrete, such as an air-void analysis and the microwave test for estimating water content, but these cannot be used as a screening test due to the time required to obtain a result (prior to discharging the concrete). Even though a w/cm ratio may not be specified under a pure performance specification, once a proposed mixture has been pre-approved for use in prequalification, measuring the water content may be an effective identity test to validate that the pre-approved mix has, in fact, been delivered to the job, within the limits of the precision of the test method.

9.17.5 Tests of mechanical properties

Standard cylinder tests are likely to play a role in performance requirements, but it is also likely that the focus will shift towards maturity calculations, in-place testing or accelerated strength tests to provide early confidence (or early warning) that the desired outcomes will be achieved. Cylinder mass may become more commonly used as an early indicator of low-strength cylinders. Abrasion tests will be more common if abrasion has been defined as an in-service exposure. Given the strong association among cracking, functionality and owner/designer satisfaction, early- and later-age volume-change tests will become more common.

9.17.6 Durability concerns

The prime concern of all the authorities reviewed is durability. Without exception, all recommend SCMs, either as additions at the concrete plant or in blended cements, as an aid to achieving durability. Many tests on the ingress of aggressive substances into hardened concrete have been developed regarding durability [60–69], but few of these are standardised and none are considered ideal [70]. The NT Build 492 rapid migration test [29,30,71,72] has achieved wide use in the Nordic countries and can be adapted to evaluate curing effects [73]. Both bulk and surface resistivity tests have merit due to their simplicity [74–77] and have been standardised (ASTM C1876 – 'Standard test method for bulk electrical resistivity or bulk conductivity of concrete' [21]). Nevertheless, it is likely that ASTM C1202 – 'Standard test method for electrical indication of concrete's ability to resist chloride ion penetration' [21] will remain a significant tool for evaluating the performance of concrete in the foreseeable future [42,52,58]. The ASTM C1202 test has been used extensively in highway specifications in Virginia, in Australia and by highway departments in Canada. The Canadian CSA A23.1-19 specification [7] uses this test to qualify concrete for environments with chloride and aggressive chemical exposures. The Ontario provincial standard specification OPSS 1350 [53] uses this test and bases acceptance on tests of cores taken from the finished structure. Testing the rate of absorption (sorptivity) is also quite useful [78–81]. ASTM C1585 – 'Standard test method for measurement of rate of absorption of water by hydraulic-cement concretes' [21] has not had extensive use in North America, but similar tests have been used for acceptance in Australia [82] and a field version has been used on some Canadian projects [83–85].

9.17.7 ASTM C457

Actual freezing and thawing tests are available for evaluating the freeze–thaw durability of concrete in place, but these are difficult to employ, expensive and time-consuming. ASTM C457 – 'Test method for microscopical determination of parameters of the air-void system in hardened concrete' [21] is likely to be important and now includes an option to use image analysis to determine the air-void parameters. This test has been extensively used by Canadian highway departments. In Ontario, Canada, acceptance is based on tests of core samples taken from the structure.

9.17.8 ASR and sulfate reactions

Several specialised tests are available for the specific durability issues relating to ASR or sulfate reactions. Given that fluid flow through porous media is a factor in both of these deteriorative mechanisms, specific ASR or sulfate testing should be coupled with permeability and transport-type tests. ASTM C1567 – 'Test method for determining the potential alkali-silica reactivity of combinations of cementitious materials and aggregate (accelerated mortar-bar method)' [21] is probably the most appropriate for job mixture ingredient qualification if the aggregates have been determined to be potentially reactive by ASTM C1293 – 'Test method for determination of length change of concrete due to alkali-silica reaction' [21] or from prior field service. Protocols for ASR testing of aggregates and mitigative measures are described in ASTM C1778 – 'Standard guide for reducing the risk of deleterious alkali-aggregate reaction in concrete' [21].

9.17.9 Need for rapid approval

A practical difficulty in implementing performance specifications is the short bidding period followed by the short time before work commences, which is common in North American contracts. Tests to determine whether concrete mixtures meet ASR, diffusion and creep requirements need very long lead times whereas tests for freeze–thaw, shrinkage and scaling take at least 3 months. Further, the longer it takes to perform a test, the more likely it is that the raw materials will have changed by the time the actual production concrete is produced. An example of a supplier's responsibility is given in Australian specification AS 3600 [25], which requires ready mixed concrete producers to determine the chloride and sulfate contents and the shrinkage of their most commonly supplied mixtures every 6 months.

9.17.10 Quality management

The properties of in-place concrete depend on both the quality of the concrete delivered to the site and the quality and workmanship of construction operations such as placing, consolidating, finishing, curing and protection from the weather. Prior to approving concrete and allowing it to be placed, a design professional may require evidence from the concrete supplier that the mixture has the potential to meet the specified criteria and may also require evidence from the contractor of their ability to install the concrete.

A performance specification, therefore, requires a quality management programme from the contractor (as the party holding contractual responsibility to the owner), which in turn will require such a programme from the concrete supplier. In view of the clarification of responsibilities inherent in any specification, but particularly in a performance specification, it must be clear that when the terms 'quality assurance' and 'quality control' are used, the definitions in ACI 116 – 'Cement and concrete terminology' apply

> **Quality assurance** – actions taken by an owner or representative to provide and document assurance that what is being done and what is being provided are in accordance with the applicable standards of good practice and following the contract documents for the work.
>
> **Quality control** – actions taken by a producer or contractor to provide and document control over what is being done and what is being provided so that the applicable standards of good practice and the contract documents for the work are followed.

It has been suggested in the European specifications that not only should independent entities, such as the North American Cement and Concrete Reference Laboratory (which inspects but does not certify), be able to certify compliance of concrete mixtures but that this could also be done by national organisations such as NRMCA or regional concrete associations. (NRMCA has a Quality Plan Guideline and Process of Certification for Quality Control projects close to completion and has a similarly long-established programme that addresses the certification of plants, trucks, plant managers, delivery and sales professionals and concrete technologists.) The Precast/Prestressed Concrete Institute in the USA has been operating a successful plant certification programme for over 30 years [86–88]. However, owner acceptance of certification by producer associations will require demonstration of a clear lack of bias. Along these lines, some programmes require that concrete suppliers report deficiencies in their operation to owners and contractors in a timely fashion. For example, in the experience of one of the authors, this was done effectively for the new parking garage at Toronto Pearson International Airport. The concrete supplier did report their errors immediately and took the necessary corrective actions. Although there was some initial resistance, in the end it did work, and everyone liked it, including the supplier.

Concerns about suppliers who may not be technically capable of producing performance concrete could be allayed by maintaining and expanding NRMCA's certification programmes. Such programmes must be, and must be perceived as, rigorously applied and maintained. This would also imply that only certified suppliers should be allowed to supply to performance-based specifications. (Similar requirements exist in Quebec and Ontario (Canada), New Zealand, UK and in other locations.)

It has been suggested in Europe that if a supplier and their concrete mixtures are certified by an acceptable third party, quality assurance testing is not necessary unless a problem is seen. This will be a hard sell in North America, at least until the use of performance specifications is a mature practice or until enough on-site tests have proven the uniformly satisfactory behaviour of the concrete.

It is to be expected that concrete suppliers will have reservations about the validity of tests, particularly when a failure to meet the specification is reported. Some jurisdictions deal with this by having a referee system that can carry out retests to satisfy all parties regarding the disputed test result (e.g. CSA A23.1 [7] and OPSS.PROV 904 [52]). Although questions may remain about how to select an impartial referee and who would pay for this service, the specifications could state that referee testing is not only at the request of the contractor (or the contractor's supplier) but also that the contractor pays for it.

9.17.11 Mixes

Typically, a concrete supplier will have a portfolio of many mixes that have been developed to meet price inquiries. Also typical of most contracts is a short bid time and a rapid construction start after award. Pretesting all, or even a significant number, of these mixes to meet anticipated performance demands would be horrendously expensive and take up to several years to complete. How concrete suppliers will cope with this aspect of performance specifications is not clear.

One possible approach is the 'family of mixes' concept put forward in the UK. These are groups of related concretes for which a reliable relation has been established and documented. Criteria are given in BS 8500 [17] for confirming that a mix truly belongs to a particular family. A recognised authority can provide certificates of conformity to confirm membership of a family of mixes. In the UK, this certification can be provided either by a wholly independent authority, such as the British Standards Institute, or the concrete industry's quality scheme for ready mixed concrete.

Although not detailed in this section, many properties of concrete can now be included in a performance specification. Many properties are measurable by an existing ASTM or other organisation's standard test method and can be specified as a minimum or maximum value. The more intractable problem now is the increasing number of durability criteria that can be specified for concrete. As new tests and criteria are developed, it will become necessary to demonstrate that mixtures in the current inventory can meet the new requirements. In some cases, current mixes will pass the new tests and in others, adjustments followed by retesting will be required. In such cases, suppliers will gather the required data and be able make adjustments as necessary. In a few cases, such as with provincial concrete specifications in Ontario, new criteria have been included for 1 or 2 years without being included as pay items, to allow suppliers to become familiar with the new tests and to develop suitable mixes.

9.18. How do we get there from here?

9.18.1 Introduction

The FHWA has published a roadmap to performance-based specifications. This mirrors similar programmes elsewhere, and these planned approaches will obviously involve many people for many hours over extended periods. The EU surely did not produce its documents without an enormous expenditure of time by representatives from 31 countries. Such organisations have and are expending huge resources on the development of their performance specifications. It would be naïve to assume that substantive changes to the philosophy for specifying and ordering concrete can occur without a well-planned and protracted programme. However, there are other examples of the successful introduction of new and different technology in the concrete industry. When high-strength concrete at or above the 70 MPa level was introduced, significant changes were needed in the design and production of concrete mixtures. In Chicago, Seattle and Toronto in North America and in other cities, the consulting, testing and ready mixed concrete industries cooperated to develop concrete mixtures and quality assurance and control procedures. Ultimately, these became established practice and were enshrined in an ACI guideline document. Similarly, at the beginning of the interest in high-performance concrete, the FHWA in the US [89] and the Concrete Canada network [51] provided the impetus, encouragement and expertise to the industry that has led to the widespread adoption of this material. As would be expected, practical difficulties with such aspects as finishing and curing have occurred, and the original specifications have been successively modified as experience has been gained. The evolution of the ACI 318 building code itself shows clear evidence of the ability of the industry to adapt and to prove the validity of good ideas. With this background and the clear examples of important changes happening around the world and closer to home in Canada and departments of transportation in various US states, we can expect progress on how we specify concrete. However, how do we get there?

9.18.2 Multi-step plan

Taylor [90] has postulated nine steps in the transition from prescription to performance. The following list of considerations and actions needed to achieve this transition builds on the Taylor plan and on NRMCA intentions.

1. Develop or modify an example performance specification (or minimally prescriptive specification) that takes advantage of the freedoms and opportunities afforded by the current ACI 318 building code. Carefully describe the conditions under which this can be recommended for use.
2. When developing this model, identify those code provisions or other aspects of standard practice that represent a barrier to taking advantage of the performance specification philosophy.
3. Draft specific change proposals for consideration by ACI 318 that would allow for advantageous use of performance specifications. Given that the cooperation and support of recognised authorities is a prerequisite to substantive changes in how concrete is specified, evaluated and accepted, it is essential that the P2P initiative be coordinated with groups such as ACI committees 318 and 301. ACI 318 'Building code requirements for structural concrete' is the bible for design professionals in the US, much of South America, the Middle East and parts of Asia, and ACI 301 – 'Specifications for concrete construction' is also widely used. It will be necessary to work with these ACI committees to assist in code, commentary and specification amendments that can expand performance-oriented options and provisions. In addition to comparing performance and prescriptive requirements, attention must be given to establishing a clearer system of exposure classifications or requirements. In common with worldwide practice, these prescriptive tables would, for specific exposures, include restrictions on type of cement or cementitious combination, minimum cementitious content, w/c or w/cm ratio and air content. Once a set of exposure classes has been agreed, the next step is to produce criteria in tabular form that are essential requirements for concrete mixtures, although, initially, some of these may be prescriptive. This would ensure that concrete mixtures based on these criteria can potentially meet the

durability requirements. Initially, the critical properties of concrete needed to meet the requirements of each exposure class may well be prescriptive, but in time, they would become performance criteria. Later, as practice changes and suitable test methods become available, any prescriptive requirements in the tables can be progressively deleted in favour of performance criteria.

4. A more comprehensive model of a performance specification can be made available as code provisions permit.
5. Industry acceptance is expected to be incremental as opportunities arise for project teams to capitalise on the advantages of performance specifications. In the first such iteration, it is probable that only a few properties of concrete will be included as performance requirements. These would be those for which proven tests with known and acceptable variability are available to check compliance. These tests would also be in common and extensive use and already considered useful by design professionals. Candidate tests and properties could include the densities of normal, heavy and light concretes, early-age compressive strength, flexural and splitting tensile strengths, permeable voids, chloride permeability, freeze–thaw resistance, de-icer scaling resistance, modulus of elasticity, shrinkage and creep. For each property, only a minimum or maximum average value needs to be specified along with a minimum or maximum for a single test result (based on the precision of the test). Although early use may incorporate only a few performance requirements, the experience gained would inform the development of code modifications.
6. There are also tests already in somewhat specialised use that have the potential for wider application in the near future. These are the rapid chloride permeability test (ASTM C1202), its alternative and more recently adopted bulk electrical resistivity test (ASTM C1876), the determination of air-void system quality by ASTM C457, and sorptivity using the procedure in ASTM C1585 – 'Standard test method for measurement of rate of absorption of water by hydraulic-cement concretes' [21] modified for in-place use [83–85]. The air permeability combined with surface resistivity test designed by Roberto Torrent [36,91] has been adopted as a Swiss standard [92] and is gaining popularity.
7. Use of performance specifications may expand due to demonstrated success and benefits, and with the incorporation of additional properties as additional tests are proven to meet accuracy and variability requirements. The procedure for bringing these tests into wider use can most easily be established by an authority such as a department of transportation. A new test can be specified for use on contracts for one or more years but for information rather than as a contractual issue. Thus, extensive field experience can be gained and the utility of the test determined. The emphasis on certification programmes is expected to increase, with attention being paid not only to concrete production and construction but also to testing facilities and testing technicians.
8. An essential part of effecting a change in mindset for stakeholders so that they adopt performance specifications will be education that not only encourages the appropriate use of performance specifications but also encourages thinking the problem all the way through.

For example, as in several early 100-year service life projects in Europe, a diffusion coefficient was specified for early concrete in two tunnel liner contracts in Ontario, Canada [93]. However, the requisite test required 120 days per specimen, although for the second contract, test and tunnel liner production was 1000 units per week. Clearly the use of a 120-day test was technically justified from the perspective of performance but logistically unmanageable. In hindsight, the use of accelerated temperature testing to increase maturity at an earlier age would have made more sense. Many more recent projects have adopted rapid fluid ingress test methods for acceptance purposes.

It will be necessary to provide clear and detailed recommendations based on the findings of field experience as it accumulates. The initial literature review [1] identified several helpful and effective examples of articles, websites, papers, reports and brochures that were paving the way for new specifications. For example,

Annex J of the current edition of the Canadian CSA A23.1, 'Guide for selecting alternatives using Table 5 when ordering concrete' [7], is a concise and clear aid. A similar annex should be part of the revision of ACI 318 [19], or similar material should be added as part of the commentary. Such guidance will play a part in the transitional thinking and education of those involved in choosing between prescription and performance or in deciding how to make the most effective blend of the two types.

The adoption of performance specifications is advancing, as they can allow the use of novel materials and mixtures to improve sustainability [94]. As already initiated by the NRMCA, the education programme should include all stakeholders. Effective use of performance specifications requires that all parties understand basic concrete technology in all phases from material acquisition to in-place service. Through education and certification that qualifies companies to bid and participate on performance projects there is a genuine opportunity to 'raise the bar'.

Acknowledgements
This chapter has been largely abstracted, with permission, from more extensive reports written for the RMC Research Foundation in the US [1,2]. It has also been updated to include reference to more recent changes, international standards and publications. The RMC Research Foundation (now the Concrete Advancement Foundation) is a non-profit organisation dedicated to continuous research and enhancing the educational opportunities of the ready mixed concrete industry.

REFERENCES

1. Bickley JA, Hooton RD and Hover KC (2006) *Preparation of a Performance-based Specification for Cast-in-place Concrete.* National Ready Mixed Concrete Association, Silver Spring, MD, USA.
2. Bickley JA, Hooton RD and Hover KC (2008) *Guide to Specifying Concrete Performance: Phase II Report of Preparation of a Performance-based Specification for Cast-in-place Concrete.* National Ready Mixed Concrete Association, Silver Spring, MD, USA.
3. Bickley JA, Hooton RD and Hover KC (2006) Performance specifications for durable concrete: Current practice and limitations. *Concrete International* **28(9)**: 51–57.
4. Hooton RD and Bickley JA (2012) Prescriptive versus performance approaches for durability design – The end of innocence? *Materials and Corrosion* **63(12)**: 1097–1101.
5. ACI (American Concrete Institute) (2010) *Innovation Task Group 8. Report on Performance-based Requirements for Concrete.* ACI, Farmington Hills, MI, USA, Report ITG8-10.
6. ACI (2014) *Performance-based Requirements for Concrete.* ACI, Farmington Hills, MI, USA, Report 329R-14.
7. CSA (Canadian Standards Association) (2019) A23.1/A23.2-19. Concrete materials and methods of concrete construction/Test methods and standard practices for concrete. CSA, Toronto, ON, Canada.
8. Bjegović D, Beushausen H and Serdar M (eds) (2014) *Proceedings of RILEM International Workshop on Performance-based Specification and Control of Concrete Durability Zagreb, Croatia.*
9. Beushausen H and Luco LF (eds) (2015) *Performance-Based Specifications and Control of Concrete Durability.* Springer, State-of-the-Art RILEM Report TC230-PSC.
10. National Ready Mixed Concrete Association (2004) *The P2P Initiative: A Shift to Performance-based Specifications for Concrete Focuses on Innovation, Quality and Customer Satisfaction.* NRMCA, Silver Spring, MD, USA, www.nrmca.org/wp-content/uploads/2020/09/P2P_Initiative.pdf.
11. FHWA (Federal Highway Administration) (2008) *An Interactive Road Map to the Future of Concrete Pavement.* FHWA, Washington, DC, USA, Publication FHWA-HRT-08-017, https://www.fhwa.dot.gov/publications/focus/08sep/03.cfm.
12. Department for Transport (2014) *Road Investment Strategy: Performance Specification.* Department for Transport, London, UK. https://assets.publishing.service.gov.uk/government/uploads/system/uploads/attachment_data/file/382814/dft-ris-performance-specification.pdf (accessed 31/08/2022).

13. Cement and Concrete Association of New Zealand (2003) *Specifying Concrete For Performance*. Cement and Concrete Association of New Zealand, Report TR-10. https://www.ccanz.org.nz/files/documents/1f93c162-37d5-4b0f-85eb-1cfb7981ab8c/TR%2010.pdf (accessed 31/08/2022).
14. ACI Committee E-1 and CRSI Committee on Engineering Practice (1928) *Joint Code: Standard Building Regulations for the Use of Reinforced Concrete*. ACI, Farmington Hills, MI, USA.
15. ACI Committee 501 (1936) 501-36T. Building regulations for reinforced concrete. ACI, Farmington Hills, MI, USA.
16. Lord AR (1928) *Handbook of Reinforced Concrete Building Design*. ACI, Detroit, MI, USA (reprinted by Portland Cement Association, Concrete Reinforcing Steel Institute and Rail Steel Bar Association).
17. BSI (2020) BS 8500-1:2015 + A2:2019 Concrete. Complementary British Standard to BS EN 206-1. Part 1. Method of specifying and guidance for the specifier. Part 2. Specification for constituent materials and concrete. BSI, London, UK.
18. Seelye EE (1966) *Data Book for Civil Engineers, Specifications and Cost*, 3rd edn. Wiley, New York, NY, USA.
19. ACI (2019) 318-19. Building code requirements for structural concrete. ACI, Farmington Hills, MI, USA.
20. CEN (2011) EN 197-1. Cement. Part 1: Composition, specifications and conformity criteria for common cements. CEN, Brussels, Belgium.
21. ASTM (2022) *Standard Test Methods and Specifications*. ASTM, West Conshohocken, PA, USA, vols. 04.01 and 04.02, www.astm.org.
22. CIA (Concrete Institute of Australia) (2022) *Concrete Durability Series. Recommended Practices*. CIA, Australia, Report Z7, https://concreteinstitute.com.au/.
23. Hover KC (2003) *National Highway Institute, Course No. 131023. Highway Materials Engineering Module VI. Portland Cement and Concrete*. Department of Transportation, FHWA, Washington, DC, Publication No. HNI 04-112.
24. CEN (2013) EN 206. Concrete. Specification, performance, production and conformity. CEN, Brussels, Belgium.
25. Standards Australia (2018) AS 3600. Concrete structures. Standards Australia, Australia.
26. Li K (2016) *Durability Design of Concrete Structures: Phenomena, Modeling, and Practice*. Wiley, Singapore.
27. Hooton RD (2019) Future directions for design, specification, testing, and construction of durable concrete structures. *Cement and Concrete Research* **Special Issue 124F**, 105827.
28. Swedish Institute for Standards (2018) *Survey on Provisions Valid in the Place of Use Used in Conjunction with the European Concrete Standard and Developing Practice*. Swedish Institute for Standards, Sweden, Technical report CEN/TR 15868. https://www.sis.se/api/document/preview/80028164/.
29. NordTest (1999) NT Build 492. Chloride migration coefficient from non-steady-state migration experiments. NordTest, Espoo, Finland.
30. Tang L, Nilsson L-O and Basheer M (2017) *Resistance of Concrete to Chloride Ingress: Testing and Modelling*. CRC Press, Taylor & Francis.
31. NordTest (1997) NT Build 355. Chloride diffusion coefficient from migration cell experiments. NordTest, Espoo, Finland.
32. Castelotte M, Andrade C and Alonso C (2001) Measurement of the steady and non-steady-state chloride diffusion coefficients in a migration test by means of monitoring the conductivity in the anolyte chamber comparison with natural diffusion tests. *Cement & Concrete Research* **31(10)**: 1411–1420.
33. AASHTO (American Association of State Highway and Transportation Officials) (2019) T358-19. Standard method of test for surface resistivity indication of concrete's ability to resist chloride ion penetration. AASHTO, Washington, DC, USA.
34. AASHTO (2021) TP 119-19. Standard method of test for electrical resistivity of a concrete cylinder tested in a uniaxial resistance test. AASHTO, Washington, DC, USA.

35. AASHTO (2021) T277-21. Standard method of test for electrical indication of concrete's ability to resist chloride ion penetration. AASHTO, Washington, DC, USA.
36. Moro F and Torrent R (2017) Durability performance of concretes made with different cement types. *Proceedings, International Conference on Advances in Construction Materials and Systems, Chennai, India*, vol. 2, pp. 437–445.
37. ACI (2020) 301-20. Specifications for concrete construction. ACI, Farmington Hills, MI, USA.
38. Powers TC, Copeland LE and Mann HM (1959) Capillary continuity or discontinuity in cement pastes. *Journal of the PCA Research and Development Laboratories* **1(2)**: 38–48.
39. United States Bureau of Reclamation (1988) *USBR Concrete Manual*, 8th edn. https://www.usbr.gov/tsc/techreferences/mands/mands-pdfs/ConcreteMan-8th_Ed-rev.pdf (accessed 30/08/2022).
40. Ruettgers A, Vidal EN and Wing SP (1935) An investigation of the permeability of mass concrete with particular reference to Boulder Dam. *Proceedings of the American Concrete Institute* **31(3)**: 382–416.
41. Grove J (2006) *Using the Air Void Analyzer for Real-Time Quality Control Adjustments in the Field*. National Concrete Pavement Technology Center, Technical briefing. https://intrans.iastate.edu/app/uploads/2018/03/ava.pdf (accessed 31/08/2022).
42. Hover KC (1994) Air content and unit weight of hardened concrete. In *Significance of Tests and Properties of Concrete and Concrete Making Materials, Manual 169C* (Klieger P and Lamond JF (eds)). ASTM, West Conshohocken, PA, USA, pp. 282–295.
43. Boyd AJ and Hooton RD (2007) Long-term scaling performance of concretes containing supplementary cementing materials. *Journal of Materials in Civil Engineering* **16(10)**: 820–825.
44. Bouzoubaâ N, Bilodeau A, Fournier B, *et al.* (2008) Deicing salt scaling resistance of concrete incorporating supplementary cementing materials: A promising laboratory test. *Canadian Journal of Civil Engineering* **35(11)**: 1261–1275.
45. Bouzoubaâ N, Bilodeau A, Fournier B, *et al.* (2011) Deicing salt scaling resistance of concrete incorporating fly ash and/or silica fume: Laboratory and field sidewalk test data. *Canadian Journal of Civil Engineering* **38(4)**: 373–382.
46. ACI Committee on Responsibility in Concrete Construction (2005) Guidelines for authorities and responsibilities in concrete design and construction. *Concrete International* **27(4)**: 33–39.
47. NordTest (1995) NT Build 443. Concrete, hardened. Accelerated chloride penetration. NordTest Build, Espoo, Finland.
48. CIA (2001) *Performance Criteria for Concrete in Marine Environments*. CIA, Australia.
49. Sprinkel MM (2004) *Performance Specification for High Performance Concrete Overlays on Bridges*. Virginia Transportation Research Council, Department of Transportation and FHWA, Charlottesville, VA, USA, Report VTRC 05-R2.
50. Bognacki C, Marsano J and Wierciszewski M (2002) High performance aeronautical pavements. *Proceedings of the Federal Aviation Administration Airport Technology Transfer Conference, Atlantic City, NJ, USA*.
51. Bickley JA and Mitchell DA (2001) *State-of-the-art Review of High-performance Concrete Structures Built in Canada: 1990–2000*. Cement Association of Canada, Ottawa, Canada.
52. Ontario Ministry of Transportation (2014) OPSS.PROV 904. High performance concrete. Ontario Ministry of Transportation, Canada.
53. Ontario Ministry of Transportation (2019) OPSS.PROV 1350. Material specification for concrete – Materials and production. Ontario Ministry of Transportation, Canada.
54. Harrison TA (1997) Performance testing of concrete for durability. *Concrete* **31(10)**: 14–15.
55. Harrison T (2003) BS EN 206-1/BS 8500. Basics: Conformity and identity testing. *Concrete* **37(9)**: 50–51.
56. Harrison T (2003) *The New Concrete Standards – Getting Started: An Introductory Guide to the New Standards for Concrete BS EN 206-1 and BS 8500*. Concrete Society and Quarry Products Association, Crowthorne, UK.

57. Kessler RJ, Powers RG, Vivas E, Paredes MA and Virmani YP (2008) Surface resistivity as an indicator of concrete chloride penetration resistance. *Proceedings of the Concrete Bridge Conference, St Louis, MO, USA*.
58. Riding KA, Thomas MDA, Hooton RD, Obla KH and Weiss WJ (2018) Performance-based specifications for concrete exposed to chlorides. *Concrete International* **40(7)**: 41–47.
59. DuraCrete (1997) *Compliance Tests State-of the-art*. DuraCrete, Report BE95-1347/R6. CUR, Gouda.
60. Baroghel-Bouny V (2002) Which toolkit for durability evaluation as regards chloride ingress into concrete? Part II: Development of a performance specification approach based on durability indicators and monitoring parameters. *Proceedings of the 3rd International Workshop, Testing and Modelling Chloride Ingress into Concrete, Madrid* (Andrade C and Kropp J (eds)). RILEM, Bagneux, France.
61. Cabrera JG, Gowripalan N and Wainwright PJ (1989) An assessment of concrete curing efficiency using gas permeability. *Magazine of Concrete Research* **41(149)**: 193–198.
62. Claisse PA, Elsayad HI and Shaaban IG (1997) Absorption and sorptivity of cover concrete. *Journal of Materials in Civil Engineering* **9(3)**: 105–110.
63. Dhir RK, Hewlett PC, Byars EA and Shaaban, IG (1995) A new technique for measuring the air permeability of near-surface concrete. *Magazine of Concrete Research* **47(171)**: 167–176.
64. Dinku A and Reinhardt HW (1997) Gas permeability coefficient of cover concrete as a performance control. *Materials and Structures* **30(7)**: 387–393.
65. El-Dieb AS and Hooton RD (1995) Water permeability measurement of high performance concrete using a high pressure triaxial cell. *Cement and Concrete Research* **25(6)**: 1199–1208.
66. Figg JW (1973) Methods of measuring the air and water permeability of concrete. *Magazine of Concrete Research* **25(85)**: 213–219.
67. Hope BB and Malhotra VM (1984) The measurement of concrete permeability. *Canadian Journal of Civil Engineering* **11(2)**: 287–292.
68. Stanish K, Hooton RD and Thomas MDA (2000) Evaluation of four short-term methods for determining the chloride penetrability in concrete. In *Water–cement Ratio and Other Durability Parameters – Techniques for Determination* (Khan MS (ed.)). ACI, Farmington Hills, MI, Special publication SP-191, pp. 81–98.
69. Streicher PE and Alexander MG (1995) A chloride conduction test for concrete. *Cement and Concrete Research* **25(6)**: 1284–1294.
70. Hooton RD (2001) Development of standard test methods for measuring fluid penetration and ion transport rates. *Materials Science of Concrete: Fluid and Ion Transport Rates in Concrete* Nov., pp. 1–12.
71. Tang L and Nilsson L-O (1991) Rapid determination of the chloride diffusivity in concrete by applying an electrical field. *ACI Materials Journal* **89(1)**: 49–53.
72. Stanish, KD, Hooton RD and Thomas MDA (2001) *Prediction of Chloride Penetration in Concrete*. Department of Transportation, FHWA, Washington, DC, USA, Report FHWA-RD-00-142.
73. Hooton RD, Geiker MR and Bentz EC (2002) Effects of curing on chloride ingress and implications of service life. *ACI Materials Journal* **99(2)**: 201–206.
74. Andrade C, Alonso C and Goni S (1993) Possibilities for electrical resistivity to universally characterise mass transport processes in concrete. In *Concrete 2000* (Dhir RK and Jones MR (eds)). Spon, Dundee, UK, vol. 2, pp. 1640–1652.
75. Millard SG, Harrison JA and Edwards JA (1989) Measurements of the electrical resistivity of reinforced concrete structures for the assessment of corrosion risk. *Journal of Non-destructive Testing* **31(11)**: 617–621.
76. Monfore GE (1968) The electrical resistivity of concrete. *Journal of the PCA Research and Development Laboratories* **10(2)**: 35–48.
77. Morris W, Moreno EI and Sagues AA (1996) Practical evaluation of resistivity of concrete in concrete cylinders using a Wenner array probe. *Cement and Concrete Research* **26(12)**: 1779–1788.

78. Balayssac JP, Detriche CH and Grandet J (1993) Validity of the water absorption test for characterising cover concrete. *Materials and Structures* **26(4)**: 226–230.
79. Hall C (1989) Water sorptivity of mortars and concretes: A review. *Magazine of Concrete Research* **41(147)**: 51–61.
80. Price WF and Bamforth PB (1993) Initial surface absorption of concrete: examination of modified test apparatus for obtaining uniaxial absorption. *Magazine of Concrete Research* **45(162)**: 17–24.
81. McCarter WJ, Ezirim H and Emerson M (1996) Properties of concrete in the cover zone: Water penetration, sorptivity and ionic ingress. *Magazine of Concrete Research* **48(176)**: 149–156.
82. Ho DWS and Chirgwin GJ (1996) A performance specification for durable concrete. *Construction and Building Materials* **10(5)**: 375–379.
83. Desouza SJ, Hooton RD and Bickley JA (1997) Evaluation of laboratory drying procedures relevant to field conditions for concrete sorptivity measurements. *Cement, Concrete, and Aggregates* **19(2)**: 92–96.
84. Desouza SJ, Hooton RD and Bickley JA (1998) A field test for evaluating high performance concrete covercrete quality. *Canadian Journal of Civil Engineering* **25(3)**: 551–556.
85. Desouza SJ, Hooton RD and Bickley JA (2000) Practical QC test program for HPC in precast concrete tunnel liners. In *Water–cement Ratio and Other Durability Parameters – Techniques for Determination* (Khan MS (ed.)). ACI, Farmington Hills, MI, Special publication SP-191, pp. 99–115.
86. Wilson CW (1968) Plant certification – A program of merit in the prestressing industry. *PCI Journal* **13(2)**: 12–18.
87. PCI (Precast/Prestressed Concrete Institute) (1999) *Manual for Quality Control for Plants and Production of Structural Precast Concrete Products,* MNL-116-99, 4th edn. PCI, Chicago, IL, USA.
88. PCI (2001) PC-4. Certified quality for people, products and performance. PCI, Chicago, IL, USA.
89. Goodspeed CH, Vanikar S and Cook RA (1996) High-performance concrete defined for highway structures. *Concrete International* **18(2)**: 62–67.
90. Taylor P (2004) Performance based specifications for concrete. *Concrete International* **26(8)**: 91–93.
91. Torrent RJ and Ebensperger L (2013) Site air-permeability test: credentials as 'durability meter'. *Proceedings of CONSEC'13, Nanjing, China*, pp. 1345–1360.
92. SIA (Swiss Society of Engineers and Architects) (2013) 262/1:2013. Concrete construction – Complementary specifications, Annex E, air-permeability on structures. SIA, Switzerland.
93. Hart AJ, Ryell J and Thomas MDA (1997) High performance concrete in precast tunnel linings; Meeting chloride diffusion and permeability requirements. *Proceedings of the PCI/FHWA International Symposium on High Performance Concrete, New Orleans*, pp. 294–308.
94. Hooton RD (2017) Obtaining durable concrete using performance specifications. *Proceedings of the International Conference on Advances in Construction Materials and Systems, Chennai, India*, vol. 1, pp. 45–54.

Soutsos M
ISBN 978-0-7277-6375-4
https://doi.org/10.1680/icehcd.63754.329

Chapter 10
Modelling and predicting the effects of deterioration mechanisms

Klaas van Breugel, Eric Schlangen, Oguzhan Copuroglu and Guang Ye
TU Delft, The Netherlands

10.1. Role of models in durability and service life design

Generally, most degradation processes in concrete are very slow. The slow rate of deterioration processes is a complicating factor for modelling. Many of these processes have multiple causes, often there are coupled chemical and physical processes, and concrete has a heterogeneous nature, all of which contribute to the complexity of modelling and predicting deterioration processes.

Depending on the particular purpose for which models are needed, one can distinguish between *science-oriented* models and *engineering* models. Even though both types of model aim to mimic reality as closely as possible, the different aims of the models may mean that they have a completely different structure. That models are developed for different purposes should be considered carefully when discussing and judging them.

Despite the aforementioned complicating factors, there is an increasing demand for reliable models able to predict deterioration and service life. Contractors are asked to build structures with service lives of 80, 100 or even 200 years, especially for infrastructure. Currently used design codes generally do not give sufficient guidance for designing concrete structures for such long service lives. Service life predictions presuppose that the rate of change of the performance of a structure can be determined reliably and accurately over time. These changes happen because of changes to the materials, that is, to the concrete or the reinforcing or prestressing steel. Changes to the concrete may be caused by ongoing hydration, in which case the strength and stiffness of the concrete increase and the material becomes denser. Those changes will generally result in an increase of the service life of concrete structures. Most feared are changes caused by degradation processes. Transport processes are involved in most forms of degradation. The rate of degradation is, therefore, related to the rate of transport processes and, hence, to the transport properties of the concrete and how these change. The accuracy of service life predictions depends on the accuracy with which the evolution of the mechanical and transport properties can be simulated over time. Moreover, the deterioration processes themselves, such as carbonation, the alkali–silica reaction (ASR) and rebar corrosion, must be understood and modelled accurately.

Once the evolution of material properties with time and degradation mechanisms are known, it is possible to predict changes of performance over time and to estimate the service life The outcome of such predictions, however, very much depends on the values of the input parameters. In reality, these values have a certain scatter. Each parameter can be described with a distribution function, for example, with a mean value and standard deviation. Thus described, the parameters can be used for making probabilistic service life

Figure 10.1 Schematic representation of probabilistic durability design [2]

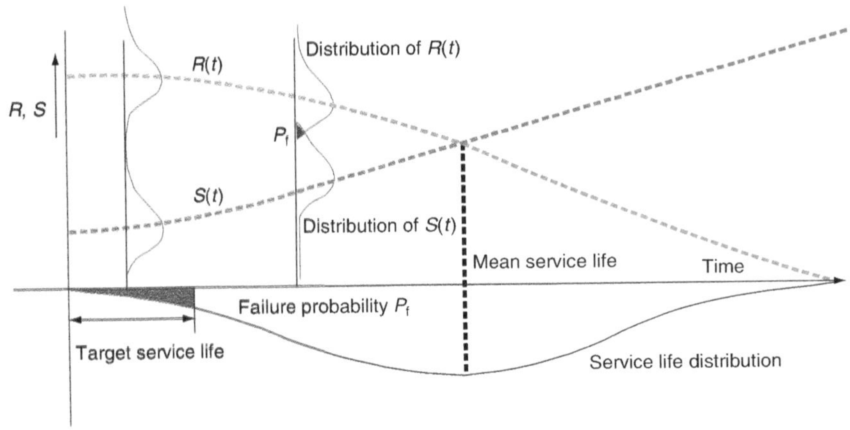

predictions [1]. In the 1990s, such a probabilistic design concept for assessing service lives was developed in the European research project DuraCrete [2,3]. The DuraCrete code follows the probabilistic design philosophy as used for structural safety. In a probabilistic service life prediction, the service life is defined as the period in which the *resistance* of a structure $R(t)$ can withstand the environmental *load* $S(t)$. The resistance $R(t)$ and load $S(t)$ are characterised by a normal distribution or another appropriate distribution function. The concept is schematically shown in Figure 10.1. The probabilities of failure for the serviceability limit state are usually in the range of a few per cent (reliability index $\beta = 1.5–2$).

As already noted, the accuracy of a service life prediction depends on the accuracy with which the evolution over time of the materials and transport properties and the deterioration processes can be simulated. The relevant length scales at which these properties and processes have to be known range from the atomic level to the mesoscopic level. The potential of some numerical models for tackling some specific durability phenomena will be briefly presented in the following sections. They give a flavour of current developments of the numerical simulation models used for studying and predicting particular durability issues.

10.2. Numerical simulation of the evolution of material properties

10.2.1 Early steps towards modelling hydration and strength development

The first steps in modelling hydration processes date back to the first decades of the 20th century. The focus was to simulate the temperature fields in hardening mass concrete structures, like concrete dams. These massive structures were prone to cracking as the concrete cooled. To predict the probability of cracking, the evolution of the modulus of elasticity and the tensile strength of the concrete had to be determined. Moreover, the effect of the time-dependent properties of the young concrete had to be known. In those days, the evolution of hydration and the evolution of material properties were often formulated as functions of time. In the absence of detailed knowledge about the relations for the evolution of material properties and the underlying microstructure, the use of time functions was the best one could do then.

In the early 1950s, models for hardening concrete were developed in which the evolution of the degree of hydration was quantified. This hydration level could be used as the basis for determining the volumes of the constituents of the cement paste. This approach was used until the late 1970s when models were developed that attempted to describe the evolution of the microstructure explicitly at the microscale. The increasing computational power of computers as well as deeper insights into hydration processes made it possible to describe the development of inter-particle contacts and the associated evolution of the mechanical and transport properties.

10.2.2 Modelling of hydration processes and microstructure development

Since the 1970s, a few types of simulation models for hydration and microstructure development have been proposed. An overview of advances of modelling up to the second decade of this century was presented by Dolado and van Breugel [4]. One of the most often quoted models is the NIST model [5]. It is a pixel-based model, each pixel representing a certain component of the cement paste to which certain chemical and physical properties are assigned. The pixels, of size 1 μm^3, can 'react' with other pixels to build a virtual microstructure.

Pioneering modelling work was performed by Jennings and Johnson [6] in the early 1970s. Later modelling work by Navi and Pignat [7] and van Breugel's HYMOSTRUC model [8,9] were inspired by Jennings and Johnson's first attempts to simulate hydration processes and microstructure development from a system of randomly distributed spaces in water, which represent cement or other powders in the paste. On reacting with water, the spheres 'grow'. Due to this growth, the spheres make mutual contact with each other to build a virtual microstructure. The formation of inter-particle contacts affects the rate of reaction of individual particles. Reactions that allow for the effects of physical interactions between particles on the rate of reaction are referred to as integrated kinetics [10]. The DUCOM concept [11] is also based on spherical particle growth but starts with a mono-sized particle system. Detailed modelling issues concern setting the multi-phase characteristics of the reactants, including the types of cement and reactive powders, the formation of a variety of reaction products and an accurate simulation of the pore structure, particularly small pores. More recently, results of studies have been published illustrating the capacity of advanced simulation models to cope with multi-phase reactions in blended cements and the formation of different reaction products [12]. These extensions of simulation models are essential for modelling chemical and chemo-mechanical deterioration processes.

After model-specific calibration, these models can simulate hydration processes and, hence, the evolution of hydration heat, quite accurately. Examples of simulated microstructures and related pore structures are presented in Figure 10.2.

Figure 10.2 Examples of virtual microstructures for cement paste: (a) NIST pixel model [5] and (b) HYMOSTRUC3D-E model of blended cement [12]. BFS, blastfurnace slag; C-S-H, calcium silicate hydrate; FA, fly ash; PC, Portland cement

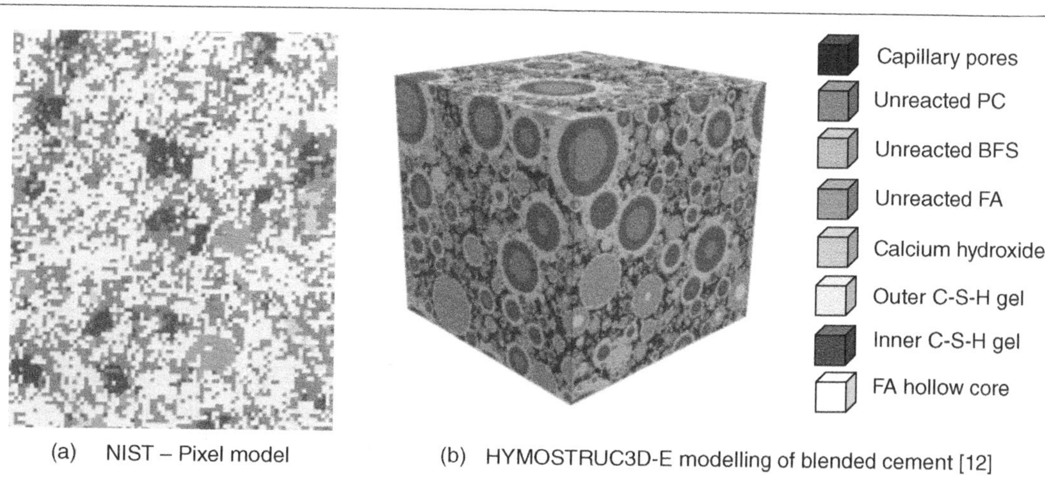

For individual concrete mixtures, isothermal or adiabatic hydration curves can be simulated, which can be input into advanced software packages that determine the temperature distribution in hardening concrete. In these software packages, which are often incorporated in concrete curing control systems, the effect of the actual curing temperature on the rate of hydration is allowed for by using the Arrhenius law. Besides the temperature distribution in hardening concrete structures, hydration fields are also obtained.

Predicting the microstructural details of a cement paste is much more complicated than predicting temperature distributions and the hydration level. Detailed simulation of the stoichiometry of the reaction products and of the pore structure, however, is essential for modelling deterioration and transportation processes [12].

10.2.3 Evolution of material properties

The evolution of the material properties is based on the evolution of the microstructure. This holds for all the material properties, such as strength, stiffness, permeability, hydration-related volume changes due to chemical shrinkage, autogenous shrinkage or thermal shrinkage, and time-dependent properties due to creep and relaxation. During hydration, all these properties change over the progress of the hydration process. To ensure consistency in the quantitative analysis of young concrete, the evolution of material properties should, preferably, be related to one single materials parameter. In modelling, therefore, the challenge is to relate the material properties to the characteristic properties of the virtual microstructure. Figure 10.3 shows an example of an attempt to relate the compressive strength of cement paste to the summarised contact area of the virtual microstructure of the paste. The growth of the stiffness and changes in the ultrasonic pulse velocity in hardening cement paste correlate well with the intensity with which the particles of a virtual microstructure are connected. Ouyang [14] showed that similarly good results could also be obtained between the compressive strength of cement pastes 'loaded' with fine fillers, such as micronised sand or limestone powder, and the simulated contact surface areas of their virtual microstructures.

Carol and Bazant [15] and Lokhorst and van Breugel [16] proposed models that explicitly relate creep and relaxation in hardening concrete to the evolution of the microstructure. All these models illustrate that researchers are at least aware that the direct coupling of material properties to the properties of the

Figure 10.3 (a) Schematic diagrams of a microstructure (left) and the contact areas between spheres (right). (b) Compressive strength as a function of summarised contact area of a virtual cement paste for mixtures with a water-to-cement (w/c) ratio ranging from 0.35 to 0.6 [13]

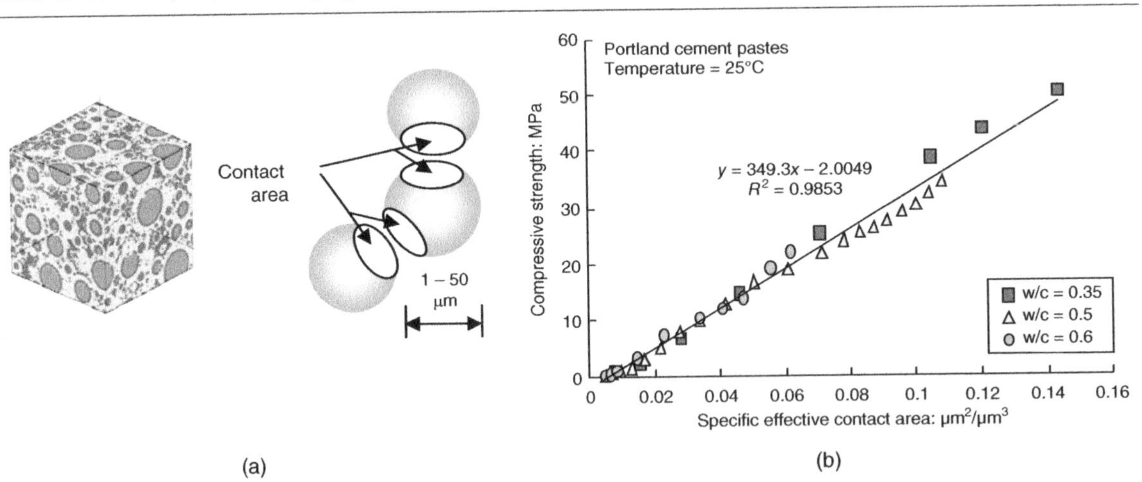

microstructure is a most challenging and, in the long-term, most promising direction for modelling the performance of materials and structures. For the time being, however, it seems more realistic to relate the evolution of material properties to the hydration level.

10.2.4 Stress development and risk of cracking in hardening concrete

Stress calculations for hardening concrete aim to predict the probability of cracking during the early lifetime of a structure. Calculating the probability of cracking requires information about the evolution of the material properties, the strain generated and the degree of restraint.

To determine the probability of cracking, both the mean values and the standard deviations of the relevant parameters – tensile stress, tensile strength or fracture energy – must be known. Basically, the scheme presented in Figure 10.4 applies, where $R(t)$ is the strength and $S(t)$ the stress. For hardening concrete, in particular, however, the strength $R(t)$ does not decrease but gradually increases with time. Parameter studies and sensitivity analyses have revealed that the degree of restraint and the time-dependent properties are very important for the stress generated and risk of cracking [16–18].

Figure 10.5 shows the results of a calculation of the evolution of the probability of cracking at the centre and surface of a hardening concrete wall. The development of the probability of cracking is given for 0.25-, 0.5- and 1.0-m-thick walls over time. These were cast on an already hardened floor slab, which restrained the deformations. The wall/slab structure could bend freely ($r_\varphi = 0$). The cracking probability was determined for the most critical cross-section, in this case about 0.5 m above the wall/slab connection. With these prevailing boundary conditions, the 1.0-m-thick wall would almost certainly crack. The 0.5-m-thick wall had a cracking probability of more than 50%. Due to the bending of the wall, there was compressive stress at the top of the wall. The reduction of the probability of cracking after reaching a maximum occurred because of the gradual decrease of the temperature differences in the cross-section.

The use of software packages to analyse the performance of hardening concrete has substantially contributed to our understanding of the state of stress and the risk of cracking in young concrete. Importantly, however, the development of modern concrete mixtures with a low water-to-binder (w/b) ratio has revealed the need for more advanced models. In the past, the performance and probability of cracking of traditional

Figure 10.4 Schematic diagram showing the development of tensile strength and stress with time in hardening concrete. The imposed thermal strains are assumed to be restrained [18]

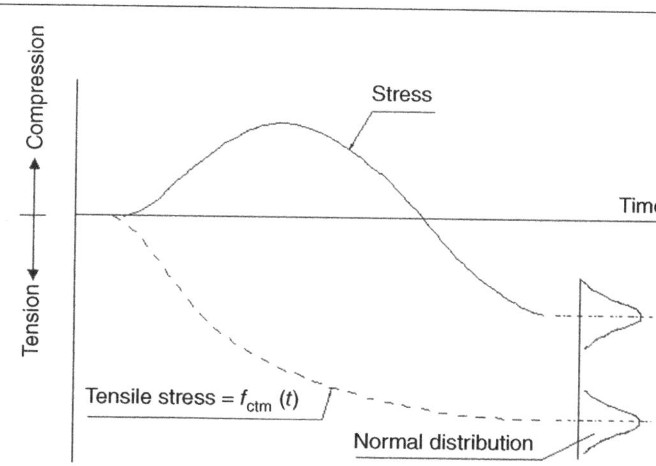

Figure 10.5 Probability of cracking in 0.25-, 0.50- or 1.0-m-thick wall as a function of time. Portland cement CEM I 52,5 R. Casting temperature 15°C. Temperature of floor slab 10°C. No rational restraint ($r\varphi = 0$) [18]

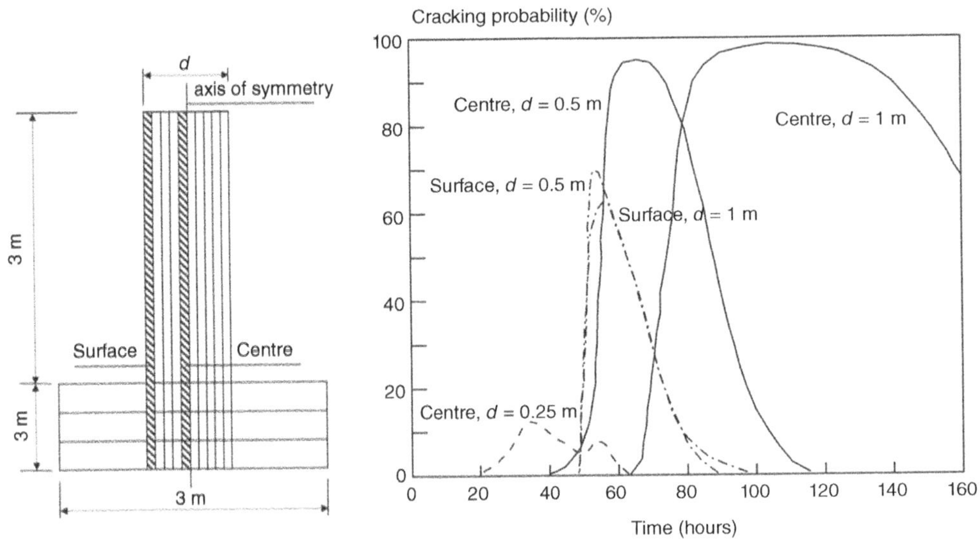

concrete mixtures could be determined satisfactorily using experience-based temperature criteria. These criteria, however, fail for concrete mixtures with a low w/b ratio that undergo substantial autogenous deformation. In concrete mixtures with w/b ratios below 0.4, the contribution of autogenous strain to the risk of cracking cannot be ignored.

10.3. Carbonation

The carbonation mechanism of concrete has been studied and described by many authors. The carbonation reaction occurs at the water/air frontier in the pore system of the cement paste. In a drying pore system, the reaction front penetrates deeper into the concrete unless the reaction product, calcium carbonate, blocks the pores and is not leached out. Calcium carbonate is formed according to the following reactions:

$$CO_2 + H_2O \rightarrow H_2CO_3 \qquad (10.1)$$

$$H_2CO_3 + Ca(OH)_2 \rightarrow CaCO_3 + 2H_2O \qquad (10.2)$$

As well as the calcium hydroxide, the C-S-H gel is also affected [19]. The major concern, however, is the consumption of the calcium hydroxide and the associated reduction of the pH, since this reduction poses a corrosion risk to the reinforcement. If the alkalinity decreases below pH 10, the steel is no longer passivated and corrosion can start over the total depassivated area, provided that all the preconditions for corrosion are fulfilled.

The most often used formula for predicting the rate of carbonation is based on the solution of Fick's second law of diffusion:

$$X = A\sqrt{t} \qquad (10.3)$$

where X is the penetration depth of the carbonation front, A is a constant and t is time. The constant A depends on the w/c ratio, the type of cement (for blended cements, A can vary quite substantially) and the hydration level. The latter depends on the curing of the concrete. In particular, early drying out of the concrete will result in enhanced carbonation rates.

Because of the many parameters that affect the actual rate of carbonation, Equation 10.3 has limited validity. The best results are obtained for temperatures of about 20°C and 65% relative humidity. Under other conditions, the carbonation rate is generally lower. When submerged, concrete will not carbonate at all.

According to Audenaert *et al.* [20], the use of blended cements affects the rate of carbonation, mainly due to associated changes in the capillary porosity and the amount of carbonatable material. Therefore, they proposed a model that accounts for the amount of carbonatable material and the capillary porosity. The model was based on work by Papadakis [21] and, to describe the pore system, the work of Powers and Brownyard [22]. Thiery *et al.* [23] proposed a more comprehensive simulation model for carbonation based on coupled CO_2–H_2O–ion transfers and chemical reactions. This model explicitly addresses the mass balances.

10.4. Chloride-induced corrosion

10.4.1 Mechanisms of chloride-ion penetration

Today, corrosion mechanisms are qualitatively well understood. Yet, it is still difficult to model rebar corrosion quantitatively. Up to now, most models have focused on the penetration of chloride ions and the numerical determination of the time needed to reach the chloride threshold at the surface of the steel. Chloride ions penetrate the concrete by *diffusion*, *convection* and *migration*. In all cases, there must be a concentration gradient if the ions are to be transported into the concrete. Temperature gradients may further complicate the transport processes. Furthermore, the binding of chloride ions to reaction products substantially affects the rate at which the chloride front penetrates the concrete.

10.4.2 Engineering models for chloride penetration

Even though several transport mechanisms play a role in the ingress of chloride ions into concrete, most models of chloride ion penetration are based on Fick's second law of diffusion. For a one-dimensional case:

$$C(x,t) = C_s - (C_s - C_i)\,\mathrm{erf}\left[\frac{x}{\sqrt{4k\,D(t)t}}\right] \tag{10.4}$$

where $C(x,t)$ is the chloride content at depth x at time t, C_s the surface chloride content, C_i the initial chloride content, k a correction factor and $D(t)$ a time-dependent diffusion coefficient. Reasons for the time dependency of this coefficient are continuing hydration, binding of chloride ions and microcracking or, conversely, self-healing of microcracks. The following has been proposed to allow for this time dependency of the diffusion coefficient (see also [24]):

$$D(t) = D_0\left(\frac{t_0}{t}\right)^n \tag{10.5}$$

where D_0 is the diffusion coefficient at reference time t_0 (e.g. 28 days) and n is the ageing coefficient. By combining Equations 10.4 and 10.5, we get

$$C(x,t) = C_s - (C_s - C_i)\,\mathrm{erf}\left[\frac{x}{\sqrt{4k\,D_0(t_0/t)^n\,t}}\right] \tag{10.6}$$

The use of Fick's second law for the transport of chloride ions into concrete has often been criticised, since it presupposes diffusion to be the decisive and, in fact, only transport mechanism, whereas in reality, other mechanisms play a role as well. Depending on the exposure conditions, the dominating transport mechanism in the outer 10 to 15 mm is either capillary suction or diffusion. The binding of chloride ions and other influencing factors mean that the diffusion coefficient must be considered as an *apparent value*. The coefficients k and n, which are found by fitting a curve, partly reflect the effect of a cocktail of influencing factors and different transport mechanisms. For offshore structures, the transport mechanisms are indicated in Figure 10.6. In the atmospheric zone, gas diffusion and water vapour diffusion occur. In the splash zone, water vapour diffusion and ionic diffusion may dominate the transport processes. In the tidal zone, water may become absorbed, whereas in the submerged zone, ionic diffusion and permeability must be considered. The model parameters have different values in each zone. Moreover, for example, the type of cement also influences the transport properties.

An example of an engineering model that has adopted Fick's second law of diffusion as the core element when analysing chloride ion profiles is the DuraCrete model [2], as mentioned in Section 10.1. That model uses an experimental value for D_0 determined for 28-day concrete samples with a rapid chloride-migration test. The surface chloride content is assumed to be independent of mix composition for simplicity, namely 3.0% m/m of cement content for marine structures [25] and 1.5% m/m for structures on land [26]. The initial chloride content in the concrete is set to 0.1% m/m.

The value of the ageing coefficient for a particular binder depends on the rate of hydration and on the extent of drying. Based on a large amount of data, n was chosen for different binders in two groups of environmental classes: very wet (XD2 and XS3) and moderately wet (XD1, XD3 and XS1) [25,27]. Table 10.1 shows the values of n used for mixtures made with different binders and for two clusters of exposure conditions.

Figure 10.6 Exposure conditions covered by model coefficients k and n in Equation 10.6

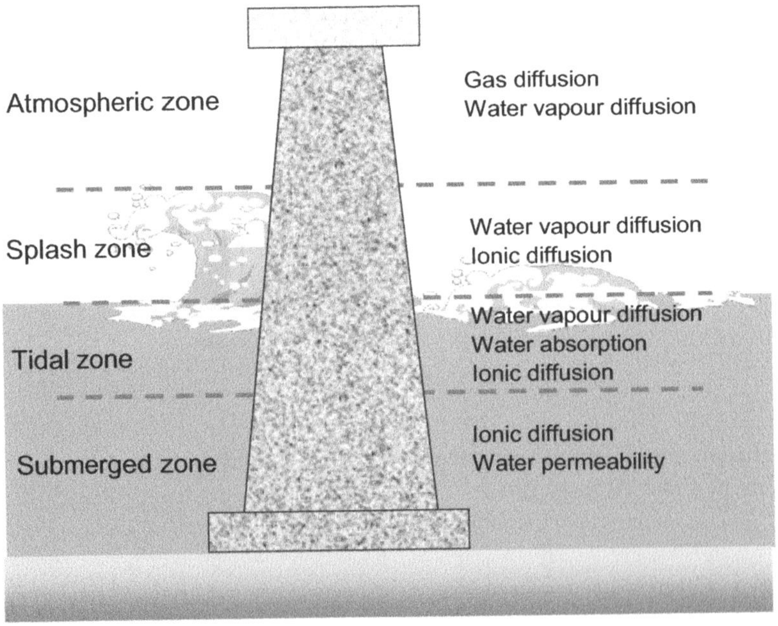

Table 10.1 Ageing coefficient *n* for different binders and different environmental classes [25,27]

Type of binder	Environmental class (NEN-EN 206)	
	Underground or the splash zone (XD2, XS3)	Above ground or the marine atmospheric zone (XD1, XD3, XS1)
CEM I	0.40	0.60
CEM I (25–50% slag), II/B-S or III/A (<50% slag)	0.45	0.65
CEM III (50–80% slag)	0.50	0.70
CEM I with 21–30% fly ash	0.60	0.70
CEM V/A (25% slag and 25% fly ash)	0.60	0.70

In an extensive field study, in which the chloride profiles in the concrete of five large marine structures were compared with simulations with the DuraCrete model, there was good agreement between the measured and simulated chloride profiles [25].

10.4.3 Durability designs in practice

One of the first large projects for which the DuraCrete concept was used was the Green Heart Tunnel in the Netherlands for the Dutch part of the high-speed train connection between Amsterdam and Paris. A 100-year service life was required for this tunnel, which has a length of more than 7 km. Construction was finished in 2006. Since that time, a monitoring system has acquired data about the real rate of degradation processes, particularly chloride ingress. More recently, the DuraCrete concept has also been used for the durability design of the Busan–Geoje Link Project, a bridge and tunnel project in South Korea [28].

To validate the DuraCrete concept, five marine structures in the Netherlands have been inspected to determine their chloride profiles. The structures were built over the period 1960–1985. Two of these structures, a quay wall and a sluice complex, are shown in Figure 10.7. Good agreement was found between the observed and measured chloride profiles [27].

Figure 10.7 Quay wall (built 1982) and a sluice complex (built in 1960), used for validating numerical predictions of chloride profiles [27]

In those cases where the cover depths and concrete quality agreed with the specification, rebar corrosion was neither predicted nor observed. Corrosion was observed in one structure due to the poor quality of the concrete and the cover depth of between 20 and 25 mm, which was much lower than specified [25].

Based on the predictive models, for example, Equation 10.6, design tables have been established from which appropriate combinations of cover depth and effective diffusion coefficients can be determined for a required service life [29]. In these tables, the exposure conditions and type of cement are also parameters. Similar tables have been presented recently by Li *et al.* [30].

10.4.4 Advanced multiscale approach

Several transport mechanisms are involved in the ingress of ions into concrete. Moreover, the transport properties of concrete can vary across the thickness of a concrete element. By using more advanced hydration and microstructural models, the evolution of the microstructure with time and across the thickness of a concrete element can be quantified explicitly. This is illustrated schematically in Figure 10.8. Instead of using an apparent diffusion coefficient with tuned model coefficients, the transport of ions can now be calculated by taking into account the actual changes of the microstructure and the related transport properties across the thickness of the concrete element.

When numerical simulations of transport processes use a finite element or discrete element mesh in which the actual transport properties are assigned to individual elements, then the model parameters that are needed in the analytical approach (Section 10.4.2) to get a good fit of actual ion profiles have no meaning anymore.

10.4.5 Detailed numerical simulation of microcracking in cement paste

The concrete in the surface zone of a concrete structure is prone to temperature- and shrinkage-induced (micro)cracking. This surface zone is important for protecting the reinforcing steel in the concrete against corrosion. How the concrete, namely, the cement paste, will crack under an imposed strain or stress has been simulated by Qian [31] using the Delft lattice model [32,33]. This model was developed in the 1990s for simulating fracture processes in heterogeneous systems, like mortars and concrete, at the *mesoscale*. Cement paste itself, however, is also a heterogeneous material, consisting of hydration products (gel) and

Figure 10.8 Development of material properties across the thickness of the concrete cover: (a) analytical model, (b) microscale model and (c) macroscale model

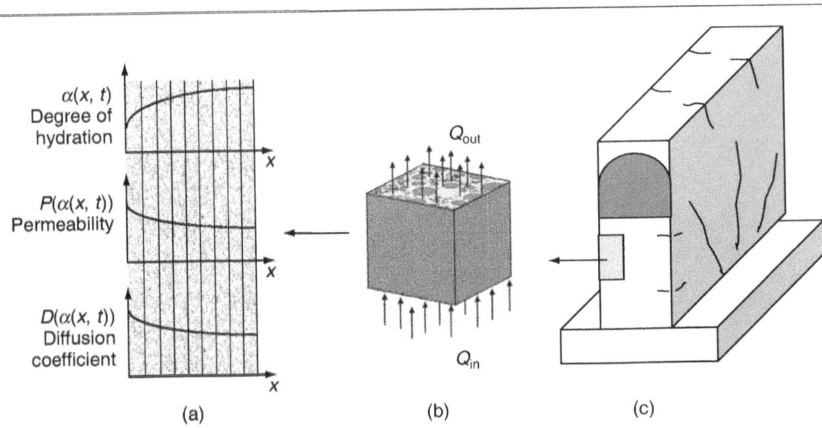

Figure 10.9 (a) Virtual 3D microstructure used to generate a lattice mesh. (b) Delft lattice model. (c) Sample of a 100 × 100 × 100 μm³ lattice system in tension [31,32]

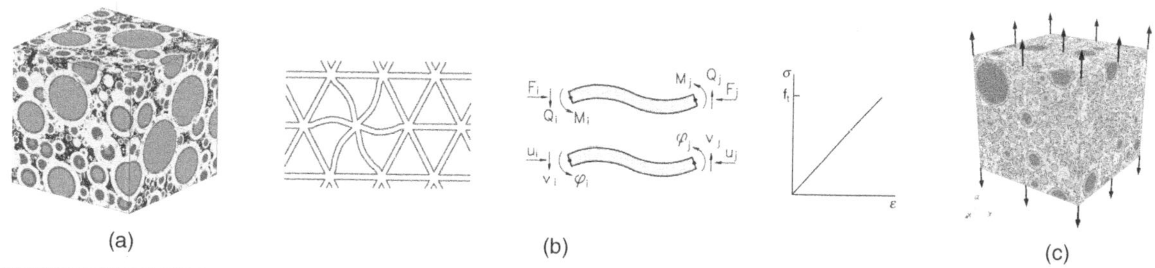

unhydrated cement cores. Qian [31] used the Delft lattice model to analyse stress patterns and the cracking of a virtual microstructure at the microscale. The cement grains were connected by beam elements, which form a three-dimensional (3D) network (Figures 10.9(a) and 10.9(b)). The mechanical properties of the elements are a function of the size of the contact areas connecting the cement particles. The virtual microstructure is then loaded in tension in a virtual test in which strain is controlled (Figure 10.9(c)). If the stress in one of the lattices exceeds its strength, this element will be 'removed' from the system, and a new equilibrium is established in a numerical iteration process.

This model has been used to analyse the effects of a variety of mixture parameters, like the fineness of the cement, the hydration level and the w/c ratio [31,32]. Figure 10.10 depicts crack patterns in a cement paste sample, w/c ratio 0.4, loaded in tension (deformation controlled) at a hydration level of 44%, 69% or 90%, respectively, for Figures 10.10(a) to 10.10(c). At the lowest hydration level, the cracks are spread throughout the volume of the paste. Cracking is more localised at the highest hydration level because fewer microcracks formed and the system was more brittle. Similar results were found in experimental studies of the tensile properties of young-age concrete by Dao *et al.* [34].

Figure 10.10 Cracking patterns in the virtual microstructure of cement paste for w/c = 0.4. Samples 100 × 100 × 100 μm³ [32]: (a) $\alpha = 44\%$, (b) $\alpha = 69\%$, and (c) $\alpha = 90\%$. The hydration level is 44%, 69% and 90%, respectively

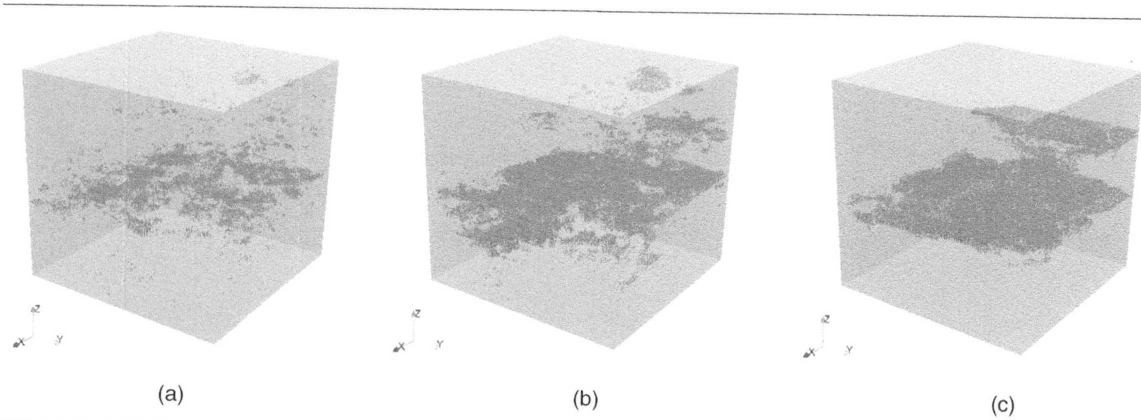

10.5. Alkali–silica reaction

10.5.1 Formation of an expanding gel

Most degradation mechanisms in concrete can be characterised by three basic mechanisms. First, the process starts with the transport of fluids or gases through the microstructure of the material. Second, there are chemical reactions among the components of the material. Third, these reactions induce expansion or shrinkage of the microstructure, leading to eigenstresses, cracking and damage of the concrete. The ASR is one of the degradation mechanisms that occur between alkali species in the pore solution and a reactive silica component in aggregate.

An alkali reacts with water:

$$Na_2O + H_2O \rightarrow 2NaOH \tag{10.7}$$

On reacting with silica, the alkali hydroxide produces an expanding gel:

$$SiO_2 \cdot n(H_2O) + 2NaOH \rightarrow Na_2SiO_3 \cdot (n+1)(H_2O) \tag{10.8}$$

Figure 10.11 shows typical gel formations and damage patterns. The ASR is often very slow, causing damage only after several years, if not decades. Therefore, accelerated testing is considered an essential tool for estimating the performance of concrete. Furthermore, it is believed that successful numerical models can help in assessing structures exposed to ASR.

10.5.2 Numerical approach for simulating ASR damage

The expanding alkali–silica gel will exert compressive and tensile stresses in the concrete. Finite element methods or a discrete lattice model can be used to simulate the effect of the expanding gel on the concrete. Finite element methods are adequate for macroscale analyses of the effect that expanding concrete structures exert on their supports or foundation. For a quantitative analysis of the effect of an expanding gel at the mesoscale, a lattice model is more appropriate. The Delft lattice model [33,36] was mentioned in Section 10.4.5.

Figure 10.11 Variation in ASR gel formation and damage: (a) gel formation in the paste (cracked in the dry state) and (b) gel formation at the interface between the aggregate and the paste (interface cracking) [35]

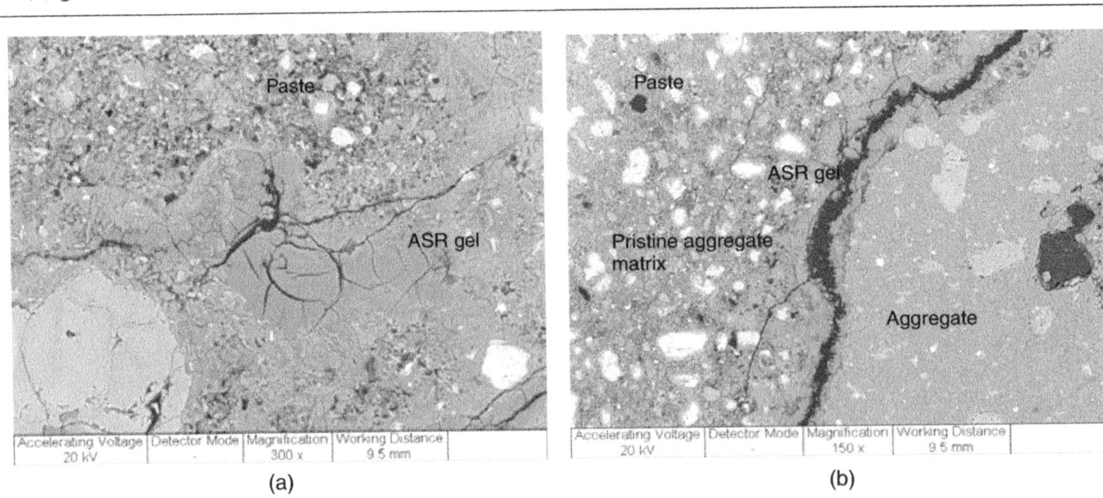

Figure 10.12 Optical micrographs of a basaltoid aggregate and surrounding paste damaged due to ASR [35]

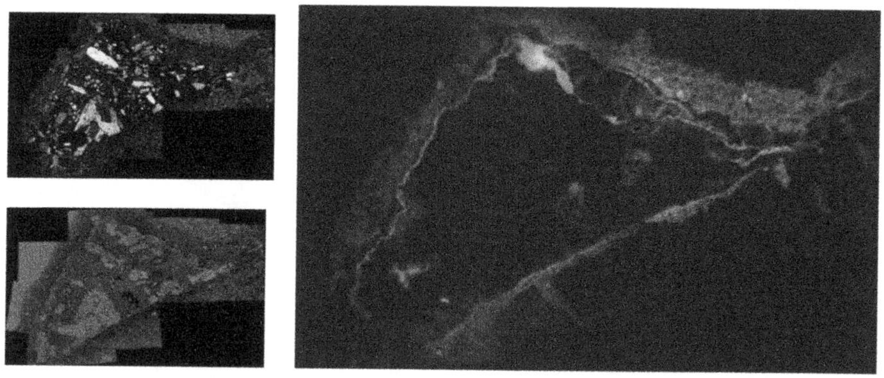

A lattice model has been used to simulate ASR damage in concrete microbars, which are often used to investigate the susceptibility of a concrete to ASR. First, concrete was prepared with aggregates of between 4.75 and 12.5 mm, a dry aggregate-to-cement ratio of 1.0 by weight and a w/c ratio of 0.33. The aggregate was a clinopyroxene-olivine andesite called basaltoid that is affected by ASR. A prismatic 40 × 40 × 160 mm microbar specimen was cured for 1 day in 80°C water and for 14 days in 80°C 1 N NaOH solution. The specimen expanded by 0.8% and showed significant cracking. Figure 10.12 shows a picture of a thin section taken from the prismatic bar.

Next, the material was discretised in the model as a lattice consisting of 42 000 small beam elements. Random values were assigned as input parameters for the beam elements in various locations for strength (and stiffness). The values were randomly chosen in the ranges given in Table 10.2. The stiffness of each element was calculated by multiplying the strength by 10 000. The sample had a single aggregate particle.

Swelling ASR gel was placed at three spots: inside the aggregate, inside the interface between the aggregate and cement matrix, and inside the cement matrix. The three locations are shown as black dots in Figure 10.13. It was assumed that the ASR gel in the three locations would swell at the same rate. Three different stages of cracking are shown in the figure. It shows that the deformations grew at the same rate in each location. These simulation results look realistic and agree well with the experimental observations in Figure 10.12.

The properties of the gel are a crucial input for a model making quantitative predictions of how much swelling and damage can be expected. Experimental studies on the expansive properties of the gel and its

Table 10.2 Strength and stiffness of components used in the lattice simulation

Material	Strength: MPa
Aggregate	7.0–13.0
Dissolved rim of aggregate	1.0–2.0
Interfacial transition zone	1.5–2.5
Cement matrix	3.0–5.0
ASR gel	1.0

Figure 10.13 ASR damage sequence simulated with a lattice model. Gel was placed in three locations, as was observed in the micrographs in Figure 10.12 [36]

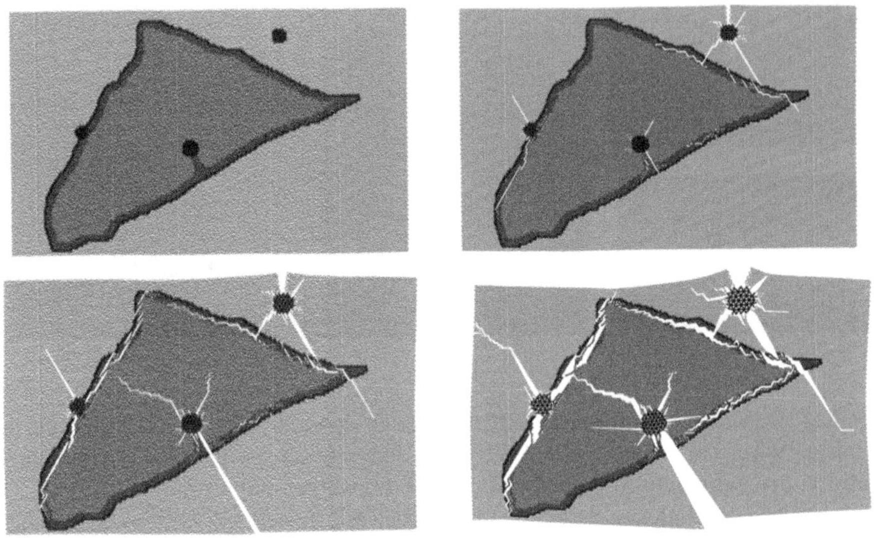

potential to exert internal stresses started in 2007 [36]. So far, even with incomplete input data, it has been possible to simulate the damage reasonably well.

10.6. Frost damage

10.6.1 Frost damage mechanisms

Freeze–thaw cycles, whether or not accompanied by additional effects due to salts penetrating the concrete, can lead to internal damage and surface scaling of concrete. It is difficult to model frost and salt scaling due to its complex physical and chemical mechanisms [37]. Plain frost action has attracted relatively more attention because of its less complicated mechanism. Setzer [38], Powers [39], Litvan [40], Pigeon and Pleau [41], Schulson [42] and many others have made significant contributions to our understanding of plain frost action. Unfortunately, similar progress has not been made for frost and salt attacks, because a single theory has not been able to explain the various phenomena [43].

An interesting theoretical explanation for the mechanism behind a frost and salt attack was proposed by Valenza and Scherer [44,45]. These researchers proposed a mechanism called *glue spalling*. According to this theory, the origin of frost and salt scaling is due to cracking of the ice/brine layer. Basically, following its formation on the concrete surface, the ice starts to shrink due to further cooling. The shrinkage creates tensile stresses in the ice. There are three consequences depending on the solute concentration of the liquid.

- Weak salt concentration (i.e. 0.1%): After the ice forms and cools, the exerted tensile stress does not exceed the tensile strength of ice. Hence, no cracking occurs, in either the ice or the concrete.
- Pessimum salt concentration (1–3%): After the ice forms on top of the concrete and then cools and shrinks, the exerted tensile stress does exceed the tensile strength of the ice and breaks it, which triggers surface scaling due to the bonding between the ice and the concrete.
- Strong salt concentration (10–20%): In this case, the ice layer is too soft to exert enough stress in the underlying cementitious material. Hence, only the ice cracks and no scaling occurs.

Figure 10.14 Schematic illustration of a cement paste: (a) under global tensile stress showing a through crack and (b) the cracks induced by the cracking of an external ice layer [46]

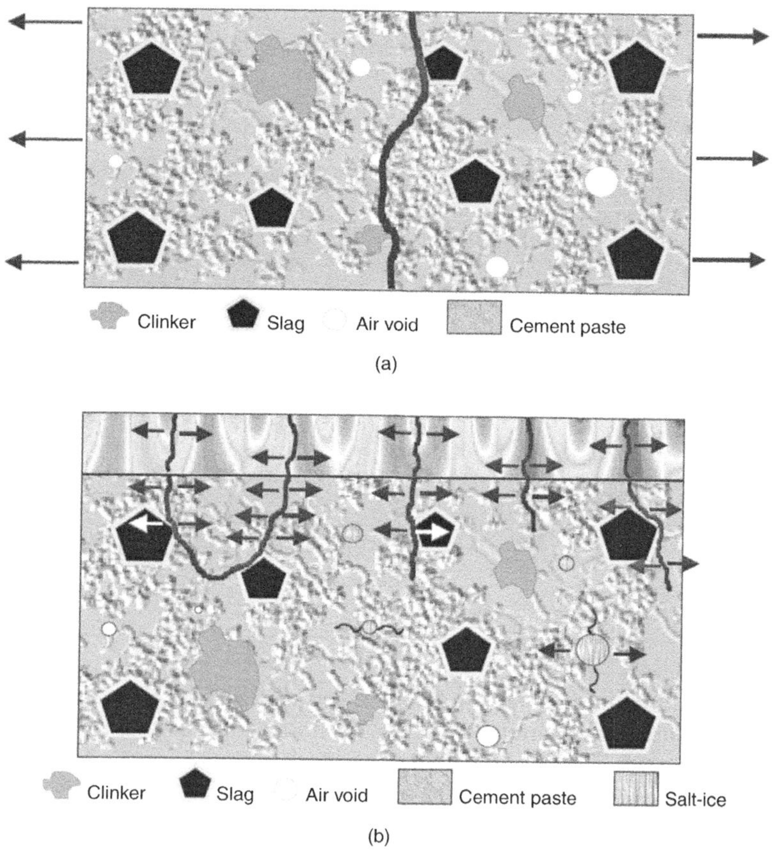

Thus, this mechanism also explains the greater amount of damage done by a pessimum salt concentration in frosty conditions, which has been unknown so far. Figure 10.14 is a schematic illustration of a cement paste specimen under external loading with a shrinking ice layer. Pure tensile loading causes a through crack, whereas the shrinking ice layer is assumed to cause scaling. Whether scaling really occurs and the severity of the damage due to the glue-spall mechanism could depend on the properties of the ice, the thickness of the ice layer and the bond between the ice and substrate.

10.6.2 Numerical modelling of glue-spall mechanism for frost and salt scaling

A lattice model, as discussed in Section 10.4.5, can be used as a numerical model of glue spalling. A typical result of a simulation is shown in Figure 10.15, which presents four subsequent stages of scaling of a mortar. An arbitrary 3-mm layer of ice on top of 12 mm of mortar is modelled. Periodic boundary conditions are assumed in the horizontal direction. At the bottom and at the top, the nodes are free. The material properties used in the simulation are listed in Table 10.3.

It is assumed in the analysis that the temperature drops down to −20°C and that the difference in thermal dilation between the concrete and the ice is 4×10^{-5}. In the analysis, the stresses are not relaxed. The first crack in the ice starts at 10% of the applied strain and cracks through at 13.5%. This correlates with a temperature change of only −2.7°C. This is somewhat less than what was found by the linear elastic calculation

Figure 10.15 Simulation of frost and salt scaling by a Delft lattice model [46]

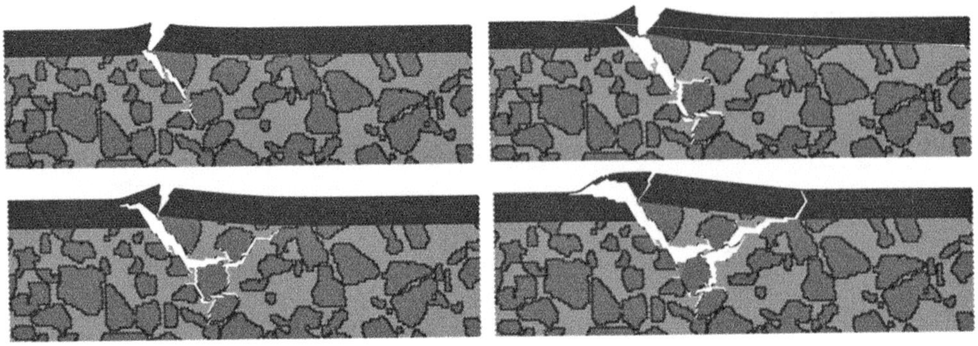

Table 10.3 Material properties used in lattice simulations of concrete with an ice layer

Material	Elastic modulus: GPa	Tensile strength: MPa
Aggregate	70	10
Matrix	25	4
Interface	10	1
Ice	10	2.5

in [45], where a decrease in temperature down to −4°C led to cracking. However, if the ice layer is on a heterogeneous microstructure, the restraint of the ice is not uniform. Due to variations in the stiffness of the supporting microstructure, stress concentrations in the uniform ice appear, which result in earlier cracking than the linear elastic calculation gives.

In practice, much of the linear elastic stress will disappear due to relaxation of the ice and also of the cement-based material under it. This was also described by Valenza and Scherer [44,45]. Due to this relaxation, the stresses generated during the overall temperature drop to −20°C are not enough to crack pure ice. They are enough, however, to crack brine ice, which is weaker than pure ice. Once the ice cracks, the crack continues in the mortar and kinks at a certain distance inside the material (Figure 10.15). It then continues until it hits the ice again, and a second crack in the ice is formed. If the simulation were to be continued, another crack in the ice would occur at another location, and the mechanism would repeat again. Of course, the cracking is influenced by the local material properties and the relaxation of the stresses.

The simulations with the lattice model revealed that the effect of the thickness of the ice layer and the pessimum effect could be simulated quite well. Note, however, that glue spalling is not the only mechanism that causes scaling. Other mechanisms, for example the ice lens theory proposed by Powers [39] and modifications of that theory, may play a role as well. Moreover, in practice, thermal eigenstresses, particularly in the surface layer of a concrete element, may also influence the cracking process.

10.7. Conclusions and prospects

The deterioration of concrete is often a very slow process in which chemical, physical and mechanical mechanisms proceed simultaneously and may interact as well. Thus, modelling is very complex. Since deterioration processes tend to start at the smallest length scales but may ultimately affect the overall structure, the need for a *multiscale modelling approach* is obvious. In the last two decades, much progress has been made in modelling hydration, developing microstructures and predicting material properties. These models enable us to determine the starting point of deterioration processes. Due to the ever increasing computational power of modern computers, we can build the tools required for modelling complex interactions and, thus, generate reliable predictions of deterioration and service life. Being able to model and simulate deterioration at the smallest possible length scales may help in designing new or modified materials with enhanced long-term performance, demonstrating that an investment in advanced material models can be worthwhile.

REFERENCES

1. Siemes A, Vrouwenvelder A and van den Beukel A (1985) Durability of buildings: A reliability analysis. *Heron* **30(3)**: 2–48.
2. DuraCrete (2000) *Probabilistic Performance Based Durability Design of Concrete Structures.* DuraCrete, Final technical report R17, Document BE95-1347/R17. CUR, Gouda, Netherlands.
3. Siemes A, Schiessl P and Rostam S (2000) Future developments of service life design of concrete structures on the basis of DuraCrete. In *Service Life Prediction and Ageing Management of Concrete Structures* (Naus D (ed.)). RILEM, Bagneux, France, pp. 167–176.
4. Dolado JS and van Breugel K (2011) Recent advances in modelling for cementitious materials. *Cement & Concrete Research* **41(7)**: 711–726.
5. Bentz DP and Garboczi EJ (1993) Digital-image based computer modelling of cement-based materials. In *Digital Image Processing: Techniques and Application* (Frost JD and Wright JR (eds)). American Society of Civil Engineers, New York, NY, USA, pp. 63–74.
6. Jennings HM and Johnson SK (1986) Simulation of microstructure development during the hydration of cement compound. *Journal of the American Ceramic Society* **69(11)**: 790–795.
7. Navi P and Pignat C (1999) Three-dimensional characterization of the pore structure of a simulated cement paste. *Cement and Concrete Research* **29(4)**: 507–514
8. van Breugel K (1991) *Simulation of Hydration and Formation of Structure in Hardening Cement-based Materials.* PhD thesis, Delft University, Delft, Netherlands.
9. Ye G (2003) *The Microstructure and Permeability of Cementitious Materials.* PhD thesis, Delft University, Delft, Netherlands.
10. van Breugel K (1984) Models for prediction of microstructural development in cement-based materials. In *The Modelling of Microstructure and Its Potential for Studying Transport Properties and Durability* (Jennings J, Kropp J and Scrivener K (eds)). NATO ASE Series, Kluwer, Dordrecht, Netherlands, vol. 304, pp. 91–106.
11. Ishida T, Mabrouk RTS and Maekawa K (2001) An integrated computational framework for performance evaluation of cementitious materials and structures under various environmental actions. In *Proceedings of the 6th International Conference on Creep, Shrinkage and Durability Mechanics of Concrete and other Quasi-brittle Materials* (Ulm FJ, Bazant Z and Wittmann F (eds)). Elsevier, Oxford, UK, pp. 511–516.
12. Gao P (2018) *Simulation of Hydration and Microstructure Development of Blended Cement.* PhD thesis, Delft University, Delft, Netherlands.
13. Sun ZH, Ye G and Shah SP (2004) Microstructure and early age properties of Portland cement pastes – Effects of the connectivity of the solid phases. *ACI Materials Journal* **102**: 122–129.
14. Ouyang X (2017) *Filler-hydrates Adhesion Properties in Cement Paste System – Development of Sustainable Building Materials.* PhD thesis, Delft University, Delft, Netherlands.
15. Carol I and Bazant ZP (1993) Solidification theory: A rational and effective framework for constitutive modelling of aging viscoelasticity. *Proceedings of the 5th International Symposium on Creep and Shrinkage of Concrete.* RILEM, vol. 22, pp. 177–189.

16. Lokhorst SJ and van Breugel K (1993) The effect of microstructural development on creep and relaxation of hardening concrete. *Proceedings of the 5th International Symposium on Creep and Shrinkage of Concrete*. RILEM, vol. 22, pp. 145–150.
17. Nilsson M (2001) *Rotational Boundary Restraint Factor*. European Research Project IPACS, Report BE96-3843/2001:69-9IPACS.
18. Eberhardt M (1993) *Probability of Cracking in Young Concrete*. MSc thesis, Delft University, Delft, Netherlands. (In German.)
19. Bijen J (2003) *Durability of Engineering Structures – Design, Repair and Maintenance*. CRC Press, Boca Raton, FL, USA and Woodhead Publishing Limited, Cambridge, UK.
20. Audenaert K, Boel V and De Schutter G (2007) Carbonation of filler type self-compacting concrete. *12th International Congress on the Chemistry of Cement*.
21. Papadakis V (2000) Effect of supplementary cementing materials on concrete resistance against carbonation and chloride ingress. *Cement and Concrete Research* **30(2)**: 291–299.
22. Powers T and Brownyard T (1946) Studies of the physical properties of hardened cement paste. *Journal of the American Concrete Institute* **43(9)**: 101–132.
23. Thiery M, Dangla P, Villain G and Platret G (2005) A prediction model for concrete carbonation based on coupled CO_2-H_2O-ions transfers and chemical reactions. *10th DBMC International Conference on Durability of Building Materials and Components, Lyon*, Paper TTI-58.
24. Maage M, Helland S, Poulsen E, Vennesland O and Carlsen JE (1996) Service life prediction of existing concrete structures exposed to marine environment. *ACI Materials Journal* **93(6)**: 602–608.
25. de Rooij MR and Polder RB (2005) *Durability of Marine Concrete Structures*. Report 215. CUR, Gouda, Netherlands. (In Dutch.)
26. Gaal GCM, Polder RB, Walraven JC and van der Veen C (2003) Critical chloride content – State of the art. In *Proceedings of the 10th International Conference on Structural Faults and Repair* (Forde MC (ed.)). Engineering Technics Press, Edinburgh, UK.
27. Polder RB and de Rooij MR (2005) Durability of marine concrete structures – Field investigations and modelling. *Heron* **50(3)**: 133–154.
28. Kim Y-J (2008) Production and construction technology of high-durable marine concrete in Busan-Geoje fixed link project. *Proceedings of the International Symposium on Advancement of Cement and Concrete Industries, Seoul*.
29. van Breugel K, Polder RB and van der Wegen GJL (2009) Service life design of structural concrete – The approach in The Netherlands. *Proceedings of the International Conference on Structural Faults and Repair, Edinburgh*.
30. Li K, Chen Z and Lian H (2008) Concepts and requirements of durability design for concrete structures: An extensive review of CCES01. *Materials and Structures* **41(4)**: 717–731.
31. Qian Z (2008) *3-D Lattice Analysis of Cement Paste*. MSc thesis, Delft University, Delft, Netherlands.
32. Schlangen E, Qian Z, Ye G and van Breugel K (2010) Materials research in progress: Simulation of cracking. *Cement* **3**: 80–85. (In Dutch.)
33. Schlangen E and Garboczi EJ (1997) Fracture simulations of concrete using lattice models: Computational aspects. *Engineering Fracture Mechanics* **57(2/3)**: 319–332.
34. Dao VTN, Dux PF and Morris PH (2009) Tensile properties of early age concrete. *ACI Materials Journal* **106(6)**: 483–492.
35. Copuroglu O and Schlangen E (2007) Modelling of effect of ASR on concrete microstructure. *Key Engineering Materials* **348–349**: 809–812.
36. Schlangen E and Copuroglu O (2007) Concrete damage due to ASR: A new method to determine the properties of the expansive gel. *Proceedings of the 6th International Conference on Fracture Mechanics of Concrete and Concrete Structures, Catania*. Taylor & Francis, London, UK.

37. Copuroglu O (2006) *Frost Salt Scaling of Cement-Based Materials with a High Slag Content.* PhD thesis, Delft University, Delft, Netherlands.
38. Setzer MJ (2001) Micro-ice-lens formation in porous solid. *Journal of Colloid and Interface Science* **243(1)**: 193–201.
39. Powers TC (1965) The Mechanisms of Frost Action in Concrete. *Stanton Walker Lecture Series on the Materials Sciences Issue 3.* University of Maryland, National Sand and Gravel Association, College Park, MD, USA.
40. Litvan GG (1973) Frost action in cement paste. *Materials and Structures* **6(4)**: 293–298.
41. Pigeon M and Pleau R (1995) Durability of concrete in cold climates. In *Modern Concrete Technology.* E&FN Spon, London, UK.
42. Schulson EM (1998) *Ice Damage to Concrete.* Cold Regions Research and Engineering Laboratory, Hanover, NH, USA, Special report 98-6.
43. Lindmark S (1999) *Mechanisms of Salt Frost Scaling of Portland Cement-bound Materials: Studies and Hypothesis.* PhD thesis, Lund University, Sweden.
44. Valenza JJ and Scherer GW (2004) Mechanism for salt scaling of a cementitious surface. *RILEM Spring Meeting 2004*, Chicago, IL, USA.
45. Valenza JJ and Scherer GW (2006) Mechanism for salt scaling. *Journal of the American Ceramic Society* **89(4)**: 1161–1179.
46. Copuroglu O and Schlangen E (2008) Modelling of frost salt scaling. *Cement and Concrete Research* **38(1)**: 27–39.

Soutsos M
ISBN 978-0-7277-6375-4
https://doi.org/10.1680/icehcd.63754.349

Chapter 11
Case Studies: Durability problems, repair strategies or proper consideration in durability design?

11.1. Introduction

Marios Soutsos, Queen's University Belfast, UK

The owners of structures are beginning to recognise the high cost of repairs. Unfortunately, it is not only the engineering achievements that are making the headlines but also collapses or failures of structures to meet their design life. Durability problems arising during the design life of a structure may be considered to be a failure in the serviceability design of the structure. The owners of infrastructure are, therefore, expecting a longer design life for new built structures, and design consultants have recognised that to achieve this longer design life, careful consideration must be given to the choice of mix proportions and the constituent materials of the concrete.

This chapter presents case studies of durability problems that have arisen in major structures to raise awareness of the extent and cost of repairs that may be needed for the structures to remain operational. Such repair costs could be avoided by identifying at the design stage the durability property required for the structure. This may not necessarily be associated with steel reinforcement corrosion. For example, improved abrasion resistance is important for durable factory floors. One hundred years or more of service life is achievable with careful selection of concrete mix proportions and constituent materials, as demonstrated by East Sea Bridge in Shanghai.

11.2. Deterioration of a dry dock

Don Wimpenny, Jacobs, UK

This dry dock facility was constructed in the Middle East between 1973 and 1979. The dock has three basins, with the widest able to take tankers up to 1 million dead weight tons (Figure 11.1). The best contemporaneous practices were employed in its design and construction, and considerable diligence was exercised during its construction. However, premature deterioration occurred to the facility within 10 years of it opening, requiring extensive remedial works. Other dry docks in hot climates have also required premature repairs [1].

11.2.1 Construction

The walls of the dock were formed from multi-celled slipformed caissons 17 m wide, 13.6 m in height and 31 m long. These cells were floated into position and ballasted with marine dredged sand. A grouted connection was made between the caissons, and then crane beams, a service gallery and a slab were cast onto the caissons to provide the dockside arms. The key requirements for the concrete caissons are listed in Table 11.1.

Figure 11.1 Cross-section through the dry dock

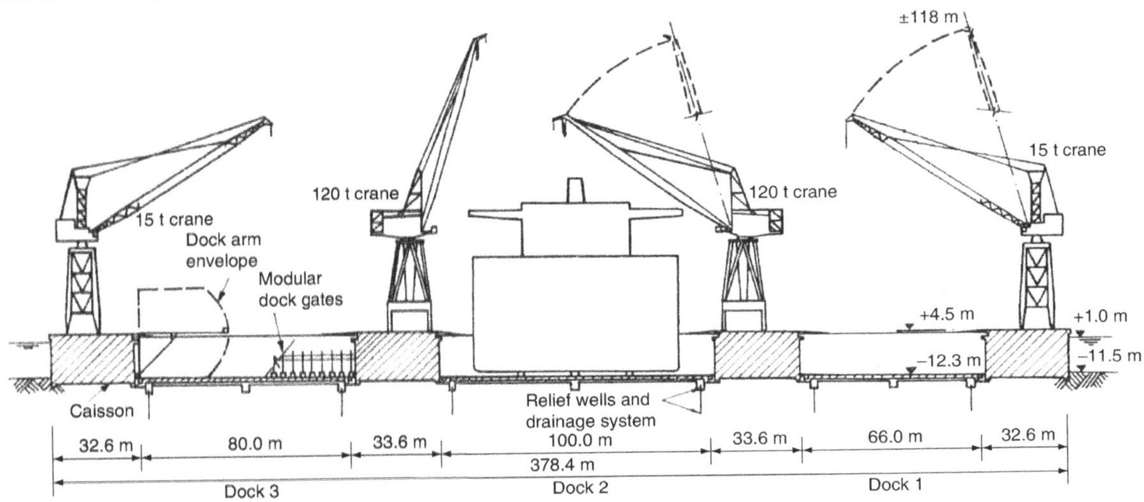

Table 11.1 Requirements for caisson concrete

Parameter	Value
Compressive strength: N/mm^2	30
Minimum cement content: kg/m^3	360
Actual cement content: kg/m^3	390
Actual w/c ratio	0.51
Target slump: mm	75
Nominal cover: mm	60

There was some uncertainty over the achievable strength and water-to-cement (w/c) ratio, so both were subject to trials. The w/c ratio used was required to be within 10% of these trial mixes. Chemical admixtures were not widely used in the early 1970s, and reticence over their use is reflected in a specification requirement that admixtures were to be used *only with the express permission of the engineer.*

Figures 11.2 and 11.3 show the dock under construction and at completion, respectively. There were significant challenges during construction, including

- workability loss at high ambient temperatures
- control of water content
- dusty gabbro aggregate
- handling of heavy reinforcement.

The absence of admixtures considerably increased the difficulties of controlling the water demand of the concrete posed by the high temperatures and dusty aggregate. Great care was exercised to monitor and control the dust content of the aggregate, but even with these precautions, the w/c ratio occasionally increased to over 0.55.

Figure 11.2 View of the casting yard for the slipformed caissons (the truck mixer provides scale)

Figure 11.3 View of the completed dock

At the time of construction, Portland cement was in short supply and of variable quality. Secondary cementitious materials, such as slag or fly ash, were not readily available within the region, and their benefits in terms of enhanced resistance to chloride ingress were not widely recognised.

11.2.2 Deterioration

The chloride profile measured in the concrete in 1980 is shown in Figure 11.4. The profile indicates that chloride levels of over 0.4% by mass of concrete had penetrated beyond the reinforcement, indicating that corrosion may have been initiated. Chloride levels of 1.0%, representing a risk of significant corrosion

Figure 11.4 Profile of chloride levels in the concrete in 1980 plotted against depth

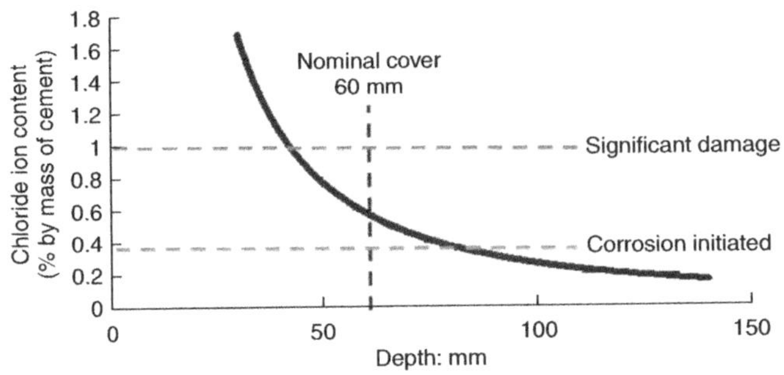

damage [2], had penetrated to beyond 40 mm depth compared to a nominal depth of cover of 60 mm. The latter represents a minimum cover of 45 mm for a typical fixing tolerance for reinforcement for in situ concrete.

By 1991, serious corrosion problems had manifest at the facility, including

- chloride ions >1% at 50–75 mm depth
- cracks up to 0.2 mm wide
- hollowness behind 20% of the caisson surface
- reinforcement pitting at cracks
- up to 50% loss of section of the reinforcement over short lengths.

Removal of the sand ballast from inside the caissons revealed deterioration to the internal surfaces in the form of hollowness under tapping, spalling of the concrete cover and corrosion of the reinforcement (Figures 11.5 and 11.6).

Figure 11.5 Deterioration inside the caisson due to areas of hollowness (H)

Figure 11.6 Exposed reinforcement showing (a) exposure in area of hollowness and (b) corrosion and cracking

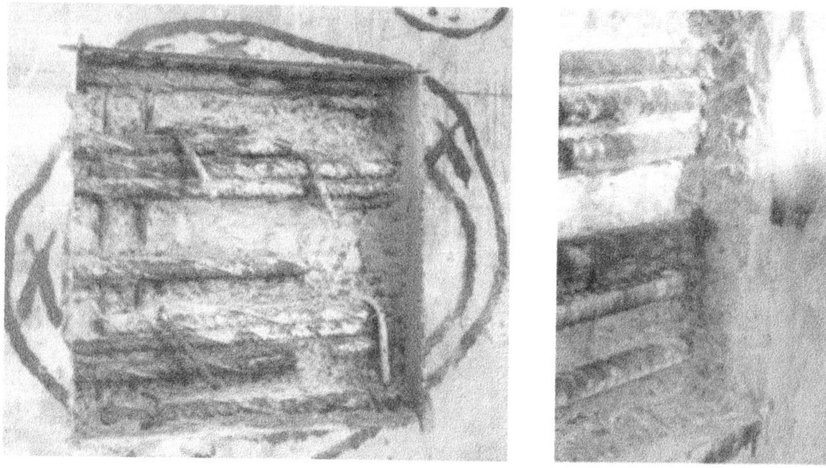

(a) (b)

The apparent chloride diffusion coefficient of the concrete was estimated as approximately 8.9×10^{-12} m²/s by solving Fick's law of diffusion for the chloride profile measured in 1980 using an in-house program called SALTS (Table 11.2). This apparent chloride diffusion coefficient is equivalent to 3.65×10^{-12} m²/s at 20°C. The latter is consistent with published values for Portland cement concrete [2].

The apparent diffusion coefficient was input into an in-house corrosion prediction model CORRPRED, which estimated the time to corrosion-related cracking (Table 11.3). The model is based on published guidance [3] and a large number of factors can be adjusted, including

- temperature and relative humidity
- effect of age on chloride diffusion coefficient
- type of reinforcement (stainless steel or carbon steel, prestressed)
- additional durability-enhancement measures (silane, corrosion inhibitors or controlled-permeability formwork).

Table 11.2 Chloride diffusion characteristics for concrete

Characteristic	Value
SALTS best-fit curve	
Assumed background level	0.15% by mass of cement
Apparent surface chloride level	2.78% by mass of cement
Apparent chloride diffusion coefficient in 1980 at service temperature (assumed 35°C)	8.9×10^{-12} m²/s
Predicted chloride diffusion coefficient at 20°C	
28 days	1.7×10^{-11} m²/s
4 years (1980)	3.65×10^{-12} m²/s
20 years	1.5×10^{-12} m²/s

Table 11.3 Output from CORRPRED analysis showing 14 years to cracking

Time to corrosion initiation and cracking

Analysis Type: DETERMINISTIC

Concrete Details		
Strength grade	31	MPa
Concrete density	2400	kg/m^3
Total cementitious content	390	kg/m^3
Binder type (pc, ggbs, pfa, sf)	PC	
Percentage Binder	0	% wt of binder
Water/binder (w/b)	0.51	
Background Chlorides	0.20	% wt of binder

Reinforcement Details		
Reinforcement Type	Carbon Steel	
Bar diameter	32	mm

Exposure Details		
Ambient Temperature	35	°C
Temperature affected Dea	Y	
Temperature affected threshold	Y	
Surface Chloride Level (% wt cem)	3.0	
Exposure Condition	3. Cyclic wet/dry	
Carbonation factor considered?	Y	
Relative Humidity (%)	65	

Coating used?	No

Additional Protective Measures	
Controlled permeability formwork?	N
Silane Impregnation	N
Integral Waterproofer	N

Diffusion Coefficient Details	
Diffusion coefficient estimated	Est by measurement
Diff Coeff (at 20°C)	3.70E-12
Age of measured value (yrs)	4.00
Age Dependant Diffusion	Y
Switch off age dependency at given date	Y
Turn off age dependency at (yrs)	20

Estimated Values (Temp adjusted)	
Dca at 1460 days	9.10E-12
Dca at 20 years	5.02E-12
Dca at 20 years	5.02E-12
Age Factor	0.37

Surface chloride level	3.00
Adjusted surface chloride level	3.00

Temp & Binder adj threshold (Ct)	0.10

Time to Serviceability Limit State (Years) (Deterministic)		
Minimum Cover Depth (mm)	Chloride threshold > 0.1 % by mass cement	Time to initiation plus cracking
45	1	14

The modelling indicates that the time to cracking at 35°C was less than 15 years. This predicted time to corrosion-related damage is consistent with the level of deterioration observed in 1991. The equivalent predicted time for temperate conditions is over 35 years.

11.2.3 Discussion

The limiting proportions and compressive strength of the concrete specified for the caisson were in accordance with the contemporaneous requirements of CP110 [4], which required a minimum cement content of 360 kg/m^3 for concrete exposed to alternating wetting and drying, to freezing while wet or to seawater.

However, as can be observed from the predictions above, the Middle East environment is much more severe than that in temperate climates, not least due to the higher ambient temperature. Furthermore, a dry dock has potentially a more severe exposure condition than a normal marine environment, as the concrete is subject to long periods of drying followed by inundation with seawater, which leads to rapid chloride ingress by capillary absorption.

There was little recognition in the 1970s of the importance of cement type and w/c ratio in achieving durability. By 1981, the recommendations for concrete, both in the UK and the Middle East, had become much more stringent in respect of cover and w/c ratio, but by this stage, the dry dock had already been constructed. Typical recommendations for the Middle East are summarised in Table 11.4 and include a requirement for reinforced concrete to use ASTM Type II, which has moderate sulfate resistance in reinforced concrete, rather than standard sulfate-resistant Portland cement. This appears to be an acknowledgement of the detrimental effect of low tricalcium aluminate levels in sulfate-resistant Portland cement on the chloride-binding capacity.

Table 11.4 Recommendations for concrete in Middle East exposure conditions [5]

Parameter	Value
Minimum cement content: kg/m³	370
Cement type	ASTM C150 Type II
Maximum w/c ratio	0.42
Nominal cover: mm	100

Table 11.5 Recommendations for concrete in a hot marine climate for a 50-year service life [6]

Parameter	Value
Minimum cementitious content: kg/m³	425
Cement type	25% Portland cement, 5% silica fume, 70% slag
Maximum water-to-cementitious (w/cm) ratio	0.34–0.38
Nominal cover: mm	75–100

More recent recommendations are listed in Table 11.5, which indicate the progressive lowering of w/c ratios with high-range water-reducing admixtures and the incorporation of one or more secondary cementitious materials.

11.2.4 Conclusions

A dry dock in a hot climate is in an extreme environment, and the best contemporaneous practice for temperate marine climates did not ensure durability. Chlorides can build up rapidly by diffusion and capillary absorption, leading to chloride-induced corrosion damage within 15 years. The corresponding time in the UK would be over 35 years. Realising durability requires a low w/cm ratio, high-range water-reducing admixtures and one or more secondary cementitious materials, such as slag or silica fume. Durability models can be used to get an accurate indication of the onset of corrosion damage.

11.3. Delayed ettringite formation in a Malaysian highway structure: Investigation and management strategy

Don Wimpenny, Jacobs, UK

The highway structure in Kuala Lumpur, comprising a 1.5-km viaduct carrying a six-lane carriageway, was opened in 2002. The structure has split decks formed from externally prestressed precast concrete box girders resting on T-shaped piers, which have an 18.7-m-long cross head that tapers from 3.5 to 2 m in depth mounted on a 3.6-m-wide octagonal column (Figures 11.7 and 11.8). The abutments are also T-shaped like the piers, but for the most part, they are buried [7].

Cracking to several piers and one abutment was reported 6 months after the viaduct was opened. A detailed survey in 2004 produced a map of the cracks, by which time some of the most severely cracked cross heads had cracks several millimetres wide. Some of the cracking could be attributed to inadequate reinforcement. However, the variability, severity and presence of lateral displacement across some cracks suggested that other mechanisms, such as an expansion reaction, may be contributing to the cracking (Figure 11.9).

Figure 11.7 View from beneath the viaduct showing the form of construction

Figure 11.8 Elevation of T-shaped piers

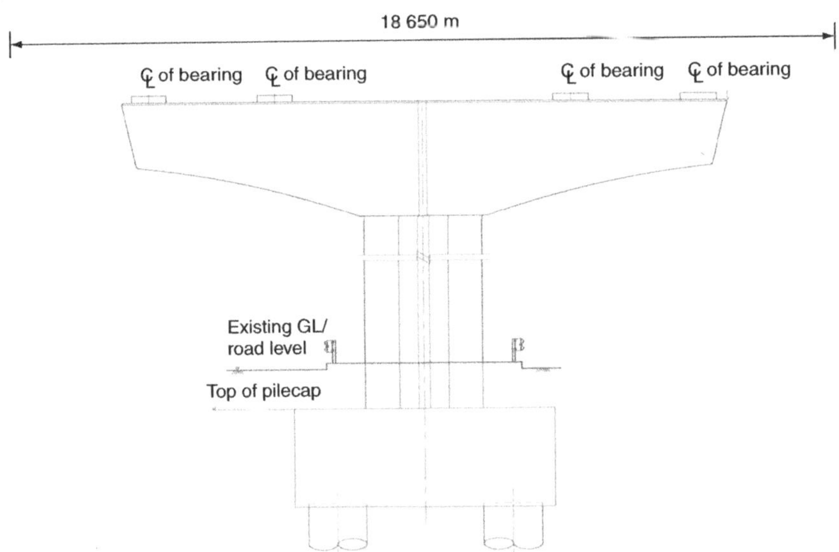

A thermal analysis using published guidance indicated temperatures in excess of 80°C could have been reached during the initial curing, leading to a risk of delayed ettringite formation (DEF).

DEF is an expansive reaction that occurs due to the delayed formation of ettringite in concrete that has been subject to high temperatures during curing (typically over 65°C). These high temperatures can be generated during steam-curing or in thicker in situ concrete members [8]. The delayed formation of ettringite in

Figure 11.9 Cracking to cross head

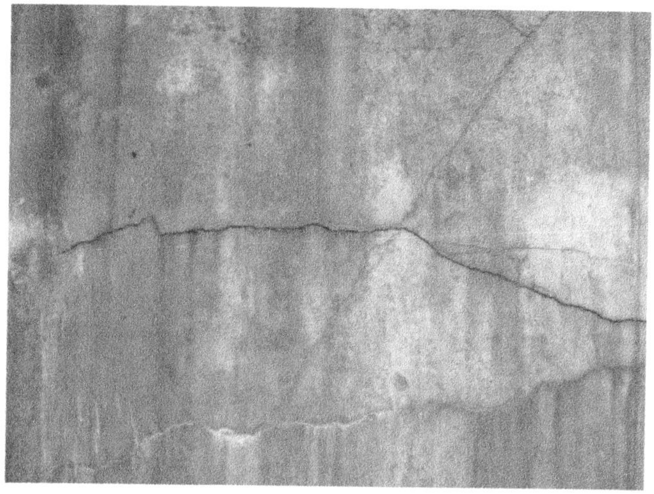

hardened concrete exposed to moisture leads to expansion and cracking. The pattern of cracking is superficially like that generated by the alkali–aggregate reaction (AAR), leading to a risk of misdiagnosis.

11.3.1 DEF management strategy

The proposed management strategy has three stages, as indicated in Table 11.6. Partial closure of the bridge led to considerable pressure to resolve the problem, which meant several of the steps occurred in parallel.

DEF was unexpected and several steps were taken to confirm the finding. Potentiometers and strain gauges were installed on selected cross heads to measure crack movement and the response of the reinforcement. These were remotely interrogated via a modem. The results indicated that there was progressive expansion that did not coincide with traffic loading but had a higher rate in the rainy season. Cores were taken for petrographic examination. Ettringite was found in cracks and rims around aggregate particles, which is a characteristic of DEF. The reaction products were confirmed using scanning electron microscopy (SEM) and microprobe analysis. Cores were subject to expansion testing using the British Cement Association (BCA) and Duggan test methods [9,10]. These confirmed damaging expansion levels of up to 0.2% at 20 days, compared to a safe limit of 0.05%. The primary cause of the expansion in the cores subjected to BCA expansion testing, petrographic techniques and electron microprobe analysis was indicated to be DEF [11].

A two-dimensional finite element analysis (FEA) was undertaken to predict the temperature in the cross head during construction (Figure 11.10). No concrete temperature records were available, except at placing (approximately 32°C), so assumptions were made about the worst case and average ambient conditions, including the solar gain. The worst case and average case analyses indicated that temperatures had been in excess of 80°C and 70°C, respectively, for over 100 hours at the core of the section. The temperatures from this analysis were used to predict the DEF expansion at different locations in the cross head, which were then used in the structural analysis.

The significant structural effects of ongoing DEF expansions were found to be

- cracking around the perimeter of the section
- loss of ultimate moment capacity due to cracking of the soffit
- failure of laps in main tension and in shear links.

Table 11.6 DEF management strategy for Malaysian highway structure

Action	Purpose
Stage 1 – Confirmation of diagnosis	
Monitor crack movement and reinforcement strain	Confirm that progressive expansion and localised yielding of the reinforcement is unrelated to live loading
Take cores from cross head for petrographic examination and SEM	Confirm that DEF is present from distribution and nature of reaction products
Take cores from cross head for Duggan expansion testing	Confirm the potential for DEF based on expansion levels
Take cores from cross head for BCA expansion testing	Confirm the occurrence of DEF based on expansion levels and petrographic examination of the specimens after expansion testing
Stage 2 – Numerical modelling	
Perform FEA of the temperature within a section during construction	Confirm the likely zone of concrete at risk of DEF
Develop a spreadsheet model to assess critical sections of the cross head	Predict the likely structural impact of DEF and determine the strengthening requirements
Perform non-linear FEA of stresses and damage based on predicted expansion	Confirm the likely structural impact of DEF and the strengthening requirements in more detail
Stage 3 – Repair	
Grout existing cracks with an epoxy-resin-based grout	Seal the existing cracks and restore the tensile strength of the concrete
Apply a waterproof coating to cross heads	Exclude ingress of rainwater and prevent water from ponding
Adopt a strengthening scheme	Apply prestressing and plate bonding to address present and possible future loss of strength

The lap failure was a particular concern, as two of the four layers of the main flexural reinforcement in the top of the cross head were lapped at the centre of the cross head with laps that were not full strength. To assess the strength of DEF-affected cross heads and identify suitable strengthening works, a spreadsheet model was developed to analyse critical sections of the cross heads from first principles. A three-dimensional non-linear FEA was also undertaken. The spreadsheet gave similar results to those of conventional section analysis software, but in addition, it also modelled the variation of DEF expansion across the cross-section, the effect of the stresses due to DEF expansion on the concrete and reinforcement, and the loss of stiffness and strength of the concrete section due to ongoing DEF. The spreadsheet model included options for modelling the tensile strength of the concrete, the stress–strain response of lapped bars and effects due to yielding of the bars and due to strengthening by plate bonding and external prestressing.

Spreadsheet and non-linear FEA models were used to predict the damage to the structure for different DEF expansion levels. There was particular concern that ongoing expansion would lead to yielding of the reinforcement and the failure of laps, potentially prejudicing the strengthening work at higher levels of expansion. At 0.1% DEF expansion, significant cracking would occur around the top and sides of the section but

Figure 11.10 (a) Section through a cross head showing the predicted temperature distribution during construction and (b) the temperature development at three locations

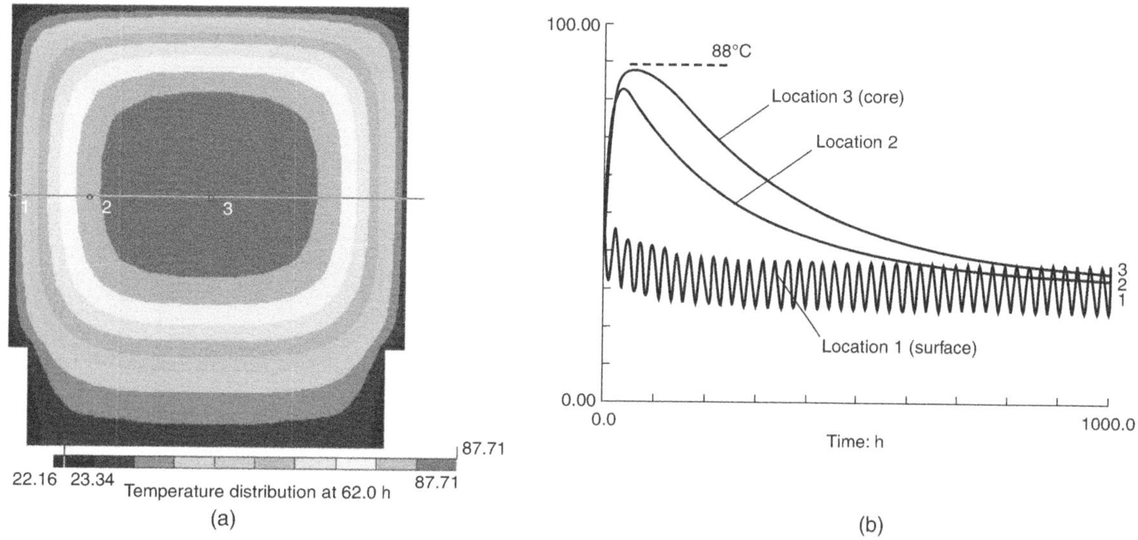

the soffit would remain intact. The effect on the ultimate moment capacity was negligible. However, as the modelled expansion increased, the moment capacity reduced due to cracking in the soffit and lap failure, which would necessitate strengthening work to reinstate the moment capacity at the root of the cantilever. The non-linear FEA produced a pattern of cracks in the cross heads over the column that very closely resembled that observed on site. The effect of continued expansion would lead to cracking developing further down the vertical face of the cross head until it extended into the soffit.

The cross heads were subject to frequent wetting due to the longitudinal gap between the split decks, leaking drainage and wind-driven rain. The rapid development of the reaction may have been promoted by the high ambient temperatures and humidity in Malaysia. The waterproofing of the structure was regarded as essential for limiting future expansion so that specifications were developed for a waterproof coating, which included requirements for

- tolerance to a damp substrate and a high relative humidity during application
- ability to transmit not less than 3g/m^2 of vapour per day out of the concrete
- transparency to permit cracks to be viewed and monitored
- a minimum elongation to failure of 100% and crack bridging of up to 2 mm width.

The strengthening scheme developed for a typical cross head is shown in Figure 11.11. Longitudinal prestressing was required to strengthen the cross head at the main lap regions, primarily by providing a relieving moment, such that if the two top layers of bars failed at the laps, then the external prestress in combination with the remaining reinforcement would be sufficient to develop the design moment. The transverse bonded steel plate straps were required to strengthen the cross head due to deficiencies in the transverse reinforcement. These straps also provided reinforcement to reduce the continued cracking from the limited and ongoing residual DEF expansion expected after waterproofing. The vertical bonded steel plate straps were required to control ongoing cracking in the weakened cross head/column connection region where the existing reinforcement has yielded and lost its stiffness at wide cracks.

Figure 11.11 Typical strengthening scheme for a cross head

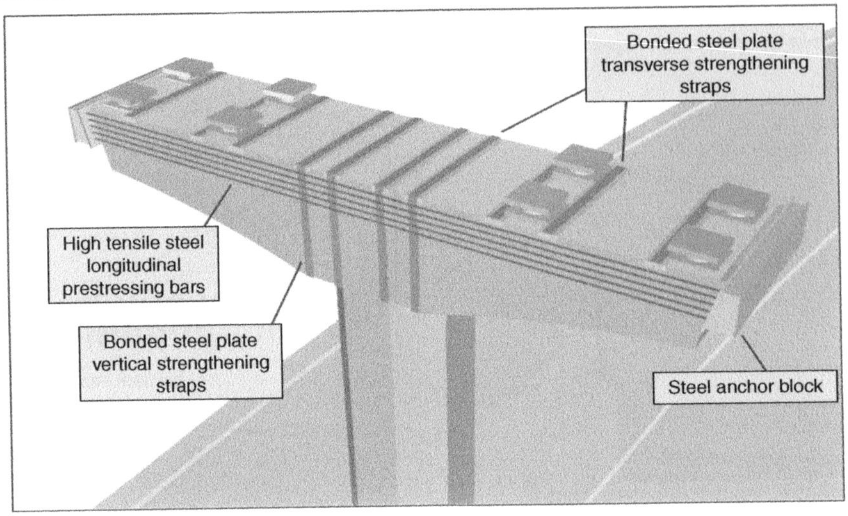

11.3.2 Conclusions

DEF can occur in both steam-cured and in situ concrete, leading to serious cracking. There are few published cases relating to bridges, which may be due to the deterioration having been diagnosed as from AAR. High temperatures during initial curing and exposure to moisture in service are critical risk factors. The structural effects of DEF can be significantly different from those of the alkali–silica reaction (ASR).

DEF investigation and management strategies must be tailored to the particular structures, but include the following elements

- removal of samples to confirm the cause of cracking
- monitoring changes to crack widths and testing expansion in cores to assess future expansion and movement and to assist in selecting repair products
- measures to exclude moisture in the form of coatings and sealing cracks
- strengthening or replacing the affected elements.

Several techniques can assist the investigation and management of DEF, including SEM to confirm the distribution of ettringite in the concrete, monitoring crack movement, and modelling temperature, stresses and deterioration.

11.4. Construction of a deepwater port

Don Wimpenny, Jacobs, UK

The construction of a deepwater port in North Africa commenced in 2004. The port has a 1.1-km-long breakwater in approximately 30 m depth of water. The original design was for a rubble mound breakwater, but an alternative, comprising reinforced concrete caissons on a 10-m-deep rubble mound, was proposed by the contractor. This alternative significantly reduced the volume of rock required but careful consideration of durability was necessary given the environment, the difficulty of undertaking any maintenance work and the 100-year service life of the facility.

11.4.1 Design

The breakwater consists of 40 caissons, approximately 35 m in height and 28 m in diameter. The caissons are principally multi-celled 'quadralobes', with external perimeter walls 0.5 m in thickness and internal dividing walls 0.4 m in thickness. There was an upper attenuation chamber to dissipate wave energy (Figures 11.12 and 11.13).

Figure 11.12 Cross-sectional elevation of a typical caisson

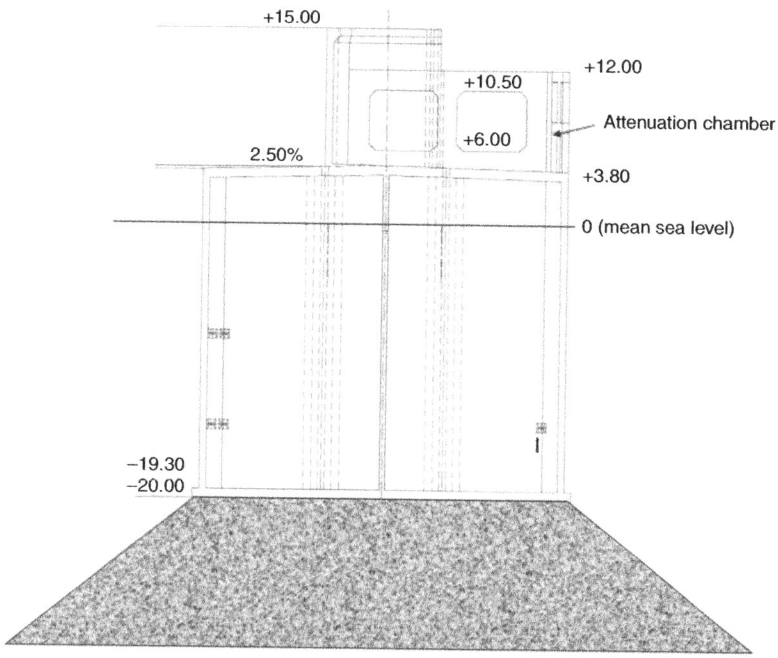

Figure 11.13 Isometric sketch and cross-sectional plan of a quadralobe caisson

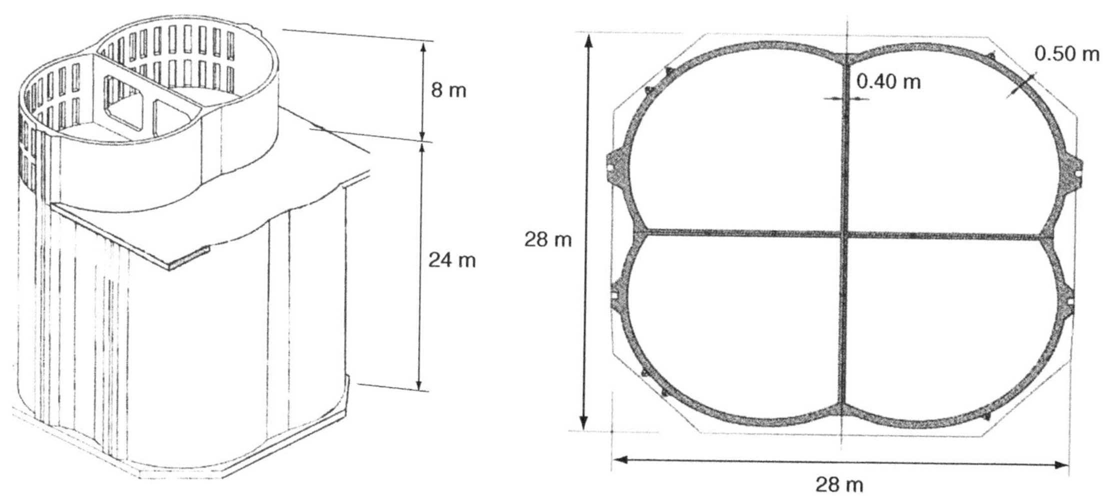

A probabilistic durability design was undertaken by the contractor using the principles outlined in the pan-European DuraCrete project [12]. This approach is like structural design in that a resistance and load are calculated with the probability of failure being defined as the probability of the resistance being less than the load. A frequency distribution is typically defined by the arithmetic mean and standard deviation for each of the variables, such as cover, to allow the failure probability to be estimated. A reliability index is calculated, which is the converse of the failure probability, using the inverse normal function. For example, probabilities of failure of 1 in 1000 and 1 in 100 give corresponding reliability indexes of 3 and 2.3. The serviceability limit state is reached when the reliability index drops below a predetermined acceptance level, which is usually between 1 and 5 depending on the cost of construction and repair, the consequences of failure, and the deterioration mechanism [13]. The proposed design was reviewed, and an independent check was made on the probabilistic design using an in-house model CORRPRED for corrosion-induced corrosion, with add-in proprietary software @RISK for statistical modelling. Figure 11.14 shows the typical output of the model (JC Knights, personal communication). Some of the key input parameters are summarised in Table 11.7. The ageing factor, which defines the rate of change of the diffusion coefficient with time, has not been included but should be selected with great care.

Figure 11.14 Reliability index for different scenarios (the dashed line indicates a failure probability of 5%)

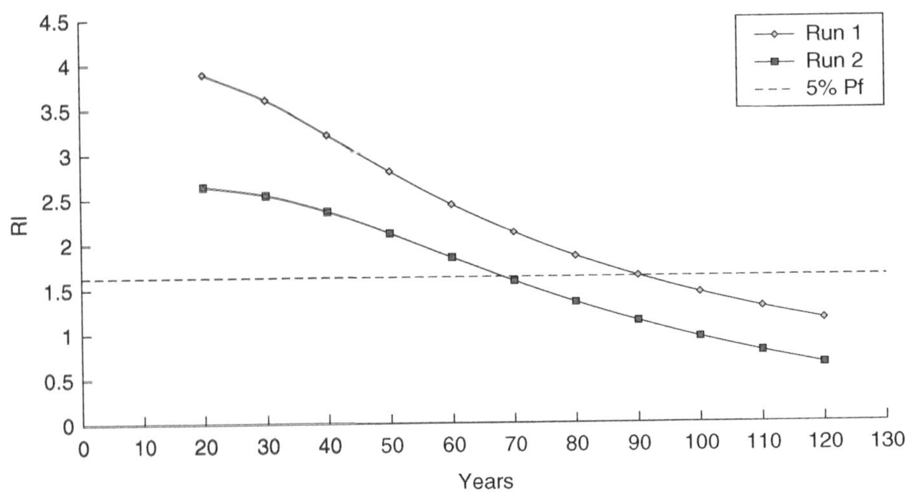

Table 11.7 Key parameters input into the model

Parameter	Mean value and (standard deviation or range)	
	Run 1	Run 2
Temperature: °C	25	25
Cover: mm	70 (55–85)	65 (50–75)
Diffusion coefficient ($\times 10^{-12}$ m²/s)	5.1 (1.1) at 28 days	1.6 (0.5) at 6 months
Threshold: %	0.6 (0.15)	0.4 (0.15)
Surface chloride level: %	4 (1)	5 (1)

Table 11.8 Durability measures

Additional durability measures	Submerged (below −4 m)[a]	Tidal and splash (above −4 m)[a]
Minimum cover: mm	70	80
Cementitious blend	20% fly ash	20% fly ash and 5% silica fume
Maximum w/c ratio	0.35	0.35
Maximum chloride migration coefficient: m^2/s	5×10^{-12}	1.5×10^{-12}
Cathodic protection	Installed	Provision for future installation, e.g. electrical continuity of reinforcement
Maximum loss of mass by wear, ASTM C1138 test	5%	5%

[a]Datum at 0 m is lowest astronomical tide.

Several potential shortcomings were identified in the proposed durability design.

- There was no allowance for construction defects.
- There was no allowance for wear.
- Overly optimistic assumptions were made, including unsubstantiated chloride diffusion characteristics of the concrete.
- There was no provision for future cathodic protection.

An addendum was inserted into the durability plan to include abrasion testing using the ASTM C1138 method and chloride migration testing using the NT Build 443 and NT Build 492 methods. NT Build 443 is a bulk diffusion test in which the chloride profile is generated by grinding a sample in depth increments of at least 1 mm. The NT Build 492 method is an accelerated chloride migration test in which an electric current is passed through a sample to promote the movement of chloride ions. However, unlike the AASHTO T277 rapid chloride migration test, the current is moderated to prevent heating of the sample and chloride penetration is assessed directly with a silver nitrate indicator [14]. A chloride migration coefficient can be estimated from the depth of penetration. The durability measures for the caissons were enhanced by making provision for cathodic protection, imposing limits on the chloride migration coefficient and requiring appropriate cementitious blends and minimum cover values, as summarised in Table 11.8.

11.4.2 Mix development

The contractor's proposed method of construction was slipforming the caisson and conventional casting of the wave attenuation chamber using self-compacting concrete. Slipforming imposes several additional requirements on the concrete compared to conventional casting. In particular, the setting time of the mix must be carefully regulated to ensure that the rate of concrete placement and slipform movement are such that the concrete has sufficient strength when it exits the bottom of the slipform so that it is self-supporting but has not stiffened so much that there is a risk of high friction and hardened plates of concrete building up at the face of the slipform or of cold joints. To regulate workability and setting, a high-range water-reducing admixture and a separate retarding admixture were used.

Table 11.9 Summary of caisson concrete

Parameter	Value	
28-day compressive strength: N/mm^2	75	
Cementitious content: kg/m^3	480	
Maximum w/c ratio	0.35	
Target slump: mm	200	
Admixtures	High range water reducer Retarder	
Cementitious type	20% pulverised fuel ash and 5% silica fume (tidal and splash zones)	20% pulverised fuel ash (submerged zone)
Approximate chloride migration coefficient by NT Build 492 method: m^2/s	3×10^{-12} at 28 days 1.5×10^{-12} at 56 days 1×10^{-12} at 91 days	6×10^{-12} at 28 days 3×10^{-12} at 56 days 2.5×10^{-12} at 91 days
Resistivity by transient electromagnetics measurement: Ω m	103 at 28 days 276 at 56 days	70 at 28 days 108 at 56 days
Wear by ASTM C1138	1.5–2.5%	N/A

Laboratory trials were undertaken for over 50 different mix designs, including the following

- seven cement types
- two fly ash sources
- two silica fume sources
- washed and unwashed sand
- three different superplasticiser and retarder combinations.

The final mixes proposed for construction are summarised in Table 11.9. Chloride migration testing indicated that both mixes met the required limits using laboratory-cast and cured specimens. Resistivity measurements were made on the concrete using the two-electrode method to give a rapid measure of concrete quality. Chloride diffusion coefficients from the NT Build 443 bulk diffusion test were slightly less than the equivalent migration values estimated from the NT Build 492 method (approximately 3.24×10^{-12} m^2/s compared to 4×10^{-12} m^2/s). In addition to the laboratory work, slipforming trials were undertaken using a slipform 3 m high and 5 m long (Figure 11.15).

11.4.3 Construction

The construction sequence involved casting the base slab for the caissons on a bespoke casting bed (Figure 11.16a). The outer walls and any cross walls were then formed by slipforming to approximately 10 m in height (Figure 11.16b). Each partially completed caisson, weighing over 3000 tonnes, was then slid onto a launching platform on the quayside on rail-mounted trolleys (Figure 11.16c). After launching, the slipforming continued from the quayside (Figure 11.16d). When the slipforming was complete, the caisson was floated into position and ballasted onto a prepared bed using sea-dredged sand. The connecting deck slab and the upper attenuation chambers were then conventionally cast to complete the breakwater (Figure 11.16e).

Figure 11.15 Slipform field trial, showing completed wall (left) and a section through the slipform (right)

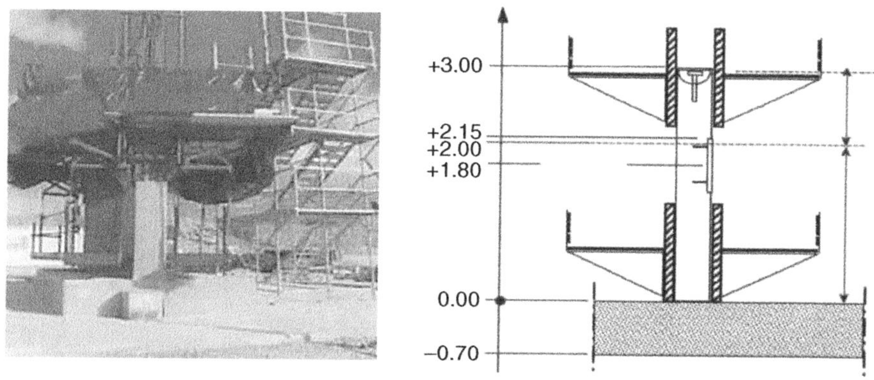

Several problems were experienced during casting, such as

- poor compaction of the concrete, especially at the top of the slipformed section
- fallout of concrete due to an excessive rate of rise of the slipform
- cracking to the walls due to wave action and early-age thermal contraction.

Prior to casting the deck slab, the open-ended caissons were vulnerable to damage due to deflection of the walls during heavy seas, so the contractor modified the method of working to address this risk. Patch repairs were undertaken to the identified areas of poor compaction and fallout. Cracks over 0.2 mm in width were injected with resin. As indicated in Section 11.4.1, the durability modelling did not allow for construction defects. However, some mitigation was provided for these defects by three forms of cathodic protection (Table 11.10).

The cathodic protection measures complicated the construction process. In addition to dealing with the cabling and the need to ensure electrical continuity of the reinforcement, the provision for cathodic protection in the upper tidal zone involved casting reservations or ducts within the concrete. The intention being that anodes would be installed in these ducts at a later stage should sensors indicate that there was a risk that corrosion was occurring. This strategy relies on the ducts being sealed from seawater ingress until they are needed and on corrosion being detected sufficiently early so that the cathodic protection system can be implemented. Figure 11.17 shows the cathodic protection system for the tidal zone installed prior to concreting.

11.4.4 Discussion

Developments in concrete technology have increased control over workability and w/c ratio and expanded the range of cement types that are resistant to chlorides. The self-compacting concrete mixes used in the caissons have a low w/c ratio and include 20% pulverised fuel ash and selective use of silica fume. These measures would be expected to increase durability and reduce the impact of poor workmanship.

Despite these measures, construction defects in the form of cracking and poor compaction occurred. These defects were not accounted for in the probabilistic durability modelling, and the repairs undertaken could not be regarded as restoring the durability to that of the parent concrete. Cathodic protection can mitigate the risk of premature deterioration associated with these features and other undisclosed defects. However, the

Figure 11.16 (a) Preparing the casting bed for the base slab of a caisson, (b) portion of a slipform being lowered into place for slipforming the walls, (c) launching a caisson, (d) completing the slipforming at the quayside and (e) caissons and deck slab in place

Table 11.10 Summary of cathodic protection measures

Location	Cathodic protection measures
Spray zone (above 4.5 m)	Reservations and ducts for future installation
Tidal zone (−4 to 4.5 m)	Ribbon anodes
Submerged zone (below −4 m)	Sacrificial anodes

Figure 11.17 Cathodic protection system showing sensors, mixed metal oxide titanium ribbons and titanium conductor

ducts where cathodic protection can be installed in the future are a potential path for chloride ingress, so that there may be an increased risk of the reinforcement corrosion that the system was intended to protect against.

Durability testing was undertaken to assess the characteristics of the concrete mix, but the results do not necessarily represent the performance of the most important concrete, that is the concrete in the actual works. The practical difficulty of slipforming large areas of exposed concrete should not be underestimated, especially for a marine facility with a long service life.

11.4.5 Conclusions

There were significant challenges in achieving a 100-year service life when using concrete caissons to form a 1.1-km-long breakwater for a deepwater port. A probabilistic durability design was proposed based on a recognised approach. However, it was unrealistic in its assumptions and, in particular, did not account for construction defects. Additional durability measures were required, including 20% fly ash, limiting the chloride migration coefficient and making provision for cathodic protection.

Difficulties encountered during construction, such as poor compaction and cracking, underlined the importance of including a provision for cathodic protection. However, this strategy had its own challenges, including the need to provide corrosion sensors and sealed ducts within the concrete to allow for timely installation of anodes.

11.5. Thaumasite sulfate attack in a UK highway structure: Investigation and management strategy

Don Wimpenny, Jacobs, UK

During bridge-strengthening works in February 1998, an attack was discovered to the foundations of a 30-year-old overbridge to the M5 in Gloucestershire. The bridge has three slender columns at each pier, 750 mm × 450 mm in cross-section and 13 m long, which rest on a spread footing approximately 5.5 m below ground level (Figure 11.18).

The thaumasite form of sulfate attack (TSA) was suspected in view of the high quality of the concrete. The Building Research Establishment confirmed the diagnosis. Prior to this discovery, the known incidence of TSA in the UK had been limited to a small number of cases in non-structural concrete exposed to cold, wet conditions. However, the potential for such an attack had been established in the laboratory, and the UK guide on sulfate attacks, BRE Digest 363, which was published in 1996, briefly referred to the phenomenon [15].

TSA is distinct from a conventional sulfate attack in that the calcium silicate hydrates react with sulfates to form calcium sulfate carbonate silicate hydrate. In the conventional reaction, calcium hydroxide and calcium aluminate hydrates form gypsum and ettringite. This difference in the form of the attack means that even concrete containing sulfate-resistant Portland cement and designed in accordance with BRE Digest 363 may be affected.

There are several risk factors in the occurrence of TSA [16]

- exposure to high levels of sulfate
- very wet, cold conditions
- source of calcium silicate (e.g. Portland cement)
- source of carbonate (e.g. limestone aggregate).

Figure 11.18 Excavation to expose the slender bridge piers and spread footing

The columns of this bridge were formed from in situ reinforced concrete cast in shutters within a cutting through undisturbed lower Lias clay. Excavated lower Lias clay was stockpiled and later used to backfill around the columns and form a 2-m embankment either side of the carriageway. Oxidation of the pyrite in this reworked clay is thought to have led to high sulfate levels developing in the ground. In addition, the excavation for the spread footing would have generated a waterlogged pit next to the structure, thus promoting TSA. The concrete used to construct the bridge included Portland cement and limestone coarse aggregate.

11.5.1 TSA management strategy

The management strategy had two stages, as indicated in Table 11.11. A condition survey was carried out in which softened areas were mapped and the depth of expansion and softening were measured to each face of the columns. Concrete cores, dust samples and fragments were subjected to petrographic examination, SEM microprobe analysis, compressive strength testing, and chloride and sulfate determination [17].

The attack was principally evident due to softening of the surface of the columns exposed during strengthening work. The distribution of the softening varied, with no attack within 1 m of ground level, partial attack in the form of patches or bands of softening up to 500 mm in size approximately 2 m below ground level and attack across the full area of all the faces below 4 m depth. The attacked concrete was observed to have a soft, white pasty layer, typically 5 mm in thickness, but no visible cracking (Figure 11.19). The concrete underneath this layer had cracks running subparallel to the surface and the coarse aggregate particles were surrounded by white 'halos' of reaction products. The frequency and size of the halos reduced with depth below the exposed surface, such that, typically at 20 mm depth, only occasional white specks of reaction products were present in pores without any physical disruption to the cement matrix, which is described as thaumasite formation rather than TSA (Figure 11.20).

Table 11.11 TSA management strategy for a UK highway structure

Action	Purpose
Stage 1 – Confirmation of diagnosis	
Take samples and conduct a logging excavation	Confirm soil type and sulfate conditions
Conduct a condition survey to map the area and depth of softening	Determine the extent of deterioration
Takes cores for petrographic examination, SEM and chemical analysis	Confirm TSA from its reaction products Determine the sulfate and chloride profiles
Stage 2 – Remedial work	
Prop up the structure	Ensure safety under dead and live loading, e.g. HGV impact
Cut and remove the existing piers	Remove deteriorated concrete and provide samples for future testing and examination
Recast new piers	Replace vulnerable concrete with a concrete mix and additional protective measures meeting the new recommendations

Figure 11.19 Appearance of concrete at depths of 5, 10 and 15 mm (left to right)

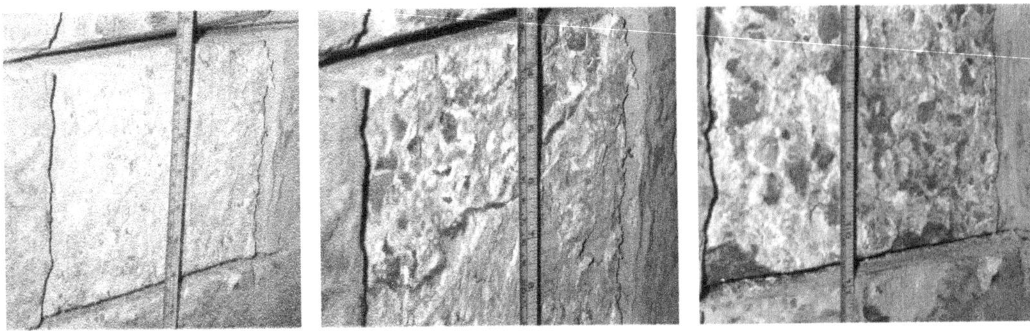

Figure 11.20 Schematic representation of TSA-affected concrete

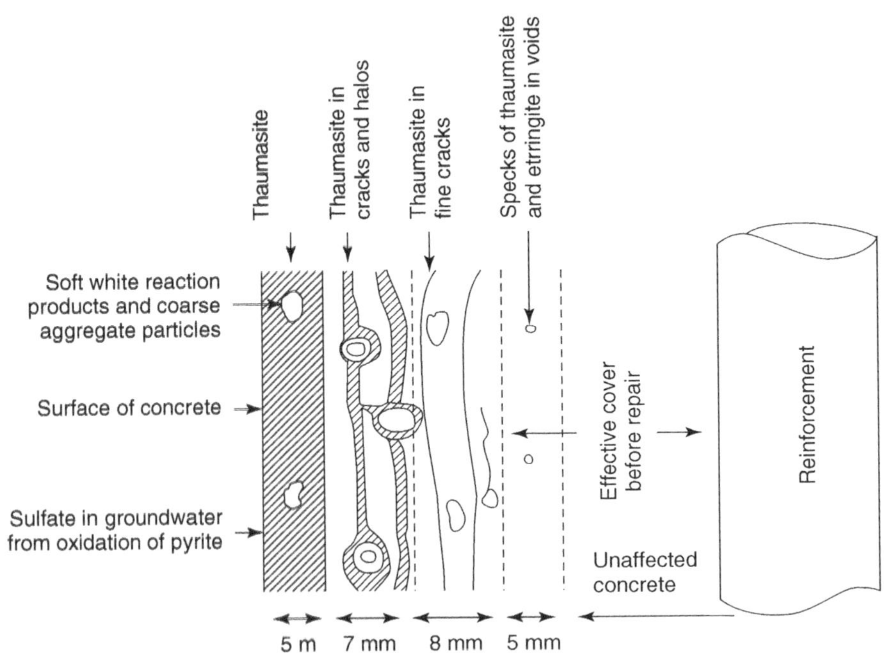

The reaction products, which were initially very soft, become hard and friable after being exposed for several days. Rust staining was present within some of the attacked areas and breakouts revealed deep pitting corrosion to the reinforcement, representing 30% loss of section, and black and green reaction products, which turned brown on exposure, characteristic of anaerobic chloride-induced corrosion (Figure 11.21).

Petrographic examination, including point counting, indicated a concrete mix containing 300–400 kg/m^3 of Portland cement and a w/c ratio of 0.48–0.57. This would meet class 1 or 2 sulfate conditions to BRE Digest 363 [15]. The major rock type in the coarse aggregate was dolomitic limestone and a fine aggregate comprising quartz/metaquartzite with 25% limestone fines. An in situ cube strength of 60.5–83.0 N/mm^2 was estimated from the proportions. The birefringent appearance of reaction products in thin sections under

Figure 11.21 Deep pitting corrosion to a reinforcing bar in TSA-affected concrete

cross polars indicated relatively pure thaumasite around the aggregate particles with little intermixing with other sulfate reaction products such as ettringite (Figure 11.22). A SEM microprobe analysis confirmed this finding. The thaumasite was associated with SiO_2/Al_2O_3 ratios of greater than 9, compared with a ratio of less than 0.4 for ettringite [18].

Sulfate and chloride determinations from depth increments of 10 mm from a 100-mm-diameter core are shown as profiles in Figure 11.23. The depth of TSA given by the depth of cracks infilled with thaumasite from the petrographic examination is also indicated. The depth of TSA approximately corresponds to the point at which sulfate values exceed approximately 5% expressed as SO_3 by mass of cement, equivalent to

Figure 11.22 Petrographic examination showing thaumasite around an aggregate particle. (Courtesy of Mike Eden)

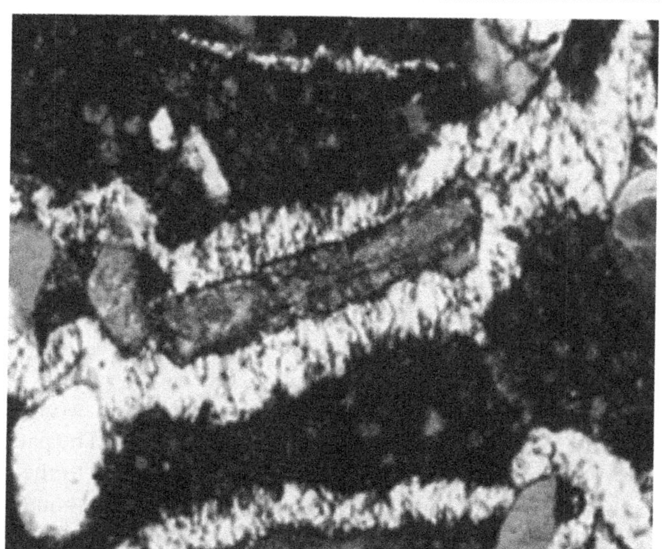

Figure 11.23 Sulfate and chloride profiles in TSA-affected concrete

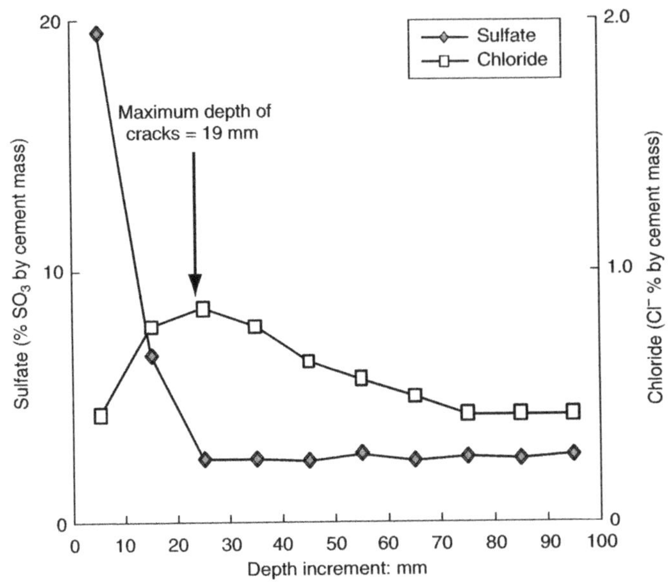

6% SO_4 by mass of cement. The chloride profile has a peak of over 0.8% chloride ions by mass of cement that is 5–10 mm deeper than the depth of TSA. This, together with the evidence of corrosion to the reinforcement, suggests that chloride-induced corrosion may be an underlying problem in areas where TSA is present.

A comparison of the distribution of attack with the results of piezometric monitoring over 18 months indicated that there was no TSA above the maximum groundwater level but full attack to concrete below the minimum water level. A partial attack at one pier coincided with the location of carrier drains through the backfill. The distribution of softening on the columns suggested that there was less softening furthest from the oncoming traffic and on faces furthest away from the carriageway. These results implied that the pattern of attack was closely associated with the presence of water, either as natural groundwater or from other sources, such as run-off and traffic spray from the carriageway. A typical pattern of softening to and expansion of the columns is indicated in Figure 11.24. Areas of expansion and softening are closely associated.

The bridge deck was propped as a precaution. The maximum net loss of cross-section of the piers at the deck was 33 mm, which is less than that associated with a risk of buckling failure to the columns. However, due to the possibility that there would be further deterioration during the remaining service life, it was decided to replace the affected lower portion of the columns and recast a new spread footing above the old base [19]. Cut samples of concrete were removed from site for further examination and testing [20].

11.5.2 Conclusions

Deterioration to piers of a 30-year-old bridge prompted a programme of investigation and repair work. Petrography in conjunction with SEM microprobe analysis confirmed TSA. The pattern of deterioration was expansion and softening of the concrete at the surface, cracking subparallel to the surface and white halos of reaction products around the aggregate particles at depth. The attack was thought to be promoted by the reworking of lower Lias clay, leading to oxidation of pyrite. The extent of attack was closely related to the availability of water.

Figure 11.24 Typical pattern of softening and expansion in columns

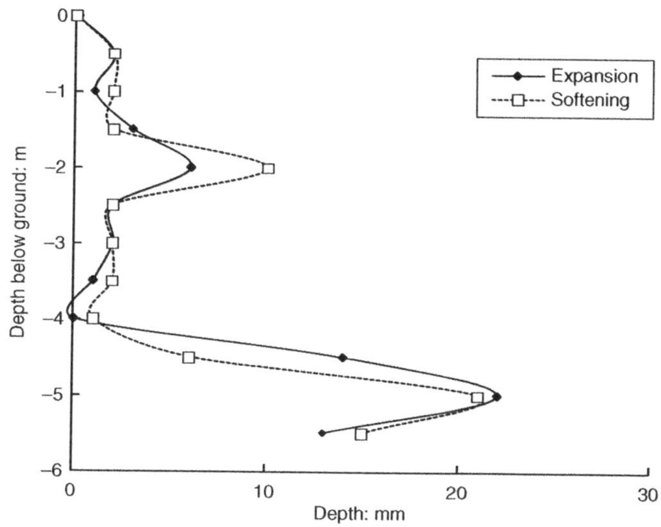

High chloride levels in the concrete and staining and pitting to the reinforcement suggest that chloride-induced corrosion may be enhanced by TSA. Uncertainty over long-term deterioration led to removal and replacement of the piers.

11.6. Installation of corrosion protection on the M4 elevated freeway, London, UK

Christopher Atkins, Mott MacDonald, UK

A radical rethink of how to install corrosion protection in reinforced concrete suffering from a chloride attack saved significant time, cost and energy and offered major health and safety improvements. Using a novel and highly innovative approach, cathodic protection was successfully installed on cross head beams supporting an elevated section of the M4 in West London.

11.6.1 Reinforcement corrosion

The M4 is a major elevated freeway into the west of London (Figure 11.25). It carries 87 000 vehicles per day into and out of the capital of the UK. It was constructed directly over an existing major highway that carries an additional 51 200 vehicles per day. Both run through a residential area with private dwellings directly adjacent to the twin-level route.

The elevated freeway is 3.5 miles long and was made with prestressed beams supported on 120 bents using two half-joints at each bent. It was constructed in 1967 and has long showed its age. Every winter, huge quantities of de-icing salts are liberally applied to prevent ice forming on the freeway. These have penetrated the leaking joints and passed into the concrete (Figure 11.26). Levels of chloride of over 5% by mass of cement (17.5 kg/m^3 or 30 lb/yd^3) are regularly recorded, with one value double this. The reinforcement is corroding, and the concrete has spalled.

11.6.2 Repair strategy

The shape of the structure meant that much of the work had to take place within lane closures. For most of the structure, lanes could be closed only at night when the traffic volumes were lowest. Moreover, because it

Figure 11.25 The M4 elevated freeway in West London

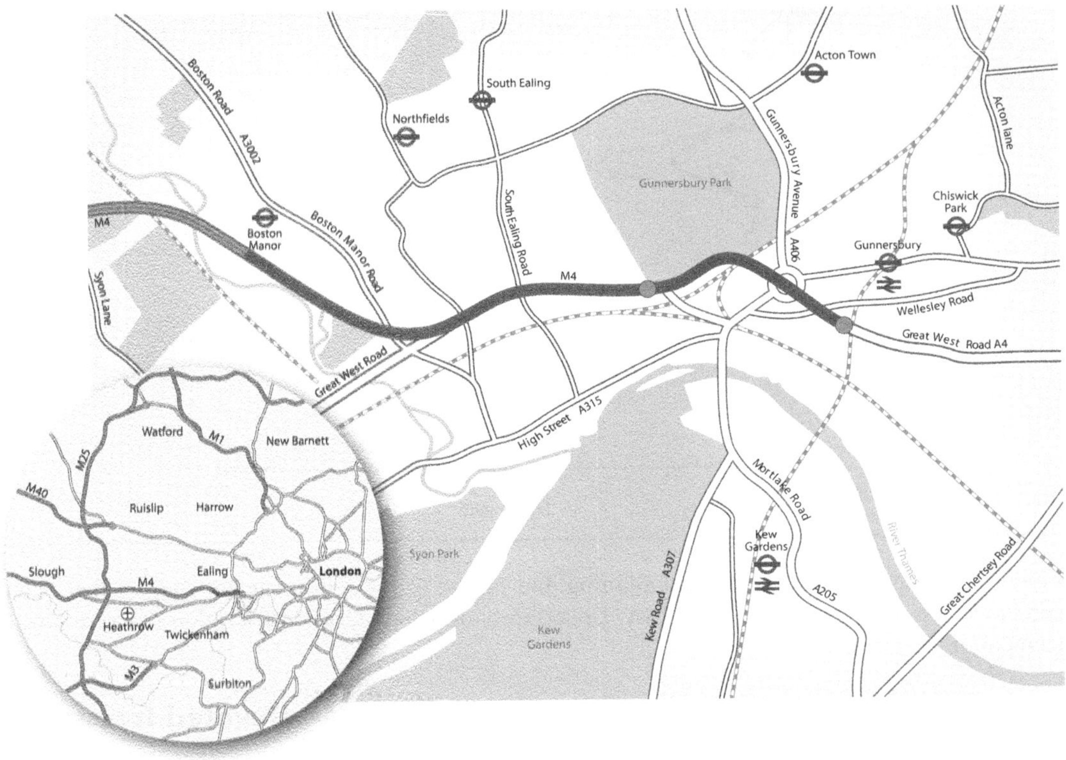

Figure 11.26 Twin half-joints contributed to the problems

was residential area, it was not acceptable to make any significant noise at night while carrying out repairs. The acceptable level of noise was roughly equivalent to loud speech.

The structure owner recognised the problems and started a series of investigations. The first priority was to confirm that there was adequate structural capacity and that the public were not at risk from an imminent collapse. Once these had been demonstrated, the owner began removing all loose concrete to prevent this falling onto vehicles (Figure 11.27). Repairs to the structure were to follow, but there was a major problem. The area of concrete most likely to be contaminated by chloride and containing the most amount of reinforcement was the area that could not be accessed due to the half-joints. A series of trials from 1995 onwards tested every available repair technique. This even included replacing the bents. Every currently available cathodic protection system was also trialled.

Cathodic protection could be applied only to the accessible areas of the structure. Since a large proportion of the piers were accessible, they could be protected using a standard mesh and overlay cathodic protection system (Figure 11.28), which successfully prevented further corrosion without needing to replace the chloride-contaminated concrete.

The half-joints also needed to be protected but could not be accessed without shutting the freeway. The only option that was worthy of further pursuit was to drill up and install anodes from the soffit. The total length of drilling would be approximately 600 feet with an estimated duration of over 30 days. If reinforcement were encountered, the hole would need to be repositioned and drilled again. Where the cross-beam ran over the pier, the holes would need to be drilled at an angle to position the anodes in the correct place. The holes would then need to have an anode inserted and be grouted from underneath. Each anode would then have to be individually wired to a primary conductor. The works would be difficult to plan and programme (Figure 11.29). The heavy and persistent traffic during the day meant that the drilling would have to be carried out at night, but the strict environmental noise limits in place for the sake of the residents meant it could not. In addition, there would be the ever present risk of having to redrill holes, which inevitably meant that the overall cost of the work was uncertain and highly variable.

Figure 11.27 For safety, all loose and delaminated material was removed

Figure 11.28 Mesh and overlay cathodic protection was installed during the day

Figure 11.29 Work had to be carried out safely without disruption to the traffic flow

The repair contract was signed in 2006. It was based around installation from the soffit. Moreover, the plan was to fix a single bent. A specialist repair contractor was appointed. After discussions with the designer, a possible alternative approach was suggested that had the potential to revolutionise the repair programme. On reviewing the reinforcement details, it was noticed that there was an area running along the cross-beam that contained no reinforcement, and it was close to the main inaccessible reinforcement. If it were possible to install anodes in this location, the risk of hitting steel was dramatically reduced, so the works became more manageable. The lengths of holes required reduced from 600 feet to 60 so the duration was shorter, which meant that the works became significantly cheaper. The neighbourhood would be less disrupted. Site safety would improve as there would be fewer people working for shorter periods. The only problem was that no one had ever attempted this before.

To see if it could be done, the specialist subcontractor obtained 60 feet of temporary concrete barriers and a coring rig. The coring started at one end and remained on course, coming out within 2 inches of where

Figure 11.30 Cantilevered gantry allows coring to proceed over the live carriageway

it was predicted. The core drifted down under gravity, but as all the cross-beams to be protected were on a camber, this was not considered to be a problem. Coring would be started at the highest end, and the natural drift would help keep it on course. Due to the success of the coring trial, it was decided to use this approach on site (Figure 11.30). The coring was successfully carried out over three night shifts.

A detailed design was produced that split the upper section of the cross head into two zones. Anodes were fixed to a titanium conductor bar at different locations to vary the current distribution in accordance with the reinforcement. Reference electrodes were also fixed to the conductor bar to allow the system to be monitored. This was then pulled through the core hole and grouted into place. The manufacturer's recommended grout did not flow sufficiently to enable its use, so a post-tensioned duct grout was employed instead.

It was still necessary to repair the cantilevered cross-beams, but this was still limited by the noise and traffic problems. The contractor came up with an innovative approach. An airport baggage-handling truck was adapted as a mobile soundproofed enclosure that could be raised into position and moved around during the night work (Figure 11.31). This meant that hydrodemolition and shotcrete reinstatement could be carried out at night without disturbing the residents.

The installation of the first pier was completed in March 2008. The estimated cost savings, assuming drilling from the soffit was possible, were 20%, reducing the cost of repair to approximately USD 600 000 per cross head. The estimated cost of replacing a cross-beam was USD 10 million. Due to the savings, another eight piers were replaced within the next 12 months. At present prices, the cost of fixing the entire viaduct fell from USD 1.2 billion to USD 72 million by employing this technique.

The repair system has proved simpler and more efficient to install than anything anyone had developed previously. The approach saves time and money, improves site safety and minimises the duration of the roadworks and the noise generated during repair. Both environmental and sustainability issues have been addressed in a positive and quantifiable manner. Finally, the project improved the appearance and feel of the neighbourhood with minimum intervention to the benefit of residents and London visitors alike. It was a truly innovative and sustainable solution to what otherwise was considered to be an intractable problem (Figure 11.32).

Figure 11.31 A modified airport vehicle became a mobile soundproof enclosure

Figure 11.32 The first four completed installations

11.6.3 Conclusions

The innovative technique for installing cathodic protection meant that the length of holes drilled was only 60 feet per beam compared to a combined length of 600 feet for a conventional approach. Moreover, the environmental footprint was reduced. Replacing a pier and cross head beam would have needed a significant amount of energy, producing a corresponding volume of carbon dioxide equivalent to that locked up by 300 acres of forest in a year. Installing cathodic protection using a conventional drilling pattern would have had almost the same carbon footprint. The single hole-through-the-middle approach used 60% less energy. The total time taken to bore each hole was 3 days, compared to 30 days or more for the conventional approach. Also, it dramatically reduced exposure to vibration for operatives. There was only limited working at height and no daytime road closures.

11.7. Systematic investigation and application of EN 1504 to a concrete repair project

John Broomfield, Broomfield Consultants, UK

11.7.1 Grade II* and Grade II listed structures

The University of East Anglia was founded in the 1960s when the main campus buildings, including the 'Teaching Wall' and the well-known 'ziggurat' residential blocks (featured on the English Heritage website) were laid out by Sir Denys Lasdun (Figure 11.33). The university is proud of its architecture, which has been supplemented by other famous architects. The Teaching Wall consists of a shallow W of reinforced concrete buildings approximately 500 m long. The runs of offices, laboratories and lecture rooms are interrupted by lift and stair 'towers' at intervals along its length. There are water tanks and plant rooms above the main building at roof level. The exposed concrete facades are a feature of the Teaching Wall and various parts of the campus, which were given grade II* and grade II listing during the work. Various sections of the Teaching Wall and other campus buildings are linked with elevated walkways.

11.7.2 Condition and situation of the structures

Broomfield Consultants were appointed as corrosion specialists by Jacobs Babtie Consultants to conduct a forensic structural engineering analysis, initially of the Biotower as phase 1. Subsequent phases would include all the other reinforced concrete structures with exposed concrete facades on the campus. Work was conducted in close collaboration with the university departments affected, as well as the Estates Department, which ran the project, English Heritage and the Norwich City planning office, which gave the planning consents for the work.

Phase I work was on the Biotower, a lift and stair tower with an air-conditioning plant room and a water tower above. Detailed investigation showed that low cover and carbonation were prevalent with some admixed calcium chloride in some 'lifts' of concrete, all leading to reinforcement corrosion. Several repair

Figure 11.33 Teaching Wall and adjacent walkways

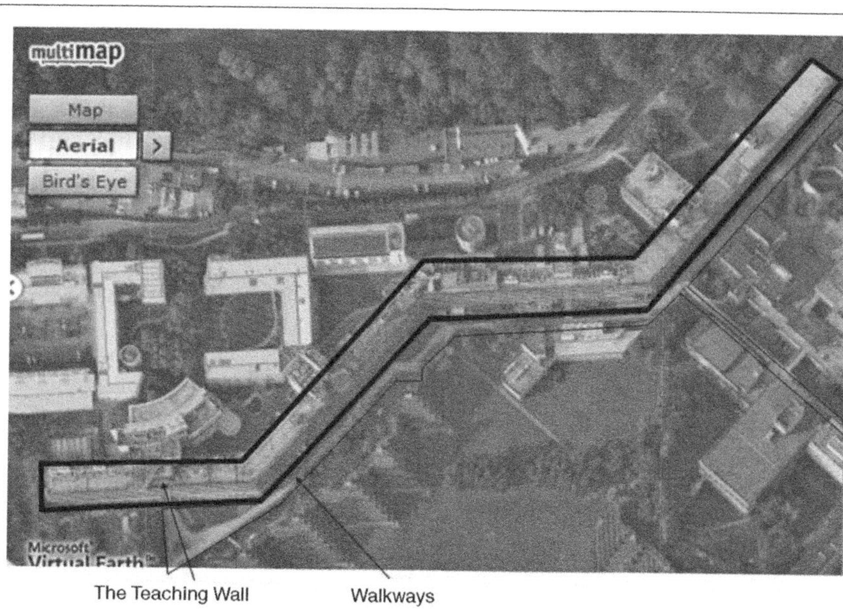

Figure 11.34 Part of the library walkway showing the Teaching Wall and towers behind. De-icing salts and leachate runs can be seen where the waterproofing and drainage has failed, allowing corrosion of the slim pier supports

options were investigated, including the possibility of cladding the facade and 'air conditioning' it to remove moisture and stop reinforcement corrosion, according to EN 1504-9 Principle 8. However, this was untried technology, and it was considered that no contractor would offer any warranties on such an installation. For that reason, impressed-current cathodic protection was applied according to EN 1504-9 Principle 10 (cathodic protection). The specification was according to BS EN 12696 – 'Cathodic protection of steel in concrete'.

Phase 2 works were on the library walkway, namely, a concrete stairway to another walkway, which had severe corrosion damage, and two further stair/lift towers in the Teaching Wall (Figure 11.34). A detailed quantitative condition survey revealed areas of concrete damage due to corrosion from carbonation. This was principally due to low cover, indifferent quality concrete and the age of the structure. Other areas were deteriorating due to de-icing salt ingress, particularly on the elevated walkways and access stairways. Using the survey data, calculations were made of ongoing chloride and carbonation ingress on a 30-year-life projection [21]. Corrosion was modelled based on Fick's law of diffusion using cover depth measurements combined with carbonation depths and chloride depth profiles [22]. This showed that other than the areas showing immediate damage, a few other areas were susceptible to future corrosion.

11.7.3 Applying the principles of EN 1504 to the rehabilitation process

EN 1504-9 Section 5.2 gives the following options.

(a) Do nothing for a certain time.
(b) Reanalyse the structural capacity.
(c) Prevent or reduce further deterioration without improving the concrete structure.
(d) Improve, strengthen or refurbish the structure.
(e) Reconstruct all or part of the structure.
(f) Demolish the structure.

Given that the structures are part of a listed site, that further deterioration could lead to health and safety problems in some areas and that the university has set aside a budget for its 'concrete preservation plan', options (c) and (d) were relevant.

The standard options for intervention on a reinforced concrete structure suffering from reinforcement corrosion are as follows.

(a) Do nothing for a certain time.
(b) Complete or partial demolition and rebuild (EN 1504-9 Principle 3.4).
(c) Patch repair of local damaged areas (EN 1504-9 Principle 3).
(d) Ingress control with coatings, membranes, sealers, water stops, enclosures or other barriers (EN 1504-9 Principles 1, 2 and 8).
(e) Impressed-current cathodic protection (EN 1504-9 Principle 10; BS EN ISO 12696).
(f) Galvanic cathodic protection (EN 1504-9 Principle 10).
(g) Electrochemical re-alkalisation (EN 1504-9 Principle 7.3; BS EN 14038-1).
(h) Electrochemical chloride extraction (EN 1504-9 Principle 7.5; BS EN 14038-2).
(i) Corrosion inhibitors (EN 1504-9 Principle 11.3).

Being part of a listed building and suffering from corrosion damage, items (a) and (b) were not feasible. Option (c) was required in some areas. Option (d) was used, but in some areas its use was constrained by the requirement to retain the board-marked finish to the concrete on the listed facades. However, control of ingress of carbon dioxide and chloride ions was required.

Thus, a proprietary architectural coating was trialled for approval by the university and by the local authority's conservation officer. This coating 'tones down' changes in concrete colour and finish and was considered ideal for minimising the visual impact of patch repairs on the board-marked finish on the concrete facades. The selected coating has anticarbonation properties and is also compatible with a silane for control of moisture and chloride ingress.

In this phase of the works, impressed-current cathodic protection, (e) in the list above, was not required on a large enough area to be cost-effective. However, given the presence of active chloride-induced corrosion, an alternative was to use galvanic anodes (f) installed in the patch repairs to minimise incipient anodes (Figure 11.35). The other electrochemical treatment techniques, (g) and (h), were not considered suitable for this project.

11.7.4 Design and specification of the work

The techniques selected, therefore, included localised galvanic cathodic protection to minimise the incipient anode effect around patches in areas of high chloride (EN 1504-9 Principle 10 CP). An example of this effect is shown in Figure 11.35. Penetrating sealers were required to keep out further chloride according to Principles 1.1 and 6.1 and to reduce moisture (Principle 8). Anticarbonation coatings were required to reduce the rate of carbonation (Principles 1.3c and 6). Renewal of the waterproofing membrane on the walkway decks was specified to keep moisture and chlorides out of the deck concrete (Principle 1.1). The membrane and improvements to drainage provided a reduction in water leakage, sheltering the walkway substructure from de-icing salt running down. These techniques were used along with conventional patch repair where required (Principle 3). A detailed analysis of the condition survey results allowed treatments to be selected for different elements of the structures, as show in Table 11.12.

Figure 11.35 Incipient anode formation around an old repair on the Biotower plant room prior to the phase I repair and impressed-current cathodic protection

Table 11.12 Treatments used in the work

EN 1504-9 Principle	Method or principle	EN standard	Elements treated	Materials used
1 – Protection against ingress 8 – Increasing concrete resistivity	Hydrophobic impregnation: Principles 1.1 and 2.1	EN 1504-2 EN 1062-3 Maximum value: $w = 0{:}035$ kg/(m^2 h$^{0:5}$)	Walkways below deck level where de-icing salts were applied and where the chloride level at the rebar is below the threshold for corrosion	Silane compatible with cosmetic coating used to tone down repairs
1 – Protection against ingress	Anticarbonation coating: Principle 1.3c	EN 1504-2 EN 1062-6 Permeability to carbon dioxide: standard deviation >50 m	Parapets on walkways above de-icing salts where chloride levels are very low	Cosmetic coating with anticarbonation properties
1 – Protection against ingress	Waterproofing membrane: Principle 1.7	Not listed in EN 1504	Walkway decks	Waterproofing system
3 – Concrete restoration 7 – Preserving or restoring passivity	Hand-applied mortar: Principle 3.1	EN 1504-3, Class R4 Compressive strength: >40MPa Adhesive bond: >2MPa	All damaged elements	Pre-bagged patch repair material
10 – Cathodic protection	Local galvanic anodes: Principle 10	Galvanic anodes not covered yet	Patch repairs with chloride levels in excess of the corrosion threshold	Zinc anodes encapsulated in a proprietary activating mortar

The following specifications were written for the job

1. A concrete repair specification based on
   a. Materials according to EN 1504-9 and 1504-3 (Class R4 structural grade repair mortar)
   b. Patch repair preparation according to EN 1504-10, Section 7 and Appendix A7
   c. Material application according to EN 1504-10, Section 8, on application, Appendix A8
   d. Testing samples on site and off site using test methods and values in EN 1504-10 and Appendices A7, A8 and A9
2. A coating specification for silane impregnation based on
   a. Materials according to EN 1504-9 and 1504-2 (1.3C for anticarbonation coating and 1.1(H) and 1.2(I) for silane impregnation for moisture or chloride ingress control)
   b. Manufacturer's literature for application
   c. Surface preparation according to EN 1504-10, Sections 7 and 8 and Appendix A8
   d. Site testing according to EN 1504-10 and Appendices A8 and A9
3. An application specification for a waterproofing membrane
   a. Lifting paving slabs
   b. Conducting repairs
   c. Repairing and improving drainage
   d. Applying a waterproofing system
   e. Replacing paving slabs.

11.7.5 Site tests

After applying coatings, cores were taken and sent for testing. Carbon dioxide permeability tests (BS EN 1062-6) gave far better than the 50-m minimum values recommended in the specifications, but the concrete alone gave starting values of 23 and 30 m.

The results of the water permeability test were

- Coated 0.03 and 0.04 kg/(m^2 $h^{1/2}$)
- Partial coated 0.05 kg/(m^2 $h^{1/2}$)
- Uncoated 0.11 and 0.12 kg/(m^2 $h^{1/2}$).

BS EN 1062-3:1998 Table 1 states

- I High >0.5 kg/(m^2 $h^{1/2}$)
- II Medium 0.1 to 0.5 kg/(m^2 $h^{1/2}$)
- III Low <0.1 kg/(m^2 $h^{1/2}$).

Given the requirements for a coating with architectural properties and that most areas of low cover were repaired, that the coated values falling in the II medium range was judged to be acceptable. Also, renewing the waterproofing and the drainage would reduce the amount of water running down the substructure, further reducing the rate of chloride ingress.

Pull-off tests on concrete patch repairs can be conducted according to EN 1542, ISO 4624 and BS 1881-201 and 1881-207, as described in EN 1504-10 A9.2 – 'Test or observation No 35'. Recommended values are given in Table A2 of EN 1504-10. In this project, pull-off tests achieved 0.8 MPa or better.

Note that the specifications where written and the work completed prior to full publication of all parts of EN 1504 and the associated test methods. Not all testing on this project was compliant with the specific CEN test mentioned but used equivalent British Standards or other tests in use at the time.

11.7.6 Conclusions

The work was successfully completed in 2007. There was minimal disruption to campus activities and both the university and the listing officer were pleased with the final finishes on the listed elevations.

In conclusion, it can be seen that concrete repair systems can be designed and their performance specified and applied using the EN 1504 set of documents along with their associated test methods following the principles of corrosion engineering, which ensures that corrosion is prevented before it initiates and it is controlled once damage has initiated.

The first critical part of any repair and refurbishment project is a condition survey, which quantifies the type and extent of damage to ensure the following.

- Only areas in need of treatment are treated.
- Appropriate treatments are selected.
- The current and future requirements of the structure are fully considered in the repair design process.

Appropriate repair systems and materials can then be selected based on the principles of EN 1504 and repair designs and specifications prepared using the product characteristics specified in EN 1504 and the associated test methods. Life-cycle costing can also be applied to ensure that the systems used are the most cost-effective ones that meet the client's requirements.

11.8. ASR on the floor of a production facility

Don Wimpenny, Jacobs, UK

11.8.1 Introduction

Blistering and delamination of an epoxy coating system to an 8000 m^2 floor was identified during installation of a clean room approximately 3 years after construction of the floor. The clean room is on the ground floor of a production facility and is intended to have a 20-year life manufacturing electronic equipment.

The floor of the clean room comprises a reinforced concrete slab 350 mm in thickness, up to approximately 119 m long in the east–west direction and 69 m wide, supported on pile caps and piles. The floor slab is reinforced top and bottom and has vertical links at its perimeter and at pile and column locations. The floor is divided into five zones A to E. To facilitate the investigation, each zone was split into 14 areas by column lines on a grid 8.4 m by 13.8 m (grid lines 5 to 19 and E to K), as illustrated in Figure 11.36. Each of the areas was further divided into eight sub-areas (S1 to S8). Manufacturing in zones A to D would generally use dry processes, whereas in zone E, there would be a wet process with chemicals. However, the wet tools would be well contained and spillages immediately removed. The clean room floor supports steel pedestals of height 1.5 m, which carry the floor of the manufacturing area. The clean room floor does not directly have traffic when in service, although air from this area circulates into the manufacturing area of the clean room.

External waterproofing beneath the floor slab comprises a bituminised membrane with a single self-adhesive face. The membrane was laid with the adhesive face down onto 50-mm-thick blinding concrete. The membrane was protected by a 20-mm screed prior to fixing the reinforcement and casting the floor slab, and it generally curtails at pile-cap locations. However, the waterproofing does not meet the intention of the specification, which was for it to be bonded to the slab not the blinding. The groundwater level is high and can be expected to occasionally approach ground level, approximately 1 m above the top surface of the slab. Groundwater passing through local breaches in the membrane can track along the underside of the slab, soaking up into it and penetrating at joints and any cracks. The joint between grid line K7 to K17 and that between grid line 19H to 19J were reported to have leaked and were subject to remedial works.

Figure 11.36 Plan of the clean room floor

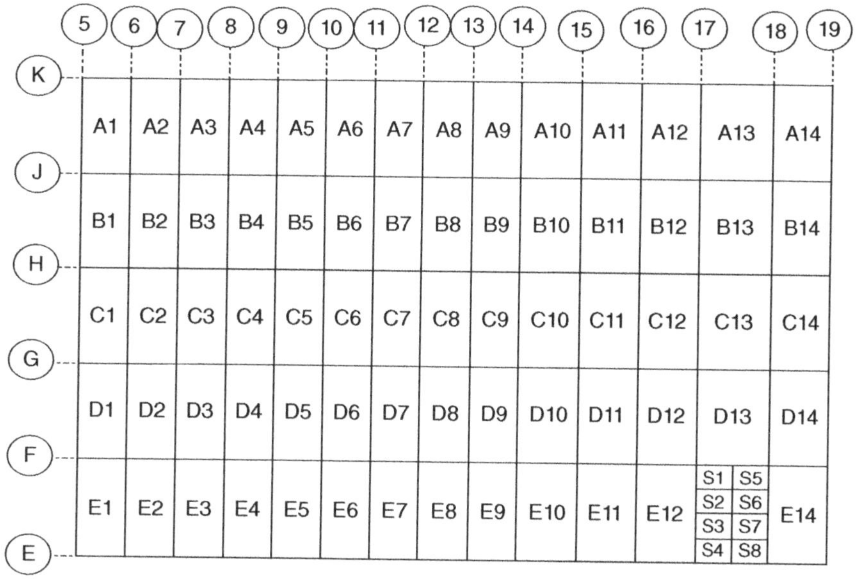

The top surface of the floor slab has an epoxy-based flooring system, which comprises four layers with a total thickness of 4 mm

- water-dispersed epoxy resin primer
- 2-mm epoxy cement floor topping, which was a temporary moisture barrier to protect the epoxy flooring until it had cured
- epoxy primer
- 2-mm epoxy chemical-resistant finish coat.

This coating system is important for sealing the concrete surface and, thereby, reducing the release of silicate particles into the clean room environment.

The floor was cast using a concrete mix with a 28-day characteristic compressive strength of 40 MPa, a total cementitious content of 415 kg/m^3 and a w/c ratio of 0.42. The mix used only Portland cement with no supplementary cementitious materials. The fine and coarse aggregates were reported to have been sourced from Vietnam and Indonesia.

Aggregate from Southeast Asia is known to be potentially reactive to alkali [23]. The reactivity may be significantly affected by discrete intrusions of particular minerals and can be ascertained only if the aggregate is tested at regular intervals. The risk of ASR can be controlled by limiting the alkali content of a mix, such as through the selection of Portland cement and supplementary cementitious materials, like silica fume, fly ash and ground granulated blastfurnace slag (GGBS).

The clean room floor was constructed in 2007, but the coating system was not applied until 2010. During the interval between construction and 2009, the area was a storage area and not subject to any air conditioning. In 2009, air conditioning was introduced to prevent condensation on the soffit of the slab to the second floor.

There was a concern about the potential impact of the blistering on the operation of the facility and, in particular, the need to meet strict limits on dust levels.

11.8.2 Investigation

11.8.2.1 Introduction

An investigation was carried out to determine the cause and extent of the failure, provide a prognosis about future deterioration and recommend corrective action. The work involved a systematic programme of inspection, sampling and testing, together with a review of construction records. The investigation had three main stages

- site investigation
- laboratory testing
- reporting and remedial recommendations.

It was important that the work be completed as quickly as possible, with minimal disruption to the fitting out of the clean room.

11.8.2.2 Site investigation

The site work involved a close inspection of the floor over 4 days to determine the extent of blistering. The available light was supplemented by a portable arc light. The blisters were tapped with a light hammer, and any that displayed hollowness were marked with indelible ink. Each grid square bounded by the column lines was divided into eight sub-areas. The number of blisters was counted within each sub-area and a photographic and video record made. The heights of representative blisters were assessed using a straightedge, spirit level and thickness gauge.

During the inspection, locations were identified for testing and sampling. The locations were chosen to provide a representative distribution across the floor area as well as to allow the investigation of any specific features of interest, for example, larger areas of delamination.

The in situ testing included

- pull-off testing in accordance with ASTM D4541-09e, the dolly pull-off method [24]
- rebound hammer checks of concrete hardness to BS EN 12504-2 [25]
- hygrometer measurement of the concrete slab within a surface-mounted enclosure in accordance with the BS 8203 method [26]
- moisture assessment by measuring the resistivity with deep probes in drilled holes.

In addition to the in situ testing, nominal 100-mm-diameter cores were taken at locations with and without blisters for laboratory examination. These were tightly wrapped in cling film and placed in polythene bags.

Some blisters were punctured to see if there was any pressure beneath them and to extract fluid from them, but the blisters were either empty or adhering tightly to the substrate. A chisel was used to break three blisters open, and concrete fragments were removed for analysis. Samples of the delaminated coating and coating attached to pull-off dollies were also removed from site for laboratory examination.

11.8.2.3 Laboratory testing

The laboratory work involved petrographic examination of cores to ASTM C856 [27] and ASTM C457 [28] to look for evidence of ASR and other deterioration, to assess the general quality of the concrete, to estimate its composition and to identify the aggregate type and any reactive particles. Following the petrographic examination, accelerated expansion testing was commenced in accordance with Appendix H of the BCA guide to the diagnosis of the alkali–silica reaction [9]. Samples of the coating (attached to cores or dollies or from delaminated areas) were also subject to petrographic examination to estimate the thickness of the layers within the coating system. SEM combined with energy-dispersive X-ray spectroscopy (EDX) was carried out on concrete fragments removed from below two blisters to determine their composition and look for evidence of contamination.

11.8.3 Findings

11.8.3.1 Site investigation

The site inspection identified over 8000 blisters, typically 20–50 mm in diameter, with a shallow height, which made them difficult to detect without the use of oblique lighting. Their distribution varied across the floor. There were 2439 blisters in zone A and 960 in zone E (Table 11.13). This may conceivably indicate differences in the concrete mix and the exposure conditions. It is notable that the highest incidences of blisters were in areas that had previously experienced an ingress of groundwater.

Six large blisters up to 1.7 × 1.0 m in size and 8 mm in height were also found in four separate areas of the floor close to columns or within an area with restricted headroom. Surface flatness measurements at the blisters gave gaps under a 1-m straightedge in excess of 4 mm. The coating in two of these large blisters had broken to reveal the concrete surface, which was generally very smooth, but in one case showed a sharp transition in texture. This transition is consistent with a change in surface preparation. It is understood that the floor was prepared by abrasive blasting, except within 1.5 m of walls and columns and other areas of restricted access, where the surface was prepared using hand grinders. These large blisters were empty; fluid would have been a useful indication of osmotic blistering [29].

The small blisters were often associated with cone-shaped pop-out failures within the concrete, which appeared to contain waxy products. Moisture was present under some blisters, but there was no evidence of fluid being expelled under pressure when the coating was punctured and there was insufficient volume to sample using a syringe. Less than 1% of the small blisters sounded hollow or displayed any evidence of cracking to the coating.

The average adhesion strength in three locations adjacent to areas of existing large blisters was less than 1.5 MPa (Table 11.14). This indicates a failure to meet the normal compliance requirement for the adhesive bonding of a coating to concrete. The failure was 100% at the interface between the concrete and the water-dispersive epoxy primer. The smooth appearance of the surface and the bond test result suggest that there had been poor preparation of the concrete surface.

The approximate thickness of the epoxy coating was measured on site at the pull-off testing locations. The overall estimated thickness of the coating was between 2.0 and 4.8 mm, compared with the 4-mm design thickness of the system (Table 11.15). Thicknesses were also measured in the laboratory.

Rebound hammer checks were carried out in two locations to verify the quality of the floor surface. In both locations tested, the floor was sound with an average rebound value in each location of over 47 (a minimum rebound value of 20 is often used as a substrate soundness criterion for coatings and repairs).

The relative humidity measured at the concrete surface within the enclosure was 77–86%. BS 8203 [26] recommends a maximum of 75% prior to application of any impervious flooring. Deep-moisture pin-type probes suggested that the moisture content increased in going from 10 to 80 mm depth below the surface (Table 11.16). The moisture content was estimated by calibrating the moisture meter readings against oven-dried values. These varied between approximately 3% and 6% at 10 mm to 6% to 14% at 80 mm.

Table 11.13 Distribution of blisters

Zone	Sub-area	Area 1	2	3	4	5	6	7	8	9	10	11	12	13	14	Total for zone
Zone A	S1	1	15	30	2	2	2	23	84	83	64	30	27	38	15	
	S2	4	16	19	7	4	6	26	26	25	41	22	23	43	15	
	S3	7	5	30	6	3	7	16	33	36	28	15	32	42	18	
	S4	14	7	12	3	4	4	5	13	10	41	40	43	35	4	
	S5	12	13	19	3	4	3	56	87	42	34	35	29	17	31	**2439**
	S6	8	24	8	3	7	6	60	53	11	24	23	29	30	14	
	S7	11	23	20	5	5	6	41	31	6	49	23	56	24	17	
	S8	9	14	14	5	0	8	11	11	25	40	36	24	27	7	
	Sub-total	**66**	**117**	**152**	**34**	**29**	**42**	**238**	**338**	**238**	**321**	**224**	**263**	**256**	**121**	
Zone B	S1	19	4	6	18	10	2	17	9	9	3	7	60	38	8	
	S2	5	8	11	5	0	2	9	7	3	20	23	26	20	12	
	S3	6	8	21	5	2	3	15	4	16	37	50	33	23	14	
	S4	11	8	5	2	1	6	14	3	10	40	38	34	16	5	
	S5	4	11	10	16	20	6	7	9	10	16	13	20	61	14	**1786**
	S6	28	18	8	4	3	1	8	4	18	17	28	20	24	26	
	S7	25	11	18	5	0	3	2	5	14	25	38	54	26	24	
	S8	17	11	16	8	9	4	11	11	12	55	76	76	14	1	
	Sub-total	**115**	**79**	**95**	**63**	**45**	**27**	**83**	**52**	**92**	**213**	**273**	**323**	**222**	**104**	
Zone C	S1	2	12	10	6	3	2	3	3	3	17	8	18	15	N/A	
	S2	22	27	7	15	10	4	6	3	7	11	28	16	32	N/A	
	S3	14	48	4	30	4	2	6	14	6	2	17	30	23	N/A	
	S4	11	50	10	47	11	3	11	3	8	8	40	36	8	N/A	
	S5	1	14	2	2	2	4	10	4	4	14	16	14	N/A	N/A	**1344**
	S6	23	27	6	2	1	5	9	2	6	25	30	24	N/A	N/A	
	S7	11	22	4	11	3	7	8	9	4	8	38	23	N/A	N/A	
	S8	3	52	10	22	2	16	10	4	6	19	55	14	N/A	N/A	
	Sub-total	**87**	**252**	**53**	**135**	**36**	**43**	**63**	**42**	**44**	**104**	**232**	**175**	**78**	**N/A**	
Zone D	S1	21	38	5	14	6	5	13	2	3	3	11	7	4	N/A	
	S2	25	45	23	21	26	12	9	8	2	6	7	10	24	N/A	
	S3	17	52	28	22	10	12	17	2	4	24	29	20	19	N/A	
	S4	15	31	12	17	7	23	22	7	6	20	24	40	20	N/A	
	S5	50	55	3	8	7	11	7	5	6	4	11	0	N/A	N/A	**1645**
	S6	44	19	17	17	12	28	6	4	0	7	5	8	N/A	N/A	
	S7	76	17	31	18	8	40	13	4	4	17	12	21	N/A	N/A	
	S8	17	7	29	6	10	42	17	3	2	35	4	28	N/A	N/A	
	Sub-total	**265**	**264**	**148**	**123**	**86**	**173**	**104**	**35**	**27**	**116**	**103**	**134**	**67**	**N/A**	
Zone E	S1	6	8	5	6	7	9	2	4	2	8	12	9	10	N/A	
	S2	2	11	7	4	4	3	8	0	6	19	12	18	19	N/A	
	S3	7	10	6	2	17	2	4	2	7	25	6	22	26	N/A	
	S4	6	14	12	10	21	20	5	3	1	41	37	34	27	N/A	
	S5	11	7	0	5	10	6	5	2	6	6	5	17	N/A	N/A	**960**
	S6	9	13	12	3	2	6	9	8	2	16	5	5	N/A	N/A	
	S7	9	6	9	4	4	3	7	0	5	11	8	13	N/A	N/A	
	S8	4	3	11	10	8	19	7	0	4	13	21	43	N/A	N/A	
	Sub-total	**54**	**72**	**62**	**44**	**73**	**68**	**47**	**19**	**33**	**139**	**106**	**161**	**82**	**N/A**	**8174**

Table 11.14 Results of pull-off tests

Location	Adhesion strength: MPa				Comment
	Result 1	Result 2	Result 3	Average	
A14/S5	2.82	2.80	2.04	2.55	
B5/S2	2.47	1.50	2.04	2.00	
C2/S4	2.59	2.67	2.55	2.60	
D6/S1	1.32	1.53	1.01	1.29	Large blister
D9/S1	0.85	1.22	1.58	1.22	Large blister
E3/S1	1.19	0.60	0.19	0.66	Large blister
E7/S7	2.11	2.22	1.05	1.79	

Table 11.15 Measured thicknesses of the coating

Location	Site measurement: mm		Laboratory measurement: mm	
	Maximum	Minimum	Maximum	Minimum
A14/S5	3.7	4.5	2.5	2.8
B5/S2	4.4	4.8	2.8	3.2
C2/S4	2.0	3.0	1.8	3.2
D6/S1	2.7	3.9	3.9	4.0
D9/S1	4.1	4.4	N/A	N/A
E3/S1	2.3	2.8	N/A	N/A
E7/S7	3.6	3.8	2.3	2.7

Table 11.16 Estimated moisture values in the concrete slab

Location	Moisture value: %			
	10 mm depth	40 mm depth	80 mm depth	120 mm depth
A2/S6	3.3	6.0	6.2	7.0
A7/S8	4.2	6.8	10	8.5
C2/S4	4.3	5.3	13.6	12.7
D9/S1	4.5	5.3	8.4	8.7
E11/S1	6.4	6.7	10.9	6.8

11.8.3.2 Laboratory testing

Petrographic examinations of core samples indicated that all were of similar concrete. There were crushed coarse and fine granite aggregates and siliceous sand in a Portland cement-based paste of moderate to low porosity. Based on the porosity of the cement paste, the equivalent w/c ratio was estimated to be in the range 0.4–0.5. This is, broadly, in line with the concrete mix design.

Although most of the coarse granite aggregate was considered to have normal alkali reactivity, in four out of five cores, the coarse aggregate comprised 5% to 11% of highly reactive tuff particles that included microcrystalline and cryptocrystalline quartz, which is frequently linked with ASR. Damp hygroscopic patches of gel were observed on the surfaces of these cores. Fine, near-surface cracks that appeared to originate from substantially deteriorated particles of coarse aggregate were observed in three of the tested samples (Figure 11.37). In two cases, the deteriorated particles were surrounded by patches of hygroscopic gel. The presence of secondary ettringite within voids in three out of five samples is consistent with moisture ingress into the concrete.

Blistering and slight reflective cracking of the floor coating coincided with localised aggregate-related expansion of the concrete directly below the coating. Although traces of gel were found within the fine near-surface cracks, it is likely that an expansive sheet of silicate minerals may also have contributed to crack development. In contrast, no cracking or ASR-related damage was observed in a core sample with a significantly lower content of reactive tuff particles. This core was taken from an area of the floor that had no visible blisters.

The measured thickness of each layer of the coating system was assessed by a petrographic examination. The total thickness is significantly below the design value for the system (Table 11.15). In three out of five samples, the coating was firmly attached to the concrete substrate. However, in the other samples, the coating had delaminated from the substrate. In one case, the concrete had a smooth upper surface with little evidence of the surface roughening or abrading expected from the specified preparation method. There was no evidence of oil or salt contamination.

Figure 11.37 Thin section through sample A14/S5 showing deteriorated tuff particle at the top and a gel-filled crack in the paste from D4 to I6. (Courtesy of GMRS)

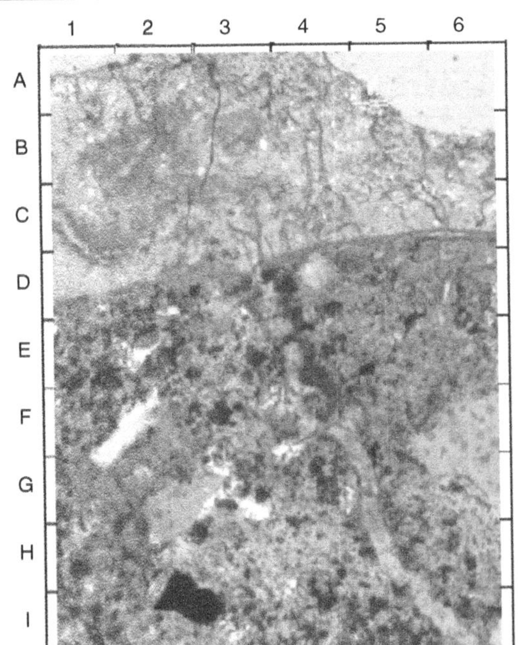

For three out of four samples, SEM identified an abundance of gel deposits associated with tuff particles in cracks. In general, the gel deposits appeared to be of recent origin, suggesting ongoing ASR. Some calcified gel deposits were indicative of reactions between the gel and the cement in the past and, possibly, ongoing ASR, even at low concentrations of alkali. The tuff particles had silica-rich compositions. Fibrous, expansive clay-grade sheet silicate particles (potentially hygrothermally altered tuff) were identified directly underneath the blistered surface of the coating.

High concentrations of silica and potassium were identified in the EDX spectra for two samples of concrete fragments collated from underneath a blister. This is consistent with the presence of an alkali–silica gel.

The average measured alkali content of the concrete, expressed as the sodium oxide equivalent (Na_2O eq), was estimated to be approximately 3.7 kg/m^3, with a range from approximately 2.5 to 5 kg/m^3. This is significantly higher than the Na_2O eq value of 2.9 kg/m^3 estimated from the reported alkali content of the constituent materials. The Na_2O eq limit recommended in BRE Digest 330 is 2.5 kg/m^3 for high-reactivity aggregate and 3 kg/m^3 for normal-reactivity aggregate [30].

Research has indicated that there is a progressive increase in expansion with the proportion of tuff particles [23]. Based on these data and the estimate of 5–11% tuff from the petrographic report, an unrestrained expansion of approximately twice that of normal concrete might be expected (vertical line on Figure 11.38).

Expansion tests on three samples exhibited similar behaviour (Figure 11.39). Rapid expansion of approximately 0.03% occurred during the first day of testing. This is most likely attributable to wetting and thermal expansion due to exposure to water at 38°C. Expansion of the concrete appears to have slowed down and largely stabilised over the monitoring period. The unrestrained expansion after 84 days was approximately 0.04% to 0.06%. Expansion values of less than 0.07% indicate that it is unlikely that there will be further expansion of the member from which they were taken [31].

11.8.3.3 Cause and remediation of the blistering

The large blisters are associated with low bond values and the presence of a smooth substrate. This is consistent with a change in surface preparation from abrasive blasting in areas close to columns and walls or

Figure 11.38 Relation between expansion and proportion of tuff particles [23]

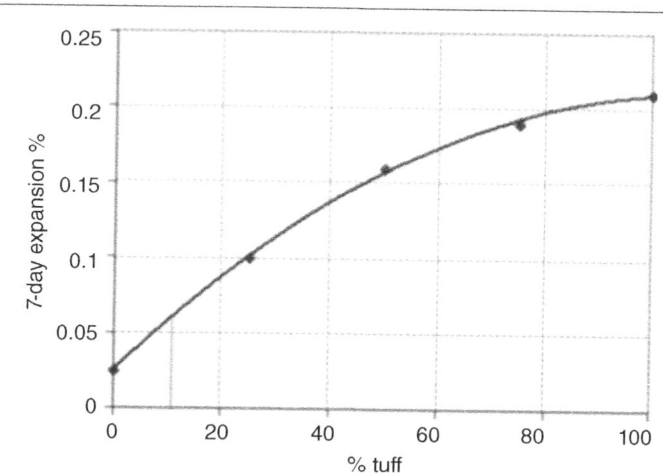

Figure 11.39 Expansion test data

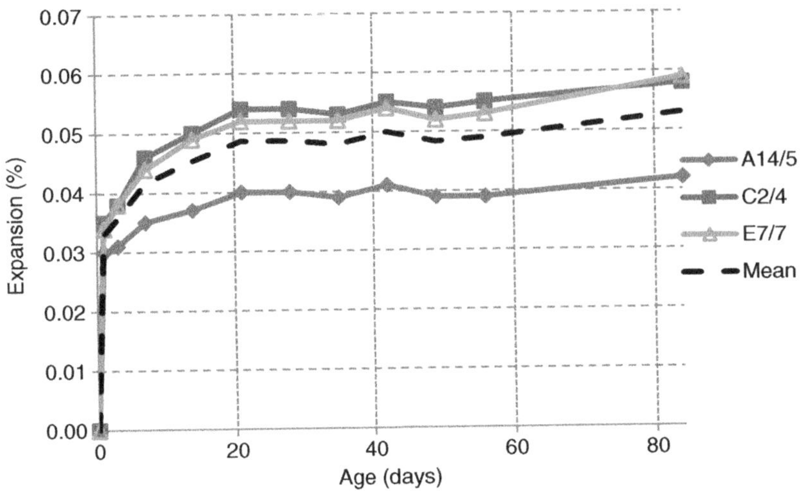

with restricted access. There was no evidence of osmotic blistering or contamination by oil or salts. The thin thickness of the coating could be a secondary factor in the risk of blistering and the rupture of the coating already observed. There is a risk that large intact blisters could rupture and expose the clean room air to silicate particles. Therefore it was recommended that these blistered areas should be repaired. The work will require preparing the concrete using vacuum abrasive blasting to remove residual primer, coating and laitance. The moisture conditions of the concrete and the bond should be tested to confirm that they are within the manufacturer's limits.

The small blisters are associated with the presence of alkali-reactive particles in the concrete leading to pop-outs at the concrete surface. The high alkali and moisture content of the concrete promotes ASR. The incidence of blisters is notably greater in areas of the floor previously subject to groundwater leakage. Because the waterproof membrane is unbonded, any water penetrating through damaged areas of the membrane can flow between it and the concrete slab and penetrate at joints and cracks.

Various remedial options were considered for the floor, including full replacement of the slab. However, the final recommendations were to monitor and patch the coating as necessary, taking into consideration the following factors.

- The small blisters were intact.
- Accelerated expansion testing suggests that further significant bulk expansion is unlikely.
- Damaging expansion is restrained by the top and bottom layers of steel, which are linked vertically at each pile location [31].
- The floor has a short service life and is not subject to direct traffic.

During the annual cleaning operation, the floor coating should be visually inspected for obvious cracks or areas that have detached, leading to exposure of the concrete substrate. Localised maintenance should be undertaken where the flooring has been breached. It is expected that temporary repairs could be undertaken using vinyl sheeting and more permanent repairs made using the epoxy resin flooring system but incorporating reinforcing strips.

Movement across the structural expansion joint should be monitored yearly using tell-tales fixed over the exposed joint.

Twenty representative blisters should be selected across the floor area for annual inspection. It was recommended that a 150-mm-diameter steel ring, 20 mm in nominal height, is installed around each blister. The ring should be fixed to the coating surface. The rings will allow blisters to be identified, provide protection to any instrumentation installed and also provide a reference for vertical displacement measurement. The horizontal strain in the flooring should be monitored by installing demountable mechanical strain gauges at 50 mm centres on the surface. The vertical strain should be monitored using the depth gauge probe of a Vernier or digital calliper to measure down from a 20-mm square steel bar placed across the ring to the surface of the coating. The ambient temperature should be measured to allow the strain values to be corrected.

The strain measurements in the coating could be supplemented by measuring the strain at selective locations using electrical resistance or vibrating-wire strain gauges fixed to the coating surface. The gauges would have wired or wireless connection to a multichannel data logger. It is envisaged that the values would be recorded weekly and manually downloaded from the data logger at quarterly intervals. The location and scope of this monitoring should be based on 1-year monitoring data for the 20 blisters.

11.8.4 Conclusions

A close inspection of the floor identified over 8000 blisters. Most of these are 20–50 mm in diameter, but four are much larger. Less than 1% of the small blisters are hollow, whereas all the large blisters are hollow. In two locations, the floor coating has failed exposing the substrate. The clean room has a strict limit on silicate particles in the air.

Pull-off testing in three areas with large blisters gave low bond values, less than the industry accepted standard of 1.5 MPa. The smooth appearance of the exposed concrete substrate in the areas of failed coating suggests that an inadequate mechanical key was developed during surface preparation.

Laboratory examination of three cores taken at the location of small blisters identified cracking in the concrete under the coating. The cracking is associated with ASR and an expansive silicate material.

The aggregate in the concrete has been identified as normally reactive, as defined by BRE Digest 330, with the exception of some highly reactive tuff particles in the coarse aggregate. Analysis of concrete fragments taken from below two blisters is consistent with the presence of alkali–silica gel.

Published information indicates that expansion increases with an increasing proportion of tuff particles. It is notable that a core taken from an area with no blisters had a very low proportion of tuff particles.

The moisture content of the floor measured by resistivity and hygrometry is higher than would be recommended for flooring. This suggests that moisture is trapped below the coating and possibly rising through the slab. Secondary ettringite that is characteristic of moisture ingress into the concrete was noted during the petrographic examination.

Expansion testing indicates that significant future bulk expansion of the concrete floor is unlikely.

Several large blisters associated with poor bonding to the concrete have been identified. The recommendation is that these should be repaired. The smaller blisters appear intact, and it is recommended that these be monitored and patch repaired.

Acknowledgement
The assistance of Dr Marcin Wieloch in preparing this case study is gratefully acknowledged.

11.9. ASR-induced service life reduction of coastal protection armour units

Jeremy Ingham, Consultant, UK

11.9.1 Introduction

In the early 1990s, 1.4 km of sloping revetment formed of concrete armour units was constructed as part of the Clacton coastal defence works in the UK (Figure 11.40). The revetment was designed for a service life of greater than 50 years, but due to severe cracking and breakages of the armour units, its entire replacement was necessary after only 22 years. Each precast concrete unit consisted of a hexagonal head on top of a central hexagonal spine on a perforated base with three legs, which connected the head to the base. The units contained no reinforcement bars. Full details of this case study can be found in a paper published in the *Proceedings of the Institution of Civil Engineers – Maritime Engineering* [32] and a brief account is included in the book *Alkali–Aggregate Reaction in Concrete: A World Review* [33]. These are summarised in the following sections.

11.9.2 The problem

Visual inspection of the concrete armour units determined that cracking was frequently present in units that were within the intertidal zone. In particular, the lower three rows of units frequently had large cracks of between 1 and 3 mm in width. These were concentrated in the head of the unit and then narrowed and terminated in the spine and legs (Figure 11.41). Conversely, units in the splash zone (i.e. above the intertidal zone) appeared to be in generally good condition with no cracking other than occasional surface crazing.

11.9.3 Investigation methodology

An investigation was undertaken to determine the causes of cracking, which involved the following four activities

- desk study of the project documentation for the design and construction stages
- site works including a comprehensive visual inspection and the collection of concrete core samples

Figure 11.40 Precast concrete armour units in the coastal defences at Clacton-on-Sea

Figure 11.41 Visual appearance of the cracked concrete armour units in the intertidal zone

- laboratory testing of the concrete core samples
- FEA to estimate the concrete temperature achieved during curing.

During the site work phase of the investigation, six concrete core samples were taken from different units by diamond core drilling (Figure 11.42). Four were taken from cracked units in the intertidal zone and two were taken from uncracked units in the splash zone. All the core samples were sent to a UKAS-accredited construction materials testing laboratory and subjected to petrographic examination (Applied Petrography Group SR2 [34] and ASTM C856 [27]), sulfate content and alkali content analysis (BS 1881-124:1988 [35]), and compressive strength testing (BS EN 12504-1:2009 [25] and BS EN 12390-3:2009 [36]).

11.9.4 Investigation findings

The desk study revealed details of the concrete mix that was specified for the armour units, as listed in Table 11.17.

Figure 11.42 Diamond drilling of a core sample from a cracked concrete armour unit in the intertidal zone

Table 11.17 Summary of the concrete mix specification for the concrete armour units

Property	Value
Strength grade	C50 (cube strength of 50 N/mm^2)
Cement type	Rapidly hardening Portland cement
Coarse aggregate type	5–20 mm natural gravel coarse aggregate consisting chiefly of quartzite and vein quartz with minor proportions of sandstone, basalt and flint
Fine aggregate type	Natural sand consisting chiefly of quartz and quartzite with minor proportions of sandstone and basic igneous rock types
Cement content	400 kg/m^3
Coarse aggregate content	1150 kg/m^3
Fine aggregate content	697 kg/m^3
Maximum w/c ratio	0.45
Calculated alkali content for mix	2.46 kg Na_2O eq/m^3

It was determined from the petrographic examination that all the concrete core samples were similar in terms of composition and that they had all been manufactured from the specified concrete mix. They were all suffering from ASR with reaction sites observed (through the microscope) in chert and quartzite particles from both the coarse and fine aggregates. The microscopic observations of ASR severity correlated well with the on-site visual inspection observations of macrocracking, with ASR being notably worse for the samples obtained from units in the intertidal range. ASR had also initiated in all samples taken from units in the splash zone, even though these units had no visible cracks.

Based on the mix design and petrographic findings, the aggregate combination for the concrete samples was classified as being of 'normal' reactivity in accordance with BRE Digest 330 [30]. For normal-reactivity aggregate combinations, BRE Digest 330 recommends that alkalis be limited to $\leq$3.0 kg Na_2O eq/m^3 (where high alkali cement is used). The calculated alkali content of the concrete mix at the time of construction (2.46 kg Na_2O eq/m^3) complied with this recommendation. An alkali content analysis of the six core samples gave results ranging from 2.25 to 3.38 with a mean value of 2.55 kg Na_2O eq/m^3 and all but one of the values were below the recommended upper limit.

The ASR observed in all core samples could not have occurred without the availability of sufficient alkali hydroxides in the pore solution of the cement paste. However, the alkali content laboratory test results were typically within the BRE Digest 330 recommended limit. This raised the following question: how were alkali levels ever sufficient to cause ASR? The answer was considered to be alkali migration caused by the action of seawater leaching through the concrete combined with the 'wicking' action of surface evaporation, which caused alkalis to migrate and accumulate in the upper sections of the unit. This appeared to have been encouraged by the geometry of the armour units and also its higher surface area to volume ratio than other designs of units (e.g. cubes or grooved cubes).

The relatively low alkali contents determined by laboratory testing during the investigation were probably due to the excess alkalis that would have existed at areas of alkali concentration being consumed by chemical binding during ASR. Owing to the alkali migration, it is probable that the alkali content did exceed the limits recommended in BRE Digest 330 in heads (and other parts) of the units at some point while they were in service.

Table 11.18 Classification of risk of DEF associated with peak temperatures that concrete reaches during curing (adapted from Quillin [8])

Risk of DEF	Peak temperature reached during curing: °C
No risk of DEF	<60
Very low risk of DEF	60–69
Low risk of DEF (if the Na_2O eq of the cement is less than 0.85%)	70–79
Increased risk of DEF	>85

The microscopic examination showed that the core samples had deposits of ettringite in voids and cracks, which were most likely caused by leaching. However, the possibility that in addition to ASR, some minor degree of DEF had occurred could not be ruled out. The sulfate content analysis gave results ranging from 3.2% to 3.7% with a mean of 3.4% by mass of cement. These values would be considered normal for Portland cement concrete.

When specifying concrete curing temperatures to prevent DEF, it is common to limit the maximum temperature to 65°C [37]. However, when investigating concrete structures that are suspected of suffering from DEF, it is typically found that much higher curing temperatures are required to cause the reaction. Quillin [8] provides a useful classification scheme for DEF risk, as shown in Table 11.18.

To investigate the possible contribution of DEF to the cracking of units, a modelling exercise was undertaken using FEA software. Further discussion of such thermal modelling techniques can be found in Sfikas *et al.* [38]. The modelling was used to estimate the maximum temperature that the precast concrete units may have reached during early-age curing (up to 3 days). Thermal simulation modelling indicated that the curing temperatures reached in the armour units were unlikely to have exceeded the temperature at which there would be a high risk of DEF (>85°C). However, they could potentially have reached temperatures where DEF is possible (70–79°C), but the associated risk would be classified as 'low'.

The thermal simulation modelling also indicated that relatively high unit core temperatures could have been generated, possibly resulting in temperature differentials with the surface temperature that could have caused some early-age thermal cracking in the units. This is likely to have caused relatively narrow cracks, which could have passed quality control inspections prior to installation, and some cracks hidden in the core of the units. It is possible that such early-age thermal cracking could have allowed some seawater to penetrate the core of the units to accelerate the initiation and early development of expansive reactions.

Compressive strength testing of the core samples indicated that they complied with the specified cube strength of 50 N/mm^2, with the mean estimated in situ cube strength being 71 N/mm^2. Note that the specimens used in the compressive strength tests were specifically selected so that they did not have any of the cracks caused by the expansive reaction, so that the overall strength of the units affected by the expansive reaction could reasonably be expected to be lower owing to the cracking.

11.9.5 Investigation outcome

The cracking observed in the armour units in the intertidal zone was interpreted as being caused by an expansive deleterious reaction. The type of expansive reaction was most likely to have been ASR alone, although the possibility of a combination of ASR and DEF (with ASR being by far the most dominant) could not be ruled out. Early-age thermal cracking may also have been a minor contributor to the deterioration.

For the units in the intertidal zone, the expansive reaction was probably initiated a few years after the structure was commissioned. In most cases of ASR, the expansive effects of the reaction could reasonably be expected to be largely complete at around 15 years after initiation, since the alkalis would then have been consumed. However, in this case, it was considered that there was a likelihood of ongoing alkali migration within the units, so that the units exposed to seawater would continue to react expansively (and deteriorate) in response to their ongoing exposure to new alkali concentrations that could generate further ASR. Hence, the cracking deterioration could reasonably be expected to be ongoing and severe.

For the units in the splash zone, the petrographic examination indicated that they had the same composition as the cracked units that were exposed to seawater and that ASR had already been initiated. It was considered that if these units were left in place, the expansive reaction would slowly proceed in them, causing expansion and cracking of lower severity than for the units in the intertidal zone.

In view of the investigation findings, the asset owner decided to remove all the armour units from the revetment. A new coastal protection scheme was constructed with fishtailed groyne structures (rock armour) and with beaches of dredged marine aggregate in between. The removed concrete armour units were crushed down and recycled, being utilised within the cores of the fishtailed groynes. The concrete material in the groynes was designed to withstand further ASR-related deterioration once exposed to seawater by blending in sound rock materials that would mitigate the effects of ASR. This approach meant that the carbon usage of project was reduced by more than 14% and provided a cost saving of approximately £550,000.

11.9.6 Durability lessons learnt

The finding of ASR in the armour units and the associated shortening of service life is surprising given that the concrete mix design complied with the current UK guidelines (BS 8500-2 [39] and BRE Digest 330 [30]) in terms of the aggregate combination used and the alkali content of the mix. This case study demonstrates the importance of considering the potential action of leaching when designing concrete structures to resist AAR.

In light of this case study, the designers of marine or coastal structures should carefully consider whether they should use concrete elements with a complex geometry or a relatively high surface area to volume ratio that could potentially promote leaching and alkali migration. If such units are selected, it would be prudent to specify a mix design that specifically resists leaching and alkali migration, and also mitigates against the risk of both ASR and DEF. Designers should also take account of other factors that could potentially accelerate expansive reactions, such as early-age thermal cracking, and seek to reduce their occurrence.

Furthermore, consideration should also be given to amending the standards for specifying concrete to improve the guidance relating to leaching and alkali migration.

11.10. Improving the abrasion resistance of factory floors

Robert C Lewis, Ferroglobe Plc, UK

11.10.1 Introduction

The textile industry has been established for many years in southern India. During that time, significant damage has been done to the floors in the factories and warehouses. Standard concrete (20 MPa) was previously used, in conjunction with a surface hardener, as the flooring material. The factory floors are now made from high-performance silica fume concrete to gain greater durability, but without surface hardeners. The design of the concrete is such that it is very economic due to the increased working lifetime, while still being produced and placed by normal labour-intensive techniques.

11.10.2 Historical perspective

The traditional method used for these large industrial floors, based on hand-batched and site-mixed concrete, has been to use the normal 20 MPa mix for ease of production and placement. When still 'green', this concrete is treated with a topping, usually a cement-bound metallic powder, which is then trowelled into the surface.

However, any damage to the surface layer will immediately allow the relatively weak concrete underneath to be attacked. Deterioration and breakdown can happen very quickly once the surface coating has been broken. Repair and reinstatement of these floor areas can directly cost a large amount of money, and the downtime and lost revenue for the production unit can also be significant.

New structures are assessed by a life-cycle analysis with the aim of achieving the best durability for the longest working time. The performance of silica fume concrete is such that most new units are built from this material (Figures 11.43 and 11.44).

Figure 11.43 New factory unit under construction

Figure 11.44 Large floor area within a factory unit

11.10.3 Improving performance by selecting the right constituents for the concrete mix

The benefits of silica fume in concrete have been known for many years in India, as it has been used by the nuclear and hydropower industries and for roads and bridges. However, such mixes are at the top end of the technology scale, used when very high strength or impermeability are required. These concretes are more expensive, although they are more economic in the long term.

The textile industry needed a mix that would provide good performance, be economic and be suitable for placement by manual labour (Figures 11.45 and 11.46). A review of the mix normally used and the method of production, placement and curing simply recommended adding about 7% silica fume to the mix. There was no increase in the cement content or other major modifications, and the water-to-binder (w/b) ratio remained at 0.40. A few trials found that this mix actually increased the final strength from 25–30 MPa, as achieved by the normal mix, to up to around 40 MPa. This was acceptable, as it meant no modifications to the bay size or joint cutting. Since silica fume concrete has no bleed water, finishing routines can be started within a few hours (Figure 11.47). This improves the rate of working, thus speeding up construction.

Figure 11.45 Hand-batching with a tumble mixer on site

Figure 11.46 Placing by hand (some larger sites use truck mixers and pumps)

Figure 11.47 Bay after a long, single pour and after screeding with a razorback

A distinct improvement was noted by the teams doing the power-floating and power-trowelling, as this mix produced a much tighter surface on the concrete, again making it easier to work (Figure 11.48). Standard curing – ponding – is frequently used. Wet hessian (sacking) is also used to great effect (Figure 11.49). Due to the 'sealed' surface of the silica fume concrete, these methods maximise the effects of curing and, thus, realise the best surface durability.

11.10.4 Long-term performance

The cotton-spinning machines are serviced by workers using hand trolleys, who collect full bobbins and replace them with fresh ones. These carts weigh up to 150 kg, when full, and run on four small plastic wheels. Each wheel has a central runner about 15 mm across. Thus, the impact area of the 150 kg is focused on four points just 15 mm wide. This is a bad enough abrasion surface when the wheels can actually roll, but once the bearings become clogged with dust and cotton fibres and the wheels drag, the damage can become significant. With the original concrete floors, damage could be seen within a couple of months as two 'tramlines' gouging into the concrete around the cotton-spinning machines.

Figure 11.48 Power-floating the surface to produce a solid, sealed finish

Figure 11.49 Wet hessian curing. Ponding can also be seen in the top right of the picture

Other damage to the floors is caused by the movement of pallet trucks and forklifts due to the weight on the wheels of these plus the lifting and dropping of pallets and other equipment and materials. A misjudged pickup with a forklift truck can mean the forks chopping into the concrete surface. Such damage would previously accelerate the breakdown of the concrete below the 'protective' surface hardener.

The new silica fume concrete has been in use for more than 16 years in several factories and has shown considerably better wear performance (Figure 11.50). In most cases, the surface of the concrete appears to be even more polished than when first placed. The paths of the trolleys can still be seen, but now the wheels clean dust from the surface and, hence, leave two highly polished lines. Close inspection of these tracks shows that there has been no damage to the surface (Figure 11.51).

The owners were very pleased with the performance of the silica fume concrete for all completed projects. Although designed as 'low tech', the durability of this silica fume concrete means that there will be a significant increase in the lifetimes of the factory and warehouse floors and, thus, will provide very large financial benefits to the owners.

Figure 11.50 Refurbishment of a worn entranceway with silica fume concrete

Figure 11.51 A completed floor in a spinning unit. The polished 'tram tracks' can be seen just in front of the machine. This picture was taken 6 months after production started and shows no surface wear

11.10.5 Conclusions

By the end of 2006, some 93 000 m^2 of flooring had been placed using this design of silica fume concrete. Twice that area is already in production. Table 11.19 shows just a selection of the work already completed.

Acknowledgements

Grateful thanks is given to GHCL, the local management of Meenakshi Mills in Madurai and Jagannata Textile Co in Coimbatore for allowing permission to visit their sites and use the information gathered.

11.11. High-strength concrete for the JJ Hospital Flyover, Mumbai, India

Robert C Lewis, Ferroglobe Plc, UK

11.11.1 Introduction

The JJ Hospital Flyover was the first project to combine strength and durability in a precast concrete with a specified strength of 75 MPa in a region that used to consider 40 MPa as the highest possible strength obtainable.

Table 11.19 List of projects where silica fume concrete has been used for industrial floors

Factory	Location	Area: m^2
GHCL Meenakshi Mills	Madurai	12 000
Gangotri Mill	Perundurai	26 000
Jagannath Textile Co	Coimbatore	9000
Divyalakshmi Textile Mills	Coimbatore	14 000
Jayajothi Textile Mills	Rajapalayam	8000
Other projects		>19 000

11.11.2 Concrete mix design requirements

Due to the location of the flyover, there was great concern over the aesthetics of the construction. Moreover, this is a very densely built-up area, and the traffic flow would be hindered if the supports of the flyover blocked much of the street below. The aim was to make the structure as slender and pleasing to the eye as possible, giving maximum room and 'air space' beneath the elevated road, as well as maximising the light to the buildings on either side.

The 'old' style of flyover, with three or four large, 1-metre-wide columns every 10 to 15 metres supporting a thick slab of concrete, would not work in this situation. The idea was for the design to go beyond previous building patterns.

Silica fume concrete had been used in India by the contractor, Gammon India Ltd, in previous projects they had already completed. After a rethink of the design in collaboration with the consultant, Dar Consultant Project Management, and the approval of Maharashtra State Road Development Corporation, it was decided to utilise the strength potential of silica fume concrete and set a new precedent for such a project.

Specifying the concrete at 75 MPa meant that the design could be radically altered to have singular, slender columns and thinner decks. The strength of the structure also meant that the spacing between the columns could be increased to over 30 metres.

11.11.3 Selection and proportioning of mix constituents

Work in the nuclear power sector in India had shown that it was possible to achieve these levels of strength with silica fume concrete. However, the scale of this project was ten times that of the work that had previously been done. It was necessary to ensure that the silica fume concrete would give the required performance when produced for such a massive precast operation.

To ensure durability, the strength, drying shrinkage, rapid chloride permeability and the initial surface absorption were all to be measured. The plastic properties of the concrete were also important. Moreover, a good surface finish was required for the aesthetics of the structure.

With a non-bleed silica fume concrete and precasting in a yard where the ambient temperature could exceed 45°C, curing, workability and slump retention were of paramount importance. It was determined that the placing temperature should not exceed 28°C. Hence, the concrete was mixed with ice-cold water, and wet hessian curing would begin as soon as the precast elements were stripped and stacked. Concrete for the precast yard had to have a slump of 80 mm at 90 minutes. For the 'on-site' casting, it was necessary to allow for 120 mm at 150 minutes, as the batching plant was some 12 km from the site.

Taking into account all these considerations, optimised mix proportions were determined after extensive trial work (Table 11.20). The results indicated that a reduction could have been made in the cement content of the mix, but other restraints meant that this was not implemented.

11.11.4 Properties of the Finished Concrete

Testing of the concrete was as extensive as the trial work. An on-site laboratory was set up to keep close track of the concrete quality. The average results are shown in Table 11.21.

11.11.5 Conclusions

It can be clearly seen from these results that the concrete achieved, and easily exceeded, the strength requirements for the construction. It is also clear that the very low permeability of the concrete will result in

Table 11.20 Silica fume concrete specifications

Parameter	Value
Characteristic strength: MPa	75
Target strength: MPa	83.2 (= 75 + (1.64 × 5))
W/cm ratio	0.269
Mix quantities	
Cement: kg	500
Silica fume: kg	50
Fine aggregate: kg	682
Coarse aggregate, 10 mm: kg	384
Coarse aggregate, 20 mm: kg	762
Water: litres	148
Admixture: litres	8.25
Laboratory cube results: MPa	
1 day	30
3 days	46
7 days	71
28 days	86

Table 11.21 Properties of the finished concrete

Parameter	Value
Compressive strength: MPa	
1 day	30
3 days	46
7 days	71
28 days	80
90 days	387
365 days	94.5
Drying shrinkage (IS 1199): %	0.00142
Rapid chloride permeability (ASTM C-1202.97): coulombs	
Cores from side segments	100
Laboratory samples	240
Initial surface absorption test (BS 1881:1970): ml/(m^2 s)	
Precast elements	0.0043
In situ concrete	0.0161
Compressive strength: MPa	
Cores from spine segments	88.5
Standard deviation	2.1

greatly improved long-term durability. The chloride permeability is the equivalent of that for latex-modified or polymer-impregnated concrete and among the lowest ever seen for silica fume concrete. The working lifetime of the flyover should easily surpass the original design expectations.

The use of silica fume concrete meant that the JJ Hospital Flyover could be built using radical design improvements, resulting in a structure that is not only functional but also pleasing to the eye, sympathetic to the environment and an elegant monument to all those involved in its construction.

11.12. High-performance concrete for Tsing Ma Bridge, Hong Kong

Robert C Lewis, Ferroglobe Plc, UK

11.12.1 Introduction

The Tsing Ma Bridge (Figure 11.52) is the centrepiece of the Lantau fixed crossing in Hong Kong. This is the link across the sea strait from the mainland at Tsing Yi to the island of Lantau and it carries both the road and rail traffic out to the island and, more importantly, to the international airport of Chek Lap Kok.

11.12.2 Durability requirements

Durability was a major design concern. The bridge would be exposed to seawater and high winds and rain and, potentially, some seismic activity. The scale of the bridge, with towers over 200 m tall and a main span of over 1400 m, meant that the authorities did not want any major maintenance costs for a long time. Hence, the design life of the finished bridge was specified at 120 years.

11.12.3 Selection of constituents and mix proportions

Construction of the towers was to be by slipform. However, the mix design required a high percentage of GGBS, some 70%. A significant amount of trial work found that this blend would not gain strength fast enough for placing, even though it met the temperature criteria for the concrete. Over 100 mix designs were tested before a suitable one was found, which was a ternary blend containing silica fume in addition to GGBS. Adding 5% of silica fume gave the early-age strengths that were required and also increased the durability of the concrete. The concrete in other parts of the bridge were also designed to use 5% silica fume in conjunction with 25% fly ash.

Figure 11.52 Tsing Ma Bridge, the centrepiece of the Lantau Link, Hong Kong

With the very tight restrictions on w/c ratio, it was noted that adding the silica fume in powder form could mean that the mixing time would be increased and that extra superplasticiser would be needed to maintain the slump. Thus, it was decided to add the silica fume in a slurry form – as an 'instant' mix with chilled water – so that the material would disperse faster.

As such, the silica fume was delivered to the site in big 1-tonne bags, which were individually broken into the 'slurry mixer'. An equal weight of chilled water was added to achieve a 50:50 blend, and the mixed suspension was pumped into a holding tank before use. Slurry was made every day when needed, so that excess material did not remain in the tank and settle out.

11.12.4 Conclusions

With a mix design of 390 kg of cementitious material comprising 30% ordinary Portland cement (OPC), 65% GGBS and 5% silica fume, the casting of the towers proceeded without problems, stopping only twice for typhoon-strength winds. Long-term testing of the concrete returned strengths of over 80 MPa (C65) and chloride diffusion coefficients of better than 10^{-14}. The project started in 1992 and was completed in time for the return of Hong Kong to China, in July 1997.

11.13. East Sea Bridge, Shanghai: Service life requirement of 100 years

Robert C Lewis, Ferroglobe Plc, UK

11.13.1 Introduction

In the last 20 years, the Shanghai shipping industry has grown along with the economy of the surrounding areas. However, the shallow water of Huangpu River prevents larger vessels from reaching the traditional ports. Thus, construction of a leading international shipping centre, Shanghai Deepwater Harbor, started in 2002. It consists of over 100 container terminals on outlying islands and includes East Sea Bridge and New Harbor City on the mainland. East Sea Bridge was to be 32.5 km long. It would start at Little Yangshan Island of Qiqu Archipelago (Zhejiang Province) and cross the north of Hangzhou Bay to New Harbor City in Luchao, Nanhui District, Shanghai (Figure 11.53). Thus, East Sea Bridge would connect the islands and the mainland and be used to transport containers, so its durability was critical. A service life of 100 years was specified for the first time in China. However, no code or guideline for such durability was available. Extensive studies were carried out at Shanghai Research Institute of Building Sciences (SRIBS) to establish scientific, practical and reliable strategies to maximise the life of the bridge.

11.13.2 Profile of East Sea Bridge

The bridge has a total length of 32.5 km and is split into three sections: a marine bridge, an island connection and a land-based bridge. The marine bridge has one major high bridge, three minor high bridges (Figure 11.54) and five sections of low bridge (Figure 11.55). The major high bridge is a cable-stayed bridge with a main span of 420 m and a vertical navigation clearance of 40 m. The minor high bridges are post-tensioned concrete cantilever bridges with main spans of 120–160 m and a navigation clearance of 17.5–25 m. The total length of the high bridges is 2.15 km. The low bridges, 22.35 km in total length, were built with precast, post-tensioned box girders, with spans from 50 to 60 m.

The island connection starts at the south end of the marine bridge at Xiaowugui (Little Turtle) Island, links four islands where the container terminals are located, and is 4.35 km long. The land-based bridge is 2.4 km long and runs from the north landing point of the marine bridge to the Old Levee of Luchao Port. These sections were built as post-tensioned continuous beams on supports.

East Sea Bridge is 31.5 m wide. It has six lanes plus an emergency lane. With a design life of 100 years [40], it needs to resist a Richter scale 7 earthquake and the '100 year' highest probable wind. It also carries a water pipeline and power and communication cables.

Figure 11.53 Axis line of East Sea Bridge

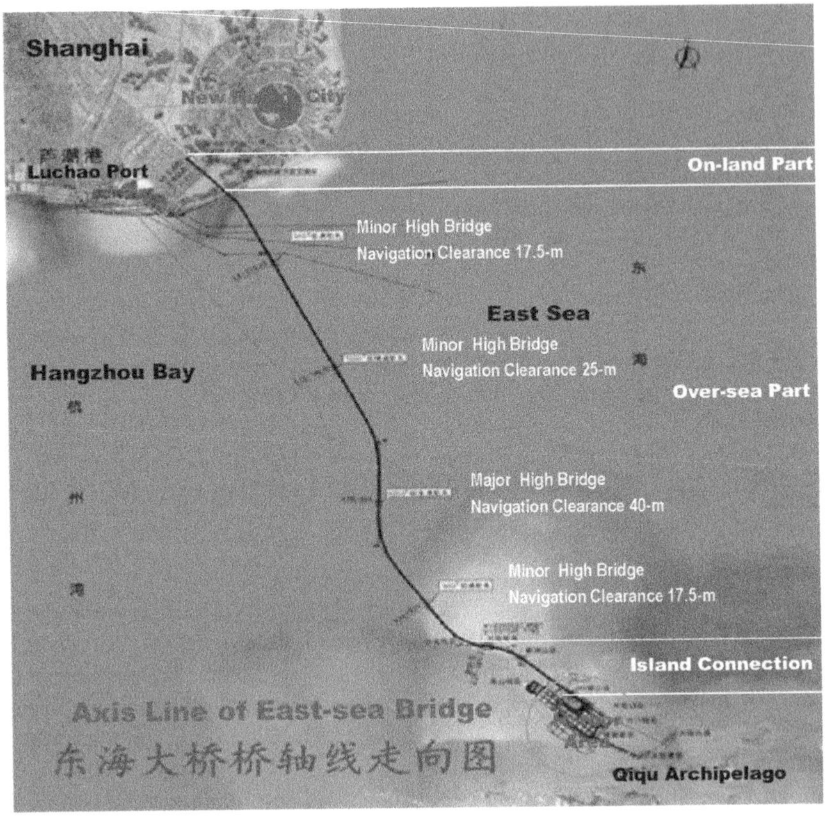

Figure 11.54 High bridge

Figure 11.55 Low bridge

11.13.3 Analysis of environmental load on structural concretes

11.13.3.1 Meteorological conditions

East Sea Bridge is on the south edge of the north subtropics. Influenced by the north-east monsoon, the area is cold in winter and warm in summer. There is high rainfall and an annual average temperature of 15.8°C. The seawater has a salt content in the range 1.0–3.2%.

11.13.3.2 Deterioration mechanisms of structural concrete

Due to the location, freeze–thaw effects and impact and abrasion from floating ice are not relevant. The following durability goals are considered important

- no expansion caused by AAR
- resistance to magnesium sulfate attack
- resistance to chloride penetration
- absence of thermally induced cracks
- resistance to carbonation.

An investigation of concrete structures near the East Sea Bridge site showed that the carbonation of concrete proceeds at a much slower rate than chloride ingress. For medium-quality concrete, the carbonation rate was about 3 mm in a decade [41]. Therefore, the major factor affecting the durability of East Sea Bridge is chloride penetration.

11.13.3.3 Technical measures for enhancing and evaluating the durability of marine concrete

Various measures to enhance durability, especially to control corrosion, were reviewed and carefully evaluated

- high-performance marine concrete (HPMC) [42–45]
- increasing the thickness of the concrete cover
- concrete surface coating
- epoxy-coated rebar [46]
- corrosion inhibitor [47]
- cathodic protection.

11.13.3.4 Durability strategy for the structural concrete of East Sea Bridge

The durability strategy had to be reliable, cost-efficient and practical. Based on the evaluation described above and on studies using local materials, HPMC was chosen. The thickness of concrete cover was increased, and for some parts, a concrete surface coating or corrosion inhibitor was to be adopted as additional protection.

Based on extensive tests at SRIBS, two blended binders, marine concretes I and II (MC I and MC II), consisting of Portland cement, silica fume, GGBS and fly ash, were formulated. These, with a carboxylic-based high-range water-reducing admixture, were used to produce the HPMC. The properties of the concrete are presented in Tables 11.22, 11.23 and 11.24.

Although the mechanical properties are like those of conventional concrete, the durability is substantially better. The Portland cement content of the HPMC (less than 40% of the total binder content) is low, leading to low heat generation. In adiabatic conditions, the temperature increase in HPMC can be reduced by 7–9°C and the peak time of heat release can be delayed for 8–16 hours, which is beneficial for reducing the thermal stress and controlling early-age cracking [2].

Table 11.22 Mechanical properties of HPMC

Concrete	Compressive strength of a 150 × 150 × 150 mm^3 cube: MPa	Split-tensile strength: MPa	Flexural strength: MPa	Axial compressive strength of a 150 × 150 × 300 mm^3 column: MPa	Modulus of elasticity: GPa
Grade 35 control	43.3	4.0	7.4	29.4	33.5
Grade 35 MC I	38.7	3.9	7.7	26.7	32.7
Grade 35 MC II	41.0	4.1	7.6	28.9	35.5
Grade 50 control	58.5	4.0	9.0	32.2	36.9
Grade 50 MC I	52.4	3.9	8.7	31.3	36.5
Grade 50 MC II	66.7	4.5	9.9	32.9	41.3

Table 11.23 Carbonation, permeability and frost resistance of HPMC

Concrete	Carbonation		Water penetration		Freeze–thaw (100 cycles)	
	Depth: mm	Strength loss: %	Water pressure: MPa	Penetration depth: mm	Mass loss: %	Loss of dynamic modulus of elasticity: %
Grade 35 control	0.30	0.63	2.5	26.3	0.9	8.1
Grade 35 MC I	0.16	0.42	2.5	7.1	0.6	6.9
Grade 35 MC II	0.16	0.46	2.5	6.5	0.6	7.2
Grade 50 control	0.25	0.50	2.5	20.5	0.7	7.2
Grade 50 MC I	0.17	0.38	2.5	6.6	0.5	6.8
Grade 50 MC II	0.14	0.37	2.5	5.4	0.4	6.4

Table 11.24 Resistance of HPMC to sulfate attack and ASR

Binder	Sulfate resistance factor	ASR expansion: %
PII cement (100%)	0.92	0.2
PII cement (40%) + MC I mineral admixture (60%)	1.25	0.05
PII cement (40%) + MC II mineral admixture (60%)	1.12	0.03

Table 11.25 Chloride-penetration resistance of HPMC

Concrete	Rapid chloride-penetration test: coulombs	Apparent diffusion coefficient[a] D_a: ($\times 10^{-12}$ m²/s)
Grade 35 control	1263	4.85
Grade 35 MC I	826	1.28
Grade 35 MC II	741	1.10
Grade 50 control	1112	4.26
Grade 50 MC I	750	1.15
Grade 50 MC II	637	0.95

[a] D_a was tested on specimens after immersion in salt water for 90 days.

The resistance of HPMC to chloride penetration has been significantly improved. The chloride diffusion coefficient was only 1/4 to 1/3 of that of conventional concrete at 90 days (Table 11.25). This will continue to decrease (i.e. increasing the resistance) with time, as studies show that the chloride diffusion coefficient could continue for 30–50 years [2,5]. Using the structural design and the concrete properties, SRIBS performed a dynamic analogue computation of chloride ingress, finding that the service life of 100 years should be ensured with this quality of concrete.

11.13.3.5 Conclusions

High-performance concrete is specified for many marine bridges with a required service life of 100–120 years. The studies and tests outlined here demonstrate the required concrete performance, thus confirming that HPMC is durable enough for marine environments.

Concrete quality control and assurance are also included in the one for East Sea Bridge, including full-scale trials and performance confirmation, a testing and inspection scheme, and wet curing requirements.

11.14. High-quality repair formulations with silica fume

Robert C Lewis, Ferroglobe Plc, UK

11.14.1 Introduction

Silica fume is an integral component of most high-quality repair formulations, in particular pre-packed materials. Silica fume is added to all the repair products mentioned in Section 5.3.4. Due to the proprietary nature of these products, the author is unable to detail performance data for them, which should be sought from the manufacturer.

The inclusion of silica fume in concrete designs is to eliminate, or at least greatly reduce, the degree of environmental attack against the concrete while also giving the performance required in the plastic and hardened states. The main use, therefore, is for 'new builds'. Silica fume concrete can be used as a designed concrete for specific repairs, such as bridge decks and overlays, and has been used to retrofit dams, after abrasion erosion damage or to increase seismic resistance. Some projects are listed below.

11.14.2 Bridge deck overlay, New York

With the enthusiastic use of de-icing salts and other chemicals to remove the ice and snow from our roads, it is not surprising that chloride attacks occur against bridge decks or that roadways easily become rutted due to chloride attacks against the reinforcing steel. In the United States alone, the Federal Highways Administration calculated in 2010 that over USD 3 trillion of road repairs were needed to bring the infrastructure up to date. Hence, cementitious blends of OPC with silica fume and of OPC, silica fume and fly ash are used to get the necessary durability.

In the repair of a bridge deck overlay in New York State (Figure 11.56), the concrete was designed to resist the penetration of chloride ions and minimise cracking. Hence, it was a triple blend with fly ash to reduce the temperature and silica fume to seal against chlorides. The basic mix design and curing requirement are listed in Table 11.26.

Figure 11.56 Bridge deck repair after abrasion and de-icing salt deterioration carried out by New York State's Department of Transportation: (a) placing and (b) curing

(a) (b)

Table 11.26 Basic mix design and curing requirement for a bridge deck overlay

Property	Value
OPC	300 kg/m^3
Fly ash	80 kg/m^3
Silica fume	25 kg/m^3
Water-to-binder ratio	0.40
Slump	75–100 mm
Air	6.5%
Wet curing	7 days

11.14.3 Kinzua Dam, Pennsylvania

Built in 1967, the stilling basin of Kinzua Dam had suffered so much abrasion erosion that it had to be replaced in 1974. The concrete for the repair comprised imported high-strength aggregates, a high-strength mix design and steel fibres. The repair lasted only 9 years before the damage was severe enough (Figure 11.57) that it had to be renewed (Figure 11.58). Local aggregates encased in a silica fume concrete matrix have so far lasted over 35 years, with the abrasion loss, noted at the 30-year inspection, giving an estimate of potentially another 100 years of service life.

Figure 11.57 Kinzua Dam: (a) damage in 1983, after only 9 years of use and (b) comparison of the high-strength concrete cored from the 1974 repair with a silica fume concrete using 'weaker' local aggregates

(a)

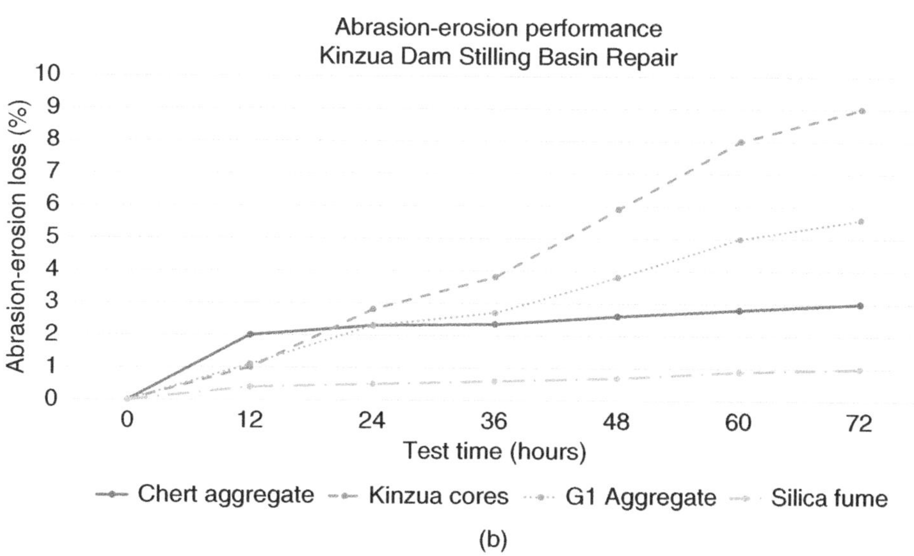

(b)

Figure 11.58 Little Rock Dam, California: (a) spraying, (b) curing and (c) return to operation

(a) (b) (c)

11.14.4 Retrofit of Littlerock Dam, California

Although still perfectly functional, the 'integrity' of the Littlerock Dam was not considered sufficient to resist the maximum potential earthquake that might hit the region. It was calculated that a flexible, high-strength layer tightly bonded to the surface of the dam would give the required safety level. The remedial work was designed with silica fume shotcrete reinforced by steel fibres, with a nominal strength of 40 MPa, sprayed to a depth of 100 mm across the face of the dam (Figure 11.58). The strengths achieved were over 50 MPa compressive and 1.0 MPa bond to the original concrete. The basic mix design was 405 kg of OPC, 41 kg of silica fume and 60 kg of steel fibres with a w/b ratio of 0.45 and a slump of 50–100 mm.

REFERENCES

1. de Almeida IR, Cordeiro TJRB and Costa JPVM (1996) The Setenave dry docks rehabilitation. *Concrete International* **18(3)**: 30–33.
2. Bamforth PB (1994) Admitting that chlorides are admitted. *Concrete* **28(6)**: 18–21.
3. Bamforth PB (2004) *Enhancing Reinforced Concrete Durability: Guidance on Selecting Measures for Minimising the Risk of Corrosion of Reinforcement in Concrete.* Concrete Society, Camberley, UK, Technical report 61.
4. BSI (British Standards Institution) (1972) CP110. The structural use of concrete. Part 1. Design, materials and workmanship. BSI, London, UK.
5. Kay EA, Fookes PG and Pollock DJ (1981) Middle East concrete 3. Deterioration related to chloride ingress. *Concrete* **15(11)**: 22–28.

6. Slater D and Sharp BN (1998) The design of coastal structures. In *Concrete in Coastal Structures* (Allen RT (ed.)). Thomas Telford, London, UK, Ch. 6.
7. Buckby RJ, White PS, Mills CA and Quillin K (2006) Severe cracking in in-situ concrete substructures due to delayed ettringite formation (DEF): Diagnosis, assessment and remediation. *Hong Kong International Conference on Bridge Engineering.*
8. Quillin K (2001) Delayed ettringite formation: In-situ concrete. Building Research Establishment (BRE), Information paper 11/01. BRE, Watford, UK.
9. British Cement Association (1992) *The Diagnosis of Alkali–silica Reaction.* BCA, Crowthorne, UK, Report 45.042.
10. Grabowski E, Czarnecki B, Gillot JE, Duggan CR and Scott JF (1992) Rapid test of concrete expansivity due to internal sulfate attack. *ACI Materials Journal* **89(5)**: 469–480.
11. Eden MA, White PS and Wimpenny DE (2007) A laboratory investigation of concrete with suspected delayed ettringite formation – A case study from a bridge in Malaysia. *11th Euroseminar on Microscopy Applied to Building Materials, Porto, Portugal.*
12. DuraCrete (1999) *Probabilistic methods for durability design.* European Union – Brite EuRam III. CUR, Gouda, Netherlands.
13. Edvardson C and Mohr L (1999) DuraCrete – A guideline for durability based design of concrete structures. *FIP Symposium, Prague.*
14. Stanish KD, Hooton RD and Thomas MDA (2000) *Prediction of Chloride Penetration in Concrete. Final Report to the Federal Highway Administration, DTFH-97-R-00022.* Department of Civil Engineering, University of Toronto, Toronto, Canada, pp. 113–145.
15. BRE (1996) *Sulfate and Acid Resistance of Concrete in the Ground.* BRE, Watford, UK, Digest 363.
16. Thaumasite Expert Group (1999) *The Thaumasite Form of Sulfate Attack: Risks, Diagnosis, Remedial Works and Guidance on New Construction.* Department for the Environment, Transport and the Regions, London, UK.
17. Floyd M and Wimpenny DE (2002) Procedures for assessing thaumasite sulfate attack and adjacent ground conditions at buried concrete structures. *Cement and Concrete Composites* **25(8)**: 1077–1088.
18. Wimpenny DE and Slater D (2002) Evidence from the Highways Agency thaumasite investigation in Gloucestershire to support or contradict postulated mechanisms of thaumasite formation (TF) and thaumasite sulfate attack (TSA). *Cement and Concrete Composites* **25(8)**: 879–888.
19. Wallace J (1999) Strengthening thaumasite-affected concrete bridges. *Concrete* **33(8)**: 28–29.
20. Slater D, Knights JC and Wimpenny DE (2002) Sulfate and chloride profiles and the visual characteristics with depth from the face in buried concrete subject to TSA. *1st International Conference on Thaumasite in Cementitious Materials.* BRE, Watford, UK.
21. Broomfield JP (2006) A web based tool for selecting repair options and life cycle costing of corrosion damaged reinforced concrete structures. *Proceedings of the 2nd International Conference on Concrete Solutions, St Malo* (Grantham MG, Jaubertie R and Lanos C (eds)). BRE, Watford, UK.
22. Broomfield JP (2007) *Corrosion of Steel in Concrete – Understanding, Investigation and Repair.* CRC Press, Taylor & Francis, London, UK.
23. Awal AS and Hussin MW (1997) The effectiveness of palm oil fuel ash in preventing expansion due to alkali–silica reaction. *Cement and Concrete Composites* **30**: 367–372.
24. ASTM (American Society for Testing and Materials) (2009) D4541-09e. Standard test method for pull-off strength of coatings using portable adhesion testers. ASTM, West Conshohocken, PA, USA.
25. BSI (2001) BS EN 12504-2. Testing concrete in structures. Non-destructive testing. Determination of rebound number. BSI, London, UK.
26. BSI (2009) BS 8203. Code of practice for installation of resilient floor coverings. BSI, London, UK.
27. ASTM (2011) C856. Standard practice for petrographic examination of hardened concrete. ASTM, West Conshohocken, PA, USA.

28. ASTM (2010) C457. Standard test method for microscopical determination of parameters of the air-void system in hardened concrete. ASTM, West Conshohocken, PA, USA.
29. FeRFA (2009) *Guide to Osmosis in Resin Flooring.* FeFRA, Guide note 2.
30. BRE (2004) *Part 2. Alkali–silica Reaction in Concrete: Detailed Guidance for New Construction.* BRE, Watford, UK, Digest 330.
31. Institution of Structural Engineers (1992) *Structural Effects of Alkali–silica reaction – Technical Guidance on the Appraisal of Existing Structures.* Institution of Structural Engineers, London, UK.
32. Ingham J, Collery D, Sfikas I and Badger M (2016) Failure of coastal protection armour units caused by alkali–silica reaction (ASR). *Proceedings of the Institution of Civil Engineers – Maritime Engineering* **169(3)**: 115–123, 10.1680/jmaen.2015.32.
33. Sims I and Poole A (2017) *Alkali–aggregate Reaction in Concrete: A World Review.* Taylor & Francis Group, London, UK.
34. Applied Petrography Group (2010) *A Code of Practice for the Petrographic Examination of Concrete.* Geological Society of London, UK, Report SR2. http://www.appliedpetrographygroup.com/reports.htm (accessed 05/09/2022).
35. BSI (1988) BS 1881-124:1988. Testing concrete. Part 124. Methods of analysis of hardened concrete. BSI, London, UK.
36. BSI (2009) BS EN 12390-3:2009. Testing hardened concrete. Part 3. Compressive strength of test specimens. BSI, London, UK.
37. Ingham JP (2012) Briefing: Delayed Ettringite formation in concrete structures. *Proceedings of the Institution of Civil Engineers – Forensic Engineering* **165(2)**: 59–62, 10.1680/feng.12.00001.
38. Sfikas I, Ingham J and Baber J (2018) Simulating thermal behaviour of concrete by FEA: State-of-the-art review. *Proceedings of the Institution of Civil Engineers – Construction Materials* **171(2)**: 59–71, 10.1680/jcoma.15.00052.
39. BSI (2015) BS EN 8500-2. Concrete. Complementary British Standard to BS EN 206. Specification for constituent materials and concrete. BSI, London, UK.
40. Shanghai Municipal Engineering Design and Research Institute (2002) *The Introduction of East-Sea Bridge Design. Science and Technology of Shanghai Construction, No. 5.* (In Chinese.)
41. Qiang X, *et al.* (2002) *Studies on High Performance Marine Concrete Applied to Shanghai Deep Water Harbor Project, Final Report for Appraisal.* (In Chinese.)
42. Jun Z (2001) Corrosion prevention of reinforcing bar of concrete structures exposed to marine environment and de-icing salt. *Highway* **4**: 52–59. (In Chinese.)
43. Dinghai H (1998) Application examples of high performance concrete with high GGBS content. *Transaction of Building Materials*, 82–86. (In Chinese.)
44. Xunjie C, *et al.* (2000) Successful application of GGBS high performance concrete to a marine structure. *Concrete* **9**: 59–61. (In Chinese.)
45. Clear C and Chapman-Andrews J (1991) GGBS in extreme environments. *Concrete* **25**: 27–30.
46. Hassan KE, Cabrera JG and Maliehe RS (2000) The effect of mineral admixtures on the properties of high-performance. *Cement and Concrete* **22(4)**: 267–271.
47. Federal Highways Administration (2000) *Materials and Methods for Corrosion Control of Reinforced and Prestressed Concrete Structures in New Construction.* Federal Highways Administration, Washington, DC, USA, Report FHWA-RD-00-081.

Soutsos M
ISBN 978-0-7277-6375-4
https://doi.org/10.1680/icehcd.63754.417

Chapter 12
Repair methods

John Broomfield
Broomfield Consultants, UK

12.1. Assessment prior to repair

Even the best designed structures can eventually need repair, either due to exceeding their design life, minor defects during construction or changes in use or environment. When considering making repairs to a structure, it is critically important to understand as much as possible about it, its condition and what has caused the deterioration. It is also critically important to understand what will be required of the structure after it is repaired, so that an appropriate, cost-effective repair is executed.

There are a huge number of destructive, semi-destructive and non-destructive tests that can be used for reinforced concrete structures. In the UK, there is a large and dynamic sector of concrete survey companies. Some are independent whereas others are part of a construction or concrete repair contracting company, a civil engineering consultancy or a university. Some do their own laboratory analysis of samples in house and are accredited by the National Measurement Accreditation Service/United Kingdom Accreditation Service (NAMAS/UKAS). Most of them have specialised teams that carry out surveys using conventional and highly specialised equipment. Using a team with expertise and regular experience with survey and non-destructive testing techniques can be very cost-effective, particularly for large investigations. An extensive list of investigation techniques has been published by the Concrete Bridge Development Group [1]. Broomfield [2] lists the national and international standards current in 2007, where available, for many of the techniques discussed, along with a detailed discussion of the merits, limitations and interpretation of techniques. NACE Standard SP0308 [3] also provides a guide to inspecting reinforced concrete structures and a comprehensive list of test methods and standards, particularly those related to reinforcement corrosion.

12.1.1 Condition survey

As stated above, it is important to carry out a sufficiently detailed condition survey to assess the present condition of the structure and how it reached that condition. Only then will it be possible to choose the most cost-effective repair so that the structure can continue to fulfil its required purpose. Access will be a major cost of any investigation and survey for most large structures. Traffic control for highway bridges is frequently the largest single cost of a survey and even of a repair. In some cases, it can be over 50% of the cost. It is, therefore, important that a survey is well planned with clear objectives and a well-briefed survey team and that it is adequately resourced. It has been suggested that about 10–20% of the repair budget should be spent on the investigation. It is often better to take more samples and do more tests than initially considered necessary rather than having to go back for more to clear up any ambiguity in the results. Table 12.1 summarises the major test techniques used in a condition survey and where to find standards and guidance on their application.

Table 12.1 Techniques for assessing concrete and relevant international standards

Method	Standards or guidance
Visual	How to recognise the alkali–silica reaction (ASR) is discussed in the manual on ASR published by the Strategic Highway Research Program [4] and in Kaetzel *et al.* [5]. The appendix of the technical guide published by the Concrete Bridge Development Group [1] has a table with 25 examples of cracks and visual defects with photographs.
Delamination	ASTM D4580 describes how to detect delamination in concrete bridge decks using a chain drag or hammers, the Delamatect system and a rotary percussion device. The standard concludes that the chain drag procedure is the most reliable of these techniques. Standard ASTM D4788 covers the infrared thermography of concrete and asphalt-covered concrete bridge decks. This requires a scanner with a minimum thermal resolution of 0.2°C. The expected temperature difference between a delaminated or debonded area and an intact area is 0.5°C. The standard is for a vehicle-mounted system. It claims that 80–90% of delamination can be found in concrete decks with or without an asphalt overlay. ASTM D6087 covers the use of ground-penetrating radar to evaluate asphalt-covered concrete bridge decks. It is designed for vehicle-mounted or manually driven systems. It claims the system is accurate in detecting delamination to within +11.2% according to a precision test on ten bridge decks in New York, Virginia and Vermont.
Cover meter	One of the few standards for cover meters is BS 1881-204. Alldred [6] discusses the accuracy of a cover meter when several rebars are close together. The technical guide published by the Concrete Bridge Development Group [1] describes cover meters, their design and performance in depth.
Phenolphthalein	There is a CEN standard on carbonation depth measurement (EN 14629:2006). The Concrete Society's Technical Report 60 [7] covers the same material as the CEN standard. BRE Digest 405 [8] specifically discusses carbonation and measuring it in Portland cement concrete. BRE Digest 444 Part II [9] and BRE IP 11/98 [10] discuss measuring carbonation in high-alumina cement concrete.
Chloride content: sampling, analysis and interpretation	CEN has produced a test method for chloride analysis, BS EN 14629:2007 – 'Products and systems for the protection and repair of concrete structures. Test methods. Determination of chloride content in hardened concrete' (draft). There are three relevant ASTM standards: ASTM C1152/C1152M-04e1 – 'Standard test method for acid soluble chloride in concrete and mortar', ASTM C1524-02a – 'Standard test method for water-extractable chloride in aggregate' and ASTM C1218/C1218M-17 – 'Standard test method for water soluble chloride in mortar and concrete'. ACI 2001 has an extensive discussion on thresholds. BRE Digest 444 [10] has a comprehensive table and a 'cloud data' graph of chloride content compared to risk based on field data.
Reference electrode (half-cell) survey	Detailed methodologies for undertaking reference electrode potential surveys and their interpretation can be found in a Concrete Society report [7], Chess and Grønvold [11] and ASTM C876-15.
Linear polarisation resistance (LPR)	The macrocell test ASTM G109 is the only standard for measuring the corrosion rate of steel in concrete (laboratory tests only). Guidance on LPR comes from the Concrete Society report [7], which has a detailed procedure and a pro forma for measuring and recording LPR-based corrosion rates. Detailed instructions are given in the manuals provided by manufacturers and suppliers of the equipment.

Table 12.1 Continued

Method	Standards or guidance
Concrete resistivity	Resistivity testing is covered in BS 1881 – 'Testing concrete. Part 201. Guide to the use of nondestructive methods of test for hardened concrete'. The Concrete Society report [7] provides a detailed procedure and a pro forma for recording resistivity measurements. The Concrete Bridge Development Group [1] also gives information and guidance. The test is based on soil resistivity, as described in ASTM G57 – 'Field measurement of soil resistivity using the Wenner four-electrode method'.
Initial surface absorption test (ISAT) for permeability	The ISAT test (BS 1881-208) is particularly useful for checking the effectiveness of coatings. Concrete Society Technical Report 50 [12] is a guide to surface treatments for protection and enhancement of concrete. RILEM Report 12 [13] goes into great detail on the various tests and transport phenomena in concrete.
Petrography	Standards for a petrographic analysis include ASTM C457-15 – 'Standard test method for microscopical determination of parameters of the air-void system in hardened concrete', ASTM C856-18e1 – 'Standard practice for petrographic examination of hardened concrete', BS 1881 – 'Testing concrete. Part 124. Methods of analysis of hardened concrete' and the Applied Petrography Group Report [14].
Radar and radiography	ASTM D4788-03 gives a standard test method for bridge decks, which can be used as guidance for other concrete elements.
Ultrasonic pulse velocity	Ultrasonic pulse velocity is documented in BS EN 12504-4:2004 – 'Testing concrete. Determination of ultrasonic pulse velocity'.
Impact echo	ASTM C597-16 – 'Standard test method for pulse velocity through concrete' and ASTM C1383-04 – 'Test method for measuring the P-wave speed and the thickness of concrete plates using the impact-echo method' cover the application of impact echo to concrete structures.
Radiography	General information regarding radiography for concrete structures can be found in BS 1881-205 – 'Concrete testing. Recommendations for radiography of concrete'. National regulations on ionising radiation should also be consulted.

12.1.2 Visual surveys and desk studies

The first part of an investigation is to check what drawings are available, along with any reports of previous investigations, repairs or maintenance relevant to the investigation. It is far easier to mark survey results on elevation drawings if available. If a structural analysis is required, then reinforcement drawings will need to be located or developed, as they will also be for the design and application of any electrochemical techniques, such as cathodic protection, electrochemical chloride extraction or re-alkalisation.

A visual survey is a critical part of an investigation. This may break down into a preliminary survey to understand the extent of the problem and how it has manifested and then a more detailed survey that identifies all relevant defects. It is normal practice to use pro formas, elevation drawings or sketches to note down all defects as well as sampling and test locations. A detailed visual survey should be done within touching distance of the surface being surveyed; however, this may not always be possible due to access limitations. In such cases, the areas that can be accessed by direct contact should be surveyed and then other areas surveyed using binoculars.

As with all aspects of a condition survey, it is important to decide how the results of a visual survey will be used. If these are to be used to develop a bill of quantities for repair, then the survey needs to quantify each area of damage to be repaired in terms of depth as well as length and breadth and the number and type of defects.

The author is not aware of any standards for visual surveys, but good examples and guidance are given by the Concrete Society [15], the Concrete Bridge Development Group [1], the American Concrete Institute [16], the Buildings Research Establishment (BRE) [17], Broomfield [2] and NACE SP0308 [3].

12.1.3 Delamination survey

A delamination survey is frequently undertaken alongside a visual survey where it is possible to touch the surface. A hammer or a chain on bridge or car park decks is usually the most cost-effective and accurate method of measurement, although other techniques, such as radar, thermography and impact echo, have been used, especially where there are overlays, such as waterproofing membranes and asphalt wearing courses on bridge decks.

12.1.4 Coring and petrography

It is common practice to extract cores for petrographic analysis and strength testing. A petrographic analysis will reveal internal problems with the concrete, such as ASR or a sulfate attack, or the presence of high-alumina cement with its risk of chemical conversion and subsequent loss of strength and increase in porosity that make it more vulnerable to carbonation. Petrography will also show the extent of voiding and poor compaction and the degree of hydration of the cement. It can be used to estimate the cement content and types and the water-to-cement ratio. There are many coring companies that can extract cores quickly and efficiently; however, this is a rather destructive technique.

12.1.5 Chloride and carbonation testing

These were discussed in detail in Section 4.4.7.5. These tests are essential for determining the correct repair for structures with corroding reinforcement, as the extent of carbonation and chloride ingress will determine the most cost-effective method of long-term corrosion control. Modelling techniques, such as those described in Chapter 10, can be used to predict future rates of deterioration and the effectiveness of applying different repair options.

12.1.6 Corrosion rate, potential mapping and concrete resistivity measurement

These techniques were also discussed in Section 4.4.7.5. Potential mapping is widely used for highway structures and car park structures, and it can be used to map anodic (corroding) areas of reinforcement. A potential survey can be misleading in the presence of excess moisture, galvanised steel or other embedded metal (such as window frames). Measurements of the local corrosion rate can be used to confirm that corrosion is occurring where mapped and to predict rates of section loss and time to cracking and spalling.

12.1.7 Other tests

Comprehensive lists of tests and their relevance have been published by the Concrete Bridge Development Group [1], Broomfield [2] and NACE SP0308 [3], and are also listed in Table 12.1.

12.2. The repair strategy

As stated in Section 12.1, a successful repair requires an understanding of the following

- the structure, including locations critical for its ongoing performance
- its present condition
- the deterioration mechanisms
- the future condition if unrepaired
- repair options, their suitability and their limitations.

European standard BS EN 1504 – 'Products and systems for the protection and repair of concrete structures' gives specifications for concrete repair materials in Parts 2 to 7. These cover the following aspects

- coatings, sealers and impregnation (Part 2)
- concrete and mortar repair materials, structural and non-structural (Part 3)
- structural bonding (Part 4)
- concrete injection materials (Part 5)
- anchoring of reinforcing bars (Part 6)
- reinforcement corrosion protection (Part 7).

Part 1 contains definitions, and Part 8 describes quality control and conformity, which is principally of interest to manufacturers to ensure their products comply with the directive.

Part 9 gives general principles for products and systems, whereas Part 10 gives information on how to specify the materials in a concrete repair performance specification. A non-mandatory appendix describes how to apply the materials and what site tests and values should be specified. Part 9 states that the basic options for protection and repair are as follows.

- Do nothing for a certain time.
- Reanalyse the structural capacity, possibly downgrading the capacity.
- Prevent or reduce further deterioration without improvement.
- Improve, strengthen or refurbish the structure.
- Undertake a partial or full reconstruction.
- Undertake partial or full demolition.

The fourth item above includes provisions relating to corrosion control for the reinforcement, controlling factors such as moisture that influence ASR or sulfate attacks, and surface protection to stop damaging agents from reaching the concrete surface.

12.2.1 Structural issues

A detailed review of structural issues is beyond the scope of this book, but any reduction of structural capacity must be assessed, whether due to deterioration or to repairs, for instance if concrete is removed. As stated in EN 1504-9, the structural impact of the repair process and how loads will be carried both during and after repair must be considered.

12.2.2 During an assessment

A condition evaluation as described here is not necessarily a structural survey. A structural engineer must be consulted if there are concerns about the capacity of the structure, either due to corrosion damage or for any other reason. Any excessive deflection of structural elements, misalignment, impact damage, excessive cracking, loss of concrete or loss of steel section will require a structural evaluation before the corrosion damage is repaired. Structural issues for concrete bridges are discussed by the Concrete Bridge Development Group [1]. For car park structures, they are discussed by the Institution of Civil Engineers [18], and for older buildings by Macdonald [19]. If there are structural issues, then the existence of reinforcement drawings becomes more important. In their absence, a careful cover meter survey with a scanning cover meter may supply sufficient information. However, it is very difficult to get detailed information about hinges, joints and other structurally important locations from cover surveys, as steel is often very congested, multilayered and sometimes buried deep below accessible surfaces.

Figure 12.1 Structural support system for an elevated section of the Midland Links Motorway (Junction 9, M6) prior to repair. This support system can be jacked up to take some of the load while the concrete is being repaired. The conductive coating for a cathodic protection system can be seen on cross heads in the background.(Courtesy of Kevin Davies, taken December 2005)

12.2.3 During repair

Removing concrete during a repair can compromise the integrity of the structure. This may limit the amount of breakout permitted during a repair. Adequate time must be allowed for the concrete to gain strength after being repaired, and there may be a requirement for propping during the repair process itself. If repairs are to be structural, then a structural engineer must design them. It is usually necessary to support the live and dead loads on the structure during repair, although the loads will go through the repairs once the support is removed. In some cases, structural support is needed if significant amounts of concrete are to be removed. A structural support system for concrete repair is shown in Figure 12.1. The above-ground structure has been tied into the foundations. BS EN 1504-3 clearly differentiates between structural and non-structural repair materials. Also note that if there has been extensive delamination of the concrete, the bond between the steel and the concrete may have been compromised.

12.3. Repair after concrete attack

There are several attack mechanisms against concrete, as covered in Chapters 3 and 4. These can be internal reactions of the constituents such as ASR or sulfate attack or they can be an external attack by acids, freeze–thaw cycles, fire, erosion or impacts. Since the aim of all repairs is to repair existing damage and control future damage, it is important to understand how far the damage has gone and how much further it may go.

12.3.1 Sulfate attack

This topic is covered more extensively in Section 4.4.3. There are several forms of sulfate attack, including external sulfate attacks, internal delayed ettringite formation and thaumasite attacks. Repairs may be structural, require replacement of lost cover to the reinforcement or provide protection from ingress of sulfates and water.

12.3.2 Alkali–silica reaction

This topic is covered more extensively in Section 4.4.6. ASR causes unsightly staining on the concrete surface. By the time it is investigated, the aggregates may have completely reacted, in which case repairs can be carried out without further concern. The repairs may, therefore, be mainly cosmetic. However, ASR-induced expansion can move bearings off pads or damage and distort other elements or adjacent structures. Repairs may be structural, requiring jacking or structural propping. Testing and analysing cores may be required to determine the extent to which further expansion could occur and the likelihood that any treatment will successfully control it.

As water is a necessary reactant, controlling moisture access by protecting the concrete surface with coatings, penetrating sealers or better drainage may reduce the reaction rate. This subject is dealt with in greater detail in Section 4.4.6.5.

12.3.3 Acid attack

Section 4.4.2 discusses the mechanism of an acid attack against concrete. Acids with a pH of less than 6.5 will attack concrete. Concrete repair, fairing coats and cementitious or other coatings may be required to repair existing damage and prevent future damage to the concrete. In some cases, the residue after an attack may have been washed away, exposing the reinforcement and leaving large aggregate residues in the system. In other cases, the attack may leave a softened layer at the surface that needs to be removed back to sound concrete before repair.

12.3.4 Fire damage

Fire may weaken concrete and, therefore, jacking or structural propping may be required during repairs, which may need to be structural. Compromised cover concrete may need to be replaced.

12.4. Physical concrete repair

12.4.1 Structural repairs

Structural repairs may include the complete or partial removal of elements of a structure or of damaged concrete while leaving the reinforcing network in place. Alternatively, they may use structural plate bonding or structural bonding of glass or carbon fibres [15,20].

12.4.2 Patch repairs

A patch repair is conventionally applied when concrete cover is damaged due to reinforcement corrosion or attack from external sources. A good-quality patch repair should use a proprietary low-shrinkage repair mortar or concrete. The material may be applied by hand, by spray or pumped into suitable shutters. The patch area must have clean surfaces free of dust and debris and squared edges.

A quantitative visual and delamination survey is required to develop the bill of quantities. When bidding for concrete patch repair work, the form of contract and the form and accuracy of the bill of quantities influence how the bidding contractor will perceive the risk to itself and, therefore, how it will price the bill of quantities. As stated in Section 12.1.2, an investigation should provide an estimate of quantities broken down into depth and area of concrete to be removed. The bill of quantities should be as accurate as possible. It must allow for ongoing deterioration between the survey and the day the contractor carries out repair, that breakouts will be larger than the area that is delaminated and any shortcomings in the methods used to measure delamination.

There are several methods for removing concrete, and the choice depends on the number, size and depth of the patches, the reinforcement layout, the specification, the budget and the contractor's preferences. If concrete is just starting to spall due to carbonation or if an electrochemical treatment such as cathodic protection is planned, then a simple repair that involves removing unsound concrete, cleaning the rebar surface, squaring the edges and putting in a sound, cementitious, non-shrinking repair material may suffice. This may also be sufficient for 'one-off' repairs, such as for fire damage or ASR that has gone to completion, or to prevent the recurrence of an external attack or damage.

Before patching carbonated concrete, any cracked and spalled concrete must be removed from around the rebar or as far as the carbonation front goes, whichever is the greater. The cementitious patch material must be chosen to ensure that the steel is back in a high-pH, alkaline environment. This will encourage the reformation of the passive layer, which stop further reinforcement corrosion.

Concrete can be removed by handheld percussion tools, hydrodemolition or, if large areas of concrete are to be removed, by milling or vehicle or robot-mounted hydrojets. With increased health and safety concerns about vibration-induced injuries, hydrodemolition is increasingly popular for reasonably extensive repairs. Hydrodemolition has several advantages over percussion tools.

- There is less noise and vibration, so there is less impact on users of the structures or the neighbours.
- It cleans the bars as it goes.
- It leaves the exposed concrete surface less damaged than percussion tools and ready for concrete application.
- It eliminates the risk of vibration injuries to operatives.

However, hydrodemolition is not always suitable because of the need to deal with the water effluent and the access requirements. Moreover, the logistics may be excessive for smaller repairs. It may need a follow-up to remove enough concrete from behind the reinforcement due to 'shadowing'.

If patching is required due to chloride-induced corrosion, then the usual specification is to remove concrete to a depth of about 25 mm behind the rebar and to ensure that all corroded steel in delaminated areas is exposed and that the rebar is cleaned to a near-white finish by removing all rust and chlorides. Removing concrete behind the reinforcement and then cleaning the rear surfaces of bars can be extremely difficult, especially if the reinforcing bars are highly congested. Thus, careful specification and selection of concrete removal techniques may be required.

Figure 12.2 shows three types of repair: a bad repair, a good chloride repair and a patch suitable for cathodic protection where protection will not be provided by the patch itself. The cut edges and faces of the concrete must be square and clean of all dust and debris in all cases.

The choice of material for a general patch repair is discussed below. More recently contractors have found it essential to roughen any saw-cut edges and slope the edges so that the breakout looks like a mix of Figures 12.2(a) and 12.2(b). Wetting or bonding agents may be used to promote adhesion between the repair and the original concrete. Bonding agents should not be used where electrochemical treatment is required, as these could create an electrically insulating layer. If an electrochemical technique is being used, then repairs need only reinstate damaged concrete. A clean rebar surface is needed, and a simple patch to rebar depth in cracked, spalled or delaminated areas will prepare the structure for the application of the anode. In some cases, the anode and overlay will fill the excavated areas in a single operation. Materials used must be compatible with the electrochemical treatment. These issues are discussed more fully in Sections 12.6–12.8.

An evaluation of 230 case histories of repairs under the CONREPNET programme found that 20% of repairs failed within 5 years, 60% failed within 10 years and 90% within 25 years [21]. The most common forms of failure were cracking and debonding of patch repairs. Corrosion and ASR were the most difficult problems to repair.

One of the reasons why corrosion is one of the most difficult problems to repair, particularly with patch repairs, is the incipient or ring anode effect. Section 4.4.7 and Figure 4.18 explain the corrosion process in terms of the formation of anodes where corrosion occurs and adjacent areas where a benign cathodic reaction occurs. Figure 12.3 shows what can happen in the adjacent concrete when an anode is patch repaired.

Figure 12.2 Patch repairs. (a) Bad. Feathered edges and poor preparation may allow breakaways. Keying will be poor. (b) Good. Concrete has been removed from behind the bar and beyond the corroded area to remove contamination and allow the restoration of passivity. Edges are squared. (c) Compatible with electrochemical treatment. It is not necessary to remove concrete beyond the corroded area for electrochemical treatment

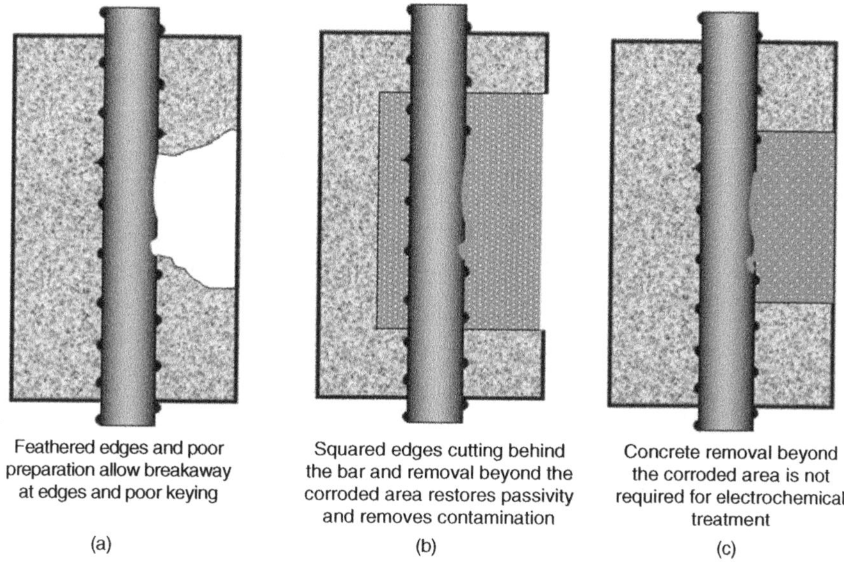

Figure 12.3 Schematic illustration of an incipient anode showing how the anode is displaced to the edge of the repair by the formation of a new cathode in the patch repair

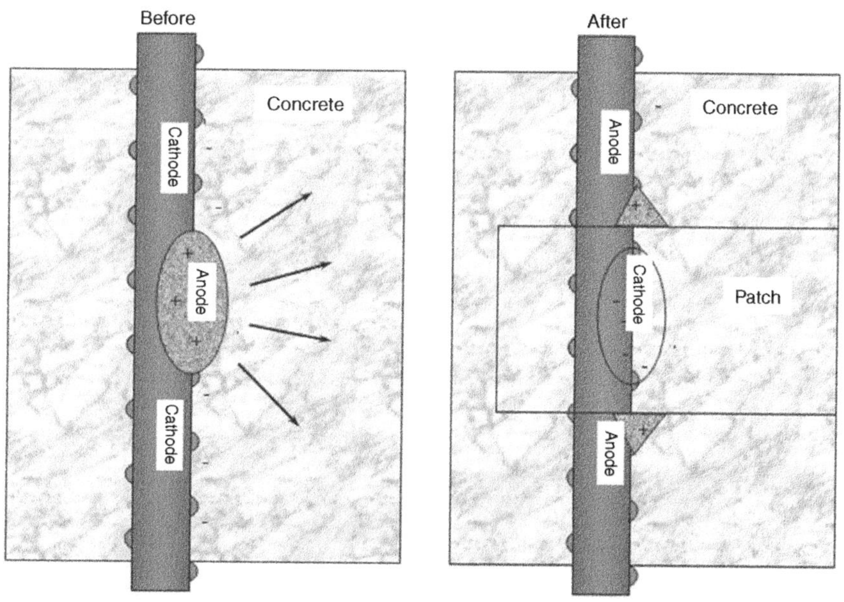

The formation of an anode forces the creation of adjacent cathodes around it, limiting the active area of corrosion. If the anode is suppressed and is turned into a cathode by patch repairing, the previously cathodic areas will become anodic, and corrosion will initiate around the repair. This problem is particularly prevalent for chloride-contaminated structures, as the concrete is generally wetter than for carbonated structures and so the lower electrical resistance of the concrete means that anodes and cathodes are well defined and separated from each other. Examples are illustrated in Figures 12.4 and 12.5.

It is, therefore, important to realise that patch repairs will only be cosmetic and short term unless the underlying cause of deterioration is fully addressed.

Figure 12.4 A poorly executed patch repair with surrounding spalling due to incipient anodes

Figure 12.5 Incipient anode showing the damaged repair (left side) and corrosion in the original concrete due to the formation of an incipient anode on an integral beam under a car park deck exposed to chlorides. The original repair appears to be well executed but could not stop the formation of an incipient anode

The UK Concrete Society has a standard specification and bill of quantities for concrete repairs [22]. A more up-to-date specification and method of measurement are available from the website of the UK Concrete Repair Association. (In the USA, the *Concrete Repair Manual* is published jointly by the American Concrete Institute and the International Concrete Repair Institute [23].)

12.5. Surface protection systems

One of the easiest ways of addressing the underlying causes of corrosion can be to apply a coating or other form of surface protection. If there is ingress of carbon dioxide, chlorides, water or oxygen, then, in principle, such barriers can stop further ingress. Obviously, a coating or barrier applied during the initial construction is more likely to be successful in keeping out contaminants than one applied once problems come to light. It is important to determine whether the application of surface protection is going to be cost-effective once deterioration has started.

12.5.1 Coatings

The construction products industry has produced a range of coatings, including penetrating sealers and impregnants. The new European standard EN 1504-2 refers to hydrophobic impregnants, impregnants and coatings. These can be used to protect against ingress, for moisture control, to increase the electrical resistivity by limiting moisture content, for physical resistance or for surface improvement.

It has been demonstrated experimentally that coatings can control corrosion in carbonated structures in outdoor conditions in the UK but not where corrosion is due to chlorides [24]. EN 1504-2 lists test methods for measuring the permeability to carbon dioxide, water and water vapour. There is no agreed European standard yet for measuring chloride ion diffusion through concrete.

Generally, anticarbonation coatings are high-build, pigmented coatings, although unpigmented, clear coatings are now being produced, and these also have good track records. Penetrating sealers, such as silanes or siloxy silanes, are hydrophobic. They line the concrete pores with a hydrophobic coating while leaving the pores open. This will minimise water and hence chloride ingress while allowing the concrete to 'breathe out' water vapour.

It is important that coatings are applied properly with the specified surface preparation, coverage rate, wet and dry film thickness, and adhesion. It is interesting to note that Sergi *et al.* [24] found that the best coatings for controlling carbonation-induced corrosion did not have a very high pull-off strength. They suggested that this allowed better bridging of cracks, which accommodated minor crack movements.

It is important that the pull-off strength of the substrate is tested prior to coating as well as after coating. A very poor-quality substrate may be unsuitable for coating so that a penetrating sealer or other measures may be required to prepare a surface suitable for coating.

12.5.2 Concrete overlays

Concrete overlays are widely used in North America for rehabilitating corrosion-damaged bare-concrete bridge decks. Shrinkage crack control, delamination and 'curling' due to differential shrinkage can be dealt with due to the considerable practical expertise of American highway contractors. Such techniques are rarely used in Europe, where waterproofing membranes and asphalt overlays are more commonly applied to bridges. Decks to be overlaid may be prepared by removing all the concrete cover behind the reinforcement or a selective area by milling, hydrodemolition, percussion tools or a combination. It may be necessary to replace damaged reinforcement or the complete top mat.

12.5.3 Waterproofing

Waterproofing is used extensively on new bridge decks in many western countries where de-icers are extensive applied. Two notable exceptions are the USA and Canada. A range of materials, in the form of either sheets or liquids, are approved for use in different countries. A synthesis of practice of waterproofing membranes in North America was published by Manning [25]. Approximately 50 systems were tested by the UK Transport Research Laboratory in a trial of membrane systems [26].

12.5.4 Barriers and enclosures

Barriers and enclosures can keep out chlorides and water. They can also be used to reduce the humidity level at the reinforcement surface below the threshold for corrosion. It is not clear from the literature what that level is, but concrete within buildings rarely corrodes, even though it is frequently carbonated to rebar depth. Also, the external facades of buildings constructed with calcium chloride added as a set accelerator may corrode, but rarely, if ever, is there internal corrosion, unless there is moisture ingress. It is, therefore, feasible to construct cladding over a building facade and ventilate it to the interior of the building to control reinforcement corrosion.

12.6. Chemical corrosion inhibitors

ISO 8044 defines a corrosion inhibitor as 'a chemical substance that decreases the corrosion rate when present in the corrosion system at suitable concentration, without significantly changing the concentration of any other corrosion agent'. This definition excludes coatings, pore blockers and other materials that affect the water, oxygen or chloride concentrations.

To protect steel in concrete, corrosion inhibitors can be admixed into the fresh concrete. Alternatively, they can be applied to the concrete surface, usually long after the concrete has hardened. They must then transport through the hardened concrete to react on the reinforcing steel surface to slow down the rate of corrosion. This difference between admixed and surface-applied inhibitors is fundamental. The precise dosage can be controlled by admixing an inhibitor during construction. Moreover, the conditions can be readily tested. The protective mechanism is in place before corrosion initiates. In contrast, when applying inhibitors to hardened concrete as part of a rehabilitation system, it is far more problematical to ensure that the dosage at the rebar is effective and to measure, test and predict the long-term performance. Also, if corrosion has started, the inhibitor must penetrate the layer of corrosion products and stop corrosion that has started, not just prevent its initiation.

There are several ways of classifying inhibitors.

- By their action. Corrosion in concrete occurs through the formation of anodes and cathodes. Corrosion inhibitors can, therefore, be
  - anodic inhibitors that suppress anodic corrosion reactions
  - cathodic inhibitors that suppress cathodic reactions
  - ambiodic inhibitors that suppress both anodic and cathodic reactions.
- By their chemistry and function
  - inorganic inhibitors such as nitrites, phosphates and other inorganic chemicals
  - organic inhibitors such as amines and other organic chemicals
  - vapour phase or volatile inhibitors, which are a subgroup of the organic inhibitors (generally amino alcohols) that have a high vapour pressure.

All these inhibitors have been widely used by corrosion engineers for many years to protect steel and other metals, such as electronic components and in the chemical process and power generation industries. In some cases, their chemistry is well understood; in other cases, less so.

12.6.1 Admixed corrosion inhibitors

Of the available materials, the most widely applied as an admixture in new concrete is calcium nitrite. This inhibitor is of proven effectiveness, provided the chloride-to-nitrite ratio stays less than about 1.8:1 [27]. A dosage rate of 10–30 litres per cubic metre of concrete is generally specified, depending on the expected maximum chloride level at the rebar. Calcium nitrite has proven efficacy and is usually the benchmark against which other inhibitors are tested. Nitrites act as setting accelerators (their original use as a concrete additive was as non-chloride setting accelerators in cold climates). They are usually sold in proprietary formulations that include a setting retarder to avoid flash setting of the mix, adjusted to local conditions.

Alternative proprietary inhibitors are generally organic with active amine and ester groups. The main problem with most organic and inorganic inhibitors is that they act as severe set retarders, In contrast, formulations with amines and esters or with nitrites do not.

The US Federal Highway Administration has expressed concerns that inhibitors may be leached out of concrete, particularly in marine exposure conditions. Some work on calcium nitrite has been undertaken to remedy this [28].

12.6.2 Applying corrosion inhibitors to hardened concrete

For many years, vapour-phase inhibitors and migrating corrosion inhibitors have been used to impregnate packaging, greases and waxes to protect steel machinery and components, particularly before use. An American company realised in the 1980s that such inhibitors might be effective in diffusing through concrete pores and protecting reinforcing steel. Several companies now offer proprietary formulations. These, too, are based on amines and esters, with amino alcohols being the main volatile component.

Amino alcohols are ambiodic, forming a film on the steel surface that blocks both anodic and cathodic reactions. They can be applied as a coating on the surface of the concrete, a 'plug' of material in a hole or admixed into repairs. The suppliers claim that these materials will move very rapidly through the air voids in the concrete, through pores and microcracks, to reach the steel and protect it.

A migrating inhibitor that is not volatile is MFP or monofluorophosphate. It relies on capillary action and other transport mechanisms to get the material down to the steel through the concrete cover. A paper has shown that it can be quite successful in carbonated concrete [29]. Calcium nitrite has also been offered in a formulation for patch repairs or overlays, with the intention that the nitrite will diffuse out of the patch and protect the steel.

12.6.3 Track record, case histories and monitoring

One issue with corrosion inhibitors is determining how effective they are. There are two aspects. Can they be detected and can levels be measured quantitatively in the concrete and at the rebar surface? Do they effectively depress the corrosion rate?

The method of detection is obviously specific to a particular active inhibiting chemical. Some tests are expensive and difficult. Much work has been undertaken to develop a test for amino alcohols, with mixed results so far. As stated earlier, it is far easier to carry out a test on admixed inhibitors than on inhibitors applied to hardened concrete. There can be dangers with under-dosing inhibitors, in that corrosion may occur very aggressively in small anodic pits. It is, therefore, important that the risk of pitting and under-dosing is fully understood by those using these materials.

The most effective method of assessing corrosion inhibitors is to measure the corrosion rate. If a corrosion inhibitor is effective, it should reduce the active corrosion rate by at least 1 and preferably 2 orders of magnitude. The author has had published a report on the long-term monitoring of structures that had corrosion inhibitors applied in 1995 [30]. Amino-alcohol-based inhibitors were applied by surface application, down holes drilled in the concrete and in repair patches to support structures. The concrete has over 1% of chloride by weight of cement cast, since calcium chloride was added to the concrete as a set accelerator. Over the years, many patch repairs have been carried out. In this trial with over 60 structures, there was only a slight reduction in the corrosion rate. The onset of cracking of treated structures occurred at approximately the same time as for untreated controls.

In a very small-scale trial with another proprietary amino-alcohol-based inhibitor, the short-term results were more encouraging, with a 91% reduction in the corrosion rate measured by LPR in an area with active corrosion and corrosion potentials. However, there has been no follow-up to this trial on a car park deck. It has low cover, which increases the likelihood that the inhibitor will reach the steel. Moreover, the BRE have done tests with calcium nitrite [31].

In the USA, corrosion inhibitors have been tested for many years. Calcium nitrite cast into concrete as an inhibitor has been widely tested in the laboratory and the field. Amino-alcohol-based inhibitors were first trialled in hardened concrete in the 1980s, with some of the most comprehensive testing done by the Strategic Highway Research Program [32,33]. A recent review of field trials on bridge decks suffering from chloride ingress showed that inhibitor treatments had no beneficial effects [34]. Similarly, a large-scale exposure trial of inhibitors on slabs also found no beneficial effects for several inhibitors [35]. The exception was calcium nitrite cast into concrete.

If coatings are applied after the inhibitor is introduced, then corrosion monitoring is harder. To take measurements of the corrosion condition, either the coating must be removed locally or (preferably) corrosion rate probes must be embedded in the concrete, as has been done on some car parks in the UK [36]. Ideally, this should be done before the inhibitor is applied, so that 'control' measurements can be taken prior to treatment. Permanent monitoring probes can effectively monitor the changes in the corrosion rate without disturbing the structure, thus ensuring that readings are made in a consistent way in a consistent location.

The biggest issues with corrosion inhibitors are for those wishing to use them for rehabilitation rather than for durable construction. The research and field trials on corrosion inhibitors reveal the following issues.

- There are very limited independent field data on the performance of corrosion inhibitors.
- The available field data are often poor, with no clear evidence of the amount of inhibitor applied, whether it has reached the rebar and whether it is reducing corrosion rates and extending the time to cracking.
- Many claims have been made about the transport of inhibitors through hardened concrete. These claims need to be independently assessed.
- There is no definitive evidence of the dosage needed for a given chloride level to achieve a given (low) corrosion rate.
- There is no quantitative data on the penetration of corrosion inhibitors into hardened concrete for different levels of cover and concrete permeability.
- More information is needed on the performance of inhibitors, particularly from well-controlled field trials and long-term corrosion monitoring.

Corrosion inhibitors are inexpensive materials that can be applied simply. However, until there is clear evidence of their envelope of effectiveness in terms of chloride level, corrosion rate reduction, dosage, cover and longevity, it will be difficult to compare the whole-life costing of an inhibitor compared to proven alternatives, such as cathodic protection. This is particularly so for inhibitors applied to hardened concrete.

Whole-life costs are, however, available from the manufacturers of inhibitors admixed into fresh concrete that show the cost-effectiveness of their inhibitor compared to some of their other products or alternative methods of improving durability in aggressive environments.

12.7. Cathodic protection

Cathodic systems can comprehensively protect against corrosion of the reinforcing steel in a concrete structure. Cathodic protection can be combined with a patch repair, as described in Section 12.4.2. The repairs need to be compatible with the passage of electricity through the concrete, so that there should not be an insulating bond coat on the steel. Nor should there be metallic fibres in repair materials, as these could short circuit the anode to the steel.

12.7.1 Impressed-current cathodic protection

The best known form of cathodic protection for steel in concrete uses an impressed current from a mains-powered transformer/rectifier to deliver a direct current along wires to a distributed anode system mounted on the concrete surface or embedded in the concrete. The level of protection needed is determined by measuring the steel potential against an embedded reference electrode (Sections 4.4.7.5 and 4.4.7.6). The arrangement is illustrated in Figure 12.6.

12.7.1.1 Cathodic protection theory

There are several ways to explain how cathodic protection works. The simplest is as follows. Consider Figure 4.18. If corrosion is occurring at an anode and generating electrons while a benign cathodic reaction is consuming electrons nearby, then installing an external anode and applying a direct current can be used to force all the steel to become anodic or cathodic. By connecting the reinforcing steel cage to the negative terminal, excess electrons are injected into the steel, thus preventing Fe^{2+} ions from being generated and encouraging the formation of hydroxide ions. Therefore, corrosion stops. Another way to understand cathodic protection is that it depresses the electrochemical potential of the anodes to below that of the cathodes so that current no longer flows between them and the corrosion is suppressed.

Figure 12.6 Schematic diagram of an impressed-current cathodic protection system

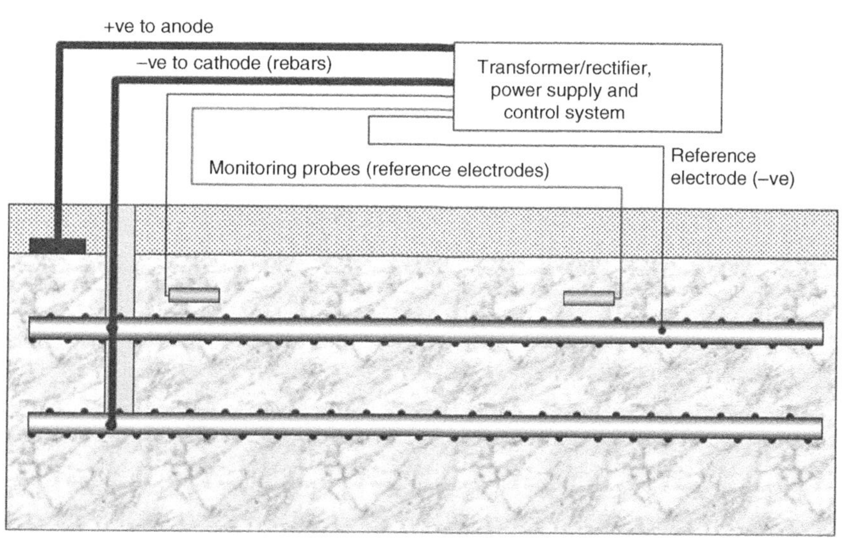

12.7.1.2 Impressed-current anodes

Several electrochemical cathodic reactions can occur on an anode surface. Most of them lead to the formation of acidic products at the anode. An impressed-current anode must, therefore, have the following characteristics.

- It must be electrically conductive so that direct current flows into the concrete.
- It must be compatible with being embedded in or mounted on concrete long term (at least 10 years and ideally 40 years or more).
- There must be low interfacial electrical resistance at the steel/concrete interface.
- It must be resistant to the acids formed at the anode/concrete interface.

The primary materials used for impressed-current anodes for steel in concrete are either mixed-metal-oxide-coated titanium or carbon-loaded composite materials. There is also a proprietary conductive ceramic that is manufactured in various configurations by one anode supplier.

A mixed-metal-oxide coating on a titanium anode lowers the interfacial resistance, allowing it to conduct electricity into the concrete. It also resists acidic reactions. The thickness and formulation of the coating determine the life of the anode in ampere-hours per square metre of anode surface area. There is a standard test for anodes embedded in concrete, which was originally designed for mixed-metal-oxide-coated titanium anodes [37].

Mixed-metal-oxide-coated titanium anodes are frequently an expanded mesh, which maximises the surface area in contact with the concrete. The mesh may be in the form of ribbons, 11–15 mm wide, which can be placed in saw cuts in concrete or spaced off from rebars when applied in a new construction. The most common application is a sheet of mesh, typically 1.2 m wide, fixed to the concrete surface with a concrete overlay applied. Another popular application is a mesh rolled into a tube and inserted into holes in the concrete. Strings of individual probes or discrete anodes are used to form a zone. In one proprietary design, each anode has a resistor, which helps to balance the current flow in the anode zone. Examples are shown in Figures 12.7 and 12.8.

Figure 12.7 A range of probe anodes. Left to right: Mixed-metal-oxide-coated titanium ribbon, mixed-metal-oxide-coated titanium tube anode, mixed-metal-oxide-coated titanium mesh probe anode with resistor, mixed-metal-oxide-coated titanium probe anode with a resistor and cap for retaining grout when installed in soffits, and two types of conductive ceramic anode. All anodes come in various diameters and lengths

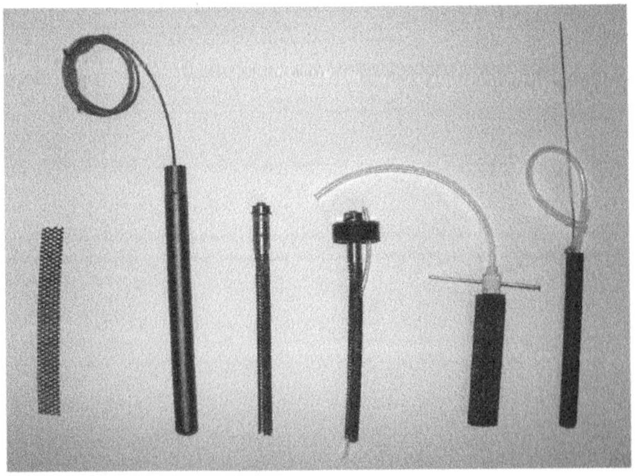

Figure 12.8 A mixed-metal-oxide-coated titanium mesh anode fixed to a concrete surface prior to applying a sprayed concrete overlay. (Courtesy of WS Atkins and the National Highways)

Carbon-based anodes are either conductive organic paint coatings with a graphite 'pigment' or two proprietary materials that use nickel-coated carbon fibres as the conductor and are encapsulated in a wet-sprayed cementitious overlay or in a conductive paint coating, as illustrated in Figures 12.9–12.11. There is a NACE test method for qualifying these anodes [38].

Thermal sprayed zinc is another anode system that can be used as an impressed-current or a galvanic cathodic protection anode (Section 12.7.2). It has the attributes of a coating (i.e. no change in loading, cross-section or clearances) but is more moisture-tolerant than carbon-loaded organic coatings. It is widely used in the USA but less so in Europe. One example is shown in Figure 12.12.

Different anode systems have different advantages and limitations.

- Conductive organic coatings are easy to apply and do not change the loads or dimensions of the structure but are moisture-sensitive and have limited resistance to wear and tear. A cosmetic topcoat can provide an attractive finish that hides any patch repairs. The system has a more limited life than mixed-metal-oxide-based anodes. They have more modest current outputs.
- Arc-sprayed zinc does not change the loads or dimensions of the structure and is more moisture-tolerant than organic carbon-loaded coatings but is more difficult to apply. It also has modest current outputs.
- Probe anodes are easy to install, do not change the loads or dimensions of the structure but require the use of percussion drills, so there are health and safety implications. They are often installed unobtrusively in surface details in buildings. They can short circuit to the reinforcing steel unless the steel location can be accurately mapped. The current output of anode zone can be tailored by altering the size and spacing of the anodes.

Figure 12.9 Conductive paint coating applied to a building in the late 1980s because calcium chloride had been admixed into the concrete on the lowest two floors during construction. (Courtesy of Vinci Technology Centre)

Figure 12.10 The same building after application of a cosmetic and protective topcoat. (Courtesy of Vinci Technology Centre)

- Ribbon anodes do not change the loads or dimensions of the structure, but they require chases in the concrete and adequate concrete cover to avoid short circuiting directly to the steel. Like probe anodes, they can be installed unobtrusively in surface details in buildings. They too can short circuit to the reinforcing steel unless the steel location can be accurately mapped. The current output can be tailored by altering the spacing of the anodes.

Figure 12.11 Applying a conductive cementitious overlay to a bridge. (Courtesy of BASF – Construction Chemicals (UK) Ltd)

Figure 12.12 Arc-sprayed zinc applied to a bridge on the M6 in Cumbria. The very top and the cantilevered ends of the leaf piers have probe anodes installed due to the higher moisture exposure

- Mesh anodes with their cementitious overlays increase the loads and dimensions of the structure. They can be extremely durable and take wear and tear very well. Patch repairs can be hidden under an overlay. However, they require excellent quality assurance and control procedures to ensure that there is a good bond between the original surface and the overlay. A very high current output can be realised by using different meshes and even multiple layers of mesh.
- A sprayed conductive mortar anode also increases the loads and dimensions of the structure. It is very durable and has excellent adhesion to the original concrete surface. Its output is not as high as a mesh but this can be increased by increasing the anode thickness.

Thus, there is a wide range of anodes suitable for different applications with different installation requirements, different durabilities, a wide range of appearances and different current outputs.

12.7.1.3 Control and monitoring systems

Impressed-current cathodic protection systems are powered by a series of small power supplies, usually applying 1 or 2 amps per zone at a few volts direct current. Each zone will have one or more embedded reference electrodes (Sections 4.4.7.5 and 4.4.7.6 and Figure 12.6). These are used to measure the electrochemical potential of the steel to determine whether there is sufficient protection for the steel in the zone.

Measurements may be made manually on a simple system of one to four small zones. More commonly, the sensors are linked to a logger that records the potential between the steel and a reference electrode automatically, switching off the current briefly to make 'iR-free' measurements. They will also take depolarisation measurements, typically once a month, where they will switch off the current for 24 hours while logging the potential decay.

The European standard for cathodic protection of steel in concrete, BS EN ISO 12696, has several optional criteria for achieving corrosion control. The most commonly applied to atmospherically exposed reinforced concrete is that the potential between the steel and a reference electrode should decay by 100 mV in less than 24 hours from the instant after switching off the system.

Control and monitoring systems are generally modular, with power supplies and voltage and current monitors slotting into a rack. They measure anode zone currents, voltages and potentials between the steel and reference electrodes. If automated, a logging system regularly records values and conducts depolarisation tests. Data can be collected by going to site, but usually it is done with a suitable telecommunications link to an office-based computer, which can also be used to adjust the current and voltage outputs once the data have been analysed.

12.7.1.4 Feasibility and conceptual design

When considering cathodic protection, a repair designer must first ensure that the cathodic protection system addresses the deterioration problem, either fully or in combination with other procedures, such as making patch repairs, applying coatings or controlling water ingress. It is essential that any adverse effects are considered. The primary one is that there is no risk of hydrogen embrittlement of any prestressing steel. Other things to consider include any coatings, sealants or other insulating materials that might impede the current flow from the anode to the cathode.

As discussed in Section 12.7.1.2, there is a wide range of anode types. More than one anode can be used on a structure, as shown in Figure 12.12.

The reinforcement dimensions and layout must be determined so that the steel surface area can be calculated and, thus, the required current density for the anode and the total current delivered by the power supplies. A structure is generally broken down into a series of independently powered and monitored anode zones, depending on its size and geometry, the steel densities within it and its exposure (e.g. submerged, tidal and splash zones on a marine bridge substructure).

Next, the location and design of connections to the reinforcement and to the anodes must be chosen, along with the number and location of reference electrodes and possibly other monitoring probes. Cable dimensions, cable runs and cable management systems have to be selected. The number, capacity and performance requirements of the control and monitoring system are also specified in the design. Some of these aspects may need adjustment once on site, but they should all be set down in design documents with suitable drawings, a quality management system, method statements, and a materials and equipment list before work starts, along with a full health and safety plan and procedures (in the UK, as required under the Construction (Design and Management) Regulations 2015).

12.7.1.5 Installation and operation

The system should be installed by a competent and experienced contractor to the designs and method statements provided. For surface-mounted anodes, surface preparation is the key to a successful, durable installation. This may require trials to achieve the required bonding and finish, as well as adhesion tests during anode application. Cable runs may be embedded in overlays, recessed into the concrete or mounted on the surface with clips, a conduit or cable trays.

In the UK, the usual requirement is for the contractor to install and commission the system, and then, after commissioning (which usually constitutes contract completion), the contractor will operate the system for 11 months, undertaking the quarterly monitoring and providing a final and full 11-month review of the system. The system is then handed over to the client, who must either be trained in how to operate the system or place a contract with a suitably qualified cathodic protection engineer to undertake the ongoing monitoring. There is a new ISO standard, BS EN ISO 15257:2017, which defines the levels of competence and certification of cathodic protection personnel. The design, installation and monitoring must be under the supervision of a level 4 certified engineer, but work can be carried out by level 2 and level 3 technicians to method statements provided by the level 4 engineer.

12.7.2 Galvanic or sacrificial-anode cathodic protection systems

12.7.2.1 Theory

Galvanic cathodic protection works as follows. Metals such as aluminium and zinc are higher in the electrochemical series than iron or steel. This means that when coupled together in a corrosive electrolyte, they will corrode preferentially, thus protecting the steel from corrosion. Consider a simple zinc–carbon cell, namely a battery. The zinc case is the anode which corrodes, whereas the carbon rod in the centre does not corrode. In reinforced concrete, we use a cruder cell made of steel and zinc. The zinc again corrodes, producing a small current at up to about 1 V DC, which protects the reinforcing steel.

12.7.2.2 Galvanic anodes

Anodes for steel in concrete are generally zinc, although there are presently two made of an aluminium–zinc–indium alloy on the market as proprietary galvanic anodes. Anodes come in the following forms.

- Coating on the surface, generally electric arc-sprayed zinc or an aluminium–zinc–indium alloy. The thermal spraying of zinc onto a bridge substructure for a galvanic anode system is shown in Figure 12.13. The coating thickness is a few tenths of a millimetre to avoid adhesion problems due to thermal expansion during application. The thickness achieved will determine the life of the system. Pure zinc can be used on marine structures, as the concrete stays wet and, therefore, is electrically conductive. A ‘humectant’ spray can be applied to decrease the electrical resistance of the concrete if applied inland in dryer conditions. The proprietary aluminium–zinc–indium systems are designed to work in marine and non-marine conditions, as the alloy produces a higher current than pure zinc.
- Zinc sheets fixed to the surface with a conductive hydrogel adhesive, as shown in Figure 12.14. The amount of zinc in the system can give lives of 25 years or more for this anode. It is important that moisture cannot get at the hydrogel, which is soluble and can leak out of inadequately sealed systems in wet conditions.

Figure 12.13 Thermal spraying of zinc onto a bridge substructure in the Florida Keys. (Reproduced with permission from the Florida Department of Transportation)

- Expanded zinc mesh in a permanent form clamped to a column and filled with cementitious grout, as shown in Figure 12.15. The amount of zinc in the system can give lives of 25 years or more for this anode. It is generally used on piles or columns in marine conditions but has been applied in a flat form to jetty soffits.
- A disk of zinc embedded in a proprietary mortar wired to any reinforcement exposed during repairs, as shown in Figure 12.16. This anode is particularly aimed at the incipient anode problem (Section 12.4.2). It can extend the life of patch repairs and reduce the ongoing risk of corrosion around a repair rather than providing structure-wide corrosion control.
- A rod of zinc embedded in a proprietary mortar inserted into a core hole in the concrete and connected in a string of other similar anodes to the steel, as shown in Figure 12.17. This is a structure-wide form of the anode shown in Figure 12.16.

Figure 12.14 Sheets of zinc hydrogel applied to a bridge substructure. (Reproduced with permission from BAC Corrosion Control Ltd)

Figure 12.15 Expanded zinc mesh anodes in a permanent form filled with cementitious grout and clamped to columns on a jetty in St Helier, Jersey

Figure 12.16 A disk of zinc embedded in a proprietary mortar wired to the reinforcement that had been exposed during repairs. (Reproduced with permission from Vector (UK) Ltd)

- The most recent development is hybrid anodes. These use an initial external impressed power supply to rapidly repel chlorides from the bar, repassivate it and stop any pitting attack. After that, a conventional galvanic anode maintains polarisation. There are two proprietary forms of hybrid anode currently on the market. The first is a galvanic anode 'charged' by an impressed current using a DC power supply of the type used for impressed current cathodic protection or can be as simple as a car battery. The second has a battery built into each probe anode. As the battery is depleted the galvanic anode maintains polarisation.

Figure 12.17 A rod of zinc embedded in a proprietary mortar being inserted into a core hole in the concrete and connected in a string of other similar anodes to the steel to form an anode string or array. (Reproduced with permission from Fosroc Ltd [20])

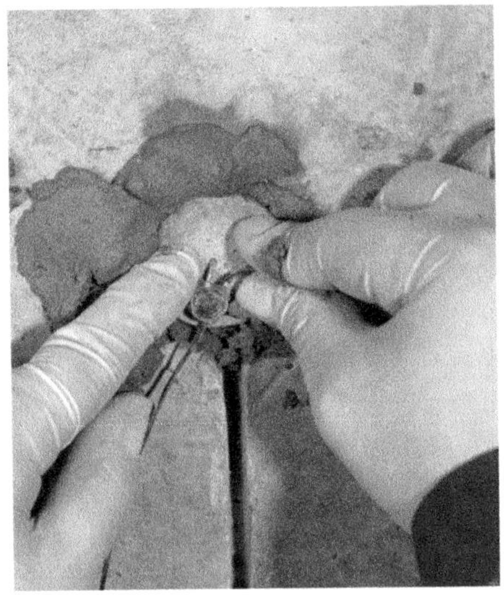

12.7.2.3 Control and monitoring systems

Generally, there is not much to control in galvanic cathodic protection systems compared with an impressed-current system. Systems for measuring current flows can be inserted in trial systems or representative locations by separating an anode or group of anodes and measuring the current flow to the steel, usually through a resistor in the circuit. It is also possible to insert a switch and use an embedded or portable reference electrode to measure the instant off and depolarisation (Section 12.7.1.3). However, the current cannot be adjusted, except with resistors if the current is too high or by adding more anodes if the current is too low.

Since galvanic anodes have finite lives and since they cannot be adjusted, structures containing galvanic anodes should be checked regularly. Monitoring can be performed as follows

- visual survey of the structure and the cathodic protection system to check for corrosion of the steel or any changes to the anodes
- potential survey of the structure to determine the activity of embedded anodes
- potential shifts of the steel (if a switch between the anode and cathode is installed)
- current flow from the anodes to the steel (if a switch or a shunt resistor between the anode and cathode is installed)
- corrosion rate measurements (if a switch between the anode and cathode is installed).

12.7.2.4 Feasibility and conceptual design

As with an impressed-current system, the choice of anode is critical. It is, then, important to ensure that there is sufficient anode material for the required design life. This can be calculated from the steel surface area and the likely current demand (measured in trials or estimated from comparable exposure conditions). For the two types of embedded anode described above, the anode supplier has tabulated the anode density against the steel surface area.

12.7.2.5 Installation and operation

Most anodes can be installed using fairly conventional techniques known to the civil engineer. The exception is if arc-spraying zinc or aluminium–zinc–indium onto concrete. A detailed specification for such an application has been written by the American Welding Society [39] based on its extensive use on bridges in the Florida Keys and other structures on the Florida coast and elsewhere in the USA.

12.8. Electrochemical chloride extraction and re-alkalisation

Electrochemical chloride extraction and electrochemical re-alkalisation are two closely related techniques that use a similar arrangement to that shown in Figure 12.6, but instead, they use a temporary anode and a high current to re-establish a passive, alkaline, chloride-free environment around the steel.

12.8.1 Theory

As established in Section 4.4.7.3, chloride ions are, effectively, a catalyst for corrosion of steel when it has accumulated at a sufficient concentration at the steel surface. We can remove chloride ions by applying a negative charge to the steel and a positive charge to an external temporary anode, which repels chloride ions from the surface and removes a certain proportion from the concrete. In addition, hydroxide ions are formed at the steel surface by the cathodic reaction, as shown in Figure 4.18. The hydroxide ions repassivate the steel surface and, in particular, the actively corroding pits, thus suppressing corrosion. As the hydroxide ion concentration rises and the chloride concentration declines, the chloride-to-hydroxide ion ratio drops below the threshold for corrosion. Additional effects include the infilling of voids at the steel surface due to the migration of ions in an electric field, leading to the precipitation of calcium hydroxide. This means that there is a reduction in initiation sites for future corrosion.

For carbonated concrete, the generation of hydroxide ions re-alkalises the concrete. Also, an alkaline electrolyte is used, which moves into the concrete under normal absorption, capillary and diffusion processes and, in some cases, by electro-osmosis (discussed in the next section).

12.8.2 Anodes

The mixed-metal-oxide expanded titanium mesh and mesh ribbon anodes described in Section 12.7.1.2 are widely used in tanks or bunded systems on decks. Plain steel mesh is also used. The steel is consumed and can stain the concrete surface. However, its consumption avoids the generation of chlorine gas at the anode and reduces acid generation in the electrolyte, which can etch the concrete surface. A proprietary system uses a shredded paper (papier mâché) spray using a suitable electrolyte. Figures 12.18 and 12.19 show examples of these anode systems. Electrolytes for electrochemical chloride extraction include tap water, calcium or other hydroxide solutions to maintain alkalinity or lithium hydroxide or lithium borate if there are concerns about ASR. For re-alkalisation, potassium carbonate is recommended, as it chemically increases the pH of the concrete at the surface while the electrochemical re-alkalisation is occurring at the steel.

12.8.3 Control and monitoring systems

Power supplies range from simple mains rectified direct current supplies to highly sophisticated monitoring and logging systems that measure and control the current flow to the systems. Voltages are kept below 50 V DC for health and safety reasons, and the currents below 4 A/m^2 of the steel surface area to minimise the effects on the concrete. The charge passed in amps per square metre of steel surface area should be logged.

12.8.4 Feasibility and conceptual design and specification

There is a US NACE Standard Practice for electrochemical chloride extraction and re-alkalisation [40]. In Europe, BS EN 14038-1 on re-alkalisation and 14038-2 on electrochemical chloride extraction have been published. Issues with electrical continuity, short circuits from the steel to the anodes, hydrogen

Figure 12.18 Cellulose fibre overlay being applied over a steel mesh for electrochemical chloride extraction. (Reproduced with permission from Makers Ltd)

Figure 12.19 Cassette-tank system using a mixed-metal-oxide-coated titanium anode. (Reproduced with permission from Makers Ltd)

embrittlement of susceptible steels and ASR are even more critical for these techniques than they are for impressed-current cathodic protection. As with impressed-current cathodic protection, the areas to be protected are divided into independently controlled zones. It is important to measure the steel surface area in each zone and ensure that zones reflect the differences in steel densities.

One major issue in the application of these systems is determining the end point. For re-alkalisation, this is fairly straightforward, as phenolphthalein can be used to measure the elevation of pH above the level where corrosion can occur (Section 4.4.7.2). However, there are risks and consequent cost implications if a contractor must achieve a given level of re-alkalisation without some limitation on the total number of amp-hours passed and, therefore, the time spent re-alkalising.

For electrochemical chloride extraction, several criteria are given in NACE SP0107-2021 [40] and in the European standard. These are the achievement of a specified level of chloride in the vicinity of the reinforcement, such as less than 0.4% chloride by mass of cement within 25 mm of the bar, or the passing of a given amount of current per square metre of steel, usually in the region of 600–1500 Ah/m^2, and achieving a chloride-to-hydroxide ratio of less than a defined level, such as 0.6. Contractors can, of course, guarantee to achieve a criterion based on the charge passed. There are risks associated with requiring the contractor to achieve given chloride concentrations or chloride-to-hydroxide ratios without giving an upper amp-hour limit. If treatment goes on longer than expected to achieve the end point criteria, this could have cost and programming implications for a project, unless current density limits are used in conjunction with a concentration criterion.

12.8.5 Installation and operation

The installation is like that for an impressed-current cathodic protection system, with checks of electrical continuity of the steel, connections to the steel, installation of the anode, checks for short circuits before and after applying the electrolyte, and then application of the current with regular monitoring, testing and sampling. There should be detailed records of all test results, before, during and on completion of the treatment.

12.9. Electro-osmosis

Electro-osmosis has already been mentioned in connection with electrochemical re-alkalisation. When a porous medium, such as concrete, contains a solution, then an electric current applied between an anode and a cathode will move the water, usually from the anode to the cathode. For pipelines in soils, this leads to drying of the cathodic protection anodes. This occurs when a compound dissolves and forms charged ions and water molecules then attach themselves to the ions. This happens more for positively charged metal ions than for negatively charged ions. Therefore, more water is carried by the positive ions towards the negative cathode.

There is proprietary technology for applying electro-osmosis to reinforced concrete. The proprietary part is the use of pulses to reduce the build-up of charged ions at the electrodes and the reinforcement, which would increase the electrical resistance. The system requires cathodes to be installed in areas where the water can be discharged, and anodes in areas where water must be removed.

Electro-osmosis has been described by McInerney *et al.* [41,42]. Several systems have been installed in the UK on a range of structures. The author was the consultant for a trial system with horizontal mixed-metal-oxide-coated titanium wire anodes and cathodes across the leaf piers, as well as cathodes in the ground to draw water away. The aim of the system was to reduce the relative humidity in the concrete to a level that would not sustain corrosion below, say, 60% [43].

The main problem that the author is aware of with such systems is that the anode will become heavily stressed if there is significant water ingress or run-down. Very high current densities can occur on the anodes, which can lead to them failing. It is, therefore, likely that the anodes will need further development before the technology can be considered fully mature for external applications where random wetting events cannot be avoided.

In addition, the pulsing process requires sophisticated electronics, which must be durable over many years. They are likely to be less reliable than simpler transformer/rectifier systems that provide a straight direct current for impressed-current cathodic protection.

Although electro-osmosis systems for drying building materials (e.g. concrete or masonry block basements) are well-known technology, at the time of writing, this technology is in its infancy when used to control corrosion in reinforced concrete structures. The issue of stray-current-induced corrosion of the reinforcement should also be considered. Some proprietary electro-osmosis systems are designed to work with the reinforcing cage. Others are not, and as such, could accelerate reinforcement corrosion.

12.10. Selection of repair techniques to achieve design life and life-cycle costs

Achieving an effective repair requires the balancing of first cost, life-cycle cost, aesthetics, effectiveness of the repair, disruption during repair and the future requirements for the structure.

12.10.1 Costs

The aim of any repair is to be cost-effective for the remaining life of the structure. A cost analysis should include the first cost and the life-cycle cost. Effectiveness encompasses what is feasible, what is aesthetically acceptable and any other constraints, such as access. Such determinations should be carried out by an engineer who is fully up to date with all the technologies being considered and with the relevant standards for their application. For instance, when considering impressed-current cathodic protection, the wide range of anodes available now makes the approach applicable to far more structures than previously, when only expanded mixed-metal-oxide-coated titanium mesh and conductive paint were available.

Obtaining cost data is difficult, particularly up-to-date prices and rates for more specialised techniques. A BRE projects website has comparative price data for 2000 for the UK for most of the techniques described in this chapter [44]. These can be used to generate first cost and life-cycle cost comparisons. The website also gives guidance on the constraints and limitations of the techniques. More up-to-date information on costs for impressed-current cathodic protection for highway structures can be found in Broomfield [45]. However, these costs were collected from UK Highways Agency projects with more complex contractual, access, and health and safety issues than many other projects, so the reported costs may be twice those of, say, a parking structure with easy access.

12.10.2 Treating the causes as well as the symptoms

A well-planned investigation of the structure and the damage to it is critical to the correct choice of repair. There is little guidance from published standards on site investigations, but guidance documents, such as NACE SP0308 [3] and those by the Concrete Society [46] and the Concrete Bridge Development Group [1], cover the techniques to be used and their application. There is also more background in books on the subject, such as that by Broomfield [2].

12.10.3 Planning investigations and conducting repairs using EN 1504

The European standard EN 1504 (Parts 1 to 10) contains much valuable information on concrete repair materials and systems and their application. It is important to understand that this standard was written under the Construction Products Directive, and therefore, it is designed to inform manufacturers of the requirements for their products to achieve CE marking and how to specify the materials properties so that comparable materials that meet the performance requirements for the project can be offered by materials suppliers. However, Parts 9 and 10 give guidance on the range of repair options available, how to specify the materials and how they should be applied on site. A systematic method of using EN 1504 would be as follows.

- EN 1504-9 summarises evaluations, covering both structural and reinforcement corrosion issues. Corrosion evaluations are covered in Section 4.4.7.5 of this book.
- EN 1504-9 summarises the options. These are covered in this chapter. EN 1504-9 discusses
  - the 'do nothing' option, with or without monitoring
  - structural issues
  - prevention or reduction of further deterioration
  - improvement, strengthening or refurbishment of all or part of the structure
  - full or partial reconstruction.
- EN 1504-9 then lists a series of factors to be considered when choosing a rehabilitation option
  - intended use of the structure
  - performance characteristics (e.g. water tightness and fire resistance)
  - repair lifetime
  - opportunities for additional repair and monitoring
  - life-cycle costing throughout the residual life of the structure
  - performance and properties of the existing substrate
  - final appearance.
- There are additional issues, such as
  - health and safety
  - environmental impact
  - appropriateness of the selected option(s)
  - conformity with the standard.
- The concrete rehabilitation materials for non-electrochemical techniques can then be listed, and the required properties selected from Table A-1 in the informative annex to EN 1504-9. This lists tests to achieve the required characteristics (bond strength, shrinkage, elasticity, etc.) from EN 1504, Parts 2 to 7.
- The production of a specification and application conditions and methods are discussed in EN 1504-10. The most useful information is in the informative annexes.
- Finally, EN 1504-9 requires the compilation of a record of the works and an inspection and maintenance procedure.

A basic procedure for conducting a repair is as follows.

- Determine the cause and extent of damage with a quantitative survey using suitable techniques that measure how much repair is needed and what causes of deterioration must be addressed.
- If the cause is structural, then look at strengthening options, such as plate bonding and other reinforcement, structural concrete repairs, partial or full replacement, etc.
- If the cause is concrete deterioration (ASR, sulfate attack, etc.), address the causes and conduct conventional concrete repairs using a first cost and life-cycle cost analysis to identify the most cost-effective solution.
- If the cause is reinforcement corrosion, again we must address the cause of deterioration. The first requirement is to determine whether it is chloride or carbonation or both, and then shortlist options.
- For all repair options, consider the following
  - required remaining life of the structure
  - required life of the treatment or the time to the next intervention
  - possible side effects of treatments
  - aesthetics

  - access issues
  - other constraints
  - available expertise and familiarity with the systems
  - ongoing maintenance and monitoring requirements.
- Once a repair option is selected, produce suitable designs and specifications using the relevant national and international standards.

12.11. Conclusion

There is a wide range of options for repairing and rehabilitating damaged reinforced concrete structures, as detailed in this chapter. No one technique is necessarily superior to the others, although some (such as corrosion inhibitors applied after corrosion initiates) have been less well tried and tested than others.

Selecting the optimum combination of repair techniques requires a well-designed and executed condition survey. This is followed by a proper evaluation of the available options by an engineer fully conversant with the techniques and experienced in their application. Although EN 1504 provides a useful systematic framework for concrete repair, it is no substitute for a knowledgeable and experienced engineer.

The design and specification of repairs are likely to be most cost-effective if the contractor is fully aware of the information available as well as any gaps in that information. The equitable sharing of risk between the contractor and client can be critical to the pricing and success of any concrete repair project.

Whichever repair and corrosion control system is applied, it is important to fully and accurately record the work done and the future maintenance requirements in an operation and maintenance manual or the health and safety file for the structure.

REFERENCES

1. Concrete Bridge Development Group (2002) *Guide to Testing and Monitoring the Durability of Concrete Structures.* Concrete Society, Camberley, UK, Technical guide 2.
2. Broomfield JP (2007) *Corrosion of Steel in Concrete. Understanding, Investigation and Repair*, 2nd edn. Taylor & Francis, London, UK.
3. AMPP (Association of Materials Protection and Performance) (2018) SP0308-2018-SG. Inspection methods for corrosion evaluation of conventionally reinforced concrete structures. AMPP, Houston, TX, USA.
4. Stark D (1991) *Handbook for the Identification of Alkali Silica Reactivity in Highway Structures, SHRP-C-315.* Strategic Highway Research Program, National Research Council, Washington, DC, USA.
5. Kaetzel L, Clifton J, Snyder K and Kleiger P (1994) *Users Guide to the Highway Concrete (HWYCON) Expert System.* National Research Council, Washington, DC, USA, SHRP report and computer program SHRP-C-406.
6. Alldred JC (1993) Quantifying the losses in cover-meter accuracy due to congestion of reinforcement. *Proceedings of the 5th International Conference on Structural Faults and Repair, Edinburgh*, vol. 2, pp. 125–130.
7. Concrete Society (2004) *Electrochemical Tests for Reinforcement Corrosion.* Concrete Society, Camberley, UK, Technical report 60.
8. BRE (Building Research Establishment) (1995) *Carbonation of Concrete and Its Effects on Durability.* BRE, Watford, UK, Digest 405.
9. BRE (2000) *Corrosion of Steel in Concrete. Part 2. Investigation and Assessment.* BRE, Watford, UK, Digest 444.

10. BRE (1998) *Assessing Carbonation Depth in Ageing High Alumina Cement Concrete.* BRE, Watford, UK, Information paper 11/98.
11. Chess P and Grønvold F (1996) *Corrosion Investigation – A Guide to Half Cell Mapping.* Thomas Telford, London, UK.
12. Concrete Society (1997) *Guide to Surface Treatments for Protection and Enhancement of Concrete.* Concrete Society, Camberley, UK, Technical report 50.
13. Kropp J and Hilsdorf HK (eds) (1995) *Performance Criteria for Concrete Durability. RILEM Report 12.* E&FN Spon, London, UK.
14. Applied Petrography Group (2010) *A Code of Practice for the Petrographic Examination of Concrete* (Eden MA (ed.)). Applied Petrography Group, Report SR2. https://www.appliedpetrographygroup.com/reports.htm.
15. Concrete Society (2003) *Strengthening Concrete Structures with Fibre Composite Materials: Acceptance, Inspection and Monitoring.* Concrete Society, Camberley, UK, Technical report 57.
16. ACI (American Concrete Institute) (2010) *Protection of Metals in Concrete Against Corrosion.* ACI, Farmington Hills, MI, USA, Report 222R-01.
17. BRE (2000) *Corrosion of Steel in Concrete. Part 2. Investigation and Assessment.* BRE Digest 444. BRE, Watford, UK.
18. Institution of Civil Engineers (2002) *National Steering Committee for the Inspection of Multistorey Car Parks. Recommendations for the Inspection, Maintenance and Management of Car Park Structures.* Thomas Telford, London, UK.
19. Macdonald S (ed.) (2003) *Concrete – Building Pathology.* Blackwells, London, UK.
20. Concrete Society (2004) *Design Guidance for Strengthening Concrete Structures Using Fibre Composite Materials.* Concrete Society, Camberley, UK, Technical report 55, 2nd edn.
21. Tilly GP (2006) Past performance of concrete repairs. *Concrete Solutions: Proceedings of the 2nd International Conference, St Malo* (Grantham MG, Jauberthie R and Lanos C (eds)). BRE Press, Watford, UK, Paper 2.
22. Concrete Society (1991) *Patch Repair of Reinforced Concrete – Subject to Reinforcement Corrosion: Model Specification and Method of Measurement.* Concrete Society, Camberley, UK, Technical report 38.
23. ACI (2008) *Concrete Repair Manual*, 3rd edn. ACI, Farmington Hills, MI, USA.
24. Sergi G, Seneviratne AMG, Maleki MT, Sadegzadeh M and Page CL (2000) Control of reinforcement corrosion by surface treatment of concrete. *Proceedings of the Institution of Civil Engineers: Structures and Buildings* **140(1)**: 85–100.
25. Manning DG (1995) *Waterproofing Membranes for Concrete Bridge Decks: A Synthesis of Highway Practice.* In *NCHRP Synthesis 220, National Cooperative Highway Research Program.* Transportation Research Board, National Research Council, Washington, DC, USA.
26. Price ARC (1989) *A Field Trial of Waterproofing Systems for Concrete Bridge Decks.* Transport Research Laboratory, Crowthorne, UK, Research report 185.
27. Virmani YP and Clemena GG (1998) *Corrosion Protection – Concrete Bridges.* Federal Highway Administration, Washington, DC, USA, Technical report RD-98-088.
28. Li L, Sagüés AA and Poor N (1999) In-situ leaching investigation of pH and nitrite concentration in concrete pore solution. *Cement and Concrete Research* **29(3)**: 315–321.
29. Raharinaivo A and Malric B (1998) Performance of monofluoro-phosphate for inhibiting corrosion of steel in reinforced concrete structures. *Proceedings of the International Conference on Corrosion and Rehabilitation of Reinforced Concrete Structures, Orlando.*
30. Broomfield JP (2000) Results of long term monitoring of corrosion inhibitors applied to corroding reinforced concrete structures. *Proceedings of Corrosion 2000, Houston*, Paper 791.

31. Treadaway KWJ and Russell AD (1968) The inhibition of the corrosion of steel in concrete. *Highways Public Works* **36(19)**: 40–41.
32. Al-Qadi IL, Prowell BD, Weyers RE, *et al.* (1993) *Concrete Bridge Protection and Rehabilitation: Chemical and Physical Techniques, Corrosion Inhibitors and Polymers.* Strategic Highway Research Program, National Research Council, Washington, DC, USA, Report SHRP-S-666.
33. Prowell BD, Weyers RE and Al-Qadi IL (1993) *Bridge Protection and Rehabilitation: Chemical and Physical Techniques. Field Validation.* Strategic Highway Research Program, National Research Council, Washington, DC, USA, Report SHRP-S-658.
34. Sohangpurwala AA, Islam M and Scannell WT (1997) Performance and long term monitoring of various corrosion protection systems used in reinforced concrete bridge structures in North America. *Proceedings of the International Conference on Repair of Concrete Structures.* Svolvaer Norway Publ, Norwegian Road Research Laboratory, Oslo, Norway, Additional papers: 9 to 19.
35. Sprinkel M and Ozyildirim C (1998) Evaluation of exposure slabs repaired with corrosion inhibitors. *Proceedings of the International Conference on Corrosion and Rehabilitation of Reinforced Concrete Structures, Florida.* Federal Highway Administration, Washington, DC, USA, Report FHWA-SA-99-014.
36. Broomfield JP, Davies K and Hladky K (1999) The use of permanent corrosion monitoring in new and existing reinforced concrete structures. *NACE Corrosion, Houston, TX, USA,* paper NACE-99559.
37. NACE (National Association of Corrosion Engineers) (2007) *Testing of Embeddable Impressed Current Anodes for Use in Cathodic Protection of Atmospherically Exposed Steel-reinforced Concrete.* NACE, Houston, TX, USA, Report TM0294-2007.
38. NACE (2007) *Test Procedures for Organic Based Conductive Coating Anodes for Use on Concrete Structures.* NACE, Houston, TX, USA, Report TM 0105-2005.
39. American Welding Society (2002) *Specification for Thermal Spraying Zinc Anodes on Steel Reinforced Concrete.* American Welding Society, Miami, FL, Report C2.20/C2.20M:2002 (ANSI).
40. NACE (2021) *Electrochemical Realkalization and Chloride Extraction for Reinforced Concrete.* NACE, Houston, TX, USA, Report SP0107-2021.
41. McInerney M, Morefield S, Cooper S, *et al.* (2002) *Electro-osmotic Pulse (EOP) Technology for Control of Water Seepage in Concrete Structures.* US Army Engineer Research and Development Center, Construction Engineering Research Laboratory, Champaign, IL, USA, Report ERDC/CERL TR-02.
42. McInerney MK, Cooper SC, Hock VF and Morefield SW (2004) Measurements of water and ion transport in concrete via electro-osmosis. *Proceedings of Corrosion 2004.* NACE, New Orleans, LA, USA, paper NACE-04351.
43. Lambert P (1997) Controlling moisture. *Construction Repair* **11(2)**: 29–32.
44. Broomfield JP (2006) A web based tool for selecting repair options and life cycle costing of corrosion damaged reinforced concrete structures. *Concrete Solutions: Proceedings of the 2nd International Conference, St. Malo* (Grantham MG, Jauberthie R and Lanos C (eds)). BRE Press, Watford, UK, pp. 505–513.
45. Broomfield JP (2008) *Budget Cost and Anode Performance Information for Impressed Current Cathodic Protection of Reinforced Concrete Highway Bridges.* Corrosion Prevention Association, Aldershot, UK, Technical note 11.
46. Concrete Society (2004) *Electrochemical Tests for Reinforcement Corrosion.* Concrete Society, Camberley, UK, Technical report 60.

Soutsos M
ISBN 978-0-7277-6375-4
https://doi.org/10.1680/icehcd.63754.449

Chapter 13
Non-destructive testing and structural health monitoring

Sreejith Nanukuttan
Queen's University Belfast, UK

Kai Yang
Chongqing University, China

PA Muhammed Basheer
University of Leeds, UK

13.1. Introduction

Reinforced concrete structures, such as bridges, roads, tunnels, dams and buildings, are a major part of civil infrastructure in all parts of the world. Reinforced concrete, as a composite construction material, is distinguished for its very long service life in comparison to other building materials. However, in reality, when concrete structures are subjected to extreme environmental conditions, they tend to deteriorate in rather an alarming rate. Therefore, maintaining the durability of concrete structures during their intended service life is a priority for civil engineers. The deterioration of porous building materials usually involves the movement of aggressive gases or liquids from the surrounding environment into the near surface zone of the porous materials, followed by physical or chemical changes to their internal structure [1,2]. For example, in some extreme environments salt and moisture or carbon dioxide penetrate the concrete cover, eventually causing corrosion of the embedded steel reinforcement. The movement of these aggressive substances occurs due to differentials in humidity, ionic concentration, pressure and temperature within the microstructure of the concrete [3]. The damage caused by deterioration can be either physical or chemical in nature or even both [4,5].

Chemical deterioration can be caused by three main types of external attack. First, there is the action of aggressive ions, such as chlorides, sulphates or other salts, originating from subsurface soil. Second, carbon dioxide from the atmosphere can penetrate the concrete, resulting in the carbonation of the concrete. Third, there is the action of acid rain [6,7]. Physical deterioration can arise due to the thermal expansion of water in concrete during 'freeze–thaw' cycles, stresses in the pore structure caused by crystallisation of salts or fatigue caused by thermal stresses [8,9]. Cracking due to deterioration of the concrete is more likely to result in non-serviceability of the structure and, if unattended, may eventually lead to its collapse [10,11].

The current situation in most developed countries is that the repair and rehabilitation costs of structures far exceed the total budget for capital development programmes. The issue with most structures is not *whether* maintenance is required but *when* and *where* to schedule it most cost-effectively. Therefore, to maintain the serviceability of a concrete structure, it is essential to monitor the performance of the structure continuously throughout its intended service life [12,13]. Figure 13.1 illustrates the usefulness of continuously monitoring

Figure 13.1 Usefulness of continuously monitoring structures

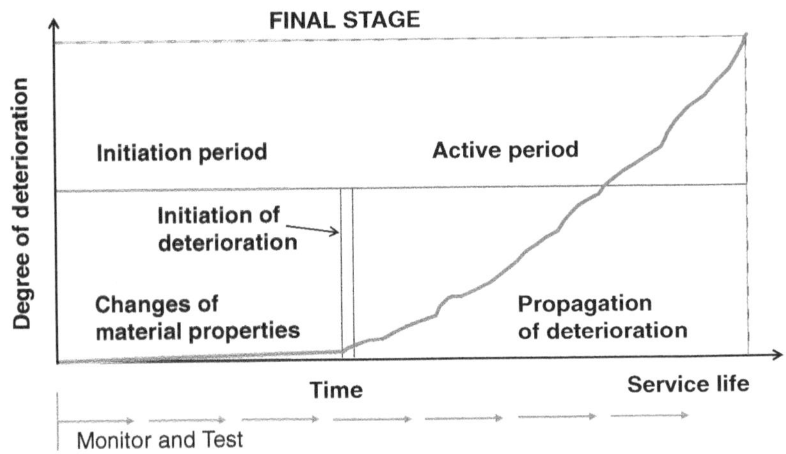

the performance of concrete structures in three different stages during their service life. In the initial stage, the material properties of the concrete are changing. The second stage is crucial, as this is when deterioration initiates. In the third stage, the deterioration propagates.

The deterioration of concrete in a structure depends not only on the quality of the mix ingredients and the manufacturing and placement of the concrete but also on the loading conditions and the surrounding environment to which it is subjected. Earlier design codes for concrete were based on prescriptive approaches, which classified the surrounding environment as mild, moderate, severe or very severe. In this prescriptive approach, the concrete was designed to achieve a particular strength, and this was believed to satisfy durability too. Thus, this mainly relied on the engineering judgement of the specifier. The present approach for the durability design of concrete in EN 206-1 [14] is based on a deterioration-specific prescriptive approach, which classifies the surrounding environment into 18 different exposure classes based on the type of deterioration mechanism and the severity of the exposure environment. In the deterioration-specific prescriptive approach, specifications for minimum binder content, maximum water-to-binder ratio, minimum concrete grade, allowable cement type and cover of concrete are based on the exposure condition and the main types of deterioration observed in the exposure environment. However, in reality, concrete structures are subjected to a combination of deterioration mechanisms and the current approach for the durability design of concrete does not take into account these multiple factors and the rapidly changing environmental conditions surrounding the structure [15,16].

In recent years, a performance-based approach for specifying the durability design of concrete has been popularly used in major construction projects all over the world [16,17]. In this approach, the expected service life of the concrete when exposed to a particular deterioration mechanism is specified quantitatively. The performance of the concrete depends on various factors and processes that are influenced by the quality of the mix ingredients, site practice when manufacturing and placing the concrete, and the surrounding exposure environment, as shown in Figure 13.2.

The conventional approach for assessing the performance of concrete in situ has been to perform mechanical and chemical testing of cores cut from the concrete structure [17]. For example, for corrosion-related deterioration, the concrete can be assessed by measuring the depth of carbonation and the chloride profile of the cover concrete [18,19]. In addition, specialised non-destructive and partially destructive techniques can be periodically used to assess the performance of concrete, depending on the type and intensity of the deterioration

Figure 13.2 Factors influencing the performance of concrete in situ [16]

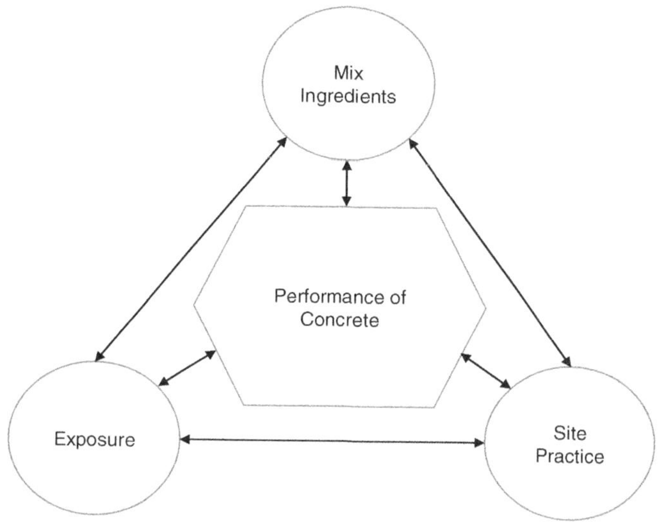

[20,21]. These techniques provide information about the condition of the concrete at the particular time and date of testing, which is useful for many purposes, including deciding on the type of maintenance needed or the type of report to be prepared. However, it can be expensive to test the condition of concrete frequently. A better understanding of the deterioration processes and, hence, the condition of concrete can be obtained by studying the temporal and seasonal changes that take place in the cover zone of the concrete [17,22]. Smart structures, which incorporate in their design sensors and actuating devices that not only can diagnose a problem but can also affect a solution to the problem, are one approach for diagnosing faults and incipient failures and for scheduling maintenance efforts most effectively and at optimum cost, thus minimising the possibility of catastrophic failure. Therefore, using sensors to monitor different parameters associated with the deterioration of concrete can be indispensable for construction projects with a performance-based specification.

13.2. Non-destructive tests for assessing mechanical properties and durability

A visual inspection is normally one of the first steps in any site investigation. Such an inspection can provide a qualified engineer with valuable information as to the cause and extent of deterioration, because almost all deterioration mechanisms eventually lead to cracking of the concrete. The ability to differentiate between different types of cracking is essential for identifying the cause of distress. Thus, the UK Concrete Society has illustrated the different types of non-structural cracks in a hypothetical structure [23], which can be used to identify the cause of distress in a structure. However, even though each form of deterioration has a definite crack pattern, some of the crack forms are similar, and hence, it is often difficult to determine the cause by visual inspection alone. Nonetheless, the results from a visual inspection are essential for selecting further identification tests.

During an investigation, many different types of test can provide information for a condition assessment and for predicting the future performance of the structure [24,25]. The test methods can be classified as follows.

- Destructive tests (DTs): In these conventional methods, samples are extracted from a structure.
- Non-destructive tests (NDTs): These do not impair the intended performance of the element or member under test.
- Partially destructive tests (PDTs): These cause localised damage to a surface zone.

In some tests, the desired property is measured directly whereas others give an estimate from a correlation with another property. The tests chosen depend on a combination of factors, such as access, cost, speed and reliability. This section briefly reviews various test methods and their application.

13.2.1 Assessing strength

It is generally recognised that due to the variability in the concrete mixing, compaction and curing conditions, the actual strength of concrete in a structure is not the same as that in a controlled test cube or cylinder cast at the time of construction and cured and tested in laboratory conditions [26]. Moreover, in some cases, problems with a concrete are not detected until it is too late, as standard testing is generally carried out at 7 and 28 days [24–26]. These limitations of conventional cube and cylinder strength tests provided the impetus to the development of non-destructive methods for assessing the strength of concrete in situ. A wide range of test techniques are available for estimating the in situ strength of concrete [25,27]. These include non-destructive and partially destructive techniques, which measure physical parameters in situ and empirically relate them to the strength properties. Some of the commonly used NDTs and PDTs for assessing the in situ strength of concrete in structures are summarised in Table 13.1 and briefly elaborated below.

Further details of all these tests can be found in Bungey *et al.* [24], Malhotra and Carino [25], Bungey and Soutsos [26] and Breysse *et al.* [27], which discuss at length the advantages and disadvantage of each test method and their practical applications. However, for completeness, the tests mentioned above are briefly described in the following sections.

13.2.1.1 Rebound hammer test

A concrete surface hardness test gives a measure of the surface hardness of concrete (up to about 30 mm deep). The rebound hammer method is easy to use and a high number of readings can be taken at little expense. The compressive strength obtained for well-controlled cubes or cylinders and the corresponding rebound values from the hammer test are reported to have a high degree of correlation under controlled conditions [25,28,29]. Similar empirical relations have also been established with other properties of concrete, such as the flexural strength and modulus of elasticity [25,39]. However, the estimated standard error in this test method varies between 10% and 25% [40]. BS EN 12504-2 [28] and ASTM C805 [29] provide detailed guidelines for the rebound hammer test, although both of these suggest that the rebound method should

Table 13.1 NDTs and PDTs for assessing the strength of concrete in structures

Test method	Intended purpose	Classification	Reference
Rebound hammer test	Surface hardness of concrete	NDT	BS EN 12504-2 [28] ASTM C805 [29]
Ultrasonic pulse velocity test	Non-uniformity and density of concrete	NDT	BS EN 12504-4 [30] ASTM C597 [31]
Pull-out tests	Pull-out strength of concrete cover	PDT	BS EN 12504-3 [32] ASTM C900 [33]
Pull-off test	Pull-off surface tensile strength of concrete	PDT	BS EN 1542 [34] ASTM C1583 [35]
Probe penetration resistance test	Probe penetration resistance of concrete cover	PDT	BS 1881-207 [36] ASTM C803 [37]
Maturity measurements	Strength development	NDT	ASTM C1074 [38]

not be considered as a substitute for other ways to determine the compressive strength of concrete, as it is useful only as a preliminary or complementary method [24,26]. This technique can be used to assess the uniformity of concrete quality, to compare a concrete with a reference statistically and to indicate changes in the characteristics of the concrete with time. The factors that influence the results of a surface hardness test are type of surface, age and compaction of concrete, surface moisture conditions, type and content of cement, type of aggregate and surface carbonation of concrete. Note that areas with honeycombing, scaling, an irregular surface or high porosity must be avoided.

13.2.1.2 Ultrasonic pulse velocity test

In the ultrasonic pulse velocity method, the time for a pulse of vibrations to travel at ultrasonic frequency through concrete between two transducers is measured [30,41]. Knowing the direct path length between the transducers, the pulse velocity through the concrete can be obtained. The velocity of ultrasonic sound in concrete is related to its quality, uniformity and strength. This method of testing concrete quality has become widely accepted all over the world due to its speed, ease of use and its non-destructive nature. Recommendations for this method are given in BS EN 12504-4 [30] and ASTM C597 [31]. Although a good correlation has been established between the ultrasonic pulse velocity and the strength of concrete, the guidelines in international standards suggest that when estimating the compressive strength of concrete, similar correlations must already have been established for the type of concrete under investigation [25]. A typical classification of concrete quality using ultrasonic pulse velocity measurements is given in Table 13.2 [42]. Although this is based on data from the 1950s, more recent work [43] suggested that the classification ranges have not changed much since then. The relation between ultrasonic pulse velocity and strength of concrete depends on the type of aggregate, cement used, water-to-cement ratio and mix proportions [24]. For a particular mix, the relation depends on age of concrete, curing conditions and the local moisture content.

The ultrasonic pulse velocity test is widely used in non-destructive evaluations of concrete strength, uniformity of concrete, depth of surface cracks and honeycombing [30,44,45]. The test method is also used for other applications, such as measuring the layer thickness, elastic modulus of concrete, heat of hydration and deterioration of the concrete [45–47]. The factors that affect the pulse velocity measured in a concrete structure are the location of the steel reinforcement, cracks and voids in the concrete and surface moisture. Any steel reinforcement along the path of the ultrasonic wave is a particular problem because the pulse velocity through steel is approximately 40% greater than that through concrete. Failure to take account of these factors can lead to erroneous estimates of the concrete strength.

13.2.1.3 Pull-out test

The compressive strength of concrete is related to the maximum tensile load applied to an insert embedded in the concrete [24–25]. The compressive strength of concrete can, thus, be estimated by obtaining

Table 13.2 Quality of concrete as a function of ultrasonic pulse velocity [42]

Ultrasonic pulse velocity: m/s	Concrete quality
>4500	Excellent
3500–4500	Good
3000–3500	Doubtful
2000–3000	Poor
<2000	Very poor

an empirical relation between the pull-out tensile load and the compressive strength of the concrete under investigation. There are two basic ways of carrying out a pull-out test, one in which a metal insert is cast into the fresh concrete at the time of construction and the other where the metal insert is fixed into a hole drilled in the hardened concrete. Thus, the former is used for assessing concrete strength in a new construction and the latter is used for existing structures. Guidelines with requirements for and the use of a pull-out test and correlations with compressive strength are given in BS EN 12504-3 [32] and ASTM C900 [33].

Unlike the rebound hammer and ultrasonic pulse velocity tests, a pull-out test subjects the concrete to a static load and produces a well-defined fracture surface in the concrete. Thus, it measures the static strength of the concrete. The main limitation of the cast-in approach is that the position and placement of formwork inserts need to be pre-planned to ensure that the reinforcement does not interfere with the conical failure zone. Similarly, when using a drilled hole, it is important to locate the reinforcement bars with a sensitive cover meter scan. Malhotra and Carino [25] have stated that a pull-out test is influenced by the ratio of mortar strength to coarse aggregate strength, the type of coarse aggregate and the maximum aggregate size. Another limitation of the pull-out technique is that, for aesthetic reasons, it is necessary to repair the damage done to the concrete surface.

13.2.1.4 Pull-off test

A pull-off test measures the in situ tensile strength of cover concrete by applying a direct tensile load through a disc bonded onto the surface. The compressive strength is estimated from the pull-off strength using empirical correlations [34]. This test method is commonly used these days for applications such as measuring the bond strength of concrete patch repairs, assessing the effects of curing and carbonation of concrete, and determining the effect of microcracking [24]. A pull-off test is carried out using two approaches, one with the metal disc directly attached to the concrete surface where the strength of the concrete is measured and the other where the concrete is cored to a depth to eliminate the influence of a carbonated layer of concrete and to measure strength at that depth. Partial cores that extend beyond the interface between a patch repair and the substrate concrete are quite extensively used for testing the bond strength of such repairs. Guidelines for this test method are available in BS EN 1542 [34] and ASTM C1583 [35]. The main advantage of a pull-off test is that it is simple and quick to perform and no pre-planning is required to avoid the reinforcement. The entire process of preparing the surface and applying the metal disc should take no more than 15 minutes. Another advantage is that the damage to the concrete surface after a test is not unduly severe [35]. The main factor that influences the pull-off strength is the type and size of the coarse aggregate in the concrete. The main limitation of this method is the rather long curing time required for most adhesives used to fix the disc onto the concrete surface. However, with recent advances in adhesives developed for the construction industry, this particular limitation is now becoming less of a problem.

13.2.1.5 Penetration resistance test

A penetration resistance test measures the depth of penetration of a probe or pin forced into hardened concrete with a specially designed gun. This depth indicates the hardness or penetration resistance of the surface concrete, which can be related to its compressive strength [25]. However, the test method is not recommended for concrete with a compressive strength above 28 MPa [48]. The main advantage of a penetration test is that the probe travels through both mortar and aggregate, so that both materials affect the penetration distance. Thus, the test can estimate the combined strength of the mortar and the aggregate. Although this test is not usually influenced by local conditions, such as texture and moisture content, it is significantly affected by the type and size of the coarse aggregate and the carbonation depth [48]. Another limitation is that the in situ concrete test member needs to be reasonably thick to withstand the large amount of energy transmitted into it during a test. It is also important to ensure compliance with the requirements of ASTM C803 [37] in relation to the use of powder-actuated driving units.

13.2.1.6 Maturity measurements

The in-place strength of concrete is estimated by measuring the effects of temperature and time (maturity) on the strength development of the concrete [38,49–50]. By establishing a relation between strength and maturity, the strength of a particular concrete mix can be expressed as a function of a maturity index [38,48]. There are two main commercial techniques for measuring the concrete maturity index in situ. One uses a disposable maturity meter, and the other uses an electronic multichannel logging device connected to embedded thermocouples or thermistors [25,49,51]. The main advantage of the maturity method is that it is a simple NDT that takes account of the temperature history within the concrete during hydration, provided sufficient moisture is present. It is particularly valuable for when construction occurs during adverse weather conditions, as knowledge of the strength development is then important. The main limitation of this approach is that the relation between strength and maturity is very specific to the particular concrete mix and curing regime. Thus, considerable expense is involved in producing unique relations for all possible mixes. Despite this limitation, the principal application of the maturity method is for monitoring the in situ strength development in relation to stripping the formwork, removing props and applying loads, including prestressing. The maturity method is particularly useful if it is used in conjunction with another NDT, for example, ultrasonic pulse velocity or surface hardness tests, because the different testing approaches complement each other very well [51–53].

13.2.1.7 Comparison and selection of test methods

The confidence with which in an situ strength assessment can be made depends on the degree of correlation with cube strength, reproducibility of test method and number of assessments made at a particular location. The effective correlation between the strength estimated by a particular test method and the standard cube or cylinder strength or the compressive strength of cores taken from a structure increases with the number of samples obtained from a particular location. Therefore, the minimum number of replicate samples recommended for each test when calculating an average strength is indicated in Table 13.3. Further, the strength ranges for the different test methods are also compared. To increase the accuracy, a combination of test techniques is proposed. For example, combinations such as a pulse velocity test and a rebound hammer test or a maturity measurement with a pull-out or pull-off test can provide a better estimate of in situ strength than when applied individually [48].

Most non-destructive methods of assessing strength have limitations, and hence, the accuracy of the strength estimates varies for each test method. It is important to understand the relative advantage and limitations of each of the test methods when estimating strength. Tables 13.4 and 13.5 briefly compare the performance, merits and limitations of various test methods applicable for both new and existing construction. This comparison is intended to help in selecting an appropriate test method or combination of methods based on the test location, effect of damage and test accuracy required.

Table 13.3 Compressive strength ranges for in situ strength test methods [48]

Test method	Range of compressive strengths: MPa	Recommended number of valid tests at a location
Rebound hammer	10–40	10
Probe penetration	10–120	3–6
Pin penetration	3–30	
Pull-out	2–130	3
Ultrasonic pulse velocity	1–70	5
Break-off or pull-off	3–50	5
Maturity	No limit	–

Table 13.4 Performance of in situ strength tests [54]

Test	Accuracy of strength estimation	Speed of test	Ease of test	Cost of testing	Damage to structure
Core	High	Slow	Moderate	Moderate	Local
Ultrasonic pulse velocity	Moderate to high	Rapid	Easy	Low	None
Surface hardness	Low	Rapid	Easy	Low	None

Table 13.5 Comparison of performance of in situ strength tests [48]

Test method	ASTM Standard	BS standard	Reliability (when estimating strength)		95% confidence limit of strength estimate in situ (BS 6089)	Ease of use
			New construction	Existing construction		
Rebound hammer	C805	12504-2	Low	Low	±20% (concrete not older than 3 months)	Easy
Penetration resistance	C803/C803M	1881-207	Low	Low	±20%	Easy
Pull-out	C900	12504-3	Moderate	Moderate	±20%	Moderate
Break-off or pull-off	C1150	1881-207	Moderate	Moderate	±20% (±15% in laboratory conditions)	Moderate
Ultrasonic pulse velocity	C597	12504-4	Moderate	N/A	±20%	Easy
Maturity	C1074	–	High	N/A	–	N/A

13.2.2 Methods for measuring permeability

The following are the main transport processes by which aggressive substances move within concrete.

- Adsorption: Molecules adhere to internal surfaces within the concrete either because of the physical forces of adhesion or because of chemical bonds.
- Absorption: Concrete takes in liquid by capillary suction, which fills the pore spaces. The rate at which liquid enters the pores is termed the sorptivity.
- Diffusion: A gas or ions penetrate the concrete under the action of a molar concentration gradient. The rate of diffusion is generally defined by a diffusion coefficient or a diffusivity value.
- Permeation: A fluid (liquid or gas) penetrates concrete under the action of a pressure gradient. The rate of transport is normally expressed by a coefficient of permeability.

By measuring the permeability of concrete, namely adsorption, absorption, diffusion and permeation, the quality of concrete against the ingress of harmful substances can be assessed [13,55,56]. As diffusion is a slow process and determining the diffusion coefficient could take several months, the flow of ions through concrete can be accelerated by applying a potential. Such tests are called migration tests [2,57]. The diffusivity of concrete determined by a migration test is the same as that obtained from a laboratory-based diffusion test [58–61]. Therefore, along with tests that assess the permeability of concrete, migration tests that can be performed on site are also included in this section. Table 13.6 summarises the permeability tests. Those tests that can be used on site are discussed further in this section.

Table 13.6 Permeability tests

Test method	Intended purpose	Classification	References
Initial surface absorption test	Absorption of the concrete cover	NDT	BS 1881-208 [62]
Figg water permeability test	Absorption in a hole drilled in the concrete cover	PDT	Figg [63]
Figg air permeability test	Air permeability in a hole drilled in the concrete cover	PDT	Figg [63]
Torrent test	Air permeability of the concrete cover	NDT	Torrent [64] Romer [65]
Autoclam air permeability test	Air permeability of the concrete cover	NDT	Basheer [13,66]
Autoclam sorptivity test	Absorption by the concrete cover	NDT	Basheer [13,66]
Autoclam water permeability test	Water permeability of the concrete cover	NDT	Basheer [13,66]
Gas permeability test using a permeability cell	Permeability of concrete	DT	Basheer [13]
Water permeability test using a permeability cell	Permeability of concrete	DT	Basheer [13]
Water absorption by immersion	Water absorption capacity of concrete	DT	Basheer [13]
BS water absorption	Short-term water absorption of concrete	DT	BS 1881-122 [67] Concrete Society [68]
DIN 1048 absorption test	Water tightness of concrete	DT	Concrete Society [68]
Capillary absorption (sorptivity) tests	Rate of water absorption of concrete	DT	Basheer [13]
Rapid chloride-permeability test	Ionic conductivity of concrete	DT	AASHTO T-277 [69]
PERMIT ion-migration test	Ionic transport of concrete	NDT	Basheer [13] Nanukuttan *et al.* [61]
Ion-migration or ion-diffusion test using diffusion cells	Ionic diffusion resistance of concrete	DT	ASTM C1202 [70]
Water vapour diffusion test	Water vapour transport of concrete	DT	Basheer [13]

13.2.2.1 Absorption tests

INITIAL SURFACE ABSORPTION TEST (ISAT)

In this test, the rate of penetration of water into a defined area of concrete (5000 mm^2) after a stated interval from the start of the test (10, 30 or 60 min) at a constant water head (200 mm height) and temperature (20°C) is measured [62]. There are specifications for preparing the test area, initial moisture condition, surface area to be in contact with water and the time at which measurements must be taken. Guidelines for the test method, calibration and sample preparation are available in BS 1881-208 [62]. Typical ISAT results reported by Concrete Society [68] are given in Table 13.7. The main advantage of this test is that it is a quick and simple in situ NDT for measuring the rate of water penetration into a concrete surface. It provides reasonably consistent results with oven-dried samples. The limitations of this test include the difficulty of ensuring a watertight seal in site conditions, the influence of the moisture condition of the concrete on the measured property and an insufficient water head.

AUTOCLAM SORPTIVITY TEST

The Autoclam Permeability System is used for the sorptivity and water and air permeability tests, giving an index for each of these properties [71]. The Autoclam sorptivity test [13,66] measures the cumulative inflow of water in the first 15 minutes from a water source 50 mm in diameter (i.e. from a base ring of internal diameter 50 mm) at an applied pressure of 0.02 bars (nearly 200 mm water head). The cumulative volume of water plotted against the square root of time gives a linear relation. The slope of the curve is the sorptivity index. For less permeable surfaces where the flow of water is low, the contact area of the water source can be increased to obtain measurable flow rates. However, the data measured must be then normalised to the standard 50-mm-diameter water source before calculating the sorptivity index. Table 13.8 lists typical values of the Autoclam sorptivity index for different levels of the protective quality of the concrete. The moisture content of the concrete surface can influence the Autoclam sorptivity index. Basheer and Nolan [3] proposed that the test should be carried out when the internal relative humidity (RH) of the concrete at 10 mm depth

Table 13.7 Typical results of an initial surface absorption test [62]

Concrete absorption	ISAT results: ml/(m^2 s)		
	Time after starting test		
	10 min	30 min	1 hour
High	>0.50	>0.35	>0.20
Average	0.25–0.50	0.17–0.35	0.10–0.20
Low	<0.25	<0.17	<0.10

Table 13.8 Performance of concrete based on the Autoclam sorptivity index [13]

Protective quality	Clam sorptivity index: m^3 × 10^{-7}/√min
Very good	≤1.30
Good	>1.30 and ≤2.60
Poor	>2.60 and ≤3.40
Very poor	>3.40

is less than 80% to eliminate this effect. The equipment is portable and easy to use on site. Fixing problems identified for the ISAT have been eliminated for the Autoclam sorptivity test.

FIGG WATER PERMEABILITY (ABSORPTION) TEST

This test involves drilling a hole 10 mm in diameter in the near surface layer of concrete to a depth of 40 mm [63,68]. After thorough cleaning, the top 20 mm of the hole is plugged with a silicon rubber plug, leaving a cavity 20 mm deep for the test. Water is admitted to this cavity at a low pressure head of 100 mm and allowed to absorb by capillarity. The time required for a certain volume of water to be absorbed is measured in seconds as an absorption index. Thus, a higher absorption index corresponds to better quality concrete. The relation between the concrete protective quality and the Figg water permeability (sorptivity) index is shown in Table 13.9. The main advantage of this test is that it is simple and relatively inexpensive compared to other tests [68]. The main disadvantage is that drilling, even at slow speed, can introduce microcracks, which may defeat the purpose of the test by altering the flow mechanism.

13.2.2.2 Permeability tests

AUTOCLAM WATER AND AIR PERMEABILITY TESTS

The Autoclam water permeability test has a procedure like that used for the Autoclam sorptivity test [66]. The main difference is the test pressure used, namely a pressure of 1.5 bar is used for the Autoclam water permeability test. The inflow of water through a test area 50 mm in diameter through a surface-mounted ring is measured at this pressure for 15 minutes. The Autoclam water permeability index ($m^3/\sqrt{min}$) is the slope of a linear plot of the cumulative inflow against the square root of time. As the inflow is measured in this test, to obtain meaningful data, the test location must be dry. The protective quality of concrete based on the Autoclam water permeability index is reported in Table 13.10.

Table 13.9 Relation between concrete protective quality and Figg water permeability (sorptivity) index [68]

Protective quality of concrete	Water permeability index	
	Measured time: s	Equivalent water absorption rate: s/ml × 10^3
Poor	<20	<2
Not very good	20–50	2–5
Fair	50–100	5–10
Good	100–500	10–50
Excellent	>500	>50

Table 13.10 Protective quality based on Autoclam water and air permeability indexes [3]

Permeability	Protective quality			
	Very good	Good	Poor	Very poor
Autoclam water permeability index: $m^3 \times 10^{-7}/\sqrt{min}$	≤3.70	>3.70 and ≤9.40	>9.40 and ≤13.8	>13.80
Autoclam air permeability index: ln(pressure)/min	≤0.10	>0.10 and ≤0.50	>0.50 and ≤0.90	>0.90

In this test, the pressure decay from 1.5 bar is measured in a test reservoir mounted on the surface of concrete over a period of 15 min. The natural logarithm of pressure is plotted against time to give a straight line graph. The slope of this graph is the Autoclam air permeability index. Although moisture influences the test results, Basheer and Nolan [3,72] have shown that the quality of concrete can be classified in terms of the Autoclam air permeability index if measurements are taken when the internal RH in a cavity in the concrete at a depth of 10 mm from the surface is less than 80%. The main advantages of this test are that it is portable, quick and simple to perform. However, like the Autoclam sorptivity test, this test provides air and water permeability indexes rather than a coefficient of permeability.

FIGG AIR PERMEABILITY TEST

This test uses a poroscope. The test set-up is like that used for the Figg water test, but instead of admitting water into the cavity, a pump is used to reduce the pressure in the cavity to 55 kPa below atmospheric pressure [63,68]. As air flows into the cavity, the pressure increases and the time taken for the pressure to increase by 5 kPa, that is to a cavity pressure of −50 kPa, is recorded using a stopwatch, and this is reported as the air permeability index. Like the Figg water test, a higher air permeability index corresponds to a less permeable concrete. Typical values of the Figg air permeability index for different levels of the protective quality of concrete are listed in Table 13.11 [68]. The advantages and disadvantages of the Figg water absorption test also apply to the Figg air permeability test.

TORRENT TEST

This test is very similar to the Schönlin air permeability test [73]. It uses a surface-mounted reservoir to apply a vacuum to the test surface [64,65]. Although a Torrent test is like a Figg air permeability test, no damage is caused because a hole is not drilled into the concrete. Air flows in one direction from the annular reservoir into a central test reservoir. Using this assumption of unidirectional flow of air, a permeability coefficient is derived from measurements taken in this test [74]. It is claimed that the electrical resistivity of concrete can be used to adjust the readings from the Torrent test to take account of the moisture variations in concrete [75].

There are several other NDT methods for evaluating the air and water permeability of concrete, details of which can be found in the published literature [68,76].

13.2.2.3 Migration tests

RAPID CHLORIDE-PERMEABILITY TEST

Whiting [77] developed an in situ rapid chloride-permeability test based on the same principle as the laboratory test (ASTM C1202 [70]), in which the charge passed during the first 6 hours after the application

Table 13.11 Relation between concrete protective quality and Figg air permeability index [68]

Protective quality of concrete	Air permeability index	
	Measured time: s	Equivalent air exclusion rating: s/ml
Poor	<30	<8
Not very good	30–100	8–25
Fair	100–300	25–75
Good	300–1000	75–250
Excellent	>1000	>250

of a potential across the specimen is used to characterise the concrete in terms of chloride penetration. A detailed description of the test instrument is available elsewhere [57,77]. Whiting suggested that the in situ test should be carried out at a potential difference of 80 V for 6 hours. The total charge passed is used as an index to characterise concrete in terms of chloride diffusivity. The disadvantages of this test are the same as those for the laboratory version.

- The test result is influenced by the conductivity of the pore solution, which can be highly variable on site.
- The temperature rise due to the high voltage (80 V) will significantly affect the charge passed during the test.
- The depth of concrete cover affects the test results. For example, if a cover of 50 mm is assumed, results may vary by as much as 25% if the actual cover varies by 25 mm from this value.

In addition, unlike the laboratory test specified in ASTM C1202 [70], it can be difficult to saturate the test area prior to carrying out the rapid chloride-permeability test on site. Although Whiting proposed a vacuum saturation technique, this was not entirely satisfactory for yielding reliable and repeatable results [78].

PERMIT ION-MIGRATION TEST

This is a unique NDT. It can determine the chloride migration coefficient of the cover concrete on site. A detailed description of the instrument, test technique and preparation required for the test area are available elsewhere [61]. Extensive use of the PERMIT ion-migration test on different concrete samples has indicated that the in situ migration coefficient derived by the test correlates well with the conventional laboratory-based steady-state diffusion and migration coefficients. The main advantage of this test is that it can provide a migration coefficient without cores having to be removed from the structure. The main disadvantage of the test is that it introduces chloride ions into the test surface. However, these ions can be effectively removed by reversing the test voltage, which may take up to 4 days. Note that the test area will be slightly stained by deposition of the ferrous by-products of the electrochemical reaction. However, the extent of staining can be reduced if this deposit is washed off immediately after the test.

13.2.2.4 Comparison and selection of test methods

The selection of an appropriate permeation test (sometimes the term penetrability test is also used) for assessing the condition of concrete in structures basically depends on the associated transport mechanism that will cause the deterioration. However, most of the test methods have common limitations, which it is essential to overcome by understanding the factors causing them. These common limitations include sensitivity to the moisture and temperature conditions of concrete, changes in the transport mechanism during the test and the influence of drilling on the values measured.

A detailed experimental study to compare different NDTs for measuring the penetrability characteristics of cover concrete and thereby detect changes in the quality of concrete was carried out by RILEM committee 189 [79]. The performance of the methods was compared for seven different concrete types. The results are summarised in Table 13.12. For five or six out of the seven cases, the tests could detect the expected variations in the penetrability of the concrete at a significant level.

13.2.3 Measuring corrosion of steel in concrete and corrosion-related concrete properties

By the time visible evidence of corrosion damage to the steel reinforcement is seen on the surface in the form of cracks and rust staining, remedial action can be very costly and often ineffective [80]. Therefore, frequent assessments of the condition of structures are essential for ensuring their serviceability, for keeping maintenance costs low and for operational safety. Different non-destructive methods used to assess the corrosion of steel in concrete as well as tests that measure corrosion-related properties of concrete are summarised in Table 13.13. Some of the most commonly used methods are briefly described in this section.

Table 13.12 Performance of the different test methods [79]

Parameter	Autoclam air permeability	Torrent air permeability	Autoclam sorptivity
Significant discrimination	5 out of 7	6 out of 7	6 out of 7
Correlation coefficient *R*[a]	0.67	0.97	0.47
Measurements per test condition	3	6	3
Duration per test condition: min	69	99	69
Visual impact: No. holes × diameter (mm)	9 × 6	0	9 × 6

[a]With a reference test for the same transport mechanism

Table 13.13 Tests for assessing corrosion of reinforcement and corrosion-related properties of concrete

Test method	Property measured	Test type	Reference
Electromagnetic cover meter	Cover to reinforcement, rebar diameter and bar clusters	NDT	BS 1881-201 [81] BSI 1881-204 [82]
Half-cell potential test	Corrosion potential (probability of corrosion)	NDT	ASTM C876 [83]
Wenner four-probe resistivity meter	Resistivity of concrete	NDT	BS 1881-201 [81]
Linear polarisation resistance (LPR) test	Rate of corrosion of the rebar	NDT	ASTM G59 [84]
Galvanostatic pulse measurement	Rate of corrosion of the rebar	NDT	Law *et al.* [85]
AC impedance analysis	Rate of corrosion of the rebar	NDT	Millard [86]
AC harmonic analysis	Rate of corrosion of the rebar	NDT	Bungey *et al.* [24] Malhotra and Carino [25]

13.2.3.1 Electromagnetic cover measurement

A cover meter survey is an NDT for measuring the depth of cover to the reinforcement in concrete structures [82]. A cover meter electromagnetically locates embedded steel by measuring the intensity of the magnetic field produced by it. This intensity can be correlated to a specific depth from the concrete surface, thus giving an estimate of the cover. The accuracy of cover meters in measuring the cover depth under normal site conditions is within ±5 mm [82]. However, for heavily reinforced members, because of the effect of the secondary reinforcement, it is not possible to measure the depth of cover accurately. Modern cover meters can locate the main and secondary reinforcement and determine the bar sizes and spacings [24]. The output of a measurement is in the form of a scan. Apart from identifying the extent of damage in low cover regions, cover meters are also used when coring, drilling or carrying out patch repairs to determine the exact location of the reinforcement. The factors influencing a cover meter test include bar diameter, spacing of bars, aggregate with magnetic properties and other electromagnetic interference on site. BS 1881-204 [82] gives recommendations and describes the principles of operation of the electromagnetic devices that are used for estimating the position, depth and size of reinforcement buried in concrete.

13.2.3.2 Half-cell potential test

The electrical half-cell potential of reinforcing steel in concrete is measured to determine any possible corrosion of the steel [87]. The corrosion potential of steel is measured against a reference electrode consisting of a metal in an electrolyte. Commonly used reference half-cells include copper in copper sulphate and silver in silver chloride among others [83]. The criteria for assessing the corrosion condition for different reference electrodes as per ASTM C876 are presented in Table 13.14. There is no equivalent British standard, but BS 1881-201 [81] gives a brief description of the limitations and applications of this method. Half-cell potential measurements are often carried out when reinforcement corrosion is suspected or evident in a structure. This low-cost test is widely used to provide iso-potential contour maps from which it is easy to identify zones of corrosion risk [88]. However, note that the potentials obtained depend on the presence of moisture. Therefore, to account for seasonal variations, consider calculating an average of many readings taken during different seasons. The test sometimes gives misleading results for re-alkalised concretes and concretes treated with organic or inorganic inhibitors [88].

13.2.3.3 Electrical resistivity method

As the corrosion of steel in concrete is an electrochemical process, measuring the electrical resistance of concrete indicates the rate at which corrosion reaction may proceed [89]. A two-point or four-point probe is often used to measure the resistivity of concrete [24,86]. The criteria for assessing the risk of corrosion using the electrical resistivity method are shown in Table 13.15. To assess the corrosion activity better, in situ resistivity measurements are often carried out along with half-cell potential surveys. The resistance of

Table 13.14 Corrosion condition from the ASTM specification for steel in concrete as measured by different standard electrodes [83]

Copper/copper sulphate: mV	Silver/silver chloride/1.0M KCl: mV	Standard hydrogen electrode: mV	Calomel electrode: mV	Corrosion condition
> −200	> −100	+120	> −80	Low (10% risk of corrosion)
−200 to −350	−100 to −250	+120 to −30	−80 to −230	Intermediate corrosion risk
< −350	< −250	−30	< −230	High (>90% risk of corrosion)
< −500	< −400	−180	< −380	Severe corrosion

Table 13.15 Empirical relation between the resistivity of concrete and the likelihood of significant corrosion [24]

Resistivity: Ω m	Likelihood of significant corrosion when steel activated (for partially saturated concrete)
>200	Low
100 to 200	Low to moderate
50 to 100	High
<50	Very high

concrete is governed by factors such as the geometry between the measuring electrodes, porosity and pore connectivity of the cover concrete, and ionic mobility in the pore system [90]. Similarly, electrical resistivity measurements are influenced by moisture, temperature, reinforcement close to the surface, maximum size of the aggregates and carbonation of concrete.

13.2.3.4 Measuring the rate of corrosion

There are several methods of measuring the true instantaneous rate of corrosion. These methods estimate the rate of corrosion using electrochemical theory. One of the most commonly used methods to determine the rate of corrosion of the steel reinforcement in concrete is the LPR test. This method does not suffer from the major drawbacks of other corrosion assessment methods as it determines the rate of corrosion directly. In LPR, a small electric current is applied to the concrete and the resulting shift in corrosion potential is measured, which is an indicator of the active corrosion in the steel. A guard ring system has been developed as an NDT of the rate of corrosion. It measures corrosion locally in a defined area. A brief overview of this technique, advantages and limitations are given in the COST Action 521 report [90]. The criteria for interpreting the damage due to corrosion using corrosion rate measurements are listed in Table 13.16. The seasonal variation in temperature and moisture conditions influences measurements of the corrosion rate.

13.2.3.5 Comparison and selection of test methods

Each of the test methods discussed above has a specific accuracy, advantages and limitations when measuring the corrosion of steel in reinforced concrete structures. Table 13.17 summarises and compares the advantages and limitations of NDT methods used for assessing steel in concrete.

Comparative tests on commercial cover meters indicate that all the instruments can measure the cover depth with 100% success at an accuracy of ±10%, provided the nominal diameter is known and the cover depth is less than 40 mm [91]. Resistivity measurements provide a good indication of the quality and density of the cover concrete, whereas half-cell potential measurements provide information about the likelihood of corrosion [88]. Neither method directly measures the corrosion rate. Although LPR techniques directly measure the corrosion rate, interpreting the data is complex and a good understanding is needed of the factors influencing the measurements. Generally, to make full use of the data from all these test methods, a deep understanding of the instrument as well as how to interpret the data in terms of the damage to the structures are essential. It is also important to understand the influence of environmental factors on the measurements as well as methods to mitigate them.

Using a combination of tests techniques can improve the accuracy and reliability of a corrosion assessment of a structure. The preliminary method of assessing corrosion activity is chosen by considering the practical

Table 13.16 Typical interpretation of corrosion rate measurements [90]

Corrosion rate: $\mu A/cm^2$	Attack penetration (loss of section of reinforcing bar): μm/year	Corrosion level
<0.1	<1	Negligible
0.1 to 0.5	1 to 5	Low
0.5 to 1	5 to 10	Moderate
>1	>10	High

Table 13.17 Advantages and limitations of NDT methods for assessment of steel [87]

Method	Advantages	Limitations
Cover meter	• Truly non-destructive technique • Lightweight, portable and easy to use • Cover depth can be estimated on site with an accuracy of ± 5 mm • Can locate reinforcing bars and estimate bar sizes and depth of cover	• Calibrations are sensitive to steel type, bar diameter and aggregate type • Accuracy of estimated cover depth is affected if the bar size is less than 10 mm or greater than 32 mm • Ability to identify individual bars is affected by probe orientation, cover depth, spacing of bars, secondary reinforcement and lap joints
Resistivity	• Simple, lightweight and easy-to-use NDT • Can be used to predict chances of reinforcement corrosion	• Correlation with the properties of the concrete depends on individual mix proportions • Influenced by moisture, salt content, temperature and water-to-cement ratio • Practical applications are restricted to comparative measurements
Half-cell potential	• NDT • Portable and lightweight equipment • Provides iso-potential contour maps indicating zones of varying degrees of corrosion risk at time of testing	• Method cannot indicate actual rate of corrosion • Electrical contact with the reinforcement must be established by drilling a small hole • Concrete must be moist • Not applicable for epoxy-coated bars • Requires an experienced operator to run the test and interpret the results
LPR	• Direct measurement of corrosion rate of steel in concrete • Lightweight and portable • Guard ring probe measures the localised corrosion rate	• Measurement is slow compared with other methods • Electrical contact with the reinforcement must be established by drilling a small hole • Cover depth must be less than 100 mm • Measurements are influenced by seasonal variations in temperature and moisture • Concrete surface must be smooth, uncracked and free from coatings • Requires an experienced operator to run the test and interpret the results

circumstances and limitations of each method. In most cases, a corrosion assessment begins with a cover meter survey, which gives a good indication of the position of the reinforcement and areas with low cover. A combination of the half-cell potential and resistivity methods represent the most effective non-destructive approach for assessing corrosion damage. However, a chemical analysis, for example of chloride content or carbonation depth, should accompany such a structural assessment.

13.2.4 Evaluating integrity

An evaluation of integrity generally deals with detecting and identifying the location of hidden cracks, voids, honeycombing and delamination in concrete structures [92,93]. These hidden defects can affect the structural performance and durability of structures. Hammer tapping and chain dragging surveys are popular and inexpensive manual methods for identifying areas where the concrete has delaminated. More sophisticated methods for the non-destructive assessment of flaws in concrete include infrared thermography, impact-echo method and radar techniques (Table 13.18). The most commonly used of these methods are described in this section.

13.2.4.1 Manual methods

There is a range of manual methods for identifying areas of delamination in concrete [24]. These methods work on the principle that when a delaminated concrete layer is struck with a metal, it produces a specific sound that indicates the presence of a hollow cavity behind the surface of the concrete cover. Thus, hand tools, such as hammers, steel rods and chain drags, are used to assess a structure for possible delamination. Steel chain drags are normally used for large horizontal areas of concrete, such as bridge decks. Although interpreting the sound produced cannot be quantified and is based on the subjective experience of the operator, this method is widely applied because it is fast, easy to use and less expensive than other sophisticated techniques. Manual tests with a hammer or steel chain are often carried out during a visual inspection of a concrete structure. Delaminated regions are marked up and other NDTs are used on the identified portions to assess the extent of damage.

13.2.4.2 Infrared thermography

In infrared thermography, the emission of thermal radiation from a surface is scanned remotely. The resulting thermal signal is represented as a visual image [100]. This technique measures the variation in surface radiance of a concrete member undergoing heating or cooling. Because there is a surface temperature differential between solid and delaminated concrete, this method is mainly used to identify delamination. Hidden cracks and voids inside the concrete can also be detected. The best thermographic observations are made when the concrete is warming up or cooling down because delaminated concrete heats or cools

Table 13.18 Integrity tests

Test method	Damage detected	Test type	Reference
Hammer tapping and chain dragging	Delamination of concrete cover	NDT	Bungey *et al.* [24] ASTM D4580 [94]
Pulse-echo method	Internal defects in concrete	NDT	Malhotra and Carino [25]
Impact-echo method	Internal defects in concrete and concrete piles	NDT	Cheng and Sansalone [95]
Acoustic emission	Internal defects in concrete elements	NDT	Eitzen and Wadley [96]
Infrared thermography	Delamination and near-surface internal defects in concrete	NDT	Milovanovic and Pecur [97]
Radar	Internal defects in concrete	NDT	Bungey *et al.* [24] Bungey and Millard [98]
Radiography	Internal flaws in concrete	NDT	BS 1881-205 [99]
Radiometry	Internal flaws in concrete	NDT	BS 1881-201 [81]

faster. The main advantage of this technique is that it can quickly scan large surface areas. Some modern systems can capture both visual and thermal images, which helps in identifying the location of defects and in providing an indication of the percentage of deterioration in the area scanned [101]. The major limitation of this test is that it is influenced by external environmental conditions, such as the air temperature, wind, shading or direct sunlight on the test surface, and surface water. It is an expensive technique in comparison with traditional methods of identifying delamination in concrete structures. Guidelines and a detailed description on using thermography for detecting delamination in concrete bridge decks are available in ASTM D4788 [102].

13.2.4.3 Impact-echo test

This method uses impact-generated stress waves – basically sound waves – that propagate through concrete and are reflected by internal flaws and external surfaces [103,104]. A stress pulse is applied to the concrete by a transducer or mechanical impact. If a transmitter is used, the method is called a pulse-echo test. If a mechanical impact is used, it is called an impact-echo test [95,105]. The location of a defect is calculated from the travel time of the pulse and the wave velocity in the medium. The main application of this technique is detecting the location and extent of flaws in concrete, such as cracks, delamination, voids, honeycombing, bonding of repairs, voids in tendon ducts and debonding in post-tensioned members [105]. It is effective in identifying flaws in structures that are inaccessible with other NDT techniques. For example, it is often used to evaluate pile foundations or other deep foundations [106]. The major limitations are that the accessibility of the structure determines the accuracy of the results and that data analysis is very time-consuming [24].

13.2.4.4 Radar methods

Radar techniques detect deterioration based on reflections of high-frequency radio waves or ultra-high-frequency microwaves due to changes in the electromagnetic properties of the material being probed [93,107]. This method is like the impact-echo technique discussed before, except that electromagnetic waves, such as radio waves and microwaves, are used instead of stress waves. Whenever a transmitted wave encounters an electromagnetic discontinuity because of deterioration or a change in the material dielectric, it is partially reflected. The patterns created by the reflected waves are received by a radar antenna [107]. The radar receiver measures the time required for the transmitted pulse to travel to a target discontinuity and for the reflected pulse to return. This time delay of a radar echo from a structural discontinuity is directly related to the depth of the fault. This truly non-destructive technique provides a rapid method of analysing defects lying below the surface of concrete structures. However, the main limitation is that both running and interpreting tests are highly specialised, which makes it an expensive technique. There are two ASTM standards specifically for this technique, namely ASTM D4748 [108] and ASTM D6087 [109].

13.3. Structural health monitoring to assess durability

13.3.1 Introduction

Sensors used to monitor the deterioration of concrete structures need to provide continuous information on the quality of the concrete cover and on the condition of the reinforcement inside the concrete [110]. For example, a concrete bridge in a marine environment is exposed to various processes of deterioration that are physical, chemical or biological in nature, as shown in Figure 13.3. The dominant deterioration mechanism varies based on the zone and location of the concrete members in the bridge. A combination of sensor types could be used to monitor the progress of such mechanisms with a view to estimating the service life of the structure. Each sensor type could monitor one or more of the associated measurands for the different mechanisms. Table 13.19 outlines the different measurands related to each deterioration type. The cyclic and seasonal variations in these measurands can be used to quantify the changes occurring in the cover concrete and thereby approximate the risk to steel reinforcement.

Figure 13.3 Deterioration processes observed in marine concrete structures

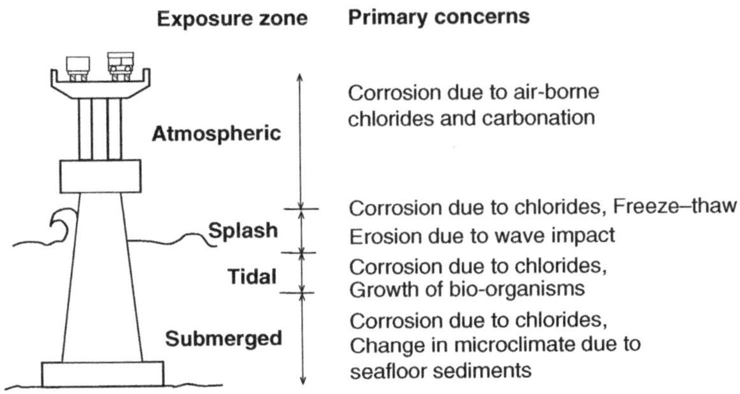

Table 13.19 Deterioration mechanisms and related measurands

Deterioration or damage type	Measurand (or parameter)	Expected duration for monitoring
Mechanical	Strain, stress, load, displacement, crack width, vibration, deformation and temperature	Short term: weeks to a month
Physical	Moisture, pressure and temperature	Medium to long term: months to years
Chemical	Chloride content or penetration, pH, carbonation depth, sulphate content or penetration, corrosion, oxidation of steel and temperature	Long term: years
Biological	Biomass, microbial or algae type, concentration, fluorescence and temperature	Information currently not available

There is a long history of monitoring stress, strain, displacement, vibration, etc., which all relate to mechanical damage. Therefore, this is a well-established field. Temperature is probably the easiest measurand to monitor and is universally monitored as a quality control parameter. Monitoring moisture is common in building physics and is becoming popular in concrete technology as well, but it needs to be done for a full cycle (1 year) or longer if any sense is to be made of the data. Biological monitoring requires bespoke sensors with very sophisticated instrumentation. This is a developing field.

The following section outlines the recent developments in sensor technology for monitoring measurands such as temperature, strain, moisture, pH and chlorides, which are crucial for assessing the durability of a concrete structure. Rather than focusing on a type of sensor and outlining its features, the section describes how some common sensor systems distinguish and quantify the measurands. Where possible, examples are also given.

13.3.2 Electrical resistance sensors

Electrical resistance sensors are both effective and inexpensive. Thin metal rods are arranged a set distance apart and have a limited surface area exposed to concrete. Numerous articles describe such sensors and cover how to calculate the formation (or shape) factor, which is a critical feature in distinguishing between an ad hoc sensor and a standardised sensor [15,111,112]. Measuring the evolution of electrical resistance is a simple and effective way to quantify any physical or chemical changes in the region near the sensors. The following aspects of concrete can change its electrical resistance

- ionic composition of the pore fluid
- pore size and connectivity
- dielectric properties of the material
- conductivity of other mix ingredients
- environmental effects such as temperature, moisture condition, carbonation, etc.

The three most commonly used sensor types in this category are discussed next.

13.3.2.1 Sensors measing the electrical resistance of the cover

A **covercrete** electrode array, as the developer termed it [15, 111,112], is a non-metallic mount with ten pairs of stainless steel electrodes, 1.6 mm in diameter and placed at different depths in the concrete. The electrodes are sleeved with heat shrink tubes except for 5 mm at the tip, as shown in Figure 13.4. The electrode array needs to be embedded in the concrete during placement. When connected to a suitable data acquisition system, real-time data on the condition of the cover can be obtained. Because of the spatial distribution of the electrodes, starting from as close as 10 mm beneath the cover surface and extending up to 70 mm from the surface, the array provides information on the ingress of moisture or salts, both as a function of time and distance. Thermistors are also attached to the mount to enable temperature profiles to be obtained concurrently.

Figure 13.4 Covercrete electrode array

The conductance (inverse of the resistance) measured across the pairs of electrodes can be presented in the following four ways.

- As-measured conductance (in siemens) as a function of time for each electrode pair.
- The as-measured conductance standardised to a reference temperature (20°C), which minimises the effects of changes in conductivity due to temperature. This is particularly important for on-site measurements.
- Relative or normalised conductance, which is the conductance standardised by dividing it by the conductance at time zero or a reference depth. This is defined as:

$$N_{Ct} = \sigma_t / \sigma_o \tag{13.1}$$

where σ_t is the standardised conductance measured using an electrode pair at time t and σ_o is the conductance measured at the same electrode pair at a datum point in time (say time zero). N_{Ct} values, thus, allow relative changes in conductance to be studied over time. The data can also be normalised by the conductance obtained at a depth of 50 mm:

$$N_{Cx} = \sigma_t / \sigma_{t,50} \tag{13.2}$$

where σ_t is as defined above and $\sigma_{t,50}$ is the conductance across a reference electrode pair at 50 mm at time t. A depth of 50 mm or greater should be chosen for the measurement electrodes, as the response at this depth will remain relatively unaffected by drying and wetting at the surface.

- Conductance can be converted to conductivity by taking into account the formation factor (shape factor) of the electrodes.

The above formalisms can also be applied to resistance. That is, instead of the normalised conductivity, it is possible to determine the normalised resistance (or resistivity). McCarter *et al.* [112] characterised the evolution of normalised conductance into four regions, as shown in Figure 13.5.

Figure 13.5 Four stages in the temporal variation of normalised conductivity for an unexposed concrete specimen [112]

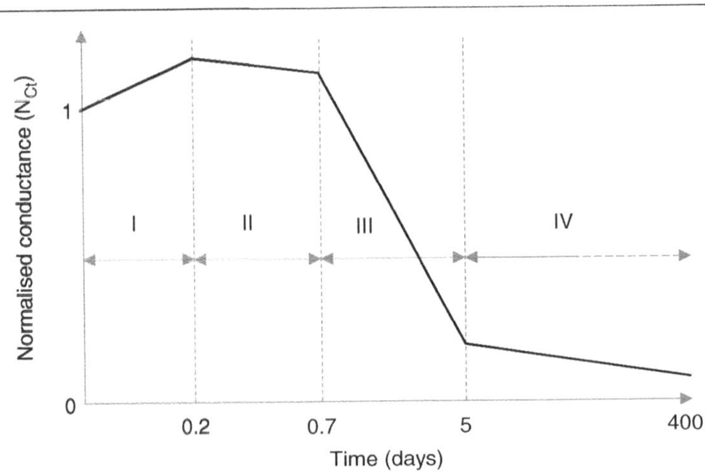

- Region I: This region is influenced by both the dissolution of the cement mineral phases, which increases the ionic concentration and the conductivity, and the growth of hydrates through which the ions are consumed so that the conductivity falls. The linear increase in conductance during this phase indicates that dissolution is dominant, but whether this occurs depend on the type of cement.
- Region II. Setting occurs during this period. More mineral phases form due to the precipitation of gel around the cement minerals, transforming the concrete from a liquid to a semi-solid phase. The conductivity gradually reduces.
- Region III: The conductivity reduces sharply as the concrete transitions to a solid phase. The continuous precipitation and crystallisation of minerals narrow the pore structure, and the increasing tortuosity restricts the movement of ions. Thus, there is a significant reduction of the conductivity over time.
- Region IV: The conductivity continues to fall but at a lower rate. New minerals form and deposit on the pore walls. The resulting narrower paths reduce the conductivity. This process can continue for a long time. If supplementary cementitious materials are used, the timescale will be even longer, as the pore structure becomes further refined due to the formation of secondary hydration products, even beyond the test period of 400 days.

Discussions on the application of this type of sensor for studying the movement of moisture in concrete can be found in Waters [113], McCarter *et al.* [114], McCarter and Garvin [115], and Rajabipour and Weiss [116]. In summary, ions flow through the concrete surrounding the electrode pair. The pore connectivity, moisture content, ionic make-up, concentration, etc. all affect the rate of flow. Electrical sensors measure the combined effects, quantifying them as the electrical resistance across the pair of electrodes. A change in the resistance, therefore, conversely indicates a change in the pore connectivity or moisture content. As water moves into the zone of influence of an electrode pair (indicated by the electric field in Figure 13.6), the electrical resistance decreases due to the increased conductance of the pore solution [116]. Profiles for the normalised resistance ($1/N_{Ct}$) for normal Portland cement concrete in a standard capillary rise test are shown in Figure 13.7. The resistance ratio decreased with time for the five electrode pairs near the surface (levels 1–5). The electrodes most affected were those nearest to the surface, that is the exposure face (levels 1 and 2).

Covercrete array sensors can also be used to study other moisture-related properties and processes in the cover zone, such as the degree of saturation and its variation, the effective porosity and the influence of cyclic wetting and drying on the pore structure. Electrical conduction in concrete occurs primarily due to the mobility of ions in the pore solution. Therefore, monitoring the electrical resistivity or conductivity

Figure 13.6 Electric field between two embedded electrodes [114]

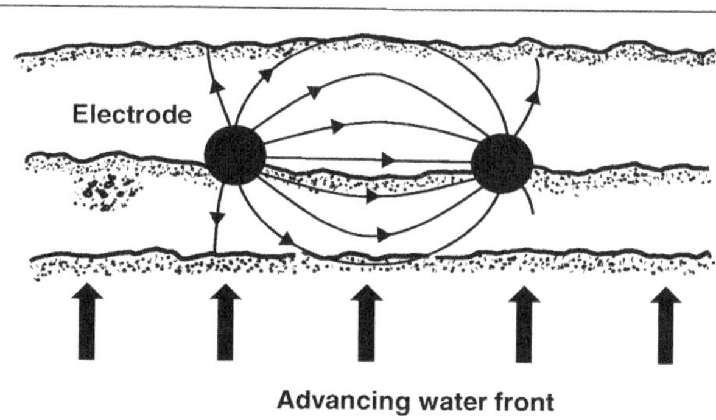

Figure 13.7 Normalised resistance for electrode pairs at different depths from the exposure face of a concrete block [114]

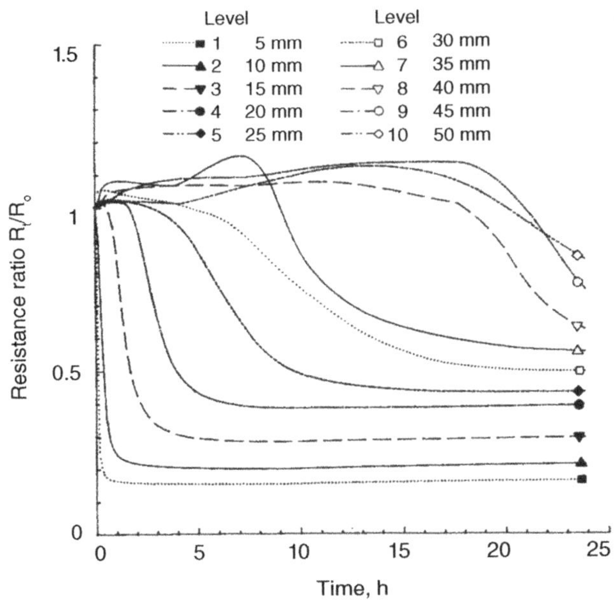

during early ages can help in understanding the hydration, setting behaviour and microstructural variations in cement-based systems. Electrical properties have been measured in studies of the hydration and microstructural evolution of cement-based material systems [112,117,118]. Multi-electrode sensors can also be used to study variations in the setting behaviour and microstructure of various types of concretes as a function of curing or hydration [119,120]. Basheer *et al.* [121] and Yoon and Chang [122] used electrical resistivity sensors to monitor the progress of carbonation. Their data show that measuring just resistance is not particularly useful for determining the progress of the carbonation front within concrete, due to the opposing effects (on resistance) brought about by the refinement of the microstructure and variations in moisture content [60].

13.3.2.2 Electrical corrosion sensors

REFERENCE ELECTRODES

As discussed earlier, monitoring the rate of deterioration of structures due to corrosion is particularly important because corrosion is the most common form of deterioration in reinforced concrete structures. Monitoring the corrosion activity of the reinforcement using sensors gives more precise information about the time when corrosion initiates and about corrosion process. Further, using a combination of sensors can provide better insights on the factors influencing corrosion.

Among the many techniques available, measuring the electrochemical potential of the steel reinforcement in concrete using reference electrodes is popular for assessing corrosion activity. The electrochemical potential of a metal in a solution is measured relative to a fixed and known potential measured by a standard reference electrode in the same electrolyte. The basic design of a reference electrode is shown in Figure 13.8. The reference electrode is embedded in the cover concrete close to the reinforcement, as shown in Figure 13.9. It measures changes in the electrochemical potential due to corrosion. The most common reference electrodes for electrochemical measurements are copper/copper sulphate ($Cu/CuSO_4$)

Figure 13.8 Reference electrode [123]

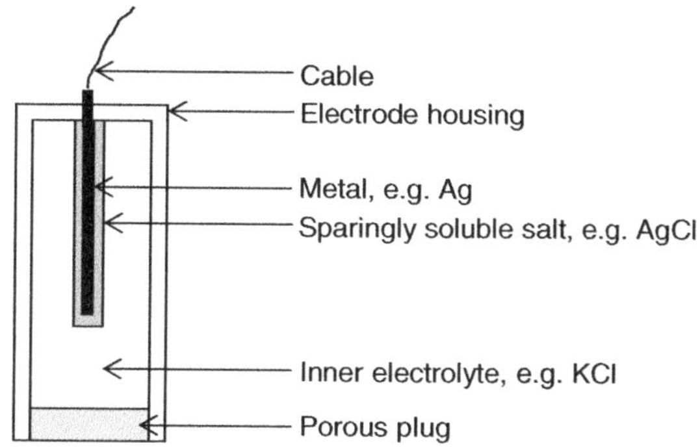

Figure 13.9 Measuring the electrochemical potential in concrete [90]

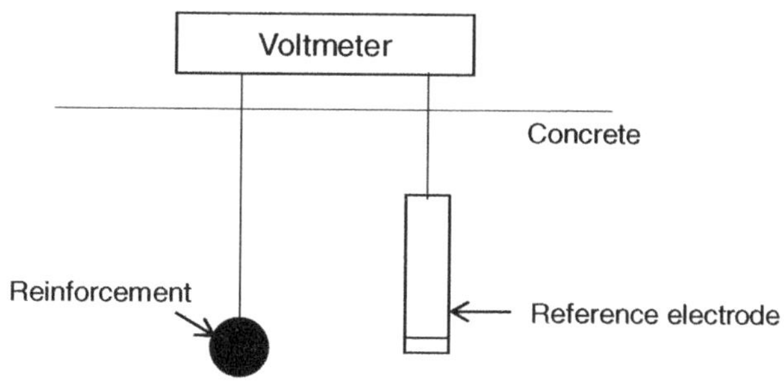

reference electrodes, standard calomel electrodes (SCEs) and silver/silver chloride (Ag/AgCl) reference electrodes. The metal mercury (Hg) in an SCE is in a sparingly soluble salt of mercury chloride (Hg_2Cl_2) contained in a saturated solution of potassium chloride (KCl), whereas with an Ag/AgCl reference electrode, potassium chloride solutions with concentrations ranging from 0.1 M to saturated solution are used [123]. Therefore, as there are different types of reference electrode, care should be taken in translating the potential obtained to the corrosion activity of steel. The potentials of various reference electrodes in reference to a normal hydrogen electrode and the criteria for assessing the corrosion condition based on the measured potentials are listed in Table 13.14.

If the reference electrode is to be embedded in the concrete, the sensor must be able to withstand the impact of fresh concrete in new construction and must be robust enough to continue operating in the aggressive environment in cover concrete. SCEs are not normally embedded in concrete, as the sensor contains a liquid metal, making it difficult to manufacture a robust design. Ag/AgCl reference electrodes have been widely embedded in concrete, as they can be manufactured with a rugged design and have a reasonably good design life of 15 years [123]. A commercial reference electrode that can be embedded in concrete uses manganese dioxide (MnO_2) in an alkaline solution. Muralidharan *et al.* [124] manufactured an embeddable manganese dioxide sensor in a three-layer design with a porous hydrated cement paste as the bottom layer, an alkaline slurry as the middle layer and manganese dioxide powder as the top layer. This embeddable manganese dioxide electrode sensor has good stability and reliability in both a simulated concrete pore solution and in concrete.

Figure 13.10 Anode-ladder system in cover concrete [87,125]

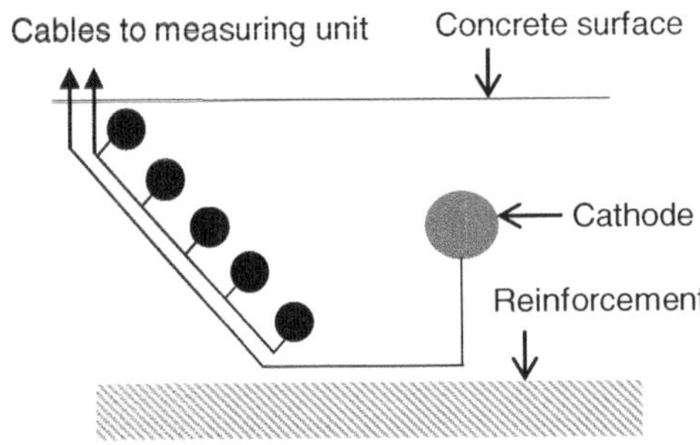

MACROCELL SENSORS

This type of sensor uses a galvanic cell consisting of an anode and a cathode [87]. The electrical current measured between the anode and cathode is proportional to the dissolution of iron at the anode due to corrosion. A change in the galvanic current gives the corrosion rate of the anodic steel in the host material. This type of sensor has been used quite widely since around 1990 to monitor corrosion in concrete.

The most popular anodic-ladder system was developed by Schiessl and Raupach [126]. It has been used to monitor the corrosion risk in new structures. It consists of a series of carbon-steel anodes mounted at different depths from the surface of the concrete, as shown in Figure 13.10. The cathode is made from stainless steel or activated titanium and is placed in the concrete near the anode. Chloride ingress and carbonation fronts are both precursors to corrosion. If either of these passes through the cover concrete and reaches the individual carbon-steel bars in the ladder, depassivation occurs, which increases the current between anode and cathode. This sensor arrangement can provide information about when corrosion of the main reinforcement will initiate and the critical depth of corrosion. Another anodic ladder system has been used in bridge decks and tunnels. It consists of a mild-steel anode and a stainless steel cathode [125] and has been widely used in concrete structures in Europe. It is available commercially.

LPR SENSORS

An LPR sensor can measure the rate of corrosion of the steel reinforcement in concrete. A small electric current is applied to the concrete and the shift in corrosion potential is measured, which is an indicator of the level of active corrosion in the steel. The working electrode is made from mild steel as it represents the actual reinforcement in the concrete. An auxiliary electrode is used to pass the current, and a reference electrode or half-cell is used to measure the potential and its changes. The LPR method with a guard ring probe for surface measurements is an established NDT for measuring the corrosion rate of the reinforcement in concrete structures. The criteria for interpreting damage due to corrosion are listed in Table 13.20.

Table 13.20 Typical interpretation of corrosion rate measurements [90]

Corrosion rate: μA/cm^2	Attack penetration (loss of section of reinforcing bar): μm/year	Corrosion level
<0.1	<1	Negligible
0.1–0.5	1–5	Low
0.5–1	5–10	Moderate
>1	>10	High

LPR sensor probes have been used to measure the corrosion rate of reinforcement in new structures [87,127,128]. An embeddable LPR sensor was developed by Broomfield *et al.* [127], as shown in Figure 13.11. The working electrode is a mild steel bar of known area to measure the corrosion rate of steel in cover concrete. Another part connects to the main reinforcement in the structure to measure the actual corrosion rate [87]. A different system [128] consists of a combination of sensors that measure the corrosion rate, temperature, chloride ion concentration and electrical resistivity in concrete structures, as shown in Figure 13.12. The sensors are contained in a porous mortar. This corrosion probe monitors various parameters related to the corrosion of the reinforcement in concrete structures.

13.3.2.3 Electrical chloride sensors

In recent years, considerable research attention has been focused on developing chloride ion sensors for monitoring concrete structures. A well-known chloride ion sensor records electrical measurements. It has silver electrodes with a silver chloride (AgCl) coating and relies on the soluble nature of AgCl [129,130].

Figure 13.11 LPR sensor for corrosion monitoring in new structures [127]

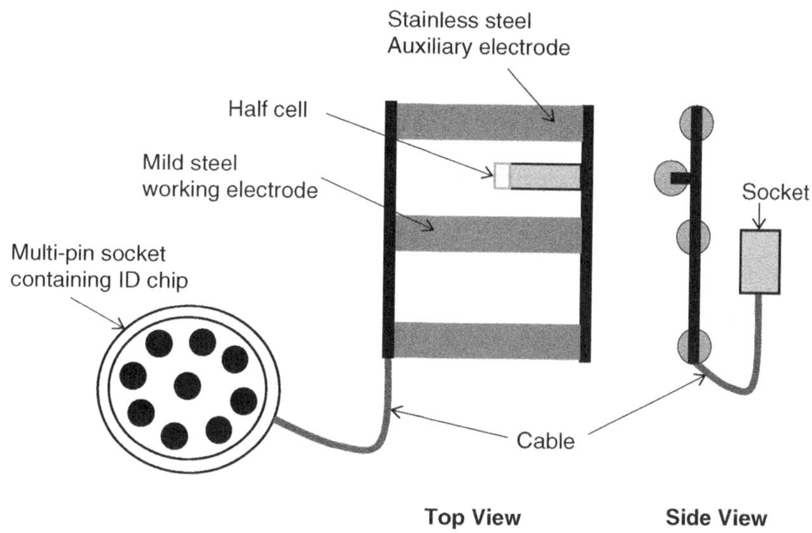

Figure 13.12 Schematic of multiple sensor probe in concrete [128]

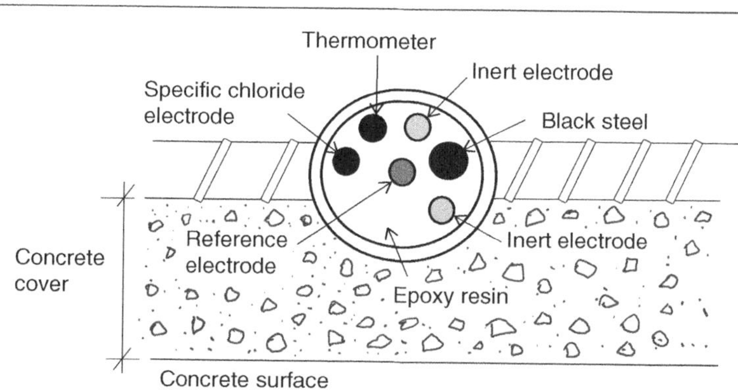

This method suffers from disadvantages, such as limited longevity of Ag/AgCl electrodes, influence of temperature and pH on measurements, and influence of other ions like bromide ions present in seawater [131]. These electrodes are insensitive to low chloride ion concentrations and another major disadvantage of this method is that Ag/AgCl electrodes must be wet so that the chloride ions are mobile at the time of measurement.

A combined chloride and resistivity sensor element has been developed and extensively used in concrete structures for monitoring free chloride concentrations and electrical resistivity [131]. The potential of an embedded chloride sensor is measured relative to a manganese dioxide (MnO_2) reference electrode [132]. Another patented chloride ion sensor has a pair of electrodes separated by a conductive polymer film that is sensitive to chloride ions. The electrical conductivity of the sensor depends on the number of chloride ions taken up by the film [133]. The polymer film (methylpyrrole) is deposited on at least one electrode of the pair, which allows measurements in wet or dry media. The electrical conductivity of the polymer film depends on the amount of chloride ions taken up by the film. Another advantage of this technique is that it can measure the chloride ion concentration at high pH values. A major drawback is that the sensor is non-reversible so that it measures only the accumulation of chloride ions.

13.3.2.4 Electrical pH and humidity sensors

Measurements of the potential values in concrete using an embedded reference electrode, as shown in Figure 13.8, depend on the moisture content, pH and chloride concentration in the vicinity of the electrode. Some researchers [134,135] have used electrolytic manganese dioxide (or calcined derivatives of this oxide) as a medium separating two conductive electrodes. The change (i.e. reduction) in the resistance of this array is influenced by the pH and relative humidity.

13.3.2.5 Integrated sensor system for monitoring corrosion in concrete

A combination of sensors is needed to monitor the causes and progress of corrosion in marine concrete structures. Therefore, an integrated sensor system was developed with three different types of sensor, namely an array of two-pin electrical resistance sensors, a temperature sensor and a corrosion sensor, as shown in Figure 13.13 [136]. The array of two-pin electrical resistance sensors monitors the moisture and ionic movement at

Figure 13.13 Septopod, an integrated corrosion sensor

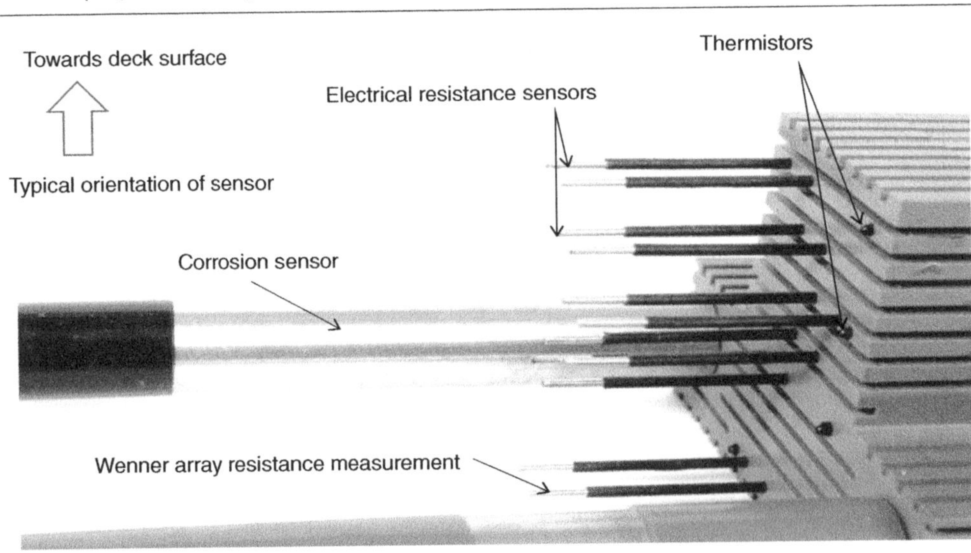

Figure 13.14 Automated set-up for remote monitoring of concrete

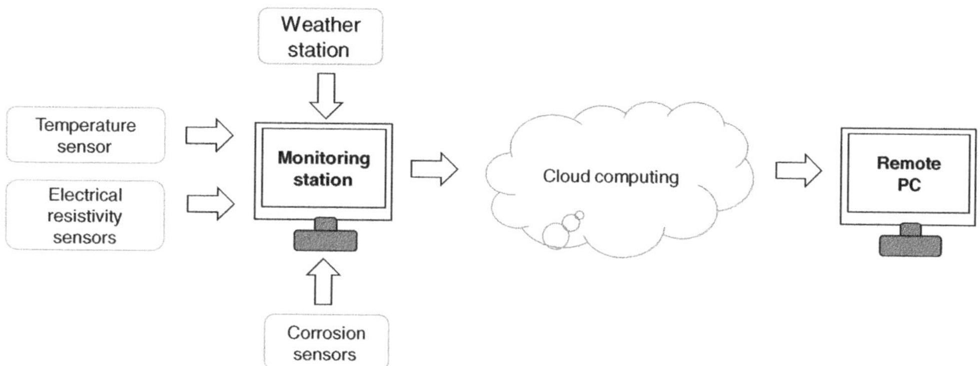

different depths in the cover zone of the concrete. The temperature sensors are placed at different depths to give the thermal gradient in the cover zone. The temperature data can also be used to predict the early-age strength of the concrete using maturity functions. In addition, the temperature measurements are useful for correcting the electrical resistivity data. The corrosion sensor in this integrated sensor probe gives an early warning before the main reinforcement in the concrete has started to corrode. The data obtained from these different sensing techniques can be combined to provide good information on the material properties of the concrete in the cover zone. Moreover, the information gathered from these sensors over time can be particularly helpful in understanding the key factors involved in the deterioration of concrete structures.

A remote monitoring set-up, as shown in Figure 13.14, can be used to monitor the performance of the concrete in a structure using an automated data acquisition system and cloud computing software [119,120]. Various sensors can be integrated to provide information about the condition of concrete in the cover zone as well as variations in the micro-environment surrounding the concrete, which can help in understanding the interrelation between climatic changes and the associated response of the cover concrete.

13.3.3 Novel fibre-optic sensor systems for monitoring concrete durability

Fibre-optic sensors have been increasingly used for structural health monitoring over the past few years. Optical fibre sensors offer numerous advantages in comparison to conventional electrical methods of monitoring. The main advantages are as follows. They are extremely small, lightweight, robust, resistant to corrosion and immune to electromagnetic noise. They can be multiplexed and have high sensitivity. Hence, they are well suited for the harsh environment in concrete structures [110,137,138].

Different sensor techniques are available for monitoring civil engineering structures. They can be classified based on which characteristics of the light, such as the intensity, wavelength, phase and polarisation, affect the parameter being measured. Fibre-optic sensors can also be classified as point-based sensors (fibre Bragg grating (FBG) sensors), long-gauge sensors (quasi-distributed sensors) and distributed sensors [139]. Fibre-optic sensors can measure extreme temperatures, such as from –200°C to 800°C if they have a silica core and up to 1500°C with a sapphire core [139]. Quasi-distributed and distributed sensors can monitor host parameters in large civil structures at multiple locations [140]. Fibre-optic sensors can measure a wide range of parameters relevant to health monitoring of civil structures, namely physical parameters (such as strain, inclination, pressure, forces, accelerations and vibrations) and temperature and chemical parameters (such as humidity, pH, and oxygen and chloride concentration). Of these, the most crucial parameters for durability are temperature, humidity, strain, pH and chloride content of the concrete, and corrosion of the rebar.

Sensors with FBG are popular for structural health monitoring of construction materials due to their better sensitivity to strain and temperature [139]. Moreover, due to their small dimensions, FBGs can be utilised as 'micro-sensors' for monitoring minute variations in the pore structure of concrete, as such changes are important in understanding aspects of the durability of the concrete. The manufacturing process, developments and application of these sensors are well documented in major books and book series [139,141,142].

13.3.3.1 FBG strain sensors

An FBG sensor has narrow gratings, which are formed by creating a periodic modulation of the refractive index of a photosensitive optical fibre using the phase-mask writing technique. The grating acts as a narrowband reflection filter that reflects a particular wavelength of light depending on the spacing or period of the grating. The reflected Bragg wavelength (λ_B) is

$$\lambda_B = 2n_{\text{eff(core)}}\Lambda \tag{13.3}$$

where $n_{\text{eff(core)}}$ is the effective refractive index of the core and Λ is the grating period. The fractional wavelength change due to strain induced on the sensor is

$$\frac{\Delta\lambda_B}{\lambda_B} = (1 - P_e)\varepsilon + [(1 - P_e)\alpha + \xi]\Delta T \tag{13.4}$$

where P_e is the photoelastic constant of the fibre (typical value for silica fibre is 0.22), ε is the strain induced on the fibre, α is the thermal expansion coefficient of the fibre and ξ is the thermo-optic coefficient of the fibre. Therefore, a shift in the wavelength of the light can be attributed to either strain or a change in the temperature based on established calibrations between wavelength and strain and between wavelength and temperature.

A single fibre containing multiple FBG sensors can be laid all along the host structure, and these can be simultaneously monitored by a single interrogator to provide measurements of strain and temperature at the different locations of the sensors. However, due to the interrelation among wavelength, strain and temperature, strain measurements must be corrected for changes in temperature and the temperature sensor must be isolated from the effects of strain.

Fibre-optic strain sensors have been widely used for structural health monitoring for more than two decades [139,143–145]. Issues related to the packaging and installation of sensors in structures have attracted considerable attention [143]. A study on methods of mounting FBG strain sensors in concrete beams suggested that packaged fibre-optic strain sensors with a mounting lug glued and screwed into the concrete specimen performed much better than sensors that were simply glued onto the surface of the concrete [146].

13.3.3.2 FBG humidity sensors

FBG sensors have also been used for measuring moisture and RH [147–150]. A moisture-sensitive polymer is coated onto a FBG sensor. The polymer swells or expands with an increase in moisture, thereby inducing a strain on the FBG. The RH is found by optically measuring the shift in wavelength caused by the expansion of the moisture-sensitive coating. A linear relation has been obtained between RH and wavelength through a calibration at standard humidity levels [148]. Further research improved the sensitivity and reliability of this technique [137,149]. Fabricating FBG humidity sensors, including the process for applying a coating of the moisture-sensitive material and the optimum coating thickness in terms of sensitivity and response time, are described in some of the references cited above. Figure 13.15 presents a typical FBG RH sensor and Figure 13.16 shows data obtained in a capillary absorption test.

Figure 13.15 Embeddable FBG probe for measuring RH and temperature [150]

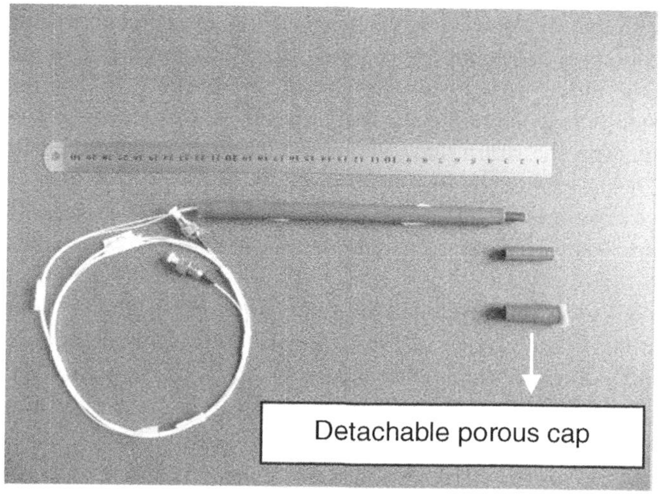

Figure 13.16 RH plotted against time for concrete in a capillary rise test as measured by a fibre-optic RH probe [150]

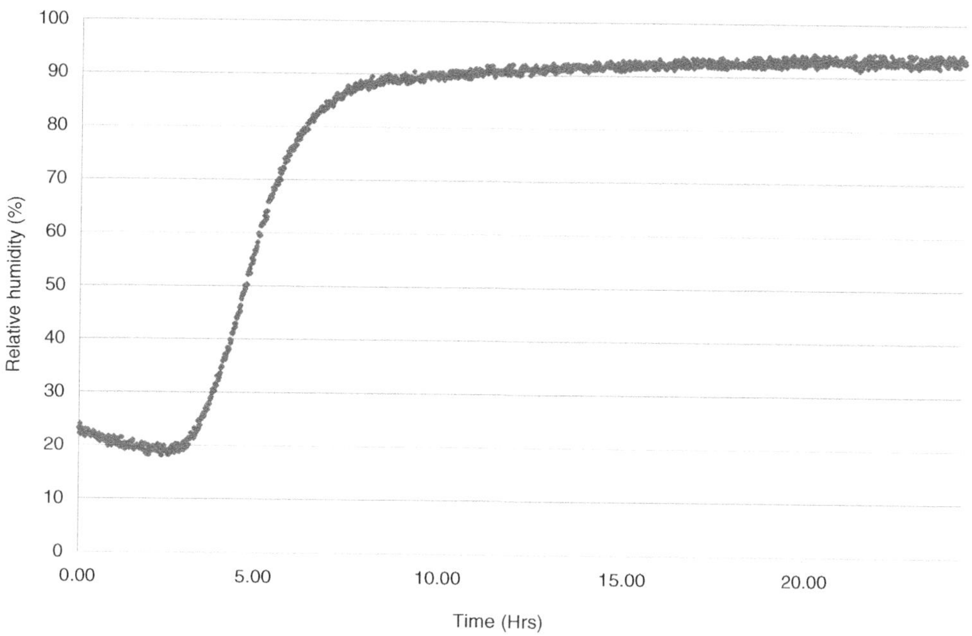

13.3.3.3 FBG corrosion sensors

Monitoring the corrosion of the reinforcement is important when assessing the durability of a concrete structure. Optical fibre sensors for monitoring steel corrosion based on different sensing techniques have been developed by many researchers [151,152]. Grattan *et al.* [152] used FBG sensors to monitor the corrosion of the rebar in concrete samples. As the reinforcement corrodes, the products of corrosion occupy a larger space than the original reinforcement, resulting in stresses in the transition layer between the rebar and the concrete. This stress was monitored using FBG strain sensors attached to the reinforcement, one on the top face of the rebar and another on the bottom face. In that work, the performance of the FBG strain sensors was compared with that of electrical resistance strain gauges embedded in separate concrete slabs subjected to similar conditions of accelerated corrosion. Strain measurements plotted over time for the two types of sensors during the onset of corrosion are shown in Figure 13.17. The results indicate that FBG strain sensors can successfully respond to the stresses caused by the onset of corrosion of the reinforcement in concrete samples. Moreover, the strain due to corrosion can be distinguished from standard strain, for example, due to loading, because once the load is removed, the strain will reduce, but the strain due to corrosion remains until the excess volume has reduced. Therefore, FBG strain sensors can show whether the reinforcement in a structure has corroded or not, although the actual corrosion rates may not be determined.

Another notable corrosion sensor was developed by Leung *et al.* [151]. The tip of an optical fibre is coated with a thin iron film. As the iron corrodes, the thickness of the coating decreases and, subsequently, the intensity of a reflected signal drops, as depicted in Figure 13.18. This fibre sensor was embedded in concrete subjected to salt ponding tests. The loss in the reflectivity of the light in the fibre-optic sensor was compared with standard macrocell current measurements, as shown in Figure 13.19.

13.3.3.4 Fibre-optic pH sensors

The requirement for a suitable pH monitoring method that is non-destructive and provides timely information on the likelihood of corrosion of the reinforcement due to carbonation of the concrete has led to the development of fibre-optic pH sensors based on optical or spectroscopic properties, such as absorbance, reflectance, fluorescence and refractive index [153,154].

Sol–gel pH sensors have potential for applications in concrete [155]. The first use of a sol–gel pH sensor was demonstrated by Badini *et al.* [156]. The absorption properties of the sol–gel are linearly related with the pH of the sample. The pH probe was constructed by coating sol–gel containing a cresol-red indicator dye

Figure 13.17 Strain measurements made by electrical resistance strain gauges (ERS) and FBG sensors during an accelerated corrosion test [152]

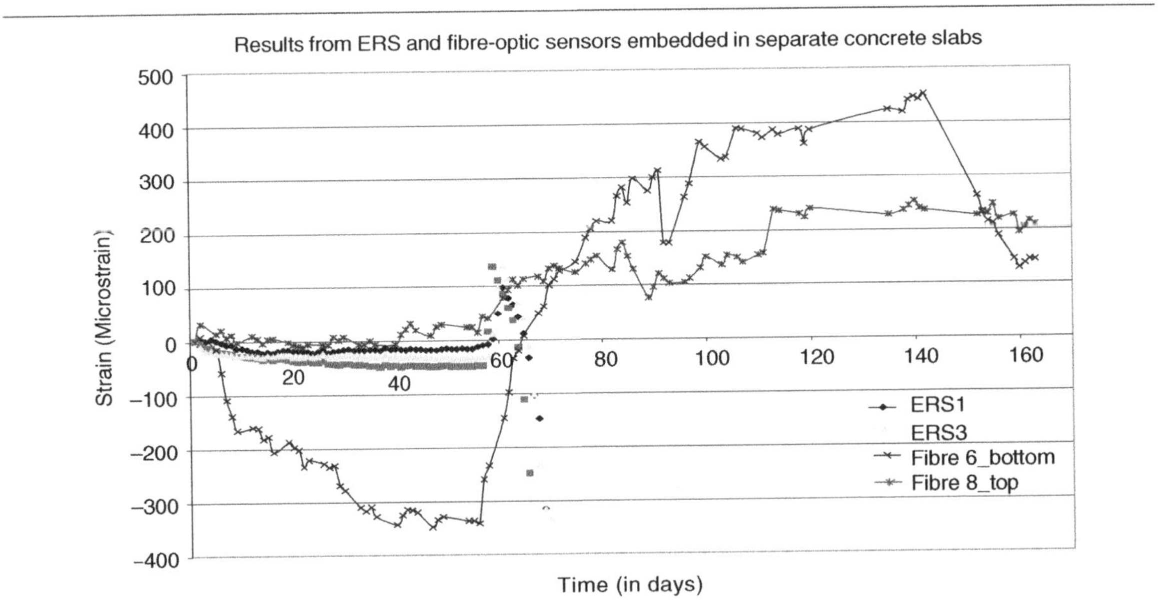

Figure 13.18 Sensing corrosion using a thin iron-film coating [151]

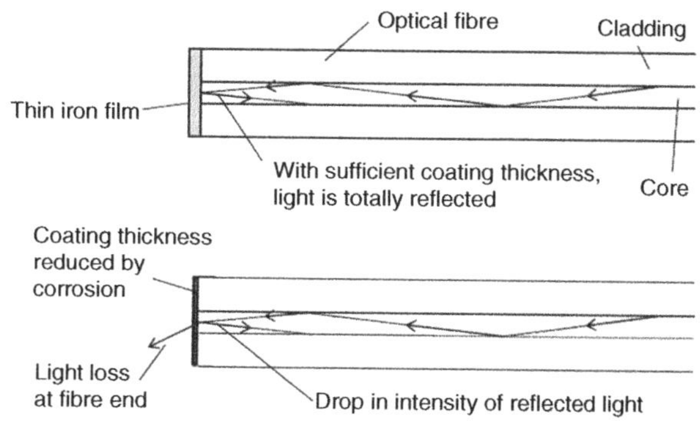

Figure 13.19 Performance of a fibre-optic sensor for corrosion monitoring compared with macrocell current measurements [151]

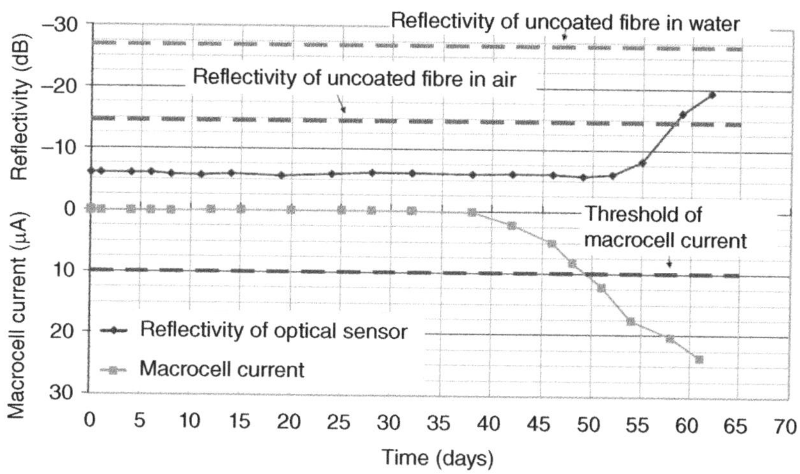

(pH range 8–13) onto a plastic-clad silica fibre with a core diameter of 600 μm. The light source was a tungsten halogen lamp, and the reflected light was analysed by a small portable spectrometer. The spectrometer, light source and pH probe are shown in Figure 13.20. In that work, the sol–gel pH probe was compared with a commercial disk-based fibre-optic probe, which had a porous matrix impregnated with an indicator dye. Both probes were embedded in mortar samples. The pH measurements obtained using the sol–gel pH probe are plotted in Figure 13.21. The probe reached a stable pH approximately 30 minutes after it was embedded in the mortar. The pH values obtained by the sol–gel probe correlated well with the apparent pH profile obtained by the digestion method [155,157].

13.3.3.5 Fibre-optic chloride sensor

Fibre-optic chloride sensors have been reported in literature. They use various properties, such as fluorescence [158], absorption and reflection [159], and changes in refractive index [160]. Most of these methods are non-reversible, and some have limited longevity at high pH ranges. In the work reported by Cosentino *et al.* [159], the sensor layer was exposed to silver nitrate ($AgNO_3$), which reacted with sodium chloride (NaCl) to form a

Figure 13.20 Fibre-optic pH sensor set-up

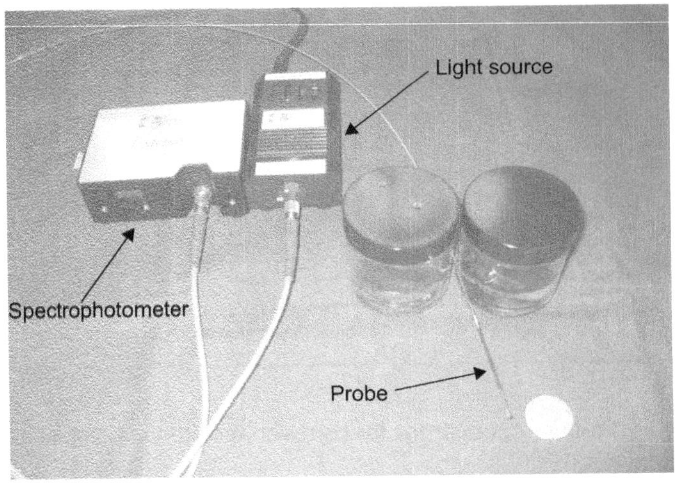

Figure 13.21 pH plotted against time in a mortar sample measured by a sol–gel pH probe

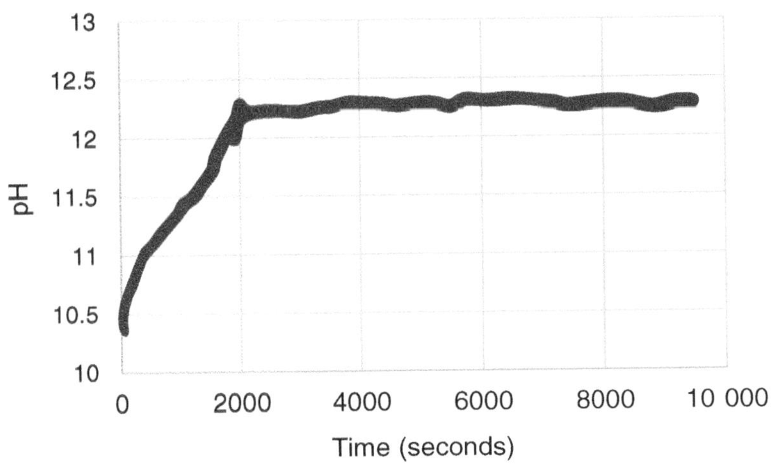

white precipitate of silver chloride (AgCl). The change in colour was sensed and related to the concentration of chlorides. Reflection and absorption sensors using silver chromate powder were also reported in the same paper. Since these methods are non-reversible, such sensors could be used for monitoring the cumulative chloride concentration at a sensor. Similar work was reported by Fuhr *et al.* [161]. The technique was modified to use dichlorofluorescein dye, which changes colour from pink to milky white depending on the concentration of positively charged silver ions in a silver nitrate solution. These sensors were embedded in various bridge deck slabs. They have also been used as chloride threshold indicators in concrete structures [162,163].

A sol–gel fibre-optic chloride sensor has been developed by McPolin *et al.* [164]. Different chloride-sensitive indicators were impregnated into the sensor, such as silver nitrate, silver chromate and fluorescein. Like the sol–gel pH sensor, sol–gel with a chloride indicator was attached to the tip of a 600-μm silica fibre using an epoxy adhesive and then covered with protective sheathing, as shown in Figure 13.22. The calibration graph

Figure 13.22 Sol–gel chloride-sensitive fibre-optic probe [164]

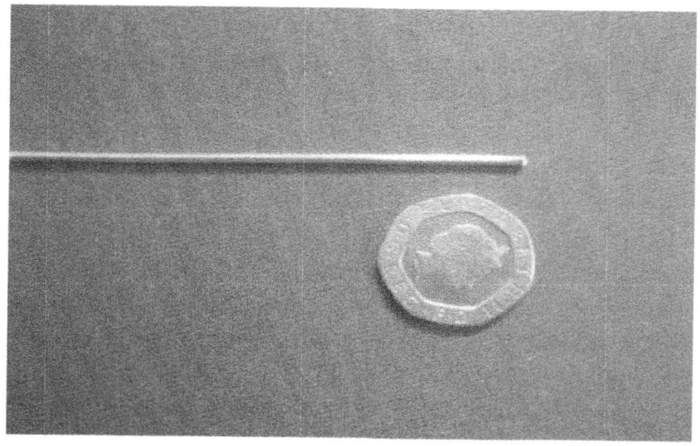

Figure 13.23 Calibration of sol–gel chloride probe with silver nitrate as an indicator [164]

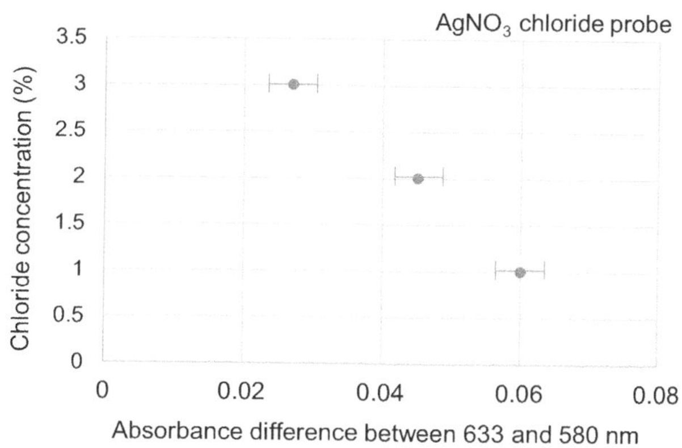

obtained with chloride concentrations of 1%, 2% and 3% and a silver nitrate indicator is shown in Figure 13.23. That investigation demonstrates the potential scope of sol–gel as an interactive membrane containing a chloride-sensitive indicator. Such sensors could be used to monitor the chloride content in concrete structures. Further research is needed on reversible chloride-sensitive indicators.

In recent research, a chloride sensor was developed based on fluorescence quenching using Lucigenin indicator dye encapsulated in sol–gel [165]. For this sensor, there was a linear correlation between intensity ratio and chloride concentration. In that study, the sensors were embedded in mortar samples at different depths and the free chloride concentrations were monitored while the mortar blocks were subjected to marine environmental conditions.

13.4. Summary

Several in situ tests for assessing the mechanical, permeation, corrosion and integrity characteristics of concrete in structures were reviewed in the first part of this chapter. The second part focused on sensing techniques, which were classified as electrical or optical. Many researchers worldwide are developing new testing and monitoring systems, and these may have major implications for buildings and civil infrastructure. It has been demonstrated that both site test methods and sensors could be used to monitor the various physical and chemical characteristics of concrete directly related to its durability. There are advantages and disadvantages with each technique, but a combination of techniques could allow a proper maintenance strategy to be developed, saving the huge sums currently spent on repair and rehabilitation and leaving more money for capital development.

REFERENCES

1. Yang K, Yang CH, Long A and Basheer PAM (2018) Use of two-pressure-head method to assess water permeability of structural concrete. *ACI Materials Journal* **115(1)**: 65–75.
2. Ma Q, Nanukuttan SV, Basheer PAM, Bai Y and Yang CH (2015) Chloride transport and the resulting corrosion of steel bars in alkali activated slag concretes. *Materials and Structures* **49(9)**: 3663–3677.
3. Basheer PAM and Nolan EA (2001) Near-surface moisture gradients and in situ permeation tests. *Construction and Building Materials* **15**: 105–114.
4. Yi Y, Zhu D, Guo S, Zhang Z and Shi C (2020) A review on the deterioration and approaches to enhance the durability of concrete in the marine environment. *Cement and Concrete Composites* **113**: 103695.
5. Amini K, Cetin K, Ceylan H and Taylor P (2019) Development of prediction models for mechanical properties and durability of concrete using combined nondestructive tests. *Journal of Materials in Civil Engineering* **31(2)**: 04018378.
6. Steindl FR, Baldermann A, Galan I, *et al.* (2019) Chemical resistance of eco-concrete – Experimental approach on Ca-leaching and sulphate attack. *Construction and Building Materials* **223**: 55–68.
7. Lauer KR (1990) Classification of concrete damage caused by chemical attack. *Materials and Structures* **23(3)**: 223–229.
8. Sun LF, Jiang K, Zhu X and Xu L (2020) An alternating experimental study on the combined effect of freeze–thaw and chloride penetration in concrete. *Construction and Building Materials* **252**: 119025.
9. Shields Y, Garboczi E, Weiss J and Farnam Y (2018) Freeze–thaw crack determination in cementitious materials using 3D X-ray computed tomography and acoustic emission. *Cement and Concrete Composites* **89**: 120–129.
10. Zhu W, François R, Zhang C and Zhang D (2018) Propagation of corrosion-induced cracks of the RC beam exposed to marine environment under sustained load for a period of 26 years. *Cement and Concrete Research* **103**: 66–76.
11. Mehta PK and Monteiro PJM (eds) (2005) *Concrete: Microstructure, Properties, and Materials*, 3rd edn. McGraw-Hill Professional.
12. Stambaugh ND, Bergman TL and Srubar WV (2018) Numerical service-life modeling of chloride-induced corrosion in recycled-aggregate concrete. *Construction and Building Materials* **161**: 236–245.
13. Basheer PAM (2001) Permeation analysis. In *Handbook of Analytical Techniques in Concrete Science and Technology: Principles, Techniques and Applications* (Ramachandran VS and Beaudoin JJ (eds)). Noyes Publications, pp. 658–727.
14. BSI (2013) BS EN 206-1:2013. Concrete. Specification, performance, production and conformity. BSI, London, UK.
15. McCarter WJ, Chrisp TM, Starrs G, *et al.* (2015) A durability performance-index for concrete: Developments in a novel test method. *International Journal of Structural Engineering* **6(1)**: 2–22.
16. Alexander M, Ballim Y and Santhanam M (2005) Performance specifications for concrete using the durability index approach. *Indian Concrete Journal* **79(12)**: 41–46.
17. McCarter WJ, Suryanto B, Taha HM, Nanukuttan S and Basheer PAM (2017) A testing methodology for performance-based specification. *Journal of Structural Integrity and Maintenance* **2(2)**: 78–88.
18. Sun D, Shi H, Wu K, *et al.* (2020) Influence of aggregate surface treatment on corrosion resistance of cement composite under chloride attack. *Construction and Building Materials* **248**: 118636.

19. Belda Revert A, De Weerdt K, Hornbostel K and Geiker MR (2018) Carbonation-induced corrosion: Investigation of the corrosion onset. *Construction and Building Materials* **162**: 847–856.
20. Breysse D, Balayssac JP, Biondi S, *et al.* (2019) Recommendation of RILEM TC249-ISC on non destructive in situ strength assessment of concrete. *Materials and Structures* **52(4)**: 52–71.
21. Galvão J, Flores-Colen I, de Brito J and Veiga MR (2018) Variability of in-situ testing on rendered walls in natural ageing conditions – Rebound hammer and ultrasound techniques. *Construction and Building Materials* **170**: 167–181.
22. Yu H, Tan Y and Feng T (2019) Study of temporal change in chloride diffusion coefficient of concrete. *ACI Materials Journal* **116(1)**: 103–112.
23. Concrete Society (2010) *Non-structural Cracks in Concrete*, 4th edn. Concrete Society, Camberley, UK, Technical report 22.
24. Bungey JH, Millard SG and Grantham M (2006) *Testing of Concrete in Structures*, 4th edn, Taylor & Francis, London, UK and New York, NY, USA.
25. Malhotra VM and Carino NJ (2004) *Handbook on Nondestructive Testing of Concrete*, 2nd edn (Malhotra VM and Carino NJ (eds)). CRC Press.
26. Bungey JH and Soutsos MN (2001) Reliability of partially-destructive tests to assess the strength of concrete on site. *Construction and Building Materials* **15(2–3)**: 81–92.
27. Breysse D, Balayssac JP, Biondi S, *et al.* (2017) Non destructive assessment of in situ concrete strength: Comparison of approaches through an international benchmark. *Materials and Structures* **50(2)**: 133, 10.1617/s11527-017-1009-7.
28. BSI (2021) EN 12504-2:2021. Testing concrete in structures. Non-destructive testing. Determination of rebound number. BSI, London, UK.
29. ASTM (2018) C805:2018. Standard test method for rebound number of hardened concrete. ASTM, West Conshohocken, PA, USA.
30. BSI (2021) EN 12504-4:2021. Testing concrete in structures. Determination of ultrasonic pulse velocity. BSI, London, UK.
31. ASTM (2016) C597:2016. Standard test method for pulse velocity through concrete. ASTM, West Conshohocken, PA, USA.
32. BSI (2005) EN 12504-3:2005. Testing concrete in structures. Determination of pull-out force. BSI, London, UK.
33. ASTM (2019) C900:2019. Standard test method for pull-out strength of hardened concrete. ASTM, West Conshohocken, PA, USA.
34. BSI (1999) EN 1542:1999. Products and systems for the protection and repair of concrete structures. Test methods. Measurement of bond strength by pull-off. BSI, London, UK.
35. ASTM (2020) C1583:2020. Standard test method for tensile strength of concrete surfaces and the bond strength or tensile strength of concrete repair and overlay materials by direct tension (pull-off method). ASTM, West Conshohocken, PA, USA.
36. BSI (2020) 1881-207:2020. Testing concrete. Part 207. Recommendations for the assessment of concrete strength by near-to-surface tests. BSI, London, UK.
37. ASTM (2018) C803:2018. Standard test method for penetration resistance of hardened concrete. ASTM, West Conshohocken, PA, USA.
38. ASTM (2019) C1074:2019. Standard practice for estimating concrete strength by the maturity method. ASTM, West Conshohocken, PA, USA.
39. Russell HG (1999) ACI defines high-performance concrete. *Concrete International* **21(2)**: 56–57.
40. Kibar H and Ozturk T (2015) Determination of concrete quality with destructive and non-destructive methods. *Computers and Concrete* **15(3)**: 473–484.
41. Carette J and Staquet S (2015) Monitoring the setting process of mortars by ultrasonic P and S-wave transmission velocity measurement. *Construction and Building Materials* **94**: 196–208.

42. Whitehurst EA (1951) Soniscope tests concrete structures. *Journal of American Concrete Institute* **47(2)**: 433–444.
43. Saint-Pierre F, Philibert A, Giroux B and Rivard P (2016) Concrete quality designation based on ultrasonic pulse velocity. *Construction and Building Materials* **125**: 1022–1027.
44. Lafhaj Z, Goueygou M, Djerbi A and Kaczmarek M (2006) Correlation between porosity, permeability and ultrasonic parameters of mortar with variable water–cement ratio and water content. *Cement and Concrete Research* **36(4)**: 625–633.
45. Zhutovsky S and Kovler K (2017) Application of ultrasonic pulse velocity for assessment of thermal expansion coefficient of concrete at early age. *Materials and Structures* **50(1)**: Art. 5.
46. Mirvalad S and Nokken M (2015) Studying thaumasite sulfate attack using compressive strength and ultrasonic pulse velocity. *Materials and Structures* **49(10)**: 4131–4146.
47. Zhou C, Li K and Han J (2012) Characterizing the effect of compressive damage on transport properties of cracked concretes. *Materials and Structures* **45(3)**: 381–392.
48. ACI (2004) *Non-destructive Test Methods for Evaluation of Concrete in Structures.* American Concrete Institute, Farmington Hills, MI, USA, Report 228.2R:2004.
49. Carino NJ and Lew HS (2001) The maturity method – From theory to application. *Proceedings of Structures Congress & Exposition, Washington, DC*, https://tsapps.nist.gov/publication/get_pdf.cfm?pub_id=860356.
50. Teixeira S, Santilli A and Puente I (2017) Demoulding vertical elements: Recommendations for apply maturity functions. *Construction and Building Materials* **145**: 392–401.
51. Soutsos M, Turuallo G, Owens KJ, *et al.* (2013) Maturity testing of lightweight self-compacting and vibrated concretes. *Construction and Building Materials* **47**: 118–125.
52. Ferreira L, Branco FG, Costa HSS, Júlio E and Maranha P (2015) Characterization of alkali-activated binders using the maturity method. *Construction and Building Materials* **95**: 337–344.
53. Benaicha M, Burtschell Y and Alaoui AH (2016) Prediction of compressive strength at early age of concrete – Application of maturity. *Journal of Building Engineering* **6**: 119–125.
54. BSI (2010) BS 6089:2010. Assessment of in-situ compressive strength in structures and precast concrete components. BSI, London, UK.
55. Elahi A, Basheer PAM, Nanukuttan SV and Khan QUZ (2010) Mechanical and durability properties of high performance concretes containing supplementary cementitious materials. *Construction and Building Materials* **24(3)**: 292–299.
56. Basheer PAM, Montgomery FR and Long AE (1994) Factorial experimental design for concrete durability research. *Structures and Buildings* **104(4)**: 449–462.
57. Tang L, Nilsson LO and Basheer PAM (2011) *Resistance of Concrete to Chloride Ingress: Testing and Modelling*. Spon Press, London, UK and New York, NY, USA.
58. Bondar D, Ma Q, Soutsos M, *et al.* (2018) Alkali activated slag concretes designed for a desired slump, strength and chloride diffusivity. *Construction and Building Materials* **190**: 191–199.
59. Wang Y, Nanukuttan S, Bai Y and Basheer PAM (2017) Influence of combined carbonation and chloride ingress regimes on rate of ingress and redistribution of chlorides in concretes. *Construction and Building Materials* **140**: 173–183.
60. Wang Y, Jiang X, Wang S, *et al.* (2019) Influence of axial loads on CO_2 and Cl^- transport in concrete phases: Paste, mortar and ITZ. *Construction and Building Materials* **204**: 875–883.
61. Nanukuttan SV, Basheer PAM, McCarter WJ, *et al.* (2015) The performance of concrete exposed to marine environments: Predictive modelling and use of laboratory/on site test methods. *Construction and Building Materials* **93**: 831–840.
62. BSI (1996) BS 1881-208:1996. Testing concrete. Recommendations for the determination of the initial surface absorption of concrete. BSI, London, UK.

63. Figg JW (1973) Methods of measuring the air and water permeability of concrete. *Magazine of Concrete Research* **25(84)**: 213–219.
64. Torrent RT (1992) A two-chamber vacuum cell for measuring the coefficient of permeability to air of the concrete cover on site. *Materials and Structures* **25(6)**: 358–365.
65. Romer M (2005) Effect of moisture and concrete composition on the Torrent permeability measurement. *Materials and Structures* **38(279)**: 541–547.
66. Basheer PAM, Long AE and Montgomery FR (1995) 'CLAM' tests for measuring in-situ permeation properties of concrete. *Nondestructive Testing and Evaluation* **12(1)**: 53–73.
67. BSI (2011) BS 1881:122:2011. Testing concrete. Method for determination of water absorption. BSI, London, UK.
68. Concrete Society (2008) *Permeability Testing of Site Concrete*. Concrete Society, Camberley, UK, Technical report 31.
69. AASHTO (American Association of State Highway and Transportation Officials) (2015) *Standard Method of Test for Electrical Indication of Concrete's Ability to Resist Chloride Ion Penetration*. AASHTO, Washington, DC, USA, Report T-277:2015.
70. ASTM (2019) C1202:2019. Standard test method for electrical indication of concrete's ability to resist chloride ion penetration. ASTM, West Conshohocken, PA, USA.
71. Nolan EA, Ali MA, Basheer PAM and Marsh BK (1997) Testing the effectiveness of commonly-used site curing regimes. *Materials and Structures* **30(1)**: 53–60.
72. Basheer PAM, Nolan EA, McCarter WJ and Long AE (2000) Effectiveness of in situ moisture preconditioning methods for concrete. *Journal of Materials in Civil Engineering* **12(2)**: 131–138.
73. Schönlin K and Hilsdorf HK (1987) Evaluation of the effectiveness of curing of concrete structures. *Concrete Durability: Katharine and Bryant Mather International Conference, Atlanta, GA*, pp. 207–226.
74. Jacobs F and Torrent RT (2009) SIA 262:2003. A step towards performance-based specifications for durability. In *Concrete in Aggressive Aqueous Environments – Performance, Testing, and Modeling*. Swiss Society of Engineers and Architects, pp. 532–539.
75. Torrent R, Denarié E, Jacobs F, Leemann A and Teruzzi T (2012) Specification and site control of the permeability of the cover concrete: The Swiss approach. *Materials and Corrosion* **63(12)**: 1127–1133.
76. Guth DL and Zia P (2001) Evaluation of new air permeability test device for concrete. *ACI Materials Journal* **98(1)**: 44–51.
77. Whiting D (1981) *Rapid Determination of the Chloride Permeability of Concrete*. Federal Highway Administration, Washington, DC, USA.
78. Wang YC, Yang K, Long A, *et al.* (2018) Effectiveness of preconditioning regimes for assessing water permeability of high performance concrete. *Cement and Concrete Composites* **94**: 126–135.
79. Romer M (2005) Recommendation of RILEM TC 189-NEC, Non-destructive evaluation of the concrete cover: Comparative test. Part I: Comparative test of 'penetrability' methods. *Materials and Structures* **38(284)**: 895–906.
80. Pedrosa F and Andrade C (2017) Corrosion induced cracking: Effect of different corrosion rates on crack width evolution. *Construction and Building Materials* **133**: 525–533.
81. BSI (1986) BS 1881–201:1986. Testing concrete. Guide to the use of non-destructive methods of test for hardened concrete. BSI, London, UK.
82. BSI (2020) BS 1881-204:2020. Testing Concrete. Part 204. Recommendations on the use of electromagnetic covermeters. BSI, London, UK.
83. ASTM (2015) C876:2015. Standard test method for corrosion potentials of uncoated reinforcing steel in concrete. ASTM, West Conshohocken, PA, USA.
84. ASTM (2014) G59-97:2014. Standard test method for conducting potentiodynamic polarisation resistance measurements. ASTM, West Conshohocken, PA, USA.

85. Law DW, Millard SG and Bungey J (2000) Galvanostatic pulse measurements of passive and active reinforcing steel in concrete. *Corrosion* **56(1)**: 48–56.
86. Millard SG (1993) Corrosion rate measurement of in-situ reinforced concrete structures. *Proceedings of the ICE. Structures and Buildings* **99(1)**: 84–88. Technical note 604.
87. Broomfield JP (2007) *Corrosion of Steel in Concrete. Understanding, Investigation and Repair.* Taylor & Francis, London, UK.
88. Elsener B (2001) Half-cell potential mapping to assess repair work on RC structures. *Construction and Building Materials* **15(2–3)**: 133–139.
89. Sanchez J, Andrade C, Torres J, Rebolledo N and Fullea J (2016) Determination of reinforced concrete durability with on-site resistivity measurements. *Materials and Structures* **50(1)**: Art. 41.
90. COST Action 521 (2003) *Corrosion of Steel in Reinforced Concrete Structures.* Publications Office of the European Union, Report EUR-20599, https://op.europa.eu/en/publication-detail/-/publication/c122c980-b071-4eb7-8ac8-fc8353a502be.
91. Luco LF (2005) Recommendation of RILEM TC 189-NEC, Non-destructive evaluation of the concrete cover: Comparative test. Part II: Comparative test of 'Covermeters'. *Materials and Structures* **38(10)**: 907–911.
92. Millard SG, Bungey J, Thomas C and Soutsos MN (1998) Assessing bridge pier scour by radar. *NDT & E International* **31(4)**: 251–258.
93. Bungey JH (2004) Sub-surface radar testing of concrete: A review. *Construction and Building Materials* **18(1)**: 1–8.
94. ASTM (2018) D4580:2018. Standard practice for measuring delaminations in concrete bridge decks by sounding. ASTM, West Conshohocken, PA, USA.
95. Cheng C and Sansalone M (1993) The impact-echo response of concrete plates containing delaminations: Numerical, experimental and field studies. *Materials and Structures* **26(5)**: 274–285.
96. Eitzen DG and Wadley HNG (1984) Acoustic emission: Establishing the fundamentals. *Journal of Research of the National Bureau of Standards* **89(1)**: 75–100.
97. Milovanovic B and Pecur IB (2016) Review of active IR thermography for detection and characterization of defects in reinforced concrete. *Journal of Imaging* **2(11)**, 10.3390/jimaging2020011.
98. Bungey JH and Millard SG (1993) Radar inspection of structures. *Structures and Buildings* **99(2)**: 173–186.
99. BSI (1986) BS 1881-205:1986. Testing concrete. Recommendations for radiography of concrete. BSI, London, UK.
100. Clark MR, McCann DM and Forde MC (2003) Application of infrared thermography to the non-destructive testing of concrete and masonry bridges. *NDT & E International* **36(4)**: 265–275.
101. Weil GJ (2004) Infrared thermographic techniques. In *Handbook on Nondestructive Testing of Concrete*, 2nd edn (Malhotra VM and Carino NJ (eds)). CRC Press, Ch. 15.
102. ASTM (2013) D4788:2013. Standard test method for detecting delaminations in bridge decks using infrared thermography. ASTM, West Conshohocken, PA, USA.
103. Liu PL, Lin LC, Hsu YY, Yeh CY and Yeh PL (2017) Recognition of rebars and cracks based on impact-echo phase analysis. *Construction and Building Materials* **142**: 1–6.
104. Azari H, Nazarian S and Yuan D (2014) Assessing sensitivity of impact echo and ultrasonic surface waves methods for nondestructive evaluation of concrete structures. *Construction and Building Materials* **71**: 384–391.
105. Carino NJ (2001) The impact-echo method: An overview. *Proceedings ASCE Structures Congress.*
106. Kim HJ, Mission JL, Dinoy PR, Kim HS and Park TW (2020) Guidelines for impact echo test signal interpretation based on wavelet packet transform for the detection of pile defects. *Applied Science* **10(7)**: 2633, 10.3390/app10072633.
107. Barrile V and Pucinotti R (2005) Application of radar technology to reinforced concrete structures: A case study. *NDT & E International* **38(7)**: 596–604.

108. ASTM (2020) D4748:2020. Standard test method for determining the thickness of bound pavement layers using short-pulse radar. ASTM, West Conshohocken, PA, USA.
109. ASTM (2015) D6087:2015. Standard test method for evaluating asphalt-covered concrete bridge decks using ground penetrating radar. ASTM, West Conshohocken, PA, USA.
110. ACI (2021) *Structural Health Monitoring Technologies for Concrete Structures*. ACI, Farmington Hills, MI, USA, Report PRC-444.2-21:2021.
111. Suryanto B, Saraireh D, Kim J, *et al.* (2017) Imaging water ingress into concrete using electrical resistance tomography. *International Journal of Advances in Engineering Sciences and Applied Mathematics* **9(2)**: 109–118.
112. McCarter WJ, Chrisp T, Starrs G, *et al.* (2013) Characterization of physio-chemical processes and hydration kinetics in concretes containing supplementary cementitious materials using electrical property measurements. *Cement and Concrete Research* **50**: 26–33.
113. Waters EH (1974) Determining the moisture content of concrete: The use of electrical resistance method. *Building Science* **9(4)**: 289–297.
114. McCarter WJ, Emerson M and Ezirim H (1995) Properties of concrete in the cover zone: Developments in monitoring techniques. *Magazine of Concrete Research* **47(172)**: 243–251.
115. McCarter WJ and Garvin S (1989) Dependence of electrical impedance of cement-based materials on their moisture condition. *Journal of Physics D: Applied Physics* **22(11)**: 1773–1776.
116. Rajabipour F and Weiss J (2007) Electrical conductivity of drying cement paste. *Materials and Structures* **40(10)**: 1143–1160.
117. Wilson JG and Whittington HW (1990) Variations in the electrical properties of concrete with change in frequency. *IEE Proceedings* **137(5)**: 246–254.
118. Wei X and Xiao L (2013) Electrical resistivity monitoring and characterisation of early age concrete. *Magazine of Concrete Research* **65(10)**: 600–607.
119. Srinivasan S, Basheer PAM, Mao JH, *et al.* (2012) Electrical based sensors and remote monitoring system for assessing the corrosion related activity in concrete at Hangzhou Bay Bridge. *Civil Structural Health Monitoring Workshop, Berlin, Germany*, pp. 1–12.
120. Srinivasan S, Basheer PAM, Mao JH, Jin WL and McCarter WJ (2013) An integrated sensing system for monitoring the corrosion activity in concrete blocks exposed at Hangzhou Bay Bridge. *2nd Conference on Smart Monitoring, Assessment and Rehabilitation of Civil Structures, Istanbul, Turkey*, pp. 1–12.
121. Basheer PAM, Russell D, Andrews R and McCarter WJ (2006) Monitoring internal changes in concrete during carbonation and chloride ingress using a covercrete electrode array. *11th Proceedings of Structural Faults and Repair, Edinburgh* (Forde MC, (ed.)). Engineering Technics Press, Edinburgh, Scotland.
122. Yoon IS and Chang CH (2020) Effect of chloride on electrical resistivity in carbonated and non-carbonated concrete. *Applied Sciences* **10(18)**: 6272, 10.3390/app10186272.
123. Myrdal R (2007) *The Electrochemistry and Characteristics of Embeddable Reference Electrodes for Concrete. European Federation of Concrete Publications 43.* Woodhead/CRC Press.
124. Muralidharan S, Ha TH, Bae JH, Ha YC and Kim DKJ (2007) A promising potential embeddable sensor for corrosion monitoring application in concrete structures. *Measurement* **40(6)**: 600–606.
125. Raupach M and Schiessl P (1997) Monitoring system for the penetration of chlorides, carbonation and the corrosion risk for the reinforcement. *Construction and Building Materials* **11(4)**: 207–214.
126. Schiessl P and Raupach M (1992) Monitoring system for the corrosion risk for steel in concrete. *Concrete International* **14(7)**: 52–55.
127. Broomfield JP, Davies H and Hladky K (2002) The use of permanent corrosion monitoring in new and existing reinforced concrete structures. *Cement and Concrete Composites* **24(1)**: 27–34.
128. Duffó G and Farina SB (2009) Development of an embeddable sensor to monitor the corrosion process of new and existing reinforced concrete structures. *Construction and Building Materials* **23(8)**: 2746–2751.

129. Atkins CP, Scantlebury JD, Nedwell PJ and Blatch SP (1996) Monitoring chloride concentrations in hardened cement pastes using ion selective electrodes. *Cement and Concrete Research* **26(2)**: 319–324.
130. Montemor MF, Alves JH, Simoes AM, *et al.* (2006) Multiprobe chloride sensor for in situ monitoring of reinforced concrete structures. *Cement and Concrete Composites* **28(3)**: 233–236.
131. COST Action 509 (1997) *Corrosion and Protection of Metals in Contact with Concrete*. Publications Office of the European Union, Report, Brussels, Belgium. https://www.cost.eu/publication/corrosion-and-protection-of-metals-in-contact-with-concrete-final-report/.
132. Bertolini L, Elsener B, Pedeferri P and Polder R (2004) *Corrosion of Steel in Concrete. Prevention, Diagnosis, Repair*, 2nd edn. Wiley, pp. 300–306.
133. Murray GM, Cain RP, Carkhuff BG and Weiskopf F (2005) *Techniques for Sensing Chloride Ions in Wet or Dry Media*. International patent application number PCT/US2003/039348.
134. Telli L, Brahimi B and Hammouche A (2000) Study of pH sensor with MnO_2 and montmorillonite based solid state internal reference. *Solid State Ionics* **128**: 255–259.
135. Xu CN, Miyazaki K and Watanabe T (1998) Humidity sensors using manganese dioxides. *Sensors and Actuators B: Chemical* **46(2)**: 87–96.
136. McCarter J, Chrisp T, Starrs G, *et al.* (2012) Developments in performance monitoring of concrete exposed to extreme environments. *Journal of Infrastructure Systems* **18(3)**: 167–175.
137. Alwis L, Sun T and Grattan KTV (2013) Optical fibre-based sensor technology for humidity and moisture measurement: Review of recent progress. *Measurement* **46(10)**: 4052–4074.
138. Enckell M, Glisic B, Myrvoll F and Bergstrand B (2011) Evaluation of a large-scale bridge strain, temperature and crack monitoring with distributed fibre optic sensors. *Journal of Civil Structural Health Monitoring* **1(1–2)**: 37–46.
139. Glisic B and Inaudi D (2007) *Fibre Optic Methods for Structural Health Monitoring*. Wiley, p. 276.
140. Inaudi D and Glisic B (2007) Distributed fibre optic sensing for long-range monitoring of pipelines. *3rd International Conference on Structural Health Monitoring of Intelligent Infrastructure, Vancouver, Canada*. https://roctest.com/wp-content/uploads/2017/03/c176.pdf.
141. Slowik V, Schlattner E and Klink T (1998) *Fiber Bragg Grating Sensors in Concrete Technology. Leipzig Annual Civil Engineering Report 3*, pp. 109–120.
142. Measures MR (2001) *Structural Monitoring with Fiber Optic Technology*. Academic Press, San Diego, CA, USA.
143. Ansari F (1997) State of the art in applications of fiber-optic sensors to cementitious composites. *Cement and Concrete Composites* **19(1)**: 3–19.
144. Inaudi D (2000) Application of civil structural monitoring in Europe using fiber optic sensors. *Progress in Structural Engineering and Materials* **2(3)**: 351–358.
145. Gebremichael YM, Li WT, Meggitt BT, *et al.* (2005) A field deployable, multiplexed Bragg grating sensor system used in an extensive highway bridge monitoring evaluation tests. *IEEE Sensors Journal* **5(3)**: 510–519.
146. Scott RH, Banerji P, Chikermane S, *et al.* (2013) Commissioning and evaluation of a fiber-optic sensor system for bridge monitoring. *IEEE Sensors Journal* **13(16)**: 2555–2562.
147. Kronenberg PK, Rastogi PK, Giaccari P and Limberger HG (2002) Relative humidity sensor with optical fiber Bragg gratings. *Optics Letters* **27**: 1385–1387.
148. Yeo TL, Sun T, Grattan KTV, *et al.* (2005) Polymer-coated fiber Bragg grating for relative humidity sensing. *IEEE Sensors Journal* **5(5)**: 1082–1089.
149. Alwis L, Sun T and Grattan KTV (2016) Fibre grating-based sensor design for humidity measurement in chemically harsh environment. *Procedia Engineering* **168**: 1317–1320.
150. Srinivasan S, Taylor SE, Basheer PAM, *et al.* (2009) Fibre optic relative humidity probe for long-term monitoring of concrete structures. *4th International Conference on Structural Health Monitoring of Intelligent Infrastructure, Zurich, Switzerland*.

151. Leung C, Wan KT and Chen LQ (2008) A novel optical fiber sensor for steel corrosion in concrete structures. *Sensors* **8(3)**: 1960–1976.
152. Grattan SKT, Basheer PAM, Taylor SE, *et al.* (2007) Fibre Bragg grating sensors for reinforcement corrosion monitoring in civil engineering structures. *Journal of Physics: Conference Series* **76**, 10.1088/1742-6596/76/1/012018.
153. Korostynska O, Arshak K, Gill E and Arshak A (2007) Review on state of the art in polymer based pH sensors. *Sensors* **7(12)**: 3027–3042.
154. Staneva D and Betcheva R (2007) Synthesis and functional properties of new optical pH sensor based on benzo[de]anthracen-7-one immobilized on the viscose. *Dyes and Pigments* **74(1)**: 148–153.
155. McPolin D, Basheer PAM, Long AE, Grattan KTV and Sun T (2007) New test method to obtain pH profiles due to carbonation of concrete containing supplementary cementitious materials. *Journal of Materials in Civil Engineering* **19(11)**: 936–946.
156. Badini GE, Grattan KTV, Sun T, Palmer AW and Tseung ACC (1989) Development of pH sensitive substrates for optical sensor applications. *Proceedings in Physics* **44**: 436–442.
157. McPolin D, Basheer PAM, Long AE, *et al.* (2005) Optical fibre sensor systems for monitoring concrete. *Concrete* **39(9)**: 54–55.
158. Fuhr PL, Huston DR, McPadden AJ and Cauley RF (1996) Embedded chloride detectors for roadways and bridges. *Proceedings of SPIE* **2719**: 229–237.
159. Cosentino P, Grossman B, Shieh C, *et al.* (1995) Fiber-optic chloride sensor development. *Journal of Geotechnical Engineering* **121(8)**: 610–617.
160. Tang JL and Wang JN (2007) Measurement of chloride ion concentration with long period grating technology. *Smart Materials and Structures* **16(3)**: 665–672.
161. Fuhr PL, MacCraith BD, Huston DR, Guerrina M and Nelson M (1997) Fiber optic chloride sensing: If corrosion's the problem, chloride sensing is the key. *Proceedings of SPIE* **3180**: 92–104.
162. Wu TG, Liu GW, Fu SG and Xing F (2020) Recent progress of fiber-optic sensors for the structural health monitoring of civil infrastructure. *Sensors* **20(16)**: 4517.
163. Fuhr PL and Huston DR (2000) Fiber optic chloride threshold detectors for concrete structures. *Journal of Structural Control* **7(1)**: 77–102.
164. McPolin D, Basheer PAM, Long AE, Grattan KTV and Sun T (2005) Obtaining progressive chloride profiles in cementitious materials. *Construction and Building Materials* **19(9)**: 666–673.
165. Laferrière F, Inaudi D, Kronenberg P and Smith IFC (2008) A new system for early chloride detection in concrete. *Smart Materials and Structures* **17(4)**: 4501–4508.

Soutsos M
ISBN 978-0-7277-6375-4
https://doi.org/10.1680/icehcd.63754.493

Chapter 14
Durability of concretes based on non-Portland cements

John L Provis
University of Sheffield, UK

14.1. Introduction

This chapter addresses the durability of concretes based on non-Portland cements, with a particular focus on four key binder types: calcium aluminate, calcium sulfoaluminate, alkali-activated and magnesia-based cements. These are discussed with respect to key characteristics related to concrete durability: mechanical performance, intrinsic stability as well as the ability to resist the ingress of aggressive species and protect the steel reinforcement. Applications in both general and specialist concreting scenarios are briefly noted.

The durability of concretes based on non-Portland cements is, in general, controlled by the same factors that control the durability of conventional concretes. For durable reinforced concretes, regardless of the binder type, it is essential to produce a material that is suitably impermeable to aggressive agents and that maintains an internal chemical environment that can protect the passive layer on mild steel reinforcing elements. For all concretes, the ability to resist relevant mechanical or chemical actions in the desired application is critical. The material must also be resistant to degradation or breakdown of its structure through internal or autogenous processes. These principles hold true regardless of the type of binder on which the concrete is based.

The comments of Neville [1] regarding concrete quality also apply to concretes based on non-Portland cement: the ingredients of a good concrete are generally the same as those of a poor concrete, but the difference is due to understanding and know-how. This refers both to realising the correct design of the mix proportioning (in formulating a binder and then using it in a concrete) and to executing and implementing the concrete.

The focus of this chapter is, therefore, directed toward identifying those cases where a concrete based on a non-Portland cement may behave differently from what would be expected of a good quality Portland-based concrete. This is by no means to say that the influence of parameters such as water-to-binder ratio, aggregate grading, compaction and finishing are not important for such materials; in fact, quite the opposite is true. However, the expected trends and good practice in these respects for conventional concretes will, in general, also hold for concretes based on non-Portland cements. This chapter will instead aim to highlight areas in which the specific chemistry of particular non-Portland cements can either offer performance advantages or raise concerns during the application of these materials. Binder types covered include calcium aluminate cements, ye'elimite-rich (calcium sulfoaluminate) cements, alkali-activated cements, magnesia-based cements and others with particular relevance under each subheading, focusing on cases where the Portland cement content is zero or very low.

14.2. Internal processes: Stability and degradation

It is clearly essential that a concrete based on a non-Portland cement has sufficient intrinsic stability – chemical and dimensional – to enable it to be effective and reliable in service. The most notorious example of a non-Portland cement failing to comply with this requirement is the strength loss leading to structural failure of some concretes based on calcium aluminate cements through a chemical degradation process known as 'conversion' [2]. This process, its consequences and the ways in which it may be mitigated have been discussed at length by many authors including [3–5]. It has been argued that calcium aluminate cement concretes formulated with sufficiently low water-to-cement ratios (below 0.40, and with a minimum cement content of 400 kg/m^3 [5]) are much less affected by strength loss due to conversion, as the water released by conversion will participate in further hydration of the remnant clinker grains and the microstructure will remain dense [6]. This approach plus designing concretes according to converted (rather than maximum) strength may allow the broader use of calcium aluminate cement in concretes, provided care is taken in the selection of materials and applications.

Cements with a high content of ye'elimite usually contain belite as a secondary phase. Tetracalcium aluminoferrite, ternesite and alite are also found in many such clinkers, and anhydrite or gypsum are added to regulate setting [7–9]. The nature and balance of the sulfate source also determines whether the cements compensate for expansion and shrinkage and whether they undergo some chemical shrinkage. Broadly (and dependent on other parameters such as the water-to-binder ratio and cement fineness), expansion occurs if sulfate availability is high and ettringite dominates the hydration product assemblage [10] (Figure 14.1). The shrinkage and creep of these sulfoaluminate cements – whether used alone or blended with Portland cement – are not well described by existing models designed and calibrated for Portland cement. Thus, further work is required so that these materials can be appropriately described in structural codes [11].

Alkali-activated cements and concretes can form alkali aluminosilicate or calcium silicate hydrate type gels as key binding phases, although the calcium silicate hydrates that do form in these binders are also extensively substituted by aluminium and alkalis [12,13]. These phases are strong, have been demonstrated in practical applications to retain and gain strength over many years in service [13,14] and have tailorable

Figure 14.1 Expansive behaviour of calcium sulfoaluminate cement pastes (65.3% ye'elimite) with a water-to-cement ratio of 0.45 and gypsum contents as marked, cured in 'ultra-pure' water at 20°C after demoulding at an age of 24 hours. (Data from [10])

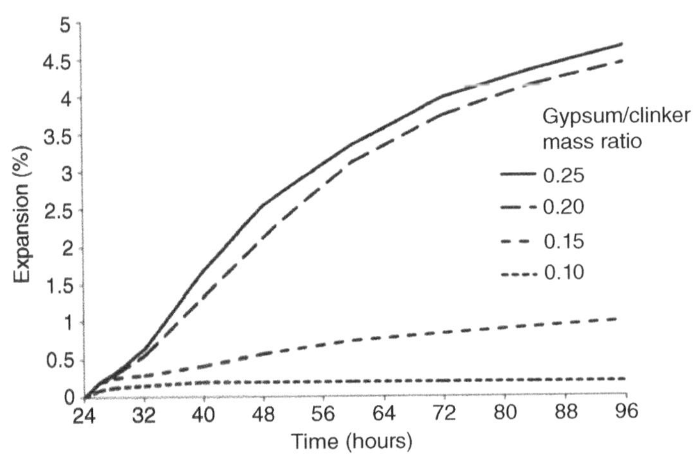

porosity based on the relative contents of high-calcium and low-calcium gels (Figure 14.2). Some binders formed by the alkali activation of blastfurnace slag (BFS) have undergone drying shrinkage [15], although effective shrinkage mitigation strategies have been developed and described [16]. Gel phases have partially crystallised to zeolites in some low-calcium alkali-activated binders subjected to warm, moist environments for prolonged periods [17], but this is not necessarily detrimental to strength retention, particularly if the water-to-binder ratio is suitably low [18]. As mentioned for sulfoaluminate cements, the relations between creep and long-term engineering properties also require further detailed investigation for alkali-activated binders.

Magnesia-based binders can be designed, depending on the chemistry of the anion that is selected to react with the magnesia, to be expansive or neutral or have some intrinsic tendency toward shrinkage. Magnesia is an expansive constituent of concretes [20], as it tends to hydrate to form brucite with a marked volume increase. This hydration can be manipulated, so that magnesia-based cements can be useful if dimensional control is necessary. Some magnesium carbonate cements are designed with nesquehonite as a key binding phase, and care is required that this phase, which is readily formed at high partial pressures of carbon dioxide during accelerated curing but can be unstable under ambient conditions with respect to hydromagnesite [21], does not then decompose while the material is in service [22]. Magnesium oxychloride (Sorel) cements are susceptible to degradation when exposed to liquid water, as the key binding phases are rather soluble by the standards of cementitious hydrates, and various modifications of the basic magnesium oxychloride chemistry have been developed and implemented to improve performance in this regard [23].

In general, alkali–aggregate reactions are not unduly problematic for all cement types discussed in this chapter except when reactive aggregates are used. This is particularly important for alkali-activated binders, whose inventory of alkalis is far beyond the usual limits imposed on Portland-based cements for use

Figure 14.2 Porosity of alkali-activated pastes formulated with different ratios of fly ash (FA) to ground granulated BFS, as determined using X-ray microtomography in [19]. All pastes were formulated with comparable water-to-binder ratios and activator contents. BFS-rich binders have a higher prevalence of calcium-rich gels, while the FA source used was very low in calcium. (Reproduced from [13])

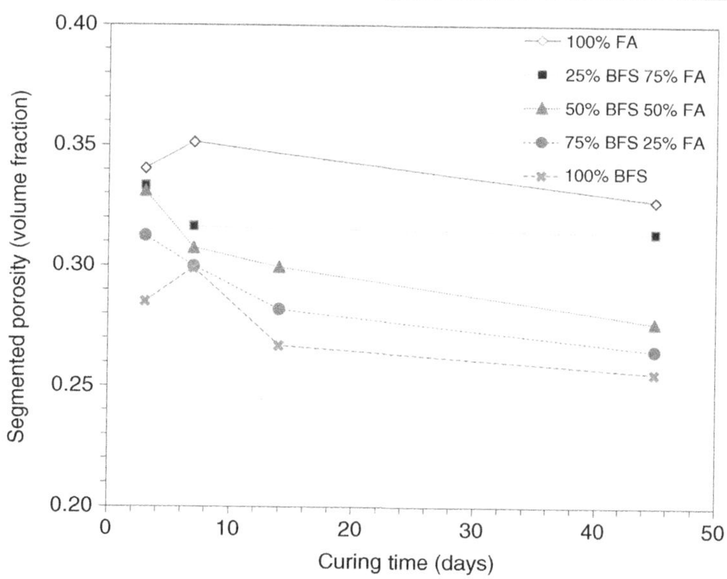

in concretes. However, the very high concentration of aluminium in alkali-activated cements appears to restrict the expansive nature of the reactions that do take place between the aggregate and these alkalis [24], meaning that, in general, this process is not as problematic as may otherwise have been anticipated simply from considering the high binder alkalinity [25,26]. The very alumina-rich cements (calcium aluminate and sulfoaluminate) are usually highly resistant to alkali–aggregate reactions if they are not blended with large quantities of Portland cement.

14.3. Transport properties and protection of steel reinforcement

A wide variety of non-Portland cementitious binders are used to produce concretes with steel reinforcement, and there are two key aspects that control the suitability of each binder type in this application: (*a*) the ability to generate an environment that passivates mild steel (in general, high pH without excessive free chloride) and (*b*) the ability to maintain steel passivity by excluding potentially damaging species (particularly chloride and carbonate) from the external environment.

Most cementitious materials (conventional and alternative) are, by nature, quite highly alkaline and so will comply with condition (*a*). This is particularly true for alkali-activated cements. For calcium aluminate [27] and sulfoaluminate [28] cements, the pore solution pH depends on the clinker alkali content (which is variable), rather than being defined intrinsically by binder chemistry or mineralogy. However, in general, condition (*a*) is very difficult to meet when using binders with near-neutral to acidic pH, such as magnesium phosphate cements [29] or carbonation-cured magnesia-based or calcium silicate cements. Sorel cements have a high free chloride content, which is generally unfavourable for durability when used in conjunction with steel elements [23]. Alinite cements, based on a chloride-containing analogue of tricalcium silicate, can suffer potentially problematic steel corrosion under conditions of cyclic immersion in sodium chloride solution, although constant or cyclic immersion in tap water did not appear to be as problematic due to the washing out of the chloride that had been released during dissolution of the alinite during clinker hydration [30].

Compliance with condition (*b*) is fundamentally dependent on the hardened material reaching a suitably low permeability, which restricts the ingress of chloride and carbonate. As mentioned in the introduction of this chapter, standard good practice for concrete mix design – particularly the selection of a low water-to-binder ratio, aided where possible with chemical admixtures – is imperative in achieving this, regardless of the type of binder used. In particular, many of the potential durability issues associated with non-Portland concretes are like those for concretes with a high volume of blended Portland-based binders [31], such as low or zero levels of portlandite to act as a pH buffer and, sometimes, also the slower development of a final, mature microstructure.

Other factors are also important for specific types of non-Portland cement, particularly those relating to the chloride-binding capacity of calcium aluminate cements [32] and alkali-activated cements [33]. Given that chloride binding tends to be strongest in layered double-hydroxide-type (LDH) phases, mixes with higher quantities of either Mg-Al (hydrotalcite-like) or Ca-Al (AFm family) LDH may have a higher chloride-binding capacity and therefore reduced chloride ingress [34]. Conversely, the ettringite that dominates calcium sulfoaluminate-based binders does not bind chloride to a comparable degree [35].

Some authors have observed that calcium aluminate cement binders are highly resistant to carbonation [27], whereas other researchers have found that they carbonate more rapidly than Portland cement [36]. Hargis *et al.* [36] demonstrated that sulfoaluminate cements carbonated more rapidly than Portland cement (Figure 14.3). For both calcium aluminate and sulfoaluminate cements, the development of a precise comparative test of the carbonation rate is hindered because the water content required for complete hydration

Figure 14.3 Carbonation depths of mortars produced from different cement types: Portland cement (PC), calcium aluminate cement (CAC) and calcium sulfoaluminate cement (M). The number immediately following the M indicates the molar ratio of calcium sulfate to ye'elimite, and the number after the underscore for all mortars indicates the water-to-binder ratio. Mortar prisms with a paste volume of 55% were carbonated at 20°C, 4% CO_2 and 57% relative humidity, after 28 days of pre-curing and conditioning. (Reproduced from [36])

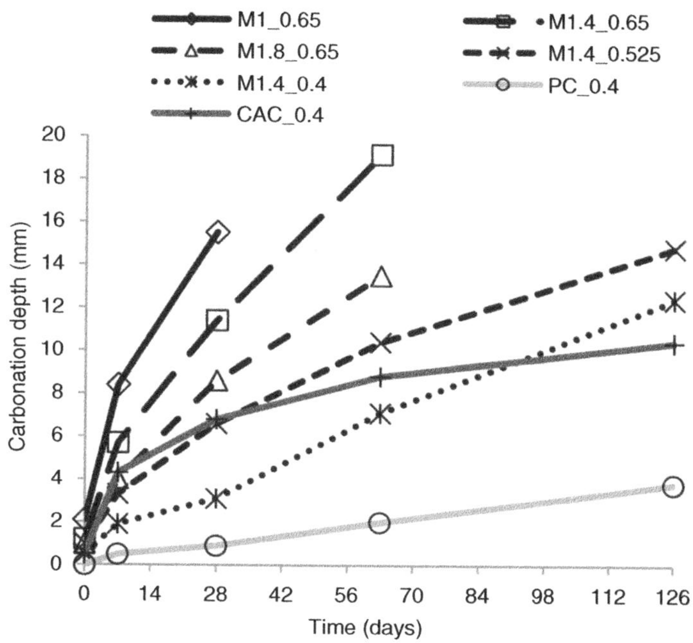

is very high for calcium aluminate cements and also for some sulfoaluminate binders that are designed to have a high ettringite content. For this reason, a comparison of different binders with an identical water-to-cement ratio may be misleading. Moreover, because the strength development profiles of the three types of binders can also be very different from each other, comparing materials designed to reach the same strength at a particular age is also challenging. Different researchers have used different factors when comparing mixes, which is likely to be a contributing reason for why different trends have been reported for different classes of binders.

The main deleterious effect due to the carbonation of concrete is the corrosion of the steel reinforcement. However, the carbonated binder may have the same or a higher strength compared to the pristine material. The complex electrochemical effects (in concretes of all types) mean that corrosion is not necessarily initiated as soon as the carbonation front reaches the steel reinforcement. This has, in particular, been observed by Dunster *et al.* for calcium aluminate cement concretes [37]. An important property of calcium sulfoaluminate cements is their strong ability to self-desiccate [38], which can increase the electrical resistivity of the material and reduce the corrosion rate of the embedded steel [39].

The water and chloride permeability parameters of alkali-activated materials are broadly comparable to those of Portland-based cements [40]. The effects of chemical binding are potentially important in influencing chloride transport [33]. Low-calcium alkali-activated materials tend to have a relatively open pore structure that enables more rapid chloride transport, whereas higher-calcium binders, which tend to form calcium silicate hydrate-type binding gels, are usually much less permeable [41]. The pore solution of alkali-activated cements tends to be highly alkaline, with hydroxide concentrations of up to 1 mol/l observed across a wide range of precursor compositions [42–45]. Interpreting the results of an analysis of the carbonation resistance of these binders requires care. Accelerated testing of alkali-activated materials often shows rapid carbon dioxide ingress [46], but under natural conditions the internal pH can remain high enough that steel is passivated even after carbonation [47]. Binders can be designed specifically to resist degradation by carbonation [48]. The role of sulfide in influencing and controlling steel corrosion (both the actual process and its measurement by electrochemical techniques) in alkali-activated binders is now starting to be better understood [49,50], and similar effects are likely to be observed in any other binder system that is blended with a high volume of ground granulated BFS as a reactive constituent, for example, supersulfated cements. The mechanisms of steel reinforcement corrosion in concretes produced from sulfide-rich cements (e.g. those depicted in Figure 14.4), either because of a pH reduction due to carbonation or induced by chloride ingress, are likely to differ very significantly from those that prevail in binders with an intrinsically oxidising chemistry, such as Portland cement [49].

It is also interesting, and potentially very valuable, to consider the use of reinforcement systems other than conventional mild steel – for example, stainless steel, polymer composites or fibre reinforcing – in partnership with non-Portland cements in concretes. This opens further possibilities for using cements with a high chloride content in high-performance reinforced structural materials.

Figure 14.4 Photographs of mild steel reinforcing bars recovered from mortar specimens after 270 days of immersion in alkaline chloride-rich solutions and Raman spectra of the surface layers for mortars produced from sodium silicate-activated BFS, a 1:1 blend of BFS with white Portland cement (wPC) and neat wPC. (Adapted from [51] under Creative Commons CC-BY licence terms)

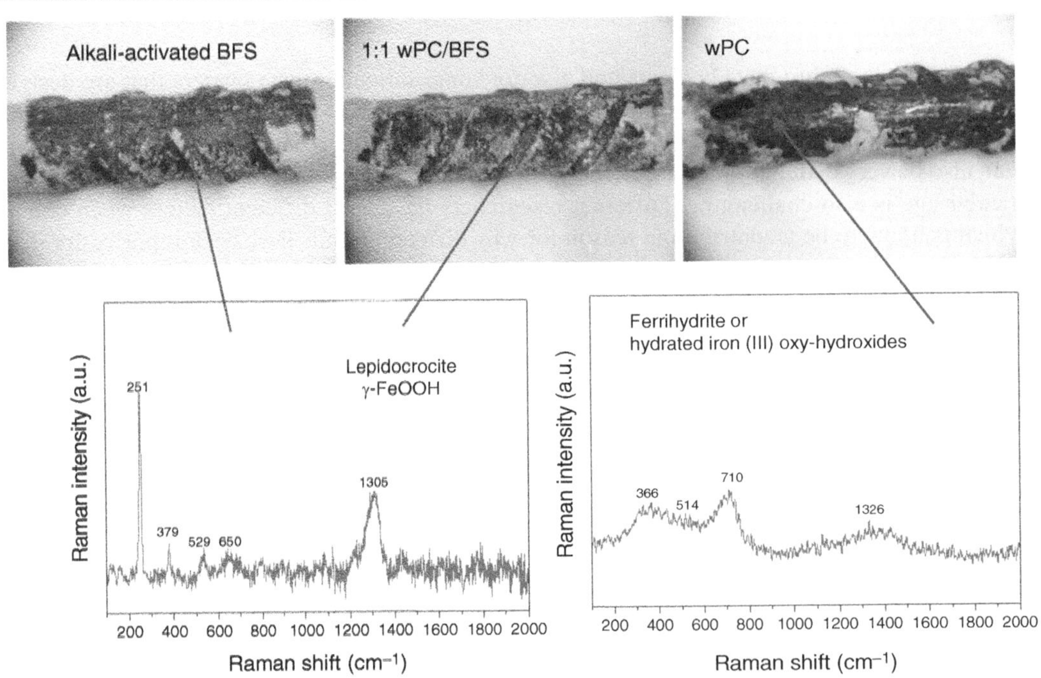

14.4. Concretes for special applications

Concretes based on non-Portland cements are often specified for use in applications that can take advantage of their specific durability characteristics. It is not possible to give an exhaustive list of all such characteristics within the context of a chapter such as this, and the reader is referred to the specialist literature for each cement type to obtain more detailed insights. However, some key areas for each major binder type are described below.

Calcium aluminate cement was historically specified due to its very rapid development of high strength, which also makes it useful in applications where mechanical forces are applied, for example, under abrasion [52]. Calcium aluminate cement concretes have also been specified for chemically aggressive environments, such as marine and sulfate-rich conditions, for sewer pipe construction or lining [5], and also in refractory applications [53], where the absence of free lime (or portlandite) is beneficial in reducing damage due to thermal cycling.

Calcium sulfoaluminate cements can be produced to be very resistant to sulfate attack due to the high content of ettringite that forms when ye'elimite is mixed with sufficient calcium sulfate. They may be specified for this reason. They may also be specified for special applications where some expansiveness is desired. Sulfoaluminate chemistry has attractive possibilities in the management of hazardous and radioactive wastes [54].

Alkali-activated materials offer particular possibilities in applications requiring service at high temperatures [55], due to the low bound water content and absence of portlandite. This class of cement can also be useful in service environments requiring high resistance to chemically aggressive conditions, for example sulfates or acids (organic and inorganic), due to their lower calcium content than other classes of cement and the possibility of achieving very good impermeability [13].

Magnesia-based cements are used in many diverse applications [23]. However, the durability of most classes of magnesia-based cement under field service environments needs further investigation, as the use of these cements in large-scale concrete production (for general or specialist application) is not yet widespread and so many aspects of behaviour and performance still require further analysis.

14.5. Mechanical properties and durability

It is assumed in standards for concretes based on Portland cement, both implicitly and explicitly, that a stronger concrete will be a more durable concrete. This is well known to be an oversimplification of a set of highly complex phenomena that control concrete durability [56]. The mismatch between strength (usually specified at 28 days and in compression) and key durability characteristics is even more marked when considering binders with diverse chemistries. However, any performance-based approach to the understanding and specification of concrete durability will almost inevitably include compressive strength as a key measurable material parameter, as do almost all prescriptive concrete standards in force at present [57].

Each type of non-Portland binder has its own specific strength development profile (some are very rapid, e.g. calcium aluminate cements [5], whereas others are somewhat slower, e.g. supersulfated cements [58]), and there are different curing conditions for optimal performance. It is important to be aware of these differences to be able to understand and optimise durability and also when designing and implementing durability tests. If a material gains maturity (and strength is a reasonable measure of this) relatively slowly, it needs to be allowed additional curing time before being exposed to potentially damaging service conditions; premature exposure can cause an excessive loss of performance. Key examples of this are processes such as carbonation and creep, which strongly depend on the microstructure of the binder and which may proceed very rapidly at the earliest ages of a material but can be much better resisted by a material that has cured for a longer duration before it is placed into service.

Moreover, durability tests that can be conducted at a very early age (e.g. ASTM C1012 [59] for sulfate resistance allows exposure to begin when a specified compressive strength is reached) can unduly penalise cements with a high early strength. Such materials may be subjected to testing in aggressive sulfate-rich environments from as early as one day after casting, when their microstructure is still far from its final mature form even though an acceptable compressive strength has already developed. This type of consideration needs to be incorporated into the design of any testing programme that aims to understand, and hence to predict or specify, the durability of concretes based on any cement whose chemistry and microstructure differ significantly from the materials for which the testing protocols were initially designed, which is almost always plain Portland cement. As an example of how this type of approach can be implemented and used, BSI PAS 8820:2016 [60] describes performance-based specifications and testing of alkali-activated cements and concretes using adaptations of standardised testing methods that are used more commonly for Portland cement. It is very possible that a similar performance-based approach to the specification of a broader set of non-Portland cements will follow in other jurisdictions in the near future, as engineers seek to design new high-performance concretes with a reduced environmental footprint compared to established materials [61,62].

14.6. Concluding remarks

The processes and phenomena that control the durability of non-Portland cements are as diverse and complex as the materials themselves and depend on both material selection and the service environment into which the materials are placed. However, a fundamental principle holds in all cases: if a good quality concrete can be manufactured and placed into service, it is wholly possible to achieve excellent performance with any of a wide range of potential binder chemistries.

REFERENCES

1. Neville AM (2011) *Properties of Concrete*, 5th edn. Pearson Education, Harlow, UK.
2. Midgley HG and Midgley A (1975) The conversion of high alumina cement. *Magazine of Concrete Research* **27(91)**: 59–77.
3. Neville AM (2009) History of high-alumina cement. Part 1: Problems and the Stone report. *Proceedings of the Institute of Civil Engineers – Engineering History and Heritage* **162(EH2)**: 81–91.
4. Neville AM (2009) History of high-alumina cement. Part 2: Background to issues. *Proceedings of the Institute of Civil Engineers – Engineering History and Heritage* **162(EH2)**: 93–101.
5. Ideker JH, Scrivener KL, Fryda H and Touzo B (2019) Calcium aluminate cements. In *Lea's Chemistry of Cement and Concrete*, 5th edn (Hewlett PC and Liska M (eds)). Butterworth Heinemann, Oxford, UK, pp. 537–584.
6. Juenger MCG, Winnefeld F, Provis JL and Ideker J (2011) Advances in alternative cementitious binders. *Cement and Concrete Research* **41(12)**: 1232–1243.
7. Glasser FP and Zhang L (2001) High-performance cement matrices based on calcium sulfoaluminate-belite compositions. *Cement and Concrete Research* **31(12)**: 1881–1886.
8. Bullerjahn F, Schmitt D and Ben Haha M (2014) Effect of raw mix design and of clinkering process on the formation and mineralogical composition of (ternesite) belite calcium sulphoaluminate ferrite clinker. *Cement and Concrete Research* **59**: 87–95.
9. Zea-Garcia JD, Santacruz I, Aranda MAG and De la Torre AG (2019) Alite-belite-ye'elimite cements: Effect of dopants on the clinker phase composition and properties. *Cement and Concrete Research* **115**: 192–202.
10. Chen IA, Hargis CW and Juenger MCG (2012) Understanding expansion in calcium sulfoaluminate–belite cements. *Cement and Concrete Research* **42(1)**: 51–60.

11. Sirtoli D, Wyrzykowski M, Riva P, *et al.* (2019) Shrinkage and creep of high-performance concrete based on calcium sulfoaluminate cement. *Cement and Concrete Composites* **98**: 61–73.
12. Bernal SA and Provis JL (2014) Durability of alkali-activated materials: Progress and perspectives. *Journal of the American Ceramic Society* **97(4)**: 997–1008.
13. Provis JL and van Deventer JSJ (eds) (2014) *Alkali-Activated Materials*. Springer/RILEM, Dordrecht, Netherlands, State-of-the-art report TC 224-AAM.
14. Provis JL and van Deventer JSJ (2019) Geopolymers and other alkali-activated materials. In *Lea's Chemistry of Cement and Concrete*, 5th edn (Hewlett PC and Liska M (eds)). Butterworth-Heinemann, Oxford, UK, pp. 779–805.
15. Ye H and Radlińska A (2016) Shrinkage mechanisms of alkali-activated slag. *Cement and Concrete Research* **88**: 126–135.
16. Ye H and Radlińska A (2017) Shrinkage mitigation strategies in alkali-activated slag. *Cement and Concrete Research* **101**: 131–143.
17. De Silva P and Sagoe-Crentsil K (2008) Medium-term phase stability of Na_2O–Al_2O_3–SiO_2–H_2O geopolymer systems. *Cement and Concrete Research* **38(6)**: 870–876.
18. Lloyd RR (2009) Accelerated ageing of geopolymers. In *Geopolymers: Structure, Processing, Properties and Industrial Applications* (Provis JL and van Deventer JSJ (eds)). Woodhead, Cambridge, UK, pp. 139–166.
19. Provis JL, Myers RJ, White CE, Rose V and van Deventer JSJ (2012) X-ray microtomography shows pore structure and tortuosity in alkali-activated binders. *Cement and Concrete Research* **42(6)**: 855–864.
20. Mo L, Deng M, Tang M and Al-Tabbaa A (2014) MgO expansive cement and concrete in China: Past, present and future. *Cement and Concrete Research* **57**: 1–12.
21. Hopkinson L, Rutt K and Cressey G (2008) The transformation of nesquehonite to hydromagnesite in the system CaO–MgO–H_2O–CO_2: An experimental FT-Raman spectroscopic study. *Journal of Geology* **116(4)**: 387–400.
22. Glasser FP, Jauffret G, Morrison J, *et al.* (2016) Sequestering CO_2 by mineralization into useful nesquehonite-based products. *Frontiers in Energy Research* **4**: 3, 10.3389/fenrg.2016.00003.
23. Walling SA and Provis JL (2016) Magnesia based cements – A journey of 150 years, and cements for the future? *Chemical Reviews* **116(7)**: 4170–4204.
24. Shi Z, Shi C, Wan S and Ou Z (2017) Effect of alkali dosage on alkali–silica reaction in sodium hydroxide activated slag mortars. *Construction and Building Materials* **143**: 16–23.
25. Tänzer R, Jin Y and Stephan D (2017) Effect of the inherent alkalis of alkali activated slag on the risk of alkali silica reaction. *Cement and Concrete Research* **98**: 82–90.
26. Winnefeld F, Gluth GJG, Bernal SA, *et al.* (2020) RILEM TC 247-DTA round robin test: Sulfate resistance, alkali–silica reaction and freeze–thaw resistance of alkali-activated concretes. *Materials and Structures* **53(6)**: Art. 140.
27. Macias A, Kindness A and Glasser FP (1996) Corrosion behaviour of steel in high alumina cement mortar cured at 5,25 and 55°C: Chemical and physical factors. *Journal of Materials Science* **31(9)**: 2279–2289.
28. Ioannou S, Paine K, Reig L and Quillin K (2015) Performance characteristics of concrete based on a ternary calcium sulfoaluminate–anhydrite–fly ash cement. *Cement and Concrete Composites* **55**: 196–204.
29. Wagh AS (2004) *Chemically Bonded Phosphate Ceramics*. Elsevier, Oxford, UK.
30. Kostogloudis GC, Kalogridis D, Ftikos C, *et al.* (1998) Comparative investigation of corrosion resistance of steel reinforcement in alinite and Portland cement mortars. *Cement and Concrete Research* **28(7)**: 995–1010.
31. Gartner E and Hirao H (2015) A review of alternative approaches to the reduction of CO_2 emissions associated with the manufacture of the binder phase in concrete. *Cement and Concrete Research* **78(A)**: 126–142.

32. Ann KY, Kim TS, Kim JH and Kim SH (2010) The resistance of high alumina cement against corrosion of steel in concrete. *Construction and Building Materials* **24(8)**: 1502–1510.
33. Ke X, Bernal SA, Hussein OH and Provis JL (2017) Chloride binding and mobility in sodium carbonate-activated slag pastes and mortars. *Materials and Structures* **50(6)**: Art. 252.
34. Jen G, Stompinis N and Jones R (2017) Chloride ingress in a belite-calcium sulfoaluminate cement matrix. *Cement and Concrete Research* **98**: 130–135.
35. Khanzadeh Moradllo M and Ley MT (2017) Comparing ion diffusion in alternative cementitious materials in real time by using non-destructive X-ray imaging. *Cement and Concrete Composites* **82**: 67–79.
36. Hargis CW, Lothenbach B, Müller CJ and Winnefeld F (2017) Carbonation of calcium sulfoaluminate mortars. *Cement and Concrete Composites* **80**: 123–134.
37. Dunster AM, Bigland DJ and Holton IR (2000) Rates of carbonation and reinforcement corrosion in high alumina cement concrete. *Magazine of Concrete Research* **52(6)**: 433–441.
38. Glasser FP and Zhang L (2001) High-performance cement matrices based on calcium sulfoaluminate–belite compositions. *Cement and Concrete Research* **31(12)**: 1881–1886.
39. Carsana M, Canonico F and Bertolini L (2018) Corrosion resistance of steel embedded in sulfoaluminate-based binders. *Cement and Concrete Composites* **88**: 211–219.
40. Ismail I, Bernal SA, Provis JL, *et al.* (2013) Influence of fly ash on the water and chloride permeability of alkali-activated slag mortars and concretes. *Construction and Building Materials* **48**: 1187–1201.
41. Gluth GJG, Arbi K, Bernal SA, *et al.* (2020) RILEM TC 247-DTA round robin test: Carbonation and chloride penetration testing of alkali-activated concretes. *Materials and Structures* **53(1)**: Art. 21.
42. Lloyd RR, Provis JL and van Deventer JSJ (2010) Pore solution composition and alkali diffusion in inorganic polymer cement. *Cement and Concrete Research* **40(9)**: 1386–1392.
43. Nedeljković M, Ghiassi B, van der Laan S, Li Z and Ye G (2019) Effect of curing conditions on the pore solution and carbonation resistance of alkali-activated fly ash and slag pastes. *Cement and Concrete Research* **116**: 146–158.
44. Song S and Jennings HM (1999) Pore solution chemistry of alkali-activated ground granulated blast-furnace slag. *Cement and Concrete Research* **29(2)**: 159–170.
45. Gruskovnjak A, Lothenbach B, Holzer L, Figi R and Winnefeld F (2006) Hydration of alkali-activated slag: comparison with ordinary Portland cement. *Advances in Cement Research* **18(3)**: 119–128.
46. Bernal SA (2015) The resistance of alkali-activated cement-based binders to carbonation. In *Handbook of Alkali-Activated Cements, Mortars and Concretes* (Pacheco-Torgal F, Labrincha JA, Leonelli C, Palomo A and Chindaprasirt P (eds)). Woodhead, Cambridge, UK, pp. 319–332.
47. Pouhet R and Cyr M (2016) Carbonation in the pore solution of metakaolin-based geopolymer. *Cement and Concrete Research* **88**: 227–235.
48. Ke X, Criado M, Provis JL and Bernal SA (2018) Slag-based cements that resist damage induced by carbon dioxide. *ACS Sustainable Chemistry & Engineering* **6(4)**: 5067–5075.
49. Mundra S, Bernal SA, Criado M, *et al.* (2017) Steel corrosion in reinforced alkali-activated materials. *RILEM Technical Letters* **2**: 33–39.
50. Mundra S and Provis JL (2021) Mechanisms of passivation and chloride-induced corrosion of mild steel in sulfide-containing alkaline solutions. *Journal of Materials Science* **56(26)**: 14783–14802.
51. Criado M and Provis JL (2018) Alkali activated slag mortars provide high resistance to chloride-induced corrosion of steel. *Frontiers in Materials* **5**: 34.
52. Scrivener KL, Cabiron J-L and Letourneux R (1999) High-performance concretes from calcium aluminate cements. *Cement and Concrete Research* **29(8)**: 1215–1223.
53. Lee WE, Vieira W, Zhang S, *et al.* (2001) Castable refractory concretes. *International Materials Reviews* **46(3)**: 145–167.
54. Ojovan MI and Lee WE (2014) Immobilisation of radioactive waste in cement. In *An Introduction to Nuclear Waste Immobilisation*, 2nd edn (Ojovan MI and Lee WE (eds)). Elsevier, Oxford, UK, pp. 205–232.

55. Rivera OG, Long WR, Weiss Jr CA, *et al.* (2016) Effect of elevated temperature on alkali-activated geopolymeric binders compared to Portland cement-based binders. *Cement and Concrete Research* **90**: 43–51.
56. Dhir RK, Jones MR, Byars EA and Shaaban IG (1994) Predicting concrete durability from its absorption. *CANMET/ACI International Conference on Concrete Durability, Nice*. American Concrete Institute, Farmington Hills, MI, USA, pp. 1177–1194.
57. Torrent R, d'Andrea R, Gonçalves A, *et al.* (2015) Prescriptive durability specifications. In *Performance-Based Specifications and Control of Concrete Durability* (Beushausen H and Fernandez Luco L (eds)). Springer/RILEM, Dordrecht, Netherlands, State-of-the-art report TC 230-PSC, pp. 19–50.
58. Moranville-Regourd M and Kamali-Bernard S (2019) Cements made from blastfurnace slag. In *Lea's Chemistry of Cement and Concrete*, 5th edn (Hewlett PC and Liska M (eds)). Butterworth-Heinemann, Oxford, UK, pp. 469–507.
59. ASTM (2018) C1012/C1012M-18b. Standard test method for length change of hydraulic-cement mortars exposed to a sulfate solution. ASTM, West Conshohocken, PA, USA.
60. BSI (2016) PAS 8820:2016. Construction materials. Alkali-activated cementitious material and concrete. Specification. BSI, London, UK.
61. ACI Innovation Task Group 8 (2010) *Performance-Based Requirements for Concrete*. ACI, Farmington Hills, MI, USA, Report ITG-8R-10.
62. ACI Innovation Task Group 10 (2018) *Alternative Cements*. ACI, Farmington Hills, MI, USA, Report ITG-10.1R-18.

Soutsos M
ISBN 978-0-7277-6375-4
https://doi.org/10.1680/icehcd.63754.505

Chapter 15
Durability of concrete made with aggregates recycled from construction and demolition waste

Rui V Silva and Jorge de Brito
Universidade de Lisboa, Portugal

15.1. Introduction

Concrete is the world's most sought-after construction material and is the basis of urban development. However, its use has significant environmental impacts, including the high incorporated energy from cement production and associated carbon dioxide emissions, the enormous consumption of natural resources [1] and the large quantity of debris produced during construction and demolition activities. This debris is called construction and demolition waste (CDW) [2]. After submitting CDW to adequate separation and beneficiation techniques, the most effective approach for maximising the valorisation of these materials is to use the resulting recycled aggregate (RA) as a partial replacement for natural aggregate (NA) in the manufacture of new construction products.

The greatest challenge in manufacturing recycled-aggregate concrete (RAC) is making a material with predictable performance so that it can be used in structural elements. In previous studies based on a meta-analysis and a systematic literature review, the authors have developed several practical rules for predicting the mechanical performance and durability of RAC [3-11]. Working from the assumptions made in those previous studies, this study presents a proposal for a performance-based design of reinforced RAC slabs and beams in conformity with Eurocode 2 or EC2 [12]. This approach allows engineers to understand the implications of a particular cross-sectional geometry of a reinforced RAC slab and adjust it so that it can guarantee the same service life and load-bearing capacity and have the same long-term deformation as a corresponding natural-aggregate concrete (NAC) slab, all within the limits established by EC2.

15.2. Durability of RAC

15.2.1 Permeability

The durability of concrete mainly depends on the porous microstructure of the material. If water penetrates the concrete, it can transport detrimental chemicals that can initiate the degradation of the concrete [13]. The transport mechanism of water in concrete, which occurs by capillary action and permeation, happens mostly through pores in the matrix [14]. Aggregates play an important role in this transport. This is especially relevant for RAC since it is likely to contain old mortar and other porous constituents capable of increasing the overall porosity of the resulting material. It is widely acknowledged that increasing the replacement level of RA leads to a higher water absorption level compared to a corresponding NAC [15-54]. The magnitude of this increase naturally depends on the constituents of the RA. The incorporation of RA with a high content of recycled-masonry aggregate (RMA) is likely to yield a concrete with a more porous structure than one containing RA with a high content of recycled-concrete aggregate (RCA) [18,29,40,44,55]. A previous study by the authors indicated that replacing coarse NA with 100% RCA increased the sorptivity by about 30% to 60% in comparison with the control NAC [56] as a result of the greater continuity of the porous structure due to the old mortar. Moreover, the increase in sorptivity may be in the region of 60% to 70% in concrete mixes containing 100% coarse RMA [23].

The size of the RA also affects the permeation characteristics of the concrete. Fine RCA usually has a high amount of old mortar, which translates into higher water absorption levels compared to a coarser counterpart [25,44,57]. For compatible particle size distributions, it has been reported that the inclusion of 100% fine RCA could result in about a 90% increase in sorptivity in comparison with a control concrete [27]. However, this trend is not typical of concrete containing coarse or fine RMA, wherein equivalent water absorption levels can be observed for the same replacement level and sometimes lower for mixes containing fine RMA [3,18,42,58]. Owing to their higher surface area, these aggregates may lead to pozzolanic reactions, thereby resulting in hydration products capable of decreasing the connectivity of the porous network [42,58-61].

Even though current methodologies for RA usually consider only their composition, the quality of their constituents should not be neglected. For RA composed solely of RCA from different sources, there is a notable scatter in those physical properties that affect the performance of concrete [3,62]. Also, the quality of RA, which is mostly governed by the quantity of adhered mortar and porous contaminants [3,63], can be controlled by the number of crushing stages, quality control of the triage process, washing stages and other beneficiation techniques in CDW recycling plants [10,64-68]. Consequently, using RA that has been subjected to more processing stages is likely to result in a concrete with a lower porosity [69]. Apart from the reduced quantity of adhered mortar, these aggregates also have a lower shape index [69], which typically results in a concrete with a lower water demand and enhanced densification [70,71].

One of the factors that has the greatest influence on the porosity of a concrete is its water-to-cement (w/c) ratio. Thus, the moisture content of the RA needs to be taken into account when formulating a RAC. Since RA absorbs more water compared to conventional NA, then if this is not compensated for, there will likely be an overall decrease in the effective w/c ratio and, thus, a less porous cementitious matrix. A concrete with such RA may absorb less water compared to a RAC made with fully compensated or pre-saturated RA [28,34,42,58].

15.2.2 Carbonation

Incorporating RA into concrete is likely to increase the volume and connectivity of the porous microstructure, thereby facilitating the diffusion of carbon dioxide, which can accelerate the corrosion of the steel reinforcement. There is a consensus that increasing the replacement level of RA leads to the carbonation progressing faster in comparison with control NAC specimens with a similar mix design [9,21,22,25,27,29,43,72-81]. The authors' previous studies have shown two things. First, there is a linear relation between carbonation depth and replacement levels, which enables the development of prediction models. Second, using 100% coarse RA will likely result in an increase of carbonation by up to 2.5 times that of a control NAC [56]. The variation in this probable increase depends on several factors related to the RA and RAC mix design. The constituent that probably has the most influence on this property is the amount of old mortar, which depends on the strength of the source concrete and the beneficiation techniques applied to the end-of-life material [9,49,79,82-86]. Concrete made with RA subjected to more processing stages is likely to have lower carbonation depths compared to a corresponding concrete made with RA that has been less processed [49,69]. In some cases, incorporating RA can enhance the performance in terms of carbonation. This improvement is attributed to the application of a different mix design [87,88] or the use of partly dry RA, both of which decrease the overall effective w/c ratio of the RAC and, thus, its porosity [89].

RA with a high RMA content will probably absorb more water than one containing more RCA. Thus, the carbonation depth of mixes made with RMA is higher [3]. Nevertheless, provided that the weighed amount of absorbed water in RA is equivalent between different mixes regardless of the composition and content of the RA, the resulting overall porosity and progression of carbonation for the mixes are likely to be comparable [29,76,80,84,85,90]. This further reinforces the need to classify RA and for certification based on physical performance rather than on composition alone, which is the norm in some specifications.

Owing to the typically greater porosity of fine RA [25,44,57], its use as a sand replacement will considerably increase the rate of carbonation. There is a 95% probability that concrete with 100% fine RA will have a carbonation depth from about 1.6 to 7.4 times that of control NAC mixes [56]. However, the diffusion dynamics of carbon dioxide in mixes with fine RMA differs from that in mixes with fine RCA, as there are opposing effects in the former. One the one hand, fine RMA may have some pozzolanicity and form additional products of hydration, thus contributing to a denser microstructure. On the other hand, this type of RA still has considerable porosity, which promotes the diffusion of carbon dioxide into the concrete. Moreover, the consumption of calcium hydroxide in the matrix also reduces the resistance to carbonation [59].

15.2.3 Chloride ion penetration

The penetration of chloride ions is one of the key factors responsible for the destruction of the passive layer on the steel reinforcement. Like carbonation, the rate of penetration depends significantly on the overall porosity and pore connectivity of the concrete, which in turn is affected by the level of substitution of NA by RA. There is a consensus that there is a linear relation between the RA level and the rate of penetration of chloride ions due to the greater porosity of RA compared to NA [21,27,45,50,69,91-125]. This effect is exacerbated by decreasing the size of the RA, which has a higher content of adhered mortar. Thus, the size of the RA has a considerable influence on the transport of chloride ions in concrete [3,126,127]. There is a 95% probability that the incorporation of 100% coarse or fine RA will result in an increase in the non-steady-state migration of chloride ions by about 45% and 77%, respectively, compared to corresponding control NAC mixes [56]. Naturally, there is some scatter involved in this increase in penetration, depending on the type, quality and moisture state of the RA. The incorporation of coarse RCA is likely to lower the rate of migration of chloride ions compared to mixes with coarse RMA [128] because of the limited connectivity of the matrix due to the low amount of water absorbed by coarse RCA and because there may also be additional physical adsorption and chemical binding of chloride ions in calcium silicate hydrate and AFm phases, respectively. Furthermore, despite the initial decline in performance, the resistance to chloride ion penetration can sometimes increase over time [45,50,113,129] due to ongoing hydration reactions at the interfacial transition zone and at the RCA. Kou and Poon [113] found that even though incorporating 100% coarse RCA led to increases of 10–21% in 28-day total charge passed, negligible differences were reported after 10 years.

As noted, even though using pozzolanic additions (e.g. fly ash) as a partial cement replacement is current practice to reduce the progression of chloride ions, the reaction between additions and most RA is likely to be non-existent due to the mostly inert nature of the latter. Thus, the interactions between these two fall outside the scope of the current study. However, note that, not only does some RA contain aluminosilicate additions that can still react with the new cementitious microstructure (i.e. because the interfacial transition zone enlarges over time), but the pozzolanicity of the RA may mean that it can mitigate or even decrease the progression of chloride ions. This occurs in some finely ground RMA [59,60,130-136]. A decrease of almost 60% was observed in the migration coefficient after the fine natural fraction was completely replaced with fine RMA from crushed ceramic bricks [59].

The mix design of the source concrete can also have a considerable effect on the transport mechanism of chloride ions [69,101,112,137]. González and Etxeberria [101] demonstrated that there was a clear increase in chloride penetration when using coarse RCA from concrete specimens with decreasing target strength. Using 100% coarse RCA from 40-, 60- and 100-MPa concrete specimens led to increases in the total charge passed of 140%, 114% and 50%, respectively. The migration of chloride ions in RAC can be further mitigated using RCA that has been subjected to further crushing stages, thereby partly removing the porous adhered mortar [69,115].

As has been stated for previous properties, the moisture level of the RA also affects the penetration of chloride ions into the concrete. The moisture level can easily be controlled during the mixing stage. Thus, mixing procedures can be developed capable of mitigating the performance loss [45,138-140]. Instead of compensating for the water absorbed by RA prior to concrete batching, which may be prone to error and impractical from an industrial standpoint [28], some have suggested adding water in different mixing stages [28,45,104], rather than introducing all the constituents into the mixer at the same time.

15.2.4 Resistance to sulfate attack

One of the most deleterious phenomena that can occur in concrete is the expansion-related damage caused by a sulfate attack against the cementitious matrix. This attack can occur due to exposure to sulfate-rich soil, industrial waste water, sulfate-rich aggregate or even excessive gypsum in the cement. This is particularly relevant for concrete with RA, as the RA may be contaminated with gypsum plaster due to inadequate construction and demolition activities [141]. The performance of RAC with gypsum-contaminated RCA is likely to decline over time due to disruption of the cementitious matrix [142-144]. Moreover, even if the RA is not contaminated in this way, its use in concrete will probably result in faster degradation when the concrete is exposed to an external source of sulfate because the old mortar also reacts with the sulfate [25,118,145].

One way to mitigate the expansion of concrete due to a sulfate attack is to reduce the amount of calcium hydroxide in the matrix using pozzolanic additions (e.g. fly ash, ground granulated blastfurnace slag or even ground ceramic-based RMA [144-147]). Another method capable of effectively increasing the resistance to a sulfate attack is to reduce the amount moisture in the RA. Using air-dried RCA in uncompensated mixes can result in expansion levels equivalent to those of a control NAC [89]. As previously explained, however, there are opposing effects when using such RCA. The microstructure is denser because of the lower effective w/c ratio, but the irrevocable increase in the total volume of the porous structure facilitates the penetration of sulfate. Moreover, using RCA made from higher strength materials [148] and with less absorbed water [149,150] can further mitigate this problem.

15.2.5 Alkali–silica reaction

Like a sulfate attack, damage caused by alkali–silica reactions is due to expansion. These reactions occur between silica-based minerals and an alkali solution in the concrete [151]. Even though there is limited information on this subject and that negligible differences have sometimes been reported [152-154], the incorporation of RCA in concrete can lead to greater expansion due to alkali–silica reactions [73,118,155,156]. This increase in expansion can occur due to the higher alkali content in the old mortar in the RCA, because of its lower soundness in comparison with NA and, thus, lower resistance [156] or because of the presence of siliceous sand in the old adhered mortar [152]. Nevertheless, although using RCA may lead to relatively higher expansion compared to corresponding NAC samples, it is likely to be of little consequence as RCA is considered to have low reactivity and pose little risk of alkali–silica reactions [157,158].

15.2.6 Resistance to freezing and thawing

There are discrepancies in the literature regarding the effect on its resistance to freezing and thawing when RA is incorporated into a concrete. Some researchers have proposed that RA has a marginal influence [25,77,159-165], whereas others have observed a decline in performance [74,80,115,166-171]. Nevertheless, the differences can be explained by the variable characteristics of RA as well as the use of chemical admixtures.

Performance is usually better with RCA from non-air-entrained concrete, since this aggregate usually absorbs less water. However, this type of RCA is likely to significantly reduce the resistance of the concrete to freezing and thawing. The old mortar in these aggregates may hold a considerable amount of water, which expands when subjected to freezing temperatures. After several freeze–thaw cycles, the repeated expansion prompts the propagation of cracks through the weaker areas of the old mortar, resulting in faster deterioration [169,172-176]. It is possible to mitigate the lower freeze–thaw resistance of concrete containing non-air-entrained RCA by submitting the RCA to further processing stages that reduce the content of old mortar [84,115,166,168,176]. Nevertheless, unsatisfactory results have still been reported in comparison with control NAC [115,176,177]. The incorporation of RCA from crushed air-entrained concrete, however, can deliver acceptable resistance to freezing and thawing, even though the RCA is generally considered to be of lower quality with regard to the other properties of concrete. Still, to ensure that there is sufficient resistance to freezing and thawing, air-entrained RCA should be used in mixes with air-entraining admixtures [176,177].

15.2.7 Resistance to abrasion

Several contradictory findings have been reported in the literature concerning the effect of RA on the abrasion resistance of concrete. Some authors have stated that the inclusion of RA has a marginal influence on abrasion resistance [16,25,27,41,178-181]. A few reports have suggested that RA may reduce the mass loss due to abrasion [23,39,182], whereas other studies have found the opposite [38,40,43,183-185]. This discrepancy may be explained by the varying replacement levels, as well as the quality, type and size of the RA. Concrete mixes containing up to 50% coarse RCA may have a negligible change in mass loss due to abrasion compared to a control NAC. In contrast, the mass loss may be up to 35% higher for mixes with 100% coarse RCA [25,38,43,161,181,183,186,187]. Using 100% coarse RMA from crushed ceramic bricks may enhance the resistance to abrasion in comparison with NAC [23], but the opposite has also been reported [184,188], such as a 75% higher mass loss [188]. Additionally, abrasion resistance may be lowered by the incorporation of fine RA in comparison with a concrete with corresponding coarser fractions [182].

Even though several properties of concrete depend on the quality of the RA, research suggests that RA may have a negligible influence on abrasion resistance. RAC specimens containing varying percentages of different contaminants (i.e. glass, tiles and bricks) comprising 10% by weight of the total aggregate had comparable rates of mass loss due to abrasion as uncontaminated specimens [189]. Nevertheless, it is possible that a high content of asphalt-based particles may increase the mass loss due to abrasion [182].

15.3. Durability design of RAC elements

The following design is based on a simply supported one-way concrete slab (Figure 15.1(a)), with a span of 4 m, a strength class of C30/37 and an exposure class of XC1/XS1 [12]. The concrete was assumed to be produced from CEM I 42.5, with a maximum w/c ratio of 0.50 and a minimum cement content of 300 kg/m^3 [190]. A simply supported beam (Figure 15.1(b)) is made from the same material and exposed to the same conditions.

In previous studies, the authors conducted meta-analyses to determine the influence of incorporating RA on several properties of concrete [3-9,11,191,192]. The results show that it is possible to predict the compressive strength loss among other properties of RAC from the characteristics of the RA. The design of the aforementioned concrete elements assumes the validity of these studies. Table 15.1 gives the strength class for each mix. Besides the control mix, two scenarios were considered (R-I and R-II), each with three types

Figure 15.1 Cross-sections of a NAC slab (a) and a beam (b)

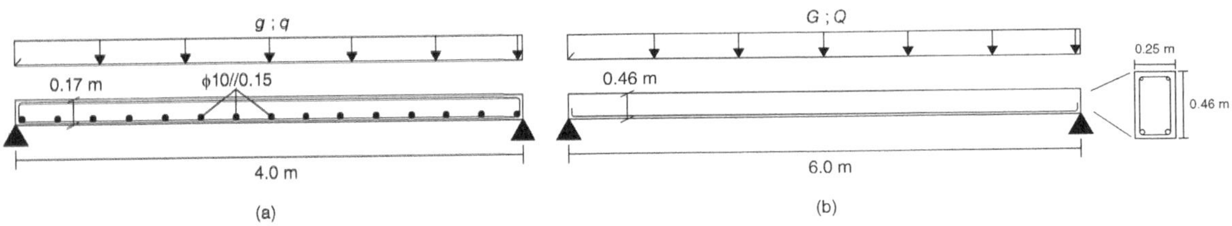

of RA (A, B and C). The characteristics of these RAs are based on the performance-based classification presented in the authors' previous study [3] and are compatible with EN 12620 [193].

The difference between scenarios R-I and R-II is the mix design of the RAC. In scenario R-I, the RAC is produced with the same design as that of the control NAC. The incorporation of RA causes a decline in the mechanical performance according to the results of the meta-analysis carried out previously by the authors [4,5,8]. For the RAC mixes in scenario R-II, there is a slight change in the effective w/c ratio (effected by maintaining the cement content and using superplasticisers) to achieve a target strength equivalent to that of the control NAC. The replacement levels of 50% for RA in class A and 30% for RA in classes B and C (Table 15.2) are based on the proposal for updating EN 206 [194], EN 12620 [193] and Eurocode 2 [12],

Table 15.1 Characteristics of concrete for all scenarios and designs of slabs and beams

Property		Scenario						
		Control	R-I			R-II		
Characteristics of concrete	Aggregate type	–	A	B	C	A	B	C
	Replacement level: %	0	50	30	30	50	30	30
	f_{ck}: MPa	32.5	29.1	28.8	27.2	36.3	36.7	38.8
	Strength class	C30/37	C25/30	C25/30	C25/30	C35/45	C35/45	C35/45
	f_{ctm}: MPa	2.9	2.56	2.56	2.56	2.9	2.9	2.9
	E_{cm}: GPa	32.8	26.8	26.8	26.8	27.9	27.9	27.9
	α	–	1.4	1.25	1.25	1.4	1.25	1.25
	$E_{c,eff}$: GPa	9.38	5.95	6.53	6.53	6.2	6.81	6.81
	c_{nom}: mm	35	40	40	40	35	35	35
Slab design	Rebar	ϕ10//0.15	ϕ10//0.15	ϕ10//0.15	ϕ10//0.15	ϕ10//0.15	ϕ10//0.15	ϕ10//0.15
	D: mm	170	190	190	190	180	180	180
	c_{nom}: mm	35	40	40	40	35	35	35
Beam design	Rebar	6ϕ20	6ϕ20	6ϕ20	6ϕ20	6ϕ20	6ϕ20	6ϕ20
	h: mm	460	520	510	510	500	500	500

f_{ck} = 28-day characteristic compressive strength, f_{ctm} = mean axial tensile strength, E_{cm} = 28-day modulus of elasticity, α = correction factor for the creep coefficient for a given time period and loading, $E_{c,eff}$ = effective modulus of elasticity (taking creep into account), c_{nom} = nominal cover; D = slab thickness and h = beam height

Table 15.2 Recommendation for the classification of RA [195]

Aggregate class	Maximum water absorption: %	Minimum oven-dry density: kg/m³	Composition according to EN 933-11 [196][a]
A	3.5	2400	Rc_{90}, Rcu_{95}, Rb_{10-}, Ra_{1-}, FL_{2-}, XRg_{1-}
B	8.5	2100	Rc_{50}, Rcu_{70}, Rb_{30-}, Ra_{5-}, FL_{2-}, XRg_{2-}
C	15	1800	Rc_{NR}, Rcu_{NR}, Rb_{NR}, Ra_{10-}, FL_{5-}, XRg_{2-}

[a]Rc = content by mass of concrete-based fragments, Rcu = concrete-based fragments + unbound aggregates, Rb = masonry-based fragments, Ra = bituminous materials, Rg = glass and X = other materials (EN 933-11 [196])

regarding the use of RA from beneficiated CDW. These specific replacement levels lead to a manageable level of performance loss, as the strength class decreases by only one level. Moreover, the expected strength loss is similar between the mixes with different RA.

In scenario R-I, the incorporation of RA results in a material with a lower strength class than that required for the assumed environmental classes. Nevertheless, the calculation demonstrates the effect that this change has on the geometry of the two elements. As expected, incorporating any of the three classes of RA would result in a similar loss of strength so that the strength class decreases from C30/37 to C25/30. Naturally, this has repercussions for the material properties that affect the structural design (e.g. lower tensile strength and modulus of elasticity and increased deformation due to creep). Since the same relation between mean axial tensile strength (f_{ctm}) and characteristic compressive strength (f_{ck}) holds for both RAC and NAC [8], it is possible to use the formulae in EC2 to estimate the expected f_{ctm}, as listed in Table 15.1 for all mixes. As expected, f_{ctm} decreases with a decrease in the compressive strength in scenario R-I.

However, in scenario R-II, where no strength loss is expected, the final strength class is the same as that of the control NAC, which is realised by slightly altering the mix design. For the given replacement levels, it was established that a mere decrease of 0.05 in the effective w/c ratio is enough to compensate for the calculated strength loss, according to existing rules of design [195,197]. Thus, in mixes with the same strength class, RAC is expected to have an equivalent tensile strength [8] but a lower modulus of elasticity with increasing replacement levels due to the added deformability of the RA in comparison with NA [5]. Furthermore, the higher creep-related deformation of RAC over time was also taken into account in the design of the structural elements. The total deformation can be calculated using the effective modulus of elasticity ($E_{c,eff}$), which was found using the correction factor (α) for the higher creep coefficient of the RAC. However, note that, due to the limited number of studies on the influence of RA on the creep behaviour of concrete, the analysis used the replacement level rather than the different RA classes [6]. For this reason, the $E_{c,eff}$ of concrete with class A is slightly lower than that for mixes with class B or C. Further research is required on the creep behaviour of RAC to understand the implications of using RA of a known quality class.

One of the other main differences of the design of the elements for the two scenarios is the cover. Both scenarios assume that the elements are going to be subjected to environmental classes XC1 and XS1. For the carbonation-based cover design, the calculations were based on the authors' previous findings concerning the close relation between the compressive strength of concrete and carbonation, which is considered a practical approach for indirectly evaluating carbonation [198–204]. Thus, a meta-analysis studied 799 mixes with different types of RA (RCA, RMA and mixed recycled aggregates) and different aggregate sizes using CEM I. These were subjected to a controlled environment with 4–5% carbon dioxide content. The coefficient of accelerated carbonation is plotted against the corresponding 28-day compressive strength in Figure 15.2(a). The power regressions for mixes with varying replacement levels and different types of

Figure 15.2 Coefficient of accelerated carbonation (a) and chloride ion migration coefficient (b) plotted against the 28-day compressive strength of concrete

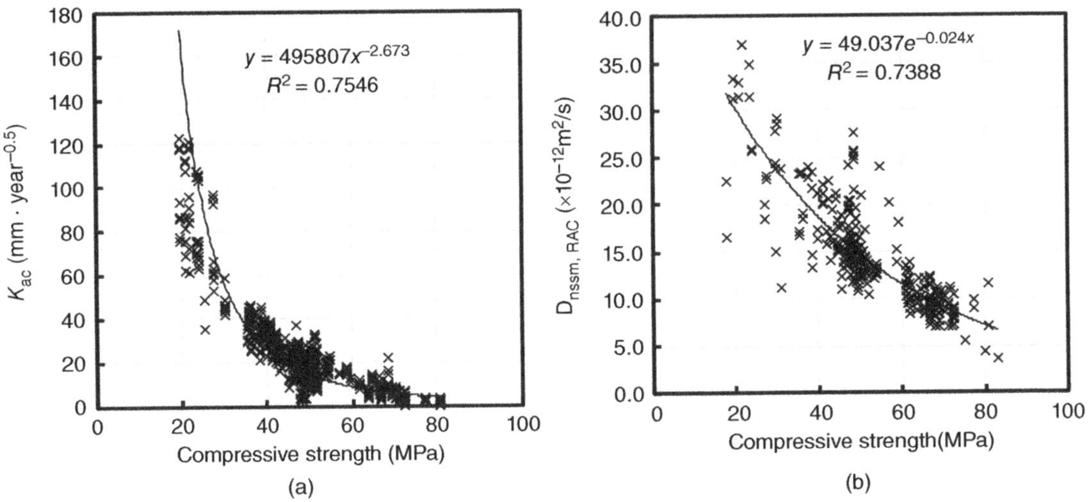

RA indicate that there is a strong correlation between the two variables. The results suggest that there was no statistically significant difference between the behaviour of RAC and NAC, regardless of the replacement level or the size or type of RA [9,205]. From a practical viewpoint, this means that RAC and NAC with similar compressive strengths will have similar carbonation rates, which agrees with previous findings [25,159–164,206–208]. Additionally, a comparison [9] of the power regression obtained from the collated values was remarkably like that proposed by LNEC E-465 specification 'Methodology for estimating the concrete performance properties allowing to comply with the design working life of the reinforced or pre-stressed concrete structures under the environmental exposures XC and XS' [202].

A parallel comparison was carried out regarding the effect of chloride ion penetration on the design of structural RAC (Figure 15.2(b)) [7]. There were statistically marginal differences between the relations for the non-steady-state chloride ion migration coefficient and the 28-day compressive strength of NAC and RAC, regardless of the size, type or quality of the RA.

The authors have developed relations for carbonation and chloride ion penetration with compressive strength. The affected depth is used to determine the cover needed for structural elements [7,9] to achieve a pre-established service life. The relation between the cover of NAC and RAC with comparable compositions can be calculated using the following equations:

$$\frac{X_{\mathrm{RAC}}}{X_{\mathrm{NAC}}} = \left(\frac{f_{\mathrm{cm,NAC}}}{f_{\mathrm{cm,RAC}}}\right)^{2.7} \quad (15.1)$$

$$\frac{X_{\mathrm{RAC}}}{X_{\mathrm{NAC}}} = \sqrt{\exp(-0.023(f_{\mathrm{cm,RAC}} - f_{\mathrm{cm,NAC}}))} \quad (15.2)$$

where X_{RAC} and X_{NAC} (mm) are the affected depths of RAC and NAC, respectively, and $f_{\mathrm{cm,RAC}}$ and $f_{\mathrm{cm,NAC}}$ are the respective 28-day compressive strengths. These equations are applicable only to concrete mixes with CEM I [190].

Table 15.1 presents the nominal cover for both elements, which are subjected to environmental classes XC1 and XS1. The decline in mechanical performance in scenario R-I is assumed to be due to the more porous cementitious microstructure than that of the control, which results in increased carbonation and chloride ion penetration. Consequently, increased cover is required to ensure the same service life as that of the control NAC (from 35 to 40 mm). On the other hand, in scenario R-II, where the same strength class was realised by altering the mix design, the original cover can be used (i.e. 35 mm) [7,9].

In all scenarios, the concrete slab and beam were designed iteratively, like the method conventionally used for NAC. After establishing the requirements for the control slab, sequential increments of 10 mm were made to the thicknesses of the slab and beam until all requirements were met for the same steel reinforcements as that obtained for the NAC elements. Table 15.1 presents the solution for the lower main steel reinforcement and the final cross-sectional thickness for the slab (*D*) and the height for the beam (*h*) in each scenario. The compressive strength loss of the concrete in scenario R-I had a marginal effect on the area of steel required to guarantee the ultimate limit state in EC2. Indeed, due to the increase in the thickness and height of the slabs and beams, respectively, a slightly lower area of steel is required to comply with the ultimate limit state. Nevertheless, the same area of reinforcements was adopted to understand what the changes would be to the geometry of the elements.

In terms of the serviceability limit state in EC2, the cross-sectional geometry required a modification due to the greater deformability of RAC. The thickness of the RAC slabs in scenario R-I was increased by 20 mm to prevent significant long-term deformation, so that the deformation would be the same as that of the NAC slab when subjected to the same loading conditions (Figure 15.1). On the other hand, in scenario R-II, an increase of only 10 mm is required. The same trend was observed for the beam design, with increases in the height of 50–60 mm and 40 mm for scenarios R-I and R-II, respectively.

15.4. Concluding remarks

Using RA as partial replacement of NA, even though it may produce concrete with comparable mechanical performance, tends to result in an overall decline in durability. This is due to the inherent porosity of RA, which also means the resulting concrete has an increased porous network. Consequently, more water is absorbed. Typically, there is increased carbonation and chloride ion penetration and decreased resistance to freezing and thawing, sulfate attacks, alkali–silica reactions and abrasion. This performance decline should, therefore, be considered when designing RAC structures.

Classifying RAs based on their physical properties, rather than their just composition, apart from being the most practical approach, means the classification is more strongly correlated with the performance of the resulting concrete products. Using the formulae derived in the authors' previous studies, it is easy to predict the behaviour of the concrete. The widespread and safe use of RAC can, thus, be made possible by adapting codes for structural concrete. The proposed method is very conservative and is flexible, covering a wider range of applications compared with existing specifications, which are severely restrictive.

Two scenarios were considered in this study. The first one used RA in concrete with the same mix design as that of the control, resulting in a decline in mechanical performance. The aim was to demonstrate the adaptability of the proposed approach. For the second scenario, by simply adjusting the w/c ratio and using superplasticisers, the mixes had the same strength class. This scenario is considered to be the most viable from a practical viewpoint, because only an adjustment to the cross-sectional geometry of the slab and beam was required to reduce the long-term deformation.

Nevertheless, there are implications in terms of environmental impact and costs. Even though the same amount of steel reinforcement is needed, the seeming benefit of incorporating RA could be offset by the need to use a greater mass of RAC to achieve equivalent performance. To overcome this, since the proposed design was based on conservative limits established by a meta-analysis of results in the literature, future research should focus on ensuring that the (lower) reliability in determining the properties of RAC is accurately taken into account when designing structures.

This is especially important for the modulus of elasticity and creep deformation, which may significantly influence the design to meet the serviceability limit state in EC2. It is likely that, by using more realistic and less conservative figures, future proposals for structural RAC design will require small changes to the cross-sectional geometry of an element to achieve the same mechanical, durability and structural performance.

REFERENCES

1. Freedonia (2012) *World Construction Aggregates*. Freedonia, Cleveland, Ohio, USA.
2. Eurostat (2016) *epp.eurostat.ec.europa.eu* (accessed 12/04/2016).
3. Silva RV, de Brito J and Dhir RK (2014) Properties and composition of recycled aggregates from construction and demolition waste suitable for concrete production. *Construction and Building Materials* **65**: 201–217.
4. Silva RV, de Brito J and Dhir RK (2014) The influence of the use of recycled aggregates on the compressive strength of concrete: A review. *European Journal of Environmental and Civil Engineering* **19(7)**: 825–849.
5. Silva RV, de Brito J and Dhir RK (2016) Establishing a relationship between modulus of elasticity and compressive strength of recycled aggregate concrete. *Journal of Cleaner Production* **112**: 2171–2186.
6. Silva RV, de Brito J and Dhir RK (2015) Comparative analysis of existing prediction models on the creep behaviour of recycled aggregate concrete. *Engineering Structures* **100**: 31–42.
7. Silva RV, de Brito J, Neves R and Dhir R (2015) Prediction of chloride ion penetration of recycled aggregate concrete. *Materials Research* **18(2)**: 427–440.
8. Silva RV, de Brito J and Dhir RK (2015) Tensile strength behaviour of recycled aggregate concrete. *Construction and Building Materials* **83**: 108–118.
9. Silva RV, Neves R, de Brito J and Dhir RK (2015) Carbonation behaviour of recycled aggregate concrete. *Cement and Concrete Composites* **62**: 22–32.
10. Silva RV, de Brito J and Dhir RK (2017) Availability and processing of recycled aggregates within the construction and demolition supply chain: A review. *Journal of Cleaner Production* **143**: 598–614.
11. Silva RV, de Brito J, Evangelista L and Dhir RK (2016) Design of reinforced recycled aggregate concrete elements in conformity with Eurocode 2. *Construction and Building Materials* **105**: 144–156.
12. CEN (Comité Européen de Normalisation) (2014) EN 1992-1-1. Eurocode 2. Design of concrete structures. Part 1-1: General rules and rules for buildings. CEN, Brussels, Belgium.
13. Eckert M and Oliveira M (2017) Mitigation of the negative effects of recycled aggregate water absorption in concrete technology. *Construction and Building Materials* **133**: 416–424.
14. Tuutti K (1982) *Corrosion of Steel in Concrete*. PhD thesis, Swedish Cement and Concrete Research Institute at the Institute of Technology, Stockholm, Sweden.
15. Dodds W, Goodier C, Christodoulou C, Austin S and Dunne D (2017) Durability performance of sustainable structural concrete: Effect of coarse crushed concrete aggregate on microstructure and water ingress. *Construction and Building Materials* **145**: 183–195.
16. Soares D, de Brito J, Ferreira J and Pacheco J (2014) Use of coarse recycled aggregates from precast concrete rejects: Mechanical and durability performance. *Construction and Building Materials* **71**: 263–272.
17. Lotfy A and Al-Fayez M (2015) Performance evaluation of structural concrete using controlled quality coarse and fine recycled concrete aggregate. *Cement and Concrete Composites* **61**: 36–43.
18. Levy SM and Helene P (2004) Durability of recycled aggregates concrete: A safe way to sustainable development. *Cement and Concrete Research* **34(11)**: 1975–1980.

19. Zega CJ and Di Maio ÁA (2011) Use of recycled fine aggregate in concretes with durable requirements. *Waste Management* **31(11)**: 2336–2340.
20. Wirquin E, Hadjieva-Zaharieva R and Buyle-Bodin F (2000) Utilisation de l'absorption d'eau des bétons comme critères de leur durabilité. Application aux bétons de granulats recyclés. *Materials and Structures* **33(6)**: 403–408.
21. Amorim P, de Brito J and Evangelista L (2012) Concrete made with coarse concrete aggregate: Influence of curing on durability. *ACI Materials Journal* **109(2)**: 195–204.
22. Buyle-Bodin F and Hadjieva-Zaharieva R (2002) Influence of industrially produced recycled aggregates on flow properties of concrete. *Materials and Structures* **35(8)**: 504–509.
23. Correia JR, de Brito J and Pereira AS (2006) Effects on concrete durability of using recycled ceramic aggregates. *Materials and Structures* **39(2)**: 169–177.
24. Debieb F, Courard L, Kenai S and Degeimbre R (2009) Roller compacted concrete with contaminated recycled aggregates. *Construction and Building Materials* **23(11)**: 3382–3387.
25. Dhir RK, Limbachiya MC and Leelawat T (1999) Suitability of recycled concrete aggregate for use in BS 5328 designated mixes. *Proceedings of the Institution of Civil Engineers – Structures and Buildings* **134(3)**: 257–274.
26. Evangelista L and de Brito J (2004) Criteria for the use of fine recycled concrete aggregates in concrete production. *International RILEM Conference on the Use of Recycled Materials in Buildings and Structures* (Vázquez E, Hendriks C and Janssen GMT (eds)). RILEM Publications, Barcelona, Spain, pp. 503–510.
27. Evangelista L and de Brito J (2010) Durability performance of concrete made with fine recycled concrete aggregates. *Cement and Concrete Composites* **32(1)**: 9–14.
28. Ferreira L, de Brito J and Barra M (2011) Influence of the pre-saturation of recycled coarse concrete aggregates on concrete properties. *Magazine of Concrete Research* **63(8)**: 617–627.
29. Gomes M and de Brito J (2009) Structural concrete with incorporation of coarse recycled concrete and ceramic aggregates: durability performance. *Materials and Structures* **42(5)**: 663–675.
30. Gomez-Soberon JMV (2002) Porosity of recycled concrete with substitution of recycled concrete aggregate: An experimental study. *Cement and Concrete Research* **32(8)**: 1301–1311.
31. Gomez-Soberon JMV (2003) Relationship between gas adsorption and the shrinkage and creep of recycled aggregate concrete. *Cement Concrete and Aggregates* **25(2)**: 42–48.
32. González-Fonteboa B and Martínez-Abella F (2005) Recycled aggregates concrete: Aggregate and mix properties. *Materiales de Construccion* **55(279)**: 53–66.
33. Henry M, Pardo G, Nishimura T and Kato Y (2011) Balancing durability and environmental impact in concrete combining low-grade recycled aggregates and mineral admixtures. *Resources, Conservation and Recycling* **55(11)**: 1060–1069.
34. Kenai S, Debieb F and Azzouz L (2002) Mechanical properties and durability of concrete made with coarse and fine recycled aggregates. *Proceedings of the International Symposium on Sustainable Concrete Construction* (Dhir RK, Dyer TD and Halliday JE (eds)). Thomas Telford, Dundee, Scotland, UK, pp. 383–392.
35. Khalaf FM and DeVenny AS (2004) Recycling of demolished masonry rubble as coarse aggregate in concrete: Review. *Journal of Materials in Civil Engineering* **16(4)**: 331–340.
36. Kou SC and Poon CS (2006) Compressive strength, pore size distribution and chloride-ion penetration of recycled aggregate concrete incorporating class-F fly ash. *Journal of Wuhan University of Technology – Materials Science Edition* **21(4)**: 130–136.
37. López-Gayarre F, López-Colina C, Serrano-López MA, García Taengua E and López Martínez A (2011) Assessment of properties of recycled concrete by means of a highly fractioned factorial design of experiment. *Construction and Building Materials* **25(10)**: 3802–3809.
38. Malešev M, Radonjanin V and Marinković S (2010) Recycled concrete as aggregate for structural concrete production. *Sustainability* **2(5)**: 1204–1225.

39. Mas B, Cladera A, Olmo T and Pitarch F (2012) Influence of the amount of mixed recycled aggregates on the properties of concrete for non-structural use. *Construction and Building Materials* **27(1)**: 612–622.
40. Poon CS and Chan DX (2006) Paving blocks made with recycled concrete aggregate and crushed clay brick. *Construction and Building Materials* **20(8)**: 569–577.
41. Poon CS and Lam CS (2008) The effect of aggregate-to-cement ratio and types of aggregates on the properties of pre-cast concrete blocks. *Cement and Concrete Composites* **30(4)**: 283–289.
42. Sadek DM and El Nouhy HA (2014) Properties of paving units incorporating crushed ceramic. *HBRC Journal* **10(2)**: 198–205.
43. Sagoe-Crentsil KK, Brown T and Taylor AH (2001) Performance of concrete made with commercially produced coarse recycled concrete aggregate. *Cement and Concrete Research* **31(5)**: 707–712.
44. Soutsos MN, Tang K and Millard SG (2011) Use of recycled demolition aggregate in precast products. Phase II: Concrete paving blocks. *Construction and Building Materials* **25(7)**: 3131–3143.
45. Tam VWY and Tam CM (2007) Assessment of durability of recycled aggregate concrete produced by two-stage mixing approach. *Journal of Materials Science* **42(10)**: 3592–3602.
46. Van Acker A (1998) Recycling of concrete at a precast concrete plant. *Proceedings of the International Symposium on Sustainable Construction: Use of Recycled Concrete Aggregate* (Dhir RK, Henderson NA and Limbachiya MC (eds)). Thomas Telford, London, UK, pp. 321–332.
47. Yang J, Du Q and Bao Y (2011) Concrete with recycled concrete aggregate and crushed clay bricks. *Construction and Building Materials* **25(4)**: 1935–1945.
48. Zaharieva R, Buyle-Bodin F, Skoczylas F and Wirquin E (2003) Assessment of the surface permeation properties of recycled aggregate concrete. *Cement and Concrete Composites* **25(2)**: 223–232.
49. Pedro D, de Brito J and Evangelista L (2015) Performance of concrete made with aggregates recycled from precasting industry waste: Influence of the crushing process. *Materials and Structures* **48(12)**: 3965–3978.
50. Kapoor MEK, Singh SP and Singh B (2016) Durability of self-compacting concrete made with recycled concrete aggregates and mineral admixtures. *Construction and Building Materials* **128**: 67–76.
51. Lopez-Uceda A, Ayuso J, Jimenez JR, *et al.* (2016) Upscaling the use of mixed recycled aggregates in non-structural low cement concrete. *Materials* **9(2)**: 91, 10.3390/ma9020091.
52. Silva YF, Robayo RA, Mattey PE and Delvasto S (2016) Properties of self-compacting concrete on fresh and hardened with residue of masonry and recycled concrete. *Construction and Building Materials* **124**: 639–644.
53. Zong L, Fei ZY and Zhang SP (2014) Permeability of recycled aggregate concrete containing fly ash and clay brick waste. *Journal of Cleaner Production* **70**: 175–182.
54. Medina C, Zhu WZ, Howind T, de Rojas MIS and Frias M (2014) Influence of mixed recycled aggregate on the physical - mechanical properties of recycled concrete. *Journal of Cleaner Production* **68**: 216–225.
55. Levy S and Helene P (2007) Durability of concrete mixed with fine recycled aggregates. In *Sustainable Construction Materials and Technologies* (Chun YM, Claisse P, Naik TR and Ganjian E (eds)). Taylor & Francis, Leiden, Netherlands, pp. 45–51.
56. Dhir RK, de Brito J, Silva RV and Lye CQ (2019) *Sustainable Construction Materials: Recycled Aggregates*. Woodhead Publishing, Cambridge, UK.
57. Padmini AK, Ramamurthy K and Mathews MS (2002) Relative moisture movement through recycled aggregate concrete. *Magazine of Concrete Research* **54(5)**: 377–384.
58. Debieb F and Kenai S (2008) The use of coarse and fine crushed bricks as aggregate in concrete. *Construction and Building Materials* **22(5)**: 886–893.
59. Vieira T, Alves A, de Brito J, Correia JR and Silva RV (2016) Durability-related performance of concrete containing fine recycled aggregates from crushed bricks and sanitary ware. *Materials & Design* **90**: 767–776.

60. Etxeberria M and Vegas I (2015) Effect of fine ceramic recycled aggregate (RA) and mixed fine RA on hardened properties of concrete. *Magazine of Concrete Research* **67(12)**: 645–655.
61. Ghernouti Y, Rabehi B, Bouziani T and Chaid R (2016) Valorization of brick wastes in the fabrication of concrete blocks. *Materiali in Tehnologije* **50(6)**: 911–916.
62. de Brito J, Evangelista L and Silva RV (2018) Concrete with multi-recycled crushed concrete aggregates. *4th International Conference on Structural Engineering, Tehran, Iran.*
63. Japan Standards Association (2012) *JIS 5023.* Recycled aggregate for concrete class L. Japan Standards Association, Tokyo, Japan. (In Japanese.)
64. de Brito J and Silva RV (2016) Current status on the use of recycled aggregates in concrete: Where do we go from here? *RILEM Technical Letters* **1**: 1–5.
65. Nagataki S, Gokce A, Saeki T and Hisada M (2004) Assessment of recycling process induced damage sensitivity of recycled concrete aggregates. *Cement and Concrete Research* **34(6)**: 965–971.
66. de Brito J and Alves F (2010) Concrete with recycled aggregates: The Portuguese experimental research. *Materials and Structures* **43(1)**: 35–51.
67. de Brito J and Robles R (2009) Methodology for the prediction of concrete with recycled aggregates properties. In *Excellence in Concrete Construction through Innovation* (Limbachiya MC and Kew HY (eds)). CRC Press/Balkema, Taylor & Francis, London, UK, pp. 477–488.
68. Van Der Wegen G and Haverkort R (1998) Recycled construction and demolition waste as a fine aggregate for concrete. *Proceedings of the International Symposium on Sustainable Construction: Use of Recycled Concrete Aggregate* (Dhir RK, Henderson NA and Limbachiya MC (eds)). Thomas Telford, London, UK, pp. 333–346.
69. Pedro D, de Brito J and Evangelista L (2014) Influence of the use of recycled concrete aggregates from different sources on structural concrete. *Construction and Building Materials* **71**: 141–151.
70. Matias D, de Brito J, Rosa A and Pedro D (2013) Mechanical properties of concrete produced with recycled coarse aggregates – Influence of the use of superplasticizers. *Construction and Building Materials* **44**: 101–109.
71. Matias D, de Brito J, Rosa A and Pedro D (2014) Durability of concrete with recycled coarse aggregates: Influence of superplasticizers. *Journal of Materials in Civil Engineering* **26**: 7.
72. Lamond JF, Campbell RL, Giraldi A, *et al.* (2001) *Removal and Reuse of Hardened Concrete.* ACI, Farmington Hills, MI, USA, Report 26.
73. Hansen TC (ed.) (1992) *Recycling of Demolished Concrete and Masonry.* E&FN Spon, London, UK.
74. Li X (2008) Recycling and reuse of waste concrete in China. Part I. Material behaviour of recycled aggregate concrete. *Resources, Conservation and Recycling* **53(1–2)**: 36–44.
75. Dillmann R (1998) Concrete with recycled concrete aggregates. *Proceedings of the International Symposium on Sustainable Construction: Use of Recycled Concrete Aggregate* (Dhir RK, Henderson NA and Limbachiya MC (eds)). Thomas Telford, London, UK, pp. 239–254.
76. Kikuchi M, Yasunaga A and Ehara E (1993) The total evaluation of recycled aggregate and recycled concrete. *Proceedings of the 3rd International RILEM Symposium on Demolition and Reuse of Concrete and Masonry, Odense, Denmark* (Lauritzen EK (ed.)). CRC Press, pp. 425–436.
77. Merlet JD and Pimienta P (1993) Mechanical and physico-chemical properties of concrete produced with coarse and fine recycled concrete aggregates. *Proceedings of the 3rd International RILEM Symposium on Demolition and Reuse of Concrete and Masonry, Odense, Denmark* (Lauritzen EK (ed)). CRC Press, pp. 400–411.
78. Rao A, Jha KN and Misra S (2007) Use of aggregates from recycled construction and demolition waste in concrete. *Resources, Conservation & Recycling* **50(1)**: 71–81.
79. Ryu JS (2002) An experimental study on the effect of recycled aggregate on concrete properties. *Magazine of Concrete Research* **54(1)**: 7–12.

80. Teranishi K, Dosho Y, Narikawa M and Kikuchi M (1998) Application of recycled aggregate concrete for structural concrete. Part 3. Production of recycled aggregate by real-scale plant and quality of recycled aggregate concrete. *Proceedings of the International Symposium on Sustainable Construction: Use of Recycled Concrete Aggregate* (Dhir RK, Henderson NA and Limbachiya MC (eds)). Thomas Telford, London, UK, pp. 143–156.
81. Ayob A, Razali ME, Alias S, Ahmad AG and Ali DSH (2017) Engineering behavior of concrete with recycled aggregate. *9th International Unimas Stem Engineering Conference, Kuching, Malaysia* (Hasan A, Khan AA, Mannan MA, *et al.* (eds)), pp. 9–16.
82. Otsuki N, Miyazato S and Yodsudjai W (2003) Influence of recycled aggregate on interfacial transition zone, strength, chloride penetration and carbonation of concrete. *Journal of Materials in Civil Engineering* **15(5)**: 443–451.
83. Pedro D, de Brito J and Evangelista L (2014) Influence of the use of recycled concrete aggregates from different sources on structural concrete. *Construction and Building Materials* **71**: 141–151.
84. Eguchi K, Teranishi K, Nakagome A, *et al.* (2007) Application of recycled coarse aggregate by mixture to concrete construction. *Construction and Building Materials* **21(7)**: 1542–1551.
85. Dosho Y (2007) Development of a sustainable concrete waste recycling system – Application of recycled aggregate concrete produced by aggregate replacing method. *Journal of Advanced Concrete Technology* **5(1)**: 27–42.
86. García-González J, Moran-Del Pozo JM, Rodriguez-Robles D, Guerra-Romero MI and Juan-Valdes A (2015) Porosity and pore size distribution in recycled concrete. *Magazine of Concrete Research* **67(22)**: 1214–1221.
87. Razaqpur AG, Fathifazl G, Isgor B, *et al.* (2010) How to produce high quality concrete mixes with recycled concrete aggregate. *2nd International Conference on Waste Engineering Management, Shanghai, China* (Xiao JZ, Zhang Y, Cheung MS and Chu RPK (eds)), pp. 11–35.
88. Fathifazl G, Abbas A, Razaqpur AG, *et al.* (2009) New mixture proportioning method for concrete made with coarse recycled concrete aggregate. *Journal of Materials in Civil Engineering* **21(10)**: 601–611.
89. Ridzuan ARM, Ibrahim A, Ismail AMM and Diah ABM (2005) Durability performance of recycled aggregate concrete. *Proceedings of the International Conference on Global Construction: Ultimate Concrete Opportunities: Achieving Sustainability in Construction* (Dhir RK, Dyer TD and Newlands MD (eds)). Thomas Telford, London, UK, pp. 193–202.
90. Katz A (2003) Properties of concrete made with recycled aggregate from partially hydrated old concrete. *Cement and Concrete Research* **33(5)**: 703–711.
91. Ho NY, Lee YPK, Lim WF, *et al.* (2015) Evaluation of RCA concrete for the construction of Samwoh Eco-Green Building. *Magazine of Concrete Research* **67(12)**: 633–644.
92. Surya M, Rao V and Lakshmy P (2015) Mechanical, durability, and time-dependent properties of recycled aggregate concrete with fly ash. *ACI Materials Journal* **112(5)**: 653–661.
93. Zhang HR and Zhao YX (2015) Integrated interface parameters of recycled aggregate concrete. *Construction and Building Materials* **101, Part 1**: 861–877.
94. de Brito J, Ferreira J, Pacheco J, Soares D and Guerreiro M (2016) Structural, material, mechanical and durability properties and behaviour of recycled aggregates concrete. *Journal of Building Engineering* **6**: 1–16.
95. Sucic A and Lotfy A (2016) Effect of new paste volume on performance of structural concrete using coarse and granular recycled concrete aggregate of controlled quality. *Construction and Building Materials* **108**: 119–128.
96. Bravo M, de Brito J, Evangelista L and Pontes J (2013) *Mechanical and Durability-related Performance of Concrete Made with Recycled Aggregates Sourced from Construction and Demolition Wastes of Portuguese Recycling Plants.* Instituto Superior Técnico, University of Lisbon, Portugal. (In Portuguese.)

97. Cartuxo F, de Brito J, Evangelista L, Jimenez JR and Ledesma EF (2016) Increased durability of concrete made with fine recycled concrete aggregates using superplasticizers. *Materials* **9(2)**: 98, 10.3390/ma9020098.
98. Corral-Higuera R, Arredondo-Rea SP, Neri-Flores MA, *et al.* (2011) Chloride ion penetrability and corrosion behavior of steel in concrete with sustainability characteristics. *International Journal of Electrochemical Science* **6(4)**: 958–970.
99. Duan ZH and Poon CS (2014) Properties of recycled aggregate concrete made with recycled aggregates with different amounts of old adhered mortars. *Materials & Design* **58**: 19–29.
100. Evangelista L (2014) *Behaviour of Concrete with Fine Recycled Concrete Aggregates. Department of Civil Engineering and Architecture*, Instituto Superior Técnico, University of Lisbon, Lisbon, Portugal, pp. 485.
101. González A and Etxeberria M (2014) Experimental analysis of properties of high performance recycled aggregate concrete. *Construction and Building Materials* **52**: 227–235.
102. Hwang JP, Shim HB, Lim S and Ann KY (2013) Enhancing the durability properties of concrete containing recycled aggregate by the use of pozzolanic materials. *KSCE Journal of Civil Engineering* **17(1)**: 155–163.
103. Kim K, Shin M and Cha S (2013) Combined effects of recycled aggregate and fly ash towards concrete sustainability. *Construction and Building Materials* **48**: 499–507.
104. Kong D, Lei T, Zheng J, *et al.* (2010) Effect and mechanism of surface-coating pozzolanics materials around aggregate on properties and ITZ microstructure of recycled aggregate concrete. *Construction and Building Materials* **24(5)**: 701–708.
105. Kou SC and Poon CS (2009) Properties of self-compacting concrete prepared with coarse and fine recycled concrete aggregates. *Cement and Concrete Composites* **31(9)**: 622–627.
106. Kou SC and Poon CS (2009) Properties of concrete prepared with crushed fine stone, furnace bottom ash and fine recycled aggregate as fine aggregates. *Construction and Building Materials* **23(8)**: 2877–2886.
107. Kou SC and Poon CS (2012) Enhancing the durability properties of concrete prepared with coarse recycled aggregate. *Construction and Building Materials* **35**: 69–76.
108. Kou SC, Poon CS and Agrela F (2011) Comparisons of natural and recycled aggregate concretes prepared with the addition of different mineral admixtures. *Cement and Concrete Composites* **33(8)**: 788–795.
109. Kou SC, Poon CS and Chan D (2004) Properties of steam cured recycled aggregate fly ash concrete. *International RILEM Conference on the Use of Recycled Materials in Buildings and Structures* (Vázquez E, Hendriks C and Janssen GMT (eds)). RILEM Publications, Barcelona, Spain, pp. 590–599.
110. Kou SC, Poon CS and Chan D (2007) Influence of fly ash as cement replacement on the properties of recycled aggregate concrete. *Journal of Materials in Civil Engineering* **19(9)**: 709–717.
111. Kou SC, Poon CS and Chan D (2008) Influence of fly ash as a cement addition on the hardened properties of recycled aggregate concrete. *Materials and Structures* **41(7)**: 1191–1201.
112. Kou SC, Poon CS and Wan HW (2012) Properties of concrete prepared with low-grade recycled aggregates. *Construction and Building Materials* **36**: 881–889.
113. Kou SC and Poon CS (2013) Long-term mechanical and durability properties of recycled aggregate concrete prepared with the incorporation of fly ash. *Cement and Concrete Composites* **37**: 12–19.
114. Lima C, Caggiano A, Faella C, *et al.* (2013) Physical properties and mechanical behaviour of concrete made with recycled aggregates and fly ash. *Construction and Building Materials* **47**: 547–559.
115. Moon DJ, Moon HY, Nagataki S, Hisada M and Saeki T (2002) Improvement on the qualities of recycled aggregate concrete containing super fine mineral admixtures. *Proceedings of the 1st fib Congress, Osaka, Japan* (Kyokai PK and Kyokai NK (eds)), pp. 113–118.

116. Pedro D (2014) *Mechanical and Durability-related Performance of Concrete Produced with Recycled Aggregates from High Performance Concrete for Use in Precast Industry.* Instituto Superior Técnico, University of Lisbon, Lisbon, Portugal. (In Portuguese.)
117. Poon CS, Kou SC and Chan D (2006) Influence of steam curing on hardened properties of recycled aggregate concrete. *Magazine of Concrete Research* **58(5)**: 289–299.
118. Shayan A and Xu AM (2003) Performance and properties of structural concrete made with recycled concrete aggregate. *ACI Materials Journal* **100(5)**: 371–380.
119. Sim J and Park C (2011) Compressive strength and resistance to chloride ion penetration and carbonation of recycled aggregate concrete with varying amount of fly ash and fine recycled aggregate. *Waste Management* **31(11)**: 2352–2360.
120. Tu T-Y, Chen Y-Y and Hwang C-L (2006) Properties of HPC with recycled aggregates. *Cement and Concrete Research* **36(5)**: 943–950.
121. Vaishali GG and Rao HS (2012) Strength and permeability characteristics of fiber reinforced high performance concrete with recycled aggregates. *Asian Journal of Civil Engineering (Building and Housing)* **13(1)**: 55–77.
122. Vieira T (2013) *Concrete Incorporating Fine Recycled Ceramic Aggregates: Durability-related Performance.* Department of Civil Engineering and Architecture, Instituto Superior Técnico, University of Lisbon, Lisbon, Portugal.
123. Pedro D, de Brito J and Evangelista L (2017) Evaluation of high-performance concrete with recycled aggregates: Use of densified silica fume as cement replacement. *Construction and Building Materials* **147**: 803–814.
124. Wang WJ, Wu J, Wang Z, Wu GZ and Yue AY (2016) Chloride diffusion coefficient of recycled aggregate concrete under compressive loading. *Materials and Structures* **49(11)**: 4729–4736.
125. Etxeberria M, Fernandez JM and Limeira J (2016) Secondary aggregates and seawater employment for sustainable concrete dyke blocks production: Case study. *Construction and Building Materials* **113**: 586–595.
126. Hobbs DW (1999) Aggregate influence on chloride ion diffusion into concrete. *Cement and Concrete Research* **29(12)**: 1995–1998.
127. Anastasiou E, Filikas KG and Stefanidou M (2014) Utilization of fine recycled aggregates in concrete with fly ash and steel slag. *Construction and Building Materials* **50**: 154–161.
128. Paine KA and Dhir RK (2010) Recycled aggregates in concrete: A performance-related approach. *Magazine of Concrete Research* **62(7)**: 519–530.
129. Olorunsogo FT and Padayachee N (2002) Performance of recycled aggregate concrete monitored by durability indexes. *Cement and Concrete Research* **32(2)**: 179–185.
130. Higashiyama H, Yagishita F, Sano M and Takahashi O (2012) Compressive strength and resistance to chloride penetration of mortars using ceramic waste as fine aggregate. *Construction and Building Materials* **26(1)**: 96–101.
131. Higashiyama H, Yamauchi K, Sappakittipakorn M, Sano M and Takahashi O (2013) A visual investigation on chloride ingress into ceramic waste aggregate mortars having different water to cement ratios. *Construction and Building Materials* **40**: 1021–1028.
132. Punmia BC, Jain AK and Jain AK (2003) *Basic Civil Engineering.* Laxmi Publications, New Delhi, India.
133. Pacheco-Torgal F and Jalali S (2010) Reusing ceramic wastes in concrete. *Construction and Building Materials* **24(5)**: 832–838.
134. Ge Z, Wang YY, Sun RJ, Wu XS and Guan YH (2015) Influence of ground waste clay brick on properties of fresh and hardened concrete. *Construction and Building Materials* **98**: 128–136.
135. Gonzalez-Corominas A and Etxeberria M (2014) Properties of high performance concrete made with recycled fine ceramic and coarse mixed aggregates. *Construction and Building Materials* **68**: 618–626.

136. Higashiyama H, Sappakittipakorn M, Mizukoshi M and Takahashi O (2014) Efficiency of ground granulated blast-furnace slag replacement in ceramic waste aggregate mortar. *Cement & Concrete Composites* **49**: 43–49.
137. Andal J, Shehata M and Zacarias P (2016) Properties of concrete containing recycled concrete aggregate of preserved quality. *Construction and Building Materials* **125**: 842–855.
138. Tam VWY, Gao XF and Tam CM (2005) Microstructural analysis of recycled aggregate concrete produced from two-stage mixing approach. *Cement and Concrete Research* **35(6)**: 1195–1203.
139. Tam VWY, Gao XF and Tam CM (2006) Quality improvement of recycled aggregate concrete. *Key Engineering Materials* **302–303**: 308–313.
140. Tam VWY, Tam CM and Wang Y (2007) Optimization on proportion for recycled aggregate in concrete using two-stage mixing approach. *Construction and Building Materials* **21(10)**: 1928–1939.
141. Barbudo A, Galvín AP, Agrela F, Ayuso J and Jiménez JR (2012) Correlation analysis between sulphate content and leaching of sulphates in recycled aggregates from construction and demolition wastes. *Waste Management* **32(6)**: 1229–1235.
142. Tovar-Rodriguez G, Barra M, Pialarissi S, Aponte D and Vazquez E (2013) Expansion of mortars with gypsum contaminated fine recycled aggregates. *Construction and Building Materials* **38**: 1211–1220.
143. Aguiar G and Selmo SMS (2004) An initial analysis of the effect of gypsum plaster in recycled mortars from C&D wastes. *International RILEM Conference on the Use of Recycled Materials in Building and Structures* (Vázquez E, Hendriks C and Janssen GMT (eds)). RILEM Publications, Barcelona, Spain, pp. 493–502.
144. Al-Attar TS, Al-Khateeb AM and Bachai AH (2006) Behavior of high performance concrete exposed to internal sulfate attack (gypsum-contaminated aggregate). In *Earth & Space: Engineering, Construction, and Operations in Challenging Environment*. American Society of Civil Engineers, 10.1061/40830(188)161.
145. Qi B, Gao JM, Chen F and Shen DM (2017) Evaluation of the damage process of recycled aggregate concrete under sulfate attack and wetting-drying cycles. *Construction and Building Materials* **138**: 254–262.
146. de Lucas EA, Medina C, Frias M and de Rojas MIS (2016) Clay-based construction and demolition waste as a pozzolanic addition in blended cements. Effect on sulfate resistance. *Construction and Building Materials* **127**: 950–958.
147. Berndt ML (2009) Properties of sustainable concrete containing fly ash, slag and recycled concrete aggregate. *Construction and Building Materials* **23(7)**: 2606–2613.
148. Gokce A, Nagataki S, Saeki B and Hisada M (2011) Identification of frost-susceptible recycled concrete aggregates for durability of concrete. *Construction and Building Materials* **25(5)**: 2426–2431.
149. Lee ST, Moon HY, Swamy RN, Kim SS and Kim JR (2005) Sulfate attack of mortars containing recycled fine aggregates. *ACI Materials Journal* **102(4)**: 224–230.
150. Lee S-T, Swamy RN, Kim S-S and Park Y-G (2008) Durability of mortars made with recycled fine aggregates exposed to sulfate solutions. *Journal of Materials in Civil Engineering* **20(1)**: 63–70.
151. Neville AM and Brooks JJ (2010) *Concrete Technology*, 2nd edn. Pearson, Harlow, UK.
152. Etxeberria M and Vázquez E (2010) Alkali silica reaction in concrete induced by mortar adhered to recycled aggregate. *Materiales de Construcción* **60(297)**: 47–58.
153. Desmyter J and Blockmans S (2000) Evaluation of different measures to reduce the risk of alkali silica reaction in recycled aggregate concrete. *Proceedings of the 11th International Conference on Alkali–Aggregate Reaction in Concrete, Canada*, pp. 603–612.
154. Jo B-W, Park S-K and Park J-C (2008) Mechanical properties of polymer concrete made with recycled PET and recycled concrete aggregates. *Construction and Building Materials* **22(12)**: 2281–2291.
155. Task Force of the Standing Committee of Concrete of Spain (2004) Draft of Spanish regulations for the use of recycled aggregate in the production of structural concrete. *International RILEM Conference*

on the Use of Recycled Materials in Building and Structures (Vázquez E, Hendriks C and Janssen GMT (eds)). RILEM Publications, Barcelona, Spain, pp. 511–525.
156. Delobel F, Bulteel D, Mechling JM, *et al.* (2016) Application of ASR tests to recycled concrete aggregates: Influence of water absorption. *Construction and Building Materials* **124**: 714–721.
157. Dhir RK, McCarthy MJ, Halliday JE and Tang MC (2005) *ASR Testing on Recycled Aggregates Guidance on Alkali Limits and Reactivity.* DTI/WRAP Aggregates Research Programme, Report STBF 13/14C.
158. McCarthy MJ, Halliday JE, Csetenyi LJ and Dhir RK (2009) *Alkali–silica Reaction Guidance for Recycled Aggregate in Concrete.* Waste & Resources Action Programme, Concrete Technology Unit, Division of Civil Engineering, University of Dundee.
159. Dhir RK and Paine KA (2004) Suitability and practicality of using coarse RCA in normal and high strength concrete. *1st International Conference on Sustainable Construction: Waste Management, Singapore*, pp. 108–123.
160. Dhir RK and Paine K (2010) Value added sustainable use of recycled and secondary aggregates in concrete. *Indian Concrete Journal* **84(3)**: 7–26.
161. Koulouris A, Limbachiya MC, Fried AN and Roberts JJ (2004) Use of recycled aggregate in concrete application: Case studies. *Proceedings of the International Conference on Sustainable Waste Management and Recycling: Challenges and Opportunities* (Limbachiya MC and Roberts JJ (eds)). Thomas Telford, London, UK, pp. 245–257.
162. Limbachiya MC (2004) Coarse recycled aggregates for use in new concrete. *Proceedings of the Institution of Civil Engineers – Engineering Sustainability* **157(2)**: 99–106.
163. Limbachiya MC, Leelawat T and Dhir RK (1998) RCA concrete: A study of properties in the fresh state, strength development and durability. *Proceedings of the International Symposium on Sustainable Construction: Use of Recycled Concrete Aggregate* (Dhir RK, Henderson NA and Limbachiya MC (eds)). Thomas Telford, London, UK, pp. 227–237.
164. Limbachiya MC, Leelawat T and Dhir RK (2000) Use of recycled concrete aggregate in high-strength concrete. *Materials and Structures* **33(233)**: 574–580.
165. Yildirim ST, Meyer C and Herfellner S (2015) Effects of internal curing on the strength, drying shrinkage and freeze–thaw resistance of concrete containing recycled concrete aggregates. *Construction and Building Materials* **91**: 288–296.
166. Salem RM, Burdette EG and Jackson NM (2003) Resistance to freezing and thawing of recycled aggregate concrete. *ACI Materials Journal* **100(3)**: 216–221.
167. Kimura Y, Imamoto K, Nagayama M and Tamura H (2004) High quality recycled aggregate concrete (HiRAC) processed by decompression and rapid release. *International RILEM Symposium on Environment-Conscious Materials and Systems for Sustainable Development, Koriyama, Japan* (Kashino N and Ohama Y (eds)). RILEM Publications, pp. 163–170.
168. Yanagibashi K, Yonezawa T, Arakawa K and Yamada M (2002) A new concrete recycling technique for coarse aggregate regeneration process. *Proceedings of the International Conference on Sustainable Concrete Construction* (Dhir RK, Dyer TD and Halliday JE (eds)). Thomas Telford, Dundee, UK, pp. 511–522.
169. Hosokawa Y (1999) Concrete products using fine aggregate recycled from waste concrete products. *8th International Conference on Durability of Building Materials and Components* (Lacasse MA and Vainer DJ (eds)). Ottawa, Canada, pp. 475–484.
170. Wu J, Jing XH and Wang Z (2017) Uni-axial compressive stress–strain relation of recycled coarse aggregate concrete after freezing and thawing cycles. *Construction and Building Materials* **134**: 210–219.
171. Tuyan M, Mardani-Aghabaglou A and Ramyar K (2014) Freeze–thaw resistance, mechanical and transport properties of self-consolidating concrete incorporating coarse recycled concrete aggregate. *Materials & Design* **53**: 983–991.

172. Zaharieva R, Buyle-Bodin F and Wirquin E (2004) Frost resistance of recycled aggregate concrete. *Cement and Concrete Research* **34(10)**: 1927–1932.
173. Kikuchi M, Dosho Y, Miura T and Narikawa M (1998) Application of recycled aggregate concrete for structural concrete. Part 1. Experimental study on the quality of recycled aggregate and recycled aggregate concrete. *Proceedings of the International Symposium on Sustainable Construction: Use of Recycled Concrete Aggregate* (Dhir RK, Henderson NA and Limbachiya MC (eds)). Thomas Telford, London, UK, pp. 55–68.
174. Bogas JA, de Brito J and Ramos D (2016) Freeze–thaw resistance of concrete produced with fine recycled concrete aggregates. *Journal of Cleaner Production* **115**: 294–306.
175. Zhang ZQ, Zhang YF, Yan CG and Liu YX (2017) Influence of crushing index on properties of recycled aggregates pervious concrete. *Construction and Building Materials* **135**: 112–118.
176. Gokce A, Nagataki S, Saeki T and Hisada M (2004) Freezing and thawing resistance of air-entrained concrete incorporating recycled coarse aggregate: The role of air content in demolished concrete. *Cement and Concrete Research* **34(5)**: 799–806.
177. Nagataki S and Iida K (2001) Recycling of demolished concrete. *5th CANMET/ACI International Conference on Recent Advances in Concrete Technology, Singapore* (Malhotra VM (ed.)). Canada Centre for Mineral and Energy Technology (CANMET), Natural Resources Canada, and ACI, pp. 1–20.
178. Guerra M, Ceia F, de Brito J and Julio E (2014) Anchorage of steel rebars to recycled aggregates concrete. *Construction and Building Materials* **72**: 113–123.
179. Lotfi S, Deja J, Rem P, *et al.* (2014) Mechanical recycling of EOL concrete into high-grade aggregates. *Resources Conservation and Recycling* **87**: 117–125.
180. Sadati S and Khayat KH (2016) Field performance of concrete pavement incorporating recycled concrete aggregate. *Construction and Building Materials* **126**: 691–700.
181. Fonseca N, de Brito J and Evangelista L (2011) The influence of curing conditions on the mechanical performance of concrete made with recycled concrete waste. *Cement and Concrete Composites* **33(6)**: 637–643.
182. Bravo M, de Brito J, Pontes J and Evangelista L (2015) Mechanical performance of concrete made with aggregates from construction and demolition waste recycling plants. *Journal of Cleaner Production* **99**: 59–74.
183. Guneyisi E, Gesoglu M, Kareem Q and Ipek S (2016) Effect of different substitution of natural aggregate by recycled aggregate on performance characteristics of pervious concrete. *Materials and Structures* **49(1–2)**: 521–536.
184. Jankovic K, Nikolic D and Bojovic D (2012) Concrete paving blocks and flags made with crushed brick as aggregate. *Construction and Building Materials* **28(1)**: 659–663.
185. Rodriguez C, Parra C, Casado G, *et al.* (2016) The incorporation of construction and demolition wastes as recycled mixed aggregates in non-structural concrete precast pieces. *Journal of Cleaner Production* **127**: 152–161.
186. Cement and Concrete Association of New Zealand (2011) *Best Practice Guide for the Use of Recycled Aggregates in New Concrete*. Cement and Concrete Association of New Zealand, New Zealand.
187. Lotfi S, Eggimann M, Wagner E, Mroz R and Deja J (2015) Performance of recycled aggregate concrete based on a new concrete recycling technology. *Construction and Building Materials* **95**: 243–256.
188. Zaetang Y, Sata V, Wongsa A and Chindaprasirt P (2016) Properties of pervious concrete containing recycled concrete block aggregate and recycled concrete aggregate. *Construction and Building Materials* **111**: 15–21.
189. Poon CS and Chan D (2007) The use of recycled aggregate in concrete in Hong Kong. *Resources, Conservation and Recycling* **50(3)**: 293–305.
190. CEN (2011) EN 197. Cement. Part 1. Composition, specifications and conformity criteria for common cements. CEN, Brussels, Belgium.

191. Silva RV (2015) *Use of Recycled Aggregates from Construction and Demolition Waste in the Production of Structural Concrete.* Department of Civil Engineering, Architecture and Georesources, Instituto Superior Técnico, Lisbon, Portugal.
192. Silva RV, de Brito J and Dhir RK (2015) Prediction of the shrinkage behavior of recycled aggregate concrete: A review. *Construction and Building Materials* **77**: 327–339.
193. CEN (2008) EN 12620. Aggregates for concrete. CEN, Brussels, Belgium.
194. CEN (2016) EN 206:2013 + A1:2016. Concrete. Specification, performance, production and conformity. CEN, Brussels, Belgium.
195. Silva RV and de Brito J (2018) *Specification for structural recycled aggregate concrete. IV International Conference Progress of Recycling in the Built Environment, Lisbon, Portugal.*
196. CEN (2009) EN 933-11. Tests for geometrical properties of aggregates. Part 11. Classification test for the constituents of coarse recycled aggregate. CEN, Brussels, Belgium.
197. Teychenné DC, Franklin RE, Erntroy HC, *et al.* (1997) *Design of Normal Concrete Mixes*, 2nd edn. Building Research Establishment, Watford, UK.
198. Bourguignon KM (2004) *Influence of Compressive Strength on the Carbonation of Concrete with Different Ground Granulated Furnace Slag Contents.* Universidade Federal do Espírito Santo, Vitória, Espírito Santo, Brazil. (In Portuguese.)
199. Kulakowski MP (2002) *Contribution to the Study in Concrete and Mortars with Active Silica.* Federal University of Rio Grande do Sul, Porto Alegre, Brazil.
200. Neves RN (2012) *Permeability to Air and Carbonation of Concrete in Structures.* Department of Civil Engineering and Architecture, Instituto Superior Técnico, University of Lisbon. (In Portuguese.)
201. Helene P (1997) *Vida útil das estruturas de concreto. IV Congresso Iberoamericano de Patologia das Construções e VI Congresso de Controle de Qualidade, Porto Alegre, Brazil.*
202. National Laboratory of Civil Engineering (2005) *Methodology for Estimating the Concrete Performance Properties Allowing to Comply with the Design Working Life of the Reinforced or Prestressed Concrete Structures under the Environmental Exposures XC and XS.* National Laboratory of Civil Engineering, Lisbon, Portugal, Report LNEC-E465. (In Portuguese.)
203. Comité Euro-International du Béton (1996) *Durable Concrete Structures.* Comité Euro-International du Béton, France.
204. Gonçalves AF, Ribeiro AB and Ferreira MJ (2007) *The new LNEC specifications on reinforced concrete durability. International RILEM Workshop on Integrated Service Life Modelling of Concrete Structures, Guimarães, Portugal*, pp. 131–139.
205. Dhir RK, de Brito J, Lynn CJ and Silva RV (2018) *Sustainable Construction Materials: Municipal Incinerator Bottom Ash.* Woodhead Publishing, Cambridge, UK.
206. Dhir RK, Dyer TD and Paine KA (2004) Dismantling barriers: Roles for research in realising markets for construction and demolition wastes. *1st International Conference on Sustainable Construction: Waste Management, Singapore*, pp. 1–22.
207. Limbachiya M, Meddah MS and Ouchagour Y (2012) Performance of Portland/silica fume cement concrete produced with recycled concrete aggregate. *ACI Materials Journal* **109(1)**: 91–100.
208. Paine KA, Collery DJ and Dhir RK (2009) Strength and deformation characteristics of concrete containing coarse recycled and manufactured aggregates. *Proceedings of the 11th International Conference on Non-conventional Materials and Technologies* (Walker P, Ghavami K, Paine K, *et al.* (eds)). Bath, UK.

Soutsos M
ISBN 978-0-7277-6375-4
https://doi.org/10.1680/icehcd.63754.525

Chapter 16
Lessons learnt by expert witnesses involved in construction disputes

Jeremy Ingham
Consultant, UK

Philip Ebbatson
HKA, London, UK

16.1. Introduction

When disputes arise relating to the design, construction or maintenance of concrete structures, experienced engineering professionals are often called upon to provide independent technical opinions and, if required, to act as expert witnesses. Claimants may seek to recover costs for remedial work, loss of revenue or reputational damage as well as their own legal expenses and costs for using experts.

Although the primary skill of an expert is their technical expertise, it is especially important that they be good communicators. Everything that they write and say (other than to other experts) must be expressed in the language of the layman and, if there are complex engineering issues, in simple steps that are easily understood. It is the expert's responsibility to ensure that their advice and evidence can be understood by all parties, their legal representatives, the judges and tribunal members.

All that an expert states in evidence must have a sound basis of one or more of

- supporting facts
- factual evidence
- sound reasoning
- technical authorities (such as codes of practice)
- the expert's own considerable knowledge and experience.

Essential attributes also include brevity, clarity and precision. In construction disputes, these can be difficult objectives because the issues can be many and complex. Clear-thinking from the outset is, therefore, advisable.

An expert will usually receive instructions from a solicitor who is acting for one of the parties involved in a dispute. This route to procuring an expert has the advantage of providing litigation privilege in certain legal jurisdictions, such as, for example, in England and Wales. Providing certain conditions are met, this protects all communications between a professional legal adviser (a solicitor or barrister) and their clients from being disclosed without the permission of the client.

As part of an expert commission, the expert will often undertake a forensic engineering investigation. Although there can sometimes be much to learn from forensic investigations, of value both in engineering practice and to industry and society, the results must often remain confidential. When disputes are either settled or decided by arbitral tribunal, the details must normally remain private, so experts are seldom at liberty to disclose what they may have learnt from cases, not even by way of confidential reporting routes, such as the Confidential Reporting on Structural Safety scheme in the United Kingdom.

16.2. The 'concrete expert'

Lawyers and their clients looking for a 'concrete expert' may not always realise that, for cracks in concrete, for example, they might need one or more of a structural engineering expert, a civil engineering expert, a materials engineering/concrete technology expert and a geotechnical engineering expert. In any situation, each specialist consulted should advise on what other expertise is likely to be needed and explain any limitations in the scope of their own.

Expert commissions usually fall into either an 'expert adviser' or an 'expert witness' role. It is important to understand the distinction between them. An expert adviser assists a party, often before a dispute has formed, as well as during the process of its resolution. In this role, the expert can assist that party's legal team in its preparation or defence of a claim by offering advice on technical issues. This will include strengths and weaknesses that will enable the legal team to help that party formulate its strongest case.

In contrast, although an expert witness is also appointed and paid by a party involved in the dispute, such an expert is required to provide independent and impartial opinions, which are addressed to the court or tribunal.

As a case progresses, an expert's role can sometimes change from that of expert adviser to expert witness. Irrespective of whether an expert adviser expects to receive instructions to act as an expert witness, honest, independent and objective opinions are essential from the outset.

Most cases in England and Wales are settled out of court. In other legal jurisdictions, litigation is more common. Since the Civil Procedure Rules came into effect in England and Wales in 1999 as part of the Woolf reforms, disputing parties have been encouraged to seek early settlement, without incurring the costs of going through the courts. The Woolf reforms have also encouraged the use of alternative dispute resolution techniques, such as mediation, expert determination and adjudication. Another option that can reduce costs and delay is for two or more parties to appoint a single joint expert rather than those parties instructing separate experts. Single joint experts are usually appointed for smaller claims. The views of an expert instructed on behalf of one party in a dispute may or may not be to that party's liking. The views of an expert instructed on behalf of both parties are highly likely to be unsupportive of at least one of them and may even be disliked by both.

Since 1 April 2013, courts in England and Wales have also had an option to streamline the litigation process by ordering opposing experts from like disciplines to give their evidence at the same time, known as concurrent evidence or 'hot-tubbing'.

Expert witnesses are more heavily regulated than expert advisers. The rules covering expert witnesses vary between jurisdictions. In England and Wales, they have a formal role under the *Civil Procedure Rules Part 35* [1], alongside the accompanying practice direction [2], with which expert witnesses must comply. Furthermore, experts should be familiar with the *Guidance for the Instruction of Experts in Civil Claims* [3]. Other jurisdictions have different rules for experts. For example, in Scotland, there are no formal rules for experts [4]. Construction is a global industry and expert witnesses involved in litigation, arbitration and certain forms of alternative dispute resolution are subject to a variety of institutional rules, procedures and codes of practice that vary depending on the location of the project and the legal jurisdiction of the construction contract. It is important that experts are familiar with the terms under which they are required to act.

Note that in March 2011, the Supreme Court of the United Kingdom removed the 400-year-old protection that gave expert witnesses immunity from suit for breach of duty whether in contract or negligence, in relation to their participation in legal proceedings [5]. It is now possible for experts who are assisting with cases in England and Wales to be sued for breach of contract or negligence.

16.3. Defining the 'durability' of concrete structures

Table 16.1 shows a selection of durability definitions from British, European, American and international concrete design standards and guidance documents. There is commonality between them in that they all require concrete elements to exhibit the following qualities if they are to be classified as durable.

- They should be designed with their service environment in mind.
- They should last for their intended working life (i.e. design or service life).
- There should be no appreciable loss of utility during the intended working life.
- There should be no need for any unforeseen maintenance or repairs.

Standards provide guidance on how to design for durability in specific service environments and for certain minimum lengths of working life. The standards impose requirements for the type of concrete mix and depth of cover to the steel reinforcement. However, clients will sometimes require a service life that is longer than allowed for in the standards, or they may expect the concrete to remain durable in a harsher environment than allowed for by the standards. In these situations, the engineer (usually a specialist materials engineer) should undertake a durability study to determine the appropriate type of concrete mix, depth of cover to the reinforcement and whether any special measures are required (e.g. the installation of a cathodic protection system). These durability studies are usually based on deterministic or probabilistic modelling and are currently relatively innovative in nature. They typically involve using predictive modelling tools to determine chloride ingress for reinforced concrete structures (and other durability parameters) and, thus, estimate the service life of a concrete structure (or structural element), following methodologies such as those in fib 34:2006 – 'Model code for service life design' [6,7].

Concrete durability requirements are specified in project drawings (e.g. design, for construction and as-built drawings) and project specifications (e.g. concrete materials and workmanship specifications). The contract requirement documents for a project may also contain concrete durability requirements, and these may be more onerous than standard requirements or, conversely, may allow exemptions from certain standard requirements. When investigating a dispute, it is important that the expert determine exactly what the employer required of the designer and what the designer required of the contractor and any subcontractors. Only then can a reliable assessment be made to determine whether the parties met their contractual obligations.

16.4. Defects

Disputes regarding the design, construction or maintenance of concrete structures are seldom (but occasionally are) due to the collapse of the structure. Rather, they typically arise from an allegation of the presence of defects of one sort or another and often, an associated serviceability failure. The expert is usually required to opine on the causes of the defects and their abundance and severity and is often required to advise on options for remediation. Table 16.2 lists the types of serviceability failure associated with concrete defects and possible consequences that impact the asset owner and users of the structure.

The following sections discuss aspects of the more common types of concrete defects that result in disputes.

Table 16.1 Definition of the term 'durability' reproduced from selected concrete design standards

Standard	Section	Durability definition
BS 8110-1:1997 Structural use of concrete. Part 1. Code of practice for design and construction [8]	3.1.5 Durability of structural concrete 3.1.5.1 General	A durable concrete element is one that is designed and constructed to protect embedded metal from corrosion and to perform satisfactorily in the working environment for the lifetime of the structure.
BS EN 1990:2001 +A1:2005 Eurocode. Basis of structural design [9]	2.4 Durability (1)P	The structure shall be designed such that deterioration over its design working life does not impair the performance of the structure below that intended, having due regard to its environment and the anticipated level of maintenance.
BS EN 1992-1-1:2004+A1:2014 Eurocode 2. Design of concrete structures. Part 1-1. General rules and rules for buildings [10]	4 Durability and cover to reinforcement 4.1 General (I)P	A durable structure shall meet the requirements of serviceability, strength and stability throughout its design working life, without significant loss of utility or excessive unforeseen maintenance (for general requirements see also EN 1990 [9]).
ISO 13823:2008 General principles on the design of structures for durability [11]	3.10 Durability	Capability of a structure or any component to satisfy, with planned maintenance, the design performance requirements over a specified period of time under the influence of environmental actions, or as a result of a self-ageing process.
fib Model code for service life design (2006) [6]	2.1.3 Design service life, durability and quality management (3)	Durability of the structure in its environment shall be such that it remains fit for use during its design service life. This requirement can be considered in one, or a combination, of the following ways: – By designing protective and mitigating systems – By using materials that, if well maintained, will not degenerate during the design service life – By giving such dimensions that deterioration during the design service life is compensated – By choosing a shorter lifetime for structural elements, which may be replaced one or more times during the design life in combination with appropriate inspection at fixed or condition dependant intervals and appropriate maintenance activities. In all cases the reliability requirements for long and short-term periods should be met.
ACI 318-14 Building code requirements for structural concrete [12]	2.3 Terminology	Durability: Ability of a structure or member to resist deterioration that impairs performance or limits service life of the structure in the relevant environment considered in design.

Table 16.2 Summary of the types of serviceability failure caused by concrete defects and their possible consequences. (Adapted from Ingham & McKibbins [13])

Serviceability failure type	Consequences
Structural failure	Fatality, injury, loss of use, restrictions on use (e.g. load restrictions)
Falling concrete	Fatality, injury, loss of use, restrictions on use (e.g. loss of access), impairing of protection to embedded steel reinforcement
Waterproofing compromised	Leaking through cracks, flooding, contamination, degradation of potable water
Loss of integrity of pavement or slab surface	Decreased ride quality for vehicles, trip hazards, increased repair frequency
Durability reduced	Reduced time to first repair, increased repair frequency, reduced overall service life, increased whole-life cost
Expansive reactions	Decreased strength, decreased durability, expansion of member
Failure of fixings	Loss of plant, equipment or cladding fixed to concrete structure, possible fatality or injury
Aesthetically unacceptable	Loss of prestige

16.4.1 Design or workmanship: Who is responsible for defects?

The expert is usually instructed to determine the causes of any significant defects and to assist with apportioning liability to the parties responsible. The term 'defect' may be defined in the contract documents. A defect is usually the result of work that was not in compliance with the contract. The expert must, therefore, have a clear understanding of what comprises a defect under the terms of the construction contract specific to the case.

Defects can be latent or patent. A latent defect is one that is concealed or has not yet arisen in the works and would not be discovered by a reasonably thorough visual inspection undertaken, for example, as part of normal building management or before a property sale. Conversely, a patent defect has become manifest in some way. Generally, a patent defect was (or should reasonably have been) known about while construction works were still underway, at practical completion of the works or during any liability period for rectifying defects.

Many disputes involve issues that could have been avoided at the design stage; these are design defects. Others tend to involve poor workmanship that could have been prevented by good construction or supervision; these are workmanship defects. In some cases, either defective or incorrect materials were used, and these are also usually classified as workmanship defects, as they are typically within the control of the party that oversees workmanship, namely the contractor. Insurers are particularly interested in whether the defects are attributable to design failings or inadequate workmanship, as the insured party may have a policy that does not provide coverage for all types of defects. For example, a construction contractor may have an insurance policy that does not cover them for their own workmanship defects. However, it is not uncommon for both design and workmanship issues to be involved in a single dispute, so that apportioning liability between the parties can be a somewhat inexact science.

For example, a recent review of professional negligence claims against structural engineers conducted by the insurer Griffiths & Armour found that an overwhelming proportion of claims arose from errors made during the design stage of the project [14]. It was also noted that designers make more mistakes during periods of rapid economic growth, as they are more likely to be overcommitted. Most innovative designs produce professional indemnity claims in one form or another. Innovative, novel or cutting-edge designs, by definition, seek to tackle architectural or engineering challenges, resulting in designs with little or no record of success.

Other causes of design inadequacy include overreliance on design codes in circumstances where they may not be adequate, for example, in very aggressive service environments or when an unusually long service life is required. In some cases, the designer has failed to undertake the design tasks required by the employer or by standards, or they may have misused or incorrectly used the design codes. In other cases, the authors have noted disputes arising from designers failing to clarify employers' requirements that are ambiguous or unclear. Furthermore, designers sometimes have difficulty in understanding, documenting or controlling clients' aspirations relating to aesthetics. The authors have observed several cases where value engineering exercises undertaken at the design stage with the aim of reducing material or programme costs have gone too far and had an unacceptably detrimental impact on durability that may not have been fully appreciated by the value engineering process.

Workmanship defects are usually due to construction operatives at the construction site who have had inadequate training or have insufficient experience in the required tasks or due to inadequate supervision of the construction workers. Less common is the use of materials that were either not of the required quality or not what were specified.

16.4.2 Forensic engineering investigations

The expert will often undertake a forensic engineering investigation to determine the causes and extent of defects. This typically starts with a desk study review and gap analysis of the available documentation. It is not uncommon for project design and quality assurance documentation to be at least partially unavailable, and this problem is usually worse for older structures. Missing documents are identified and a request for information is made to determine whether any of these missing documents can be obtained. If there are gaps in the information, it is likely that more investigative work will need to be undertaken during the subsequent site work stages to fill the gaps as far as possible.

Following the initial desk study, a walkover site visit is usually undertaken for reconnaissance purposes and to get an initial indication of the defects present. This may be followed by carefully planned site work investigations involving a visual inspection, concrete sampling, on-site testing and subsequent laboratory testing of the samples. This is likely to include drilling of concrete core samples (Figure 16.1) for testing the

Figure 16.1 Obtaining a concrete core sample from a working concrete structure by diamond core drilling

compressive strength, chemical analysis and petrographic examination. These tests may be supplemented by a range of indirect on-site tests to assess the uniformity of the concrete between elements, such as rebound hammer tests, ultrasonic pulse velocity tests and various other forms of non-destructive testing. Furthermore, the type, diameter, location and depth of steel reinforcement bars are often estimated by a combination of direct measurement from breakouts and non-destructive testing using a cover meter or a Ferroscan™ system. The expert must be fully conversant with the concrete testing techniques used in the investigation and fully aware of their respective levels of accuracy and the factors that can make the results inaccurate or unreliable.

Estimating the strength of concrete by crushing core samples taken from an old structure is subject to a significant degree of error. Moreover, attempting to estimate whether the concrete would have complied with the construction specification at the time of construction is fraught with difficultly. These difficulties are discussed further in Section 16.4.4.

Furthermore, note that the accuracy of the British standard methods for a chemical analysis of hardened concrete samples [15] have recently been called into question following inaccurate results reported by a Concrete Society laboratory trial [16,17]. In the decades since the methods were introduced, there have been considerable developments in concrete technology and changes in concrete construction practice, which the tests do not appear to have kept pace with. In particular, the now common use of mineral additions (e.g. fly ash and ground granulated blastfurnace slag) and chemical admixtures together with changes in cement composition present considerable (in some cases, possibly insurmountable) challenges for the chemical analysis methods in BS 1881-124 [15]. The implications of this inaccuracy in the BS 1881-124 [15] methods have been discussed in detail in a paper by Ingham *et al.* [18]. The Concrete Society [19] has recommended that the methods in BS 1881-124 [15] continue to be used until new, more accurate test methods have been developed. Furthermore, for cases in dispute, the Concrete Society recommends that an increased number of BS 1881-124:2015 [15] tests be undertaken and that the results be cross-checked using the alternative methods listed in Table 16.3.

Table 16.3 Alternative analysis methods for hardened concrete. (From the Concrete Society [19])

Concrete property	Clauses in BS 1881-124 [20] 'Wet' chemical analysis method	Alternative analysis method with potentially improved (or at least equivalent) accuracy
Cement content	Clauses 5.4 and 5.5	Microscopic analysis by point count of concrete slice to SR2 published by the Applied Petrography Group [21]
Chloride content	Clause 10.2 (in commercial laboratories the analysis is typically undertaken by potentiometric titration)	The Volhard method also given in BS 1881-124 [20] Instrumental chemical analysis by X-ray fluorescence spectroscopy to ISO 29581-2 [22]
Sulfate content	Clause 10.3	Instrumental chemical analysis by X-ray fluorescence spectroscopy to ISO 29581-2 [22]
Alkali content	Clause 10.4 (in commercial laboratories, the analysis is typically undertaken by atomic absorption spectroscopy)	Inductively coupled plasma with atomic emission spectrophotometry (ICP-AES) or flame photometry
Water-to-cement ratio	Clause 7	Fluorescence microscopy of thin section specimens to NT Build 361-1999 [23]

Once the information that can be obtained from all stages of the investigation is available, the expert will collate and interpret it and provide a report in an appropriate format. The interpretation will typically include a limited commentary on suitable repair options to remedy the defects but not a design for remedial works or tender documents. Prior to (and in anticipation of) litigation, the report may comprise a technical advice note, which (depending on the legal jurisdiction) may be subject to litigation privilege and non-disclosable to any party than the one that commissioned the expert. Reports for litigation and arbitration are subject to formal presentational standards and not subject to litigation privilege.

16.4.3 Cracks as defects

Concrete is prone to cracking (Figure 16.2), and specifications usually contain clauses detailing what degree of cracking is permissible, for example by specifying a maximum permissible crack width. Nevertheless, unexplained or unexpected cracking is one of the most common causes of disputes relating to concrete structures.

Concrete typically cracks where the tensile forces exceed the relatively low tensile strength of concrete. The range of causes of concrete cracking are summarised in Table 16.4. They can be divided into two broad groups as follows

- structural cracks: arising from external loading or self-weight
- non-structural cracks: arising from all other causes including exposure to and interaction with the environment or by the nature of the constituent materials of the concrete.

16.4.4 Low concrete strength

Allegations of contractors providing concrete with 'low strength' (i.e. concrete with a lower strength than they were contractually required to provide) feature commonly in disputes. It is important to understand what the contract required. For example, was there a minimum compressive strength that had to be achieved or was there a minimum 'characteristic strength' that had to be achieved at 28 days after casting the concrete? Characteristic strength is a statistical concept. It is defined in BS EN 206:2013 [25] as a 'value of

Figure 16.2 View of an unexpected crack in a concrete element caused by inadequate design. The crack is 1 mm wide and had been injected with resin as a repair

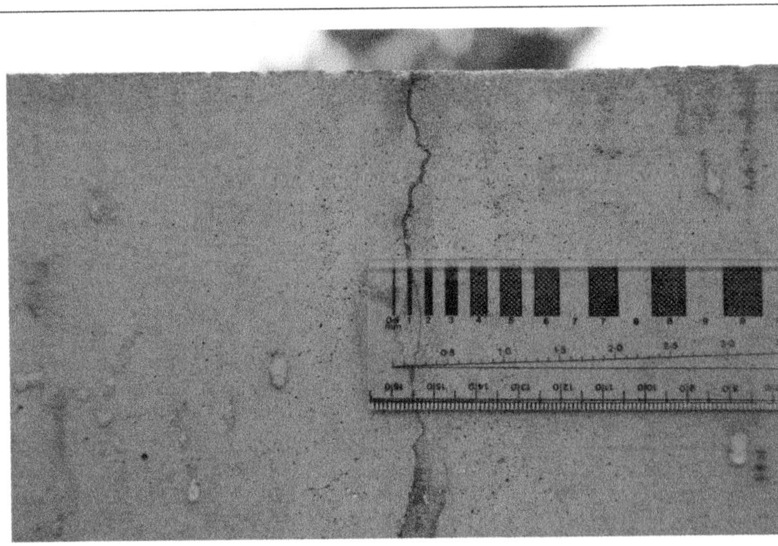

Table 16.4 Types of cracks in concrete. (Reproduced from Ingham & McKibbins [13] and after Concrete Society TR 22 [24])

Timing of crack formation	Classification of origin of cracking	Mechanism causing cracking
After hardening	Physical	Shrinkable aggregates
		Drying shrinkage
		Crazing
	Chemical	Corrosion of reinforcement
		Alkali–aggregate reaction
		Delayed ettringite formation
		Cement carbonation
	Thermal	Freeze–thaw cycles
		External seasonal temperature variations
		Early thermal contraction (with either external restraint or internal temperature gradients)
	Structural	Accidental overload
		Creep
		Design loads
Before hardening	Thermal	Early frost damage
	Plastic	Plastic shrinkage
		Plastic settlement
	Constructional movement	Formwork movement
		Subgrade movement

strength below which 5% of the population of all possible strength determinations of the volume of concrete under construction, are expected to fall'.

In a dispute, the first part of an investigation is an audit of the construction quality assurance records relating to the 7-day and 28-day cube sample testing strengths. This is a common way that contractors demonstrate compliance at the time of construction, although it is increasingly common practice to rely on the accreditation of the batching plant of the concrete supplier company. Providing that the records are not lacking and that they are not suspected of having been falsified, this should be sufficient to demonstrate that the contractor supplied concrete of the specified strength.

If no cube test results are available from the construction stage (quality assurance records), then it is possible to attempt to estimate strength by sampling cores from the structure while in-service using diamond drilling and testing the specimens removed. BS EN 13791:2007 [26] and the complementary standard BS 6089:2010 [27] provide an internationally accepted procedure for assessing the in situ compressive strength of concrete. However, note that the results will be less accurate than cube test data from the time of construction. The reasons for the relative inaccuracy of core test results are discussed in an article by Neville [28] and are addressed below.

The first difficulty with testing cores is that concrete strength will vary depending on where the core sample is taken from. The outer 50 mm of an element is expected to exhibit lower strength and is unrepresentative of the inherent concrete strength of the element (as a whole). Strength test specimens should not include this outer unrepresentative concrete. Section 5.5 of BS 6089:2010 [27] specifically requires that cores for crushing strength tests do not include concrete from with 50 mm of any surface. Furthermore, most concrete will, when placed, display a degree of sedimentation. This involves gravity drawing dense materials towards the bottom of the lift, which displace the lighter air and water upwards. Hence, slightly increased quantities of air and water are likely to be trapped in the concrete near to the top of the lift, leading to a reduction in strength. Consequently, this part of the element is also regarded as unrepresentative. BS 6089:2010 [27] requires that where the depth of lift is not more than 1.5 m, core samples must not contain concrete within 50 mm or the top 20% of the lift (whichever is the higher amount). If the depth of lift is 1.5 m or more, BS 6089:2010 [27] requires that core samples should not contain concrete from the top 300 mm of the lift.

To illustrate where unrepresentative concrete is found, Figure 16.3 is a schematic diagram of a column. It is based on separate publications by Dewer [29] and Somerville [30], who have both adopted the term 'heartcrete' to denote the portion of a column containing representative concrete that is suitable for core sampling as part of an in situ strength assessment.

Figure 16.3 Schematic illustration of a concrete column (after Dewer [29] and Somerville [30]). Core test pieces for compressive strength testing must be taken from the heartcrete to ensure that the concrete tested is representative of the actual strength of the column

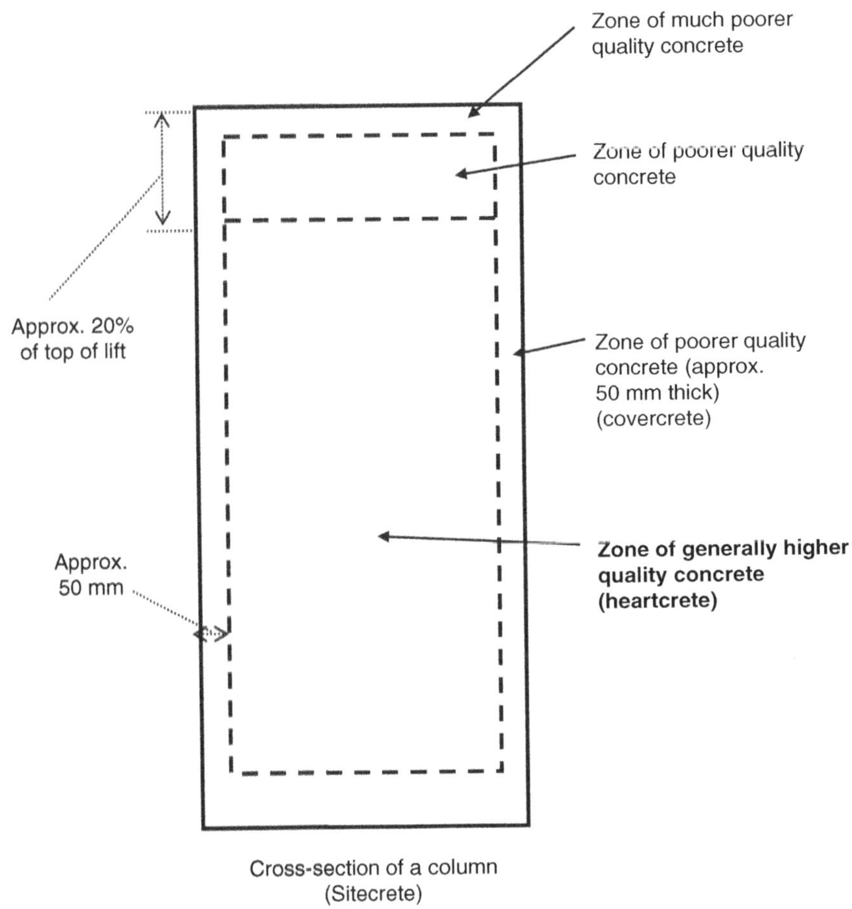

The second difficulty with core strength tests is that no matter how carefully core sampling is undertaken, the drilling process will always induce some damage in a retrieved core sample. Neville [28] states that

> It may be worth repeating that, however carefully performed, the coring operation causes some damage to the core: there may be a weakening of the bond of large particles of aggregate near the surface of the core, and there may be a dynamic effect of drilling, which induces some fine cracks.

Hence, if even carefully drilled core samples have some damage, it follows that incorrectly drilled cores may be greatly damaged. Any subsequent strength testing of a damaged sample will be inaccurate and unreliable. Furthermore, the potential for drilling damage is thought to be greater for old brittle concretes, as internal cracking of the core may be aggravated by the loss of the confining effect of the surrounding body of concrete [31].

A further difficulty is that to compare 'in situ concrete strengths' determined from core testing with cube test strengths from a project specification, it is necessary to calculate 'estimated in situ cube strengths'. This is an inherently inaccurate procedure, as numerical factors used in the conversion calculations are based on assumptions and best guess estimates. A laboratory study by the Concrete Society [32] entitled *In Situ Concrete Strength: An Investigation into the Relationship Between Core Strength and Standard Cube Strength*, demonstrated the relatively poor correlation between the two parameters. As an illustration of this, Grantham states in Section 1.3.1 of his textbook *Concrete Repair: A Practical Guide* [33] that

> The relationship between core tests and cube tests is very complex, depending on the type of member, its curing, the type of concrete, orientation and a range of other factors. *Core testing often results in posing more questions than it answers.* [Emphasis added.]

And also

> In the author's view, attempting to correlate the core and original cube test, however, is so fraught with difficulties that it is not recommended.

Furthermore, a recent comprehensive experimental study by Khoury *et al.* [34] reviewed the approaches in various national and international codes (including BS EN 12504-1:2009 [35]) for the conversion used to estimate the in situ concrete cube strength. All the code provisions studied seemed to be unreliable for predicting the in situ concrete cube strength from the results of core tests. The conclusion of the paper states that

> Comparison between the actual concrete cube strength and the estimated values using different approaches recommended by many codes is unsatisfactory.

Another issue with core testing is that it does not consider the stress history of the structure. A paper by Yip [36] demonstrated that the compressive load history of the concrete before coring could result in reductions in the measured strength of up to 30% due to internal microcracking. Consequently, the compressive strength tests obtained from core compression tests will tend to underestimate the true in situ capacity in these situations. This effect may be quite significant, even at relatively low stress levels, and adds considerably to interpretation uncertainties.

Considering the above, it is important to recognise that 'low' core strength test results do not necessarily indicate that poor quality concrete was supplied to the construction site nor that the concrete was inadequately cured. It is critical that such results are further investigated to determine why the concrete gave a low strength test result. It is important that the core sampling and testing are reviewed to determine whether they were correctly conducted and whether the results can be relied upon. Further testing of the concrete core samples by petrographic examination (using microscopes) and chemical analysis is recommended to determine whether any mix design or curing-related causes of low strength can be identified. If these are eliminated and it is confirmed that the core tests were conducted correctly, then this would point towards the concrete having been damaged in-service, for example because of structural underdesign.

Concrete strength is controlled by many factors, including

- the water-to-cement ratio
- the aggregate-to-cement ratio
- the physical properties of the aggregate
- the air-void content
- microcracking.

These factors can all be estimated by a petrographic examination. Petrography typically assesses the parameters listed in Table 16.5 to eliminate each of the common causes of low strength systematically.

A common cause of low-strength concrete is the addition of too much mixing water, so that the water-to-cement ratio of the concrete mix is higher than specified. The water-to-cement ratio can be estimated by a concrete petrographer using several different microscopy methods. The results can be cross-checked, thus increasing the reliability of the estimate. The first method is to count the number of residual cement grains in a given area, the second is to examine the portlandite crystallites in the cement matrix and the third (most commonly used) utilises fluorescent light microscopy to compare the relative fluorescence of a thin section of concrete from samples with reference specimens. The intensity of the fluorescence emitted from the cement matrix is proportional to the capillary porosity and the original water-to-cement ratio of the concrete mix.

Another important concrete parameter that can be investigated by petrographic examination is whether the concrete has been adequately cured. This is done by examining the abundance of relict cement grains in the concrete matrix. Microscopic features such as the abundance, size and hydration condition of the relict cement grains allow the petrographer to estimate the degree of curing of the cement paste and opine on whether it is a cause of low concrete strength.

A petrographic examination is also eminently suited to investigating cracks in concrete. The crack patterns can be examined in detail by a concrete petrographer using fluorescent light microscopy and their cause determined.

Table 16.5 Some of the causes of low concrete strength that can be determined by petrographic examination. (From Concrete Society TR 71 [37])

Possible cause of low strength	Parameters assessed by a petrographic examination
Too much water added to concrete	Water-to-cement ratio by fluorescence microscopy
Low cement content	Cement content by modal analysis
Inadequate curing	Microscopic estimation of the degree of hydration
Unsuitable aggregate	Hand specimen and microscopic estimation of coarse and fine aggregate properties
Poor aggregate–paste bonding	Microscopic examination of the aggregate/cement paste interface
Excessive content of fines	Microscopic estimation of fines content in the total aggregate combination
Excessive air content	Determination of air-void parameters
Deleterious reactions	Microscopic screening for leaching, alkali–aggregate reactions, sulfate attack, etc.

This method can also be used to examine the abundance and morphology of microcracking to determine the degree of structural damage from, for example, overloading of the structure. In combination with structural analysis, this can be used to determine whether inadequate building design contributed to cracking.

A petrographic examination of samples of hardened concrete is conducted in accordance with BS 1881-211:2016 [38] or ASTM C856-18 [39]. Further information regarding concrete petrography can be found in Technical Report 71 published by the Concrete Society [37], which provides a good introduction. More comprehensive information is provided in the textbooks by Poole and Sims [40] and Ingham [41].

16.4.5 Waterproofing defects

Underground structures, such as basements, have become increasingly popular in new developments. Underground construction offers advantages, as it is not limited by height restrictions for buildings and can be used to increase the total usable floor area in the same footprint. Despite the availability of comprehensive design guidance, there are still numerous instances of water ingress to concrete basements [42] and other underground structures, such as tunnels. Confirming this, *Reducing the Risk of Leaking Substructure: A Client's Guide* [43], published by the Institution of Civil Engineers, indicates that half of all basements leak at some point during their construction. The proportion that leak post-construction is significant. Although many failures are due to poor workmanship or unexpected events [44], some occur because of design decisions. Table 16.6 summarises common causes of water ingress to underground structures.

The authors regularly encounter disputes related to basement waterproofing systems that have either been incorrectly designed or inadequately installed and, thus, leak unacceptably. A common design shortcoming

Table 16.6 Common causes of water ingress to underground concrete structures and associated contributory factors involving inadequate design and workmanship

Cause of water ingress	Possible design failure	Possible workmanship failure
Through cracking of the concrete	Failure to control early-age thermal shrinkage cracking with an adequate reinforcement design, specification of concrete mix, placement restrictions and adequate curing	Failure to comply with the designer's specification and good practice for curing
Honeycombing of concrete	Unnecessarily congested design that makes placement and compaction difficult, e.g. excessive reinforcement or a tall and narrow formwork configuration	Inadequate compaction or a failure to stop grout loss through formwork joints
Open construction joints	Failure to specify adequate water bars and other waterproofing measures	Failure to install properly specified water bars and other waterproofing measures. Cold joints allowed to form between concrete pours
Open interfaces between concrete elements	Failure to specify adequate waterproofing at interfaces	Failure to install waterproofing products as per project specification or system manufacturer's requirements
Water ingress through waterproofing membrane	Wrong type of system specified	Failure to install waterproofing system as per project specification or system manufacturer's requirements

is where the designer elects to use an unbonded waterproofing membrane to tank an exterior basement wall instead of a fully bonded type. If such a membrane is damaged during construction, groundwater can penetrate and track for considerable distances along the external face of the basement wall. When the groundwater encounters a through crack or a construction joint, ingress can occur through the concrete wall. This would either not have happened or have been considerably reduced had a fully bonded membrane system been specified. Associated workmanship defects include any damage to the membrane during construction. The contractor should always protect a membrane against such damage.

16.4.6 Reinforcement corrosion

Reinforcement corrosion is the most common form of deterioration of reinforced concrete structures and is a common cause of disputes. This type of corrosion causes defects that range from aesthetic issues, such as rust staining, through to serviceability loss and, in the worse cases, structural collapse. An example of collapse involved a cantilevered reinforced concrete canopy above the entrance to a residential building (Figure 16.4). The canopy fell, seriously injuring three people [45]. The most common serviceability failure associated with reinforcement corrosion involves spalling cover concrete falling at height from structures, which presents a danger of impact injury to any users of the structure.

Reinforcement corrosion is relatively predictable and should be considered by the designer. Recommended design approaches are set out in standards. If reinforcement corrosion initiates earlier or proceeds faster than anticipated, there are two main causes.

- *Design inadequacy:* Failing to appreciate the severity of the service environment, in terms of exposure to chloride ions or the speed of carbonation of the cover concrete, can lead to a consequent failure to design sufficiently durable concrete.
- *Workmanship inadequacy:* This includes a failure to install the specified minimum depth of cover to the reinforcement. It also includes providing concrete of a lower compressive strength class than the designer specified or failing to adequately place, compact and cure the concrete in the manner prescribed in the designer's materials and workmanship specification.

Figure 16.4 Reinforced concrete building canopy that collapsed, injuring three people. The inset shows the heavily corroded condition of the steel reinforcement. (Reproduced from Ingham & Leek [45])

16.4.7 Floor slabs

Disputes relating to floor slabs are relatively common, especially the ground floor slabs of warehouses. Most claims relate to unexpected cracking, but there are also frequent disputes regarding joints, surface flatness, alleged voids in the slab, surface dusting, surface 'pop-outs' and unacceptable aesthetic appearance of the upper surface.

Common design-related causes of slab cracking include

- specifications that do not prevent the use of concrete mixes with high shrinkage properties
- specifying the mesh reinforcement or joints incorrectly
- specifying thin overlaid unbonded overslabs (polythene separated) onto an existing concrete base during refurbishment. In a warehouse with forklift loading, any curl of the overslab will be crushed, snapping the concrete panel edges. Such cracking is likely to be more abundant in aisles.

Common workmanship-related causes of slab cracking include

- failure to level the sub-base, so that the slab is too thin in places
- polythene on the sub-base without properly taped joints, which can stick up and form stress raisers.

Other types of workmanship-related failures can occur during the curing and finishing of concrete floors. These include inadequate curing, which causes a dusty surface because of excessive bleeding. Early severe drying (e.g. due to wind blowing over the slab) can weaken the top surface, leading to dusting and crazing. Other disputes have involved wetting and reworking of concrete in the plastic state or the use of a dry-shake surface finish (which is used to improve abrasion resistance). Dry-shake materials are factory-prepared blends of cement and hard fine aggregates (and sometimes admixtures and pigments), applied dry as a thin layer and trowelled into the fresh concrete base. Claims can relate to the uneven appearance of a finished floor, as excessive trowelling of the dry shake can mix it, in places, into the underlying base concrete, reducing the intended durability benefit and being aesthetically unacceptable to the owner (Figure 16.5).

Figure 16.5 Top surfaces of six 100-mm-diameter core samples taken from the concrete ground floor slab of a warehouse, showing an unacceptably variable dry-shake surface finish. (Reproduced from Ingham [41])

16.5. Conclusions and recommendations

The following recurrent themes have been identified as contributing to the causes of disputes involving the failure of design or workmanship in construction projects involving concrete structures.

- Inadequate 'employers requirements' specifications that are unclear or ambiguous regarding what the employer or designer required.
- Difficulty in understanding, documenting or controlling clients' aspirations relating to aesthetics and cleaning procedures in use.
- Poor design by inappropriately qualified or inexperienced designers. Under-resourcing or inadequate checking by the design team during the design stage.
- Over-reliance on design codes by designers in circumstances where they may not be adequate.
- Misuse or incorrect use of design codes by designers.
- Failure by the designer to adequately undertake the design tasks required by the employer or by standards.
- Value engineering to save materials or programme costs that go too far or a failure by the designers to adequately check the impacts of such changes.
- Innovative, novel or cutting-edge designs that go wrong because no warnings are available from past precedent.
- Inadequately trained or inexperienced construction workers employed by contractors or subcontractors.
- Failure to supervise construction workers adequately during construction (by any party with a supervisory role).
- Use of materials by the contractor that were either not of the required quality or not what were specified by the designer.

The authors offer the following recommendations with the aim of reducing the number and severity of future disputes involving concrete structures.

- Employers should ensure that tender and project specifications are adequately detailed, whether they are prescriptive in nature or performance-based specifications, so that contactors are in no doubt as to what is required.
- Employers, designers and contractors should involve appropriate technical experts from the earliest project stages onwards (e.g. starting at concept design stage), with the aim of saving significant time and cost on projects by identifying and mitigating technical risks. Thus, disputes are less likely to occur during or after construction.
- When looking for technical experts, employers, designers and contractors should carefully check that candidates have demonstrable relevant academic and professional qualifications and suitable project experience applicable to the specific design task.
- Designers should undertake early risk reviews for design tasks and compile a commercial and technical risk register to assist with identifying design decisions that could result in a dispute. Technical specialists should be involved in compiling the risk register and devising mitigation measures. This should be a 'live' document that is updated throughout the project and after regular risk review meetings.
- The due diligence activities that employers undertake when awarding design or build contracts should extend to all subcontractors and not just to the main designer and contractor. The employer is recommended to commission independent checks of the design and to appoint an independent tester as part of a comprehensive suite of quality assurance activities during construction.
- Engineering and construction professionals and academics should expend additional effort towards preventing disputes by forensically studying the reasons that disputes arise and seeking to implement targeted improvements to design and construction practice.

REFERENCES

1. MoJ (Ministry of Justice) (2014) *Ministry of Justice Rules and Practice Directions for Civil Courts. Part 35. Experts and Assessors*. MoJ, London, UK. http://www.justice.gov.uk/courts/procedure-rules/civil/rules/part35 (accessed 06/09/22).
2. MoJ (2014) *Ministry of Justice Rules and Practice Directions for Civil Courts, Practice Direction 35. Experts and Assessors*. MoJ, London, UK. http://www.justice.gov.uk/courts/procedure-rules/civil/rules/part35/pd_part35 (accessed 06/09/22).
3. Civil Justice Council (2014) *Guidance for the Instruction of Experts in Civil Claims*. Civil Justice Council, London, UK. https://www.judiciary.uk/wp-content/uploads/2014/08/experts-guidance-cjc-aug-2014-amended-dec-8.pdf (accessed 06/09/22).
4. Wassermann D (2015) Hints and pitfalls: The successful expert witness. *Proceedings of the Institution of Civil Engineers – Forensic Engineering* **168(3)**: 134–139, 10.1680/feng.14.00010.
5. *Jones* v. *Kaney* (2011) UKSC 13; (2011) 2 W.L.R. 823; (2011) 2 All E.R. 671; (2011) B.L.R. 283; 135 Con. L.R. 1; (2011) 119 B.M.L.R. 167; (2011) C.I.L.L. 3037; (2011) 14 E.G. 95 (C.S.); (2011) 108(15) L.S.G. 19; (2011) 161 N.L.J. 508; (2011) 155 (13) S.J.L.B. 30.
6. fib (International Federation for Structural Concrete) (2006) Model code 34. Model code for service life design. Sprint-Digital-Druck, Stuttgart, Germany.
7. Sfikas IP and Ingham JP (2016) Service life design of concrete structures using probabilistic modelling tools: Statistical analysis of input parameters. *Concrete Solutions: Proceedings of the 6th International Conference on Concrete Repair, Thessaloniki, Greece* (Grantham MG, Papayianni I and Sideris K (eds)). CRC Press, Taylor & Francis, London, UK, pp. 437–446.
8. BSI (1997) BS 8110-1:1997. Structural use of concrete. Code of practice for design and construction. BSI, London, UK.
9. BSI (2001) BS EN 1990:2001+A1:2005. Eurocode. Basis of structural design. BSI, London, UK.
10. BSI (2004) BS EN 1992-1-1:2004+A1:2014. Eurocode 2. Design of concrete structures. General rules and rules for buildings. BSI, London, UK.
11. ISO (International Organization for Standardization) (2008) ISO 13823:2008. General principles on the design of structures for durability. ISO, Geneva, Switzerland.
12. ACI (American Concrete Institute) (2014) ACI 318-14. Building code requirements for structural concrete. ACI, Farmington Hills, MI, USA.
13. Ingham JP and McKibbins L (2013) Briefing: Concrete structures affected by cracking. *Proceedings of the Institution of Civil Engineers – Forensic Engineering* **166(1)**: 3–8, 10.1680/feng.12.00018.
14. Powell T (2015) Trends and influencing factors in professional indemnity claims. *Proceedings of the Institution of Civil Engineers – Forensic Engineering* **168(3)**: 109–116, 10.1680/feng.15.00009.
15. BSI (2015) BS 1881-124. Testing concrete. Methods for analysis of hardened concrete. BSI, London, UK.
16. Ingham J and Barnes R (2015) Briefing: Testing times for concrete structures. Is BS 1881-124 still fit for purpose? *Proceedings of the Institution of Civil Engineers – Construction Materials* **168(6)**: 253–258.
17. Ingham JP and Barnes R (2016) Briefing: Doubts remain on the reliability of concrete chemical analysis methods. *Proceedings of the Institution of Civil Engineers – Civil Engineering* **169(CE3)**: 103.
18. Ingham JP, McKibbins L and Barnes R (2016) Chemical analysis of concrete for management of civil engineering structures. *Infrastructure Asset Management* **3(3)**: 106–119.
19. Concrete Society (2014) *Analysis of Hardened Concrete: A Guide to Test Procedures and Interpretation of Results*, 2nd edn. Concrete Society, Camberley, UK, Technical report 32.
20. BSI (1988) BS 1881-124. Testing concrete. Methods for analysis of hardened concrete. BSI, London, UK.
21. Applied Petrography Group (2010) *A Code of Practice for the Petrographic Examination of Concrete*. Geological Society of London, London, UK, Report SR2, http://www.appliedpetrographygroup.com/reports.htm (accessed 06/09/2022).

22. ISO (2010) ISO 29581-2. Cement. Test methods. Part 2. Chemical analysis by X-ray fluorescence. ISO, Geneva, Switzerland.
23. Nordic Innovations Center (1999) NT Build 361. Concrete, hardened. Water-cement ratio, 2nd edn. Nordic innovations Center, Oslo, Norway.
24. Concrete Society (2010) *Non-structural Cracks in Concrete*, 4th edn. Concrete Society, Camberley, UK, Technical report 22.
25. BSI (2013) BS EN 206:2013+A1:2016. Concrete. Specification, performance, production and conformity. BSI, London, UK.
26. BSI (2007) BS EN 13791. Assessment of in-situ compressive strength in structures and pre-cast concrete components. BSI, London, UK.
27. BSI (2010) BS 6089. Assessment of in-situ compressive strength in structures and precast concrete components. Complementary guidance to that given in BS EN 13791. BSI, London, UK.
28. Neville A (2001) Core tests: Easy to perform, not easy to interpret. *Concrete International* **23(11)**: 59–68.
29. Dewer JD (1988) *Manual of Ready Mixed Concrete*. Blackie Academic & Professional, London, UK.
30. Somerville G (2008) *Management of Deteriorating Concrete Structures*. Taylor & Francis, Abingdon, UK and New York, NY, USA.
31. Bungey JH, Millard SG and Grantham MG (2006) *Testing of Concrete in Structures*, 4th edn. Taylor & Francis, Abingdon, UK.
32. Concrete Society (2004) *In Situ Concrete Strength: An Investigation into the Relationship between Core Strength and Standard Cube Strength*. Concrete Society, Camberley, UK, Project report 3, Ref CS 126.
33. Grantham MG (ed.) (2011) *Concrete Repair: A Practical Guide*. Taylor & Francis, Abingdon, UK.
34. Khoury S, Aliabdo AA and Ghazy A (2014) Reliability of core test – Critical assessment and proposed new approach. *Alexandria Engineering Journal* **53(1)**: 169–184.
35. BSI (2009) BS EN 12504-1. Testing concrete in structures. Cored specimens. Taking, examining and testing in compression. BSI, London, UK.
36. Yip WK (1993) Estimating the potential strength of concrete with prior load history. *Magazine of Concrete Research* **45(165)**: 301–308.
37. Concrete Society (2010) *Concrete Petrography: An Introductory Guide for the Non-specialist*. Concrete Society, Camberley, UK, Technical report 71.
38. BSI (2016) BS 1881-211:2016. Testing concrete. Procedure and terminology for the petrographic examination of hardened concrete. BSI, London, UK.
39. ASTM (2018) ASTM C856-18. Standard practice for petrographic examination of hardened concrete. ASTM, West Conshohocken, PA, USA.
40. Poole AB and Sims I (2015) *Concrete Petrography: A Handbook of Investigative Techniques*, 2nd edn. CRC Press, UK.
41. Ingham JP (2011) *Geomaterials under the Microscope. A Colour Guide*. Manson Publishing, London, UK.
42. Barnes RA (2015) Investigation of water ingress to concrete basements. *Proceedings of the Institution of Civil Engineers – Forensic Engineering* **166(3)**: 107–115, 10.1680/feng.13.00006.
43. ICE (Institution of Civil Engineers) (2009) *Reducing the Risk of Leaking Substructure: A Client's Guide*. ICE, London, UK.
44. Lazarus D (2013) Water, water everywhere: Ingress in underground structures. *Proceedings of the Institution of Civil Engineers – Forensic Engineering* **166(4)**: 189–197, 10.1680/feng.13.00011.
45. Ingham JP and Leek D (2017) Forensic engineering of construction materials: lessons learnt from disputes. *Proceedings of the Institution of Civil Engineers – Forensic Engineering* **170(1)**: 33–44.

FURTHER READING

Bond C, Solon M, Harper P and Davies G (2007) *The Expert Witness: A Practical Guide*, 3rd edn. Sweet & Maxwell, London, UK.

Buckingham P (2009) The effective use of experts and expert witnesses. In *Forensic Engineering: From Failure to Understanding* (Neale BS (ed.)). Thomas Telford, London, UK, pp. 544–553.

Choat R (2012) Seven rules for success as an expert. *Proceedings of the Institution Civil Engineers – Forensic Engineering* **165(1)**: 3–12, 10.1680/feng.2012.165.1.3.

Clancy B (2016) Briefing: Reflections on the role of the expert witness. *Proceedings of the Institution of Civil Engineers – Forensic Engineering* **169(2)**: 31–43, 10.1680/jfoen.16.00002.

Horne R and Mullen J (2013) *The Expert Witness in Construction*. Wiley-Blackwell, London, UK.

Institution of Structural Engineers (2003) *Expert Evidence: A Guide for Expert Witnesses and Their Clients*, 2nd edn. Institution of Structural Engineers, London, UK.

Saunders S (2016) Briefing: How civil engineers can be better expert witnesses. *Proceedings of the Institution of Civil Engineers – Forensic Engineering* **170(1)**: 6–7, 10.1680/jfoen.16.00021.

Soutsos M
ISBN 978-0-7277-6375-4
https://doi.org/10.1680/icehcd.63754.545
Emerald Publishing Limited:

Index

Locators in *italics* refer to figures and tables.

Printed and bound by CPI Group (UK) Ltd, Croydon, CR0 4YY
15/09/2026
14962491-0002